CINNOLINES AND PHTHALAZINES

Supplement II

This is the Sixty-Fourth Volume in the Series

THE CHEMISTRY OF HETEROCYCLIC COMPOUNDS

THE CHEMISTRY OF HETEROCYCLIC COMPOUNDS

A SERIES OF MONOGRAPHS

EDWARD C. TAYLOR and PETER WIPF, *Editors*

ARNOLD WEISSBERGER, *Founding Editor*

CINNOLINES AND PHTHALAZINES

Supplement II

D. J. Brown

Research School of Chemistry
Australian National University
Canberra

AN INTERSCIENCE PUBLICATION

JOHN WILEY & SONS, INC.

Published by John Wiley & Sons, Inc., Hoboken, New Jersey.
Published simultaneously in Canada.

Library of Congress Catalog Card Number: 96-6182

Includes index.
ISBN-13 978-0-471-48587-2
ISBN-10 0-471-48587-X

Classification Number: QD401.F96

Printed in the United States of America

10 9 8 7 6 5 4 3 2 1

Dedicated to my revered colleagues
Wilfred L. F. Armarego and Gordon B. Barlin,
both prolific contributors to
heterocyclic research and authors
of books in this series

The Chemistry of Heterocyclic Compounds Introduction to the Series

The chemistry of heterocyclic compounds is one of the most complex and intriguing branches of organic chemistry, of equal interest for its theoretical implications, for the diversity of its synthetic procedures, and for the physiological and industrial significance of heterocycles.

The Chemistry of Heterocyclic Compounds has been published since 1950 under the initial editorship of Arnold Weissberger, and later, until his death in 1984, under the joint editorship of Arnold Weissberger and Edward C. Taylor. In 1997, Peter Wipf joined Prof. Taylor as editor. This series attempts to make the extraordinarily complex and diverse field of heterocyclic chemistry as organized and readily accessible as possible. Each volume has traditionally dealt with syntheses, reactions, properties, structure, physical chemistry, and utility of compounds belonging to a specific ring system or class (e.g., pyridines, thiophenes, pyrimidines, three-membered ring systems). This series has become the basic reference collection for information on heterocyclic compounds.

Many broader aspects of heterocyclic chemistry are recognized as disciplines of general significance that impinge on almost all aspects of modern organic chemistry, medicinal chemistry, and biochemistry, and for this reason we initiated several years ago a parallel series entitled General Heterocyclic Chemistry, which treated such topics as nuclear magnetic resonance, mass spectra, and photochemistry of heterocyclic compounds, the utility of heterocycles in organic synthesis, and the synthesis of heterocycles by means of 1,3-dipolar cycloaddition reactions. These volumes were intended to be of interest to all organic, medicinal, and biochemically oriented chemists, as well as to those whose particular concern is heterocyclic chemistry. It has, however, become increasingly clear that the above distinction between the two series was unnecessary and somewhat confusing, and we have therefore elected to discontinue *General Heterocyclic Chemistry* and to publish all forthcoming volumes in this general area in *The Chemistry of Heterocyclic Compounds* series.

Dr. Des J. Brown is again to be applauded and profoundly thanked for another fine contribution to the literature of heterocyclic chemistry. This volume on *Cinnolines and Phthalazines* covers both ring systems for the period 1973–2004, with a comprehensive compilation and discussion of the literature of the 31 years that have elapsed since the Singerman and Patel supplemental review (Volume 27 in this series) on *Cinnolines and Phthalazines* in 1973. It must be noted with admiration that many of the books in this series that have come to be regarded as classics in heterocyclic chemistry (*The Pyrimidines, The Pyrimidines Supplement I, The Pyrimidines Supplement II, Pteridines, Quinazolines Supplement I, The*

Quinoxalines Supplement II, and The Pyrazines Supplement I) are also from the pen of Dr. Brown.

Department of Chemistry
Princeton University
Princeton, New Jersey

EDWARD C. TAYLOR

Department of Chemistry
University of Pittsburgh
Pittsburgh, Pennsylvania

PETER WIPF

Preface

Cinnolines and phthalazines have been reviewed twice in this *Chemistry of Heterocyclic Compounds* series: first by J. C. E. Simpson as part of Volume 5 in 1953 and subsequently in a supplementary way by G. M. Singerman and N. R. Patel, respectively, as part of Volume 27 (edited by R. N. Castle) in 1973. The present *Second Supplement* seeks to build on those solid foundations by covering the literature of both systems for the period 1972–2004, inclusive. In doing so, it seemed advisable to make several changes in format to conform with the treatments of related diazines and benzodiazines in more recent volumes of the series. Thus primary syntheses have been collected for the first time into a single chapter for each system; nucleus-reduced derivatives are no longer treated as separate systems but have been integrated into appropriate cinnoline or phthalazine chapters; the content of each chapter has been expanded to embrace a family of derivatives rather than a single type; and the myriad scattered tables of cinnoline or phthalazine derivatives have been replaced by a single alphabetical table (for each system) that lists almost all defined simple derivatives described before 2005 (including those listed in the earlier reviews). In view of the foregoing and other changes in format, the supplementary nature of the present volume has been maintained by numerous cross-references (e.g., *H* 11 or *E* 250) to corresponding pages in Simpson's original volume (*Hauptwerk*) or Castle's update (*Ergänzungswerk*), respectively.

The chemical nomenclature used in this supplement follows current IUPAC recommendations [*Nomenclature of Organic Chemistry, Sections A–E, H* (J. Rigaudy and S. P. Klesney, eds., Pergamon Press, Oxford, 1970)] with one important exception. In order to keep "cinnoline" or "phthalazine" as the principal part of each name, those groups that would normally qualify as principal suffixes but are not attached directly to the nucleus are rendered as prefixes. For example, 2-carbamoylmethyl-1,4(2*H*, 3*H*)-phthalazinedione is used instead of 2-(1,4-dioxo-1,2,3,4-tetrahydrophthalazin-2-yl) acetamide. Secondary, tertiary, and quaternary amino substituents are also rendered as prefixes. Ring systems are named according to the *Chemical Abstracts* Service recommendations [*Ring Systems Handbook* (eds. anonymous, American Chemical Society, Columbus, Ohio, 2003 edition and supplements)]. In preparing this supplement, patent literature has been virtually ignored in the belief that useful factual information therein usually appears later in the regular literature.

Throughout this book, an indication such as 0°C→85°C (within parenthesized reaction conditions) means that the reaction commenced at the first temperature and was completed at the second; in contrast, an indication such as 25–35°C means that the reaction was conducted somewhere within that range. Terms such as "more recent literature" invariably refer to publications within the period 1972–2004.

I am greatly indebted to the Dean of the Research School of Chemistry, Professor Denis Evans, for the provision of postretirement facilities within the school; to the librarian, Mrs Joan Smith, for kindly assistance in all library matters; and to my wife, Jan, for her patient encouragement and practical help during the years of writing.

Research School of Chemistry,
Australian National University, Canberra

DES J. BROWN

Contents

CINNOLINES AND PHTHALAZINES

Supplement II

This is the Sixty-Fourth Volume in the Series

THE CHEMISTRY OF HETEROCYCLIC COMPOUNDS

CHAPTER 1

Primary Syntheses of Cinnolines

The primary synthesis of cinnolines or hydrocinnolines may be done by cyclization of carbocyclic substrates already bearing appropriate substituents; by cyclocondensation of carbocyclic substrates with acyclic synthons that provide one or more of the ring atoms needed to complete the cinnoline system; by similar processing of pyridazine substrates; or by rearrangement, ring expansion, ring contraction, degradation, or modification of suitable derivatives of other heterocyclic systems. Typical pre-1972 examples in each category may be found from the cross-references to Simpson's volume[906] (e.g., *H* 16) or to Singerman's volume[907] (e.g., *E* 62) that appear at some section headings. Some pre- and post-1972 primary syntheses have also been reviewed elsewhere.[903–905,908–916]

1.1. FROM A SINGLE CARBOCYCLIC SUBSTRATE (*H* 3, 6, 16, 46; *E* 18, 22, 62, 70, 190, 255)

Such syntheses are subdivided according to whether the N1–C8a, N1–N2, N2–C3, C3–C4, or C4–C4a bond is formed during the procedure to afford a cinnoline.

1.1.1. By Formation of the N1–C8a Bond

This process usually involves cyclization of *o*-halogeno-α-hydrazonoacetophenones or related substrates, as illustrated in the following examples.

Using *o*-Halogeno-α-hydrazonoacetophenones as Substrates

α-Ethoxycarbonyl-2,4,5-trifluoro-α-hydrazonoacetophenone (**1**) gave ethyl 6,7-difluoro-4-oxo-1,4-dihydro-3-cinnolinecarboxylate (**2**) (dioxane, reflux,

Cinnolines and Phthalazines: Supplement II, The Chemistry of Heterocyclic Compounds, Volume 64, by D.J. Brown

16 h: 70%).[414]

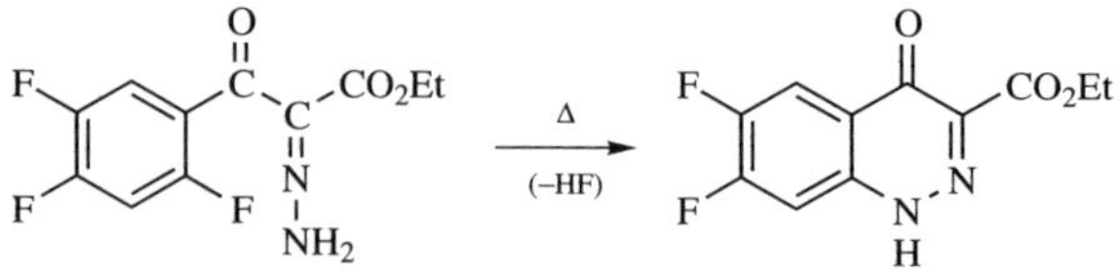

(1) (2)

2,4-Dichloro-α-ethoxycarbonyl-5-fluoro-α-(*p*-fluorophenylhydrazono)acetophenone (**3**, R = C_6H_4F-*p*) gave ethyl 7-chloro-6-fluoro-1-*p*-fluorophenyl-4-oxo-1,4-dihydro-3-cinnolinecarboxylate (**4**, R = C_6H_4F-*p*) (K_2CO_3, 18-crown-6, Me_2NCHO, 100°C, 1 h: 94%); analogs likewise.[617]

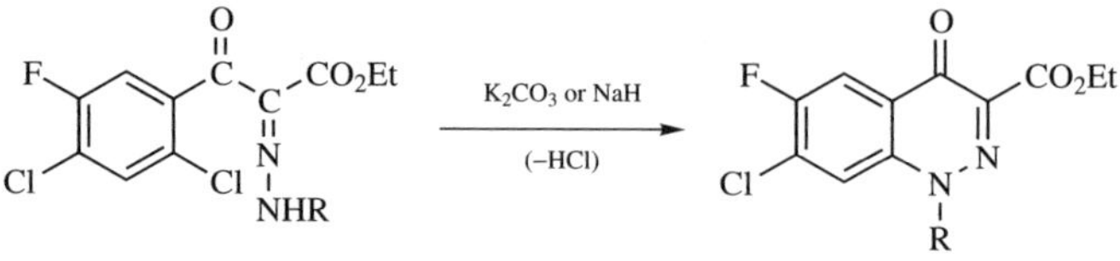

(3) (4)

2,4-Dichloro-α-ethoxycarbonyl-5-fluoro-α-(methylhydrazono)acetophenone (**3**, R = Me) gave ethyl 7-chloro-6-fluoro-1-methyl-4-oxo-1,4-dihydro-3-cinnolinecarboxylate (**4**, R = Me) (NaH, dioxane, 10°C→90°C, 15 min: 93%).[414]

α-Ethoxalyl-2,3,4,5,6-pentafluoro-α-(*p*-methoxyphenylhydrazono)acetophenone (**5**) gave 3-ethoxalyl-5,6,7,8-tetrafluoro-1-*p*-methoxyphenyl-4(1*H*)-cinnolinone (**6**) (Et_3N, $CHCl_3$, reflux, 4 h: 89%).[781]

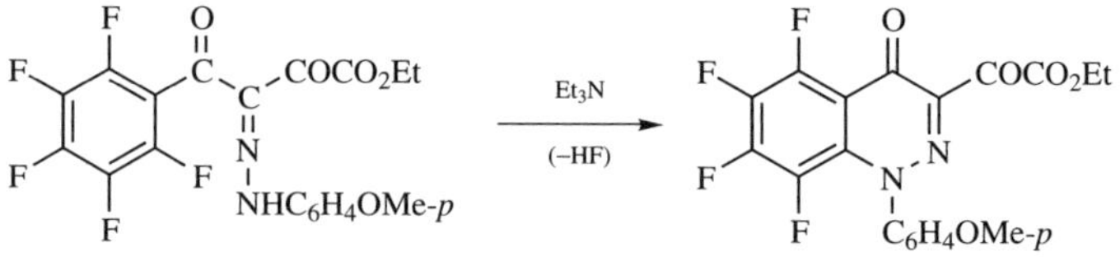

(5) (6)

Also other examples.[333,359,765]

Using Related Substrates

1-Nitro-2-[2-(phenylhydrazono)acetoacetyl]benzene (**7**) gave 3-acetyl-1-phenyl-4(1*H*)-cinnolinone (**8**) (Na_2CO_3, EtOH, H_2O, reflux, 1 h: 92%); several analogs likewise.[140]

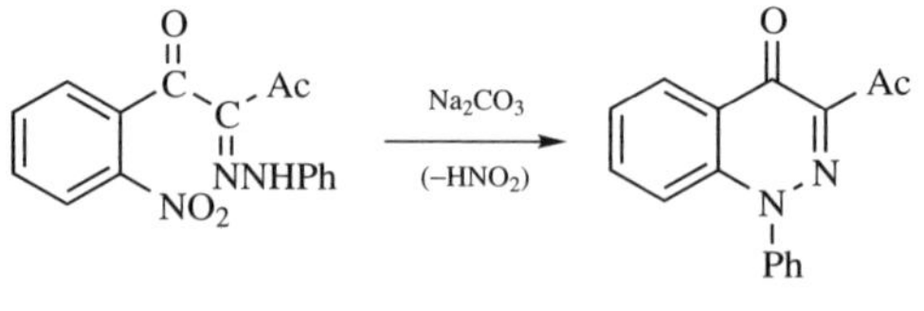

(7) (8)

2-(2-Fluoro-5-nitrophenyl)-*N'*,*N'*-dimethylacetohydrazide (**9**) gave 1,1-dimethyl-6-nitro-1,4-dihydrocinnolin-1-ium-3-olate (**10**) (K_2CO_3, H_2O, reflux, 3 h: 82%); analogs likewise.[717]

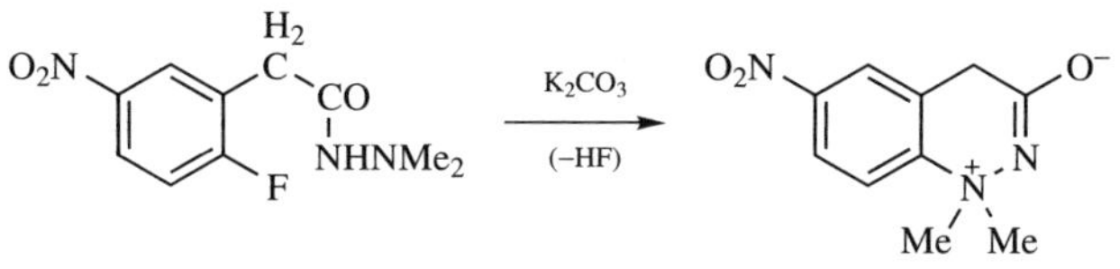

K_2CO_3 (−HF)

(**9**) (**10**)

6-Methoxy-α-*p*-methoxyphenyl-β-(*p*-nitrophenylazo)styrene (**11**) gave 6-methoxy-4-*p*-methoxyphenyl-1-*p*-nitrophenyl-1,2-dihydrocinnoline (**12**) (AcOH, reflux, 3 h: 90%);[639] the identities of this product and analogs made similarly are not fully established; they could be isomeric indole derivatives.[639]

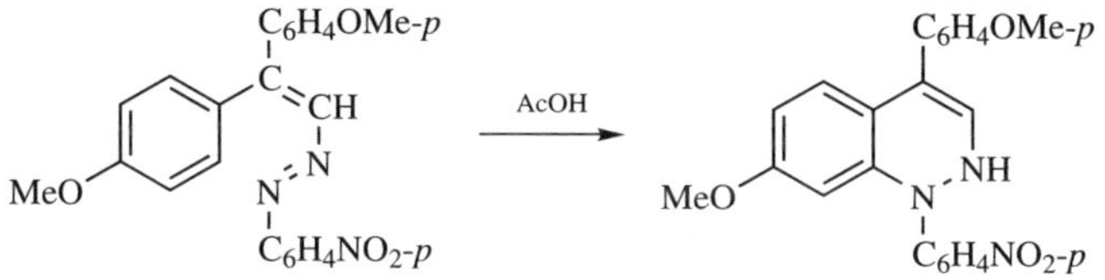

AcOH

(**11**) (**12**)

α-Diazo-α-ethoxycarbonyl-2,4,5-trifluoroacetophenone (**13**) gave ethyl 6,7-difluoro- 4-oxo-1,4-dihydro-3-cinnolinecarboxylate (**14**) (Bu_3P, dioxane, 20°C → reflux, 5 h: 60%); analogs likewise.[414]

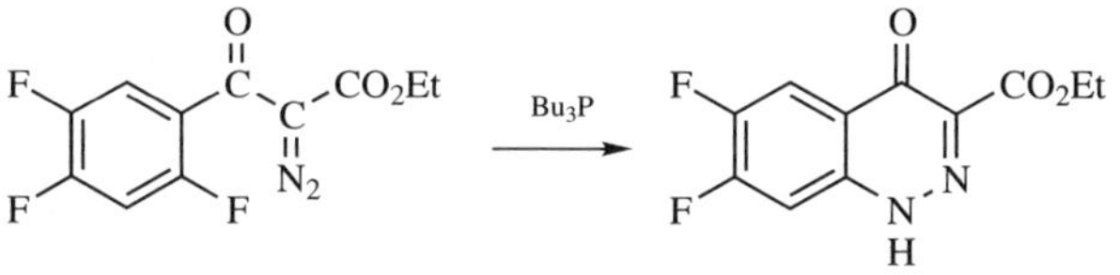

Bu_3P

(**13**) (**14**)

Also other examples.[324,325]

1.1.2. By Formation of the N1–N2 Bond

This unusual procedure is represented by the controlled electroreduction of appropriate dinitro alcohols followed by aerial oxidative cyclization under basic conditions; these processes are illustrated in the following examples.

1-(1-Hydroxy-2-nitropropyl)-2-nitrobenzene (**15**, R = Me) gave 3-methylcinnoline (**17**), via the unisolated intermediate (**16**, R = Me) ([H], pH 5, 0°C; then K_2CO_3↓, open to air, 12 h: 59%; for details, see original).[747]

In contrast, 1-(α-hydroxy-β-nitrophenethyl)-2-nitrobenzene (**15**, R = Ph) likewise gave a separable mixture of 3-phenylcinnoline (**18**) and its 1-oxide (**19**) (47% and 22%, respectively).[747]

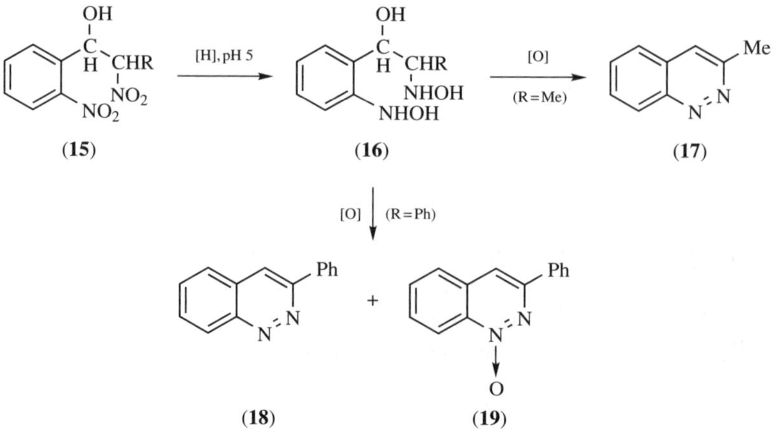

(**15**) (**16**) (**17**)

(**18**) (**19**)

1,4-Dihydrocinnoline (45%), cinnoline (65%), and 3,3-dimethyl-3,4-dihydrocinnoline (12%) have been made somewhat analogously;[64,341,983] also benzo[*c*]cinnoline and its 5-oxide.[1035]

1.1.3. By Formation of the N2–C3 Bond

Several types of substrate may be used for this cyclization, as illustrated by the following broadly classified examples.

Using Derivatives of *o*-Ethylphenylhydrazine as Substrates

3-Benzyloxy-*N′*,*N′*-di-*tert*-butoxycarbonyl-6-(2,2-dimethoxyethyl)-4-methoxyphenylhydrazine (**20**) gave either di-*tert*-butyl 7-benzyloxy-3,6-dimethoxy-1,2,3,4-tetrahydro-1,2-cinnolinedicarboxylate (**21**) (TsOH, MeOH, 20°C, 16 h: 72%) or di-*tert*-butyl 7-benzyloxy-6-methoxy-1,2-dihydro-1,2-cinnolinedicarboxylate (**22**) (TsOH, dioxane, 20°C, 16 h: 80%).[494]

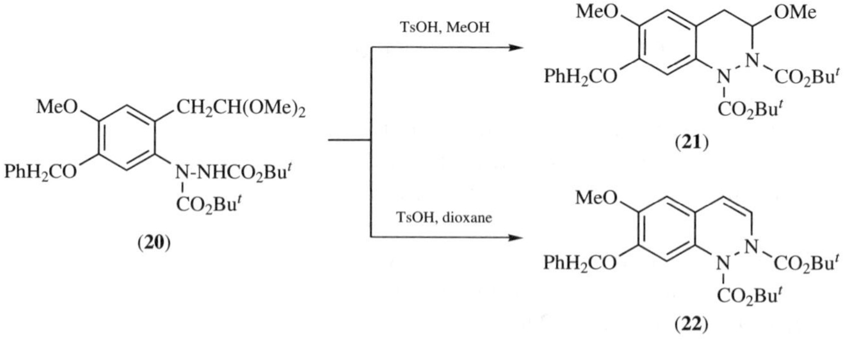

(**20**) (**21**) (**22**)

1-Ethyl-1-methoxycarbonylmethyl-2-(*o*-propylphenylhydrazono)cyclohexane (**23**) (prepared in situ) gave 4a-ethyl-2-(*o*-propylphenylhydrazono)-4,4a,5,6,7,8-hexahydro-3(2*H*)-cinnolinone (**24**) (20% H_2SO_4, reflux, 30 min: 26% after separation from another product); analogs likewise.[212]

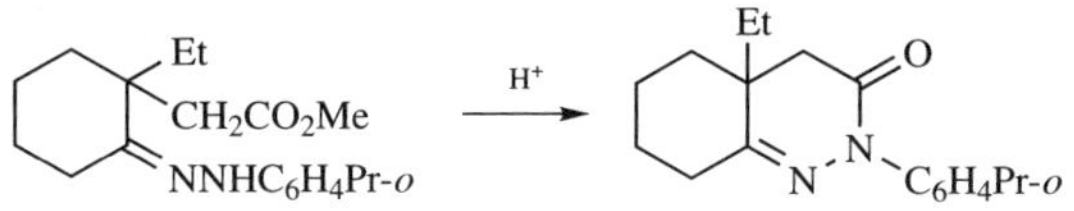

(**23**) (**24**)

3′-Oxo-5-phenylhydrazono-3′,5′,7′,8′,8a′-hexahydrospiro[cyclohexane-1,1′(2′*H*)-naphthalene]-2′,2′,4′-tricarbonitrile (**25**) gave 3-oxo-2-phenyl-2,3,5,6,7,8-hexahydro-4-cinnolinecarbonitrile (**26**), a reaction said to involve attack by NH at the carbonyl group and loss of cycloalkylidenemalononitrile as shown [$HN(CH_2)_5$, EtOH, 35°C, 1 h: 18%]; analogs likewise.[954]

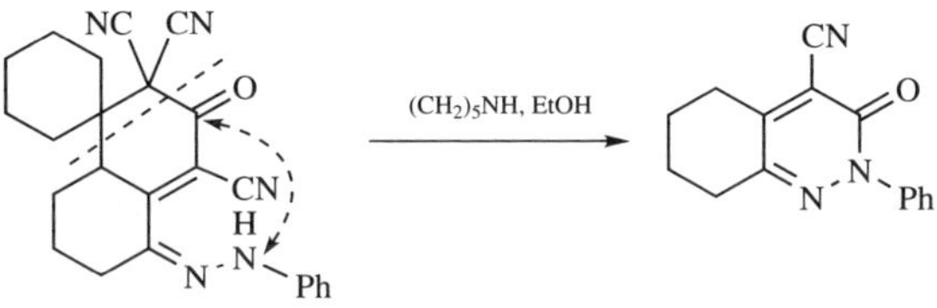

(**25**) (**26**)

Also other examples.[233,383]

Using Derivatives of 2-Ethylazo- or 2-Ethylazoxybenzene

2-Carboxymethyl-4′-methoxyazobenzene (**27**) gave 2-*p*-methoxyphenyl-3(2*H*)-cinnolinone (**28**) (ClOCCOCl, CH_2Cl_2, 20°C, 15 min: 95%).[83]

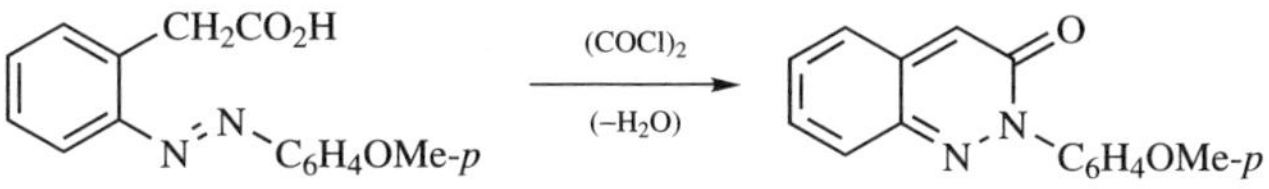

(**27**) (**28**)

2-Propionyl-*NNO*-azoxybenzene (**29**) gave 3-methoxy-3-methyl-2-phenyl-2,3-dihydro-4(1*H*)-cinnolinone (**31**), possibly via the intermediate (**30**) (MeONa, MeOH, 20°C, 4 h: 22% after separation from another product).[741]

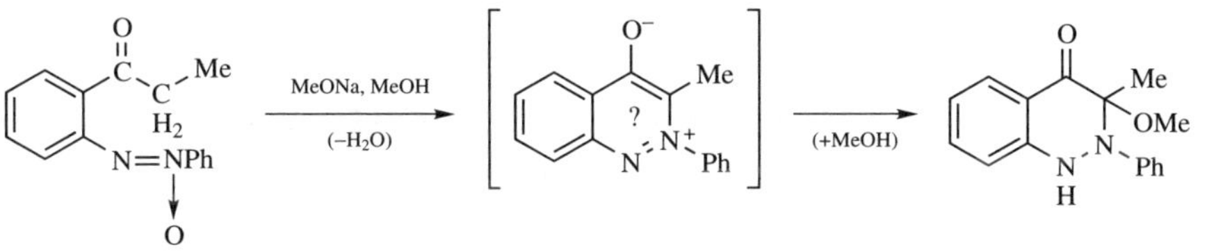

(**29**) (**30**) (**31**)

Also other examples.[83,722]

Using Derivatives of *o*-Ethylphenyltriazene as Substrates

4-Methoxy-2-(3,3-tetramethylenetriazeno)acetophenone (**32**) was converted into the sodium salt of ethyl 2-[4-methoxy-2-(3,3-tetramethylenetriaz-1-eno)benzoyl]acetate (**33**) [NaH, $OC(OEt)_2$, THF, reflux; substrate↓ during 8 h: crude] and thence into ethyl 7-methoxy-4-oxo-1,4-dihydro-3-cinnolinecarboxylate (**34**) with loss of pyrrolidine (neat F_3CCO_2H, 0°C, 12 h: 83% overall).[387]

(**32**) → $OC(OEt)_2$ → (**33**) → F_3CCO_2H → (**34**)

o-(3,3-Diethyltriaz-1-eno)phenylacetylene (**35**, R = H) gave cinnoline (**36**, R = H) ($C_6H_4Cl_2$-*o*, 200°C, sealed, 12 h: 99%; the stoichiometry is unclear).[817,819]

(**35**) → 200 °C → (**36**) + EtN=CHMe (?)

3-Chloro-6-(3,3-diethyltriaz-1-eno)phenylacetylene (**35**, R = Cl) likewise gave 6-chlorocinnoline (**36**, R = Cl) (96%)[816,817,819] and other 6-substituted analogs were made similarly.[817,819]

1.1.4. By Formation of the C3–C4 Bond

Although several procedures within this category have been reported, none has been developed to any extent. However, the following examples may point toward useful general syntheses.

o-(Nitromethylenehydrazino)benzaldehyde (**37**) gave 3-nitrocinnoline (**38**) (1,4-diazabicyclo[2.2.2]octane, H_2O, 60°C, 3 h: 86%).[370]

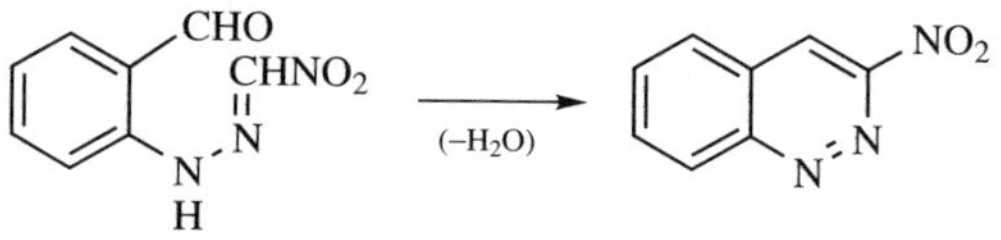

(**37**) (**38**)

3-Chloro-6-(methyl-*ONN*-azoxy)benzophenone (**39**) gave 6-chloro-4-phenylcinnoline 2-oxide (**40**) (KOH, H_2O, EtOH, reflux, 10 min: 72%).[11]

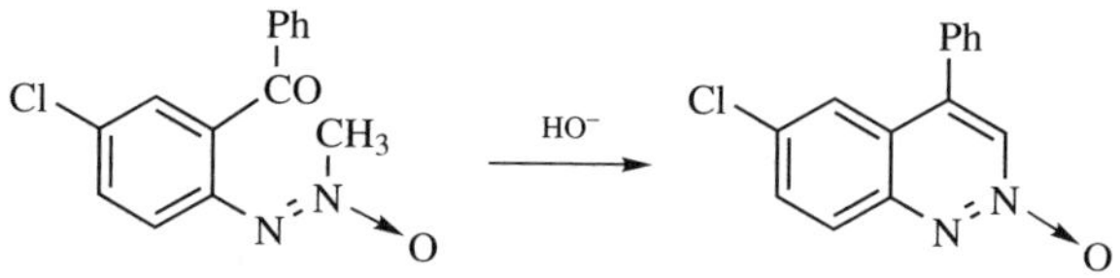

(**39**) (**40**)

N′-Benzylidene-*o*-trifluoromethylphenylhydrazine (**41**) gave 3-phenyl-4-cinnolinamine (**42**) [$NaN(SiMe_3)_2$ (4 equiv), THF (tetrahydrofuran), −78° → 20°C, 4 h: 68%); several substituted-phenyl analogs were made similarly, and a mechanism was proposed.[93]

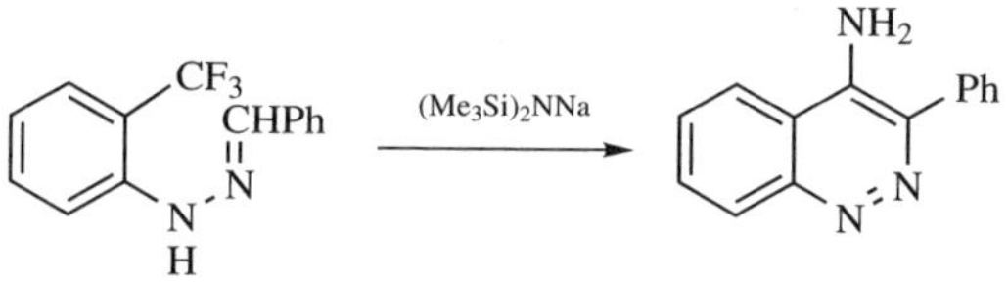

(**41**) (**42**)

1-Bromo-3-methoxalyl-4-(α-methoxycarbonyl-α-triphenylphosphoranylidenemethylazo)benzene (**43**) gave dimethyl 6-bromo-3,4-cinnolinedicarboxylate (**44**) (PhMe, reflux, 48 h: 46%);[43] several related processes have been reported, but all gave unsatisfactory yields.[26,43,515]

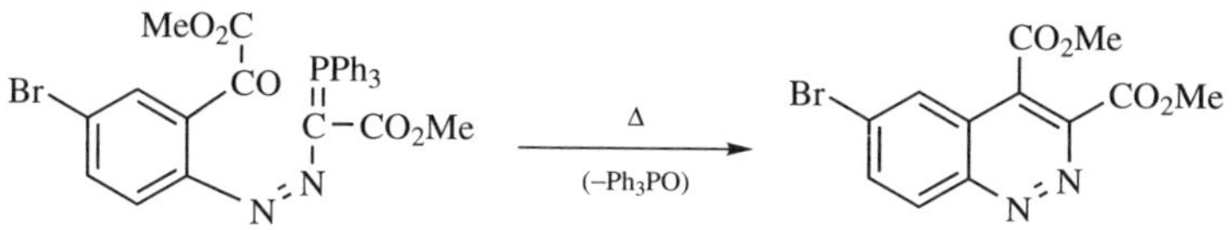

(**43**) (**44**)

1.1.5. By Formation of the C4–C4a Bond

This is a frequently used synthesis with wide applicability. The required substrates, such as diethyl α-phenylhydrazonomalonate, are easily made by coupling a benzenediazonium salt with an activated methylene synthon, and subsequent cyclization can be done in several ways. An interesting study on the regioselectivity of such cyclizations has been presented.[739] The following examples are classified according to the terminal groups that actually take part in the ring closure.

Using (Alkoxycarbonylmethylene)hydrazinobenzenes as Substrates

N′-(1-Ethoxycarbonylacetonylidene)hydrazinobenzene (**45**, R = H) gave 3-acetyl-4(1*H*)-cinnolinone (**46**, R = H) ($AlCl_3$, PhCl, 100°C, 1 h: 65%); several *p*-substituted analogs were made similarly.[619,620]

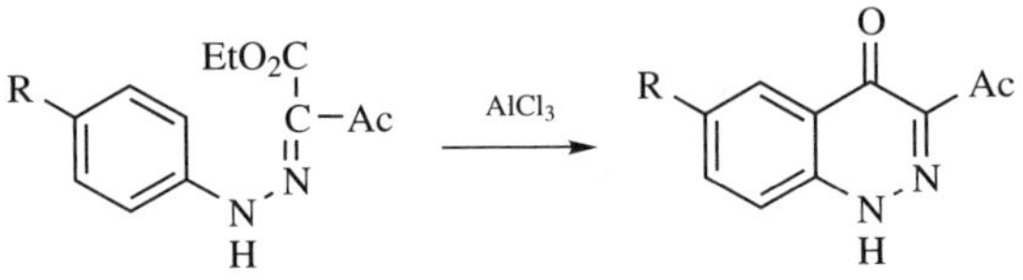

(**45**) (**46**)

1-[*N*′-(Diethoxtcarbonylmethylene)hydrazino]-4-(*p*-nitrophenylthio)benzene(**47**, R = $SC_6H_4NO_2$-*p*) gave ethyl 4-oxo-1,4-dihydro-3-cinnolinecarboxylate (**48**, R = $SC_6H_4NO_2$-*p*) ($AlCl_3$, PhCl, reflux, 1 h: 69%);[795] several 6-substituted analogs were made similarly.[663,795]

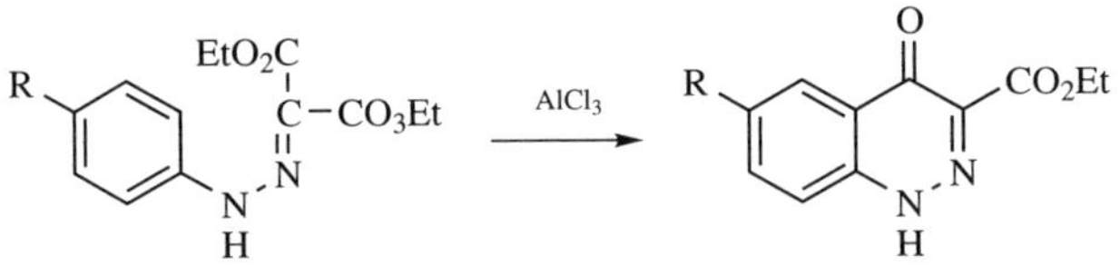

(**47**) (**48**)

o-(α-Cyano-α-ethoxycarbonylmethylene)toluene (**49**) gave 8-methyl-4-oxo-1,4-dihydro-3-cinnolinecarbonitrile (**50**) ($AlCl_3$, PhCl, reflux, 1 h: 71%); analogs likewise.[505]

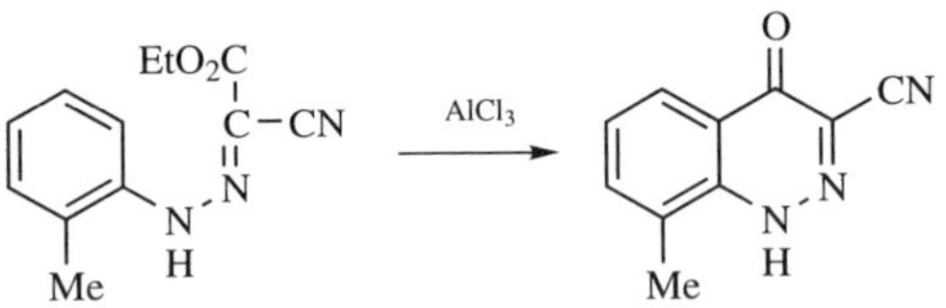

(**49**) (**50**)

Note: The evident preference in the foregoing examples for ring closure to involve an ester rather than an acyl or cyano group is important.

Using (Carboxymethylene)hydrazinobenzenes as Substrates

1-[(Carboxymethylene)hydrazino]-4-nitrobenzene (**51**) gave 6-nitro-4(1*H*)-cinnolinone (**52**) (P_2O_5, H_3PO_4, 65° → 135°C, 90 min: 56%); the 8-nitro isomer (47%) was made similarly.[497]

P_2O_5, H_3PO_4

(**51**) (**52**)

Using (Chloroformylmethylene)hydrazinobenzenes as Substrates

1-[Di(chloroformyl)methylene]hydrazino-4-ethylbenzene (**53**) gave 6-ethyl-4-oxo-1,4-dihydro-3-cinnolinecarboxylic acid (**54**) ($TiCl_4$, PhCl, 100°C, 6 h; then 20°C, 10 h: 81%; note hydrolysis of the chloroformyl group during workup).[782]

$TiCl_4$

(**53**) (**54**)

4-[Di(chloroformyl)methylene]hydrazino-3-nitroanisole (**55**) gave 6-methoxy-8-nitro-4-oxo-1,4-dihydro-3-cinnolinecarboxylic acid (**56**) ($TiCl_4$, $PhNO_2$, 100°C, 24 h: 51%).[27]

$TiCl_4$

(**55**) (**56**)

Also other examples.[1034]

Using (Cyanomethylene)hydrazinobenzenes as Substrates

(α-Cyanobenzylidene)hydrazinobenzene (**57**) gave 3-phenyl-4-cinnolinamine (**58**) ($AlCl_3$, PhMe, reflux, 1 h: 95%); several analogs bearing substituents on the carbocyclic ring were made similarly.[940]

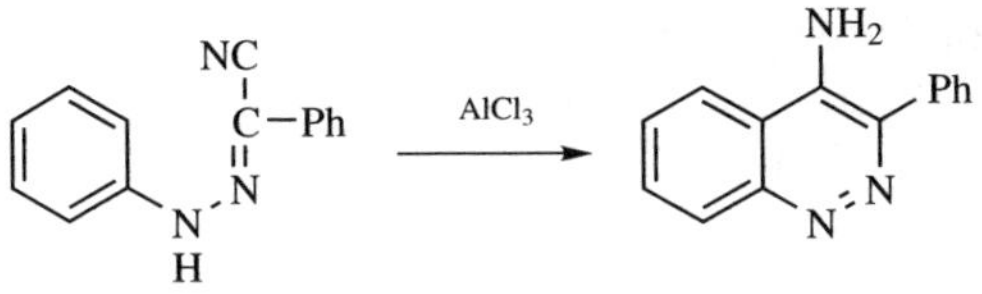

(57) (58)

(Dicyanomethylene)hydrazinobenzene (**59**) gave 3-(*p*-methylbenzoyl)-4-cinnolinamine (**62**), arising from a Friedel–Crafts reaction of the expected product (**60**) with solvent via intermediate (**61**) ($AlCl_3$, PhMe, reflux, 3 h: 32%);[487] many substituted-phenyl analogs of the substrate (**59**) likewise gave only appropriate derivatives of the product (**62**).[815,830] However, similar treatment of 1-[(dicyanomethylene)hydrazino]-2,4-dimethylbenzene did give a separable mixture of 4-amino-6,8-dimethyl-3-cinnolinecarbonitrile (**63**) and 6,8-dimethyl-3-(*p*-methylbenzoyl)-4-cinnolinamine (33% and 47%, respectively).[830,cf. 677,1011]

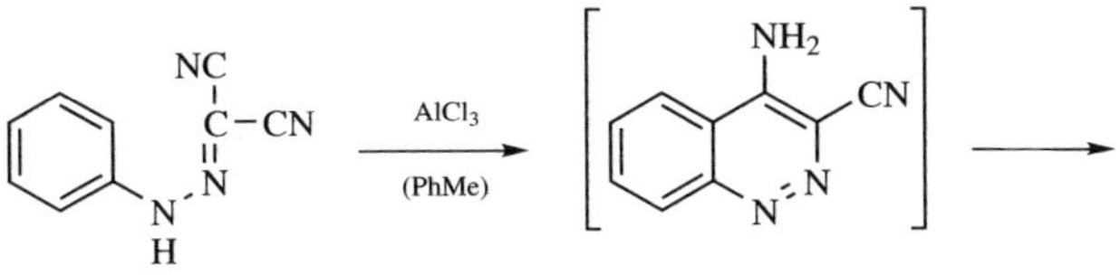

(59) (60)

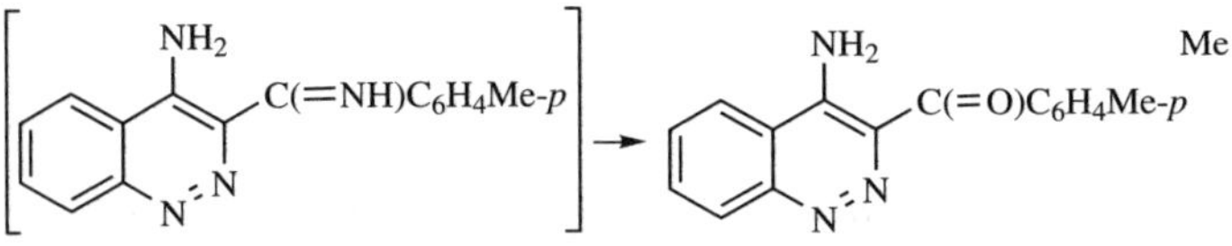

(61) (62) 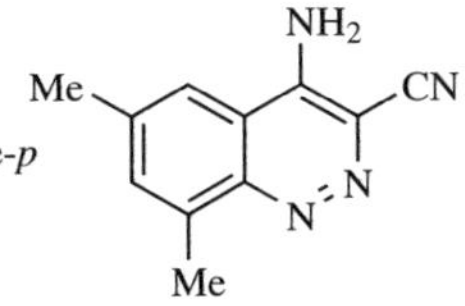 (63)

N′-(α-Carbamoyl-α-cyanomethylene)hydrazinobenzene (**64**) gave only 4-amino-3-cinnolinecarboxamide (**65**) ($AlCl_3$, PhCl, reflux, 1 h: 50%);[487] many analogs were made similarly.[487,502,503,556,667,670]

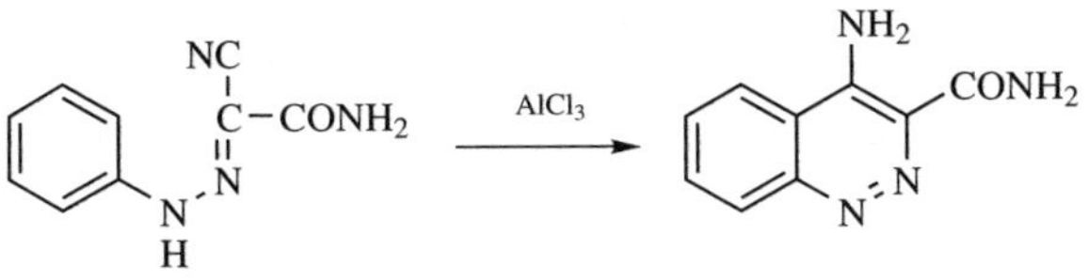

(64) (65)

Using (Acylmethylene)hydrazinobenzenes as Substrates

Note: As might be expected, such substrates with a terminal aldehydo group (formyl) appear to undergo cyclization more readily than do those with a terminal ketonic group (e.g., acetyl or benzoyl).

N′-(α-Benzoyl-α-formylmethylene)hydrazinobenzene (**66**, R = H) gave 3-benzoylcinnoline (**67**, R = H) (96% H_2SO_4, 100°C, 4 min: 60%); *N′*-[α-formyl-α-(*p*-methoxybenzoylmethylene]hydrazinobenzene (**66**, R = OMe) gave 3-*p*-methoxybenzoylcinnoline (**67**, R = OMe) (P_2O_5, H_3PO_4, 110°C, 9 min: 55%); several analogous products were made by each procedure.[824] The kinetics of such cyclizations have been studied.[818]

OHC C−C(=O)C_6H_4R-*p* N N H —H_2SO_4 (R=H); P_2O_5, H_3PO_4; (R = Me) (−H_2O)→ C(=O)C_6H_4R-*p* N N

(**66**) (**67**)

N′-(α-Formyl-α-phenylthiomethylene)hydrazinobenzene (**68**) gave 3-phenylthiocinnoline (**69**) [P_2O_5, H_3PO_4, 80°C (exothermic), 10 min: 18%]; analogs likewise.[518]

OHC C−SPh N N H —P_2O_5, H_3PO_4→ SPh N N

(**68**) (**69**)

1-[α-Acetyl-α-(phenylhydrazono)methyl]benzotriazole (**70**) gave 3-(benzotriazol-1-yl)-4-methylcinnoline (**71**) [HN(CH_2)$_5$, xylene, reflux, 1 h: 80%].[835]

Me O=C N N C−N N N H —HN(CH_2)$_5$→ Me N N N N N

(**70**) (**71**)

Also other examples.[637]

Using Miscellaneous Substrates

N-[2-Ethoxycarbonyl-2-(*m*-methoxyphenylhydrazono)ethylidene]pyrrolidinium tetrafluoroborate (**72**) gave ethyl 7-methoxy-3-cinnolinecarboxylate (**73**) (MeCN, reflux, 60 h: 63%).[292,cf. 570]

BF_4^- $\overset{+}{N}$ CH C−CO_2Et N N H MeO —Δ [−(CH_2)$_4$NHBF_4]→ CO_2Et N N MeO

(**72**) (**73**)

N′-(α-Trifluoromethylbenzylidene)hydrazinobenzene (**74**, R = H, X = F) and potassium bis(trimethylsilyl)amide gave 3-phenyl-4-cinnolinamine (**75**, Q = NH_2) (THF, −78°C, 4 h: ~70%); analogs likewise using the same or different amides.[85]

(**74**) → $(Me_3Si)_2NK$ or $(NC)_2C{=}C(CN)_2$ → (**75**)

The related substrate, [1-(*N*′-methyl-*N*′-phenylhydrazono)ethyl]benzene (**74**, R = Me, X = H), with tetracyanoethylene, gave 3-phenyl-4-cinnolinecarbonitrile (**75**, Q = CN) (MeCN, reflux, A, 8 h: 60%); analogus were made similarly, but the mechanism and categorization remain uncertain).[91]

Methyl 4-[(2-ethoxycarbonylethylidene) hydrazino]-3-(methanesulfonyloxy)-benzoate (**76**) gave methyl 3-ethoxycarbonyl-1,2-dihydro-6-cinnolinecarboxylate (**77**) (P_2O_5, H_3PO_4, 80°C, 75 min: 22% after separation from another product).[827]

(**76**) → P_2O_5, H_3PO_4 (−MsOH) → (**77**)

1.2. FROM A CARBOCYLIC SUBSTRATE AND ONE SYNTHON (*H* 6, 17, 47; *E* 20, 63)

Of the 10 possible subcategories within this major category, no less than 5 (in which the synthon would supply N1, C3, C4, N1 + N2 + C3, or N2 + C3 + C4) appear to be unrepresented in the 1972–2004 literature. However, the remaining subcategories are of considerable importance, as indicated in the following subsections.

1.2.1. When the Synthon Supplies N2 of the Cinnoline

This synthesis always involves diazotization of an *o*-ethylaniline derivative followed by spontaneous cyclization.[265] The following examples, classified according to the type of substrate, illustrate the procedures employed. Fused cinnolines have been made likewise.[831,832,834]

Using *o*-Aminoacetophenones as Substrates

Note: Such substrates naturally produce 4(1*H*)-cinnolinone or its derivatives.

o-Aminoacetophenone (**78**, R = H) gave 4(1*H*)-cinnolinone (**80**, R = H) by spontaneous cyclization of the intermediate diazonium salt (**79**, R = H) (HCl, H_2O, $NaNO_2$↓ slowly, < 5°C, 45 min; then 0°C, 1 h; then 80°C, 48 h: 35%);[509] 2-amino-4-(pyridin-4-yl)acetophenone (**78**, R = pyridin-4-yl) gave 7-(pyridin-4-yl)-4(1*H*)-cinnolinone (**80**, R = pyridin-4-yl) (HCl, $NaNO_2$↓ slowly, < 2°C, 45 min; then 0°C, 2 h; then 20°C, 12 h: 77%);[45] analogs likewise.[828]

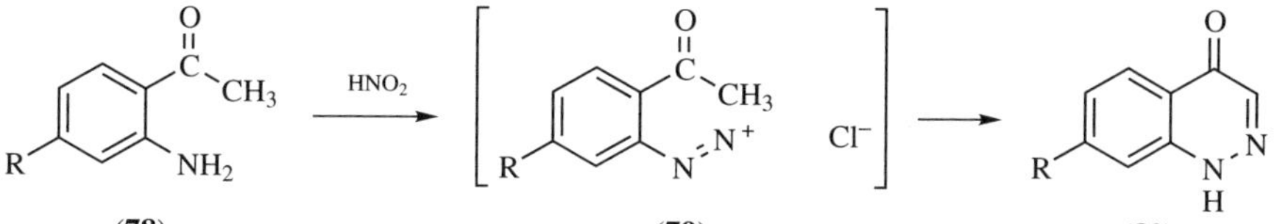

(**78**) (**79**) (**80**)

o-(5-Methoxyvaleryl)aniline gave 3-(2-methoxypropyl)-4(1*H*)-cinnolinone (**81**) (HCl, H_2O, $NaNO_2$↓ slowly, < 5°C, 40 min; then 20°C, 3 days: ~40%).[693]

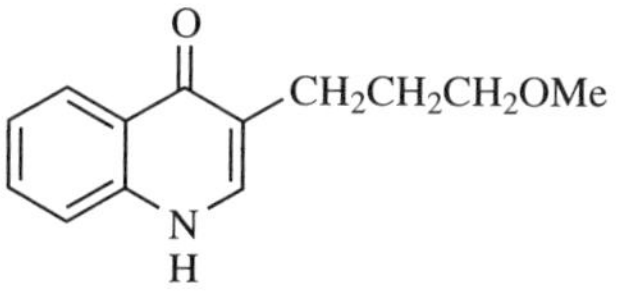

(**81**)

2-(Cyclohexylacetyl)-4,5-dimethoxyaniline gave 3-cyclohexyl-6,7-dimethoxy-4(1*H*)-cinnolinone (**82**) (HCl, H_2O, $NaNO_2$↓ slowly, −5°C, 30 min; then 0°C, 1 h; then 70°C, 4 h; ~30%);[20] *o*-(pyridin-2-ylacetyl)aniline *N*-oxide gave 3-(*N*-oxidopyridin-2-yl)-4(1*H*)-cinnolinone (**83**) (HCl, H_2O, $NaNO_2$↓ slowly, 5°C; then 20°C, 20 min: 63%).[13]

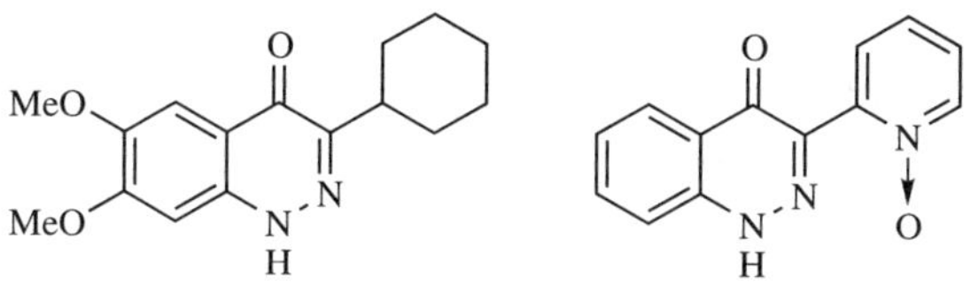

(**82**) (**83**)

o-Sulfoacetyl)aniline, as the crude sodium salt (**84**, R = ONa), gave 4-oxo-1,4-dihydro-3-cinnolinesulfonic acid as its sodium salt (**85**, R = ONa) (HCl, H_2O, $NaNO_2$↓ slowly, −5°C; then 20°C, 12 h: 64%); 4-oxo-1,4-dihydro-3-cinnolinesulfonanilide (**85**, R = NHPh) (19%), 3-phenylsulfonyl-4(1*H*)-cinnolinone (**85**, R = Ph) (62%), and other such analogs were made essentially by the same process.[485]

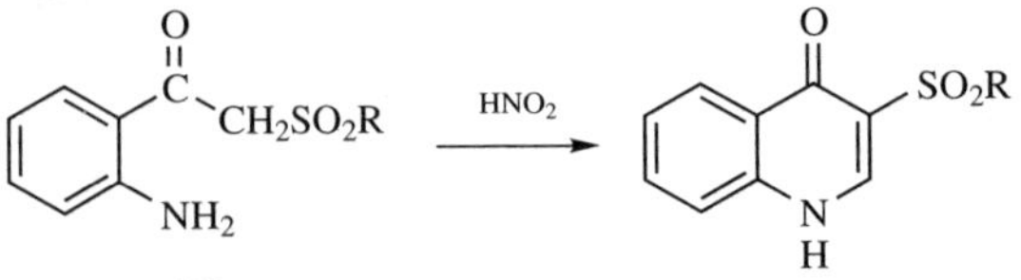

(**84**) (**85**)

o-[(Triphenylphosphoranylidene)acetyl]aniline (**86**) gave 3-triphenylphosphoranylidene-3,4-dihydro-4-cinnolinone (**87**) ($C_5H_{11}ONO$, HCl, EtOH, 0°C → 20°C, ~1 h: 91%) and thence 4(1*H*)-cinnolinone (**88**) (NaOH, MeOH, reflux, 2 h: 97%); several analogs likewise.[335]

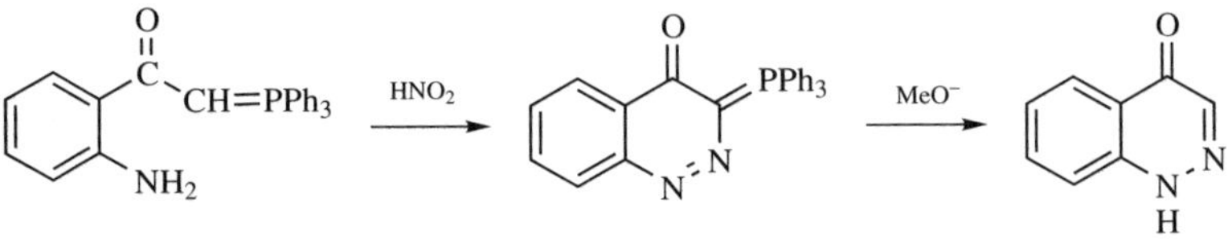

(**86**) (**87**) (**88**)

Using *o*-Aminostyrenes as Substrates

o-Amino-α-phenylstyrene (**89**) gave 4-phenylcinnoline (**90**) (MePrCHONO, Ac_2O, PhH, 80°C, 20 h: 53%).[486]

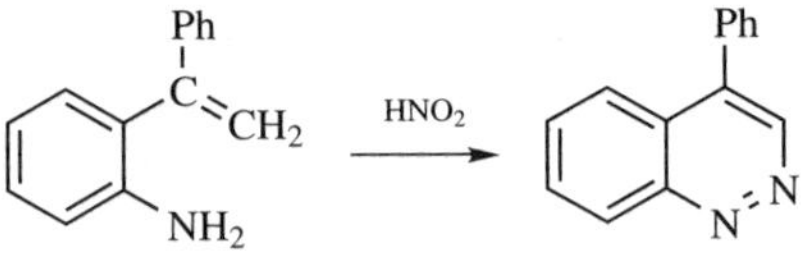

(**89**) (**90**)

2-Amino-5-chloro-β-methyl-α-phenylstyrene (**91**) gave 6-chloro-3-methyl-4-phenylcinnoline (**92**) (HCl, H_2O, $NaNO_2$↓ slowly, 0°C; then 4°C, 64 h: > 47%).[290]

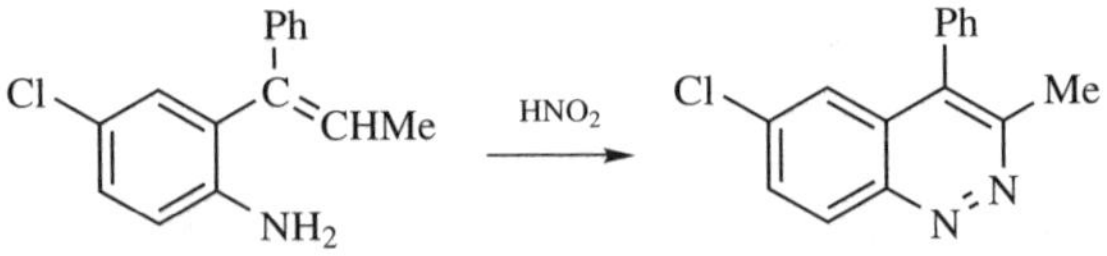

(**91**) (**92**)

Also other examples.[272,637,711]

Using *o*-Aminophenylacetylenes as Substrates

1-(*o*-Aminophenyl)-2-phenylacetylene (**93**) gave 4-bromo-3-phenylcinnoline (**94**, X = Br) (47% HBr, $NaNO_2\downarrow$ slowly, −15°C, 10 min; then 28°C, ~15 min: 86%; note the addition of HBr to the triple bond prior to cyclization); the same reaction using 36% HCl afforded 4-chloro-3-phenylcinnoline (**94**, X = Cl) (41%); and several bromo and chloro analogs were made similarly.[388,492]

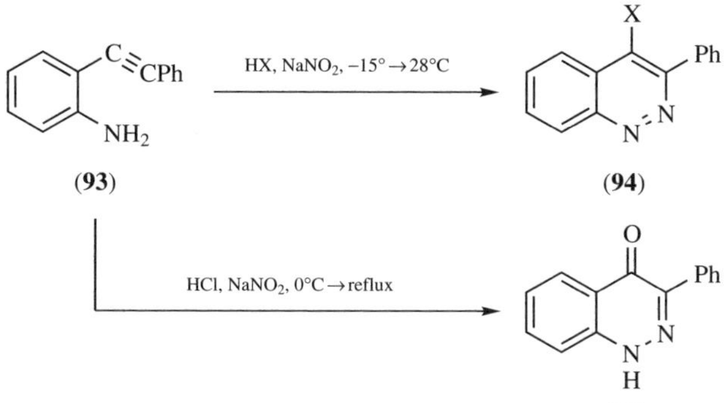

HX, $NaNO_2$, −15° → 28°C

(**93**) (**94**)

HCl, $NaNO_2$, 0°C → reflux

(**95**)

In contrast, the same substrate (**93**) under less gentle conditions gave only 3-phenyl-4(1*H*)-cinnolinone (**95**), presumably via the initial product (**94**, X = Cl) (36% HCl, $NaNO_2\downarrow$ slowly, 0°C, 2 h; then reflux, 1 h: 82%).[585]

1-(*o*-Aminophenyl)-2-trimethylsilylacatylene (**96**) gave 4(1*H*)-cinnolinone (**97**) (6M HCl, $NaNO_2\downarrow$ slowly, < 0°C, 30 min; then reflux, 3 h: 73%).[822]

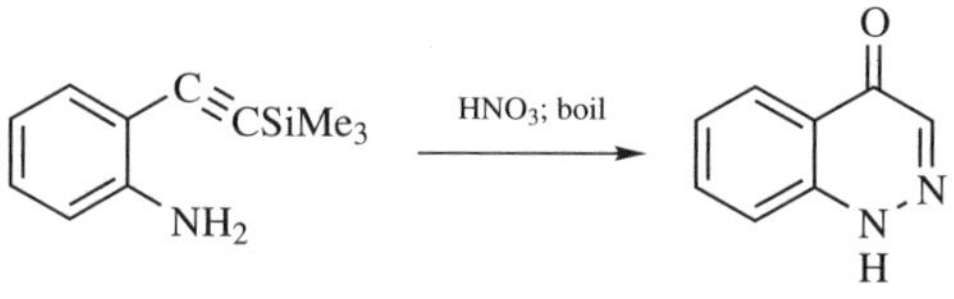

HNO_3; boil

(**96**) (**97**)

1.2.2. When the Synthon Supplies N1 + N2 of the Cinnoline

This type of synthesis has proved particularly useful for the preparation of partially nucleus-reduced cinnolines, although regular aromatic cinnolines have also been so made. The N–N fragment is easily supplied by a hydrazino, diazo, or azo synthon. The use of such synthons with convenient substrates is illustrated in the following examples. Fused cinnolines have been made similarly.[833]

2,3,5-Trimethyl-6-phenacyl-1,4-benzoquinone (**98**) gave 5,7,8-trimethyl-3-phenyl-6(2*H*)-cinnolinone (**99**) ($H_2NNH_2\cdot H_2O$, trace AcOH, PhMe, 20°C,

18 h: ~25%).[706]

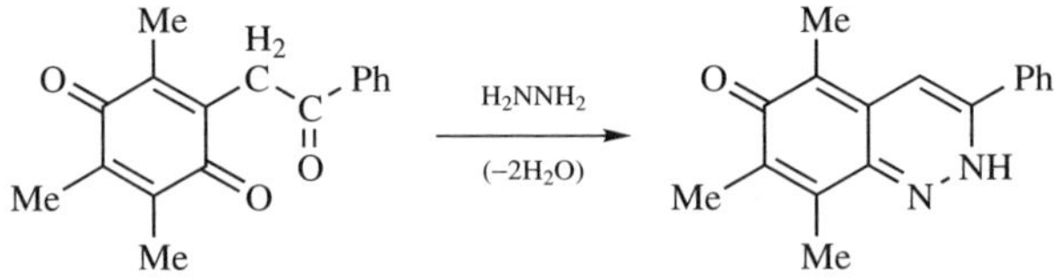

(98) (99)

2-(Ethoxycarbonylmethyl)cyclohexane (**100**) gave 4,4a,5,6,7,8-hexahydro-3(2*H*)-cinnolinone (**101**) ($H_2NNH_2 \cdot H_2O$, EtOH, reflux, 1 h: 72%) and thence 5,6,7,8-tetrahydro-3(2*H*)-cinnolinone (**102**) ($CuCl_2$, MeCN, reflux, 1 h: 87%);[600] minor variations in the foregoing reaction produced lower yields.[925]

In contrast, 2-(α-carboxy-α-morpholinomethyl)cyclohexanone, as its morpholinium salt (**103**), gave 5,6,7,8-tetrahydro-3(2*H*)-cinnolinone (**102**) directly; oxidation was provided by spontaneous loss of morpholine ($H_2NNH_2 \cdot H_2O$, EtOH, reflux, 4 h: 75%).[952]

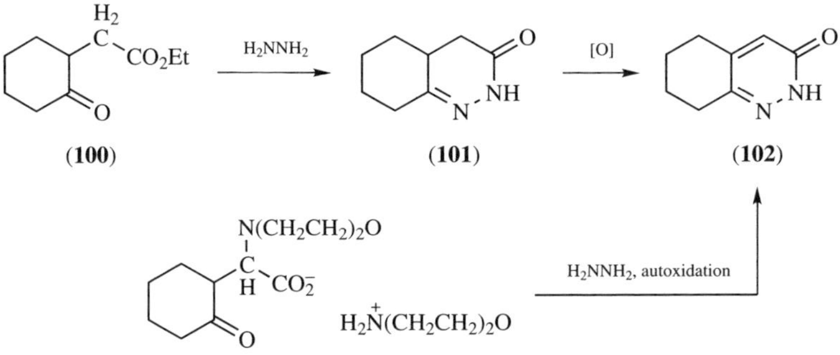

(100) (101) (102)

(103)

5,5-Dimethyl-2-phenacyl-1,3-cyclohexanedione (**104**) (formulated as the corresponding enol) gave 7,7-dimethyl-1,3-diphenyl-4,6,7,8-tetrahydro-5(1*H*)-cinnolinone (**105**) (H_2NNHPh, EtOH, reflux, 20 h: ~60%);[656] analogs likewise.[139,186,656]

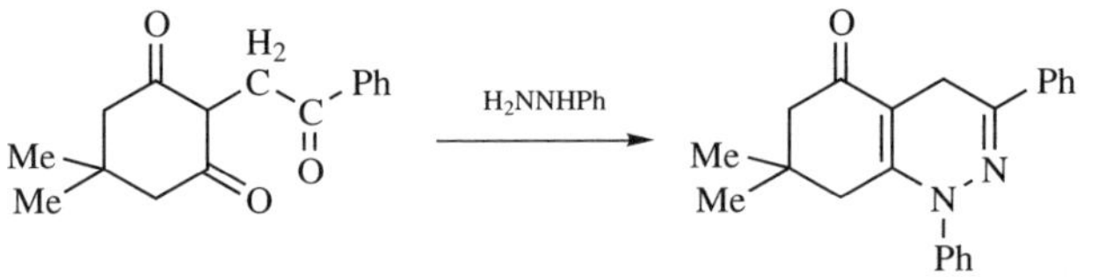

(104) (105)

2-Phenacyl-1,3-cyclohexanedione (**106**) gave 3-phenyl-4,6,7,8-tetrahydro-5(1*H*)-cinnolinone (**107**; unisolated) and thence 3-phenyl-5,6,7,8-tetrahydro-5-cinnolinone (**108**) by oxidation (H_2NNH_2, EtOH, 20°C, 30 min; then dichlorodicyanobenzoquinone↓, reflux, 30 min: 78%).[284]

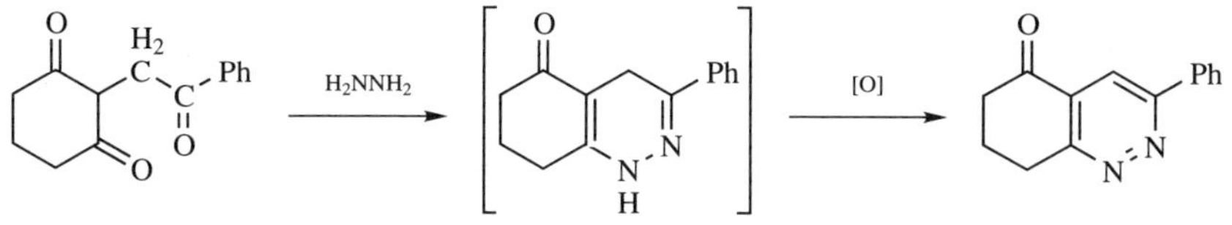

(**106**) (**107**) (**108**)

2-Carboxymethyl-5-phenylcyclohex-5-enone (**109**) gave 7-phenyl-4,4a,5,6-tetrahydro-3(2*H*)-cinnolinone (**110**, R = H) (neat $H_2NNH_2 \cdot H_2O$, reflux, 4 h: 92%) or 2-methyl-7-phenyl-4,4a,5,6-tetrahydro-3(2*H*)-cinnolinone (**110**, R = Me) (H_2NNHMe, EtOH, reflux, 16 h: 65%); analogs likewise.[517]

(**109**) (**110**)

2-(*m*-Benzyloxyphenyl)-2-(carboxymethyl)cyclohexanone (**111**) gave 4a-(*m*-benzyloxyphenyl)-2-cyclopropylmethyl-4,4a,5,6,7,8-hexahydro-3(2*H*)-cinnolinone (**112**) [$H_2NNHCH_2(CH_2)_2$, PhH, reflux, 6 h: 70%]; analogs likewise.[185,572,962]

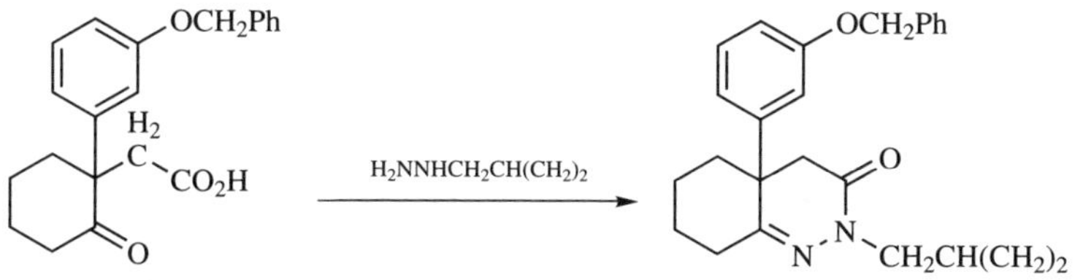

(**111**) (**112**)

2,5,9-Decanetrione (**113**, R = Me) gave 3,5-dimethyl-5,6,7,8-tetrahydrocinnoline (**114**, R = Me) ($H_2NNH_2 \cdot H_2O$, AcOH, 60°C, 1 h: 89%); the 3-ethyl, 3-propyl, and other such homologs (**114**, R = Et, Pr, etc.) were made similarly.[129]

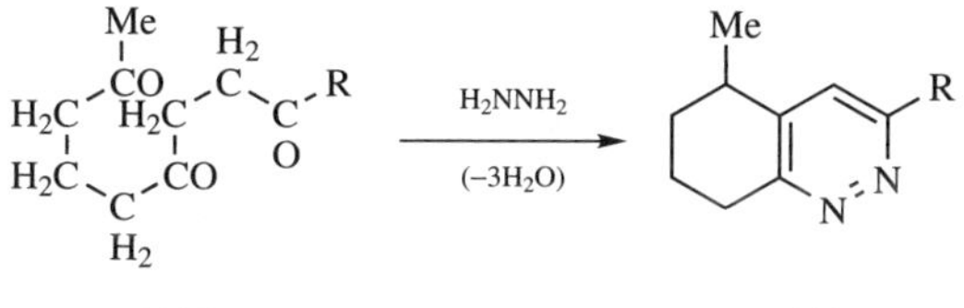

(**113**) (**114**)

A kinetic study of the reaction of styrene (**115**) with diethyl azodicarboxylate (2 mol) to give diethyl 4-(*N,N'*-diethoxycarbonylhydrazino)-1,2,3,4-tetrahydro-1,2-cinnolinedicarboxylate (**116**) showed the reaction to be first-order with respect to each reactant and to be suppressed strongly in the presence of a radical inhibitor.[746]

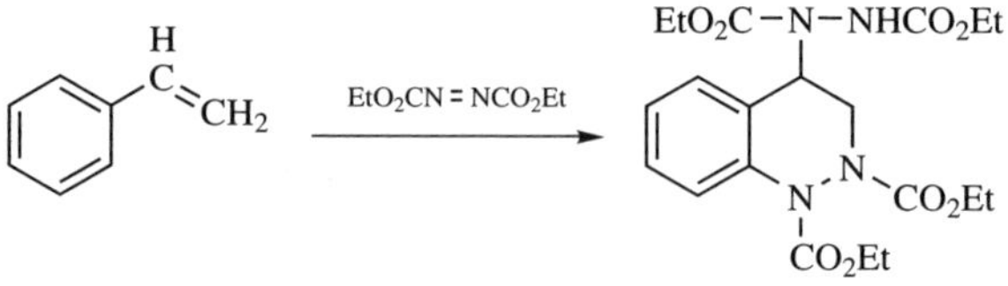

(**115**) (**116**)

Also other examples.[40,65,361,973,978]

1.2.3. When the Synthon Supplies N2 + C3 of the Cinnoline

This type of synthesis is rarely used but is illustrated in the following examples.

4-(1,3-Dioxan-2-yl)-2-nitro-α-(*p*-tolylsulfonyl)toluene (**117**) and 2-nitropropane (**118**) gave 2,3-dimethyl-7-(1,3-dioxan-2-yl)-3,4-dihydrocinnoline 1/3-oxide (**119**) [NaOH, H_2O, reflux, > 2 days [until substrate invisible on thin-layer chromatography (tlc)]: 40%; mechanism obscure].[529]

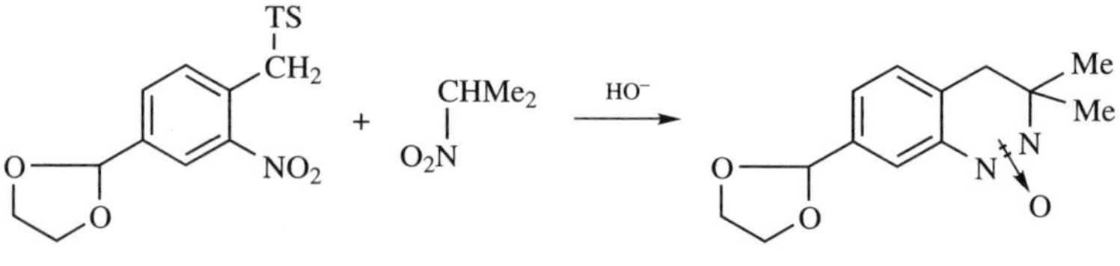

(**117**) (**118**) (**119**)

o-Chloro-*N*-(*o*-nitrobenzylidene)aniline (**120**) gave 4-(*o*-chloroanilino)-3-methoxycinnoline 1-oxide (**121**) (KCN, MeOH, reflux, 3 h: 55%).[709] Different substitution patterns in the substrate (**120**) led to many analogous products but all in lower yields;[72,194,709] a detailed mechanism has been proposed.[709]

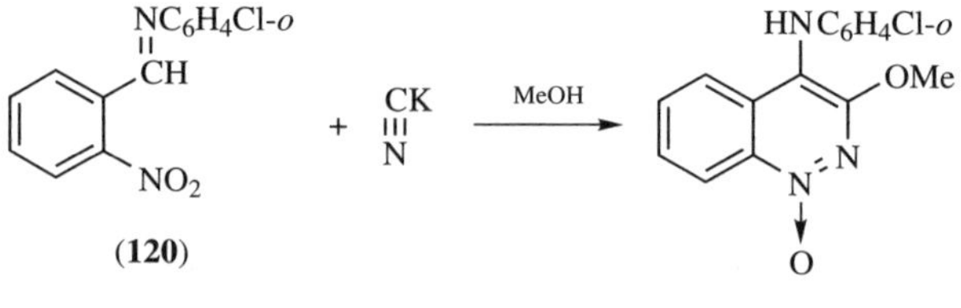

(**120**) (**121**)

1.2.4. When the Synthon Supplies C3 + C4 of the Cinnoline

A variety of substrate and synthon types may be used for this synthesis, resulting in its widespread use. The oxidation levels of the products depend on those of both reactants. For pragmatic reasons, the examples that follow are classified broadly according to the degree of unsaturation in the synthon that supplies C3 + C4 of the cinnoline produced.

Using Ethane Derivatives as Synthons

Note: Not surprisingly, simple ethane derivatives are inactive as synthons, but activated derivatives of acetaldehyde, acetic acid, or acetonitrile are ideal for this purpose.

3,5-Di-*tert*-butyl-1,2-benzoquinone mono(phenylhydrazone) (**122**) and 2-(triphenylphosphoranylidene)acetaldehyde gave a dihydrocinnoline formulated as 5,7-di-*tert*-butyl-2-phenyl-1,2-dihydro-3-cinnolinol (**123**) (EtOH, Et_3N, reflux, 2 days: 48%).[390]

$Ph_3P{=}CHCHO$

(**122**) (**123**)

5,5-Dimethyl-2-phenylhydrazono-1,3-cyclohexanedione (**124**) and methyl 2-(triphenylphosphoranylidene)acetate gave a product formulated as 6,6-dimethyl-2-phenyl-5,6,7,8-tetrahydro-3,8(2*H*)-cinnolinedione (**125**) (PhMe, reflux, 15 h: 70%); analogs likewise.[383]

$Ph_3P{=}CHCO_2Me$

(**124**) (**125**)

p-Nitrophenylhydrazine (**126**) and phenacyltriphenylarsonium bromide gave 6-nitro-3-phenyl-1,2-dihydrocinnoline (**127**) (neat $PhNMe_2$, reflux, 4 h: 56%); several analogs were made similarly.[651]

$Ph_3\overset{+}{As}CH_2C({=}O)Ph \quad Br^-$

(**126**) (**127**)

2-Phenylhydrazino-1,3-cyclohexanedione (**128**) and malononitrile (2 mol) gave 3-amino-8-dicyanomethylene-2,8-dihydro-4-cinnolinecarbonitrile (**129**) [HN $(CH_2)_5$, EtOH, 100°C, 40 min: 51%: aerial (?) oxidation]; analogs likewise.[533]

$2 \times H_2C(CN)_2$, [O]

(**128**) (**129**)

Using Ethylene Derivatives as Synthons

m-Methoxybenzenediazonium tetrafluoroborate (**130**) and ethyl 3-morpholino-isocrotonate (**131**) gave ethyl 7-methoxy-4-methyl-3-cinnolinecarboxylate (**132**) (MeCN, 20°C, 1 h; then reflux 24 h: 57%).[292]

(**130**) (**131**) (**132**)

A solution of the substrate, triphenyldiazenium perchlorate (**134**), was prepared by electrochemical oxidation of triphenylhydrazine (**133**) in acetonitrile containing lithium perchlorate;[60,736] this solution and an excess of methoxy-ethylene (methyl vinyl ether) gave 4-methoxy-1,2-diphenyl-1,2,3,4-tetrahydrocinnoline (**135**) (MeCN, 20°C, 8 h: 96%).[736] Analogous products were made similarly,[60,61,69,71,442,443,730,736,786] and some were oxidized to the dihydro analogs; for example, 1-methyl-4-phenyl-1,2,3,4-tetrahydrocinnoline (**136**) gave 1-methyl-4-phenyl-1,4-dihydrocinnoline (**137**) (Et_2O, $O_2\downarrow$, 24 h: 90%).[730]

Ph_2NNHPh —[O]→ [(**134**)] —$MeOCH=CH_2$→ (**135**)

(**133**) (**134**) (**135**)

O_2

(**136**) (**137**)

Using Acetylene Derivatives as Synthons

Azobenzene (**138**) and diphenylacetylene (**139**) (2 mol) gave 8-(1,2-diphenylvinyl)-2,3,4-triphenyl-2,3-dihydrocinnoline (**140**) [reactants mixed together at 85°C; $Co(N_2)(PPh_3)_3\downarrow$ portionwise (gas↑); then 85°C, 2 h: 70%].[722,749] Such reactions have been explored in some detail.[722,749–751]

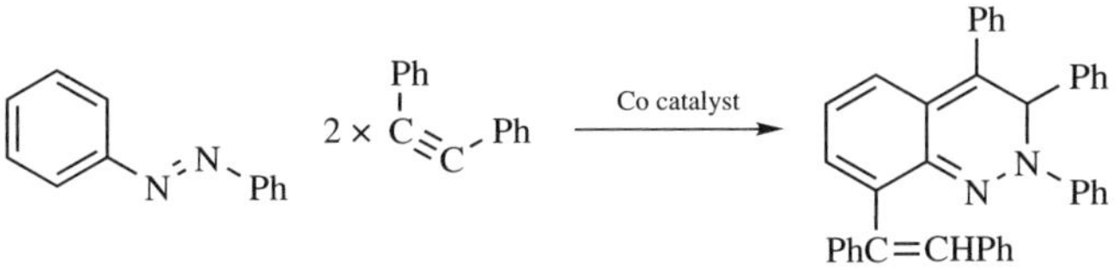

Co catalyst

(**138**) (**139**) (**140**)

4,4′-Dimethylazobenzene, as its Pd complex (**141**), and diethylacetylene gave 3,4-diethyl-6-methyl-2-*p*-tolylcinnolin-2-ium tetrafluoroborate (**142**) ($AgBF_4$, $MeNO_2$, N_2, 20°C, 4 h: 81%); analogs, such as 3,4-dimethoxycarbonyl-2-phenylcinnolin-2-ium tetrafluoroborate, were made in a broadly similar way.[768,769]

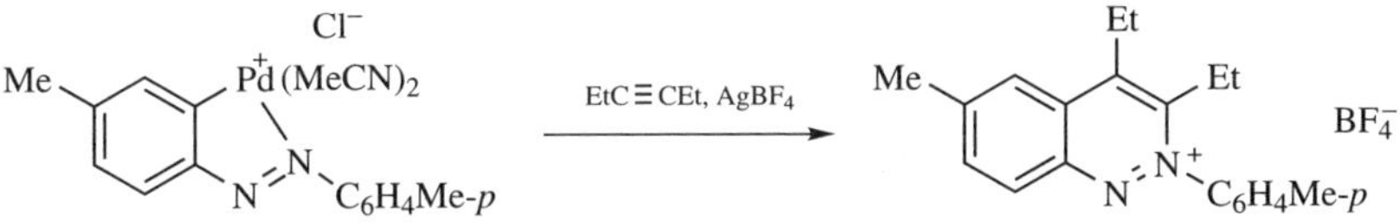

$EtC{\equiv}CEt$, $AgBF_4$

(**141**) (**142**)

3-(*N*,*N*′-Dimethylhydrazino)cyclohex-2-en-1-one (**143**) and methyl propiolate gave 1,2-dimethyl-4-methylene-1,4,5,6,7,8-hexahydro-3,5(2*H*)-cinnolinedione (**144**) (PhMe, reflux, 6 h: 8% after chromatographic separation from two other products).[127]

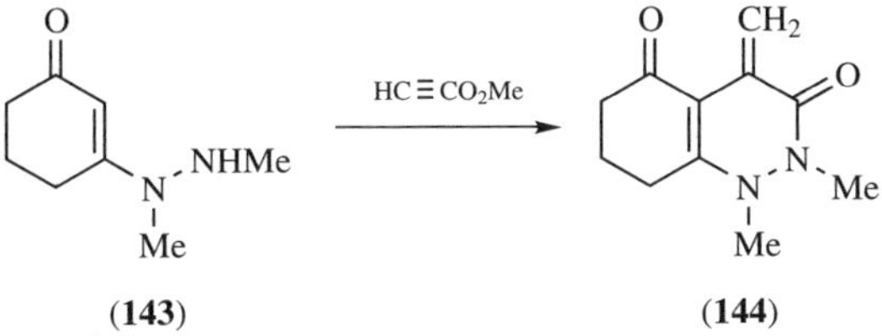

$HC{\equiv}CO_2Me$

(**143**) (**144**)

1.2.5. When the Synthon Supplies N1 + N2 + C3 + C4 of the Cinnoline

Nearly all examples in this category employ cyclohexane rather than benzene derivatives as substrates and accordingly afford partially reduced cinnolines, as illustrated here.

5,5-Dimethyl-1,3-cyclohexanedione (dimidone, **145**) and ethyl 2-(*p*-tolylhydrazono)acetoacetate (**146**) gave ethyl 4,7,7-trimethyl-5-oxo-1-*p*-tolyl-1,5,6,7-tetrahydro-3-cinnolinecarboxylate (**147**) (neat $AcONH_4$, 170°C, 30 min: 80%);[534] the same substrate (**145**) and benzil monohydrazone (**148**) gave 7,7-dimethyl-3,4-diphenyl-6,7-dihydro-5(1*H*)-cinnolinone (**149**) (or tautomer) (Et_3N, EtOH, reflux, 2 h: 80%).[673]

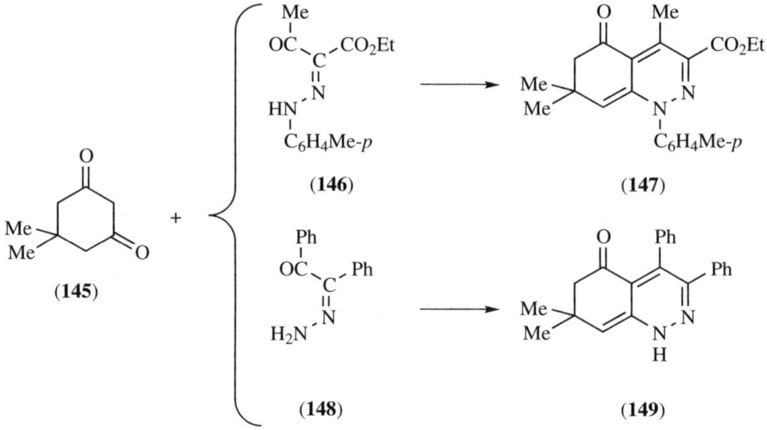

(**145**) (**146**) (**147**) (**148**) (**149**)

1-Morpholinocyclohex-1-ene (**150**) and ethyl (2,2,2-trichloro-1-phenylethylidene)hydrazinecarboxylate (**151**) gave a separable mixture of ethyl 4-chloro-3-phenyl-1,5,6,7-tetrahydro-1-cinnolinecarboxylate (**152**, $R = CO_2Et$) and 4-chloro-3-phenyl-1,5,6,7-tetrahydrocinnoline (**152**, R = H) (or tautomer) ($EtPr^i_2N$, CH_2Cl_2, N_2, reflux, 5 h: 44% and 9%, respectively).[86,123,cf. 287]

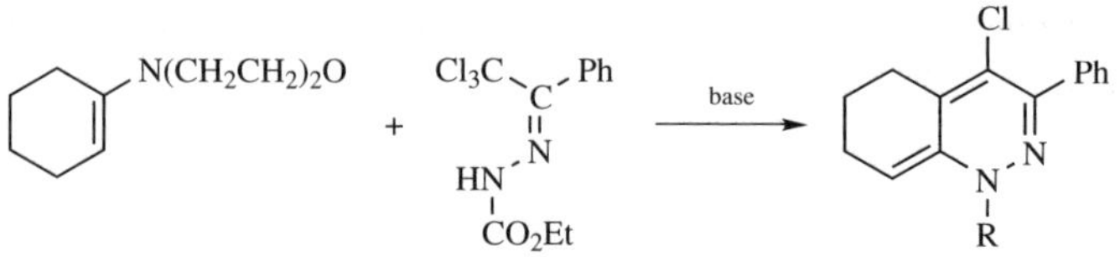

base

(**150**) (**151**) (**152**)

Also other examples.[516,548,602,618]

1.3. FROM A PYRIDAZINE SUBSTRATE

This potentially wide area of primary synthesis appears to be represented by only two types in which appropriate pyridazine substrates undergo cyclocondensation with synthones that supply either C6 + C7 or C6 + C7 + C8 of the resulting cinnolines. Examples follow.

1-*m*-Chlorophenyl-4-methyl-6-oxo-1,6-dihydro-3,5-pyridazinedicarbonitrile (**154**, $R = C_6H_4Cl$-*m*) and α-benzylidenemalononitrile (**153**) gave 8-amino-2-

m-chlorophenyl-3-oxo-6-phenyl-2,3-dihydro-4,7-cinnolinedicarbonitrile (**155**, R = C_6H_4Cl-*m*) [trace $HN(CH_2)_5$, EtOH, reflux, 1 h: 38%]; two analogs likewise.[297]

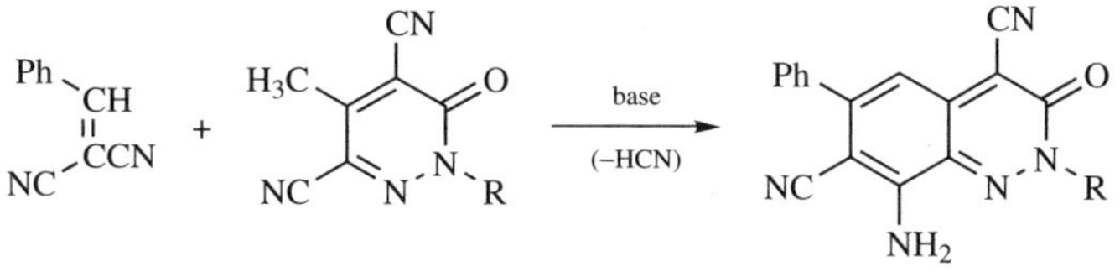

(**153**) (**154**) (**155**)

The related substrate, 4-methyl-6-oxo-1-phenyl-1,6-dihydro-3,5-pyridazinedicarbonitrile (**154**, R = Ph), and α-benzylidenemalononitrile (**153**) gave 8-amino-3-oxo-2,6-diphenyl-2,3-dihydro-4,7-cinnolinedicarbonitrile (**155**, R = Ph) [$HN(CH_2)_5$, pyridine, reflux, 4 h: 75%];[535] analogs likewise.[67,535]

Ethyl 5-cyano-1-*o*-methoxyphenyl-4-methyl-6-oxo-1,6-dihydro-3-pyridazinecarboxylate (**157**) and diethyl 3-oxoglutarate (**156**) gave ethyl 4-cyano-6-ethoxycarbonylmethyl-8-hydroxy-2-*o*-methoxyphenyl-2,3-dihydro-7-cinnolinecarboxylate (**158**) (AcOH, dioxane, reflux, 8 h: 79%); one analog likewise.[618]

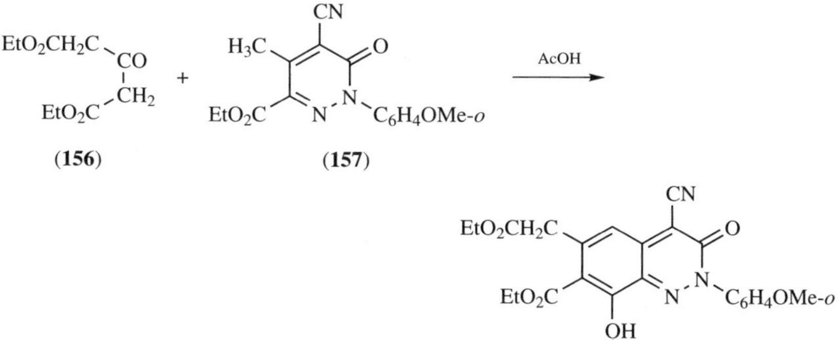

(**156**) (**157**)

(**158**)

4,5-Dibenzylidene-4,5-dihydro-3,6(1*H*,2*H*)-pyridazinedione (**160**) and ethyl acetoacetate (**159**) gave ethyl 4-benzylidene-3,7-dioxo-5-phenyl-1,2,3,4,4a,5,6,7-octahydro-6-cinnolinecarboxylate (**161**) [synthon (**159**), EtONa, EtOH, 20°C, 1 h; then substrate (**160**)↓, reflux, 3 h: 60%].[969]

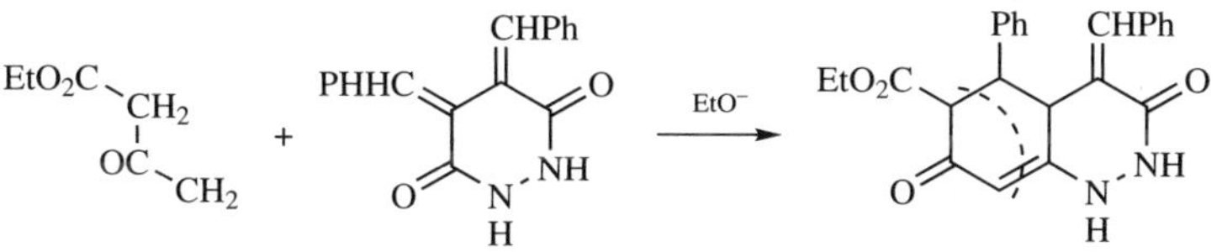

(**159**) (**160**) (**161**)

Note: Fused cinnolines may also be made from pyridazine substrates.[977]

1.4. FROM OTHER HETEROMONOCYCLIC SUBSTRATES

The formation of cinnolines from heteromonocyclic systems other than pyridazine is rare. However, at least three such systems have been so used, as illustrated in the following examples.

1,2-Diazete Derivatives as Substrates

2-Acetyl-1,4,4-triphenyl-1,2-diazetidin-3-one (**162**) rearranged into 2-acetyl-4,4-diphenyl-1,4-dihydro-3(2*H*)-cinnolinone (**163**) (neat F_3CCO_2H: > 95%; no further details).[720]

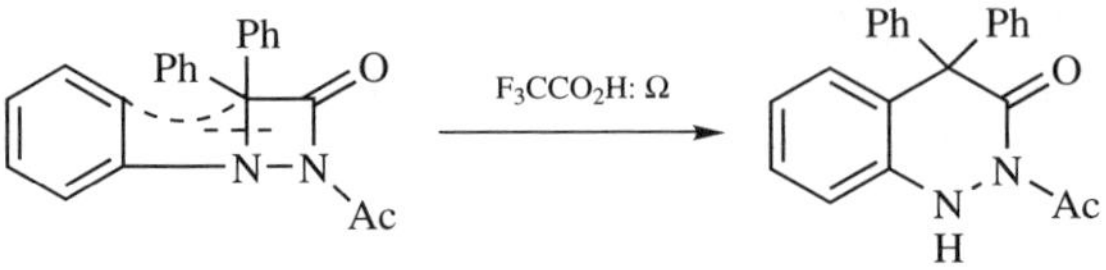

F_3CCO_2H: Ω

(**162**) (**163**)

Furan Derivatives as Substrates

o-[Bis(5-methylfuran-2-yl)methyl]aniline (**164**) gave 3-acetonylidenemethyl-4-(5-methylfuran-2-yl)cinnoline (**165**) (Me_3SiCl, $Me_2CHCH_2CH_2ONO$, MeCN, 20°C, 15 min: 79%; a logical mechanism via a diazonium intermediate was suggested); several analogs were made similarly.[823]

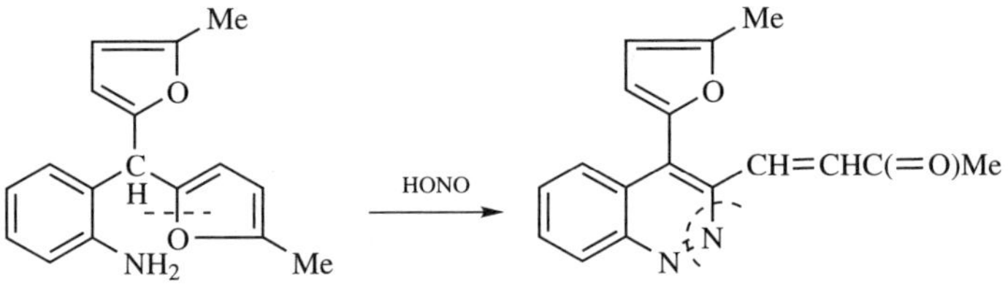

HONO

(**164**) (**165**)

1,2,4,5-Tetrazine Derivatives as Substrates

3-(1,1-Dicyanohex-5-ynyl)-6-morpholino-1,2,4,5-tetrazine (**166**) underwent loss of nitrogen and recyclization to give 3-morpholino-5,6,7,8-tetrahydro-8,8-cinnolinedicarbonitrile (**167**) (xylene, 132°C, 8 h: 58%); its 4-phenyl derivative was made similarly.[953]

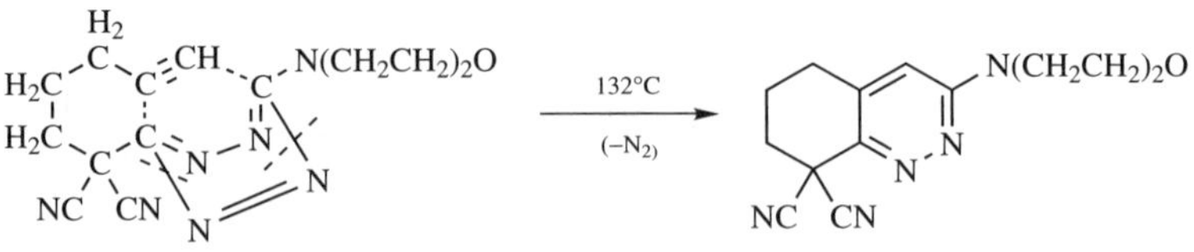

132°C ($-N_2$)

(**166**) (**167**)

1.5. FROM HETEROBICYCLIC SUBSTRATES

Several heterobicyclic systems have been used to make cinnolines, but only one has been so employed to any extent. The following examples illustrate the variety of reactions involved.

1-Benzazocine Derivatives as Substrates

3,4,5,6-Tetrahydro-1-benzazocine-2,6(1*H*)-dione (**168**) gave 3-(2-carboxyethyl)-4(1*H*)-cinnolinone (**169**) ($MeOCH_2CH_2OMe$, trace H_2O, BuONO, HCl gas↓, 25°C, 5 min; then stirred, 25°C, 12 h; then suspension of crude diazonium intermediate, 100°C, 5 min: 61%); a dozen analogs, substituted in the phenyl ring, were made similarly.[213]

HONO

$CH_2CH_2CO_2H$

(**168**) (**169**)

Benzofuran Derivatives as Substrates

2-Acetoxy-2-methyl-2,3,4,5,6,7-hexahydrobenzofuran-3-one (**170**) with hydrazine gave 3-methyl-5,6,7,8-tetrahydro-4(1*H*)-cinnolinone (**171**) (EtOH, 20°C, 12 h: 85%) or with methylhydrazine gave a separable mixture of 1,3-dimethyl-5,6,7,8-tetrahydro-4(1*H*)-cinnolinone (**172**) and 2,3-dimethyl-5,6,7,8-tetrahydrocinnolin-2-ium-4-olate (**173**) (EtOH, 0°C → 20°C, 12 h: 67% and 15%, respectively); the related substrate, 2-methoxy-2-methyl-

H_2NNH_2

($-AcOH$, $-H_2O$)

$MeHNNH_2$

$MeHNNH_2$

(**170**) (**171**) (**172**) (**173**) (**174**)

2,3,4,5,6,7-hexahydrobenzofuran-3-one (**174**), with methylhydrazine, also gave a separable mixture of the products **172** and **173** but in approximately reverse proportion (6% and 58%, respectively).[577]

7a-Morpholino-2,4,5,6,7,7a-hexahydrofuran-2-one (**175**) gave 5,6,7,8-tetrahydro-3(2*H*)-cinnolinone (**176**) ($H_2NNH_2 \cdot H_2O$, EtOH, 20°C → reflux, 4 h: 80%).[55]

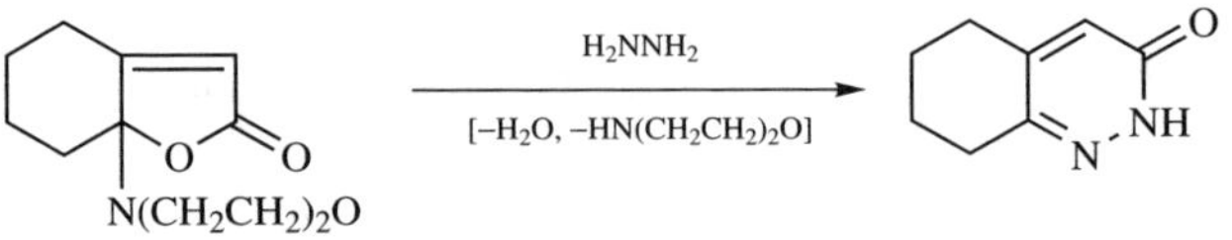

Also other examples.[30]

1,2,3-Benzotriazine Derivatives as Substrates

4-Methylene-3,4-dihydro-1,2,3-benzotriazine (**177**) rearranged into 4-methylaminocinnoline (**178**) (98% H_2SO_4, AcOH, 37°C → 55°C, 5 min: 12–30%).[627]

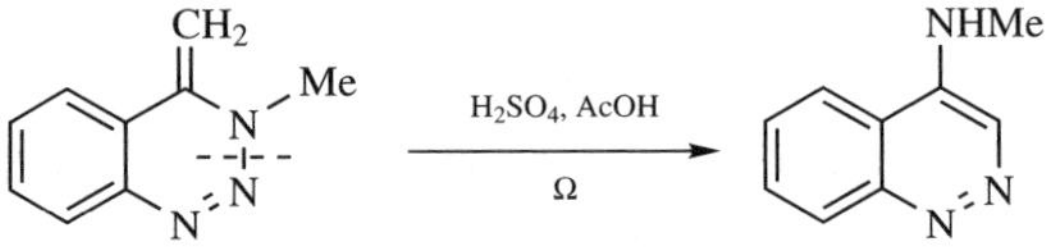

Indazole Derivatives as Substrates

6-Chloro-2-diethylamino-3-(tosylhydrazonomethyl)-2*H*-indazole (**179**) was converted into its crude sodio derivative and thence into 6-chlorocinnoline (**180**) (NaH, THF, 20°C, 15 min, evaporation; then $C_6H_4Cl_2$-*o*↓, 200°C, 12 h: 51%).[816]

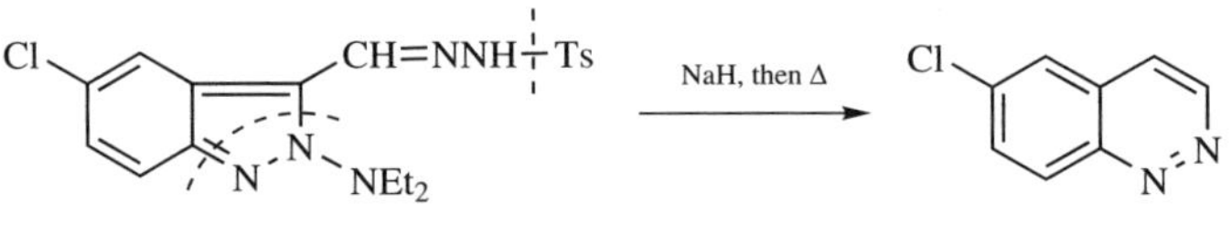

Indole Derivatives as Substrates

2-Methyl-1-indolamine (**181**) in methanolic hydrogen chloride gave a separable mixture of 3-methyl-1,4-dihydrocinnoline (**182**) and 3-methylcinnoline (**183**) (3%HCl/MeOH, reflux, 14 h: 56% and 24%, respectively; presumably some of the dihydro product suffered aerial oxidation during workup);[604] the same substrate (**183**) in methanolic hydrogen chloride containing nitrobenzene gave only the aromatic product (**183**) (3% HCl/MeOH, $PhNO_2$, reflux, 42 h; 92%);[605] analogous products of both types were made similarly.[604,605]

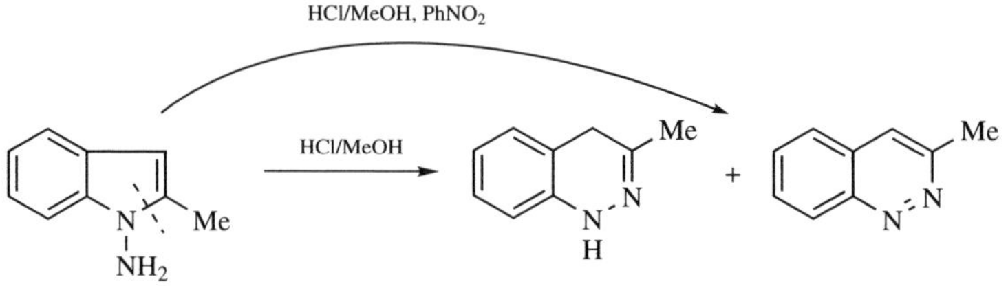

HCl/MeOH, $PhNO_2$

HCl/MeOH

(**181**) (**182**) (**183**)

1-Benzylamino-2-indolinone (**184**) underwent oxidative rearrangement into 2-benzyl-3(2*H*)-cinnolinone (**185**) (Bu^tOCl, PhH, 20°C until substrate gone: 76%);[42] analogs likewise.[38,42] Lead tetracetate has also been used for such reactions, but it appears to be less effective.[38,187,373]

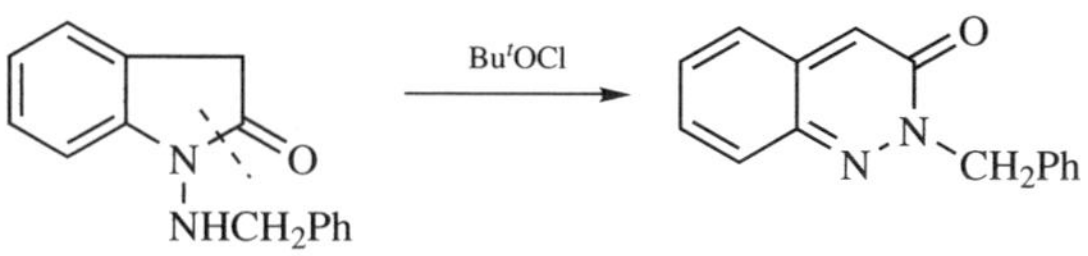

Bu^tOCl

(**184**) (**185**)

4,6-Dimethyl-2,3-indolinedione (4,6-dimethylisatin, **186**) gave 5,7-dimethyl-3,4(1*H*,2*H*)-cinnolinedione (**187**) ($NaNO_2$, HCl, 15°C, 5 min; then $SnCl_2\downarrow$, 0°C, 1 h: 80%); several analogs likewise.[809]

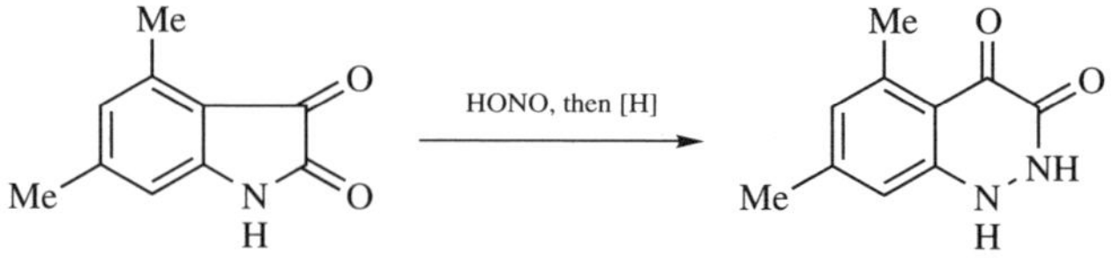

HONO, then [H]

(**186**) (**187**)

4-Hydroxyimino-2-methyl-2-morpholino-2,3,4,5,6,7-tetrahydro-3a*H*-indole 1-oxide (**188**) gave 5-hydroxyimino-3-methyl-5,6,7,8-tetrahydrocinnoline (**189**) ($H_2NNH_2 \cdot H_2O$, AcOH, H_2O, reflux, 15 min: 85%).[744]

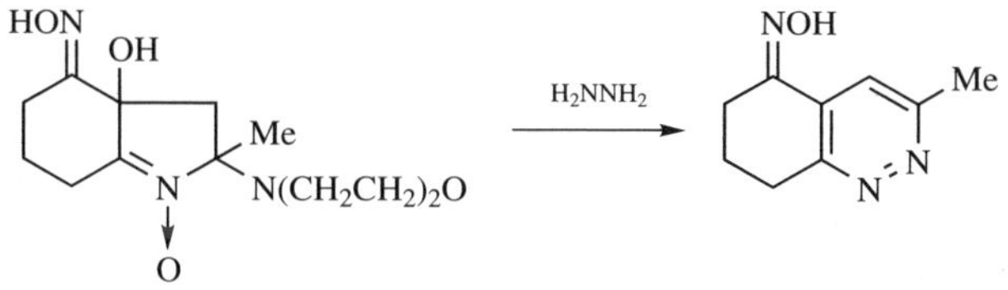

H_2NNH_2

(**188**) (**189**)

Also other examples.[2,cf. 812,325]

1.6. FROM HETEROPOLYCYCLIC SUBSTRATES

A few heterotri- to heteropentacyclic substrates have been used to prepare cinnolines, but, to date, such procedures are more of interest than utility. The following examples illustrate the processes involved.

[1]Benzopyrano[2,3-*c*]cinnoline Derivatives as Substrates

10-Bromo-2,3-dimethoxy-12*H*-[1]benzopyrano[2,3-*c*]cinnolin-12-one (**190**) underwent hydrolytic ring fission to 4-(3-bromo-6-hydroxybenzoyl)-6,7-dimethoxy-3(2*H*)-cinnolinone (**191**) (NaOH, H_2O, no further details: > 60%).[227]

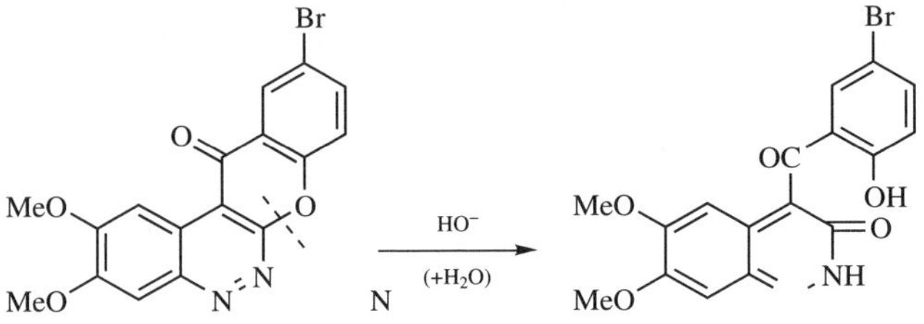

(**190**) (**191**)

Cyclopropa[*c*]cinnoline Derivatives as Substrates

Ethyl 1,1-dimethyl-7b-phenyl-1a,7b-dihydro-[1*H*]-cyclopropa[*c*]cinnoline-1c-carboxylate (**192**) underwent rearrangement into ethyl 4-isopropenyl-4-phenyl-1,4-dihydro-3-cinnolinecarboxylate (**193**) (AcOH, reflux, 30 min: 55% after separation from another product).[702]

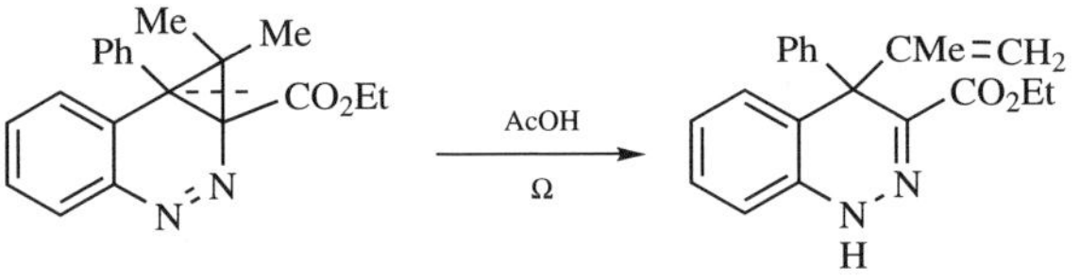

(**192**) (**193**)

Diethyl 1-phenyl-1a,7b-dihydro-[1*H*]-cyclopropa[*c*]cinnoline-1,1a-dicarboxylate (**194**) gave a separable mixture of ethyl 4-(α-ethoxycarbonylbenzyl)-1,4-dihydro-3-cinnolinecarboxylate (**195**) and ethyl 3-cinnolinecarboxylate (**196**) (AcOH, reflux, 30 min: ~40% and ~5%, respectively, after separation from another product).[702]

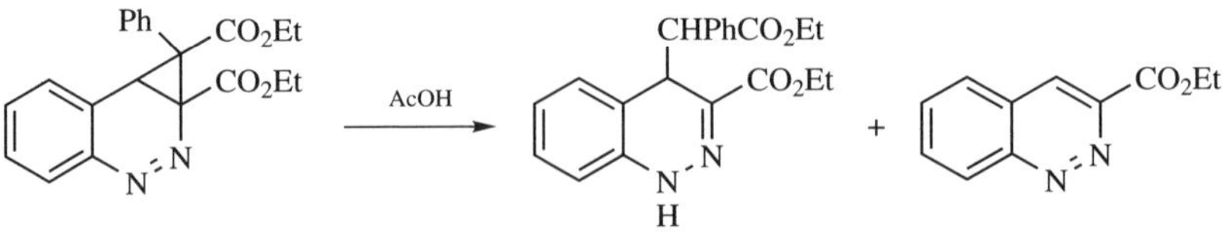

(**194**) (**195**) (**196**)

Isoxazolo[3,4-*c*]cinnoline Derivatives as Substrates

Isoxazolo[3,4-*c*]cinnolin-1(3*H*)-one 5-oxide (**197**, R = H) or its 3-methyl derivative (**197**, R = Me) underwent reductive cleavage to give 3-amino- (**198**, R = H) or 3-methylamino-4-cinnolinecarboxylic acid 1-oxide (**198**, R = Me), respectively ($H_2NNH_2 \cdot H_2O$, EtOH, reflux, ? h: 98% or 83%, respectively); analogs likewise.[147]

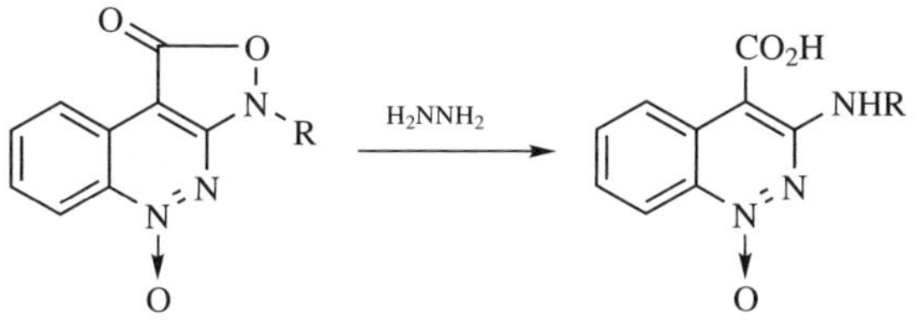

H_2NNH_2

(**197**) (**198**)

Naphtho[2′,1′:5,6]pyrano[2,3-*c*]cinnoline Derivatives as Substrates

2,3-Dimethoxy-14*H*-naphtho[2′,1′:5,6]pyrano[2,3-*c*]cinnolin-14-one (**199**) gave 4-(1-hydroxy-2-naphthoyl)-6,7-dimethoxy-3(2*H*)-cinnolinone (**200**) (NaOH, H_2O, no details: > 60%).[227]

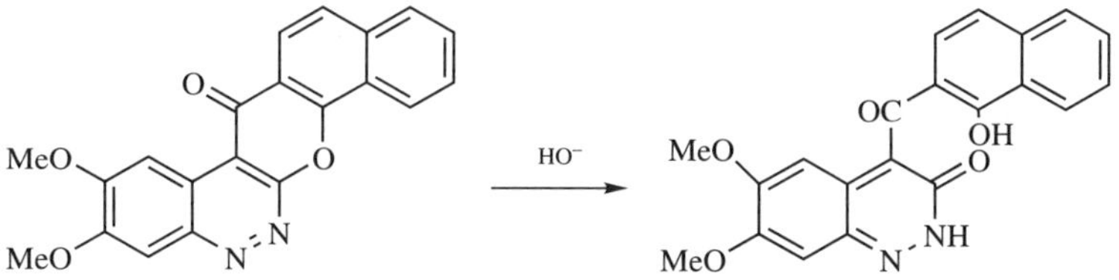

HO^-

(**199**) (**200**)

Pyrazolo[3,4-*c*]cinnoline Derivatives as Substrates

3*H*-Pyrazolo[3,4-*c*]cinnolin-1(2*H*)-one 5-oxide (**201**) underwent ring fission with loss of N_2 to give 4-cinnolinecarboxylic acid 1-oxide (**202**) (NaOCl, NaOH, H_2O, 20°C, 30 min: > 95%); analogs likewise.[146]

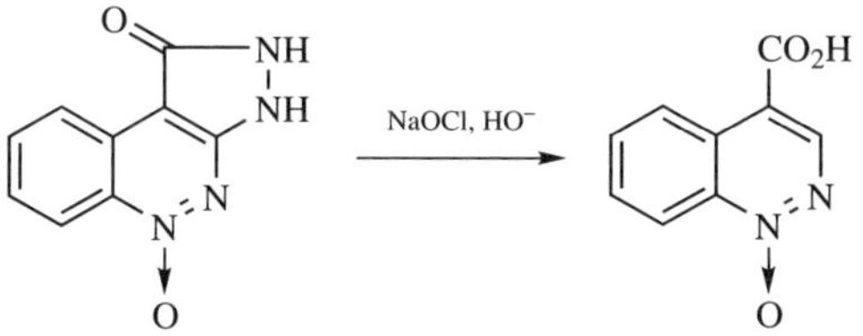

NaOCl, HO^-

(**201**) (**202**)

Pyrido[4,3,2-*de*]cinnoline Derivatives as Substrates

3,8,8-Trimethyl-4,7,8,9-tetrahydro-5*H*-pyrido[4,3,2-*de*]cinnolin-5-one 6-oxide (**203**) isomerized into 4-carboxymethyl-3,6,7-trimethyl-5-cinnolinamine (**204**)

(P_2O_5, H_2PO_4, 135°C, 7 h: 85%; structure consistent with spectra).[656]

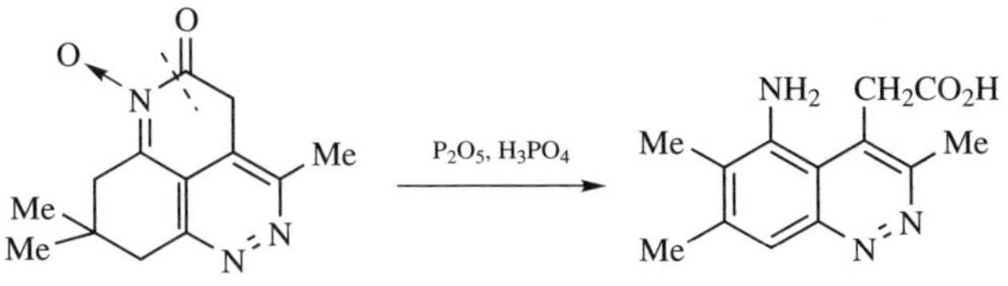

P_2O_5, H_3PO_4

(203) (204)

1.7. GLANCE INDEX TO TYPICAL CINNOLINE DERIVATIVES AVAILABLE BY PRIMARY SYNTHESES

This glance index may assist in the choice of a primary synthesis for a required cinnoline derivative; such syntheses are based on aliphatic, carbocyclic, or heterocyclic substrates with or without ancillary synthons. In using the index, it should be borne in mind that products broadly analogous to those formulated may often be obtained by minor changes to the substrates and/or synthons involved.

Section	Typical Products
1.1.1 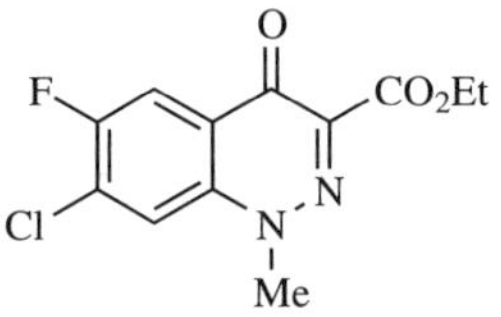 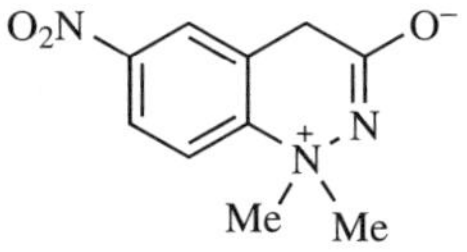	
1.1.2	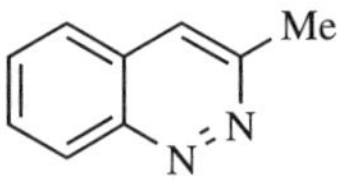
1.1.3 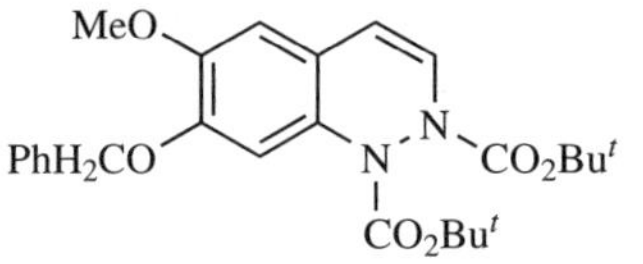 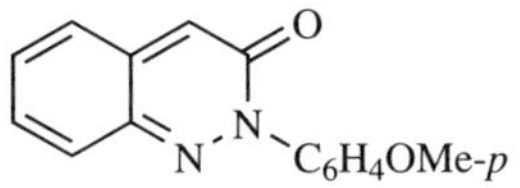	
1.1.4 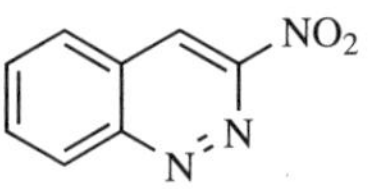 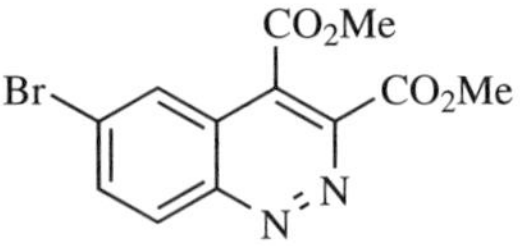	
1.1.5 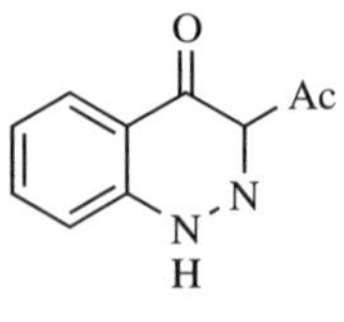 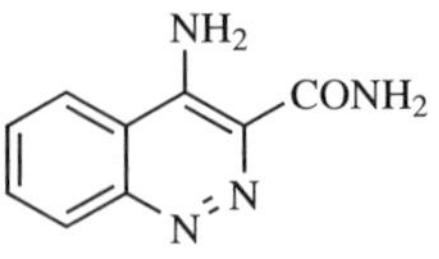	

Section	Typical Products
1.2.1	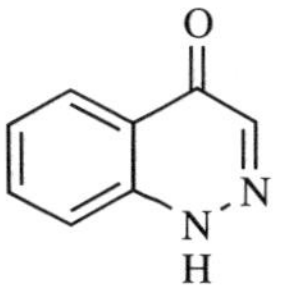 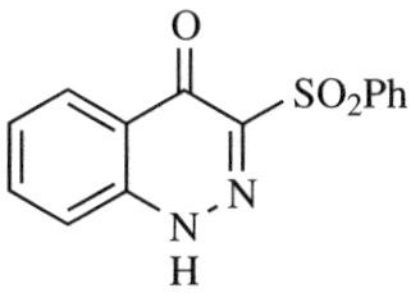 O; N; N; H | O; SO_2Ph; N; N; H
1.2.2	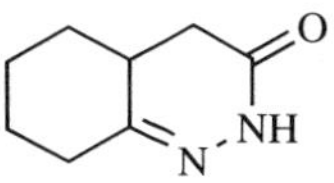 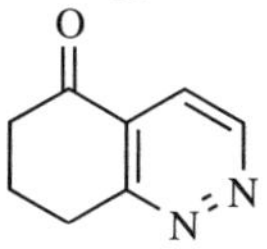 O; NH; N | O; N; N
1.2.3	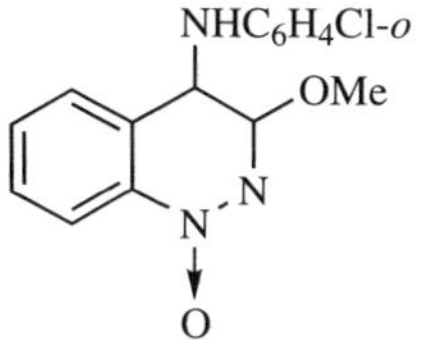 NHC_6H_4Cl-*o*; OMe; N; N; O
1.2.4	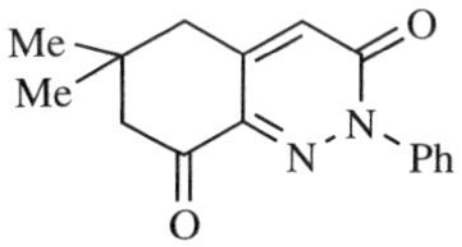 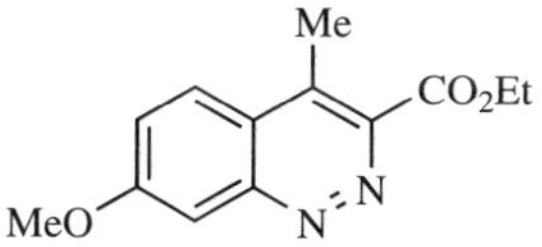 Me; Me; O; N; N; Ph; O | Me; CO_2Et; MeO; N; N
	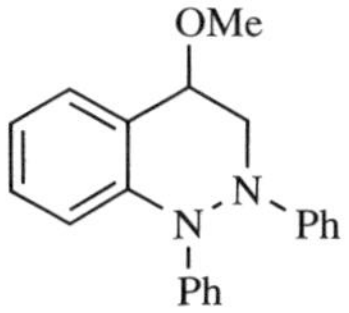 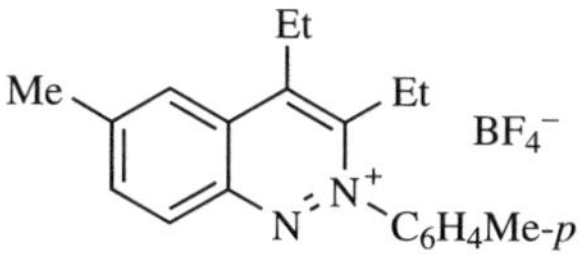 OMe; N; Ph; N; Ph | Et; Me; Et; BF_4^-; N^+; N; C_6H_4Me-*p*
1.2.5	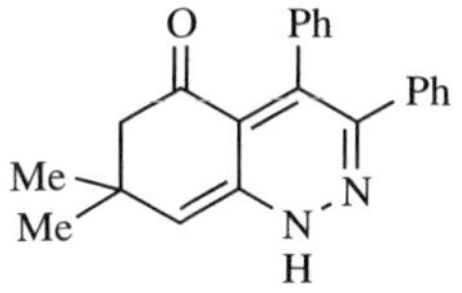 O; Ph; Ph; Me; Me; N; N; H
1.3	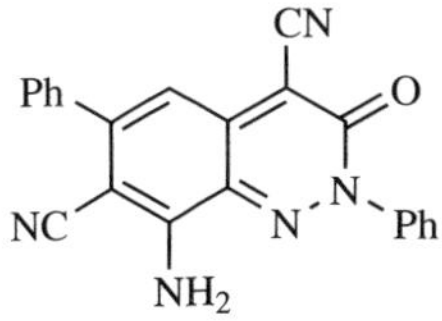 CN; Ph; O; N; NC; N; Ph; NH_2
1.4	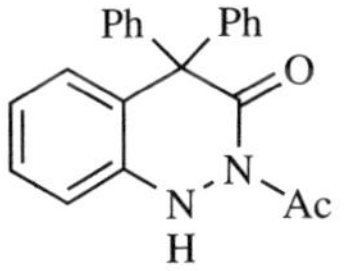 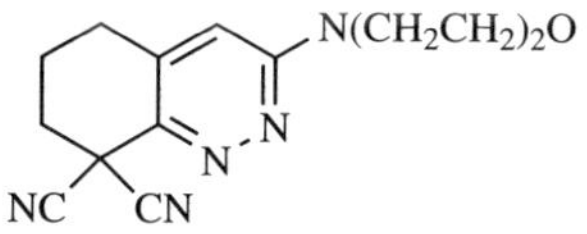 Ph; Ph; O; N; N; Ac; H | $N(CH_2CH_2)_2O$; N; N; NC; CN

(*Continued*)

Section	Typical Products	
1.5	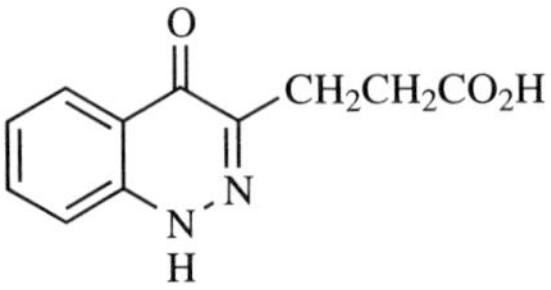 O $CH_2CH_2CO_2H$ N N H	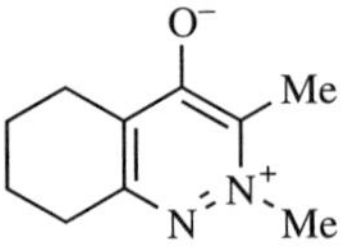 O^- Me N^+ N Me
	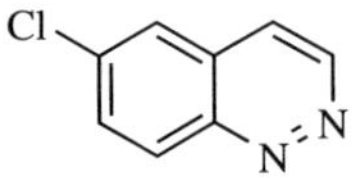 Cl N N	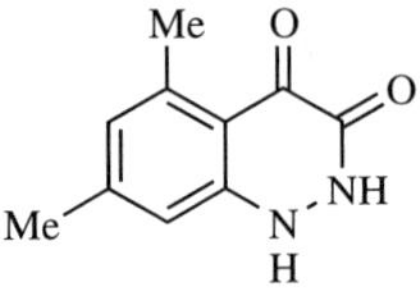 Me O O NH Me N H
1.6	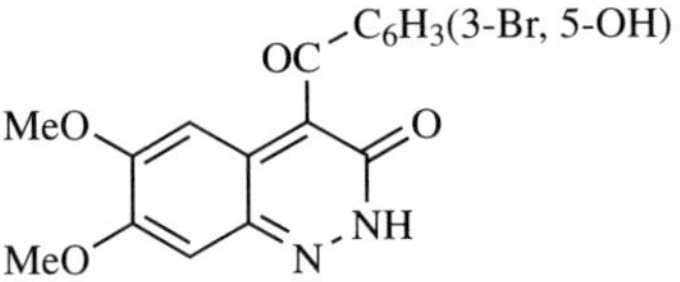 OC–C_6H_3(3-Br, 5-OH) MeO O NH MeO N	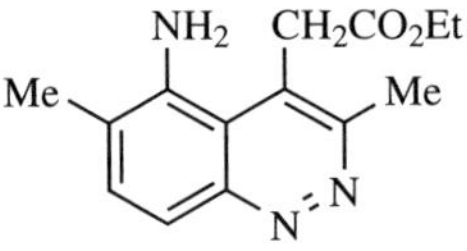 NH_2 CH_2CO_2Et Me Me N N

CHAPTER 2

Cinnoline, Alkylcinnolines, and Arylcinnolines (*H* 4, 6, 46; *E* 1, 18, 300)

This chapter covers information (reported during the period 1972–2004) on the preparation, physical properties, and reactions of cinnoline and its *C*-alkyl, *C*-aryl, *N*-alkyl, and *N*-aryl as well as their respective nucleus-reduced analogs. In addition, it includes methods for introducing alkyl or aryl groups (substituted or otherwise) into cinnolines already bearing substituents and reactions specific to the alkyl or aryl groups in such compounds. For simplicity, the term *alkylcinnoline* in this chapter is intended to cover alkyl-, alkenyl-, alkynyl-, cycloalkyl-, and aralkylcinnolines; likewise, *arylcinnoline* includes both aryl- and heteroarylcinnolines.

Since the appearance of Simpson's review[906] and Singerman's update[907] within this series, several brief to detailed reviews of general cinnoline chemistry have appeared,[181,552,903–905,908–916]

2.1. CINNOLINE (*H* 4, 46; *E* 1, 300)

2.1.1. Preparation of Cinnoline and Hydrocinnolines

Although cinnoline is costly to purchase, only two new preparative routes have been developed: a good but somewhat inconvenient primary synthesis[817] (see Section 1.1.3) and the oxidation of 1,4-dihydrocinnoline (MnO_2, 3% HCl/MeOH; or chloranil, PhH; no details for either procedure).[604] In addition, reductive procedures leading to unsubstituted tetrahydro-, hexahydro-, and octahydrocinnolines have been reported.[53]

5,6,7,8-Tetrahydro-3(2*H*)-cinnolinone (**1**) gave mainly 4,4a,5,6,7,8-hexahydro-3(2*H*)-cinnolinone (**2**) ($LiAlH_4$, PhH, reflux, 12 h) or mainly 2,3,4,4a,5,6,7,8-octahydrocinnoline (**4**) ($LiAlH_4$, PhH, reflux, 60 h), in each case accompanied by small amounts of 5,6,7,8-tetrahydrocinnoline (**3**) and 4,4a,5,6,7,8-hexahy-

Cinnolines and Phthalazines: Supplement II, The Chemistry of Heterocyclic Compounds, Volume 64, by D.J. Brown

drocinnoline (**5**); the products were all isolated, albeit with appreciable loss.[53]

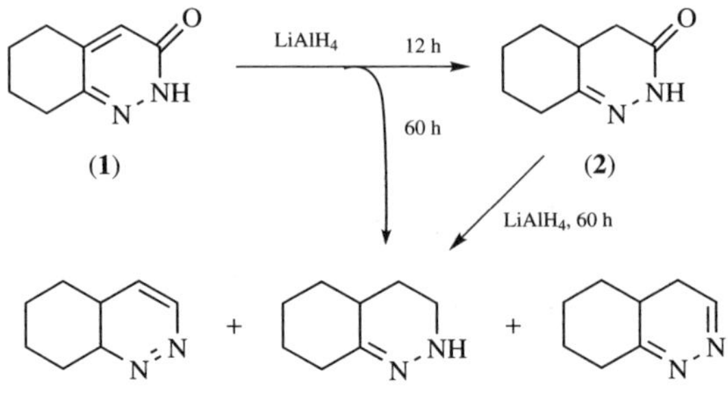

In contrast, 5,6,7,8-tetrahydro-3(2*H*)-cinnolinethione (**6**) gave a highly unstable octahydrocinnoline (**7**) that underwent aerial oxidation during treatment with picric acid to afford only 5,6,7,8-tetrahydrocinnoline (**8**) as picrate (Raney Ni, EtOH, reflux, 2 h; then picric acid↓: 15%).[53]

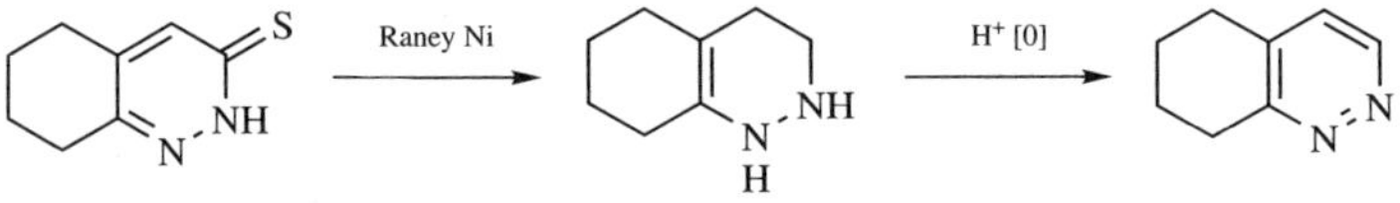

2.1.2. Physical Properties of Cinnoline

Any new or revised physical data for cinnoline and its salts or complexes may be found under "cinnoline" in the Appendix (Table A.1) at the end of this book. More notable studies on the physical properties of cinnoline or reduced cinnoline are indicated briefly here.

Aromaticity. New aromaticity indices for cinnoline and related heterocycles have been derived from 12 weighted experimental or theoretical data.[527]

Complexes. Anomalous variations in the absorption spectra for cinnoline in nonpolar solvents on temperature change appears to result from three phenomena: hydrogen bonding interactions, microcrystalline complex formation at low temperatures, and facile photoadduct formation.[506] An X-ray analysis of the complex, cinnoline.2$ZnCl_2$, has been reported.[1030]

Electron Spin Resonance. A study has been made of the ESR spectra for radicals derived by photolysis of cinnoline and related substrates.[806]

Energy Calculations. Because cinnoline is insufficiently stable for experimental study, the total and bond separation energies for cinnoline have been calculated for comparison with those of related heterocycles.[683] Highest occupied molecular orbital (HOMO) energies have been used to calculate pK_a values for cinnoline and other benzodiazines.[813]

Nuclear Magnetic Resonance Studies. The effects of a variety of 8-substituents on the ^{1}HNMR spectra of cinnoline have been studied.[289] A ^{13}CNMR study has indicated that both the 1- and 2-protonated species are present in cinnolinium salts.[757] The ^{15}NNMR chemical shifts for cinnoline and a variety of other heterocyclic systems have been reported and discussed.[152]

Ultraviolet/Visible Spectra. A useful compilation of UV spectra for cinnolines and some related systems has been prepared.[84] The UV spectra of cinnoline and other azanaphthalenes have been calculated[798] and discussed in some detail.[763]

2.1.3. Reactions of Cinnoline

Only a few reactions of unsubstituted cinnoline have been reported since 1972. They are illustrated here.

Deuteration

The deuteration of cinnoline by D_2O over Pt/asbestos at 170–220°C has been studied.[329]

Reissert-Type Additions

Cinnoline (**9**) gave 1,2-dibenzoyl-1,2-dihydro-4-cinnolinecarbonitrile (**10**) [Me_3SiCN (2 mol), BzCl (2 mol), $AlCl_3$, CH_2Cl_2, 20°, 36 h: ~40%] and thence 4,4′-bicinnoline (**11**) (NaOH, H_2O, EtOH, reflux, 30 min: ?%).[23]

The same substrate (**9**) gave ethyl 4-allyl-1,4-dihydro-1-cinnolinecarboxylate (**12**, R = Et) ($Bu_3SnCH_2CH{=}CH_2$, $ClCO_2Et$↓ dropwise, CH_2Cl_2, 0°C, 2 h: 67%) or 1-chloroethyl 4-allyl-1,4-dihydro-1-cinnolinecarboxylate (**12**, R = CHClMe) ($ClCO_2CHClMe$, likewise: 46%).[431,588]

[0]

Me3SiCN, BzCl

HO⁻

CN

Bz

Bz

(**9**) (**10**) (**11**)

$Bu_3SnCH_2CH{=}CH_2$, $ClCO_2R$

$CH_2CH{=}CH_2$

CO_2R

(**12**)

Oxidation

Cinnoline (**9**) gave 4,4′-bicinniline (**11**) [($AgMnO_4 \cdot$pyridine$_2$), $(H_2NCH_2)_2$, 5°C→20°C, 5 days: ~3%].[993]

Nitration

Cinnoline (**13**) gave a separable mixture of 5- (**14**) and 8-nitrocinnoline (**15**) (fuming HNO_3, 98% H_2SO_4, −5°C; substrate↓ slowly; then →20°C, 1 h: 23% and 29%, respectively, after separation).[217,cf. 724,1005]

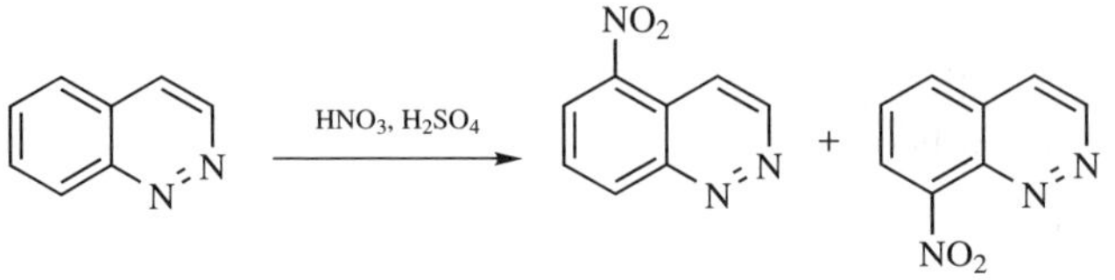

(**13**) (**14**) (**15**)

An effort has been made to extrapolate nitration rates for 130 substrates (including cinnoline) in HNO_3–H_2SO_4–H_2O at 25°C and H_0 −6.6.[684]

Photolysis on Colloidal Silver

The photolysis of cinnoline absorbed on colloidal silver has been studied by Raman spectral means; the final product was formulated as the *N*,*N*′-disilver derivative (**16**) of *o*-(2-aminovinyl)aniline.[820]

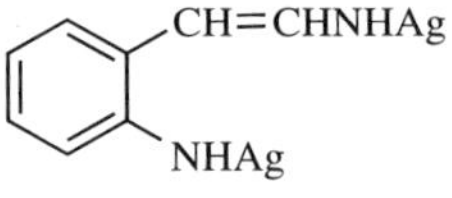

(**16**)

2.2. ALKYL- AND ARYLCINNOLINES (*H* 6, 13, 39; *E* 18)

This section covers both *C*- and *N*-alkyl/arylcinnolines as well as alkyl/arylcinnolinium salts. For many years it was believed that there were no naturally occurring cinnolines, but 4(or 3)-methylcinnoline has now been identified as a minor component of the volatiles from okra (*Hibiscus esculentus* L.).[794] Unlike alkyl/aryl derivatives of diazines and the other benzodiazines, alkyl/arylcinnolines have attracted relatively little attention. However, X-ray analyses have been reported for 4-methylcinnoline,[630] 6-chloro-2-*p*-chlorophenyl-8-(*trans*-1,2-diphenylvinyl)-3,4-diphenyl-2,3-dihydrocinnoline,[722,749] and two analogs thereof;[749] the MS (mass spectral) fragmentation of 1-methyl-4,4-diphenyl-1,2,3,4-tetrahydrocinnoline has been studied,[756] and a few preparative methods and reactions have been reported, as illustrated in the subsections that follow.

2.2.1. Preparation of Alkyl- and Arylcinnolines

Many alkyl/aryl cinnolines have been made by *primary syntheses* (see Chapter 1), and several diverse preparative procedures are illustrated in the following classified examples.

By Alkylation of Metallated Substrates

Reductive metallation of 3-phenylcinnoline gave the dihydro dianion, formulated for simplicity as 3-phenyl-1,4-disodio-1,4-dihydrocinnoline (**17**) (Na, THF, A, 20°, 12 h: not isolated), and subsequent treatment with benzyl chloride (2.5 equiv) gave a separable mixture of 4-benzyl- (**18**, R = CH_2Ph) and 1,4-dibenzyl-3-phenyl-1,4-dihydrocinnoline (**19**, R = CH_2Ph) (−78°C, 4 h: 51% and 33%, respectively); a similar reaction using methyl iodide gave only 1,4-dimethyl-3-phenyl-1,4-dihydrocinnoline (**19**, R = Me) (67%).[767]

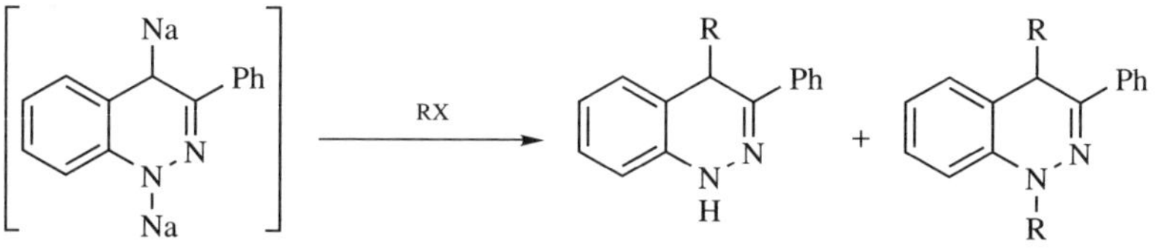

(**17**) (**18**) (**19**)

In much the same way, 3,4-diphenylcinnoline (**20**) underwent metallation and subsequent alkylation to give 4-(4-chlorobutyl)-3,4-diphenyl-1,4-dihydrocinnoline (**21**) [Na, THF, A, 20 h; then $Cl(CH_2)_4Cl$↓, −78°C, 3 h: 82%]; analogs likewise.[759]

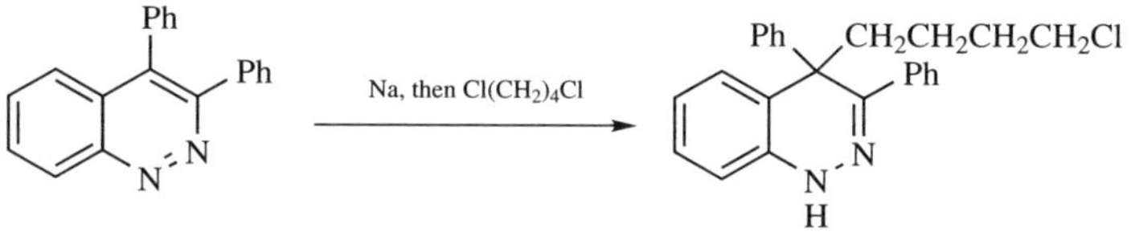

(**20**) (**21**)

In contrast to the foregoing examples, 4-chlorocinnoline underwent nonreductive metallation to 4-chloro-3-lithiocinnoline (**23**) [$LiPr^i_2N$ (made in situ), THF, A, −75°C, 30 min: not isolated]. Subsequent treatment with methyl iodide (3.5 equiv) gave 4-chloro-3-methylcinnoline (**22**) (−75°C, 2 h: 86%); with acetaldehyde gave 4-chloro-3-(1-hydroxyethyl)cinnoline (**24**, R = Me) (1 h: 89%); or with benzaldehyde (5 equiv) gave 4-chloro-3-

(α-hydroxybenzyl)cinnoline (**24**, R = Ph) (4 h: 88%).[307]

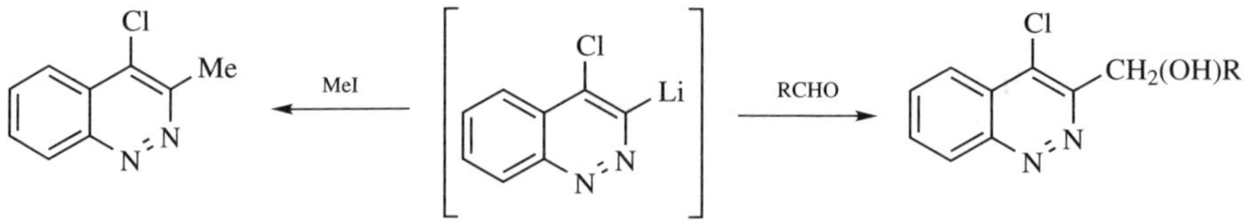

(**22**) (**23**) (**24**)

When a solution of 3-lithio-4-methoxycinnoline (**25**) was treated with methyl iodide in much the same way, a separable mixture of 4-methoxy-3-methyl- (**26**) and 3-ethyl-4-methoxycinnoline (**27**) resulted (34% and 24%, respectively, presumably by α-metallation of some of the first product (**26**) by remaining substrate (**25**) and subsequent methylation.[307]

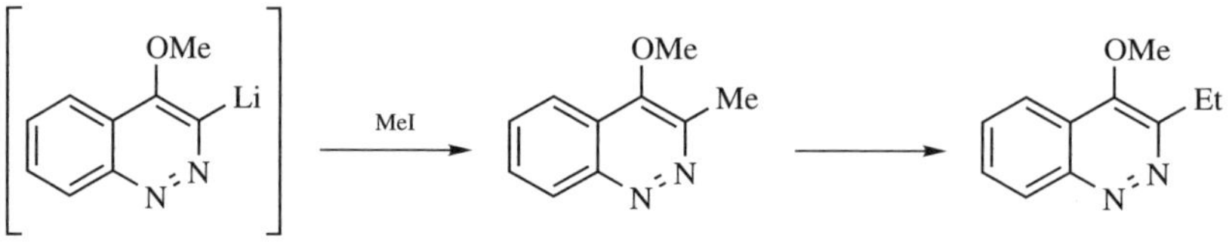

(**25**) (**26**) (**27**)

From Halogenocinnolines

4-Chlorocinnoline (**28**) and *p*-(trifluoromethyl)phenylboronic acid in the presence of a Pd catalyst gave 4-[*p*-(trifluoromethyl)phenyl]cinnoline (**29**) [$Pd(PPh_3)_4$, K_2CO_3, H_2O, EtOH, $MeOCH_2CH_2OMe$, N_2, reflux, 44 h: 85%]; many analogs likewise.[822]

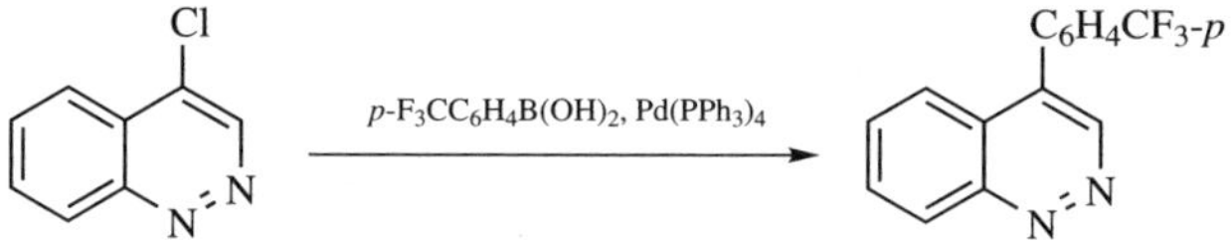

(**28**) (**29**)

4-Iodo-3-methoxycinnoline (**30**) gave 4-(α-hydroxybenzyl)-3-methoxycinnoline (**31**) [Li powder (2 mol), PhCHO (1 mol), THF, A, 20°C, ultrasound, 30 min: 64%; a Barbier reaction].[821]

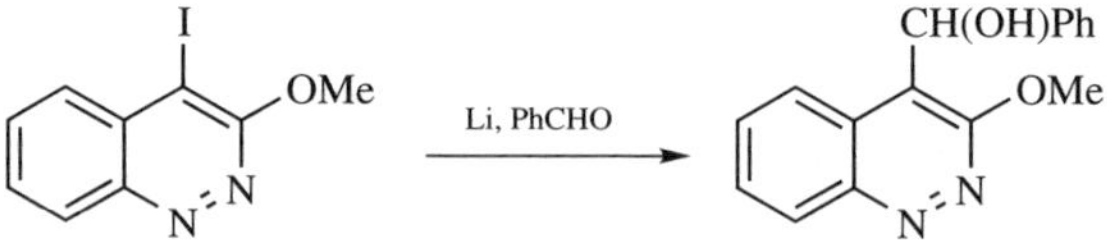

(**30**) (**31**)

3-Iodocinnoline (**32**) and phenylacetylene gave 3-phenylethynylcinnoline (**33**) [PhC≡CH, CuI, $(Ph_3P)_2PdCl_2$, neat Et_2NH, N_2, 20°C, 4 h: 45%];[293] similar treatment of 3-bromo-4-chlorocinnoline (**34**) in the presence of diethylamine, piperidine, or triethylamine afforded 4-diethylamino-3-phenylethynylcinnoline (**35**, R = NEt_2) (37%), 3-phenylethynyl-4-piperidinocinnoline [**35**, R = $N(CH_2)_5$] (64%), or 4-chloro-3-phenylethynylcinnoline (**35**, R = Cl) (29%), respectively; it seems clear that 3-alkanelysis is preferred to 3-aminolysis.[293]

PhC≡CH, CuI, $(Ph_3P)_2PdCl_2$, Et_2NH

(**32**) (**33**)

PhC≡CH, CuI, $(Ph_3P)_2PdCl_2$, Et_2NH or $HN(CH_2)_5$

(**34**) (**35**)

3-Bromocinnoline (**36**) gave 3-(3-hydroxy-3-methylbut-1-ynyl)cinnoline (**37**) [HC≡$CCMe_2OH$, $(Ph_3P)_2PdCl_2$, CuI, neat Et_2NH, N_2, 20°C, 15 h: 50%]; analogs likewise.[338]

HC≡$CCMe_2OH$, CuI, $(Ph_3P)_2PdCl_2$, Et_2NH

(**36**) (**37**)

3-Iodo-4(1*H*)-cinnolinone (**38**) and styrene gave 3-styryl-4(1*H*)-cinnolinone (**39**) [PhCH=CH_2, $(Ph_3P)_2PdCl_2$, Et_3N, MeCN, sealed, 150°C, 5 h: 52%; note severe conditions relative to those in foregoing examples.].[293]

PHCh=CH_2, $(Ph_3P)_2PdCl_2$, Et_3N

(**38**) (**39**)

Also other examples.[695,774]

From Cinnolinecarbonitriles or Alkoxycinnolines

Note: Alkanelysis of a cyano or alkoxy group has been done with activated methylene reagents in the presence of sodium amide, potassium cyanide, or sodium hydride.

4-Cinnolinecarbonitrile (**40**, Q = CN) gave 4-(dicyanomethyl)cinnoline (**41**, R = CN) [$H_2C(CN)_2$, KCN, Me_2SO, 100°C, 1 h: 66%], 4-(α-cyanobenzyl)-cinnoline (**41**, R = Ph) ($PhCH_2CN$, $NaNH_2$, PhMe, reflux, 3 h: 60%), or 4-(α-cyano-α-ethoxycarbonylmethyl)cinnoline (**41**, R = CO_2Et) (EtO_2CCH_2CN, $NaNH_2$, PhMe, reflux, 7 h: 72%).[934]

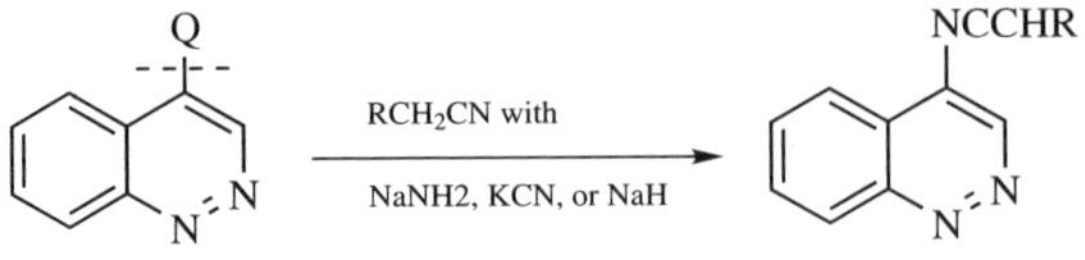

Q — NCCHR

RCH_2CN with NaNH2, KCN, or NaH

(**40**) (**41**)

4-Methoxycinnoline (**40**, Q = OMe) gave 4-(dicyanomethyl)cinnoline (**41**, R = CN) [$H_2C(CN)_2$, NaH, dioxane, 20°C; 30 min; then substrate↓, reflux, N_2, until no substrate visible on tlc: 59%] or 4-(α-cyanobenzyl)cinnoline (**41**, R = Ph) ($PhCH_2CN$, THF, reflux, 30 min; then substrate↓ and as above: 62%).[586]

By Direct *C*-Arylation

3(2*H*)-cinnolinone (**42**, R = H) gave 4-(*p*-dimethylaminophenyl)-3(2*H*)-cinnolinone (**42**, R = $C_6H_4NMe_2$-*p*) ($PhNMe_2$, AcOH, reflux, 3 h: 42%).[441]

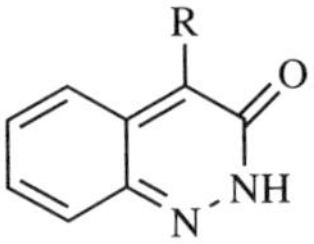

(**42**)

From Acylcinnolines

Note: *N*-Acylated hydrocinnolines have been reduced to the corresponding *N*-alkylhydrocinnolines.

4a-(*m*-Benzyoxyphenyl)-2-cyclopropylmethyl-1-phenylacetyldecahydrocinnoline (**43**) gave 4a-(*m*-benzyloxyphenyl)-3-cyclopropylmethyl-1-phenethyldecahydrocinnoline (**43a**) ($LiAlH_4$, THF, ?°C, 4 h: 75%);[964] related products were made similarly.[962,964]

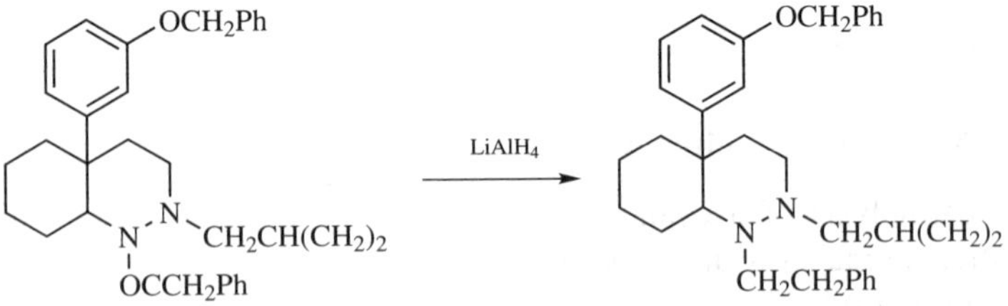

$LiAlH_4$

(**43**) (**43a**)

From Other Alkylcinnolines

Note: Modification of existing alkyl may be done in a number of ways; some are exemplified here.

1-Methyl-3-phenylethynyl- (**44**) gave 1-methyl-3-phenethyl-4(1*H*)-cinnolinone (**45**) (EtOH, H_2, Pd/C, until gas uptake ceases: 80%).[293]

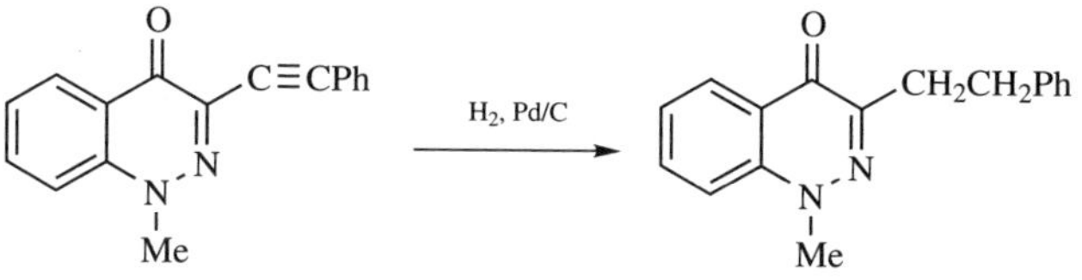

(**44**) (**45**)

3-(3-Hydroxy-3-methylbut-1-ynyl)-4-phenoxycinnoline (**46**) gave 3-ethynyl-4-phenoxycinnoline (**47**) (NaOH, PhMe, reflux, 7 h: 61%).[338]

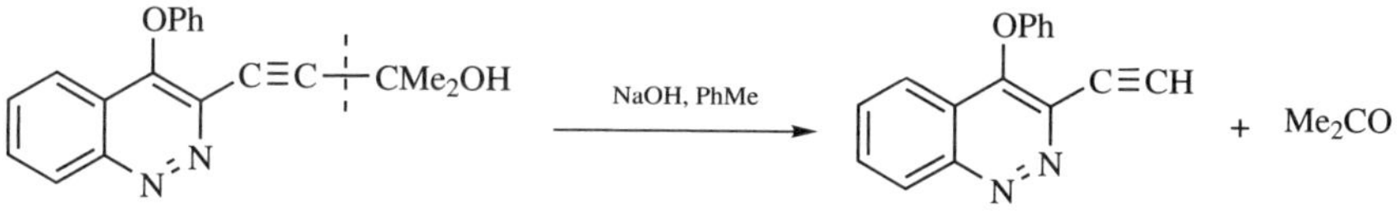

(**46**) (**47**)

2,3-Dimethylcinnolinium iodide (**49**, R = Me) gave 3-(2-anilinovinyl)-2-methylcinnolinium iodide (**48**) (neat PhN=CHNHPh, 90°C, 2 min: 60%); the homologous substrate, 2-ethyl-3-methylcinnolinium iodide (**49**, R = Et), gave 2-ethyl-3-*p*-dimethylaminostyrylcinnolinium iodide (**50**) (*p*-$Me_2NC_6H_4CHO$, neat Ac_2O, reflux, 5 min: ?%); analogs likewise.[636,959]

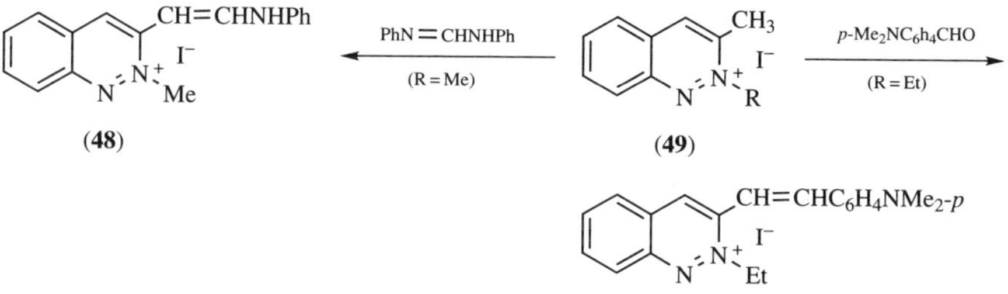

(**48**) (**49**) (**50**)

By Quaternization

Note: This process is often used but seldom described. A typical example is given here.

3-Methylcinnoline gave 2,3-dimethylcinnolinium iodide (**49**, R = Me) (neat MeI, sealed, 100°C, 1 h: ~65%).[636]

2.2.2. Reactions of Alkyl- and Arylcinnolines

Reactions that involve alkyl/aryl groups attached to cinnolines, or that involve modification of the cinnoline nucleus of alkyl/arylcinnolines bearing no other groups, are exemplified here.

Nitration

4-Methylcinnoline (**51**) gave 4-(nitromethyl)cinnoline (**52**) (substrate, KNH_2, liquid NH_3; then $PrONO_2$↓, <−33°C, 5 min: 88%); the NMR suggests that the product exists as an equilibrium mixture in which the zwitterionic form (**52b**) predominates over the regular form (**52a**).[95]

Me

KNH_2, NH_2, $PrONO_2$

CH_2NO_2

$CH=NO_2^-$

(**51**) (**52**) (**52b**)

Halogenation

4-Methylcinnoline (**53**) gave 4-(trichloromethyl)cinnoline (**54**) (NaOCl, H_2O, N_2, 20°C, 7 days: 77%).[291]

Me NaOCl CCl_3

(**53**) (**54**)

3,4-Dimethylcinnoline (**56**) gave 4-chloromethyl-3-methylcinnoline (**55**) (*N*-chlorosuccinimide, Bz_2O_2, CCl_4, reflux, 30 min: 40%) or 4-dibromomethyl-3-methylcinnoline (**57**) (substrate, AcOH, Br_2↓ dropwise, 20°C: 50%).[291]

CH_2Cl Me N-chlorosuccinimide Me Me Br_2 $CHBr_2$ Me

(**55**) (**56**) (**57**)

Oxidative Reactions

3,4-Dimethylcinnoline (**58**) gave 3-methyl-4-cinnolinecarbaldehyde (**59**) {SeO_2, AcOH, reflux, 4 h: 26%;[291] $Mn(OAc)_3$, AcOH, Ac_2O, reflux, 15 min: 43%;[711]

or $[PhSe(=O)]_2O$, PhCl, reflux, N_2, 30 min: 28% after separation from 3-methyl-4-(phenylselenomethyl)cinnoline (**60**) (30%)}.[711]

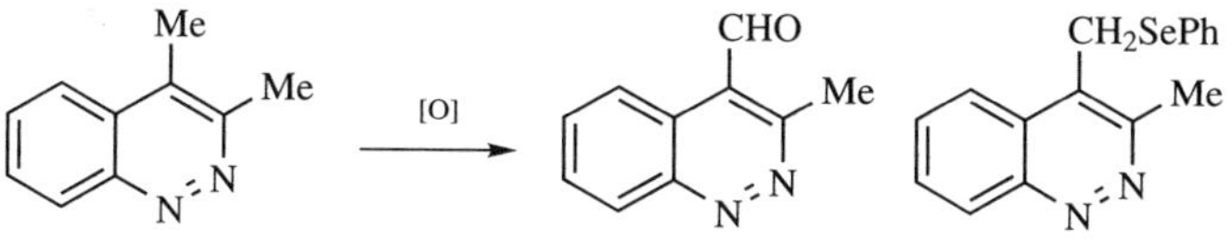

(**58**) (**59**) (**60**)

3-Phenylcinnoline (**61**) gave 6-phenyl-3,4-pyridazinedicarboxylic acid (**62**) [$KMnO_4$, H_2O, reflux (?), 3 h: 54%; note survival of the phenyl group, even under such vigorous oxidative conditions].[929]

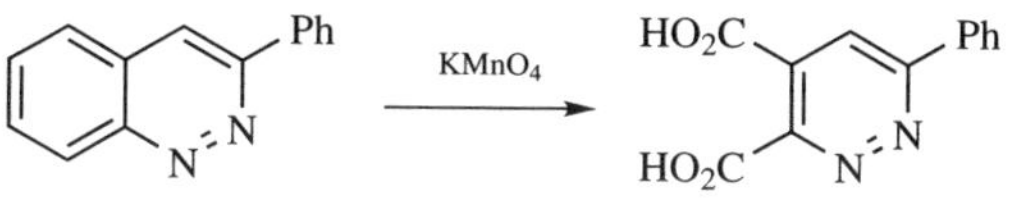

(**61**) (**62**)

Cyclization Reactions

2-Methylcinnolinium iodide (**63**) with ethyl *p*-nitrophenylacetimidate gave a single tricyclic product (**64**), the analysis and spectra of which were consistent with either 2-ethoxy-4-methyl-3-*p*-nitrophenyl-4*H*-pyrrolo[3,2-*c*]cinnoline (**64a**) or the isomeric 2-ethoxy-4-methyl-1-*p*-nitrophenyl-4*H*-pyrrolo[2,3-*c*]cinnoline (**64b**) (EtOH, 20°C, 5 h: 68%).[739]

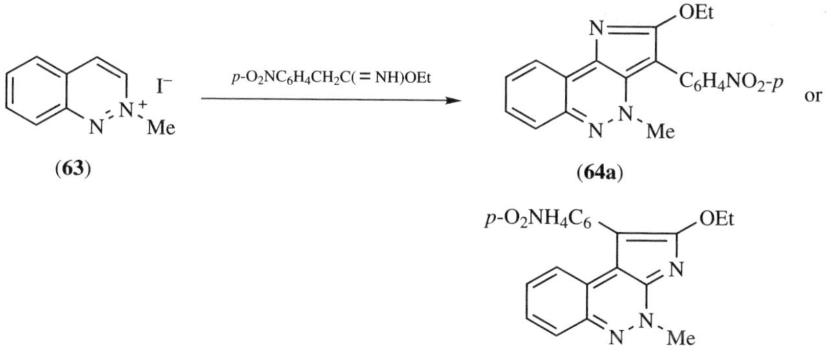

(**63**) (**64a**) (**64b**)

4-Chloro-3-phenylethynylcinnoline (**66**) with methylhydrazine gave 3-benzyl-1-methyl-1*H*-pyrazolo[4,3-*c*]cinnoline (**65**) (EtOH, 20°C, 12 h: 39%), but with phenylhydrazine it gave 1-anilino-2-phenyl-1*H*-pyrrolo[3,2-*c*]cinnoline (**67**) (EtOH, reflux, 15 min: 30% as hydrochloride).[293]

(**65**) ← $MeHNNH_2$ (−HCl) — (**66**) — $PhHNNH_2$ (−HCl) → (**67**)

Also other examples.[979]

Ring Contraction

3,4-Dimethyl-2-phenyl-2,3-dihydrocinnoline (**68**) rearranged into 1-anilino-2,3-dimethylindoline (**69**) (AcOH, BuOH, 100°C, 5 min: ?%).[750]

(**68**) → H^+, Ω → (**69**)

Also another minimally described example.[829]

CHAPTER 3

Halogenocinnolines (*H* 29; *E* 121)

An extraordinary paucity of fresh data on all aspects of halogenocinnolines in the 1972–2004 literature is reflected in the brevity of treatment accorded here to these important compounds.

Halogeno substituents at the 3- or 4-position of cinnoline are appreciably activated by N2 and N1, respectively; those at positions 5–8 or an extranuclear position have activities only marginally better than those in corresponding carbocyclic compounds. There seems to be little difference in reactivity of a fluoro, chloro, bromo, or iodo substituent at the same position.

X-ray analysis of 8-(*trans*-1,2-diphenylvinyl)-5,7-difluoro-2,3,4-triphenylcinnoline has been reported.[540]

3.1. PREPARATION OF HALOGENOCINNOLINES (*H* 29; *E* 121)

Many halogenocinnolines, especially those with halogeno substituents on the carbocyclic ring or at extranuclear positions, have been made by *primary syntheses* (see Chapter 1), and some extranuclear halogenocinnolines have been made by *passenger alkylations* (see Section 2.2.1) or by *direct halogenation* (see Section 2.2.2). Other important preparative procedures are illustrated in the following examples.

By Direct Halogenation

4(1*H*)-Cinnolinone (**1**, R = H) gave 3-iodo-4 (1*H*)-cinnolinone (**1**, R = I) (substrate, AcONa, AcOH, then ICl↓ dropwise, 100°C: 95%).[293]

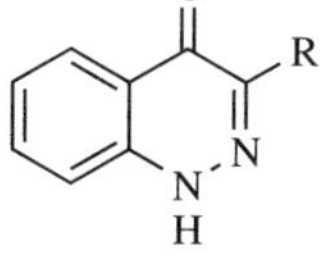

O
R
N
N
H

(**1**)

Cinnolines and Phthalazines: Supplement II, The Chemistry of Heterocyclic Compounds, Volume 64, by D.J. Brown

Ethyl 1,4-dihydro-3-cinnolinecarboxylate (**2**) gave ethyl 4-bromo-3-cinnolinecarboxylate (**3**) (substrate, AcOH, then Br_2↓ dropwise, 20°C → reflux, 30 min: 60%; note additional dehydrogenation).[54]

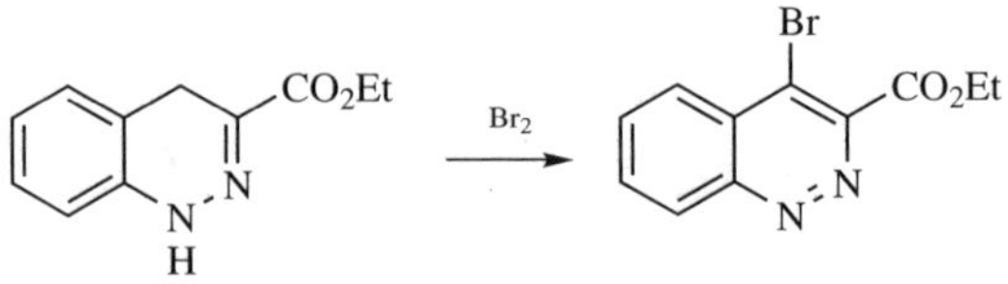

(**2**) (**3**)

1-Ethyl-7-(pyridin-4-yl)-4(1*H*)-cinnolinone (**4**, R = H) gave 3-bromo-1-ethyl-7-(pyridin-4-yl)-4(1*H*)-cinnolinone (**4**, R = Br) (substrate, AcOK, AcOH, reflux, then Br_2/AcOH↓ during 2 h: 76%).[45]

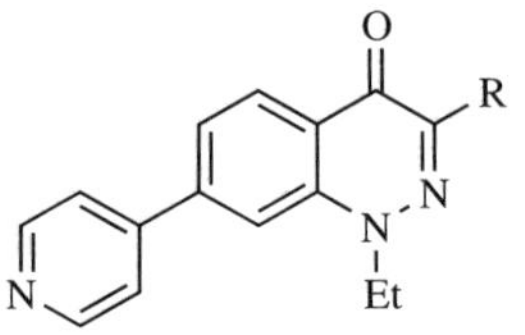

(**4**)

3-Cyclohexyl-6,7-dimethoxy-4(1*H*)-cinnolinone (**5**, R = H) gave 8-bromo-3-cyclohexyl-6,7-dimethoxy-4(1*H*)-cinnolinone (**5**, R = Br) (*N*-bromosuccinimide, Bz_2O_2, $CHCl_3$, reflux, 5 h: ~30%).[20]

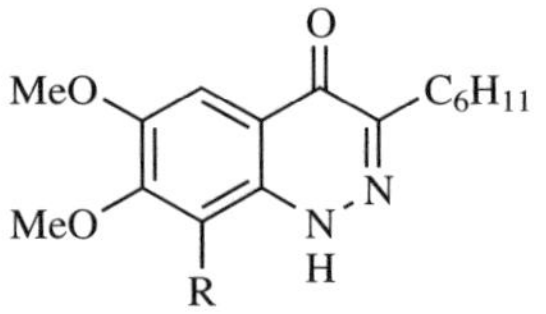

(**5**)

3(2*H*)-Cinnolinone (**6**, R = H) with chloramine (NH_2Cl) gave 4-chloro-3(2*H*)-cinnolinone (**6**, R = Cl) (CH_2Cl_2–Et_2O, 20°C, 12 h: 18% after separation from other products).[687]

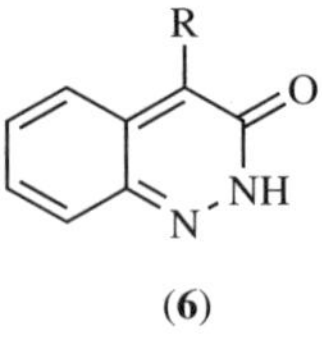

(**6**)

3-Acetyl-4(1*H*)-cinnolinone (**7**, Q = R = H) gave 3-acetyl-6-bromo-4(1*H*)-cinnolinone (**7**, Q = Br, R = H) (substrate, AcOH, Br_2↓ dropwise, 20°C:

80%) or 3-bromoacetyl-4(1*H*)-cinnolinone (**7**, Q = H, R = Br) (*N*-bromosuccinimide, CCl_4, reflux, 6 h: 75%).[619]

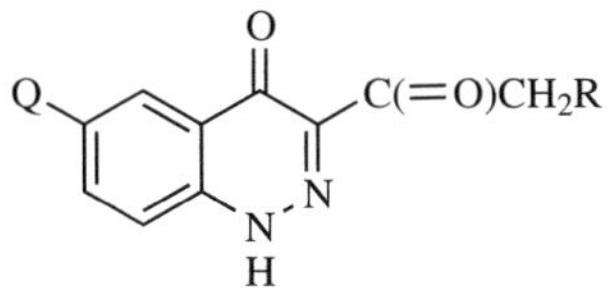

(**7**)

The fused substrate, quinoxalino[2,3-*c*]cinnoline (**7a**, R = H), gave 10-chloroquinoxalino[2,3-*c*]cinnoline (**7a**, R = Cl)(substrate, HCl gas, $CHCl_3$, 20°C, 1 min: deep blue precipitate; this dry solid, $CHCl_3$, 4M NaOH, shaken until orange: 75%; a rational mechanism, involving addition and aerial oxidation, is suggested).[710]

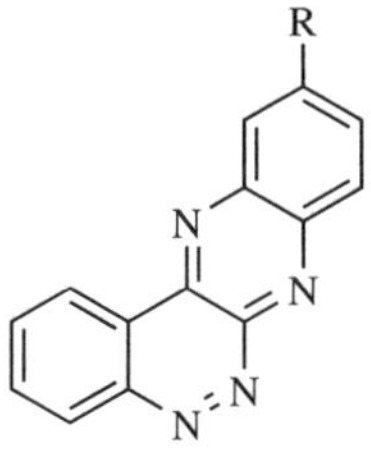

(**7a**)

By Halogenation via Metallo Derivatives

4-Chlorocinnoline (**8**, R = Cl) gave 4-chloro-3-iodocinnoline (**9**, R = Cl) [substrate, $LiNPr^i_2$ (made in situ), THF, −75°C, 30 min; then $I_2\downarrow$, −75°C, 2 h: 70%]; 4-methoxycinnoline (**8**, R = OMe) likewise gave 3-iodo-4-methoxycinnoline (**9**, R = OMe) (77%);[307] and 4-chlorocinnoline (**8**, R = Cl) likewise (but with an excess of lithiating agent and of iodine) gave a separable mixture of 4-chloro-3,8-diiodocinnoline (**10**, X = I) (56%) and 4-chloro-3-iodocinnoline (**10**, X = H) (37%).[307] 4-Iodo-3-methoxycinnoline (73%) was made in the same way as its isomer.[1010]

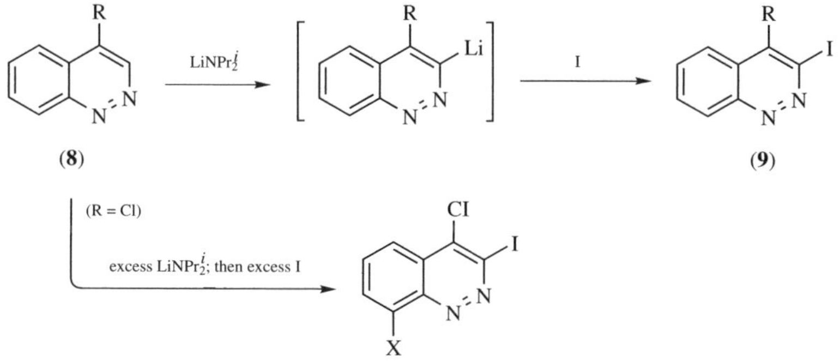

(**8**) → $LiNPr^i_2$ → [Li intermediate] → I → (**9**)

(R = Cl) excess $LiNPr^i_2$; then excess I → (**10**)

In contrast, 6,7-dimethoxy-4-*p*-methoxyphenylcinnoline (**11**, R = H) gave 8-iodo-6,7-dimethoxy-4-*p*-methoxyphenylcinnoline (**11**, R = I) [substrate, *N*-lithio-2,2,6,6-tetramethylpiperidine (made in situ), THF, −78°C, 1 h; then I↓, −78°C, 2 h: 84%; the selective 8-lithiation and halogenation may be due to steric hindrance at the 3-position by the massive 4-aryl substituent?].[822]

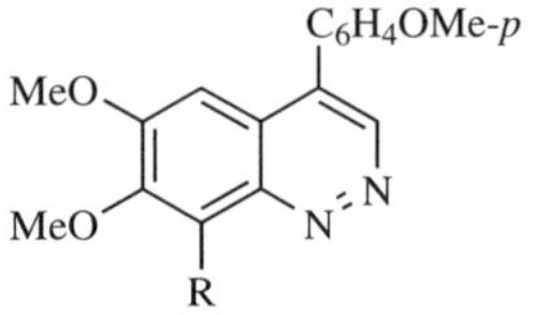

(**11**)

From Tautomeric Cinnolinones

Note: This is the most widely used method for making 3- or 4-chlorocinnolines; phosphoryl chloride (with or without a tertiary base), or a Vilsmeier reagent is usually employed.

4(1*H*)-Cinnolinone (**12**) gave 4-chlorocinnoline (**13**) ($POCl_3$, pyridine, PhCl, reflux, 1 h: 77–87%).[307,822,1012]

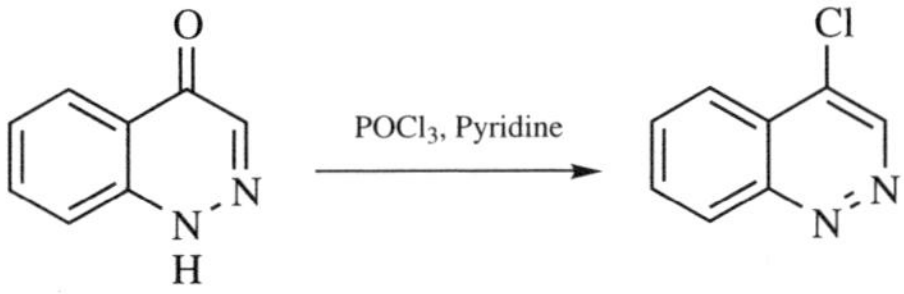

(**12**) (**13**)

6-Methoxy-8-nitro-4(1*H*)-cinnolinone (**14**) gave 4-chloro-6-methoxy-8-nitrocinnoline (**15**) (neat $POCl_3$, 80°C, 12 min: 98%).[27]

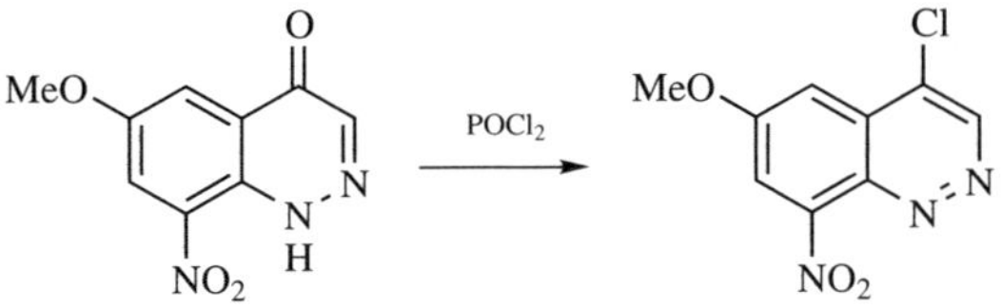

(**14**) (**15**)

3-Cyclohexyl-6,7-dimethoxy-4(1*H*)-cinnolinone gave 4-chloro-3-cyclohexyl-6,7-dimethoxycinnoline (**16**) (neat $POCl_3$, reflux, 30 min: ∼60%);[20] 4-oxo-1,4-dihydro-3-cinnolinecarbonitrile similarly gave 4-chloro-3-cinnolinecarbonitrile (**17**) [140°C → 120°C (bath temperature), 10 min: 50%].[695]

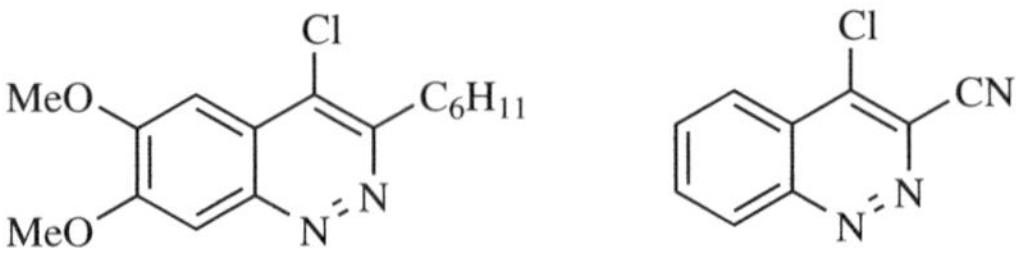

(**16**) (**17**)

3(2*H*)-Cinnoline (**18**) gave 3-chlorocinnoline (neat $POCl_3$, reflux, 10 days: 91%);[1010, cf. 307] the reduced substrate, 5,6,7,8-tetrahydro-3(2*H*)-cinnolinone (**19**), likewise gave 3-chloro-5,6,7,8-tetrahydrocinnoline (neat $POCl_3$, 100°C, until substrate dissolved: 20%).[952]

(**18**) (**19**)

6-Methoxy-7-(2-methoxyethoxy)-4(1*H*)-cinnolinone (**20**) gave 4-chloro-6-methoxy-7-(2-methoxyethoxy)cinnoline (**21**) ($SOCl_2$, Me_2NCHO, 80°C, 2 h: 74%).[828]

$SOCl_2$, Me_2NCHO (Vilsmeier reagent)

(**20**) (**21**)

Also other examples.[15,293]

3.2. REACTIONS OF HALOGENOCINNOLINES (*H* 29; *E* 131)

The *alkanelysis of halogenocinnolines* has been discussed in Section 2.2.1. Not all the other possible reactions of halogenocinnolines are represented in the more recent literature, but most of the important reactions are illustrated in the following classified examples.

Hydrogenolysis

Note: Direct reduction of a halogenocinnoline may result in simple hydrogenolysis or in the formation of a bicinnoline; indirect routes may be employed also.

4-Chloro-6-methoxy-8-nitrocinnoline (**22**) gave 6-methoxy-8-cinnolinamine (**23**) [Pd/C, H_2 (1 atm), EtOH, trace HCl, 20°C: 84%; note the concomitant reduction of the nitro group).[27]

[H]

(**22**) (**23**)

4-Chloro-3-cinnolinecarbonitrile (**24**) gave 4,4′-bicinnoline-3,3′-dicarbonitrile (**25**) ($Pd/BaSO_4$, H_2, EtOH, Et_3N, ?°C: 39%).[695] For a comparison, 3-bromocinnoline (**26**) gave 3,3′-bicinnoline (**27**) [Et_3N, $(Ph_3P)_2PdCl_2$, MeCN, sealed, 150°C, 5 h: 81%].[293]

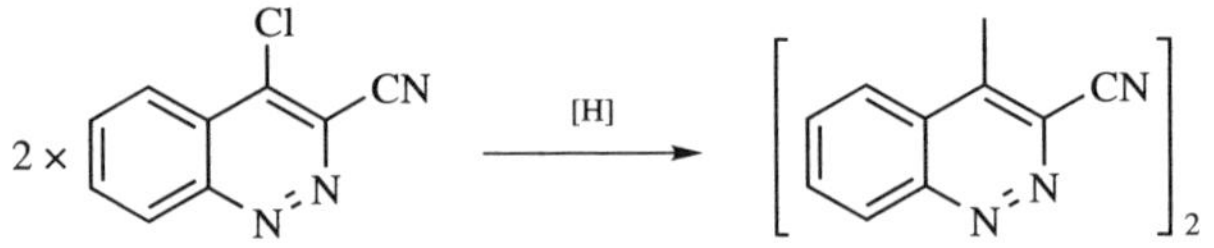

(**24**) (**25**)

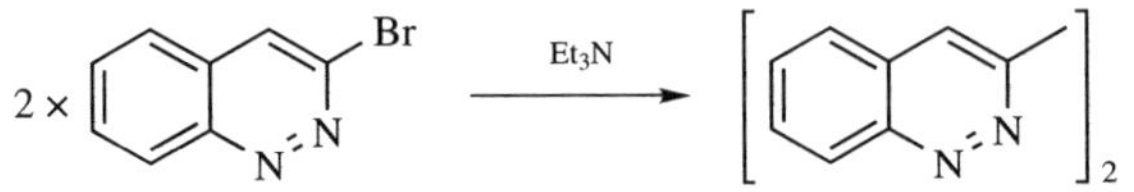

(**26**) (**27**)

4-Chloro-3-iodocinnoline (**28**) gave crude 3-iodo-4-(*N*′-tosylhydrazino) cinnoline (**29**) ($TsNHNH_2$, $CHCl_3$, 20°C, 1 week: uncharacterized) and thence 3-iodocinnoline (**30**) (5M Na_2CO_3, reflux, 90 min: 72% overall; note selective aminolysis of the 4-halogeno substituent).[293]

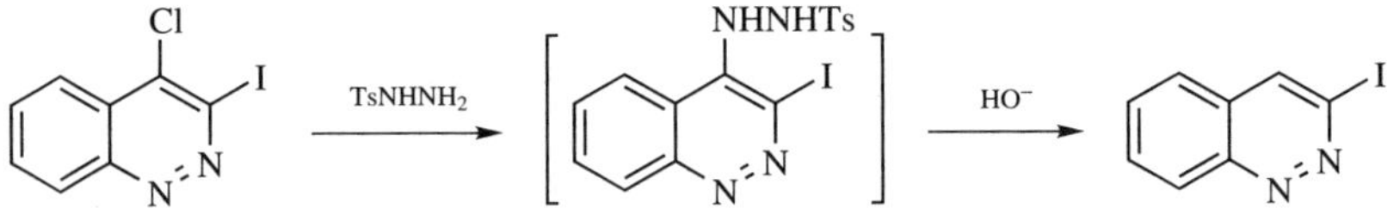

(**28**) (**29**) (**30**)

Aminolysis of 3-Halogenocinnolines

3-Chloro-5,6,7,8-tetrahydrocinnoline (**31**) gave 3-hydrazino-5,6,7,8-tetrahydrocinnoline (**32**) (neat $H_2NNH_2 \cdot H_2O$, reflux, 2 h: 71%).[952]

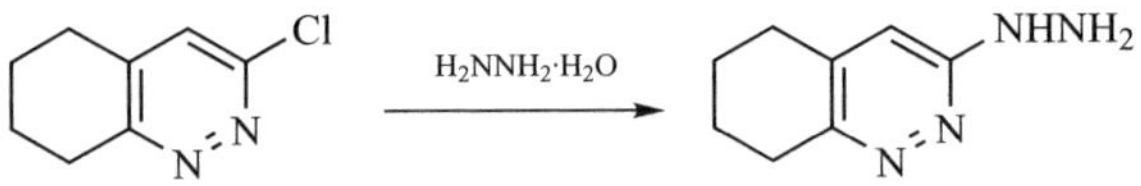

(**31**) (**32**)

Aminolysis of 4-Halogenocinnolines

4-Chlorocinnoline (**34**) gave 4-cinnolinamine (**33**, R = H) (NH_3 gas, THF, sealed, 140°C, 12 h: 84%)[307] or 4-(*p*-bromobenzylamino)cinnoline (**33**, R = $CH_2C_6H_4Br$-*p*) (*p*-$BrC_6H_4CH_2NH_2 \cdot HCl$, K_2CO_3, trace KI, Me_2NCHO, reflux, 4 h: ~60%);[420] analogs likewise.[420]

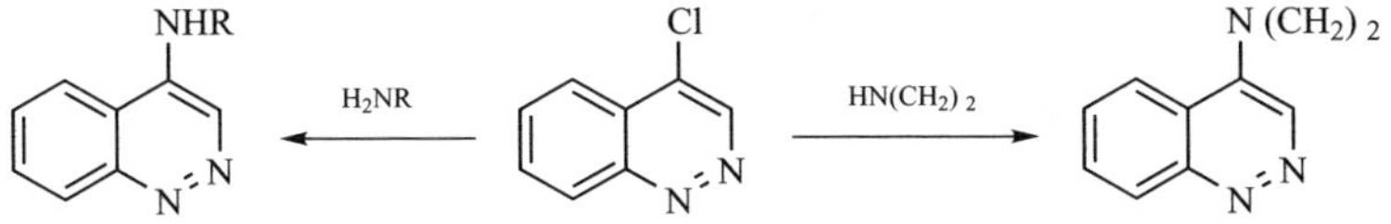

(33) (34) (35)

The same substrate (**34**) gave 4- (aziridin-1-yl)cinnoline (**35**) [$HN(CH_2)_2$, Et_3N, PhH, 50°C, 4 days: 64%], analogs likewise.[17]

4-Chloro-3-methoxycinnoline 1-oxide (**36**) gave 3-methoxy-4-*p*-toluidinocinnoline 1-oxide (**37**) ($H_2NC_6H_4Me$-*p*, NaH, PhH, reflux, 2 h; then substrate↓, reflux, 6 h: 38%).[709]

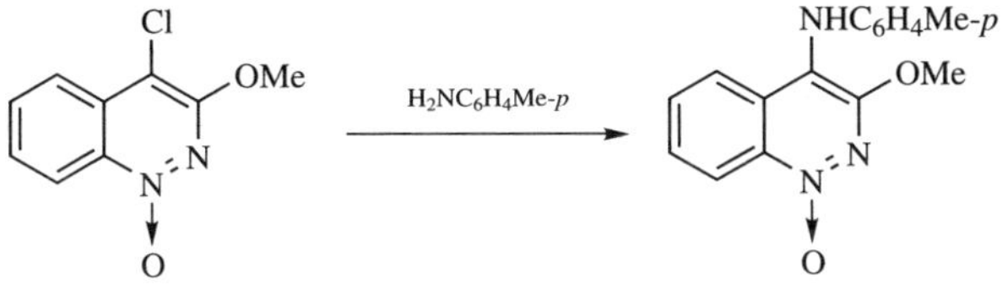

(36) (37)

Also other examples.[236,281,293,772,828,993,1012]

Aminolysis of 5-Halogenocinnolines

1-*p*-Chlorophenyl-5-fluoro- (**38**) gave 1-*p*-chlorophenyl-5-dimethylamino-4-oxo-1,4-dihydro-3-cinnolinecarboxylic acid (**39**) [substrate (as K salt), $Me_2NH \cdot HCl$, K_2CO_3, H_2O, reflux, 36 h: 94%); analogs likewise.[765]

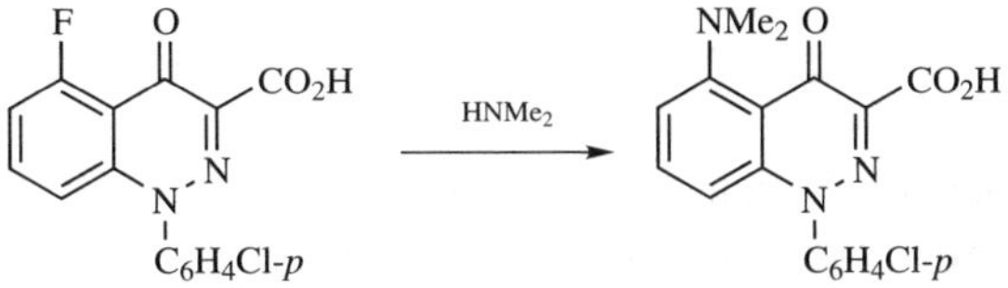

(38) (39)

Aminolysis of 7-Halogenocinnolines

7-Chloro-6-fluoro-1-methyl-4-oxo-1,4-dihydro-3-cinnolinecarboxylic acid (**40**, R = Me) gave 6-fluoro-1-methyl-4-oxo-7-(piperazin-1-yl)-1,4-dihydro-3-cinnolinecarboxylic acid (**41**, R = Me) [$HN(CH_2CH_2)_2NH$, pyridine, 85°C, 45 min: 91%]; analogs likewise.[416]

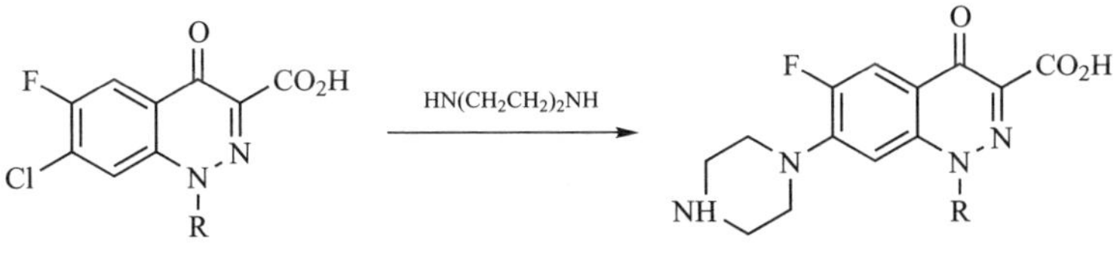

(40) (41)

7-Chloro-6-fluoro-1-*p*-fluorophenyl-4-oxo-1,4-dihydro-3-cinnolinecarboxylic acid (**40**, R = C_6H_4F-*p*) gave 6-fluoro-1-*p*-fluorophenyl-4-oxo-7-(piperazin-1-yl)-1,4-dihydro-3-cinnolinecarboxylic acid (**41**, R = C_6H_4F-*p*) [HN(CH_2CH_2)$_2$NH, pyridine, 100°C, until no substrate on tlc (> 5 h): 89%]; analogs likewise.[617]

Azidolysis

4-Chlorocinnoline (**42**) gave 4-azidocinnoline (**43**) (NaN_3, H_2O, EtOH, 95°C, 3 h: 72%).[402]

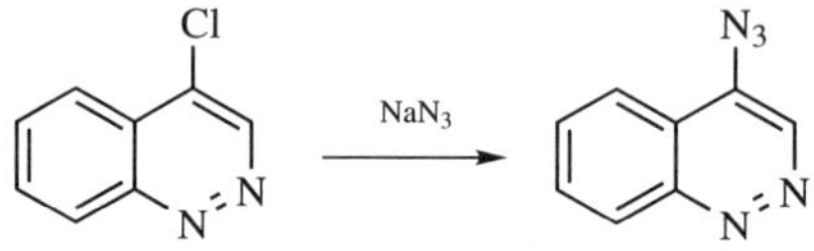

(**42**) $\xrightarrow{NaN_3}$ (**43**)

Cyanolysis

3-Bromo-4(1*H*)-cinnolinone (**44**) gave 4-oxo-1,4-dihydro-3-cinnolinecarbonitrile (**45**) (CuCN, pyridine, reflux, 16 h: 92%).[695]

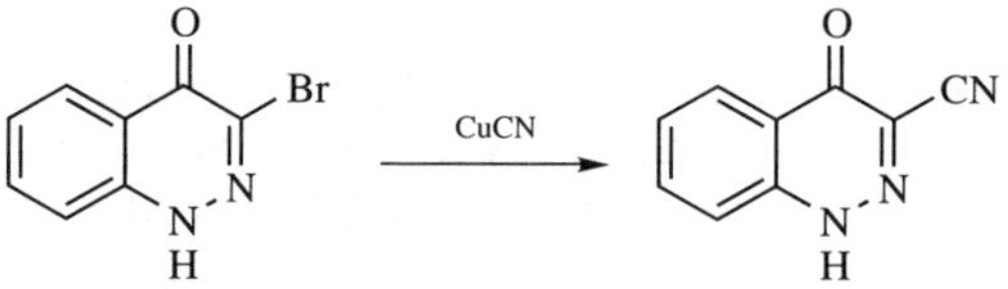

(**44**) $\xrightarrow{CuCN}$ (**45**)

3-Bromo-1-phenyl-7-(pyridin-4-yl)-4(1*H*)-cinnolinone (**46**, R = Br) gave 4-oxo-1-phenyl-7-(pyridin-4-yl)-1,4-dihydro-3-cinnolinecarbonitrile (**46**, R = CN) (CuCN, Me_2NCHO, reflux, 18 h: crude nitrile), characterized by conversion into 4-oxo-1-phenyl-7-(pyridin-4-yl)-1,4-dihydro-3-cinnolinecarboxylic acid (**46**, R = CO_2H) (50% H_2SO_4, 110°C, 12 h: 33% overall).[45]

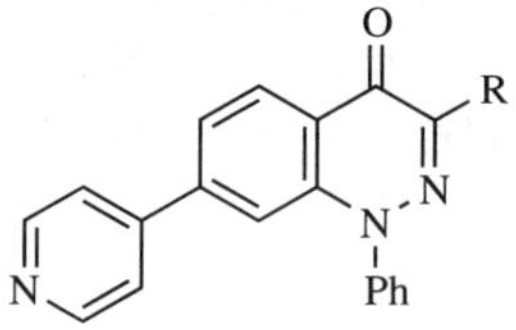

(**46**)

3-Bromo-4-chlorocinnoline (**47**) gave 3,4-cinnolinedicarbonitrile (**49**) via tolylsulfonyl intermediates (*p*-$MeC_6H_4SO_2Na$, Me_2NCHO, N_2, 0°C, 40 min; then KCN↓, 5-10°C, 4 h: 82%); one such intermediate, 4-*p*-tolylsulfonyl-3-

cinnolinecarbonitrile (**48**) (33%) was isolated by stopping the reactions after 2 h.[291]

p-$MeC_6H_4SO_2Na$, KCN

(**47**) (**48**) (**49**)

Also other examples.[485]

Conversion into Acylcinnolines

4-Chlorocinnoline (**50**) gave 4-benzoylcinnoline (**51**) [PhCHO, p-MeC_6H_4-SO_2Na, 1,3-dimethylimidazolium iodide (catalyst), Me_2NCHO, 80°C, 10 min: 73%; when the sulfinate was omitted, the yield was 39%: this suggests that the reaction proceeds better via an intermediate sulfone]; several substituted-benzoyl analogs were made similarly.[601]

PhCHO, p-$MeC_6H_4SO_2Na$, NaH, catalyst

(**50**) (**51**)

Hydrolysis

Note: The hydrolysis of nuclear halogenocinnolines is seldom intentional, but hydrolysis of di(or tri) halogenomethyl derivatives is a useful route to carbaldehydes or carboxylic acids, respectively.

4-Chloro-3-cinnolinecarbonitrile (**52**) underwent hydrolysis of both substituents to give 4-oxo-1,4-dihydro-3-cinnolinecarboxylic acid (**53**) (4M NaOH, reflux, 2.5 h: ~65%).[695]

HO^-

(**52**) (**53**)

4-Trichloromethylcinnoline (**54**) gave 4-cinnolinecarboxylic acid (**55**) (4M NaOH, ?°C, ? h: ?%).[291]

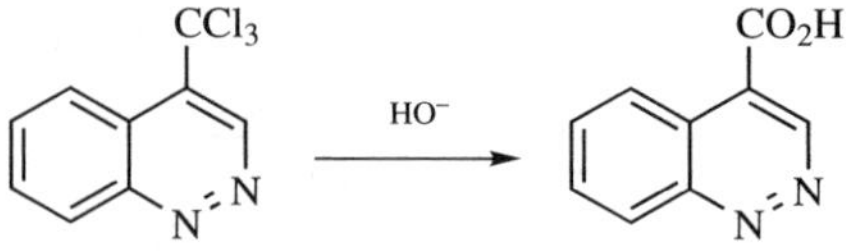

(**54**) (**55**)

4-Dichloromethyl-3-methylcinnoline (**57**), prepared in situ from 3,4-dimethylcinnoline (**56**) (NaOCl, H_2O, 20°C, 7 days: not isolated), likewise gave 3-methyl-4-cinnolinecarbaldehyde (**58**) (4M NaOH, no details: 16% overall).[291]

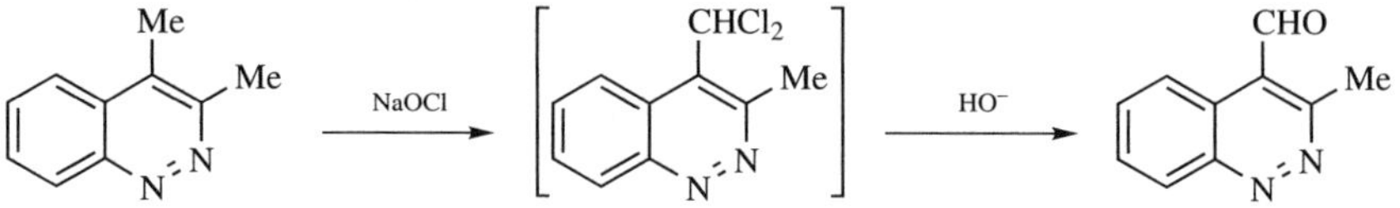

(**56**) (**57**) (**58**)

Alcoholysis or Phenolysis

3-Chlorocinnoline (**59**) gave 3-methoxycinnoline (**60**) (MeONa, MeOH, sealed, 120°C, 4 h: 93%).[307]

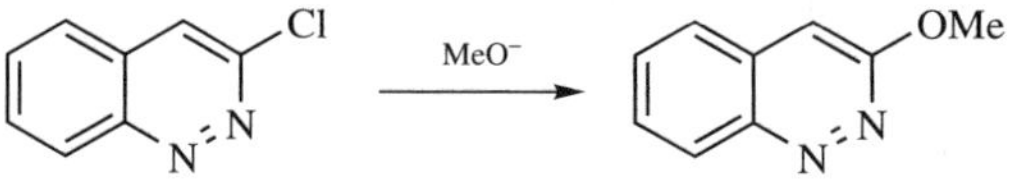

(**59**) (**60**)

1-*p*-Chlorophenyl-5-fluoro-4-oxo-1,4-dihydro-3-cinnolinecarboxylic acid (**61**, R = F) gave 1-*p*-chlorophenyl-4-oxo-5-propoxy-1,4-dihydro-3-cinnolinecarboxylic acid (**61**, R = OPr) (PrONa, PrOH, dioxane, hot → 20°C, 16 h: 95%); analogs likewise.[765]

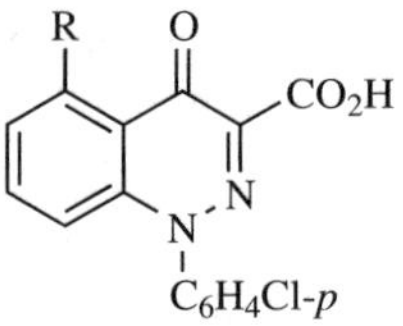

(**61**)

Also other examples.[162]

Thiolysis

Note: The direct thiolysis of halogenocinnolines with sodium hydrogen sulfide appears to be unrepresented, but thiolysis with thiourea, via an uncharacterized thiouronio salt, has been used.

3-Bromo-4-chloro-6,7-dimethoxycinnoline (**62**) underwent selective reaction with thiourea to give the 4-thiouronio salt (**63**) (EtOH, reflux, 1 h: crude) and thence 3-bromo-6,7-dimethoxy-4(1*H*)-cinnolinethione (**64**) (Na_2CO_3, H_2O, 20°C: 80% overall).[635]

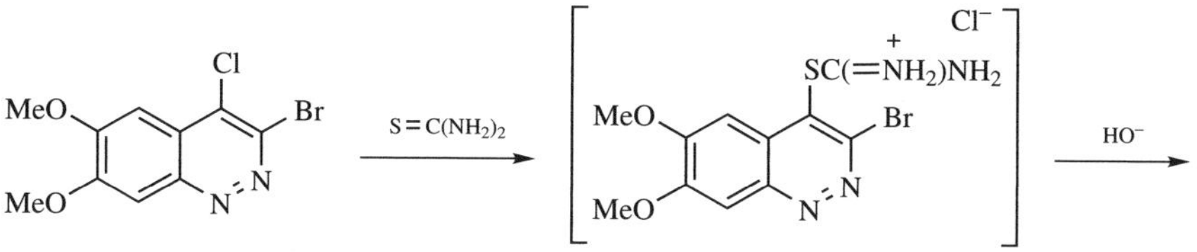

(**62**) (**63**)

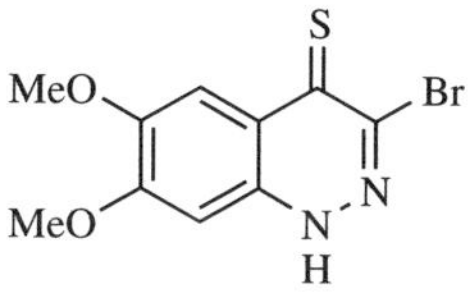

(**64**)

In much the same way, 4-chloro-3-cinnolinecarbonitrile (**65**) gave 4-thioxo-1,4-dihydro-3-cinnolinecarbonitrile (**66**) [$S{=}C(NH_2)_2$, MeOH, reflux, 10 min: ~90%; no alkaline treatment needed).[695]

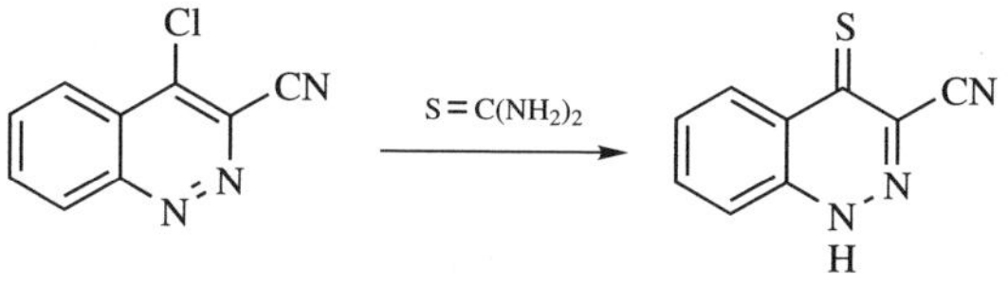

(**65**) (**66**)

Alkane- or Arenethiolysis

4-Chloro- gave 4-*tert*-butylthiocinnoline [Bu^{t}SLi (made in situ), THF, substrate↓, 0°C → 66°C, 2.5 h: 93%]; homologs prepared somewhat similarly.[1010]

4-Iodo-3-methoxycinnoline (**67**) gave 3-methoxy-4-phenylthiocinnoline (**68**) (Li powder, PhSSPh, THF, A, 20°C, ultrasound, 30 min: 24%).[821]

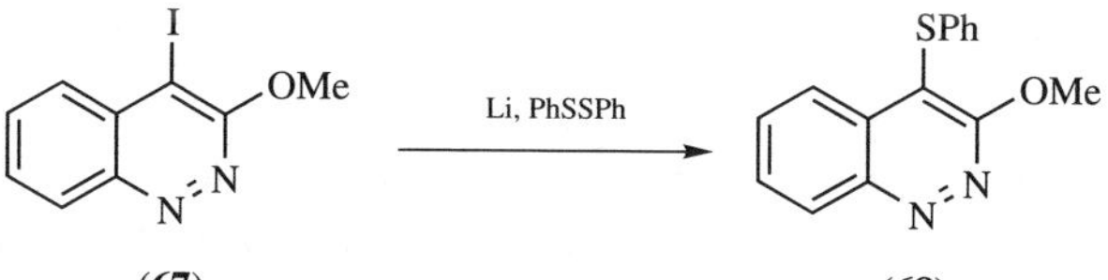

(**67**) (**68**)

Arenesulfinolysis

4-Chlorocinnoline (**69**, R = H) gave 4-*p*-tolylsulfonylcinnoline (**70**, R = H) (*p*-$MeC_6H_4SO_2Na$, Me_2SO, 55°C, 15 min: 80%).[937]

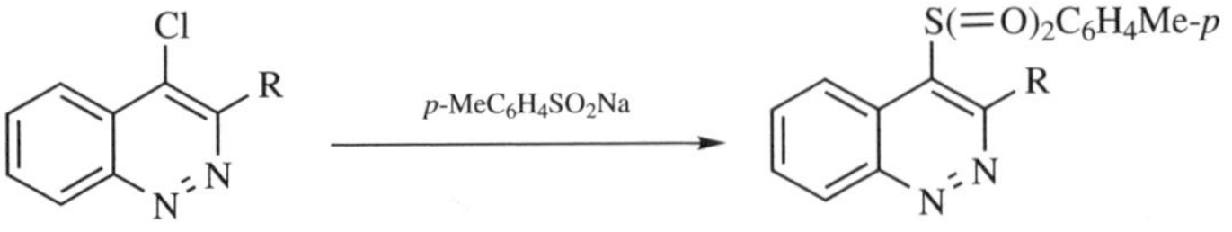

(**69**) (**70**)

4-Chloro-3-cinnolinecarbonitrile (**69**, R = CN) gave 4-*p*-tolylsulfonyl-3-cinnolinecarbonitrile (**69**, R = CN) (*p*-$MeC_6H_4SO_2Na$, Me_2NCHO, 0°C → 5°C, 1 h: ?%).[291]

Pyrolysis

Hexachlorocinnoline (**71**) gave hexachlorophenylacetylene (**72**) (vapor over SiO_2, 770°C, vacuum: 37%, after separation from an unidentified product).[688]

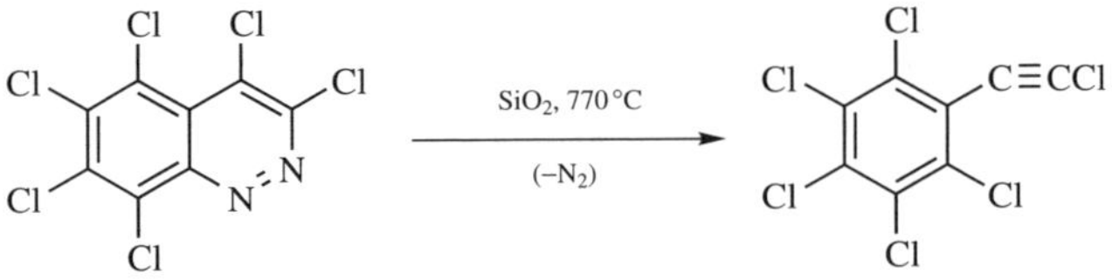

(**71**) (**72**)

Cyclization Reactions

Note: A few random examples of cyclizations involving halogenocinnolines are given here.

3-Iodo-4(1*H*)-cinnolinone (**73**) with phenylacetylene in the presence of appropriate catalysts gave 2-phenylfuro[3,2-*c*]cinnoline (**74**) [$(Ph_3P)_2PdCl_2$, CuI, $OP(NMe_2)_3$, Et_3N, N_2, 20°C, 6 h: 49%]; analogs likewise.[293]

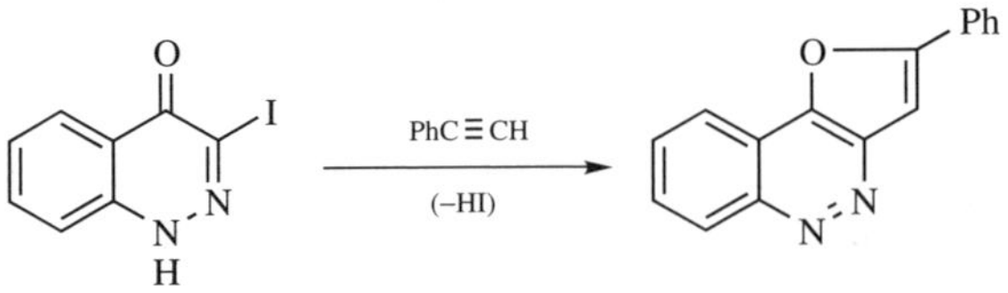

(**73**) (**74**)

4-Chloro-3-cinnolinecarbonitrile (**76**) with methyl 2-mercaptoacetate gave methyl 3-aminothieno[3,2-*c*]cinnoline-2-carboxylate (**75**) (Na_2CO_3, EtOH, reflux, 4 h: ~90%) or with methylhydrazine gave 1-methyl-1*H*-pyrazolo [4,3-*c*]cinnolin-3-amine (**77**) ($MeHNNH_2$, EtOH, reflux, 2 h: 53%; analogs likewise); the foregoing products were formulated as their 3-imino tautomers.[695]

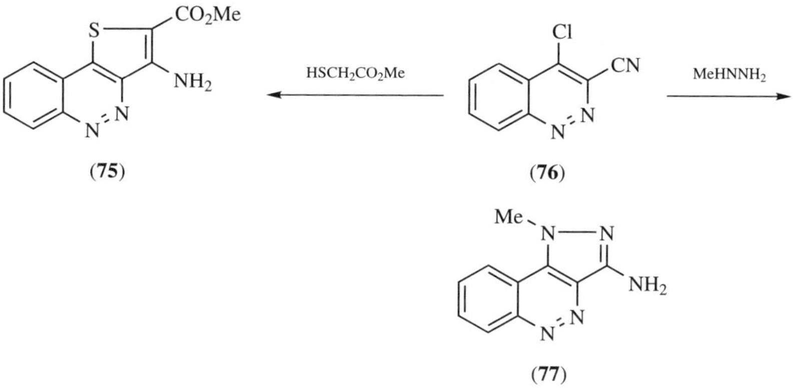

(**75**) (**76**) (**77**)

3-Bromo-4-chloro-6,7-dimethoxycinnoline (**78**) gave 7,8-dimethoxythiazolo[4,5-*c*]cinnolin-2-amine (**79**) [S=C(NH_2)$_2$, EtOH, reflux, 5 h: 36%]; analogs likewise.[635]

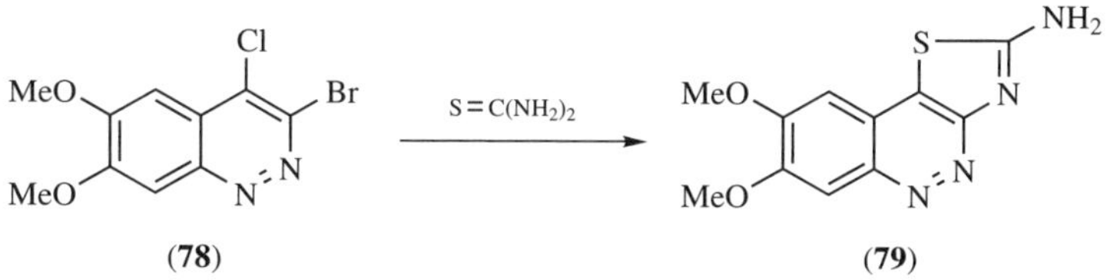

(**78**) (**79**)

1,2-Bis(2-bromoacetyl)-2,3-dihydro-4(1*H*)-cinnolinone (**80**) gave 3-phenyl-2,3,4,5,7,8-hexahydro-1*H*-[1,2,5]triazepino[1,2-*a*]cinnoline-1,5,8-trione (**81**) ($PhNH_2$, K_2CO_3, MeCN, reflux, ? h: ?%); analogs likewise (for details, see original).[475]

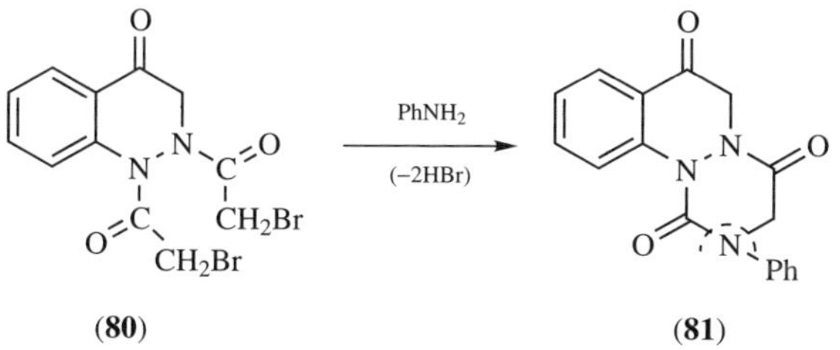

(**80**) (**81**)

4-[*N*-(2-Dimethylaminoethyl)-*o*-iodobenzamido]cinnoline (**82**) gave 11-(2-dimethylaminoethyl)isoquino[4,3-*c*]cinnolin-12(11*H*)-one (**83**) [substrate,

$Pd(OAc)_2$, $P(C_hH_4Me\text{-}o)_3$, Ag_2CO_3, Me_2NCH), reflux, 30 min: 15%]; analogs likewise.[1012]

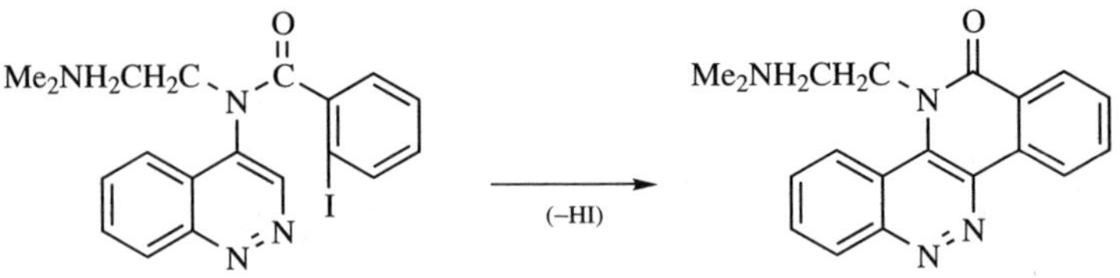

(–HI)

(**82**) (**83**)

CHAPTER 4

Oxycinnolines (*H* 16, 29, 48; *E* 62, 150, 273)

The term *oxycinnoline* includes compounds such as the tautomeric cinnolinone (**1**); its nontautomeric *N*-methylated (**2** and **3**), *O*-methylated (**4**), or *O*-acetylated derivative (**5**); the extranuclear hydroxymethylcinnoline (**6**, R = H) and its *O*-methylated (**6**, R = Me) or *O*-acetylated derivative (**6**, R = Ac); the cinnoline *N*-oxide (**7**); and the cinnolinequinone (**8**). Not all such categories are represented in the 1972–2004 literature.

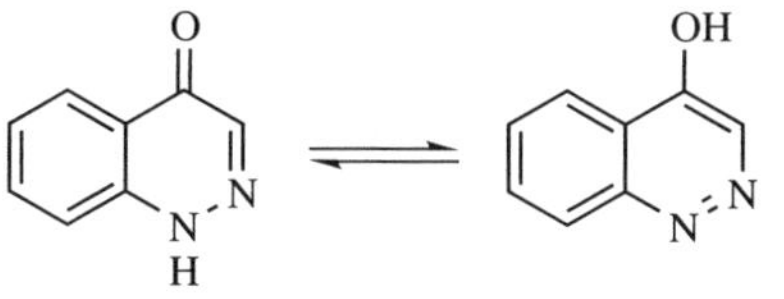

(**1**) 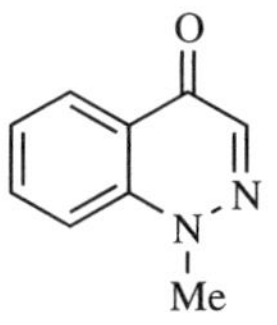 (**2**)

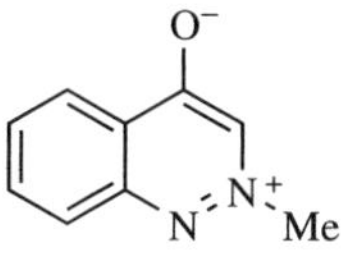

(**3**) 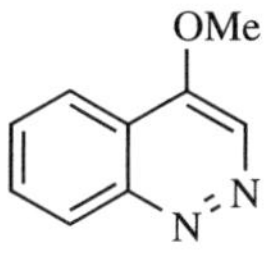 (**4**) 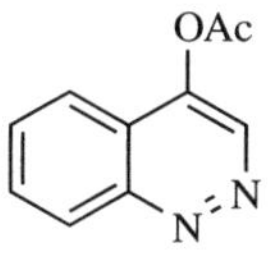 (**5**)

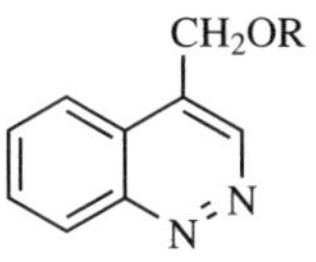

(**6**) 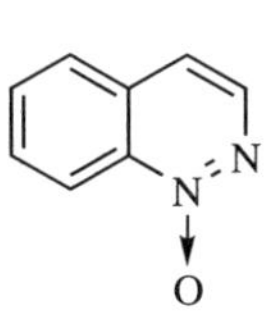 (**7**) 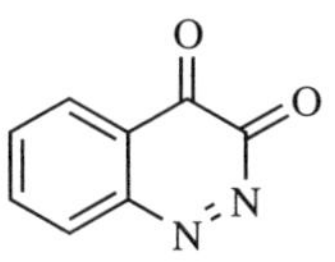 (**8**)

4.1. TAUTOMERIC CINNOLINONES (*H* 16; *E* 61)

Although there is little doubt that 4(1*H*)- (1) and 3(2*H*)-cinnolinone exist predominantly as such rather than as the tautomeric 4- and 3-cinnolinols, respectively, this issue has received further theoretical,[761,807] nuclear quadrupole resonance,[807] and X-ray crystallographic study.[807] The tautomeric states of compounds such as 5(1*H*)-cinnolinone (**9**) are still open to doubt; however, for pragmatic reasons, such compounds are also referred to as *cinnolinones* in this book.

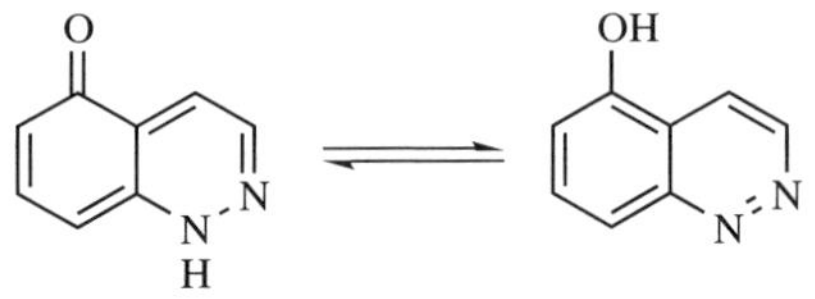

(**9**)

X-Ray crystallographic analyses have been reported for 4(1*H*)-cinnolinone (**10**, R = H),[807] 6-chloro-4(1*H*)-cinnolinone (**10**, R = Cl),[634] 8-chloro-2,3-dihydro-4(1*H*)-cinnolinone (**11**, R = H),[632] 8-chloro-2-chloroacetyl-2,3-dihydro-4(1*H*)-cinnolinone (**11**, R = $COCH_2Cl$),[632] 7,7-dimethyl-3-phenyl-4,6,7,8-tetrahydro-5(1*H*)-cinnolinone (**12** or a tautomer),[219] and 2-ethoxycarbonylmethyl-2,3-dihydro-4(1*H*)-cinnolinone;[633] also 6-fluoro- and 7-methyl-4-oxo-1,4-dihydro-3-cinnolinecarboxylic acid.[981]

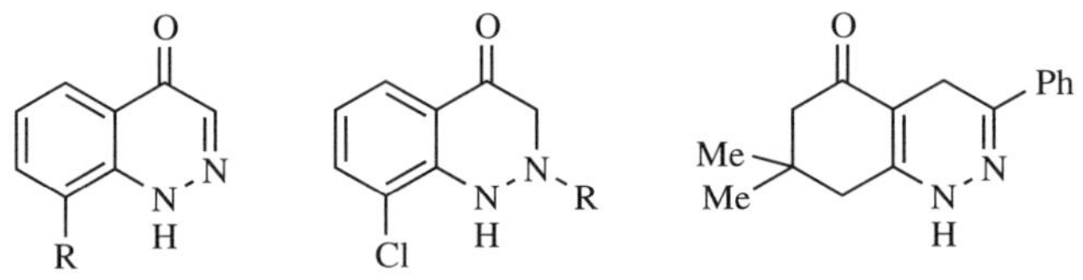

(**10**) (**11**) (**12**)

4.1.1. Preparation of Tautomeric Cinnolinones (*H* 16; *E* 62)

Most such cinnolinones have been made by *primary synthesis* (see Chapter 1), a few by *hydrolysis of halogenocinnolines* (see Section 3.2), and others using procedures illustrated in the examples below; it is noteworthy that several useful preparative routes, such as hydrolysis of alkylthiocinnolines or other types of thiocinnoline, appear to be unrepresented in the 1972–2004 literature.

From Cinnolinamines

4-Amino-3(2*H*)-cinnolinone (**13**) gave 3,4(1*H*,2*H*)-cinnolinedione (**14**) (diazotization and subsequent hydrolysis: 60%; for details, see original).[772]

Alkaline hydrolysis has also been used.[554]

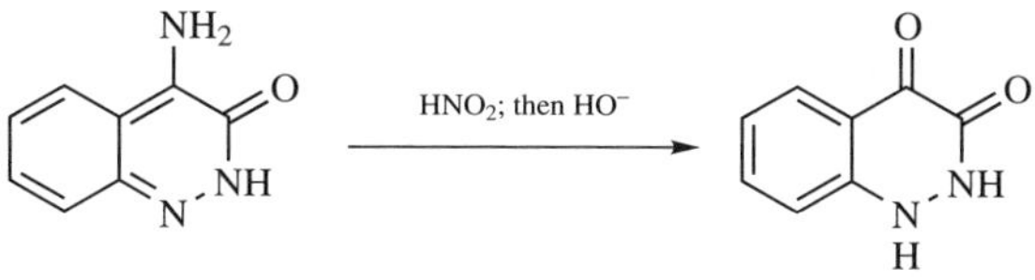

(13) → HNO_2; then HO^- → (14)

From Alkoxy- or Aryloxycinnolines

Note: This transformation can be done by hydrolysis or (in the case of a benzyloxy substrate) by reduction.

3-Ethynyl-4-phenoxycinnoline (**15**) gave 3-ethynyl-4(1*H*)-cinnolinone (**15a**) (HCl, H_2O, EtOH, reflux, 10 min: 72%); 3-phenylethynyl-4(1*H*)-cinnolinone (~50%) was made similarly.[293]

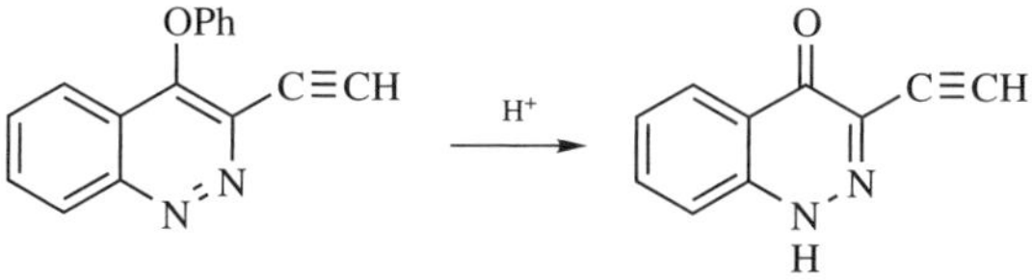

(15) → H^+ → (15a)

Di-*tert*-butyl 7-benzyloxy-6-methoxy-1,2-dihydro-1,2-cinnolinedicarboxylate (**16**) gave di-*tert*-butyl 6-methoxy-7-oxo-1,2,3,7-tetrahydro-1,2-cinnolinedicarboxylate (**16a**) (formulated as the 7-hydroxy tautomer) [H_2 (1.3 atm), Pd/C, AcOEt, 20°C, 4 h: 70%]; also an analog likewise.[494]

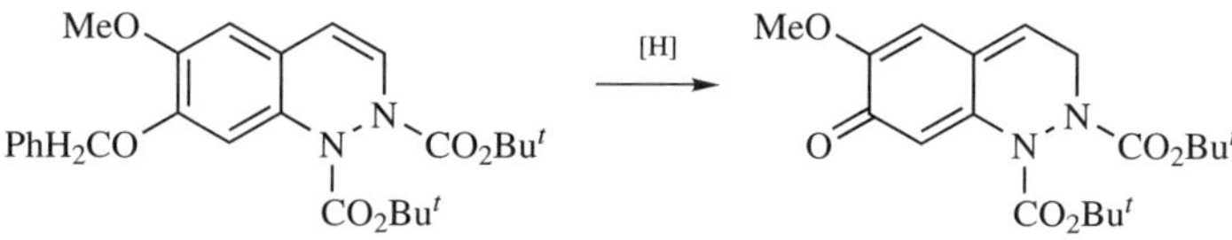

(16) → [H] → (16a)

From Nontautomeric Cinnolinones

3-(Triphenylphosphoranylidene)-3,4-dihydro-4-cinnolinone (**17**) gave 4(1*H*)-cinnolinone (**18**) (NaOH, MeOH, H_2O, reflux, 2 h: 97%); several analogs likewise.[335]

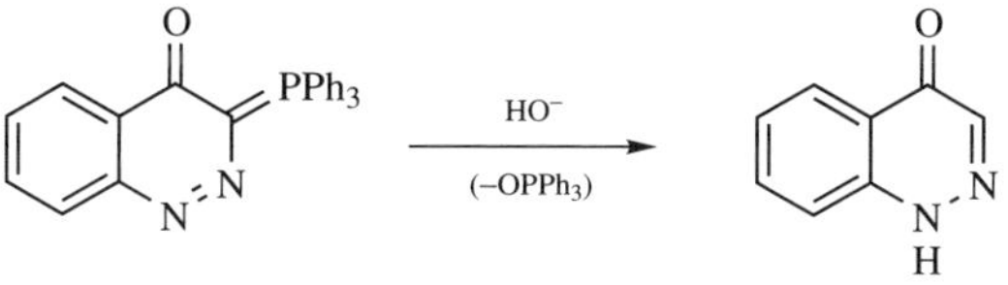

(17) → HO^- ($-OPPh_3$) → (18)

4.1.2. Reactions of Tautomeric Cinnolinones (*H* 23; *E* 80)

The conversion of *tautomeric cinnolinones into halogenocinnolines* has been covered in Section 3.1. Other reactions are discussed in the subsections that follow.

4.1.2.1. Alkylation of Tautomeric Cinnolinones

Tautomeric cinnolinones usually undergo *N*-alkylation to afford regular *N*-alkylcinnolinones, zwitterionic *N*-alkylcinnoliniumolates, or a mixture of both; *O*-alkylation to afford alkoxycinnolines is rare but not unknown. These possibilities are illustrated in the following classified examples.

Formation of Only Regular *N*-Alkylcinnolinones

3(2*H*)-Cinnolinone (**19**) gave 2-methyl-3(2*H*)-cinnolinone (**20**, R = Me) (substrate, NaOH, H_2O, EtOH, then $Me_2SO_4\downarrow$, mild exotherm, 90 min: 80%;[37] or CH_2N_2, Et_2O, 30 min: 71%)[373] or 2-benzyl-3(2*H*)-cinnolinone (**20**, R = CH_2Ph) (substrate, KOH, EtOH, reflux, then $PhCH_2Cl\downarrow$ dropwise, reflux, 12 h: 62%).[42]

(**19**) (**20**)

7-Phenyl-5,6-dihydro-3(2*H*)-cinnolinone (**21**, R = H) gave 2-ethoxycarbonylmethyl-7-phenyl-5,6-dihydro-3(2*H*)-cinnolinone (**21**, R = CH_2CO_2Et) (EtO_2-CCH_2Br, K_2CO_3, AcMe, reflux, 24 h: 40%).[517]

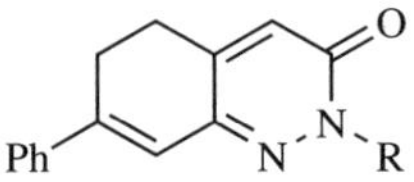

(**21**)

Ethyl 7-methoxy-4-oxo-1,4-dihydro-3-cinnolinecarboxylate (**22**, R = H) gave ethyl 1-ethyl-7-methoxy-4-oxo-1,4-dihydro-3-cinnolinecarboxylate (**22**, R = Et) [EtI, K_2CO_3 (1 equiv), AcMe, reflux, 8 h: 78% (as hydriodide salt); then Et_3N, CH_2Cl_2, 20°C, 48 h: 91% (as base)].[387]

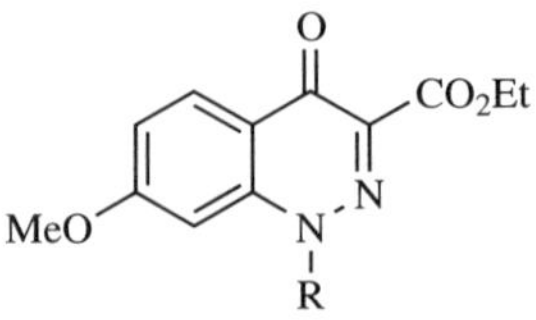

(**22**)

3-Methylsulfonyl-4(1*H*)-cinnolinone (**23**, R = H) gave 1-methyl-3-methylsulfonyl-4(1*H*)-cinnolinone (**23**, R = Me) (substrate, NaH, Me_2NCHO, 20°C, 1 h; then MeI↓ dropwise, 20°C, 2 h: 81%).[485]

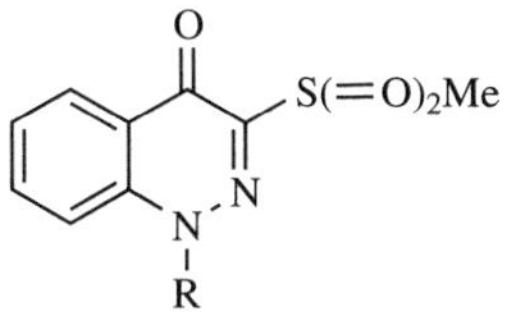

(**23**)

3-Bromo-7-(pyridin-4-yl)-4(1*H*)-cinnolinone (**24**, R = H) gave 3-bromo-1-ethyl-7-(pyridin-4-yl)-4(1*H*)-cinnolinone (**24**, R = Et) (substrate, NaH, Me_2NCHO, 20°C, 30 min; then EtI↓ dropwise, 20°C, 2 h: 76%).[45]

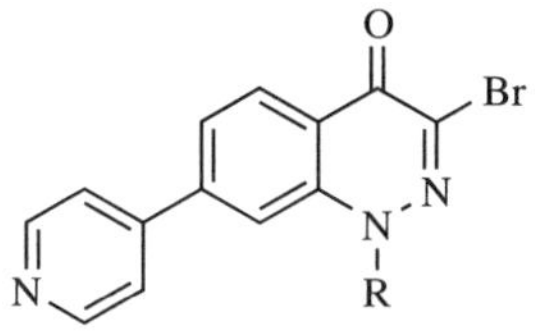

(**24**)

3-Triphenylphosphoranylidene-3,4-dihydro-4-cinnolinone (**25**) gave 1-methyl-3-triphenylphosphonio-4(1*H*)-cinnolinone iodide (**26**) (neat MeI, reflux, ? h: > 95%) and thence 1-methyl-4(1*H*)-cinnolinone (**27**) (NaOH, MeOH, H_2O, 20°C, <10 min: 80%); analogs similarly.[334]

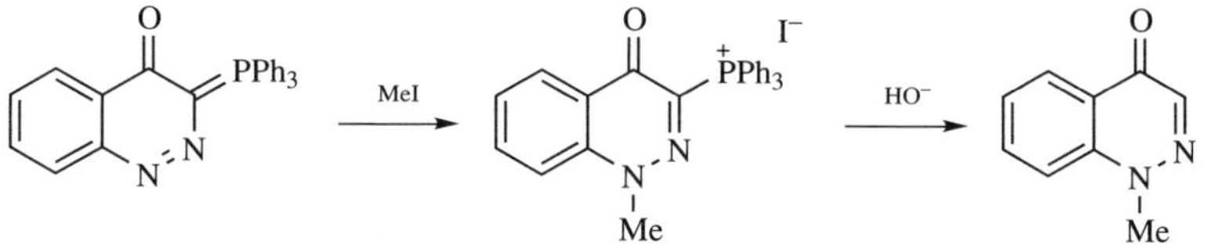

(**25**) (**26**) (**27**)

Also other examples.[54,213,509]

Formation of Only *N*-Alkylcinnoliniumolates

6-Chloro-4(1*H*)-cinnolinone (**28**) gave 6-chloro-2-methyl-4-oxo-1,4-dihydrocinnolinium *p*-toluenesulfonate (**29**) (TsOMe, kerosene, 170°C, 4 h: 80%) and thence 6-chloro-2-methylcinnolin-2-ium-4-olate (**30**) (H_2O, NaOH↓ to pH 8: 40%).[690]

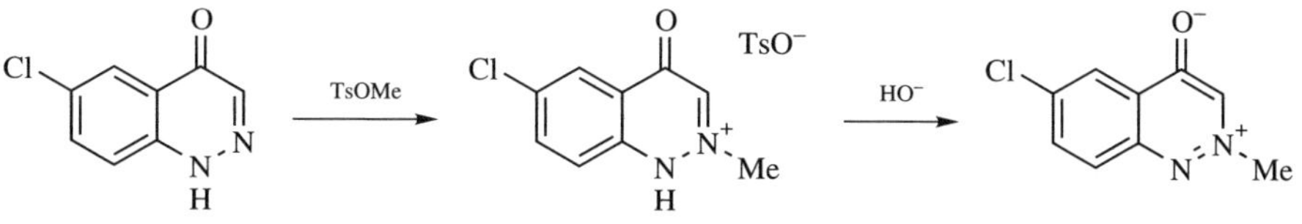

(**28**) (**29**) (**30**)

Formation of Both *N*-Alkylated Cinnolinones and Cinnoliniumolates

Ethyl 6,7-difluoro-4-oxo-1,4-dihydro-3-cinnolinecarboxylate (**31**) gave a separable mixture of ethyl 6,7-difluoro-1-methyl-4-oxo-1,4-dihydro-3-cinnolinecarboxylate (**32**) and 3-ethoxycarbonyl-6,7-difluoro-2-methylcinnolin-2-ium-4-olate (**33**) (substrate, K_2CO_3, Me_2NCHO, 80°C, 30 min, then Me_2SO_4↓, 80°C, 30 min: 62% and 9%, respectively); several pairs of isomers were made similarly.[416]

(**31**) $\xrightarrow{Me_2SO_4,\ K_2CO_3}$ (**32**) + (**33**)

6-Bromo-4-oxo-1,4-dihydro-3-cinnolinecarboxylic acid (**34**) gave a separable mixture of 6-bromo-1-methyl-4-oxo-1,4-dihydro-3-cinnolinecarboxylic acid (**35**, R = Me) and 6-bromo-2-methyl-4-oxo-1,4-dihydrocinnolin-2-ium-3-carboxylate (**36**, R = Me) (Me_2SO_4, KOH, H_2O, 27°C, 30 min: 42% and 43%, respectively, prior to final purification of each); the latter product (**36**, R = Me) underwent decarboxylation easily to afford 6-bromo-2-methylcinnolin-2-ium-4-olate (**37**, R = Me) (recrystallization from Me_2NCHO: 33% overall).[12] Ethylation of substrate (**34**) required more vigorous conditions and afforded a separable mixture of 6-bromo-1-ethyl-4-oxo-1,4-dihydro-3-cinnolinecarboxylic acid (**35**, R = Et) and 6-bromo-2-ethylcinnolin-2-ium-4-olate (**37**, R = Et) directly (EtI, EtOH, reflux, 6 h: 19% and 18%, respectively, after final purification of each).[12]

(**34**) → (**35**) + (**36**)

(**36**) $\xrightarrow{(-CO_2)}$ (**37**)

Also other examples.[30,577]

Formation of Both Alkoxycinnolines and *N*-Alkylated Cinnolinones

1-Methyl-6-nitro-1,4-dihydro-3(2*H*)-cinnolinone (**38**) gave a separable mixture of 3-methoxy-1-methyl-6-nitro-1,4-dihydrocinnoline (**39**) and 1,2-dimethyl-

6-nitro-1,4-dihydro-3(2*H*)-cinnolinone (**40**) (MeI, K_2CO_3, AcMe, reflux, 8 h: 26% and 68%, respectively, after chromatographic separation).[717]

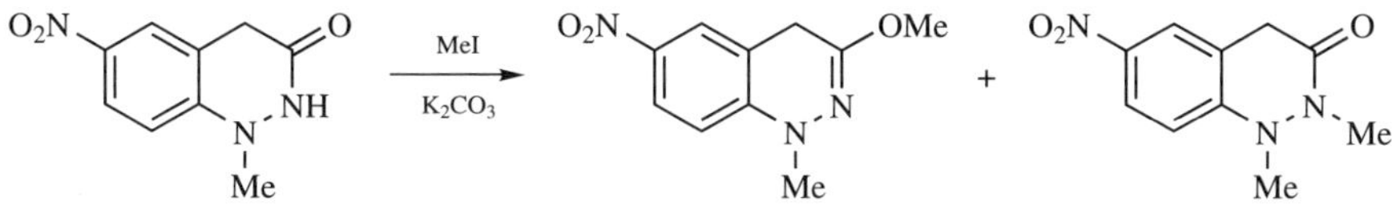

(**38**) (**39**) (**40**)

4.1.2.2. *Other Reactions of Tautomeric Cinnolinones*

Several other important reactions of tautomeric cinnolinones are represented in the more recent literature, as illustrated in the following classified examples.

Thiation

Ethyl 4-oxo-1,4-dihydro-3-cinnolinecarboxylate (**41**, R = CO_2Et) gave ethyl 4-thioxo-1,4-dihydro-3-cinnolinecarboxylate (**42**, R = CO_2Et) (P_2S_5, trace $NaHCO_3$, MeCN, reflux, 4 h: 75%).[54]

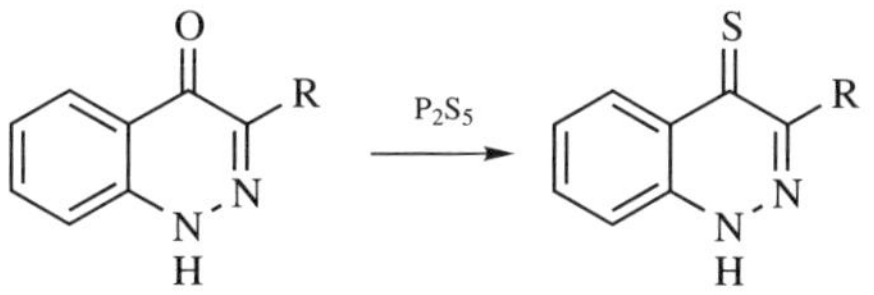

(**41**) (**42**)

4-Oxo-1,4-dihydro-3-cinnolinecarbonitrile (**41**, R = CN) gave 4-thioxo-1,4-dihydro-3-cinnolinecarbonitrile (**42**, R = CN) (P_2S_5, trace $NaHCO_3$, MeCN, reflux, 20 h: 55%).[695]

5,6,7,8-Tetrahydro-3(2*H*)-cinnolinone (**43**, X = O) gave 5,6,7,8-tetrahydro-3(2*H*)-cinnolinethione (**43**, X = S) (P_2S_5, pyridine, reflux, 3 h: 63%).[952]

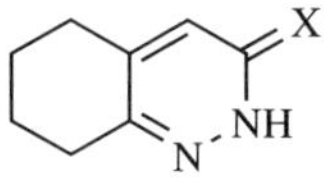

(**43**)

Aminolysis (Indirect)

7,7-Dimethyl-3-phenyl-4,6,7,8-tetrahydro-5(1*H*)-cinnolinone (**44**) gave the oxime, 5-hydroxyimino-7,7-dimethyl-3-phenyl-5,6,7,8-tetrahydrocinnoline (**45**) [$H_2NOH.HCl$, pyridine, 100°C, 16 h: ~95%; note concomitant aerial (?) oxidation] and thence 6,7-dimethyl-3-phenyl-5-cinnolinamine (**46**) [98% H_2SO_4, 120°C, 10 min: > 95% (or in lower yields by using P_2O_5–H_3PO_4 or $POCl_3$); by rearrangement with loss of H_2O];[139,656] alternatively, the same substrate (**44**) was oxidized to 7,7-dimethyl-3-phenyl-5,6,7,8-tetrahydro-5-cinnolinone (**47**) (TsCl, pyridine, 100°C, 16 h: 75%; mechanism obscure) and thence by a Schmidt-type rearrangement, with loss of water, into 6,7-dimethyl-3-phenyl-5-cinnolinamine (**46**) (substrate, 98% H_2SO_4, 20°C, then NaN_3↓ slowly, 20°C, 2 h: 75%).[656]

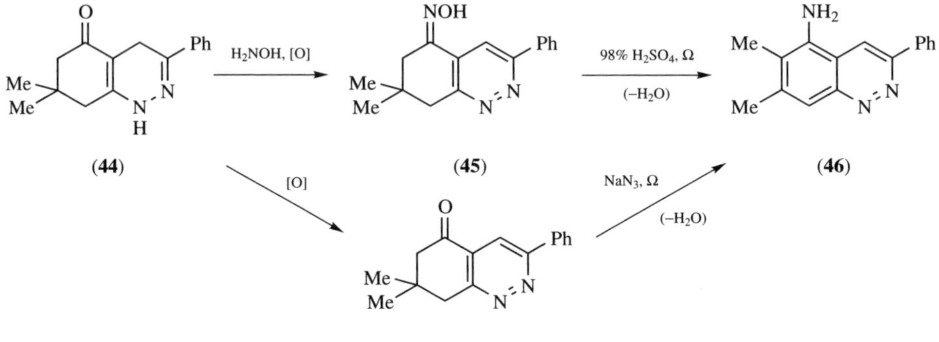

(**44**) (**45**) (**46**) (**47**)

O- or *N*-Acylation

Note: Unlike alkylation, acylation of a tautomeric cinnolinone usually affords an acyloxycinnoline rather than an *N*-acylated cinnolinone.

4-Oxo-1,4-dihydro-3-cinnolinecarbonitrile (**48**) gave 4-acetoxy-3-cinnolinecarbonitrile (**49**) (neat Ac_2O, 95°C, 30 min; 70%); several analogs likewise.[505]

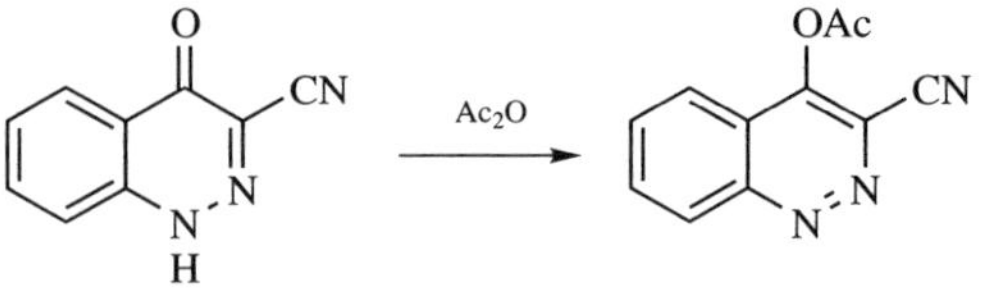

(**48**) (**49**)

5,7,8-Trimethyl-3-phenyl-6(2*H*)-cinnolinone (**50**) gave 6-acetoxy-5,7,8-trimethyl-3-phenylcinnoline (**51**) (no details except NMR that does not preclude *N*-acetylation).[706]

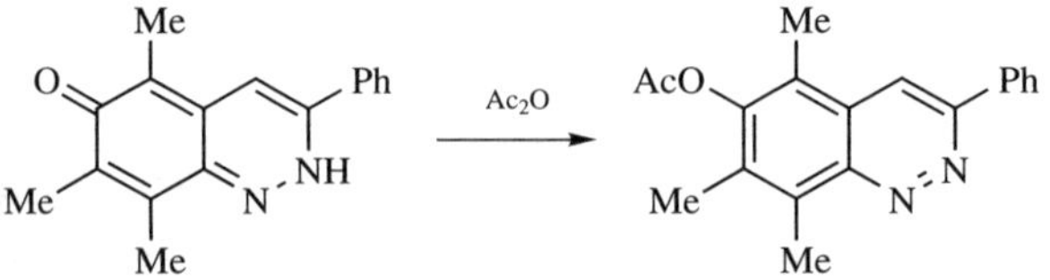

(**50**) (**51**)

6,7-Dimethyl-2,3-dihydro-4(1*H*)-cinnolinone (**52**, R = Me) gave 2-chloroacetyl-6,7-dimethyl-2,3-dihydro-4(1*H*)-cinnolinone (**53**) ($ClCH_2COCl$, H_2O,

0°C, ? h: > 50%) and thence 1,2-bis(chloroacetyl)-6,7-dimethyl-2,3-dihydro-4(1*H*)-cinnolinone (**54**) ($ClCH_2COCl$, Et_3N, see original for details); 2,3-dihydro-4(1*H*)-cinnolinone (**52**, R = H) with 3-chloropropionyl chloride gave 2,3,5,6-tetrahydro-1*H*-pyrazolo[1,2-*a*]cinnoline-3,6-dione (**55**) as the final product (PhH, reflux, 10 h: 54%);[476] also another *N*-acylation.[15]

$ClCH_2COCl$, H_2O (R = Me); $ClCH_2COCl$, Et_3N; $ClCH_2CH_2COCl$, PhH (R = H)

(**52**) (**53**) (**54**) (**55**)

Nuclear Oxidation or Reduction

A typical example[284] of the oxidation of a hexahydro- to a tetrahydrocinnolinone has been given in Section 1.2.2.

3(2*H*)-Cinnolinone (**56**) gave 1,4-dihydro-3(2*H*)-cinnolinone (**57**) (Zn dust, HBr, EtOH, reflux, 5 min: 56%;[37] or Zn dust, 10% H_2SO_4, AcOEt, 25°C, 45 min: 78%).[6]

[H]

(**56**) (**57**)

4(1*H*)-Cinnolinone (**58**) gave 2,3-dihydro-4(1*H*)-cinnolinone (**59**) (Zn dust, AcOH, EtOH, ?°C, ? h: > 70%); many analogs likewise (55–80%).[360]

[H]

(**58**) (**59**)

Cyclocondensations

3-Acetyl-4-oxo-1,4-dihydro-6-cinnolinesulfonamide (**60**) gave 3-methyl-1*H*-pyrazolo[4,3-*c*]cinnoline-8-sulfonamide (**61**) ($H_2NNH_2 \cdot H_2O$, EtOH, reflux, 3 h: 93%);[663] analogs likewise.[619,663,795]

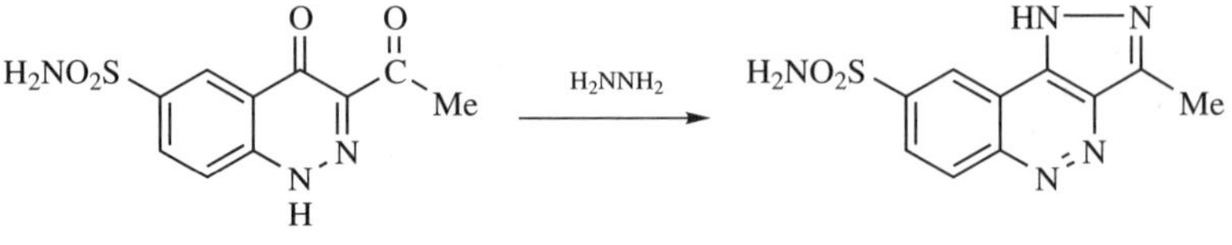

(**60**) (**61**)

7,7-Dimethyl-3,4-diphenyl-6,7-dihydro-5(1*H*)-cinnolinone (**62**) with α-benzylidenemalononitrile gave 2-amino-5,5-dimethyl-4,9,10-triphenyl-5,6-dihydro-4*H*-pyrano[2,3-*f*]cinnoline-3-carbonitrile (**63**) (Et_3N, EtOH, reflux, 1 h: 60%).[673]

Miscellaneous Reactions

7,7-Dimethyl-3,4-diphenyl-6,7-dihydro-5(1*H*)-cinnolinone (**62**) underwent hydrazinolysis (≡ hydrazone formation) to give 5-hydrazono-7,7-dimethyl-3,4-diphenyl-1,5,6,7-tetrahydrocinnoline (**64**) ($H_2NNH_2.H_2O$, EtOH, reflux, 1 h: 70%).[673]

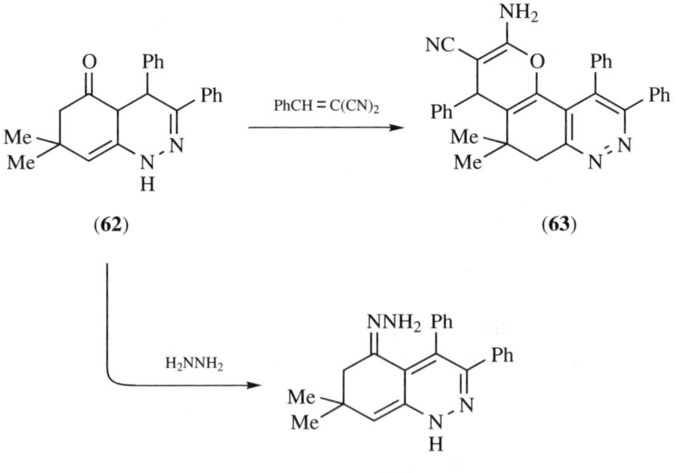

(**62**) (**63**)

(**64**)

3-Acetyl-4-oxo-1,4-dihydro-6-cinnolinesulfonamide (**65**) with diazotized aniline gave 3-acetyl-4-oxo-1-phenylazo-1,4-dihydro-6-cinnolinesulfonamide (**66**) (substrate, Na_2CO_3, H_2O, EtOH, then PhN_2Cl solution ↓ slowly, < 5°C, 15 min: 70%).[663]

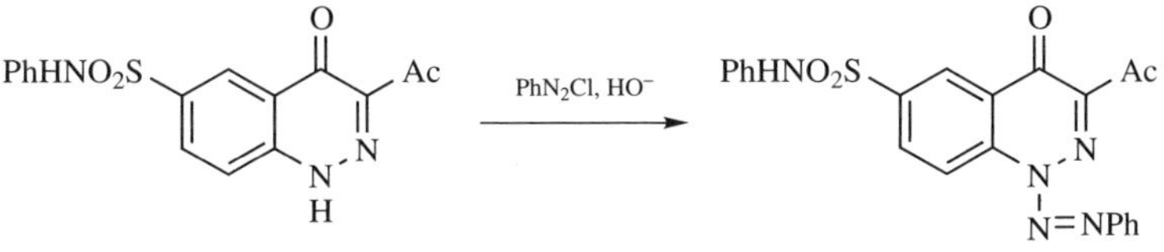

PhHNO$_2$S, Ac, O, N, N, H — PhN_2Cl, HO^- → PhHNO$_2$S, Ac, O, N, N, N=NPh

(**65**) (**66**)

1,4-Dihydro-3(2*H*)-cinnolinone (**67**) rearranged to 1-amino-2-indolinone (10% HCl, 100°C, 15 min: 85%).[6]

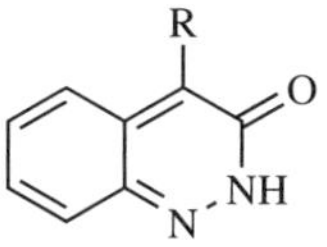

R, O, NH, N

(**67**)

3(2*H*)-Cinnolinone (**67**, R = H) with lead tetracetacte in acetic acid did not undergo *O*- or *N*-acetylation but oxidative *C*-acetoxylation to afford 4-acetoxy-3(2*H*)-cinnolinone (**67**, R = OAc) (for details, see original).[187]

4.2. OTHER OXYCINNOLINES

There is so little additional information on all other oxycinnolines in the literature of 1972–2004 that the preparation and reactions of each type can be covered briefly as notes and/or any available examples in the following classified list.

Extranuclear Hydroxycinnolines: Preparation

Note: Many such hydroxycinnolines have been made by *primary synthesis* (see Chapter 1) and some by passenger *hydroxyalkylation* (see, e.g., Section 2.2.1) or as illustrated here.

The *dimerization* of 3-cinnolinecarbaldehyde (**68**) by a benzoin-type isomerization gave 1,2-di(cinnolin-3-yl)-1,2-ethylenediol (**69**) (trace KCN, EtOH, N_2, reflux, 15 min: 32%; formulation of the product as an enediol is based on spectral evidence).[785]

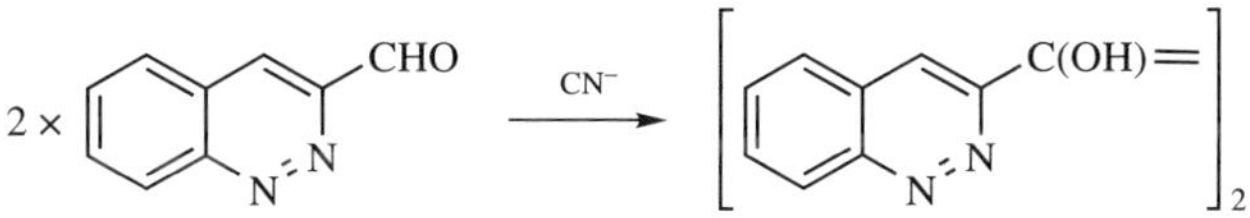

2 × CHO, N, N — CN^- → [C(OH)=, N, N]$_2$

(**68**) (**69**)

The *reductive debenzylation* of 4a-*m*-benzyloxyphenyl-1-cyclopropylmethyl-2-methyldecahydrocinnoline (**70**, R = CH_2Ph) gave 1-cyclopropylmethyl-4a-

m-hydroxyphenyl-2-methyldecahydrocinnoline (**70**, R = H) (H_2, Pd/C, HCl, EtOH: 84%; see original for more details);[964] 1-acetyl-4a-*m*-benzyloxyphenyl-2-cyclopropylmethyldecahydrocinnoline gave 1-acetyl-2-cyclopropylmethyl-4a-*m*-hydroxyphenyldecahydrocinnoline (H_2, Pd/C, EtOH: 76%);[962] and analogs were made similarly.[962,964]

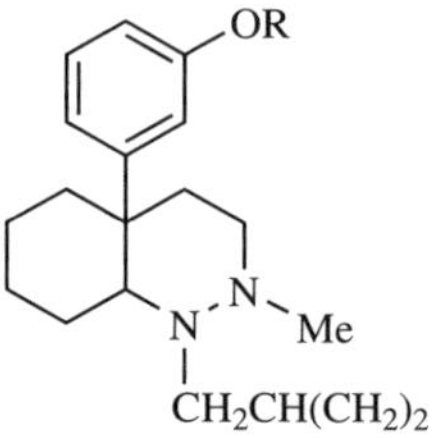

(**70**)

The *reduction* of 3,4-cinnolinedicarbaldehyde (**71**, R = CHO) gave 4-hydroxymethyl-3-cinnolinecarbaldehyde (**71**, R = CH_2OH) ($NaBH_4$, EtOH, H_2O, reflux, 30 min: 89%).[711]

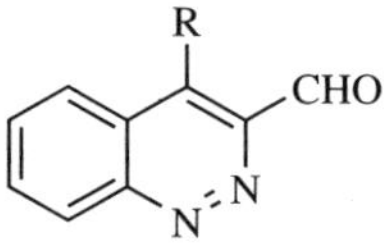

(**71**)

Extranuclear Hydroxycinnolines: Reactions

Note: An example of *dehydroxylation* has been given in Section 2.2.1; no examples of *halogenolysis* appear to have been reported.

1,2-Di(cinnolin-3-yl)-1,2-ethylenediol (**72**) underwent *oxidation* to give di(cinnolin-3-yl)glyoxal (**73**) (Me_2NCHO, air↓, 80°C, until yellow: 70%).[785] Also other oxidations.[711]

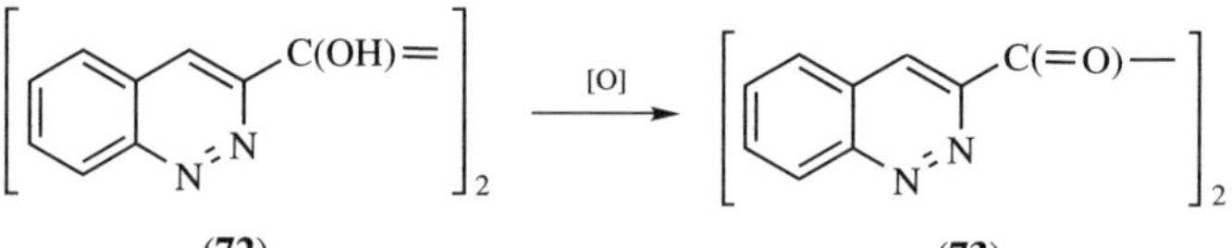

(**72**) (**73**)

1-Acetyl-4a-*m*-hydroxyphenyl-2-phenethyldecahydrocinnoline (**74**, R = H) underwent *acylation* to give 4a-*m*-acetoxyphenyl-1-acetyl-2-phenethyldecahydrocinnoline (**74**, R = Ac) (Ac_2O, 5 h: 74%; see original for more detail);

analogs likewise.[962]

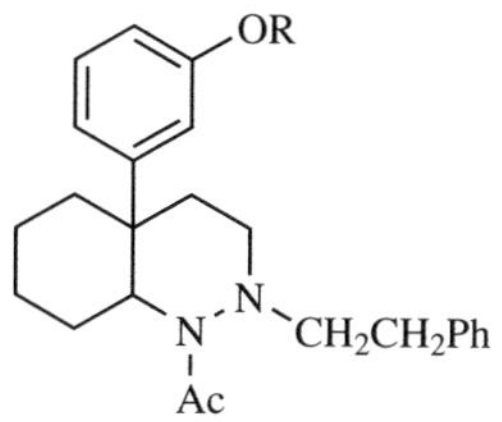

(74)

Cinnolinequinones

Note: Efforts to make 3,4-cinnolinequinone (**75**) failed;[187,571,772] other cinnolinequinones appear to be unknown also.

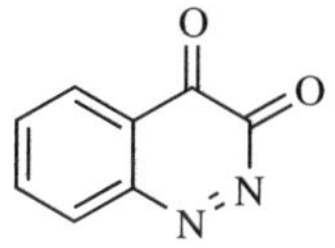

(75)

Alkoxy- or Aryloxycinnolines: Preparation (*H* 31, 32; *E* 150)

Note: The formation of these ethers by *primary synthesis* Chapter 1), by *alcoholysis or phenolysis of halogenocinnolines* (Section 3.2), and by *alkylation of tautomeric cinnolinones* (Section 4.1.2.1) has been covered already. Other routes appear to be unrepresented.

Alkoxy- or Aryloxycinnolines: Reactions (*H* 33; *E* 156)

Note: Examples of *alkanelysis* (Section 2.2.1) and *conversion into tautomeric cinnolinones* (Section 4.1.1) have been given already.

4-Phenoxy-3-phenylethynylcinnoline (**76**) underwent *aminolysis* to give 4-methylamino-3-phenylethynylcinnoline (**77**, R = Me) ($MeNH_2$, EtOH, reflux, 10 min: 83%) or 4-ethylamino-3-phenylethynylcinnoline (**77**, R = Et) ($EtNH_2$, likewise: 59%).[293]

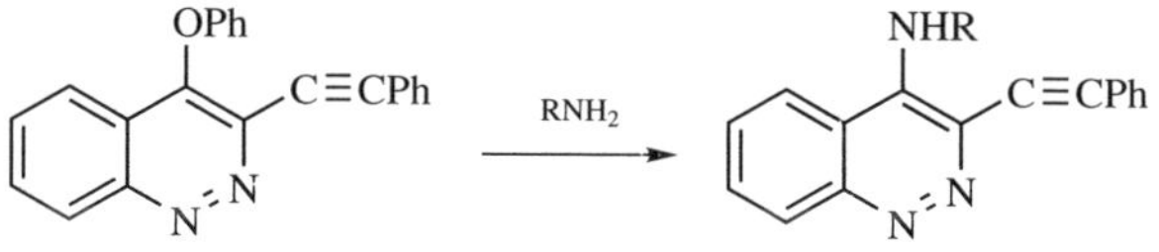

(76) $\xrightarrow{RNH_2}$ (77)

3-Bromo-4-phenoxycinnoline (**78**, R = OPh) likewise gave 4-anilino-3-bromocinnoline (**78**, R = NHPh) (neat $PhNH_2$, 175°C, 20 min: 83%; note the remarkable survival of the bromo substituent);[293] also other such aminolyses.[236,1012]

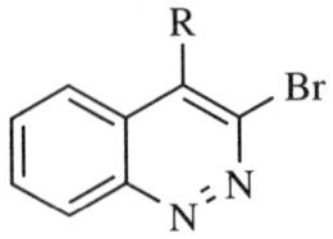

(78)

Di-*tert*-butyl 3,6-dimethoxy-7-oxo-1,2,3,4,4a,7-hexahydro-1,2-cinnolinedicarboxylate (**79**) underwent *dealkoxylation* by loss of methanol to give di-*tert*-butyl 6-methoxy-7-oxo-1,2,3,7-tetrahydro-1,2-cinnolinedicarboxylate (**80**) (F_3CCO_2H, $CHCl_3$, 0°C, 1 h, then 20°C, 2 h: 64%; both cinnolines were formulated as their 7-hydroxy tautomers).[494]

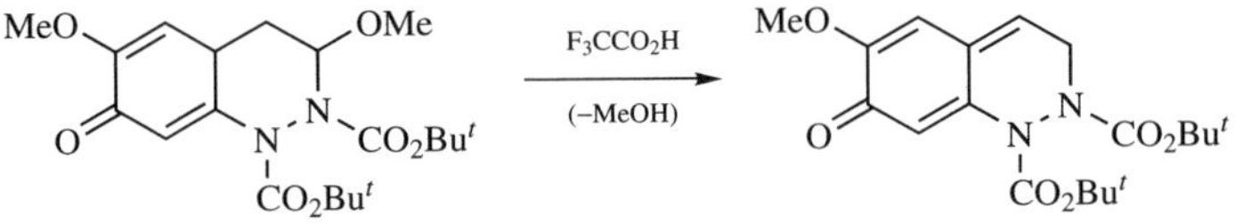

F_3CCO_2H (−MeOH)

(79) (80)

Nontautomeric Cinnolinones and Cinnoliniumolates: Preparation

Note: A few such products have been made by *primary synthesis* (see Chapter 1) but most, by *alkylation of tautomeric cinnolinones* (see Section 4.1.2.1). Minor routes are exemplified here.

4-Cinnolinecarbonitrile (**81**) with *Grignard reagents* gave 1-phenyl-4(1*H*)-cinnolinone (**82**, R = Ph) [substrate, PhH, PhMgBr (made in situ in Et_2O)↓ slowly, then reflux, 1 h: 6%; PhLi similarly: 3%) or 1-methyl-4(1*H*)-cinnolinone (**82**, R = Me) (MeMgI, similarly: 5%).[399]

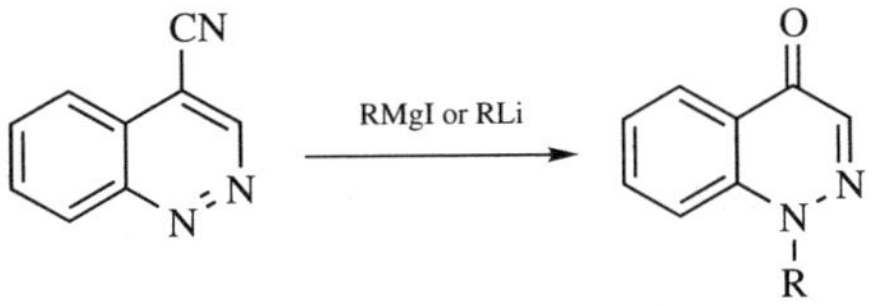

RMgI or RLi

(81) (82)

1-Methyl-6-nitro-1,4-dihydro-3(2*H*)-cinnolinone (**83**) underwent *direct oxylation* to afford 1-methyl-6-nitro-3,4(1*H*,4*H*)-cinnolinedione (**84**) (H_2O_2, $NaHCO_3$, H_2O, 20°C, 3 h: 81%).[717]

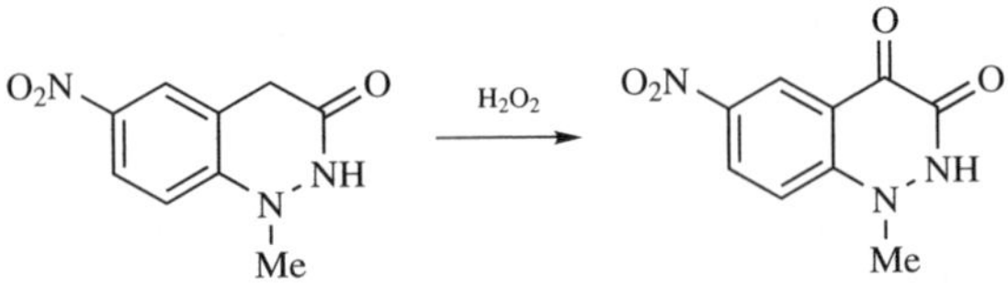

H_2O_2

(83) (84)

Ethyl 1-methyl-4-thioxo- underwent *virtual hydrolysis* to give ethyl 1-methyl-4-oxo-1,4-dihydro-3-cinnolinecarboxylate (MeONa, MeOH, MeI, reflux, 5 h: $>95\%$: mechanism unknown).[54]

Nontautomeric Cinnolinones and Cinnoliniumolates: Structure and Reactions

Note: The structure and bonding of 2-methyl-3(2*H*)-cinnolinone (**85**), 1-methyl-4(1*H*)-cinnolinone (**86**, R = H), 1-ethoxycarbonylmethyl-4(1*H*)-cinnolinone (**86**, R = CO_2Et), 2-methylcinnolin-2-ium-4-olate (**87**, R = H), and 2-ethoxycarbonylmethylcinnolin-2-ium-4-olate (**87**, R = CO_2Et) have been examined by X-ray analysis, nuclear quadrupole resonance, and ab initio calculations;[480,807,808] X-ray analyses for 1-[2-(hydrazinocarbonyl)ethyl]- (**88**, R = $NHNH_2$) and 1-{2-[(pyrrolidin-1-yl)carbonyl]ethyl}-4(1*H*)-cinnolinone [**88**, R = $N(CH_2)_5$] have been reported.[550]

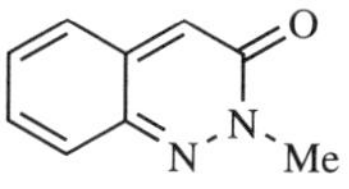

(**85**) 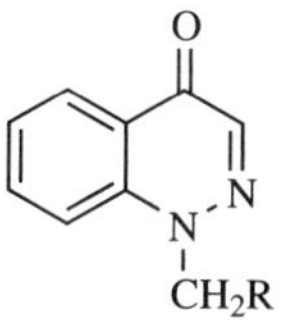 (**86**) 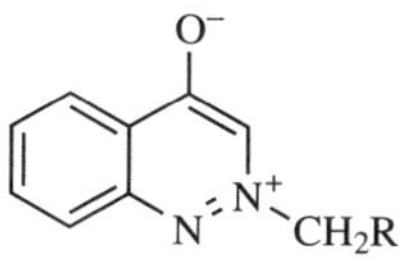 (**87**)

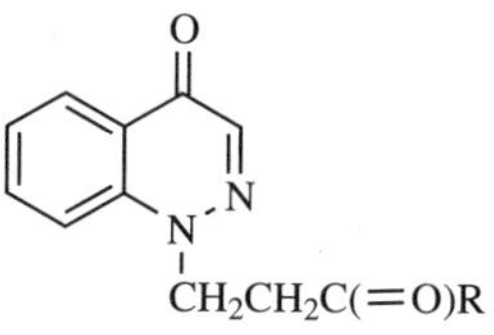

(**88**)

Ethyl 1-methyl-4-oxo- (**89**) underwent *thiation* to give ethyl 1-methyl-4-thioxo-1,4-dihydro-3-cinnolinecarboxylate (**90**) (P_2S_5, MeCN, trace $NaHCO_3$, reflux, 4 h: 70%).[54]

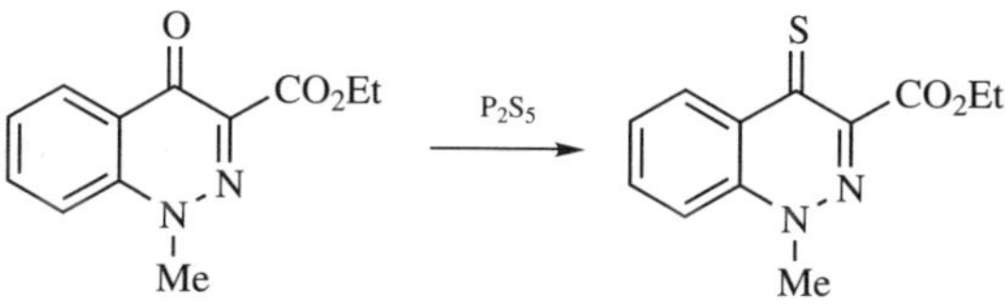

(**89**) (**90**)

2-Methyl-1,4-dihydro-3(2*H*)-cinnolinone (**91**) underwent *nuclear oxidation* to afford 2-methyl-3(2*H*)-cinnolinone (**92**) (Bu^tOCl, PhH, 20°C, 30 min: 93% as hydrochloride);[38] the latter (**92**) underwent *nuclear reduction* to the dihydrocinnolinone (**91**) (substrate, Zn dust, EtOH, reflux, 3M H_2SO_4↓ during 20 min, then reflux, 10 min: 96%).[37]

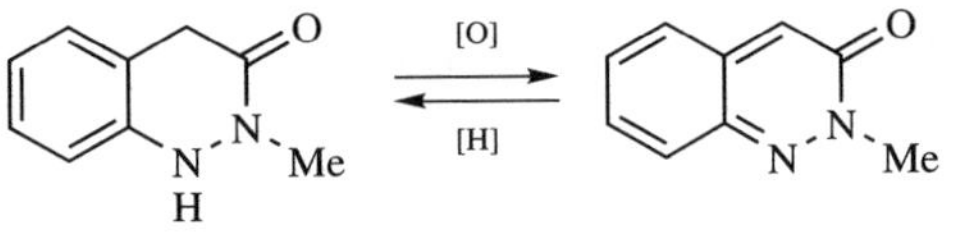

(**91**) ⇌ (**92**) ([O] / [H])

4a-*m*-Benzyloxyphenyl-2-cyclopropylmethyl-4,4a,5,6,7,8-hexahydro-3(2*H*)-cinnolinone (**93**) underwent *reductive deoxygenation* and *nuclear reduction* to give 4a-*m*-benzyloxyphenyl-2-cyclopropylmethyldecahydrocinnoline (**94**) ($LiAlH_4$, dioxane, reflux, 6 h: 51% as hydrochloride).[185]

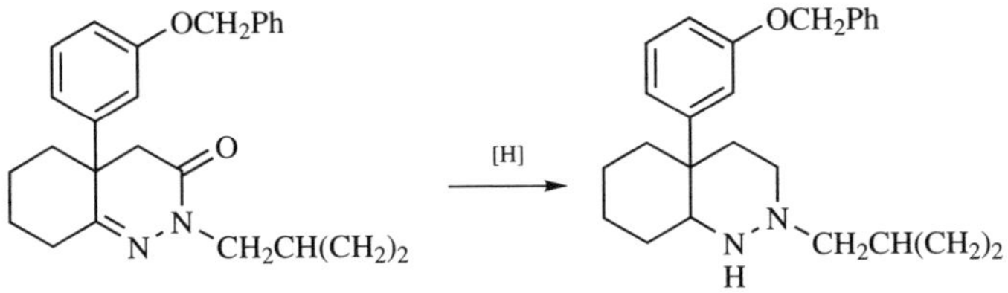

(**93**) → (**94**) ([H])

2-Methyl-3(2*H*)-cinnolinone (**95**) underwent *reductive ring contraction* to furnish 2-indolinone (**96**, R = H) (red P, HI: 18%);[373] in contrast, 2-methyl-1,4-dihydro-3(2*H*)-cinnolinone (**97**) underwent *ring contraction by rearrangement* to give 1-methylamino-2-indolinone (**96**, R = NHMe) (H_2SO_4, EtOH, N_2, reflux, 1 week: ?%).[37]

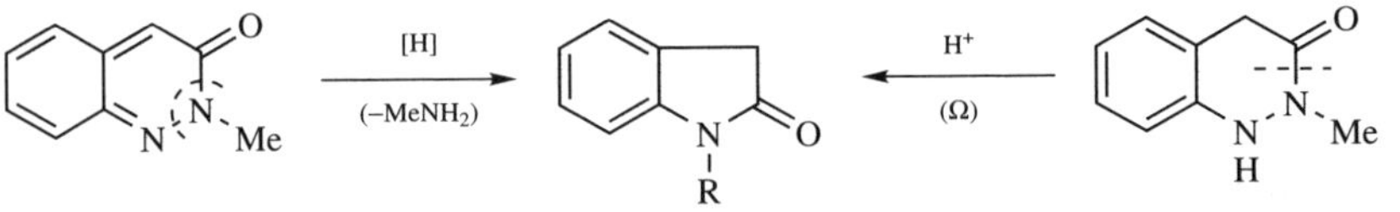

(**95**) → (**96**) ← (**97**) ([H], (−$MeNH_2$); H^+, (Ω))

6-Chloro-2-methylcinnolin-2-ium-4-olate (**98**) underwent *cycloadduct formation* with acetylene derivatives. With dimethyl acetylenedicarboxylate (**99**, Q = R = CO_2Me) it gave dimethyl 7-chloro-10-methyl-5-oxo-4,5-dihydro-1,4-imino-1*H*-1-benzazepine-2,3-dicarboxylate (**100**, Q = R = CO_2Me) (PhH, reflux, 12 h: 66%); with diphenylacetylene (**99**, Q = R = Ph) it gave 7-chloro-10-methyl-2,3-diphenyl-4,5-dihydro-1,4-imino-1*H*-1-benzazepin-5-one (**100**, Q = R = Ph) (ClC_6H_4Cl-*o*, reflux, 36 h: 30%); and with phenylacetylene

(**99**, Q = Ph, R = H) it gave a single product, formulated on NMR evidence as 7-chloro-10-methyl-2-phenyl-4,5-dihydro-1,4-imino-1*H*-1-benzazepin-5-one (**100**, Q = Ph, R = H) (xylene, reflux, 24 h: 60%).[690]

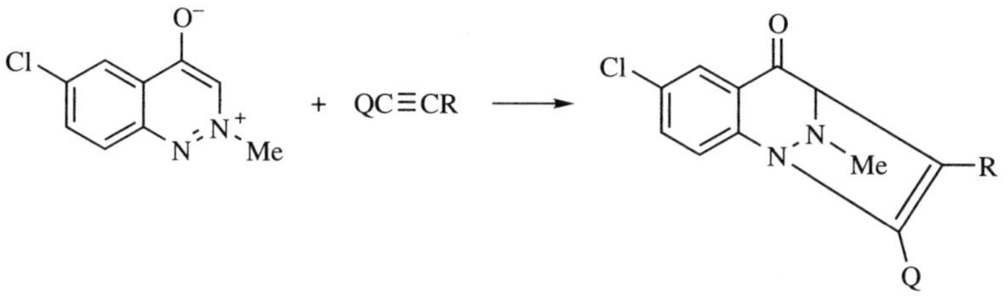

(**98**) (**99**) (**100**)

2-Methylcinnolin-2-ium-4-olate (**101**) underwent *photolytic rearrangement* to afford 3-methyl-4(3*H*)-quinazolinone (**102**) (EtOH, *hν*, reflux, 5 h: 80%); analogs likewise.[693]

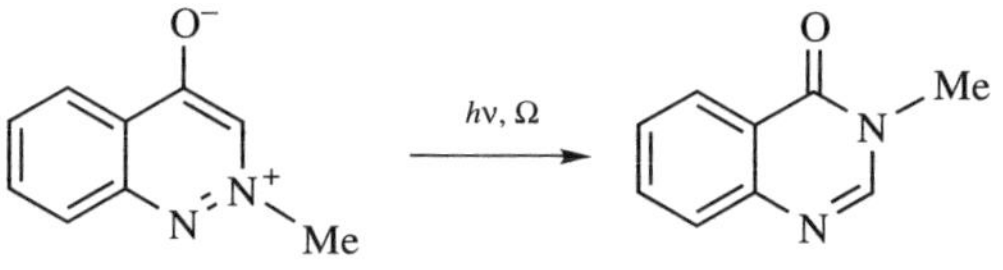

(**101**) (**102**)

Cinnoline *N*-Oxides: Preparation (*E* 272)

Note: Cinnoline *N*-oxides may be prepared directly by *primary synthesis* (see Chapter 1) or by *N-oxidation* of an existing cinnoline with a peroxy reagent, as exemplified here.

4-Azidocinnoline (**103**) gave a separable mixture of 4-azidocinnoline 1-oxide (**104**) and the 2-oxide (**104a**) (*m*-$ClC_6H_4CO_3H$, $CHCl_3$, 0°C → 20°C, 20 h: 21% and 53%, respectively).[402] See also Section 6.1.

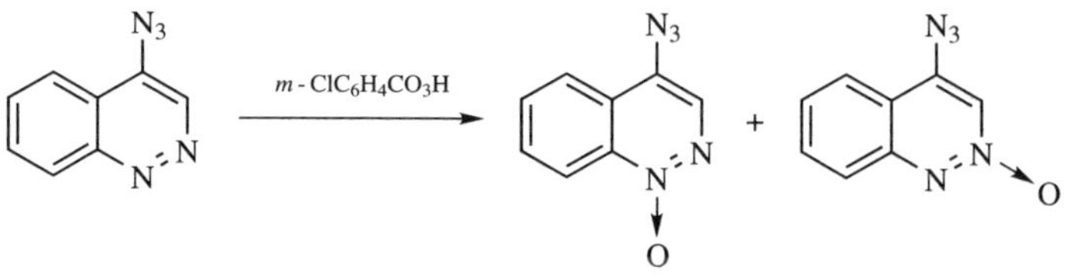

(**103**) (**104**) (**104a**)

6-Chloro-4-phenylcinnoline gave a separable mixture of 6-chloro-4-phenylcinnoline 1-oxide (**105**) and the 2-oxide (**105a**) (*m*-$ClC_6H_4CO_3H$, CH_2Cl_2,

20°C, 90 min: ~5% and ~40%, respectively).[11]

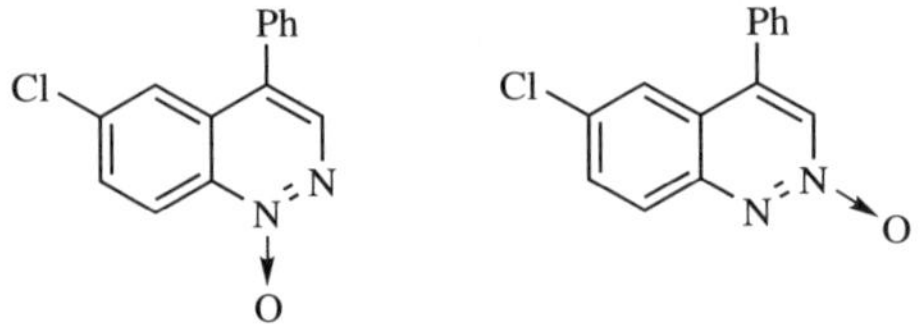

(**105**) (**105a**)

See also Section 6.1.

Cinnoline *N*-Oxides: Properties and Reactions

Note: The ^{14}N NMR spectra of cinnoline 1- and 2-oxides have been measured for comparison with those of related diazine oxides.[760]

3-Methoxy-4-(*o*-nitroanilino)cinnoline 1-oxide (**106**) underwent *deoxygenation* to give 3-methoxy-4-(*o*-nitroanilino)cinnoline (**107**) (PCl_3, $CHCl_3$, reflux, 1 h: 90%).[694] Also other examples.[147]

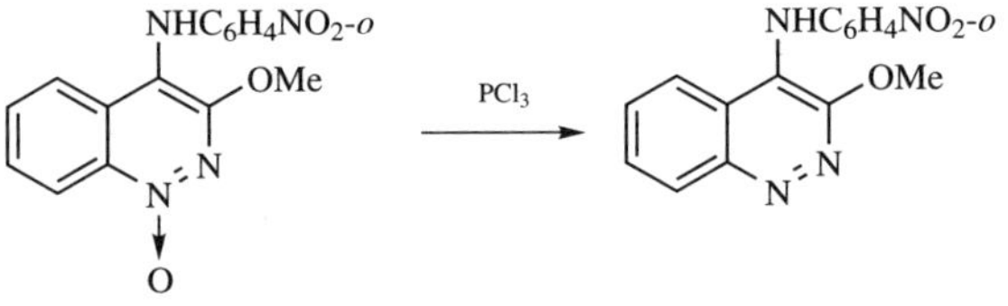

$NHC_6H_4NO_2$-*o* OMe PCl_3 $NHC_6H_4NO_2$-*o* OMe

(**106**) (**107**)

Cinnoline 1-oxide (**108**, R = H) resisted *photolysis*,[138] but 4-methylcinnoline 1-oxide (**108**, R = Me) gave a separable mixture of the deoxygenated product, 4-methylcinnoline (**109**), and 3-methyl-2,1-benzisoxazole (**110**) (MeOH, *h*ν, N_2, 84 h: 42% and 11%, respectively; also minor products); similar treatment of 4-methylcinnoline 2-oxide (**111**) also gave 4-methylcinnoline (**109**) (58%) but accompanied by 3-methyl-1*H*-indazole (**112**) (25%) and minor products.[137]

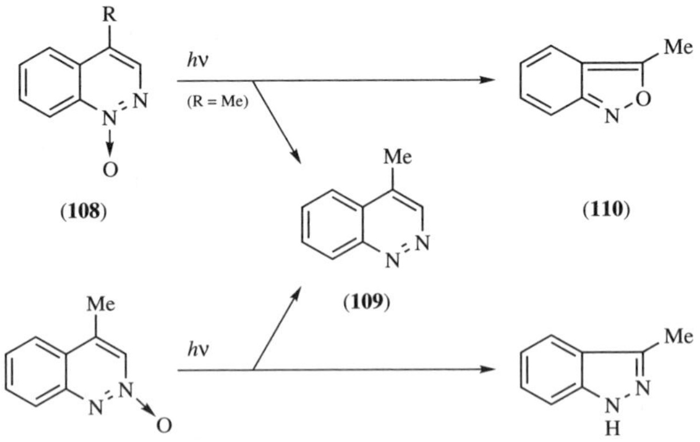

*h*ν (R = Me) *h*ν

(**108**) (**109**) (**110**) (**111**) (**112**)

Cinnoline 2-oxide (**113**) underwent *deoxidative cine-amination* with primary or secondary amines to afford, for example, 3-propylaminocinnoline (**114**, R = NHPr) (neat $PrNH_2$, reflux, 25 h: 53%) or 3-(pyrrolidin-1-yl)cinnoline [**114**, R = $N(CH_2)_4$] [$HN(CH_2)_4$, reflux, 60 h: > 95%].[993]

amine
($-H_2O$)

(**113**) (**114**)

CHAPTER 5

Thiocinnolines (*E* 170)

The term *thiocinnoline* includes any cinnoline bearing a sulfur-containing substituent that is joined directly or indirectly to the nucleus through its sulfur atom: cinnolinethiones (tautomeric and nontautomeric), alkylthiocinnolines, alkylsulfinylcinnolines, alkylsulfonylcinnolines, cinnolinesulfonic acids (and derivatives), dicinnolinyl sulfides and disulfides, and the like. So little information on thiocinnolines has emerged from the 1972–2004 literature that this chapter is necessarily brief.

5.1. CINNOLINETHIONES (*E* 170)

Preparation

Note: A few tautomeric cinnolinethiones have been made by the *thiolysis of halogenocinnolines* (see Section 3.2) and both tautomeric and nontautomeric cinnolinethiones, by *thiation of cinnolinones* (see Sections 4.1.2.2 and 4.2).

Reactions

Note: The *virtual hydrolysis* of a nontautomeric cinnolinethione to the corresponding cinnolinone has been exemplified in Section 4.2.

Ethyl 4-thioxo-1,4-dihydro-3-cinnolinecarboxylate (**2**, R = H) underwent *desulfurization* to give ethyl 1,4-dihydro-3-cinnolinecarboxylate (**1**, R = H) (Raney Ni, EtOH, reflux, 2 h: 25%); ethyl 1-methyl-4-thioxo-1,4-dihydro-3-cinnolinecarboxylate (**2**, R = Me) likewise gave ethyl 1-methyl-1,4-dihydro-3-cinnolinecarboxylate (**1**, R = Me) (40%).[54]

Ethyl 4-thioxo-1,4-dihydro-3-cinnolinecarboxylate (**2**, R = H) underwent *S-alkylation* to give ethyl 4-methylthio-3-cinnolinecarboxylate (**3**) (MeI, MeONa, MeOH, 20°C, 1 h: 70%);[54] 3-bromo-6,7-dimethoxy-4(1*H*)-cinnolinethione (**5**) behaved similarly with alkyl halides to give 4-allylthio-3-bromo-

Cinnolines and Phthalazines: Supplement II, The Chemistry of Heterocyclic Compounds, Volume 64, by D.J. Brown

6,7-dimethoxycinnoline (**4**, R = $CH_2CH{=}CH_2$) ($ClCH_2CH{=}CH_2$: 82%; no details) and 3-bromo-6,7-dimethoxy-4-methylthiocinnoline (**4**, R = Me) (MeI: 82%; no details), but with dimethyl sulfate it apparently underwent *N-alkylation* to afford 3-bromo-6,7-dimethoxy-1-methyl-4(1*H*)-cinnolinethione (**6**) (no details: 80%).[635] Also other *S*-alkylations.[695]

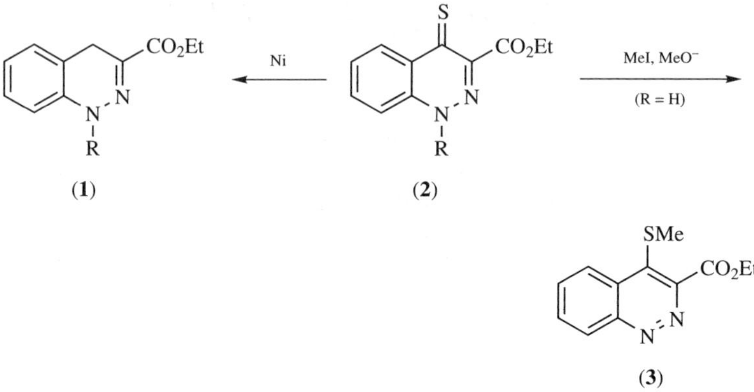

(**1**) (**2**) (**3**)

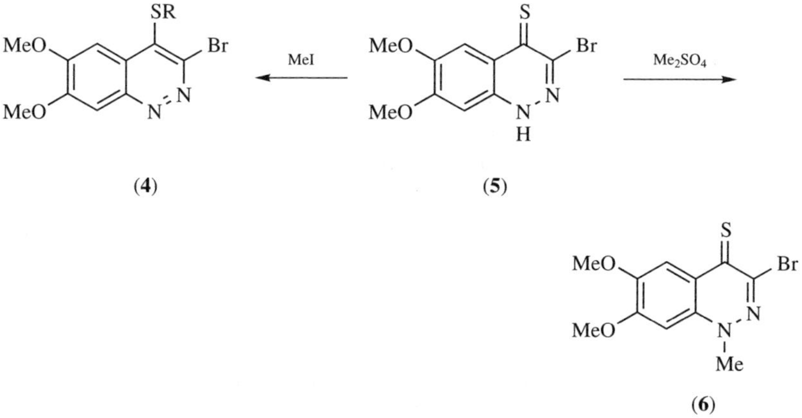

(**4**) (**5**) (**6**)

5,6,7,8-Tetrahydro-3(2*H*)-cinnolinethione (**7**) underwent *aminolysis* to afford 3-hydrazino-5,6,7,8-tetrahydrocinnoline (**8**) (neat $H_2NNH_2.H_2O$, reflux, 2 h: 86%);[952] ethyl 4-thioxo-1,4-dihydro-3-cinnolinecarboxylate (**9**) behaved similarly, but with concomitant aminolysis of the ester group, to give 4-amino-3-cinnolinecarboxamide (**10**, R = H) [NH_3 gas, EtOH, sealed 100°C (?), 18 h: 80%] or 4-hydrazino-3-cinnolinecarbohydrazide (**10**, R = NH_2) ($H_2NNH_2.H_2O$, EtOH, reflux, until $H_2S\uparrow$ ceased: 90%);[54] and ethyl 1-methyl-4-thioxo-1,4-dihydro-3-cinnolinecarboxylate gave the nontautomeric product, 4-imino-1-methyl-1,4-dihydro-3-cinnolinecarboxamide (NH_3, EtOH, sealed, 120°C, 10 h: 70%).[54]

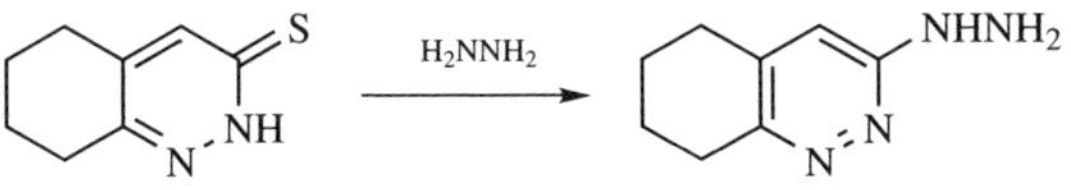

(7) $\xrightarrow{H_2NNH_2}$ (8)

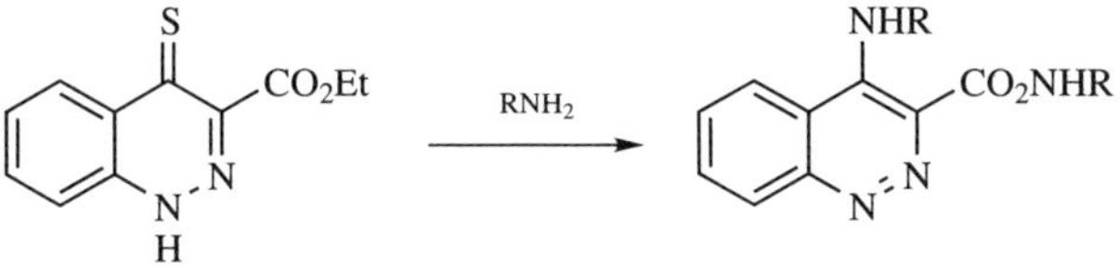

(9) $\xrightarrow{RNH_2}$ (10)

5.2. OTHER THIOCINNOLINES (*E* 171)

Alkylthiocinnolines: Preparation

Note: Alkylthio- and arylthiocinnolines have been made by *alkane- or arenethiolysis of halogenocinnolines* (see Section 3.2) or by *S-alkylation of cinnolinethiones* (see Section 5.1). In addition, (substituted-alkyl)thiocinnolinones have been made *from alkylsulfinylcinnolines*, as illustrated here.

3-Methylsulfinyl-4(1*H*)-cinnolinone (**12**) gave 3-acetoxymethylthio-4(1*H*)-cinnolinone (**11**) (neat Ac_2O, reflux, 4 h: 65%; by a Pummerer-type[970] reaction and additional acetylation) or 4-chloro-3-chloromethylthiocinnoline (**13**) (neat $SOCl_2$, reflux, 6 h: 92%; note additional chlorolysis).[15]

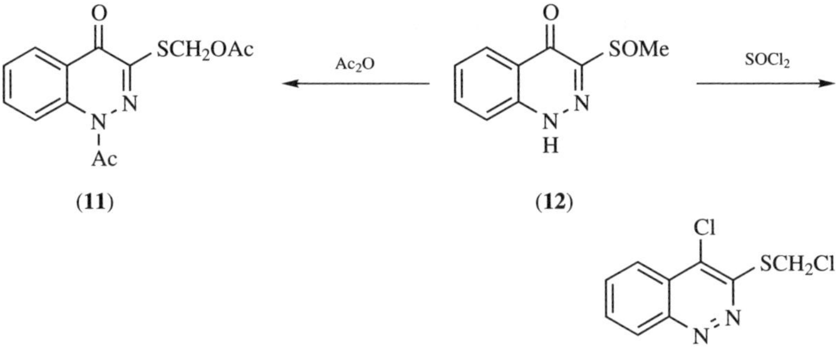

(11) $\xleftarrow{Ac_2O}$ (12) $\xrightarrow{SOCl_2}$ (13)

Alkylthiocinnolines: Reactions

Note: Of all the possible reactions of alkylthiocinnolines, only *oxidation* has been used more recently, as illustrated here.

3-Acetyl-6-(*p*-nitrophenylthio)-4(1*H*)-cinnolinone (**14**) gave 3-acetyl-6-(*p*-nitrophenylsulfonyl)-4(1*H*)-cinnolinone (**14a**) (substrate, AcOH, 30% H_2O_2↓ dropwise, 20°C, < 6 days: 43%).[795]

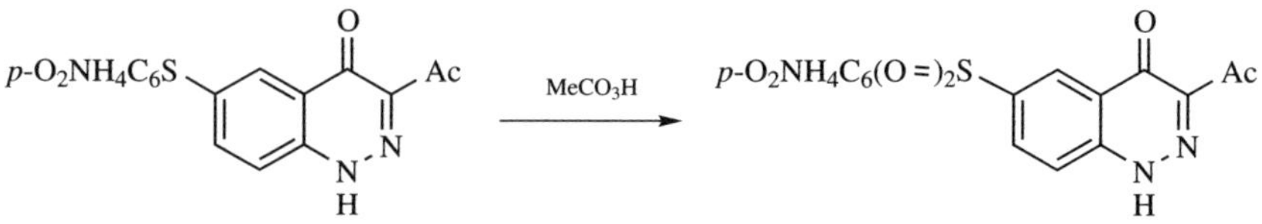

p-O2NH4C6S, Ac, O, N, N, H — MeCO3H → p-O2NH4C6(O=)2S, Ac, O, N, N, H

(**14**) (**14a**)

4-*tert*-Butylthio- (**15**) gave 4-*tert*-butylsulfinyl- (**15a**) [substrate, CH_2Cl_2, −10°C, N_2; *m*-$ClC_6H_4CO_3H$ (1 mol) in CH_2Cl_2↓ dropwise, 1 h: 89%] or 4-*tert*-butylsulfonylcinnoline (**15b**) (substrate, AcOH, H_2O, 20°C, $KMnO_4$ in H_2O↓, 30 min: 59%); analogs somewhat similarly.[1010]

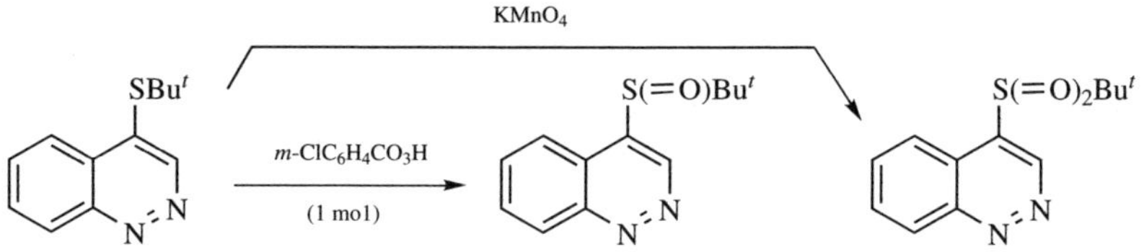

KMnO4; SBu^t — m-ClC6H4CO3H (1 mol) → S(=O)Bu^t; S(=O)2Bu^t

(**15**) (**15a**) (**15b**)

Alkylsulfinyl- and Alkylsulfonylcinnolines: Preparation

Note: Examples have been given already for the formation of the sulfones by *primary synthesis* (Section 1.2.1), by *arenesulfinolysis of halogenocinnolines* (Section 3.2), and by *oxidation of alkylthiocinnolines* (this section).

Alkylsulfinyl- and Alkylsulfonylcinnolines: Reactions

Note: The conversion of such a *sulfoxide into (substituted-alkyl)thiocinnolines* has been exemplified already in this section. The *cyanolysis* of such sulfones is illustrated here.

1-Ethyl-3-methylsulfonyl-4(1*H*)-cinnolinone (**16**, Q = H) gave 1-ethyl-4-oxo-1,4-dihydro-3-cinnolinecarbonitrile (**17**, R = H) (KCN, Me_2NCHO, 120°C, 2 h: 80%); and 7-chloro-1-ethyl-3-methylsulfonyl-4(1*H*)-cinnolinone (**16**, Q = Cl) gave 1-ethyl-4-oxo-1,4-dihydro-3,7-cinnolinedicarbonitrile (**17**, R = CN)

[KCN, Me_2NCHO, 100°C, 2.5 h: 85% (crude)].[485]

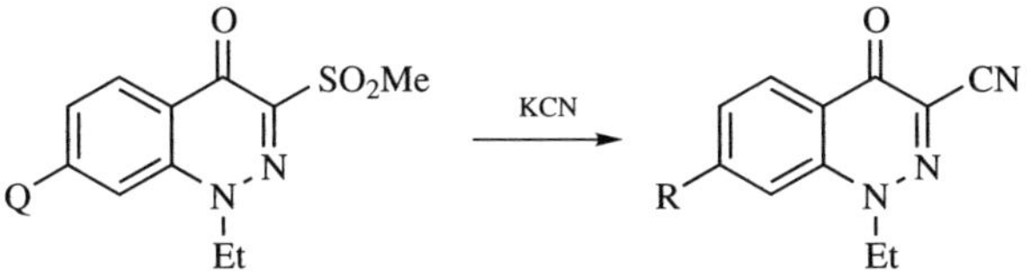

(16) KCN → (17)

Also other examples using unisolated sulfones as substrates.[291,933,937]

Cinnolinesulfonic Acids and Derivatives

Note: The *primary syntheses* of a cinnolinesulfonic acid and of several sulfonamides and sulfonanilides[485] have been covered in Section 1.2.1.

CHAPTER 6

Nitro-, Amino-, and Related Cinnolines (*H* 35; *E* 87, 207)

This chapter summarizes the meager more recent information on cinnolines that bear nitrogenous groups joined to the nucleus through their nitrogen atoms: nuclear and extranuclear nitro, nitroso (no data), amino, hydrazino, and arylazo derivatives.

6.1. NITROCINNOLINES

The preparation and reactions of nitrocinnolines are covered briefly in the following notes and examples.

Nitrocinnolines: Preparation (*E* 188)

Note: The formation of nitrocinnolines by *primary synthesis* (Chapter 1) and by *nitration* (Sections 2.1 and 2.2.2) has been noted already. An additional example of nitration is given here.

4-Oxo-1,4-dihydro-3-cinnolinecarboxylic acid gave 6-nitro-4-oxo-1,4-dihydro-3-cinnolinecarboxylic acid (**1**) (96% H_2SO_4, 0°C; substrate↓ slowly; 96% HNO_3↓; 85°C, 50 min: 67%).[1034]

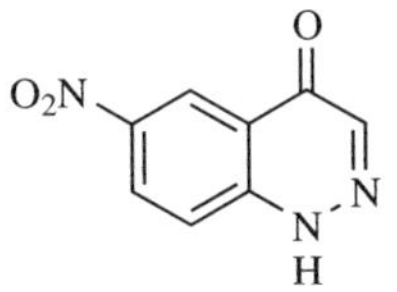

(**1**)

Nitrocinnolines: Reactions (*E* 193)

Note: The *reduction*, *azidolysis*, and *oxidative ring contraction* of nitrocinnolines are illustrated in the following examples.

Cinnolines and Phthalazines: Supplement II, The Chemistry of Heterocyclic Compounds, Volume 64, by D.J. Brown

8-Nitrocinnoline (**2**) gave 8-cinnolinamine (**2a**) (H_2, PtO_2, EtOH, 20°C, 2 h: 72%;[217] $SnCl_2$, HCl, 50°C, 10 min: 90%;[217] or $TiCl_3$, AcOH, H_2O, 11°C, 7 min: 92%);[403] similar treatment of 5-nitrocinnoline (**3**) gave either 5-cinnolinamine (**4**) (H_2, PtO_2, EtOH, 20°C, 2 h: 88%)[217] or 1,4-dihydro-5-cinnolinamine (**4a**) ($TiCl_3$, 26°C: 84%).[403]

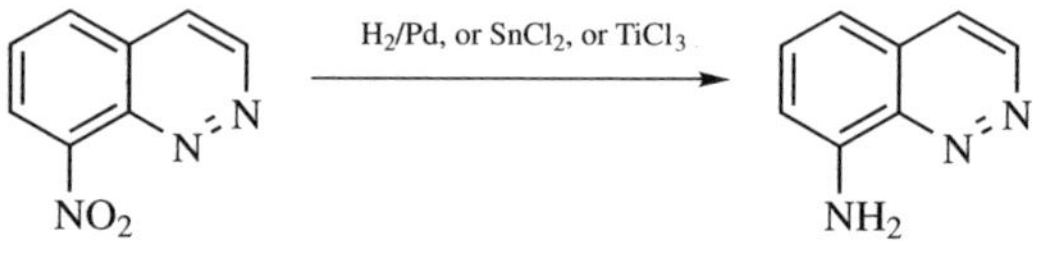

H_2/Pd, or $SnCl_2$, or $TiCl_3$

(**2**) (**2a**)

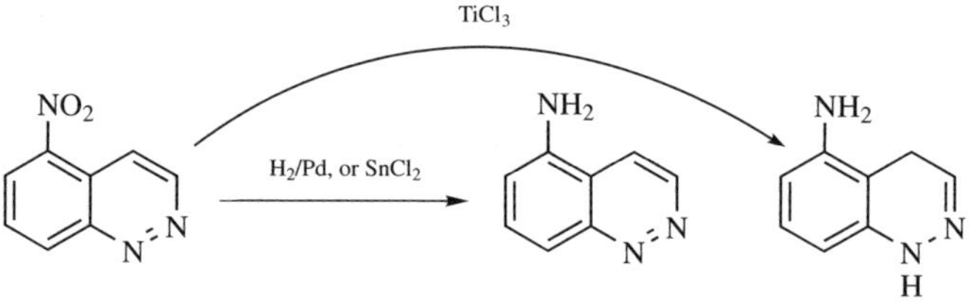

$TiCl_3$

H_2/Pd, or $SnCl_2$

(**3**) (**4**) (**4a**)

3-Nitro-4-cinnolinamine (**5**, R = NO_2) gave 3,4-cinnolinediamine (**5**, R = NH_2) ($SnCl_2.2H_2O$, 10M HCl, 100°C, 2 h: 96%).[370,cf. 971]

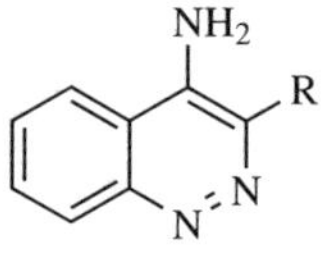

(**5**)

4-Chloro-6-methoxy-8-nitrocinnoline (**6**) gave 6-methoxy-8-cinnolinamine (**7**) [H_2 (1 atm), Pd/C, EtOH, trace HCl, 20°C: 84%; note concomitant dechlorination].[27]

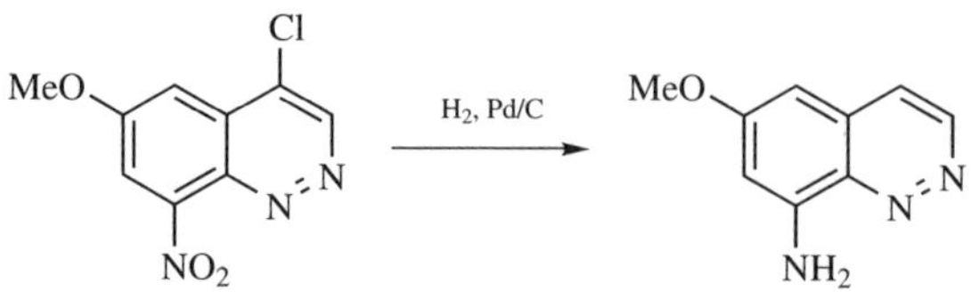

H_2, Pd/C

(**6**) (**7**)

4-Nitrocinnoline 1-oxide (**8**) gave 4-azidocinnoline 1-oxide (**9**) (NaN_3, no details: 81%).[402]

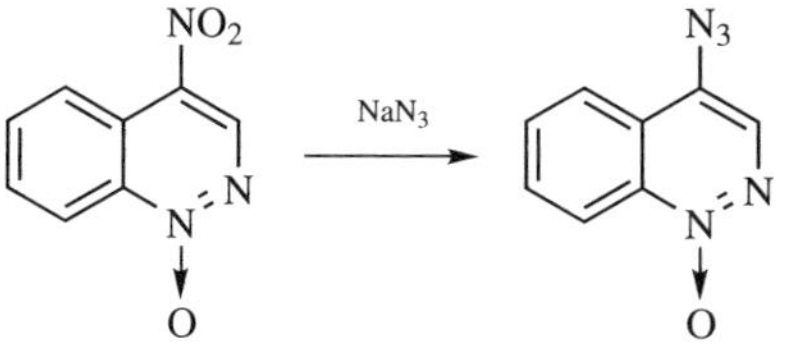

(**8**) (**9**)

8-Nitrocinnoline (**10**) gave 7-nitroindazole (**11**) and some 8-nitrocinnoline 2-oxide (**12**) (30% H_2O_2, AcOH, 65°C, 8 h: 33% and 17%, respectively); 5-nitrocinnoline behaved somewhat similarly to afford 4-nitroindazole (14%) and a mixture of 5-nitrocinnoline 1- and 2-oxide (43%).[724]

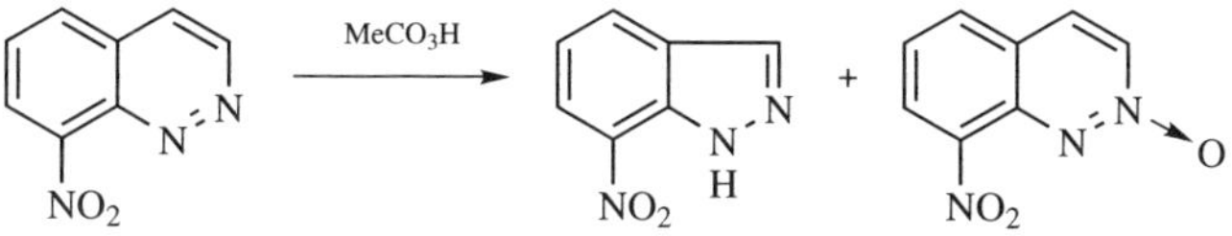

(**10**) (**11**) (**12**)

Also solid-phase reductions with Bu_4NHS.[1034]

6.2. AMINOCINNOLINES AND RELATED COMPOUNDS (*H* 35; *E* 207)

Even this important group of cinnolines is only sparsely represented in the 1972–2004 recent literature. The X-ray analyses for 4-amino-3-cinnolinecarboxylic acid (**13**) (as its sodium salt tetrahydrate)[539] and the nontautomeric imine, 4-imino-1-methyl-1,4-dihydro-3-cinnolinecarboxamide (**14**) indicate that both are broadly planar because of strong hydrogen bonding between the carbonyl and amino/imino groups.

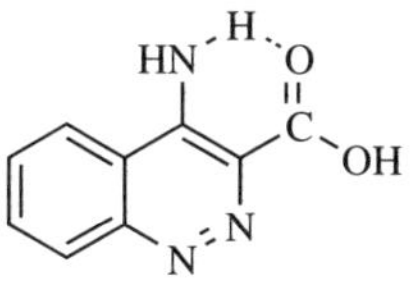

(**13**) 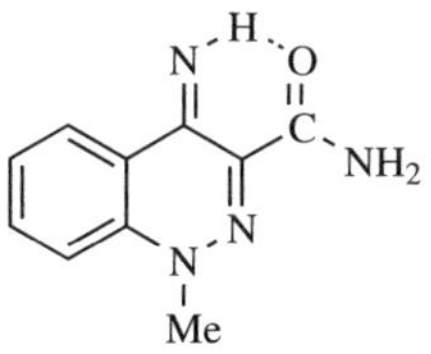 (**14**)

6.2.1. Preparation of Amino-, Hydrazino-, and Arylazocinnolines

The following routes to aminocinnolines have been covered already: by *primary synthesis* (Chapter 1), by *aminolysis of halogenocinnolines* (Section 3.2), by indirect *aminolysis of tautomeric cinnolinones* (Section 2.1.2.2), by *aminolysis of alkoxycinnolines* (Section 4.2), by *deoxidative cine-amination of cinnoline N-oxides* (Section 4.2), by *aminolysis of tautomeric or nontautomeric cinnolinethiones* (Section 5.1), and by the *reduction of nitrocinnolines* (Section 6.1); also arylazocinnolines by *primary synthesis* (Chapter 1).

The remaining approaches to aminocinnolines and arylazocinnolines are exemplifies here.

Direct Amination

3-Nitrocinnoline (**15**) gave 3-nitro-4-cinnolinamine (**16**) (substrate, $H_2NOH \cdot HCl$, EtOH, KOH in EtOH↓ slowly, 27°C → 50°C, 1 h: ~55%).[370,cf. 971]

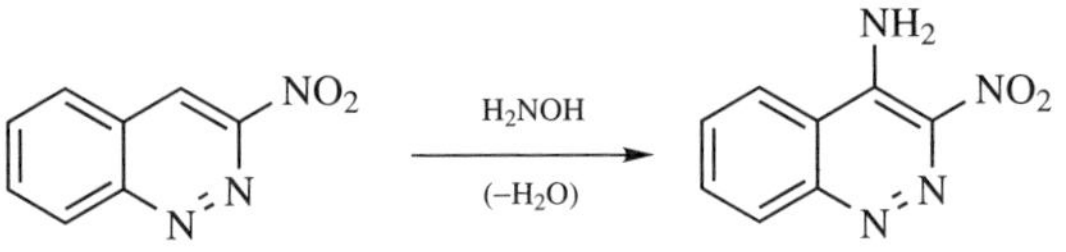

H_2NOH ($-H_2O$)

(**15**) (**16**)

3(2*H*)-Cinnolinone (**17**) gave 2-amino-3(2*H*)-cinnolinone (**18**) (substrate, NaOH, H_2O, H_2NOSO_3H↓, 50–55°C: 22%).[687]

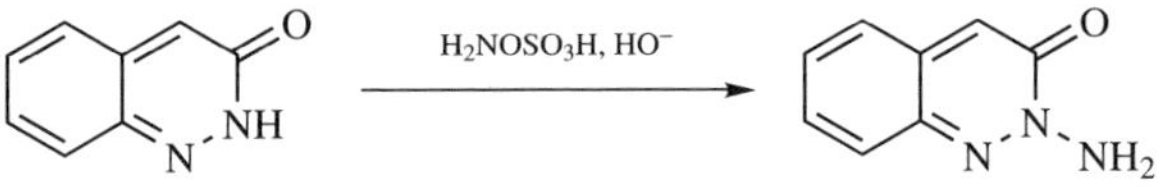

H_2NOSO_3H, HO^-

(**17**) (**18**)

3-Phenylcinnoline (**19**) gave 1-amino-3-phenylcinnolin-1-ium mesitylenesulfonate (**19a**) [substrate, CH_2Cl_2, 2,4,6-$Me_3C_6H_2S(=O)_2ONH_2$ in CH_2Cl_2↓ dropwise, 0°C; then 20°C, 10 min: 65%).[5]

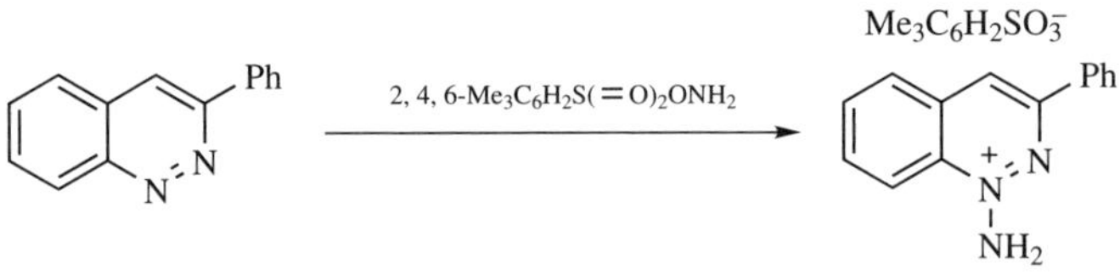

2, 4, 6-$Me_3C_6H_2S(=O)_2ONH_2$

$Me_3C_6H_2SO_3^-$

(**19**) (**19a**)

Aminolysis of Cinnolinecarbonitriles

3,4-Cinnolinedicarbonitrile (**20**) with ammonia or benzylamine gave 4-amino-3-cinnolinecarbonitrile (**20a**, R = H) (substrate, trace $CuSO_4$, MeOH, THF; NH_3 gas↓, 20°C, 30 min: 59%) or 4-benzylamino-3-cinnolinecarbonitrile (**20a**, R = CH_2Ph) ($PhCH_2NH_2$, THF, 20°C, 12 h: 72%), respectively; in contrast, the same substrate (**20**) with dimethylamine gave 3-dimethylamino-

4-cinnolinecarbonitrile (**20b**) (Me_2NH, MeOH, 20°C, 2 h: 60%; steric hindrance to 4-aminolysis by the secondary amine was suggested as a reason for the changed regioselectivity).[291]

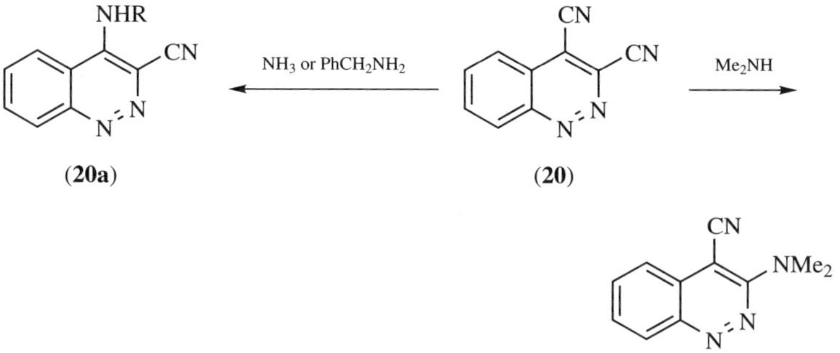

NHR, CN, N, N; NH_3 or $PhCH_2NH_2$; CN, CN, N, N; Me_2NH

(**20a**) (**20**)

CN, NMe_2, N, N

(**20b**)

Azo Coupling

3-Acetyl-4(1*H*)-cinnolinone (**21**) unexpectedly gave 3-acetyl-1-phenylazo-4(1*H*)-cinnolinone (**22a**) [substrate, Na_2CO_3, EtOH, H_2O; PhN_2Cl (fresh solution)↓ slowly, < 5°C: 95%]; analogous substituted-phenylazo analogs likewise.[620] However, the product (**22a**) was later reformulated as 3-phenylazo-4(1*H*)-phthalazinone (**22b**).[982]

O, Ac, N, N, H; PhN_2Cl; O, Ac, N, N, N=NPh; or; O, N=NPh, N, N, H

(**21**) (**22a**) (**22b**)

Azido- to Azocinnolines

4-Azidocinnoline 1-oxide (**23**) gave 4,4′-azocinnoline 1,1′-dioxide (**24**) [PhH, *hν* (sunlight), 5 h: 29%].[402]

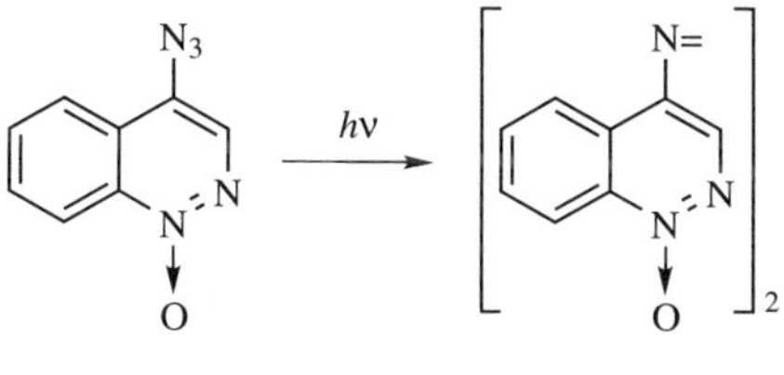

N_3, N, N, O; *hν*; N=, N, N, O, 2

(**23**) (**24**)

6.2.2. Reactions of Amino- and Hydrazinocinnolines

The direct or indirect *hydrolysis* of cinnolinamines to cinnolinones has been covered in Section 4.1.1.

The Co(II), Ni(II), Fe(II), and Cu(II) complexes of 4-amino-3-cinnolinecarboxylic acid and 4-amino-7-chloro-6-fluoro-3-cinnolinecarboxylic acid have been prepared and characterized.[556]

Other reactions are exemplified here.

N-Deamination

2-Amino-3(2*H*)-cinnolinone (**25**) gave 3(2*H*)-cinnolinone (**26**) (substrate, AcOH, $NaNO_2$ in H_2O↓ dropwise, 20°C: 96%).[687]

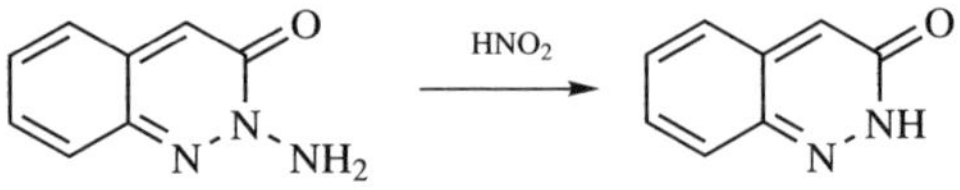

HNO_2

(**25**) (**26**)

N-Alkylidenation

4-Amino- (**27**, X = H_2) gave 4-benzylideneamino-7-chloro-6-fluoro-3-cinnolinecarboxamide (**27**, X = CHPh) (PhCHO, trace AcOH, EtOH, reflux, 3 h: 60%; analogs likewise).[670]

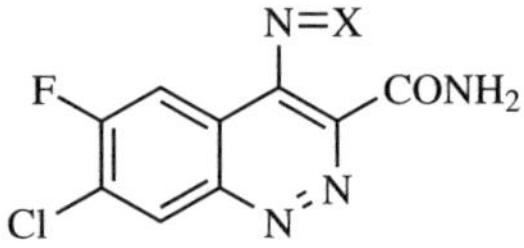

(**27**)

3-Hydrazino-5,6,7,8-tetrahydrocinnoline (**28**, X = H_2) gave 3-isopropylidenehydrazino-5,6,7,8-tetrahydrocinnoline (**28**, X = CMe_2) (AcMe, reflux, 10 min: 80%).[952]

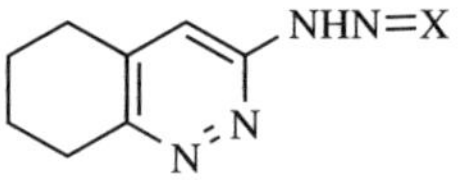

(**28**)

N-Acylation

3-Benzoyl-6,8-dimethyl-4-cinnolinamine (**29**, R = H) gave 4-acetamido-3-benzoyl-6,8-dimethylcinnoline (**29**, R = Ac) (neat Ac_2O, reflux, 5 min: 85%);[830] analogs were made similarly,[815,830] sometimes in admixture with diacetylated products.[815]

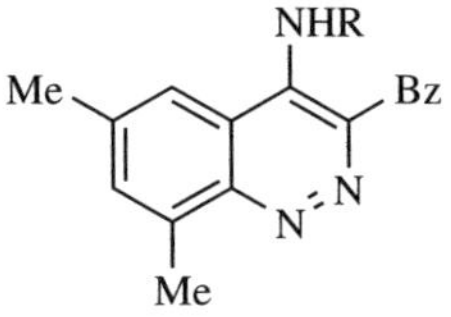

(29)

4-Cinnolinamine (**30**) gave a separable mixture of 4-pivalamidocinnoline (**31**) and 1-pivaloyl-4-pivaloylimino-1,4-dihydrocinnoline (**32**) (substrate, Et_3N, THF, 0°C; $Bu^tCOCl\downarrow$ slowly; then 20°C, 24 h: 32% and 16%, respectively, after separation).[307]

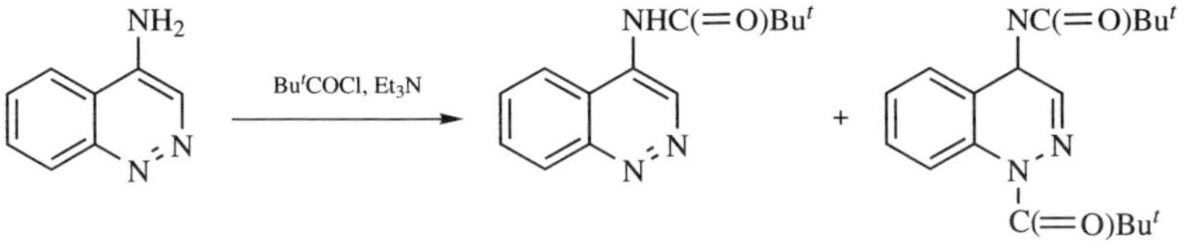

(30) (31) (32)

1-Amino-3-phenylcinnolin-1-ium mesitylenesulfonate (**33**) gave 3-phenylcinnolin-1-ium-1-benzamidate (**34**) (neat BzCl, 95°C, 3 h: 75%).[5]

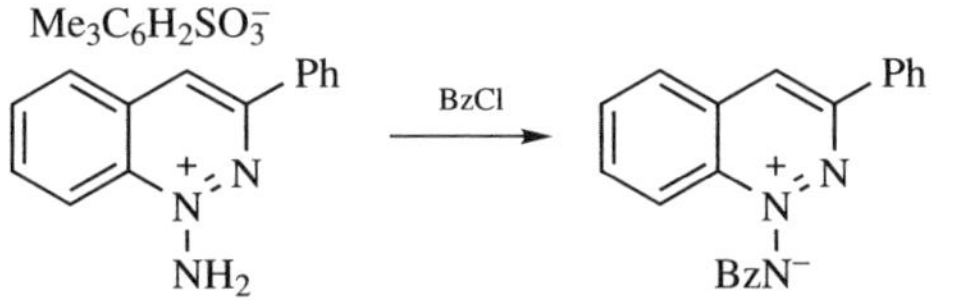

(33) (34)

4a-*m*-Benzyloxyphenyl-2-methyldecahydrocinnoline (**35**, R = H) gave 4a-*m*-benzyloxyphenyl-1-cyclopropanecarbonyl-2-methyldecahydrocinnoline [**35**, R = C(=O)CH$(CH_2)_2$] [$(H_2C)_2$CHCOCl, Et_3N, $ClCH_2CH_2Cl$: 74%; for further details, see original];[964] also many analogous acylations.[962,964,1037]

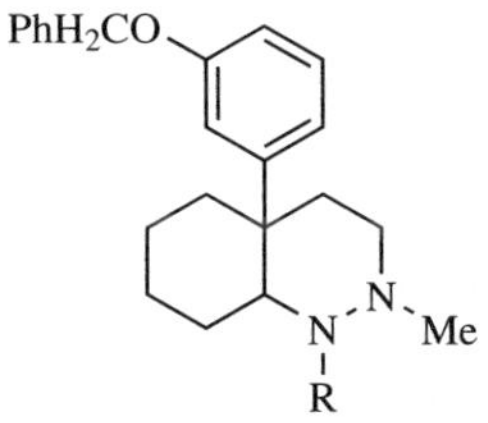

(35)

Cyanolysis (Indirect)

5-Cinnolinamine (**36**) underwent a Sandmeyer reaction to afford 5-cinnolinecarbonitrile (**37**) ($NaNO_2$, 50% H_2SO_4, −2°C; KCN + CuCN↓, 65°C, 10 min: 37%); 8-cinnolinamine did not so react.[217]

HNO_2; CuCN

(**36**) (**37**)

Cyclization Reactions

Note: Cinnolinamines and their derivatives have been used widely as substrates for further cyclocondensations. Some typical examples are given here.

3,4-Cinnolinediamine (**38**) with glyoxal gave pyrazino[2,3-*c*]cinnoline (**39**) (H_2O, 100°C, 15 min: 71%).[370]

OHCCHO

(**38**) (**39**)

3-Amino-4-cinnolinecarboxylic acid 1-oxide (**40**) with acetic anhydride gave 3-methyl-1*H*-[1,3]oxazino[4,5-*c*]cinnolin-1-one (**41**) (neat Ac_2O, reflux, ? h: 96%).[147]

Ac_2O

(**40**) (**41**)

4-Amino-3-cinnolinecarboxamide (**43**) gave pyrimido[5,4-*c*]cinnolin-4(3*H*)-one (**42**) [neat $HCONH_2$, reflux, 1 h: 60%;[667] or $HC(OEt)_3$, AcOH, reflux, 2 h: 82%],[487] pyrimido[5,4-*c*]cinnoline-2,4(1*H*,3*H*)-dione (**44**, X = O) [neat $O=C(NH_2)_2$, fused at 200°C, 15 min: ~30%],[503] or 2-thioxo-1,2-dihydropyrimido [5,4-*c*] cinnolin-4(3*H*)-one (**44**, X = S) [EtOC(=S)SK, Me_2NCHO, reflux, 2 h: ~75%).[502,503]

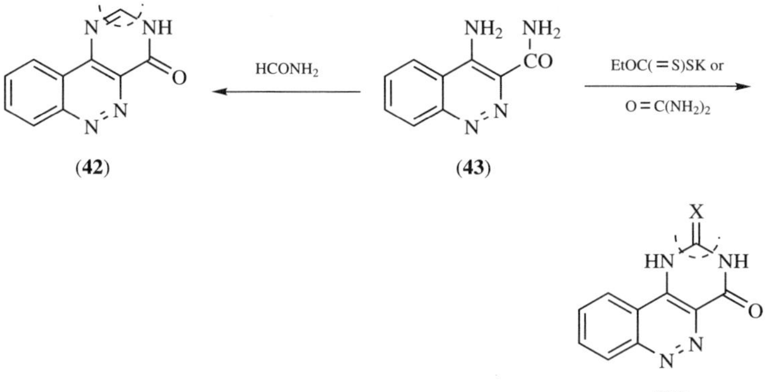

(**42**) (**43**) (**44**)

4-Amino-3-cinnolinecarbonitrile (**45**) gave pyrimido[5,4-*c*]cinnoline-2,4(1*H*, 3*H*)-dithione (**46**) [EtOC(=S)SK, Me_2NCHO, reflux, 2 h: ~70%]; analogs likewise.[677]

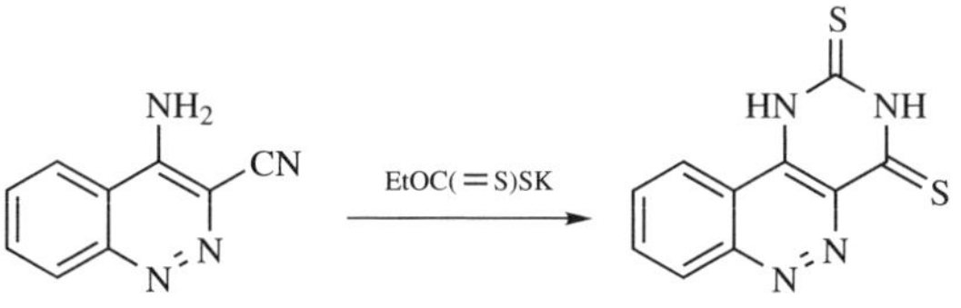

(**45**) (**46**)

4-Acetamido-3-benzoyl-6,8-dimethylcinnoline (**47**) gave 7,9-dimethyl-4-phenyl-pyrido[3,2-*c*]cinnolin-2(1*H*)-one (**48**) (K_2CO_3, Me_2NCHO, "heated," 2 h: 82%);[830] analogs likewise.[815,830]

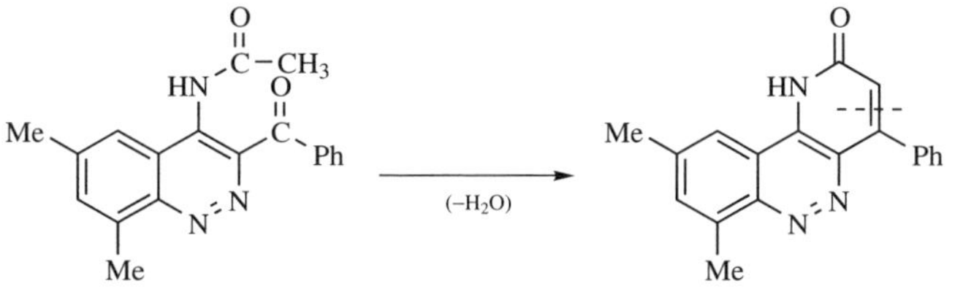

(**47**) (**48**)

3-Hydrazino-5,6,7,8-tetrahydrocinnoline (**49**) gave 6,7,8,9-tetrahydro-1,2,4-triazolo[4,3-*b*]cinnoline (**50**) (neat HCO_2H, reflux, 5 h: 79%).[952]

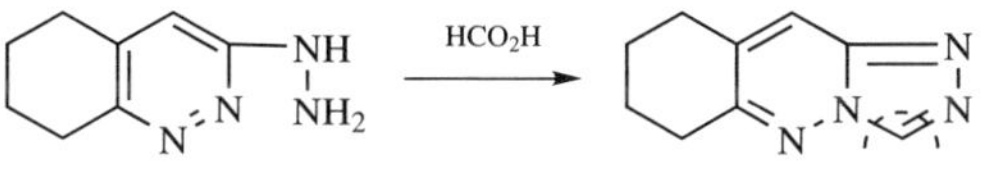

(**49**) (**50**)

Also other examples.[185,293,950]

Transamination

4-(2-Aminoethyl)aminocinnoline (**51**) underwent an atypical transamination to 4-cinnolinamine (**52**) (substrate, KNH_2, liquid NH_3, $-60°C$; then $KMnO_4\downarrow$, $-60°C \rightarrow 20°C$, 5 h: ~7%).[993]

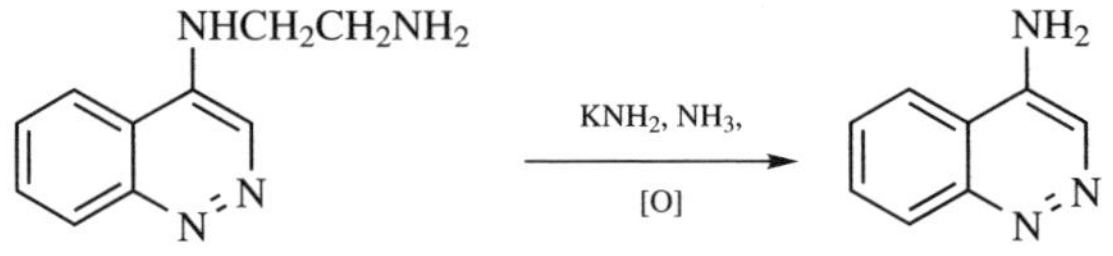

$NHCH_2CH_2NH_2$ — KNH_2, NH_3, [O] → NH_2

(**51**) (**52**)

Miscellaneous Reactions

4-(2-Dimethylaminoethylamino)cinnoline (**53**, R = H) gave 4-[*N*-(2-dimethylaminoethyl)-*o*-iodobenzamido]cinnoline (**53**, R = Bz) [substrate, Et_3N, CH_2Cl_2; *o*-IC_6H_4COCl (made in situ) in $CH_2Cl_2\downarrow$; 50°C, 18 h: 33%]; analogs likewise.[1012]

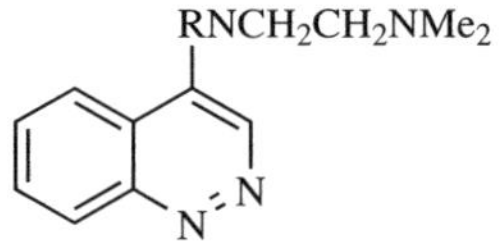

$RNCH_2CH_2NMe_2$

(**53**)

3-Benzoyl-4-cinnolinamine (**54**) gave 3-benzoyl-4-[*N'*-phenyl(thioureido)]cinnoline (**55**) (PhNCS, trace Et_3N, EtOH, reflux, 20 h: 66%).[1011]

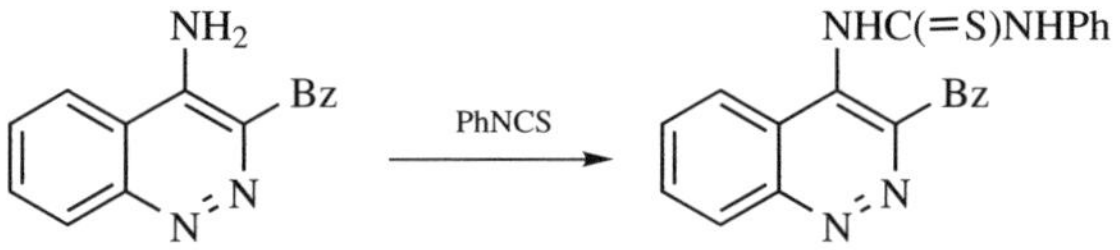

NH_2, Bz — PhNCS → NHC(=S)NHPh, Bz

(**54**) (**55**)

CHAPTER 7

Cinnolinecarboxylic Acids and Related Derivatives (*H* 11; *E* 250)

This chapter deals with nuclear and extranuclear cinnolinecarboxylic acids and the corresponding carboxylic esters, acyl halides, carboxamides, carbohydrazides, carbonitriles, and carbaldehydes, and the ketonic acylketones. To avoid repetition, the interconversion of these cinnoline derivatives are discussed only at the first opportunity: for example, the esterification of cinnolinecarboxylic acids is covered as a reaction of cinnolinecarboxylic acids rather than as a preparative route to carboxylic esters, simply because the section on acids precedes that on esters. To avoid any confusion, appropriate cross-references have been included.

7.1. CINNOLINECARBOXYLIC ACIDS (*H* 11; *E* 250)

7.1.1. Preparation of Cinnolinecarboxylic Acids

The formation of cinnolinecarboxylic acids by *primary synthesis* (see Chapter 1) and by *hydrolysis of trihalogenomethylcinnolines* (see Section 3.2) have been covered already; the *oxidative routes* from alkylcinnolines, hydroxyalkylcinnolines, or cinnolinecarbaldehydes appear to be unrepresented in the 1972–2004 literature; the remaining approaches are illustrated in the following classified examples.

By Carbonation of a Metallated Cinnoline

4-Chlorocinnoline (**1**) gave its 2-lithio derivative (**2**) [$LiNPr^i_2$ (made in situ), THF, substrate↓, −75°C] and thence, by carbonation of the reaction mixture,

Cinnolines and Phthalazines: Supplement II, The Chemistry of Heterocyclic Compounds, Volume 64, by D.J. Brown

4-chloro-3-cinnolinecarboxylic acid (**3**) (solid CO_2↓, −75°C, 2 h: 57% overall).[307]

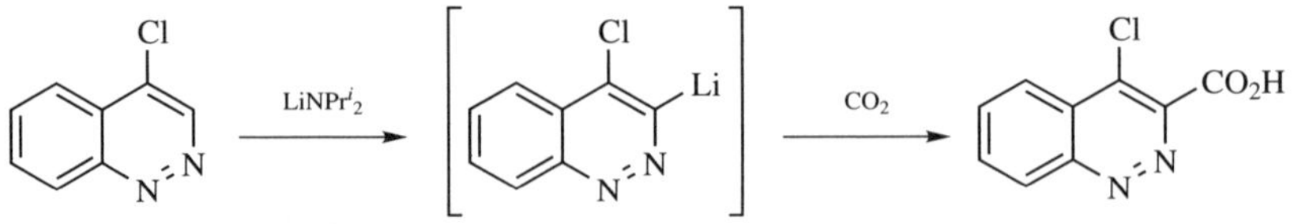

(**1**) (**2**) (**3**)

By Hydrolysis of Cinnolinecarboxylic Esters

Note: Such deesterification has been done by acidic or alkaline hydrolysis or by the use of boron tribromide, as illustrated here.

Methyl 1-*p*-chlorophenyl-5-fluoro-4-oxo-1,4-dihydro-3-cinnolinecarboxylate (**4**, R = Me) gave 1-*p*-chlorophenyl-5-fluoro-4-oxo-1,4-dihydro-3-cinnolinecarboxylic acid (**4**, R = H) (HCl, H_2O, dioxane, reflux, 4 h: 72%).[765] Also other acidic hydrolyses.[416]

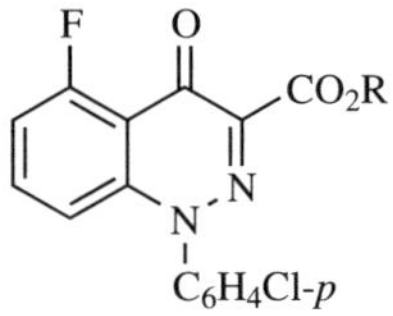

(**4**)

Ethyl (**5**, R = Et) or methyl 7-chloro-6-fluoro-1-*p*-fluorophenyl-4-oxo-1,4-dihydro-3-cinnolinecarboxylate (**5**, R = Me) gave 7-chloro-6-fluoro-1-*p*-fluorophenyl-4-oxo-1,4-dihydro-3-cinnolinecarboxylic acid (KOH, MeOH, H_2O, 50°C → 20°C, 2 h: 96% or 82%, respectively).[617]

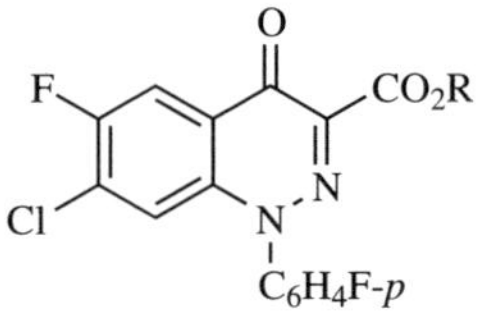

(**5**)

Ethyl 1-ethyl-7-methoxy-4-oxo-1,4-dihydro-3-cinnolinecarboxylate (**6**, R = Et) as its hydriodide gave 1-ethyl-7-methoxy-4-oxo-1,4-dihydro-3-cinnolinecarboxylic acid (**6**, R = H) (BBr_3, −78°C → 20°C: 40%).[387]

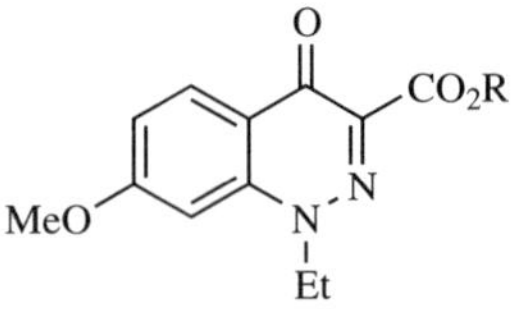

(**6**)

Also other examples.[54]

By Hydrolysis of Cinnolinecarboxamides

Note: These hydrolyses may be done in acid or alkali.

4-Amino-7-chloro-6-fluoro-3-cinnolinecarboxamide (**7**, R = NH_2) gave 4-amino-7-chloro-6-fluoro-3-cinnolinecarboxylic acid (**7**, R = OH) (KOH, EtOH, reflux, 3 h: 60%);[670] analogs likewise.[556]

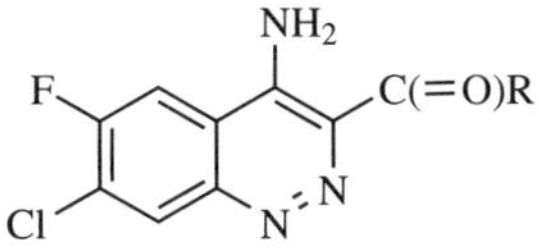

(**7**)

4-Amino-7-methoxy-3-cinnolinecarboxamide (**8**, R = NH_2) gave 4-amino-7-methoxy-3-cinnolinecarboxylic acid (**8**, R = OH) (KOH, EtOH, reflux, 3 h: 50%; or HBr, AcOH, reflux, 3 h: 66%).[487]

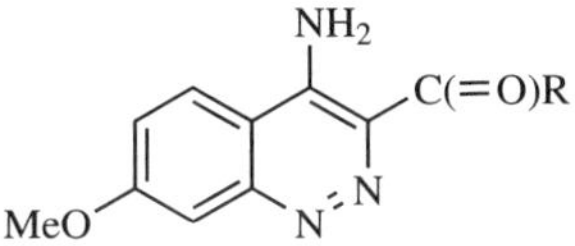

(**8**)

By Hydrolysis of Cinnolinecarbonitriles

Note: Hydrolysis of cinnolinecarbonitriles has been done under acidic or alkaline conditions.

1-Ethyl-4-oxo-1,4-dihydro-3-cinnolinecarbonitrile (**9**, Q = H) gave 1-ethyl-4-oxo-1,4-dihydro-3-cinnolinecarboxylic acid (**10**, R = H) (10M HCl, AcOH, reflux, 16 h: 92%); 1-ethyl-4-oxo-1,4-dihydro-3,7-cinnolinedicarbonitrile (**9**, Q = CN) likewise gave 1-ethyl-4-oxo-1,4-dihydro-3,7-cinnolinedicarboxylic acid (**10**, R = CO_2H) (3 h: 51%).[485]

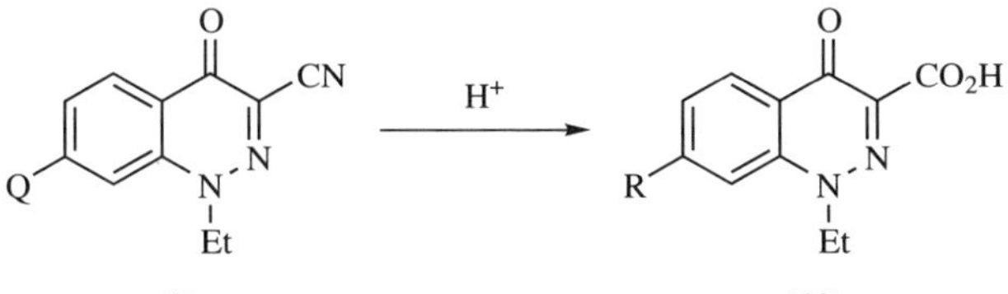

H^+

(**9**) (**10**)

3,4-Cinnolinedicarbonitrile (**11**, R = CN) gave 3,4-cinnolinedicarboxylic acid (**11**, R = CO_2H) (10M HCl, reflux, 4 h: 53%).[291]

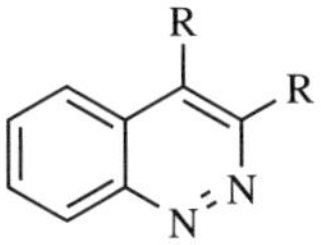

(**11**)

4-Chloro-3-cinnolinecarbonitrile (**12**) gave 4-oxo-1,4-dihydro-3-cinnolinecarboxylic acid (**13**) (4M NaOH, reflux, 2.5 h: ~55%).[695]

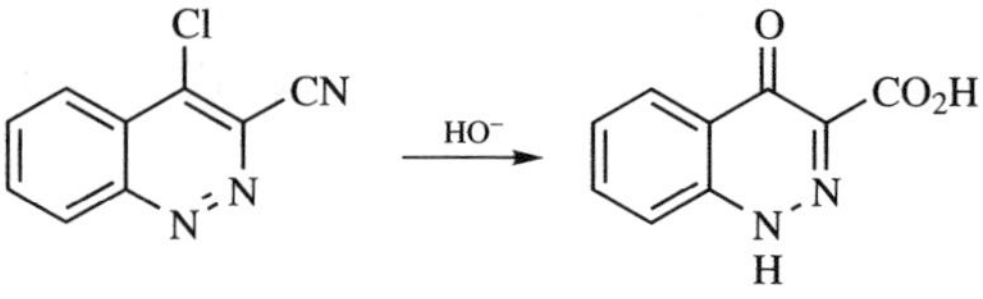

(**12**) (**13**)

8-Amino-2-*m*-chlorophenyl-3-oxo-6-phenyl-2,3-dihydro-4,7-cinnolinedicarbonitrile (**14**, R = CN) gave 8-amino-2-*m*-chlorophenyl-7-cyano-3-oxo-6-phenyl-2,3-dihydro-4-cinnolinecarboxylic acid (**14**, R = CO_2H) (NaOH, EtOH, reflux, 5 h: 94%; note selective hydrolysis of the 4-cyano group).[297]

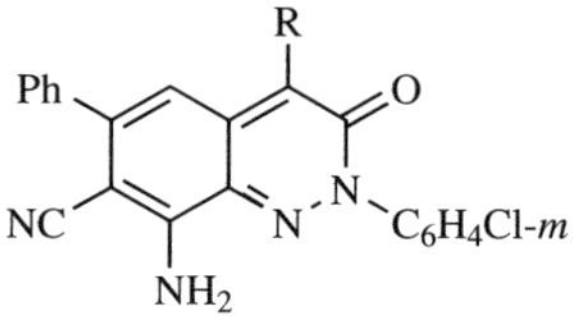

(**14**)

Also other examples.[535]

7.1.2. Reactions of Cinnolinecarboxylic Acids

Cinnolinecarboxylic acids undergo several useful reactions, illustrated briefly by the following classified examples gleaned from the 1972–2004 literature.

Decarboxylation

6-Methoxy-8-nitro-4-oxo-1,4-dihydro-3-cinnolinecarboxylic acid (**15**, R = CO_2H) gave 6-methoxy-8-nitro-4(1*H*)-cinnolinone (**15**, R = H) (Ph_2CO, ~190°C, 90 min: > 95%).[27]

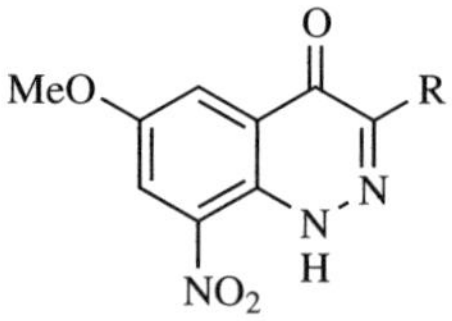

(**15**)

6-Bromo-4-oxo-1-propyl-1,4-dihydro-3-cinnolinecarboxylic acid (**16**, R = CO_2 H) gave 6-bromo-1-propyl-4(1*H*)-cinnolinone (**16**, R = H) (neat substrate,

330°C, 50 s: 23%).[12]

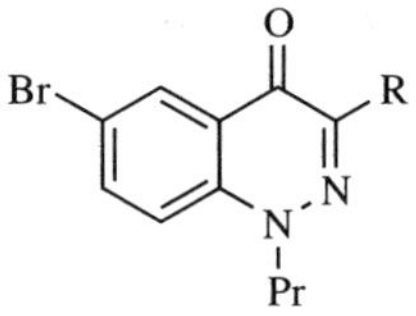

(16)

3-Amino-4-cinnolinecarboxylic acid 1-oxide (**17**) gave 3-cinnolinamine (**18**) ($Na_2S_2O_4$, Me_2NCHO, H_2O, reflux; no further details).[147]

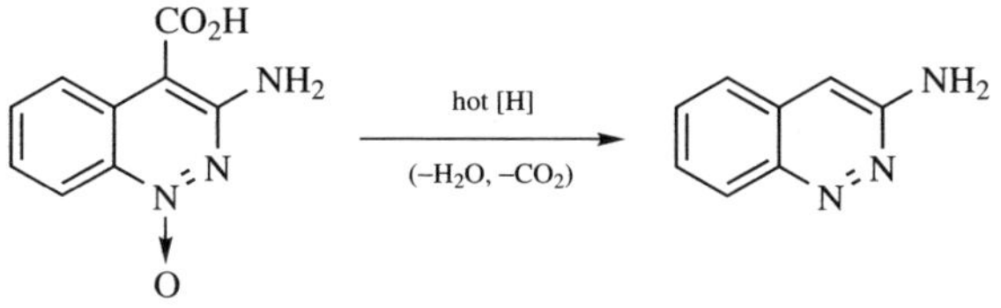

hot [H] ($-H_2O$, $-CO_2$)

(17) (18)

Also other examples.[554]

Esterification

Note: This may be done directly or via an acyl halide. Both procedures are illustrated here.

1-Methyl-4-oxo-1,4-dihydro-3-cinnolinecarboxylic acid (**19**) gave ethyl 1-methyl-4-oxo-1,4-dihydro-3-cinnolinecarboxylate (**20**) (EtOH, 95% H_2SO_4, reflux, 18 h: 80%).[54]

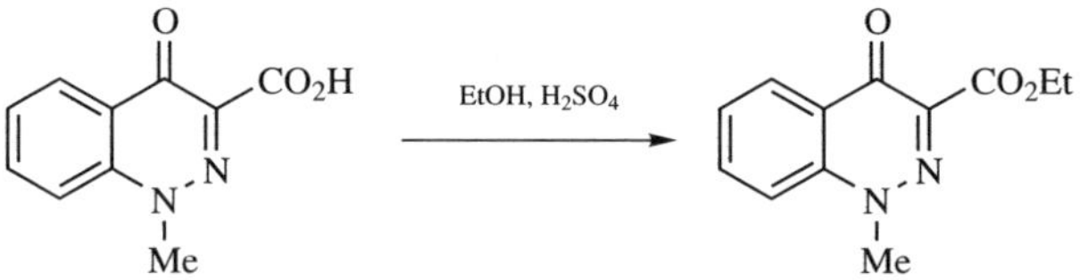

EtOH, H_2SO_4

(19) (20)

4-Oxo-1,4-dihydro-3-cinnolinecarboxylic acid (**21**) gave 4-oxo-1,4-dihydro-3-cinnolinecarbonyl chloride (**22**) (neat $SOCl_2$, reflux, 4 h: crude uncharacterized) and thence ethyl 4-oxo-1,4-dihydro-3-cinnolinecarboxylate (**23**) (EtOH, 20°C, substrate↓ portionwise; then reflux, 10 min: 70% overall).[54]

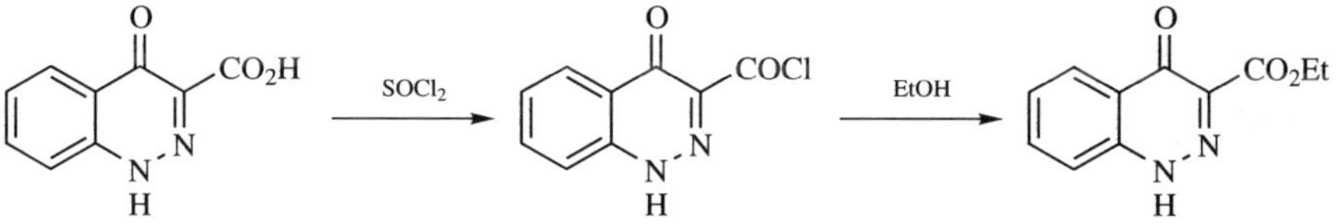

$SOCl_2$ EtOH

(21) (22) (23)

3-(2-Carboxyethyl)-6-ethyl-4(1*H*)-cinnoline (**24**, R = H) gave 6-ethyl-3-(2-methoxycarbonylethyl)-4(1*H*)-cinnolinone (**24**, R = Me) (MeOH, reflux, HCl gas↓, 30 min; then reflux, 12 h: 75%);[213] many analogs likewise.[213,782]

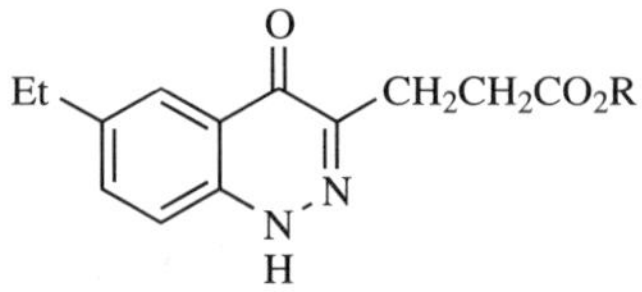

(**24**)

6,7,8-Trifluoro-1-methyl-4-oxo-1,4-dihydro-3-cinnolinecarboxylic acid (**25**, R = H) gave ethyl 6,7,8-trifluoro-1-methyl-4-oxo-1,4-dihydro-3-cinnolinecarboxylate (**25**, R = Et) (substrate, Et_3N, $CHCl_3$, −5°C; $ClCO_2Et$↓, 20 min; then EtOH↓, 20°C → reflux, 1 h: 83%).[416]

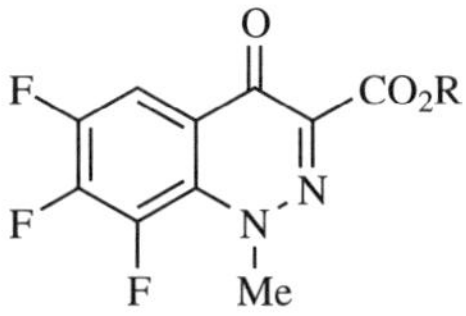

(**25**)

Conversion into Cinnolinecarboxamides

Note: Such conversions may be done directly but usually via an acyl halide or ester; examples of the first two procedures are given here.

4-Cinnolinecarboxylic acid (**26**) and 1-benzyl-4-methylperhydro-1,4-diazepin-6-amine (**27**) in the presence of 1,1′-carbonyldiimidazole gave *N*-(1-benzyl-4-methylperhydro-1,4-diazepin-6-yl)-4-cinnolinecarboxamide (**28**) [substrate (**26**), 1,1′-carbonyldiimidazole, Me_2NCHO, 20°C, 30 min; then amine (**27**)↓, 20°C, 18 h: 60%, as an oxalate].[436]

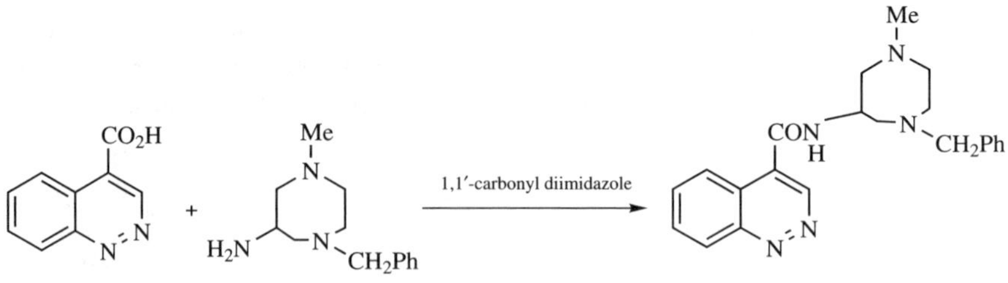

1,1′-carbonyl diimidazole

(**26**) (**27**) (**28**)

4-Thioxo-1,4-dihydro-3-cinnolinecarboxylic acid (**29**) gave 4-thioxo-1,4-dihydro-3-cinnolinecarbonyl chloride (**30**) ($SOCl_2$, $CHCl_3$, Me_2NCHO, reflux until clear: 40% crude) and thence 4-thioxo-1,4-dihydro-3-cinnolinecarboxamide (**31**) (NH_4OH, 20°C → reflux, 15 min: 15%).[54]

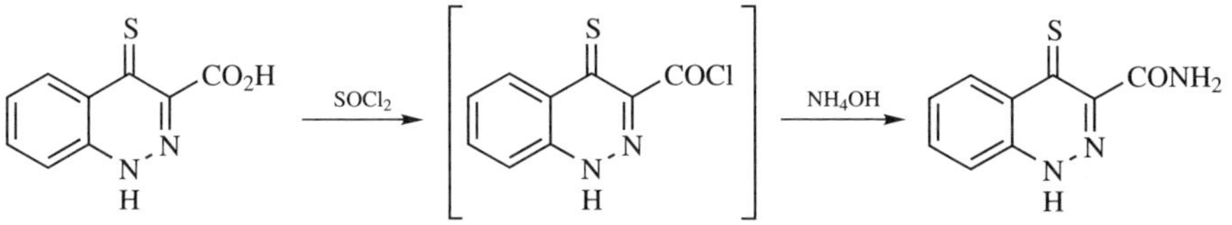

(**29**) —$SOCl_2$→ [(**30**)] —NH_4OH→ (**31**)

Also other examples.[433,1034]

7.2. CINNOLINECARBOXYLIC ESTERS (*E* 254)

All more recently used preparative routes to cinnolinecarboxylic esters and some of their reactions have been covered already. Cross-references and any additional examples are given here.

Preparation

Note: Routes to these esters include *primary synthesis* (see Chapter 1), *Reissert-type additions to cinnoline* (see Section 2.1.3), and *esterification of cinnolinecarboxylic acids* (see Section 7.1.2); also *passenger introductions* (e.g. alkoxycarbonylalkylations in Section 4.1.2.1).

Reactions

Note: *Reduction to hydroxymethylcinnolines* remains unrepresented, *hydrolysis to cinnolinecarboxylic acids* has been covered in Section 7.1.1, and *conversion into cinnolinecarboxamides or carbohydrazides* is illustrated here.

Ethyl 4-thioxo-1,4-dihydro-3-cinnolinecarboxylate (**32**) gave 4-amino-3-cinnolinecarboxamide (**33**) [NH_3 gas, EtOH, sealed, 100°C (?), 18 h: 80%; note additional aminolysis].[54]

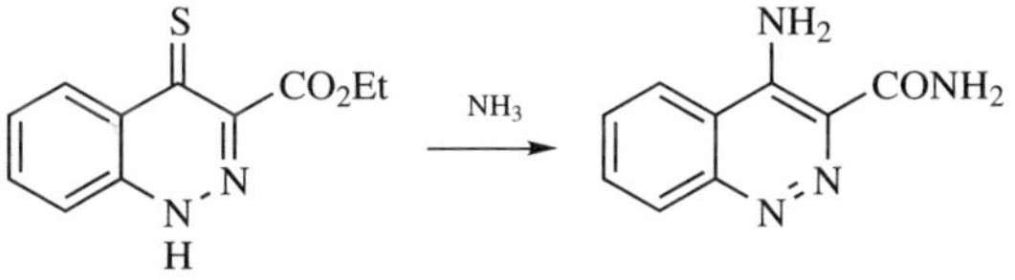

(**32**) —NH_3→ (**33**)

Ethyl 4-cinnolinecarboxylate (**34**) gave 4-cinnolinecarbohydrazide (**35**) (neat $H_2NNH_2 \cdot H_2O$, 20°C → 115°C, 90 min: > 95%).[616]

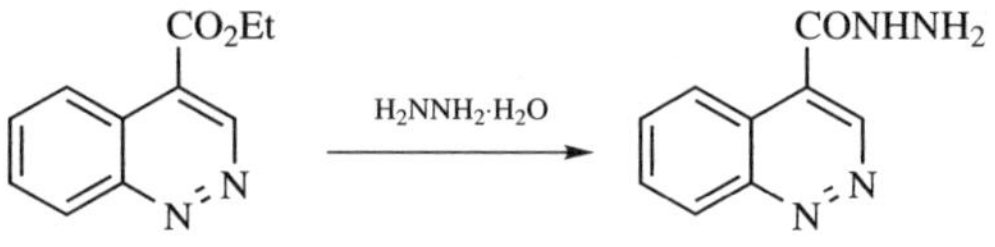

(**34**) (**35**)

Also other examples.[54,554]

7.3. CINNOLINECARBOXAMIDES AND CINNOLINECARBOHYDRAZIDES (*E* 253)

Most preparative routes and reactions for these amides and hydrazides have been covered; cross-references and any additional data are given in the following paragraphs.

Preparation

Note: Most cinnolinecarboxamides and carbohydrazides (both nuclear and extranuclear) have been made by *primary synthesis* (see Chapter 1), *from cinnolinecarboxylic acids* (directly or indirectly via cinnolinecarbonyl halides) (see Section 7.1.2), or *from cinnolinecarboxylic esters* (see Section 7.2). The remaining route, *by hydrolysis or thiolysis of cinnolinecarbonitriles*, is illustrated here.

3,4-Cinnolinedicarbonitrile (**36**) gave 3,4-cinnolinedicarboxamide (**37**) (95% H_2SO_4, 20°C, 12 h: > 95%; the limited hydrolytic capacity of this medium ensures that hydrolysis is controlled and does not proceed to the carboxylic acid stage).[291]

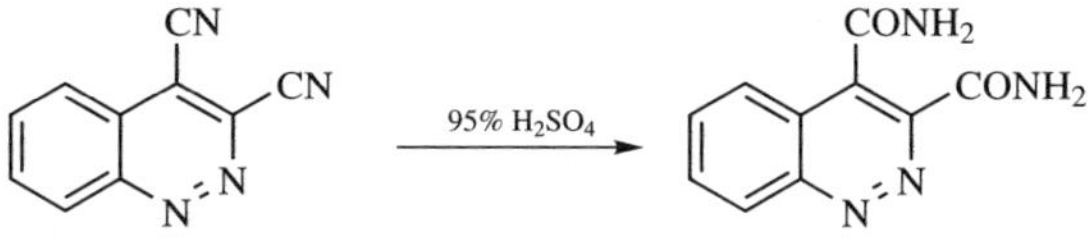

(**36**) (**37**)

4-Amino-3-cinnolinecarbonitrile (**38**) gave 4-amino-3-cinnolinecarbothioamide (**39**) (substrate, Et_3N, pyridine, Me_2NCHO, $H_2S\downarrow$ to saturation, 20°C, 12 h: 70%).[487]

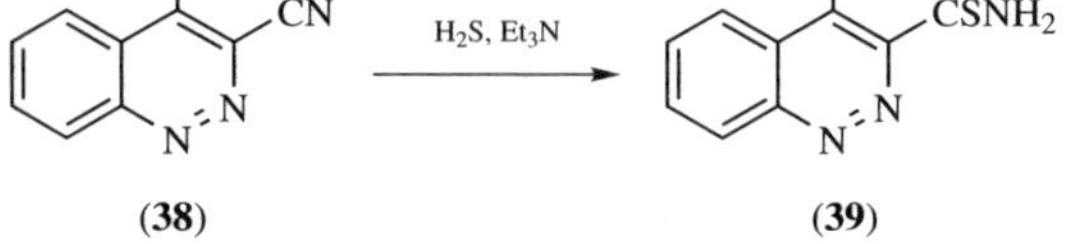

(**38**) (**39**)

See also Section 7.4 for a Radziszewski-type controlled hydrolysis of a cinnolinecarbonitrile.

Reactions

Note: The *hydrolysis of cinnolinecarboxamides to cinnolinecarboxylic acids* has been covered in Section 7.1.1. One important reaction of such amides and one esoteric reaction of such hydrazides are illustrated here.

4-Amino-3-cinnolinecarboxamide (**40**) underwent *dehydration* to give 4-amino-3-cinnolinecarbonitrile (**41**) (neat $POCl_3$, $Me_3N\downarrow$, $< 20°C \rightarrow$ reflux, 2 h: 78%); analogs likewise.[677]

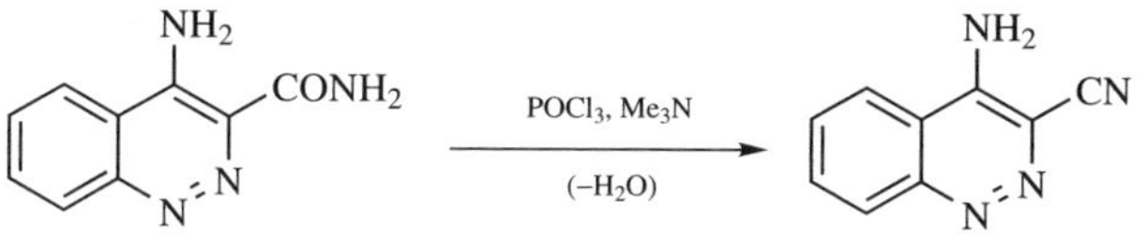

(**40**) $\xrightarrow[(-H_2O)]{POCl_3,\ Me_3N}$ (**41**)

4-Cinnolinecarbohydrazide underwent *N′-heteroarylation* by 7-chloro-5-*o*-chlorophenyl-2*H*-1,4-benzodiazepine-2-thione to give 7-chloro-5-*o*-chlorophenyl-2-[*N′*-(4-cinnolinecarbonyl)hydrazino]-1*H*-1,4-benzodiazepine (**42**) with loss of H_2S (BuOH, N_2, reflux, 6 h: 82%).[616]

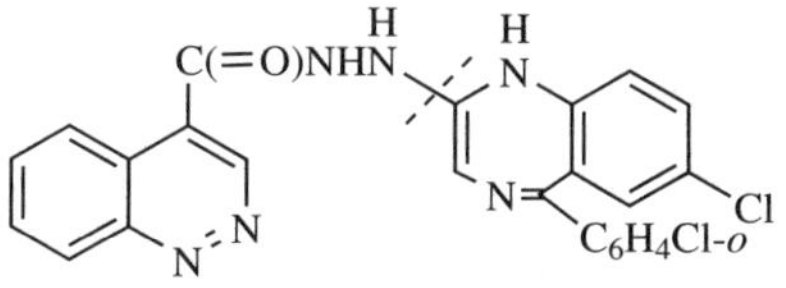

(**42**)

7.4. CINNOLINECARBONITRILES (*E* 262)

The preparative routes to cinnolinecarbonitriles and most of their reactions have been discussed earlier in this book. Cross-references and any additional data follow.

Mass Spectra

The MS of 4-cinnolinecarbonitrile and its 3-methyl, 3-ethyl, 3-propyl, 3-isopropyl, and 3-butyl derivatives have been studied in detail.[401]

Preparation

Note: More recently used routes to these nitriles have been covered already: by *primary synthesis* (Chapter 1), by *cyanoalkanelysis of alkoxycinnolines*

(Section 2.2.1), by *cyanolysis of halogenocinnolines* (Section 3.2), by *cyanolysis of alkylsulfonylcinnolines* (Section 5.2), by *Sandmeyer reactions on cinnolinamines* (Section 6.2.2), and by *dehydration of cinnolinecarboamides* (Section 7.3).

Reactions

Note: Reactions already discussed include *aminolysis* (Section 6.2.1), *hydrolysis to cinnolinecarboxylic acids* (Section 7.1.1), and *controlled hydrolysis or thiolysis to carboxamides or carbothioamides*, respectively (Section 7.3). Further minor reactions are illustrated here.

4-(α-Cyanobenzyl)cinnoline (**43**) underwent *oxidative decyanation* to afford 4-benzoylcinnoline (**43a**) (substrate, NaH, THF, 5 min; $O_2\downarrow$ until colorless: 94%; a rational mechanism was proposed).[586]

CH(CN)Ph → (NaH, O_2) → C(=O)Ph

(**43**) (**43a**)

4-Amino-5,7-dimethyl-3-cinnolinecarbonitrile (**44**) underwent *amine addition* to give 4-amino-*N*-isopropyl-5,7-dimethyl-3-cinnolinecarboxamidine (**44a**) ($Pr^i NH_2$, MeOH, reflux, 5 h: 46%); analogs likewise.[1011]

Me, NH_2, CN → (Pr^iNH_2) → Me, NH_2, C (=NH)NHPr^i

(**44**) (**44a**)

4-Cinnolinecarbonitrile (**45**) underwent *acylalkanelysis* by acetophenone carbanion to give 4-phenacylcinnoline (**46**, R = Ph) ($NaNH_2$, PhH, reflux, 4 h: 47%); acetone carbanion behaved similarly to afford 4-acetonylcinnoline (**46**, R = Me) (30 min: 28%) but diethyl ketone gave 4-ethylcinnoline (**48**), presumably by spontaneous deacylation of the intermediate (**47**) (10 min: 65%).[933]

4-(1-Acetylacetonyl)-3-cinnolinecarbonitrile (**49**) underwent *intramolecular cyclization* to give 1-acetyl-2-methylpyrido[3,4-*c*]cinnolin-4(3*H*)-one (**51**), presumably via the amide (**50**) [0.5M NaOH, $KMnO_4$ (or H_2O_2), 100°C, 2 h: 38%; perhaps the oxidizing agent was added to mimic Radziszewski conditions for controlled hydrolysis of the cyano group?].[695]

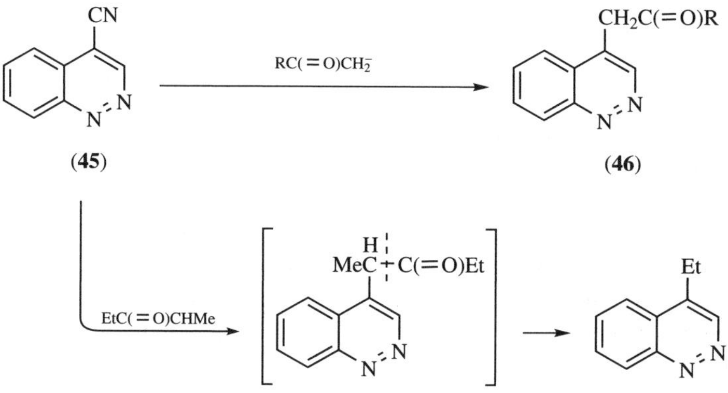

(45) (46) (47) (48)

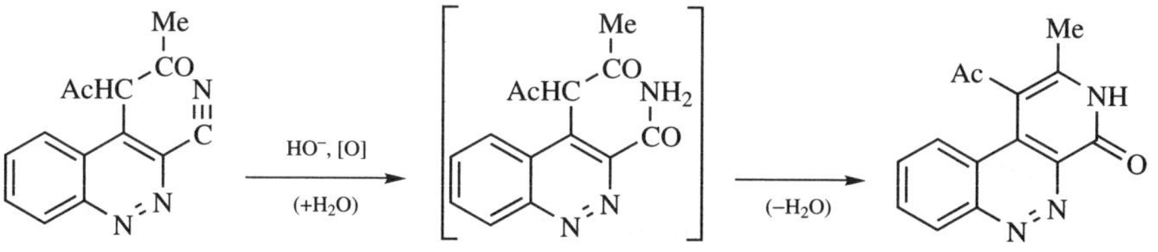

(49) (50) (51)

7.5. CINNOLINE ALDEHYDES AND KETONES (*E* 256, 259)

Most of the more recently reported data on these *C*-acylcinnolines and *N*-acylated hydrocinnolines have been discussed previously in this book, as indicated in the following paragraphs.

Preparation

Note: These acylcinnolines have been made by *primary synthesis* (see Chapter 1), *Reissert-type addition to cinnoline* (Section 2.1.3), *oxidation of alkylcinnolines* (Section 2.2.2), *displacement of halogeno substituents* (Section 3.2), *hydrolysis of dihalogenomethylcinnolines* (Section 3.2), *N-acylation of tautomeric cinnolinones* (Section 4.1.2.2), *oxidation of extranuclear hydroxycinnolines* (Section 4.2), or as illustrated here.

4-Methoxycinnoline was converted into 3-lithio-4-methoxycinnoline (**52**) and thence by *direct formylation* into 4-methoxy-3-cinnolinecarbaldehyde (**53**)

[$LiNPr^i_2$ (made in situ), substrate↓, THF, −75°C, 30 min; then HCO_2Et↓, −75°C, 2 h: 57%].[307]

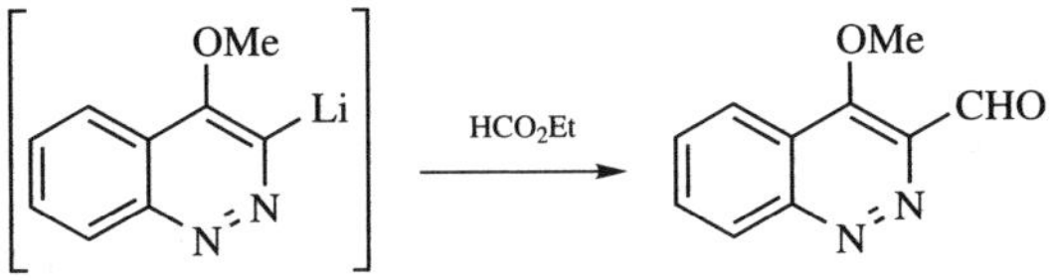

(52) (53)

4-Methylcinnoline (**54**) underwent *reductive acetylation* to give 1,2-diacetyl-4-methyl-1,2-dihydrocinnoline (**55**) [electrolytic, Ac_2O, Me_2NCHO; a physicochemical study without isolation); 4-phenylcinnoline behaved similarly].[941]

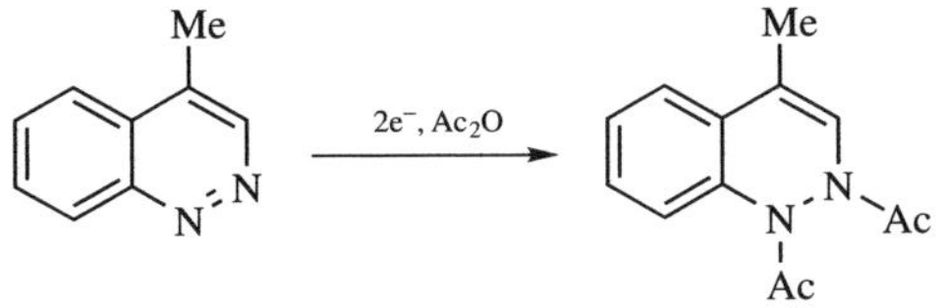

(54) (55)

Reactions

Note: These reactions of acylcinnolines have been covered already: *reduction to alkylcinnolines* (Section 2.2.1), some *cyclocondensations* (Section 4.1.2.2), and *reduction to extranuclear hydroxycinnolines* (Section 4.2). Their *oxidation to cinnolinecarboxylic acids* appears to be unrepresented in the more recent literature. Some other reactions are illustrated in the examples that follow.

The equilibrium between 4-cinnolinecarbaldehyde (**56**) and its hydrate, 4-dihydroxymethylcinnoline (**57**), has complicated a study of the reaction of the aldehyde (**56**) with hydroxyl ion.[101]

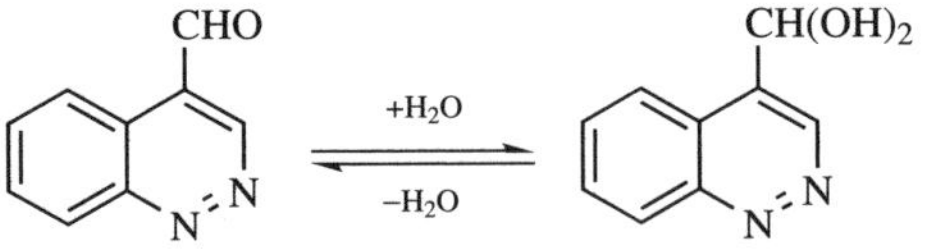

(56) (57)

3-Acetyl-4(1*H*)-cinnolinone (**59**) underwent cyclocondensation with malononitrile to give 4-methyl-2-oxo-1,2-dihydropyrido[3,2-*c*]cinnoline-3-carbonitrile (**58**) (EtOH, trace piperidine, reflux, 2 h: 65%) or, with ethyl cyanoacetate, to give 4-methyl-2-oxo-2*H*-pyrano[3,2-*c*]cinnoline-3-carbonitrile (**60**) (neat reactants, 165°C, 5 h: 70%).[619] The same substrate (**59**) gave its *thiosemicarbazone* (**61**) ($H_2NNH_2CSNH_2$, AcOH, reflux, 2 h: 70%), which underwent *cyclization* with ethyl 2-chloropropionate to afford 3-[1-(5-methyl-4-oxothiazolidin-2-ylidenehydrazono)ethyl]-4(1*H*)-cinnolinone (**62**) (AcONa, EtOH, reflux, 4 h: 65%).[619]

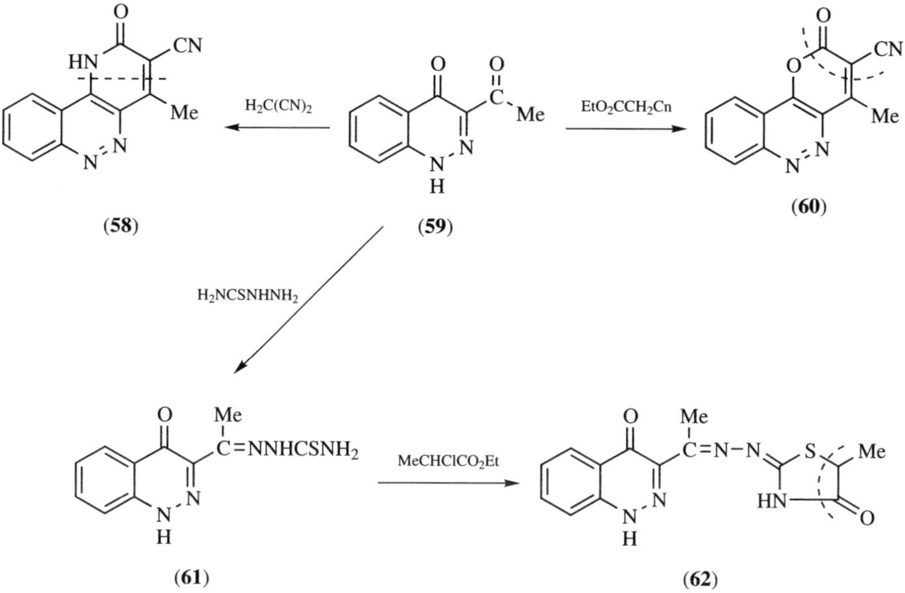

(**58**) (**59**) (**60**)

(**61**) (**62**)

3-Benzoylcinnoline (**63**) underwent flash vacuum pyrolysis to afford a mixture in which phenanthrene (**64**), anthracene (**65**), phenylacetylene (**66**), and diphenylacetylene (**67**) were identified (quartz tube, 900°C, 0.02 mmHg: 30%, 8%, 36%, and 6% respectively); substituted-benzoyl and analogous substrates were treated similarly with broadly similar results.[825]

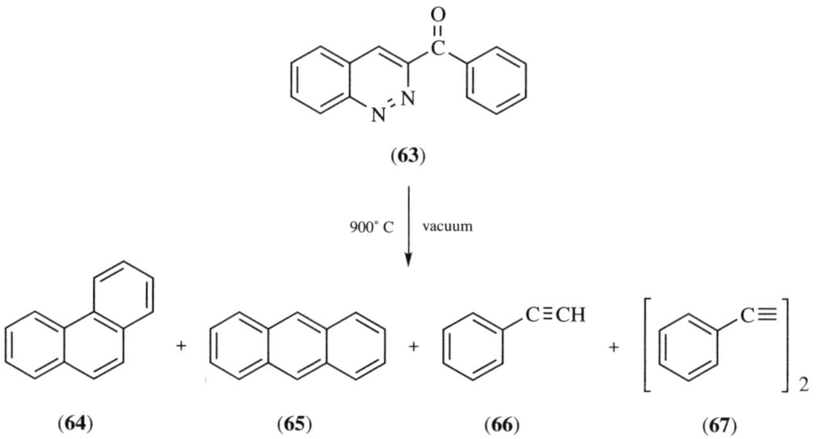

(**63**)

(**64**) (**65**) (**66**) (**67**)

3-Benzoyl-4-cinnolinamine (**68**) gave a separable mixture of the oxime, 3-[(hydroxyimino)benzyl]-4-cinnolinamine (**69**), and the product of its cyclization, 3-phenylisoxazolo[4,5-*c*]cinnoline (**70**) ($H_2NOH \cdot HCl$, EtOH, reflux, 20 h: 61% and 20%, respectively); independently, thermolysis of the oxime (**69**) gave the tricyclic product (**70**) (neat substrate, 150°C, vacuum, 3 h: 50%).[1011]

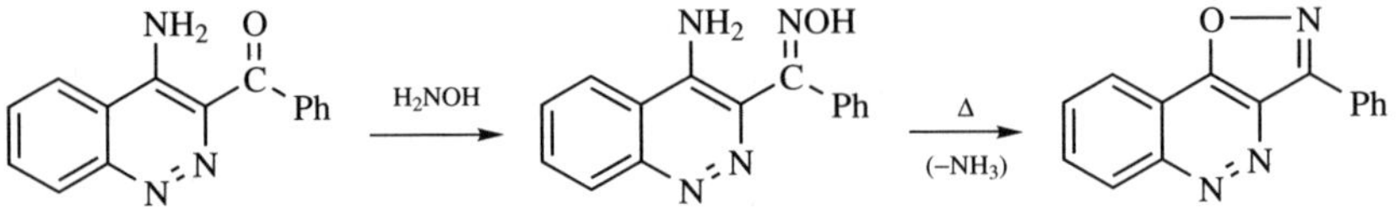

(**68**) $\xrightarrow{H_2NOH}$ (**69**) $\xrightarrow[(-NH_3)]{\Delta}$ (**70**)

CHAPTER 8

Primary Syntheses of Phthalazines

In much the same way as synthesis of their cinnoline counterparts (see Chapter 1), the primary synthesis of phthalazines (or hydrophthalazines) may be done by cyclization of benzene (or cyclohexane) derivatives already bearing appropriate substituents, by cyclocondensation of benzene (or cyclohexane) derivatives with acyclic synthons that provide one or more of the ring atoms needed to produce the phthalazine system, by analogous processing of other carbocyclic or pyridazine substrates, or by modification of other heterocyclic substrates in various ways. Typical pre-1972 examples in each category of synthesis may be found from cross-references to Simpson's volume[906] (e.g., *H* 72) or to Singerman and Patel's volume[907] (e.g., *E* 333) that appear in some section headings. A variety of pre- and post-1972 syntheses have also been reviewed elsewhere.[903–905,908–916]

8.1. FROM A SINGLE BENZENE DERIVATIVE AS SUBSTRATE (*H* 69, 72, 85, 107; *E* 342, 379)

Such syntheses of phthalazines may be done by formation of the C1–C8a, C1–N2, or N2–N3 bond. There appear to be no authenticated examples of the last mechanism in the 1972–2004 literature.

8.1.1. By Formation of the C1–C8a Bond

This bond formation has been achieved, for example, by simple isomerization of the substrates, by dehydrogenation of substrates bearing an ω-aryl groups, or by dehydration of substrates bearing conventional leaving groups. Examples in each category follow.

Cinnolines and Phthalazines: Supplement II, The Chemistry of Heterocyclic Compounds, Volume 64,
by D.J. Brown

By Isomerization of Substrates

m-(Isopropylidenehydrazinomethyl)phenol (**1**) gave 1,1-dimethyl-1,2,3,4-tetrahydro-6-phthalazinol (**2**) (10M HCl, reflux, 1 h: 58% as hydrochloride); several analogs likewise.[686]

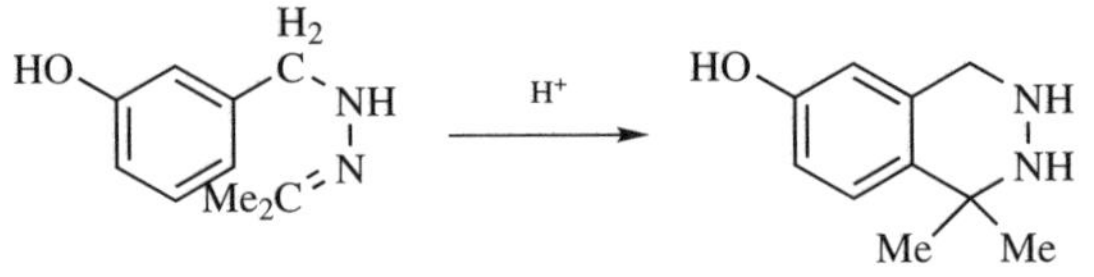

(**1**) (**2**)

By Dehydrogenation of Substrates with ω-Aryl Groups

Note: This can occur either by aerial (?) oxidation with retention of the aryl group or by oxidative loss of arene.

N,N′-Bis(2,4-dimethylbenzylidene)hydrazine (**4**) gave 1-(2,4-dimethylphenyl)-5,7-dimethylphthalazine (**3**) ($AlCl_3$, Et_3N, 190°C, 1 h: 60%) or 5,7-dimethylphthalazine (**5**) ($AlCl_3$, $AlBr_3$, 190°C, 30 min: 65%);[80] many analogs were made similarly.[80,208,220,343,352,353,456,857]

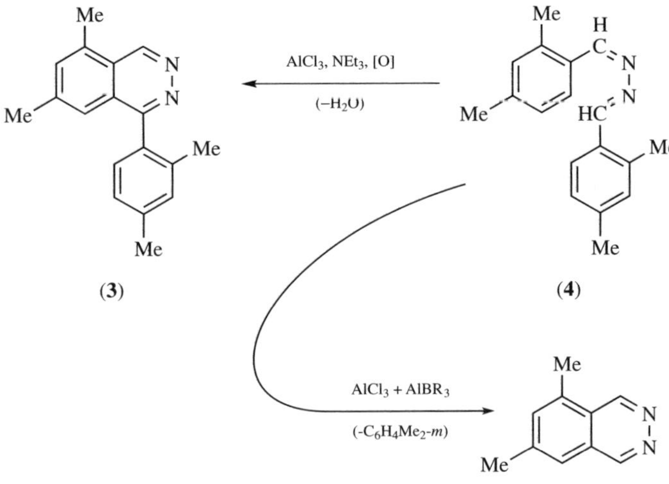

(**3**) (**4**) (**5**)

By Cyclization of Substrates Bearing Conventional Leaving Groups

Note: In these examples, the substrate undergoes cyclization by loss of water, an alcohol, ammonia, acetic acid, or the like.

o-[(Acetylhydrazono)methyl]nitrobenzene (**6**) gave 1-methyl-5-nitrophthalazine (**7**) (MeOH, *h*ν, 7.5 h: 50%); analogs likewise.[330,701]

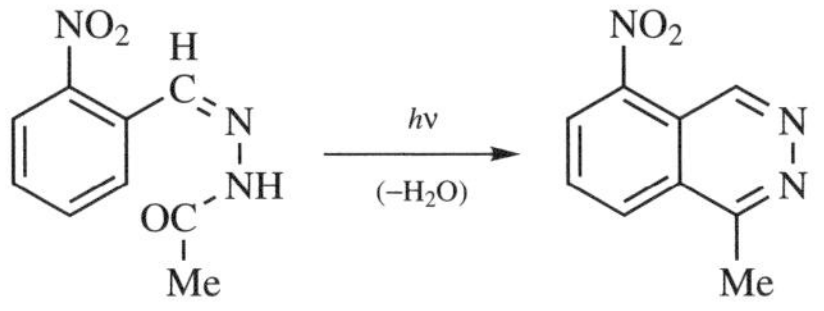

(**6**) (**7**)

1-[(Benzoylhydrazono)methyl]-3,4-dimethoxybenzene (**8**) gave 6,7-dimethoxy-1-phenylphthalazine (**9**) ("polyphosphate ester," $CHCl_3$, reflux, 10 h: 78%); analogs likewise.[902]

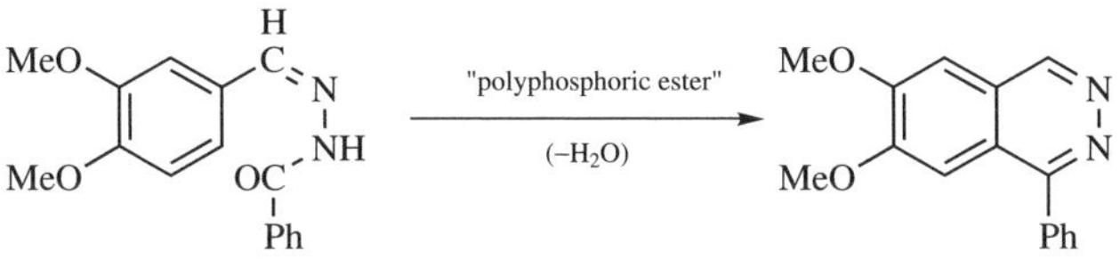

(**8**) (**9**)

N-Benzyl-*N*,*N*′-dimethoxycarbonyl-*N*′-(methoxymethyl)hydrazine (**10**) gave dimethyl 1,2,3,4-tetrahydro-2,3-phthalazinedicarboxylate (**11**) ($TiCl_4$, CH_2Cl_2, −78°C → 20°C, 3 h: >95%).[82]

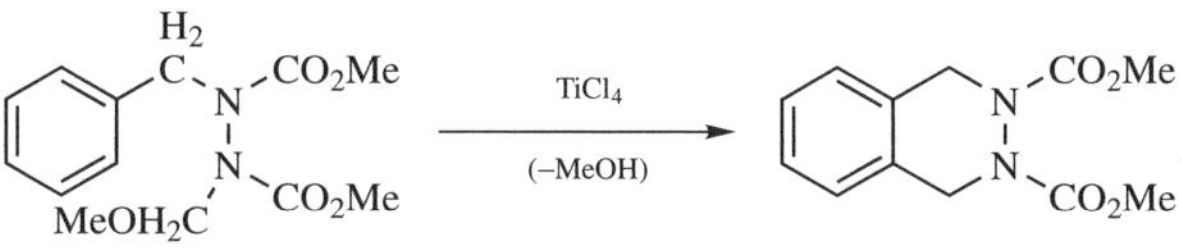

(**10**) (**11**)

N′-(α-Aminophenethylidene)benzohydrazide (**12**) gave 4-phenoxymethyl-1(2*H*)-phthalazinone (**13**) (TsOH, PrOH, reflux, 6 h: 86%; or 96% H_2SO_4, AcOH, 20°C, 30 min: >95%); analogs likewise.[336]

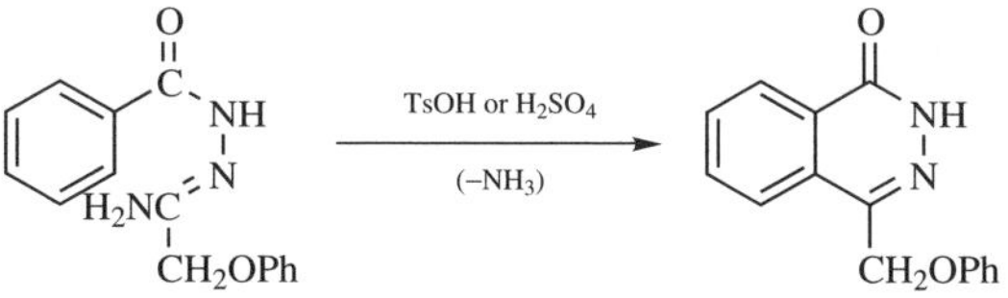

(**12**) (**13**)

Methyl 2-acetoxy-2-(*N*-allyloxycarbonyl-*N*,*N*′-dibenzylhydrazino)acetate (**14**) gave methyl 2-allyloxycarbonyl-3-benzyl-1,2,3,4-tetrahydro-1-phthalazinecarboxylate (**15**) ($SnCl_4$, CH_2Cl_2, −78°C, 15 min; then 20°C, >5 h: 91%).[302]

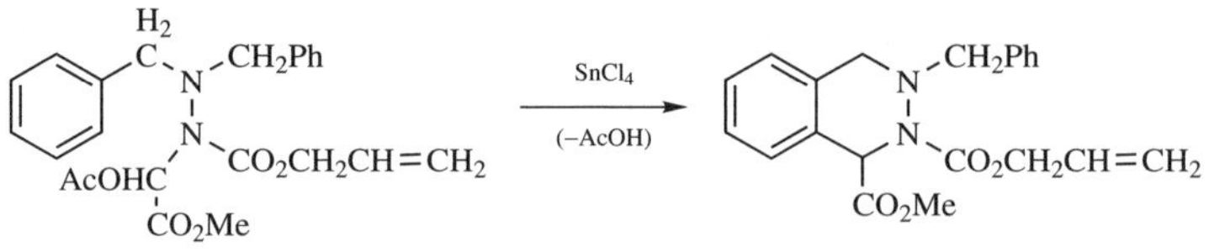

(14) (15)

Also other examples,[28,175] including the formation of fused phthalazines.[980]

8.1.2. By Formation of the C1–N2 Bond

This type of synthesis has not been explored systematically, but the following examples indicate its considerable potential.

o-[α-(2,4-Dinitrophenylhydrazono)benzyl]benzoic acid (**16**) gave 2-(2,4-dinitrophenyl)-4-phenyl-1(2*H*)-phthalazinone (**17**) (AcONa, Ac_2O, reflux, 3 h: 62%); analogs likewise.[459]

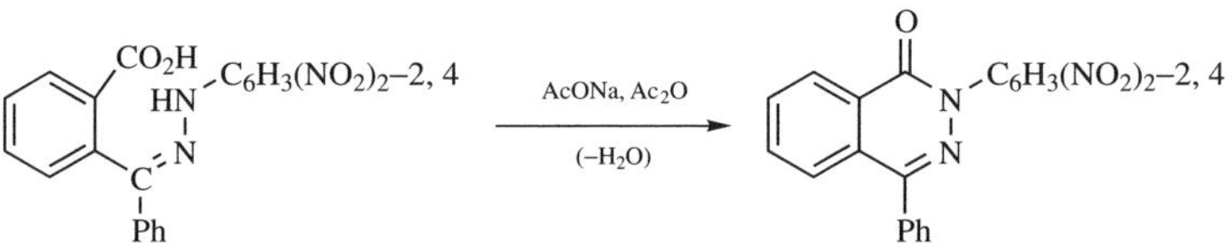

(16) (17)

o-(*p*-Nitrophenylhydrazonomethyl)benzaldehyde (**18**) gave 2-*p*-nitrophenyl-1(2*H*)-phthalazinone (**20**) via the intermediate shown [$Hg(OAc)_2$, AcOH, reflux, 30 min: 91%; analogs likewise);[372] the same substrate (**18**) (2 molecules) gave 1-(2-*p*-nitroanilino-3-oxoisoindolin-1-yl)-2-*p*-nitrophenyl-1,2-dihydrophthalazine (**19**) (HCl, H_2O, dioxane, reflux, 40 min: 75%; structure confirmed by X-ray analysis).[375]

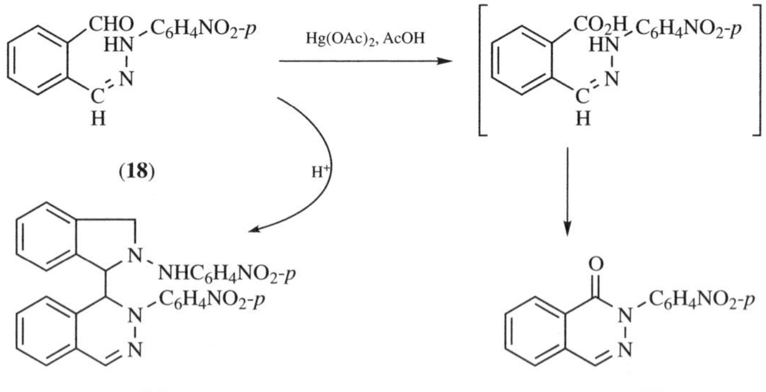

(18) (19) (20)

o-[α-(Tosylhydrazono)benzyl]benzophenone (**21**) gave 1,4-diphenylphthalazine (**22**) (EtOH, reflux, briefly: 76%).[81]

EtOH, Δ
(−TsOH)

(**21**) (**22**)

Oxidation of *o*-bis(tosylhydrazonomethyl)benzene (**23**) gave phthalazine (**24**) [$Pb(OAc)_4$, CH_2Cl_2, 20°C, 2 h: 81% as its *p*-toluenesulfonate salt].[372]

$Pb(OAc)_4$

(**23**) (**24**)

Also other examples.[230,477,645,851]

8.2. FROM A BENZENE DERIVATIVE AS SUBSTRATE AND ONE SYNTHON

In such a phthalazine synthesis, the synthon could supply C1, N2, C1 + N2, N2 + N3, C1 + N2 + N3, or C1 + N2 + N3 + C4. Of these possibilities, the second, third, and last are unrepresented (at least in the literature under review), and of the remaining three categories only that in which the synthon supplies N2 + N3 has been used widely.

8.2.1. Where the Synthon Supplies C1 of the Phthalazine

This rare type of synthesis is exemplified in the conversion of *o*-bromobenzophenone tosylhydrazone (**25**) into *o*-lithiobenzophenone lithiotosylhydrazone (**26**) (treatment with MeLi and BuLi sequentially at −78°C in THF) and subsequent treatment of the reaction mixture with oxalyl chloride (−78°C → 20°C) to give 4-phenyl-2-tosyl-1(2*H*)-phthalazinone (**27**) in 65% overall yield.[81]

$(-COCl)_2$

(**25**) (**26**) (**27**)

8.2.2. Where the Synthon Supplies N2 + N3 of the Phthalazine (*H* 98, 140; *E* 329, 376, 446)

This synthesis has been used very extensively. The synthon that supplies N2 + N3 is always hydrazine, a hydrazine derivative, or a closely related entity. Accordingly, the discussion that follows is subdivided according to the type of substrate involved. This must be an *o*-disubstituted benzene (or cyclohexane) derivative bearing any two of the following substituents: alkyl (usually bearing an appropriate leaving group), aldehydo (formyl), keto (acyl other than formyl), carboxy, halogenoformyl, alkoxycarbonyl, carbamoyl (or related substituents), and cyano. Of the 34 possible combinations, at least half have been used in the 1972–2004 recent literature, as illustrated in the subsections that follow. Please note that *o*-benzenedicarboxylic anyhydrides (phthalic anhydrides) and lactones of *o*-hydroxymethylbenzoic acids (phthalides) are considered as derivatives of isobenzofuran and that *o*-benzenedicarboximides (phthalimides) and lactims of *o*-(aminomethyl)-benzoic acids are considered as derivatives of isoindole; accordingly, their extensive use as substrates is covered in Sections 8.6.6 and 8.6.7, respectively.

8.2.2.1 Using 1,2-Dialkylbenzenes as Substrates

This type of substrate affords 1,2,3,4-tetrahydrophthalazines. All the available examples that follow employ α-bromoalkyl substrates.

o-Bis(bromomethyl)benzene (**28**) with *N,N'*-diacetylhydrazine (**29**, R = Ac) (as its monopotassium salt) gave 2,3-diacetyl-1,2,3,4-tetrahydrophthalazine (**30**, R = Ac) [synthon, warm Me_2NCHO, substrate (**28**)↓, dropwise during 3 h; then reflux, 3 h: 69%; homologs likewise];[924] the same substrate (**28**) with *N*-acetyl-*N'*-(*tert*-butoxycarbonyl)hydrazine (**29**, R = CO_2Bu^t) gave *tert*-butyl 3-acetyl-1,2,3,4-tetrahydro-2-phthalazinecarboxylate (**30**, R = CO_2Bu^t) [NaH, Me_2NCHO, synthon (**29**, R = CO_2Bu^t)↓ slowly, 0° → 20°C, 30 min; then substrate (**28**)↓ dropwise, 0° → 20°C, 24 h: 52%].[309]

CH_2Br CH_2Br + HN(Ac)–HN(R) —EtOK or NaH→ (tetrahydrophthalazine with N-Ac, N-R)

(**28**) (**29**) (**30**)

1,2,4,5-Tetrakis(bromomethyl)brnzene (**31**, R = H) gave di-*tert*-butyl 6,7-bis(bromomethyl)-1,2,3,4-tetrahydro-2,3-phthalazinedicarboxylate (**32**, R = H) [substrate, $(\text{—HNCO}_2Bu^t)_2$ (0.33 equiv), Me_2NCHO, 65°C; then NaH↓, 25 min: 41%];[326] the analogous substrate, 1,2,3,4-tetrakis(bromomethyl)-3,6-dimethoxybenzene (**31**, R = OMe) with the same synthon likewise gave di-*tert*-butyl

6,7-bis(bromomethyl)-5,8-dimethoxy-1,2,3,4-tetrahydro-2,3-phthalazinedicarboxylate (**32**, R = OMe) (50%).[327]

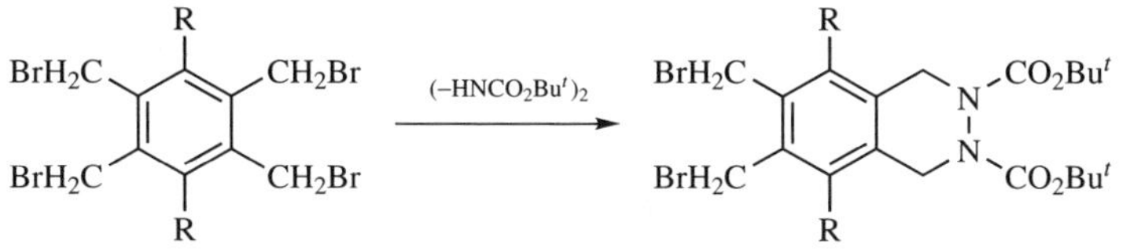

(**31**) (**32**)

o-Bis(α-bromo-α-methoxycarbonylmethyl)benzene (**33**) gave dimethyl 2-phenyl-1,2,3,4-tetrahydro-1,4-phthalazinedicarboxylate (**34**) ($PhNHNH_2$, PhH, reflux, 24 h: 33% after separation from other products).[7]

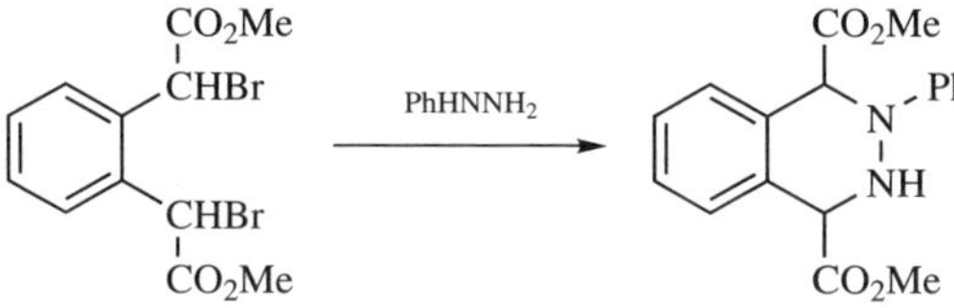

(**33**) (**34**)

8.2.2.2. *Using 1-Alkyl-2-ketobenzenes as Substrates*

This rarely used synthesis is illustrated by the cyclocondensation of *o*-methylbenzophenone (**35**) with diethyl azodicarboxylate to afford diethyl 1-hydroxy-1-phenyl-1,2,3,4-tetrahydro-2,3-phthalazinedicarboxylate (**36**) (PhH, *h*ν, N_2, 20°C, 7 days: 73%).[510]

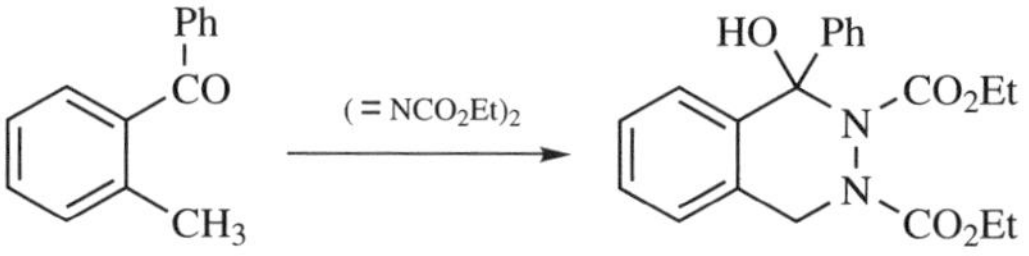

(**35**) (**36**)

8.2.2.3. *Using 1-Alkyl-2-halogenoformylbenzenes as Substrates*

This type of cyclocondensation affords a 3,4-dihydro-1(2*H*)-phthalazinone as illustrated in the reaction of 2-bromomethyl-5-nitrobenzoyl chloride (**37**) with *N,N′*-diacetylhydrazine to give 2,3-diacetyl-7-nitro-3,4-dihydro-1(2*H*)-phthalazinone (**38**, R = Ac) [$(-HNAc)_2$, NaH, dioxane, 85°C, 1 h; then substrate↓, 15°C → 85°C, 1 h: ~37%] and thence 7-nitro-3,4-dihydro-1(2*H*)-phthalazinone (**38**, R = H) (10M HCl, 20°C, 5 h: ~50%).[919]

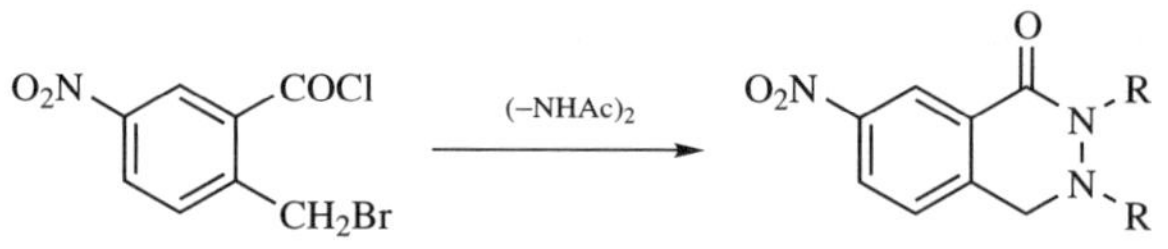

(37) (38)

8.2.2.4. *Using 1,2-Dialdehydobenzenes as Substrates*

Such substrates are usually phthalaldehydes. These give aromatic phthalazines with hydrazine but (partially reduced) phthalazinols with substituted hydrazines, as illustrated in the following examples.

Phthalaldehyde (**39**, Q = R = H) gave phthalazine (**40**, Q = R = H) ($H_2NNH_2 \cdot H_2O$, EtOH, 0°C; ethanolic substrate↓ during 2 h, N_2, 0°C; then 0°C → 20°C, 3 h: >95%);[374,cf. 100] 4-methoxyphthalaldehyde (**39**, Q = OMe, R = H) gave 6-methoxyphthalazine (**40**, Q = OMe, R = H) in a broadly similar way.[888,cf. 100]

4,5-Dimethoxyphthalaldehyde (**39**, Q = R = OMe) gave 6,7-dimethoxyphthalazine (**40**, Q = R = OMe) (substrate, EtOH, warm; $H_2NNH_2 \cdot H_2O$↓ dropwise; then reflux, 5 h: 93%).[948]

Phthalaldehyde (**39**, Q = R = H) gave 2-phenyl-1,2-dihydro-1-phthalazinol (**41**) ($PhNHNH_2 \cdot HCl$, H_2O, reflux, 2 h: 96%)[145] and thence *trans*-2,2′-diphenyl-1,1′-bi(1,2-dihydrophthalazin-1-ylidene) (**42**) (AcOH, MeCN, reflux, 2 h: "good yield");[145,cf. 144] analogs were made similarly and the structure of *trans*-2,2′-di-*p*-bromophenyl-1,1′-bi(1,2-dihydrophthalazin-1-ylidene) was confirmed by X-ray analysis.[145]

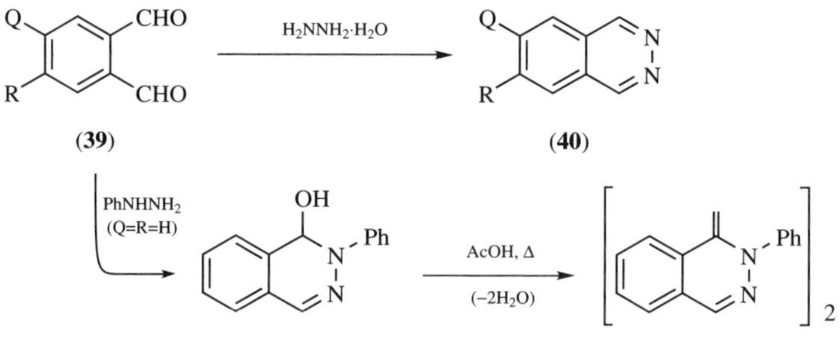

(39) (40)

(41) (42)

Also other examples.[584]

8.2.2.5. *Using 1-Aldehydo-2-ketobenzenes as Substrates*

This synthesis naturally gives a 1-alkyl(or aryl)phthalazine, as illustrated in the following examples.

3,4-Dichloro-6-(diacetoxymethyl)-4′-nitrobenzophenone (**43**, an acylal of the corresponding aldehyde) gave 6,7-dichloro-1-*p*-nitrophenylphthalazine (**44**) ($H_2NNH_2 \cdot H_2O$, EtOH, reflux, 3 h: 91%).[848]

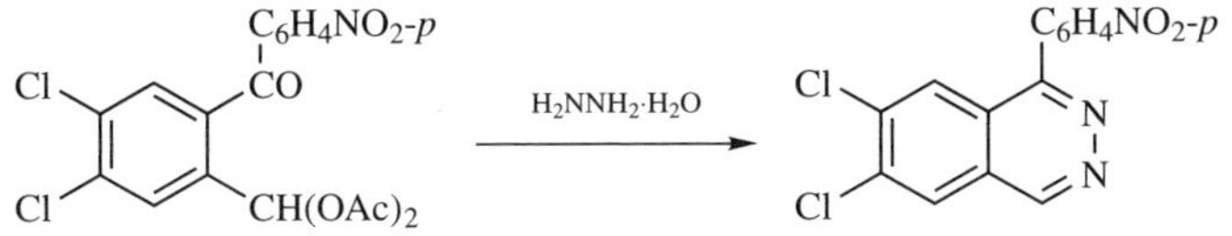

$H_2NNH_2 \cdot H_2O$

(**43**) (**44**)

2-Benzophenonecarbaldehyde (**45**) with benzylhydrazine dihydrochloride gave 2-benzyl-1-phenylphthalazinium chloride (**46**) (K_2CO_3, THF, CH_2Cl_2, <5°C → 20°C, 90 min: 49%).[119]

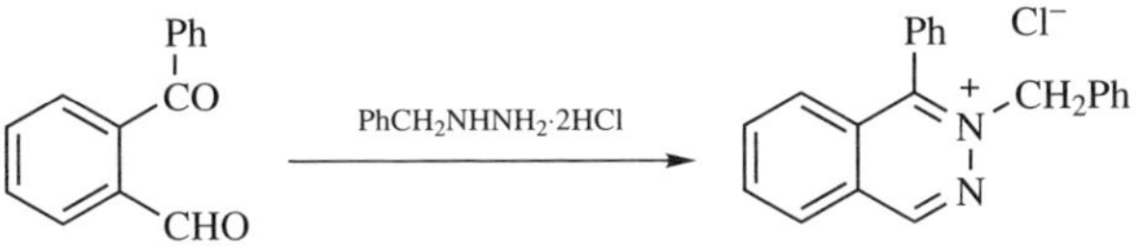

$PhCH_2NHNH_2 \cdot 2HCl$

(**45**) (**46**)

8.2.2.6. *Using 1-Aldehydo-2-carboxybenzenes as Substrates*

These substrates give 1(2*H*)-phthalazinones as illustrated in the following examples.

Phthalaldehydic acid (**47**) gave 1(2*H*)-phthalazinone (**48**, R = H) ($H_2NNH_2 \cdot H_2O$, 20°C, 12 h: 70%).[690]

The same substrate (**47**) gave 2-(3-oxo-1,3-dihydroisobenzofuran-1-yl)-1(2*H*)-phthalazinone (**49**) (neat $H_2NNH_2 \cdot H_2O$, polyphosphoric acid, 50°C → 125°C, 9 h: 50%) and thence 1(2*H*)-phthalazinone (**48**, R = H) ($H_2NNH_2 \cdot H_2O$, EtOH, reflux, 2 h: 73%).[754]

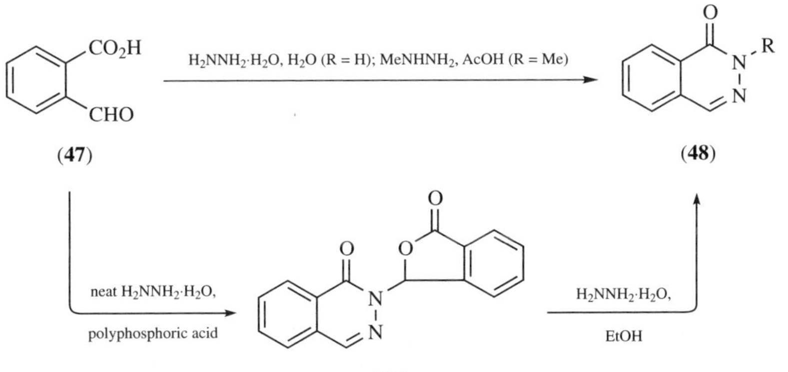

$H_2NNH_2 \cdot H_2O$, H_2O (R = H); $MeNHNH_2$, AcOH (R = Me)

(**47**) (**48**)

neat $H_2NNH_2 \cdot H_2O$, polyphosphoric acid

$H_2NNH_2 \cdot H_2O$, EtOH

(**49**)

The same substrate (**47**) gave 2-methyl-1(2*H*)-phthalazinone (**48**, R = Me) ($MeNHNH_2$, AcOH, reflux, N_2, 18 h: 81%).[623]

Also other examples.[214,258]

8.2.2.7. *Using 1-Aldehydo-2-alkoxycarbonylbenzenes as Substrates*

Although seldom used, such substrates also give 1(2*H*)-phthalazinones. Thus diethyl 4-formyl-2,6-dimethylisophthalate (**50**) with hydrazine hydrate gave ethyl 5,7-dimethyl-4-oxo-3,4-dihydro-6-phthalazinecarboxylate (**51**) (EtOH, reflux, 3 h: 72%).[783]

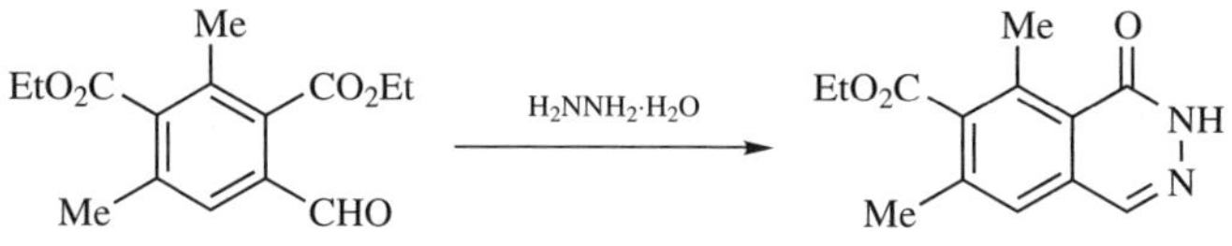

$H_2NNH_2{\cdot}H_2O$

(**50**) (**51**)

8.2.2.8. *Using 1-Aldehydo-2-cyanobenzenes as Substrates*

Such substrates clearly should afford 1-phthalazinamines, but the only reported examples underwent concomitant or subsequent hydrolysis under the conditions of cyclocondensation to give the corresponding phthalazinones; 3-formylphthalonitrile (**52**) gave 4-oxo-3,4-dihydro-5-phthalazinecarbonitrile (**53**) ($H_2NNH_2 \cdot H_2O$, 12M HCl, PhH, reflux, 6 h: 66%).[615]

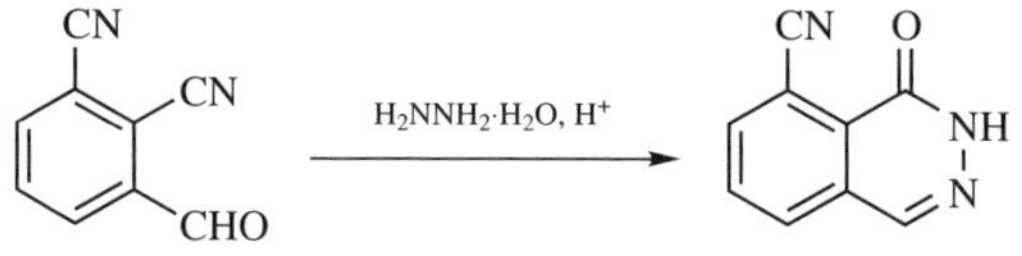

$H_2NNH_2{\cdot}H_2O$, H^+

(**52**) (**53**)

8.2.2.9. *Using 1,2-Diketobenzenes as Substrates*

This synthesis affords 1,4-dialkyl(or aryl)phthalazines, as illustrated in the following examples.

2-Benzoylbenzophenone (**54**) gave 1,4-diphenylphthalazine (**55**) [$H_2NNH_2 \cdot H_2O$, EtOH, reflux, 7 h: ~85%;[315] likewise but 20°C, 17 h: 67%;[707] or $(Ph_3P{=}N{-})_2$,

THF, N_2, reflux, 14 h: 56%].[125]

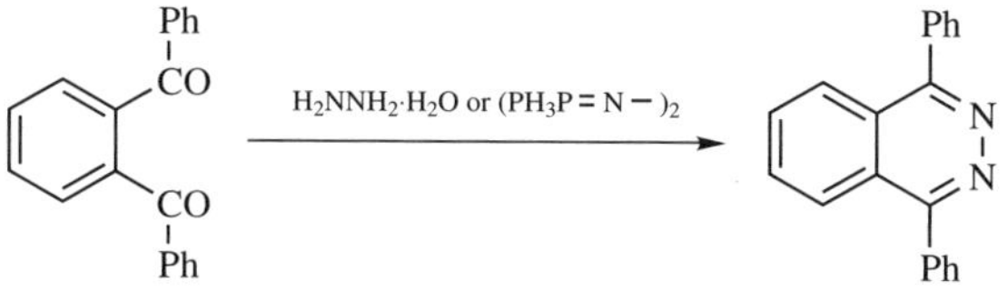

(54) → (55) [$H_2NNH_2 \cdot H_2O$ or $(PH_3P=N-)_2$]

Appropriately substituted *o*-benzoylbenzophenones and hydrazine hydrate likewise gave 1,4-bis(3,4-dimethoxyphenyl)-6,7-dimethoxyphthalazine (**56**) (EtOH, reflux, 1 h: 80%),[789] 1,4-bis-*p*-chlorophenyl-6,7-dimethoxyphthalazine (**57**) (similarly: 89%),[385] and other such products.[226,385]

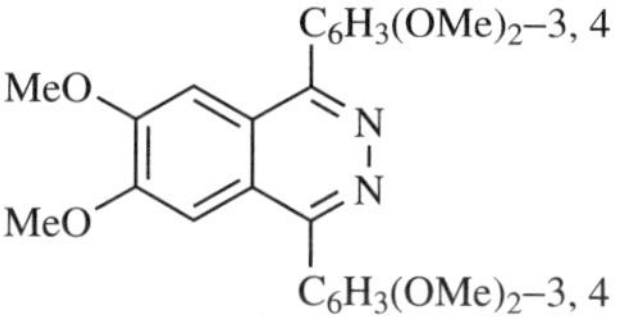

(56) 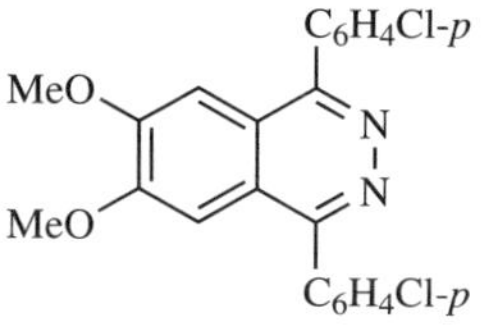 (57)

o-Bis(trifluoroacetyl)benzene (**58**) gave 1,4-bis(trifluoromethyl)phthalazine (**59**) ($H_2NNH_2 \cdot 2HCl$, pyridine, 85°C, 26 h: 73%).[779]

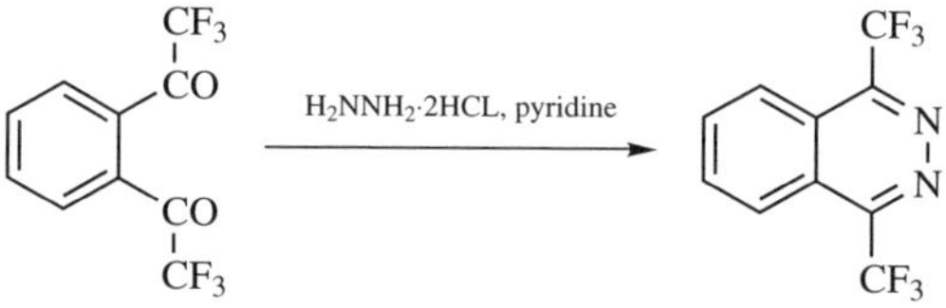

(58) → (59) [$H_2NNH_2 \cdot 2HCL$, pyridine]

o-(Bromoacetyl)benzophenone (**60**) gave 1-bromomethyl-4-phenylphthalazine (**61**) ($H_2NNH_2 \cdot H_2O$, Et_2O, −20°C, substrate↓ dropwise; then −20°C, 2 h: 55%; presumably the especially gentle conditions are necessary to avoid hydrazinolysis of the bromomethyl substituent).[49]

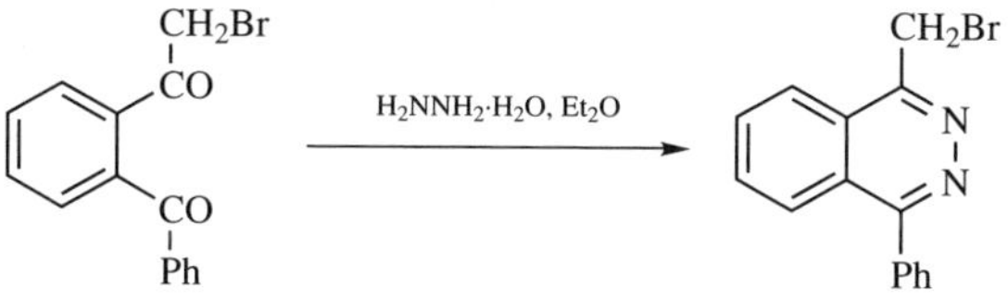

(60) → (61) [$H_2NNH_2 \cdot H_2O$, Et_2O]

Also other examples.[16,128,322,366,942]

Note: Such completions of the phthalazine ring have also been used in quite complicated systems to give, for example, the so-called 4,7,14,17-tetraphenyl [2.2](5,8)phthalazinophane (**62**) ($H_2NNH_2 \cdot H_2O$, Me_2NCHO, 80°C, 10 h: 80%) from the appropriate tetrabenzoyl precursor.[984]

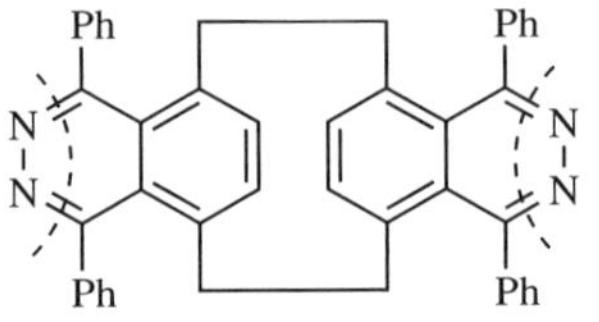

(**62**)

8.2.2.10. *Using 1-Keto-2-carboxybenzenes as Substrates*

These substrates afford 4-alkyl(or aryl)-1(2*H*)-phthalazinones. The examples that follow are so numerous that they have been divided into groups according to the type of hydrazine used as synthon.

With Unsubstituted Hydrazine as Synthon

o-Benzoylbenzoic acid (**63**, Q = R = H) gave 4-phenyl-1(2*H*)-phthalazinone (**64**, Q = R = H) ($H_2NNH_2 \cdot H_2O$, EtOH, reflux, 5 h: 92%).[278]

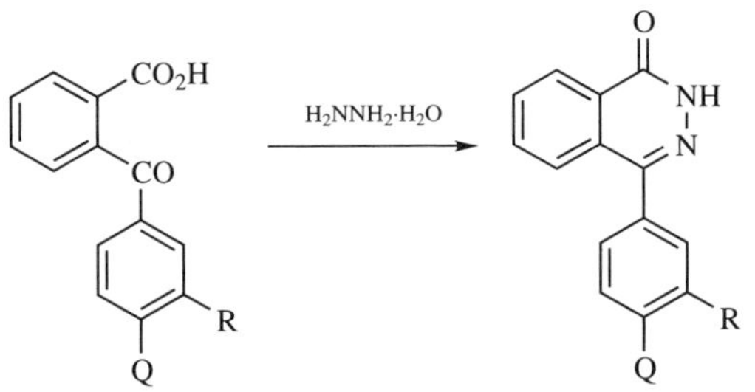

$H_2NNH_2 \cdot H_2O$

(**63**) (**64**)

o-(4-Hydroxy-3-isopropylbenzoyl)benzoic acid (**63**, Q = OH, R = Pr^i) gave 4-(4-hydroxy-3-isopropylphenyl)-1(2*H*)-phthalazinone (**64**, Q = OH, R = Pr^i) ($H_2NNH_2 \cdot H_2O$, BuOH, reflux, 2 h: >95%);[130] analogs likewise.[132]

2-Benzoylcyclohexanecarboxylic acid (**65**, R = H) gave 4-phenyl-4a,5,6,7,8,8a-hexahydro-1(2*H*)-phthalazinone (**66**, R = H) ($H_2NNH_2 \cdot H_2O$, EtOH, reflux, 4 h: 90%; analogs likewise);[873] 2-benzoyl-5-phenylcyclohexanecarboxylic acid (**65**, R = Ph) gave 4,7-diphenyl-4a,5,6,7,8,8a-hexahydro-1(2*H*)-phthalazinone (**66**, R = Ph) ($H_2NNH_2 \cdot H_2O$, PhH, reflux, 3 h: 43%).[589]

R CO2H H2NNH2·H2O CO Ph (65) → R O NH N Ph (66)

o-(6-Ethylnicotinoyl)benzoic acid (**67**) gave 4-(6-ethylpyridin-3-yl)-1(2*H*)-phthalazinone (**68**) ($H_2NNH_2 \cdot H_2O$, EtOH, reflux, 2 h: 73%); analogs likewise.[432]

CO2H H2NNH2 CO N Et (67) → O NH N N Et (68)

2-(5-Ethyl-2-thenoyl)-3,4,5,6-tetrahydrobenzoic acid (**69**) gave 4-(5-ethylthien-2-yl)-5,6,7,8-tetrahydro-1(2*H*)-phthalazinone (**70**) ($H_2NNH_2 \cdot H_2O$, EtOH, reflux, 6 h: 40%).[430]

CO2H H2NNH2·H2O CO S Et (69) → O NH N S Et (70)

Also many other examples.[214,231,244,268,273,279,348,351,404,430,469,649,650,668,727,735,743,792,872,926,951,1038]

With Simple Alkyl- or Arylhydrazines as Synthons

o-Benzoylbenzoic acid (**71**) with methyl- or *tert*-butylhydrazine gave 2-methyl- (**72**, R = Me) or 2-*tert*-butyl-4-phenyl-1(2*H*)-phthalazinone (**72**, R = Bu^t) (PhMe, reflux, H_2O removal, <4 h: 93% or 67%, respectively).[119,cf. 447]

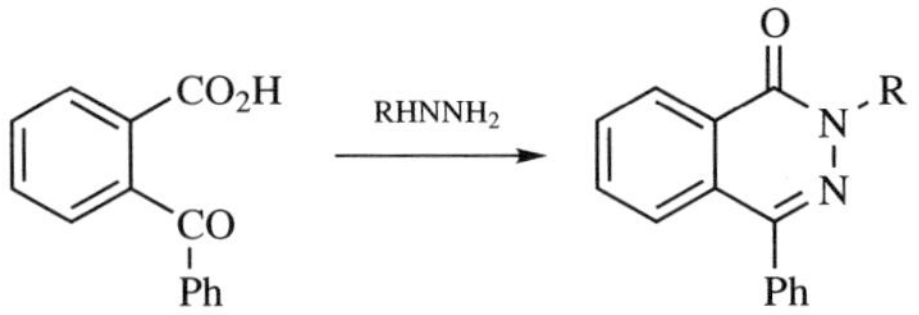

(71) (72)

o-Acetylbenzoic acid (**73**) gave 4-methyl-2-(2-methylallyl)-1(2*H*)-phthalazinone (**74**) ($H_2NNHCH_2CMe{=}CH_2$, AcOH, reflux, 3 h: 65%).[465]

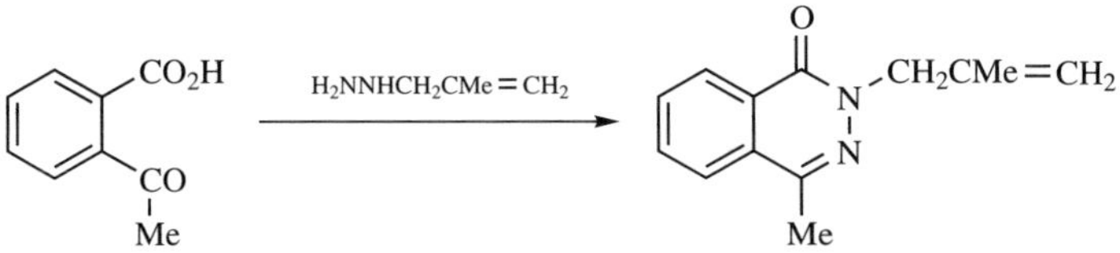

(73) (74)

o-(4-Chloro-2-hydroxybenzoyl)benzoic acid (**75**) gave 4-(4-chloro-2-hydroxyphenyl)-2-phenyl-1(2*H*)-phthalazinone (**76**) (neat $PhNHNH_2$, 125°C, 3 h: ?%).[294]

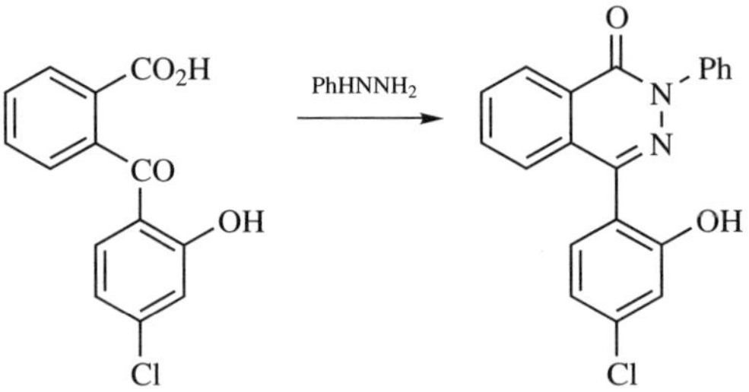

(75) (76)

cis-2-*p*-Toluoylcyclohexanecarboxylic acid (**77**) gave *cis*-2-phenyl-4-*p*-tolyl-4a,5,6,7,8,8a-hexahydro-1(2*H*)-phthalazinone (**78**) ($PhNHNH_2$, PhMe, reflux, <2 h: 62%); the corresponding *trans*-substrate likewise gave the *trans*-product (46%).[758]

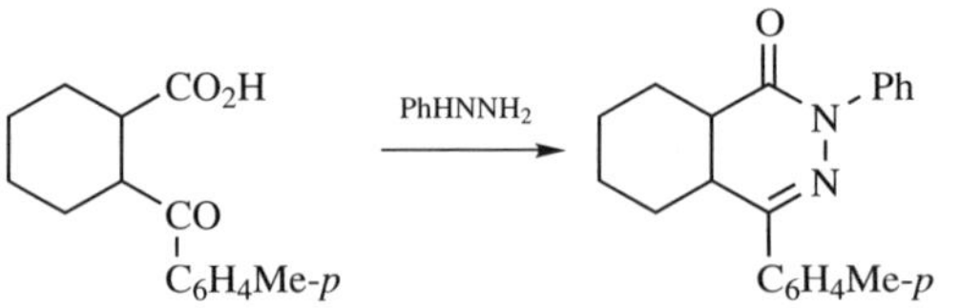

(77) (78)

Also other examples.[130,241,342,411,446,499,869]

With Functionally *C*-Substituted Alkyl- or Arylhydrazines as Synthons

o-Acetylbenzoic acid (**79**) and (2-hydroxyethyl)hydrazine gave 2-(2-hydroxyethyl)-4-methyl-1(2*H*)-phthalazinone (**80**) (KOH, H_2O, 95°C, 2 h: 66%).[737]

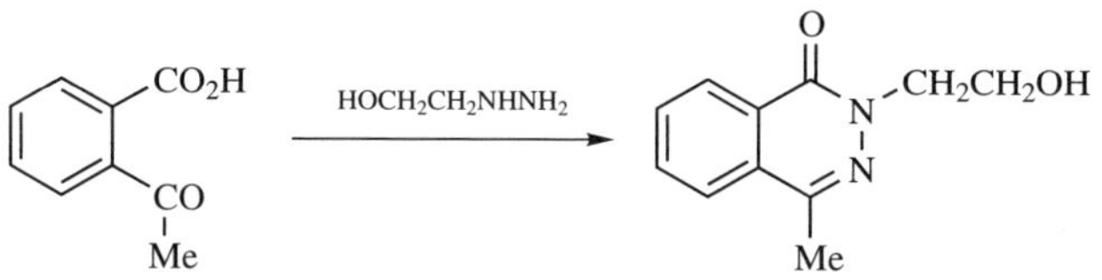
CO2H
CO
Me
HOCH2CH2NHNH2
O
N
CH2CH2OH
N
Me

(**79**) (**80**)

2-*p*-Toluoylbenzoic acid (**81**) with *p*-nitrophenylhydrazine gave 2-*p*-nitrophenyl-4-*p*-tolyl-1(2*H*)-phthalazinone (**82**) (HCl, EtOH, H_2O, reflux, 3 h: 73%).[459]

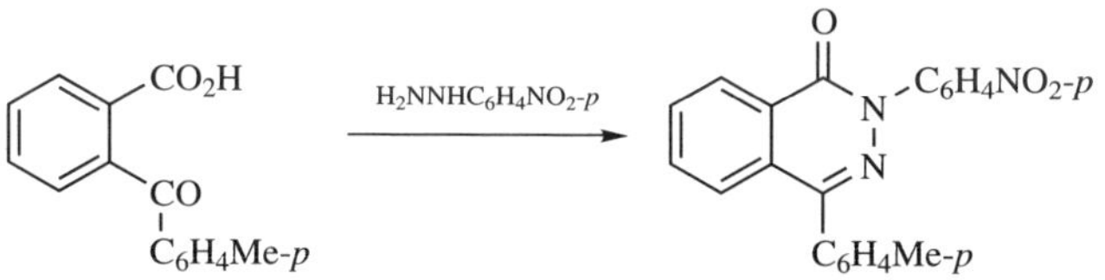
CO2H
CO
C6H4Me-p
H2NNHC6H4NO2-p
O
N
C6H4NO2-p
N
C6H4Me-p

(**81**) (**82**)

2-(3,4-Diethoxybenzoyl)-1,2,3,6-tetrahydrobenzoic acid (**83**) with *p*-hydrazinobenzoic acid gave 2-*p*-carboxyphenyl-4-(3,4-diethoxyphenyl)-4a,5,8,8a-tetrahydro-2(1*H*)-phthalazinone (**84**) (pyridine, reflux, 5 h: 71%).[894]

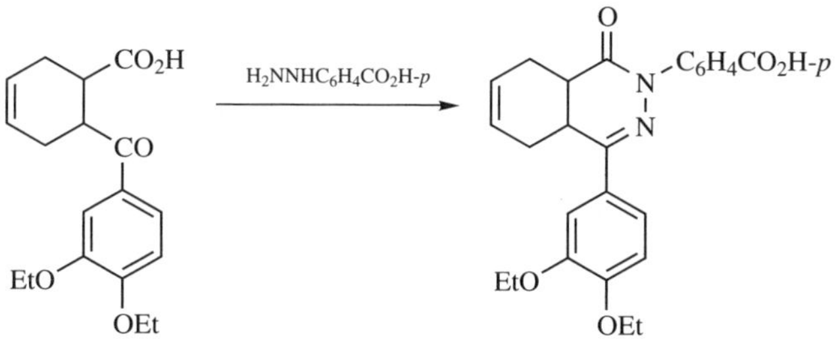
CO2H
CO
EtO
OEt
H2NNHC6H4CO2H-p
O
N
C6H4CO2H-p
N
EtO
OEt

(**83**) (**84**)

o-Acetylbenzoic acid (**85**) with 3-hydrazino-6-*p*-tolylpyridazine gave 4-methyl-2-(6-*p*-tolylpyridazin-3-yl)-1(2*H*)-phthalazinone (**86**) (EtOH, reflux, 3 h: 80%)[644] or with 8-hydrazinoquinoline gave 4-methyl-2-(quinolin-8-yl)-1(2*H*)-phthalazinone (**87**) (Et_3N, EtOH, H_2O: 95%; structure confirmed by X-ray analysis).[896]

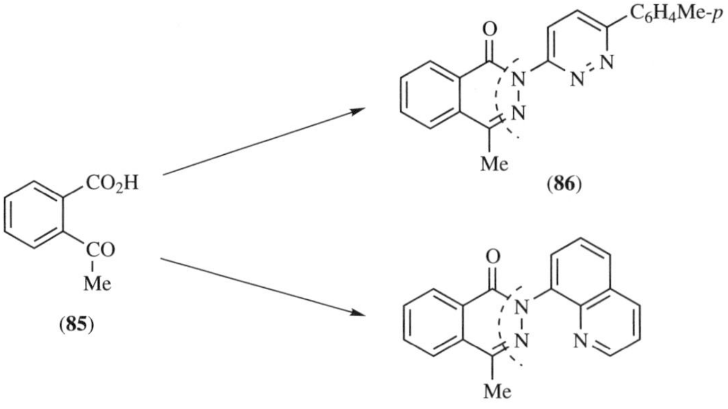

(85)

(86)

(87)

Also other examples.[345,531,951,972,985]

With Hydrazides or the Like as Synthons

Note: Most of the examples that follow have been gleaned only from abstracts and therefore lack indications of conditions and yields.

o-(4-Chloro-3-methylbenzoyl)benzoic acid (**88**) with semicarbazide or thiosemicarbazide gave 4-(4-chloro-3-methylphenyl)-1-oxo-1,2-dihydro-2-phthalazinecarboxamide (**89**, X = O) or the corresponding carbothioamide (**89**, X = S), respectively.[363,471]

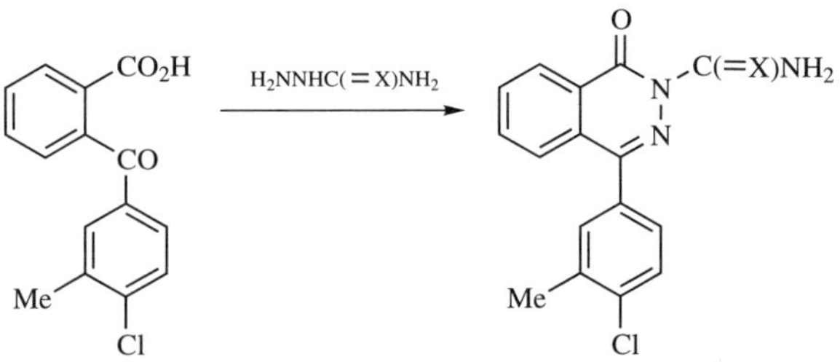

$H_2NNHC(=X)NH_2$

(88) (89)

2,3,4,5-Tetrachloro-6-(2,4-dimethylbenzoyl)benzoic acid with ethyl carbazate gave ethyl 5,6,7,8-tetrachloro-4-(2,4-dimethylphenyl)-1-oxo-1,2-dihydro-2-phthalazinecarboxylate (**90**).[473]

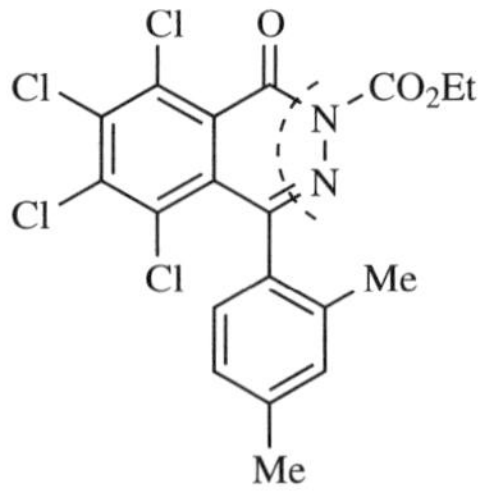

(90)

o-(2,4-Dichlorobenzoyl)benzoic acid with acetohydrazide or benzenesulfonohydrazide gave 2-acetyl- (**91**, R = Ac) or 2-benzenesulfonyl-4-(2,4-dichlorophenyl)-1(2*H*)-phthalazinone (**91**, R = SO_2Ph), respectively.[257]

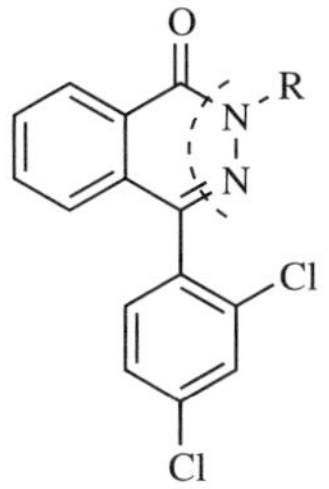

(**91**)

o-Acetylbenzoic acid (**92**) with thiosemicarbazide gave 4-methyl-1(2*H*)-phthalazinone (**94**), presumably via hydrolysis of the thioamide (**93**) (Na_2CO_3, H_2O, reflux, 20 h: ~70%).[24]

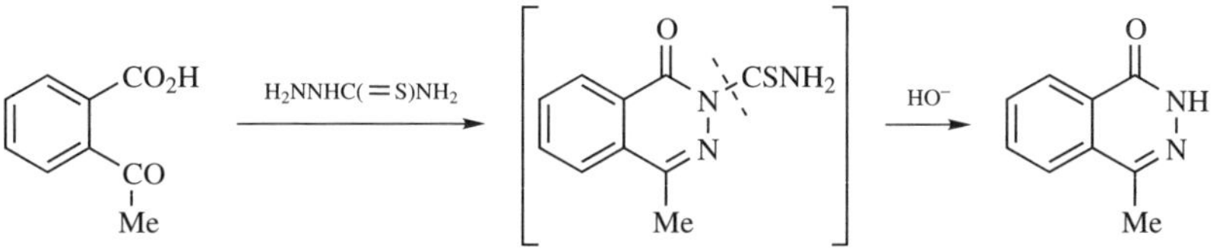

(**92**) (**93**) (**94**)

Also other examples.[258,470,621,868]

8.2.1.11. *Using 1-Keto-2-alkoxycarbonylbenzenes as Substrates*

Like the corresponding carboxylic acids, these substrates give 4-alkyl(or aryl)-1(2*H*)-phthalazinone. However, the esters are seldom used simply because the acids are usually made more easily. An example of the use of keto esters is the condensation of ethyl 2-(2-methoxy-1-naphthoyl)benzoate (**95**) with hydrazine to afford 4-(2-methoxynaphthalen-1-yl)-1(2*H*)-phthalazinone (**96**) (neat $H_2NNH_2 \cdot H_2O$, reflux, 4 h: 94%).[371]

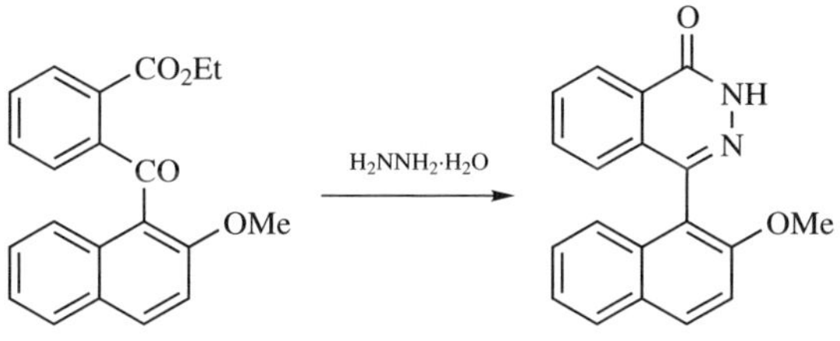

(**95**) (**96**)

8.2.2.12. Using 1-Keto-2-carbamoylbenzenes or the Like as Substrates

These substrates, like their acid or ester counterparts, furnish 4-alkyl(or aryl)-2(1*H*)-phthalazinones. Although no keto amides have been so used in the more recent literature, some closely related *o*-acylbenzoyl azides have been employed. Thus *o*-benzoylbenzoyl azide (**97**) with hydrazine or phenylhydrazine gave 4-phenyl-1(2*H*)-phthalazinone (**98**, R = H) (PhH, reflux, 2 h: 60%) or 2,4-diphenyl-1(2*H*)-phthalazinone (**98**, R = Ph) (likewise: 65%), respectively; analogs were made similarly.[638]

CON_3, CO, Ph —$RNHNH_2$→ O, N–R, N, Ph

(**97**) (**98**)

8.2.2.13. Using 1,2-Dicarboxybenzenes (Phthalic Acids) as Substrates

Such substrates furnish 1,4(2*H*,3*H*)-phthalazinediones, as illustrated in the following examples.

3-Nitrophthalic acid (**99**) with hydrazine gave 5-nitro-1,4(2*H*,3*H*)-phthalazinedione (**100**) [$(HOCH_2CH_2OCH_2\text{-})_2$: ?%;[546] or >85%;[203] for details of each procedure, see original papers].

NO_2, CO_2H, CO_2H —H_2NNH_2→ NO_2, O, NH, NH, O

(**99**) (**100**)

Phthalic acid (**101**) with *N*-(1,2-diphenylethylidene)-*N*′-methylhydrazine gave 2-methyl-3-(α-phenylstyryl)-1,4(2*H*,3*H*)-phthalazinedione (**102**) (neat PCl_3, 20°C, 8 h: 68%); a dozen analogs likewise.[563]

CO_2H, CO_2H —$MeNHN{=}CPhCH_2Ph$ PCl_3→ O, N–CPh=CHPh, N–Me, O

(**101**) (**102**)

8.2.2.14. *Using 1,2-Bis(chloroformyl)benzenes (Phthaloyl Chlorides) as Substrates*

Like the corresponding phthalic acids, these substrates afford 1,4(2*H*,3*H*)-phthalazinediones as illustrated in the examples that follow.

Phthaloyl chloride (**103**) with *N*,*N*′-dibenzylhydrazine gave 2,3-dibenzyl-1,4(2*H*,3*H*)-phthalazinedione (**104**) [$(PhCH_2NH\text{—})_2$ (made in situ), pyridine, THF, N_2, 20°C; then substrate/THF↓ dropwise (exothermic), 20°C, 24 h: 23%).[3]

The same substrate (**103**) with *N*,*N*′-bis[(*m*-acetoxymethyl)phenyl]hydrazine gave 2,3-bis[(*m*-acetoxymethyl)phenyl]-1,4(2*H*,3*H*)-phthalazinedione (**105**) (synthon, $PhNMe_2$; substrate↓, N_2, exothermic, 70°C, 5 h: 83%)[134] or with *o*-fluorobenzohydrazide gave 2-(*o*-fluorobenzoyl)-1,4(2*H*,3*H*)-phthalazinedione (**106**) (synthon, Et_3N, Me_2NCHO, substrate↓, exothermic to 70°C, 90 min).[780]

COCl, COCl (**103**) —$PhCH_2NHNHCH_2Ph$→ (**104**) (CH_2Ph, CH_2Ph)

(**103**) —$(\text{–}NHC_6H_4CH_2OAc\text{-}m)_2$→ (**105**) ($C_6H_4CH_2OAc$-*m*, $C_6H_4CH_2OAc$-*m*)

(**103**) —*o*-$FC_6H_4CONHNH_2$→ (**106**) ($C(=O)C_6H_4F$-*o*, NH)

8.2.2.15. *Using 1,2-Dialkoxycarbonylbenzenes (Phthalic Esters) as Substrates*

Such substrates have been used more widely than the two preceding types to give 1,4(2*H*,3*H*)-phthalazinediones. Examples follow:

Dimethyl 4-amino-5-nitrophthalate (**107**) gave 6-amino-7-nitro-1,4(2*H*,3*H*)-phthalazinedione (**108**) ($H_2NNH_2 \cdot H_2O$, Et_3N, MeOH, reflux, 2 h: 90%); analogs similarly.[4]

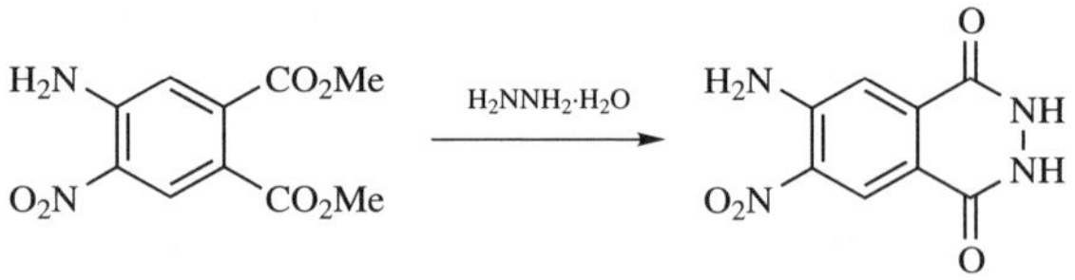

(107) (108)

Dimethyl 3-hydroxy-6-methoxyphthalate (**109**) gave 5-hydroxy-8-methoxy-1,4(2*H*,3*H*)-phthalazinedione (**110**) ($H_2NNH_2 \cdot H_2O$, EtOH, reflux, 12 h: 84%); analogs likewise.[715]

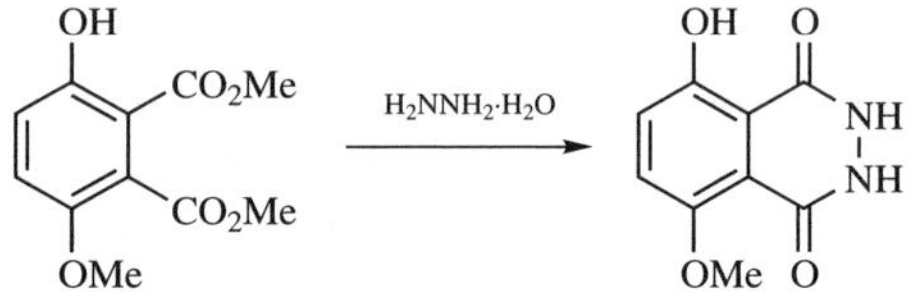

(109) (110)

Diethyl phthalate (**111**) gave 2-phenyl-1,4(2*H*,3*H*)-phthalazinedione (**112**) ($PhNHNH_2$, EtONa, EtOH, reflux, 22 h: 52% after purification).[103]

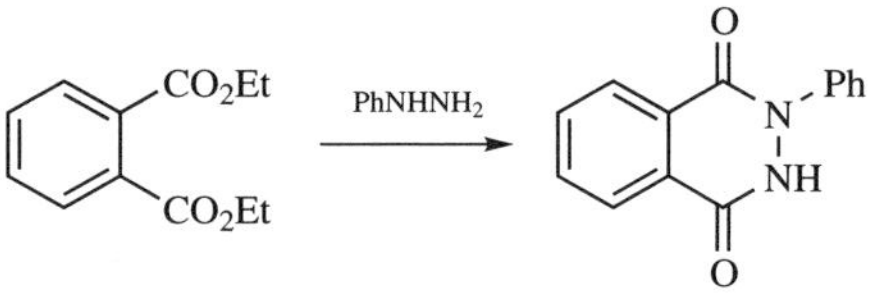

(111) (112)

In an unpredictable but interesting way, dimethyl phthalate (**113**, R = Me) with *N*,*N*-dimethylhydrazine gave 2-methyl-1,4(2*H*,3*H*)-phthalazinedione (**114**) (neat reactants, −30°C, He↓, then sealed; 125°C, ~18 h: 56%) plus gaseous products, identified as N_2, CH_4, Me_3N, and NH_3; in contrast, similar treatment of diethyl phthalate (**113**, R = Et) gave a separable mixture of the same product (**114**) and 2,3-dimethyl-1,4(2*H*,3*H*)-phthalazinedione (**115**) (22% and 16%, respectively, plus N_2, CH_4, $EtNMe_2$, and NH_3; a mechanistic explanation was offered).[1]

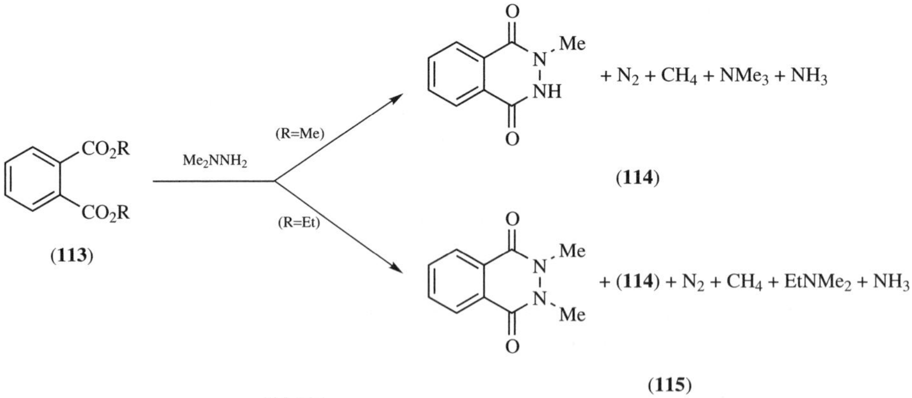

(113) (114) (115)

Also other examples.[328,797]

8.2.2.16. Using 1-Alkoxycarbonyl-2-cyanobenzenes as Substrates

The cyclocondensation of these substrates with hydrazines is represented by only one example. Methyl *o*-cyanobenzoate (**116**) with methylhydrazine gave 4-amino-2-methyl-1(2*H*)-phthalazinone (**117**) (MeONa, MeOH, 20°C, 12 h: 24%).[34]

CO_2Me, CN; $MeNHNH_2$ →; O, N, Me, N, NH_2

(**116**) (**117**)

8.2.2.17. Using 1,2-Dicyanobenzenes (Phthalonitriles) as Substrates

The use of phthalonitriles as substrates has not been developed to any extent, but existing examples are interesting. Thus phthalonitrile (**118**) with an excess of hydrazine gave 1,4-dihydrazinophthalazine (**120**), perhaps by transamination of the initial product (**119**) (AcOH, dioxane, reflux, 3 h: 83%;[275] or AcOH, H_2O, dioxane, reflux, 3 h: 45% as dihydrochloride);[804] however, it does appear possible (under subtly different conditions) to isolate phthalazinediamine (**119**).[176]

CN, CN; $H_2NNH_2 \cdot H_2O$ →; [NH_2, N, N, NH_2] →; $NHNH_2$, N, N, $NHNH_2$

(**118**) (**119**) (**120**)

8.2.3. Where the Synthon Supplies C1 + N2 + N3 of the Phthalazine

This type of cyclocondensation is represented only by a somewhat specialized example. Thus methyl *p*-benzoquinone-2-carboxylate (**121**) with *N*-acetonylidene-*N′*-methylhydrazine (**122**) gave 4-acetyl-5,8-dihydroxy-2-methyl-1(2*H*)-phthalazinone (**123**) [or its tautomer, 4-acetyl-8-hydroxy-2-methyl-1,5(2*H*,3*H*)-phthalazinedione] (MeOH, 20°C, 12 h: 50%).[890]

O, CO_2Me, O + HNMe, N, HC, Ac → (−MeOH); OH, O, N, Me, N, OH, Ac

(**121**) (**122**) (**123**)

8.3. FROM OTHER CARBOCYCLIC DERIVATIVES AS SUBSTRATES

Both carbomonocyclic (other than benzene derivatives, already discussed) and carbobicyclic substrates have been used sparingly (with or without ancillary synthons) for the primary synthesis of phthalazines. The examples that follow are grouped under the parent name of the system used as substrate.

Bicyclo[4.2.0]octanes as Substrates

2,3,4,5-Tetrachlorobicyclo[4.2.0]octa-1(6),2,4-triene-7,8-dione (**124**) with hydrazine hydrochloride gave 5,6,7,8-tetrachloro-1(2*H*)-phthalazinone (**125**, R = H) (EtOH, reflux, 24 h: 71%) or with phenylhydrazine gave 5,6,7,8-tetrachloro-2-phenyl-1(2*H*)-phthalazinone (**125**, R = Ph) (H_2SO_4, EtOH, reflux, 2 h: 79%);[288] the same substrate (**124**) was converted into its monophenylhydrazone (**126**) ($PhNHNH_2$, H_2SO_4, EtOH, reflux, briefly: 46%)[999] and thence into the product (**125**, R = Ph) (HCO_2H, 100°C, 2 h: 83%).[288]

Cl Cl Cl Cl O O (**124**) $RNHNH_2$, H^+ Cl O Cl Cl Cl N R N (**125**)

$PhNHNH_2$ HCO_2H, Δ

Cl Cl Cl Cl O NHNHPh (**126**)

Bicyclo[4.2.0]octa-1(6),2,4-triene-7,8-dione (**127**) with azobenzene gave 2,3-diphenyl-1,4(2*H*,3*H*)-phthalazinedione (**128**) (CH_2Cl_2, C_6H_{12}, *hν*, 9 h: 72%).[150]

O O (**127**) PhN = NPh O N Ph N Ph O (**128**)

Indenes as Substrates

2,2-Diphenyl-1,3-indanedione (**129**) gave 4-diphenylmethyl-1(2*H*)-phthalazinone (**130**) (neat $H_2NNH_2 \cdot H_2O$, reflux, 15 min: 92%); several analogs likewise.[899,995,1013]

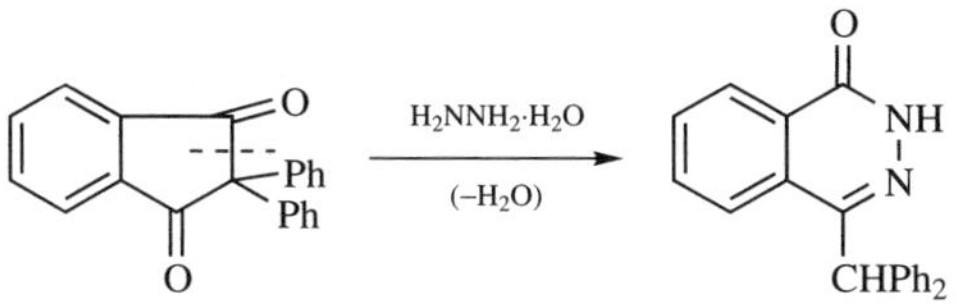

(**129**) (**130**)

4-(3-Hydroxy-1-oxo-1*H*-inden-2-yl)pyridine (**131**) gave 4-(pyridin-4-ylmethyl)-1(2*H*)-phthalazinone (**132**) (neat $H_2NNH_2.H_2O$, 110°C, 8 h: 82%); analogs likewise.[870]

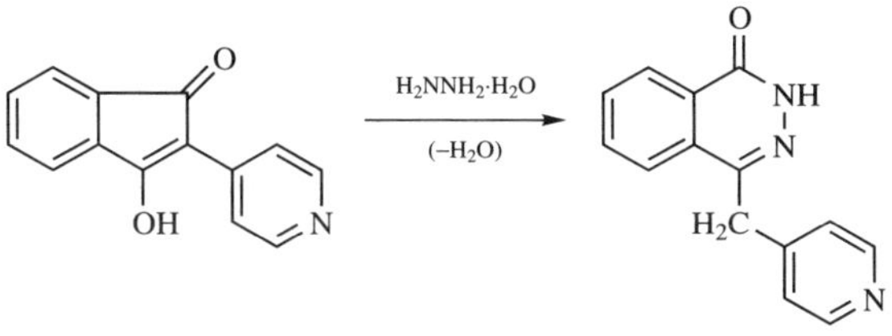

(**131**) (**132**)

Methyl 2-oxo-1-indanecarboxylate (**133**) with *p*-nitrobenzenesulfonyl azide gave methyl 2-[*N*-(*p*-nitrobenzenesulfonyl)carbamoyl]-3,4-dihydro-1-phthalazinecarboxylate (**134**) (or tautomer) (Et_3N, THF, 0°C → 20°C, 1 h: crude) and thence methyl 4-[*N*-(*p*-nitrobenzenesulfonyl)carbamoyl]-1-phthalazinecarboxylate (**135**) (tetrachloro-1,2-benzoquinone, PhH, reflux, 10 min: ?%); structure (**135**) was confirmed by X-ray analysis and a synthetic mechanism was proposed.[124]

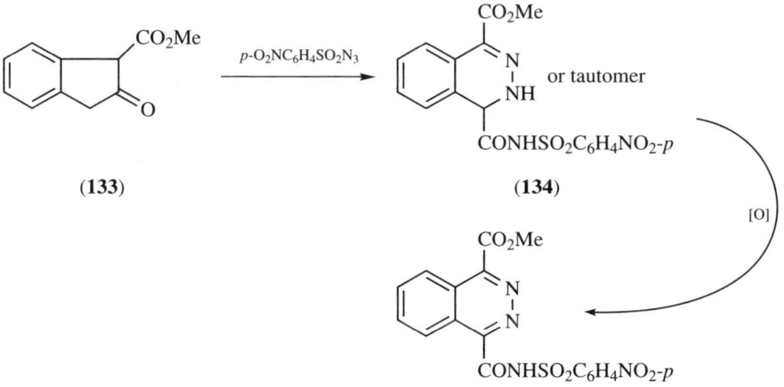

(**133**) (**134**) (**135**)

Also other examples.[24,1013]

Cycloheptanes as Substrates

3-Acetyl-2-chlorocycloheptatrien-1-one (**136**) or the isomeric substrate, 2-acetyl-7-chlorocycloheptatrien-1-one (**138**), underwent condensation and rearrangement with hydrazine to give 4-methyl-1(2*H*)-phthalazinone (**137**) ($H_2NNH_2 \cdot H_2O$, MeOH, reflux, 2 h: 29 or ?%, respectively); substrate (**138**) with methylhydrazine gave 2,4-dimethyl-1(2*H*)-phthalazinone (**139**) (MeOH, reflux, 2 h: 53%).[575]

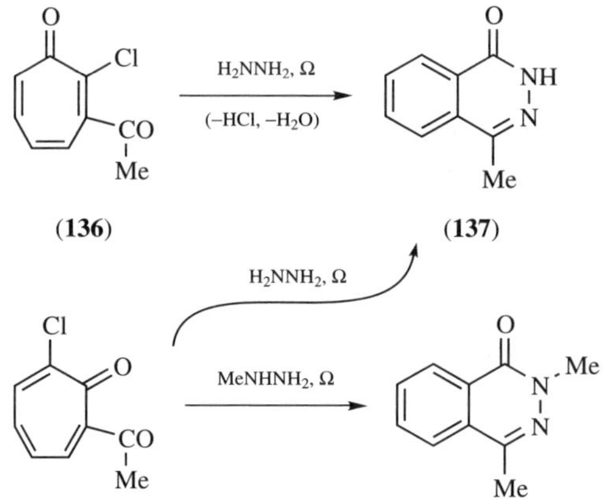

(**136**) →[H_2NNH_2, Ω; (−HCl, −H_2O)] (**137**)

(**138**) →[H_2NNH_2, Ω] (**137**)

(**138**) →[$MeNHNH_2$, Ω] (**139**)

8.4. FROM PYRIDAZINE DERIVATIVES AS SUBSTRATES

Pyridazine substrates have been used in at least four ways to synthesize phthalazines: with one synthon to supply C6, with one or two synthons to supply C6 + C7, or with one synthon to supply C5 + C6 + C7 + C8 of the product. The following classified examples illustrate these procedures.

Using a Synthon to Supply C6 of the Phthalazine

Ethyl 5-cyano-4-*p*-methoxystyryl-6-oxo-1-phenyl-1,6-dihydro-3-pyridazinecarboxylate (**140**) with malononitrile gave ethyl 5-amino-6-cyano-7-*p*-methoxyphenyl-4-oxo-3-phenyl-3,4-dihydro-1-phthalazinecarboxylate (**141**) reactants, Na, dioxane, reflux, 4 h: 70%); several analogs likewise.[534]

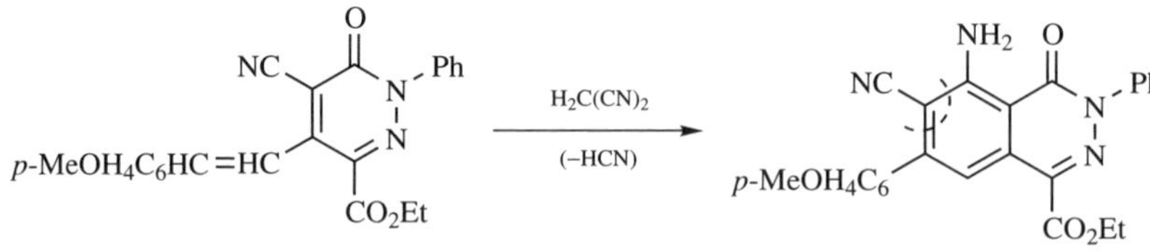

(**140**) →[$H_2C(CN)_2$; (−HCN)] (**141**)

Also other somewhat analogous examples.[66,364,81,526]

Using One Synthon to Supply C6 + C7 of the Phthalazine

Ethyl 5-cyano-6-methyl-4-oxo-3-phenyl-3,4-dihydro-1-pyridazinecarboxylate (**142**) and α-benzylidenemalononitrile gave ethyl 5-amino-6-cyano-4-oxo-3,7-diphenyl-3,4-dihydro-1-phthalazinecarboxylate (**143**) (reactants, piperidine, EtOH, reflux, 3 h: 70%;[489] analogs likewise)[468,489] or with ethyl 2-cyanoacetate gave ethyl 5-amino-6-cyano-7-hydroxy-4-oxo-3-phenyl-3,4-dihydro-1-phthalazinecarboxylate (**144**) (similar conditions: 75%).[526]

(**142**) —PhCH = C(CN)$_2$ (–HCN)→ (**143**)

(**142**) —EtO_2CCH_3CN (–EtOH)→ (**144**)

Diethyl 4,5-pyridazinedicarboxylate (**145**) with diethyl succinate gave diethyl 5,8-dioxo-2,3,5,8-tetrahydro-6,7-phthalazinedicarboxylate (**146**) (NaH, THF, reactants↓ dropwise, reflux, 6 h: 10%).[179,697]

(**145**) —$EtO_2CCH_2CH_2CO_2Et$ (–2EtOH)→ (**146**)

Also other examples.[378,474,530,618–674,810,969]

Using Two Synthons to Supply C6 + C7 of the Phthalazine

Ethyl 5-cyano-6-methyl-4-oxo-3-phenyl-3,4-dihydro-1-pyridazinecarboxylate (**147**) with 3-thiophenecarbaldehyde and malononitrile gave ethyl 5-amino-6-cyano-4-oxo-3-phenyl-7-(thien-3-yl)-3,4-dihydro-1-phthalazinecarboxylate (**148**) (reactants, trace piperidine, EtOH, reflux, <30 min: 85%); analogs likewise.[66]

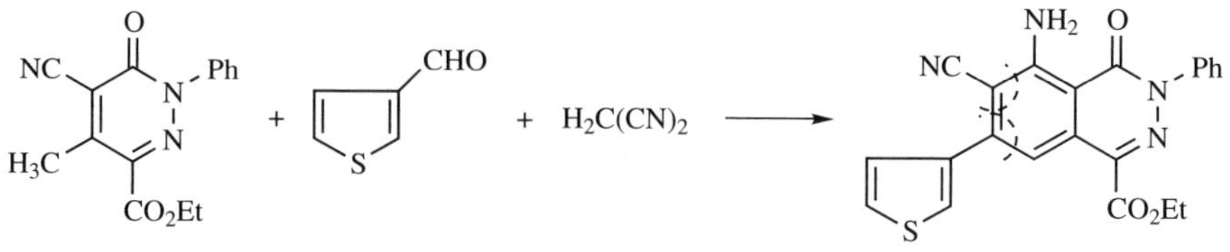

(147) (148)

4-Cyano-6-ethoxycarbonyl-5-methyl-1-phenylpyridazin-1-ium-3-olate (**149**) with formaldehyde and malononitrile gave 5-amino-6-cyano-1-ethoxycarbonyl-2-phenylphthalazin-2-ium-4-olate (**150**) (reactants, trace piperidine, Me_2NCHO, reflux, 6 h: ?%).[520]

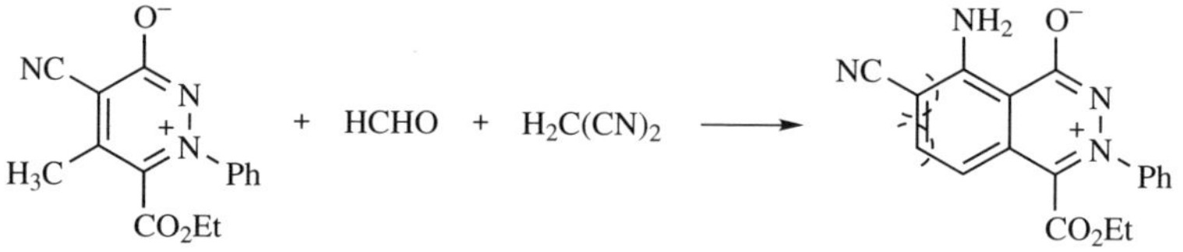

(149) (150)

Also other examples.[384]

Using a Synthon to Supply C5 + C6 + C7 + C8 of the Phthalazine

2-Methyl-5-phenylsulfonyl-3(2*H*)-pyridazinone (**152**, R = SO_2Ph) underwent Diels–Alder addition by 2,3-dimethyl-1,3-butadiene (**151**) to give 2,6,7-trimethyl-4a-phenylsulfonyl-4a,5,8,8a-tetrahydro-1(2*H*)-phthalazinone (**153**, R = SO_2Ph) (reactants, PhH, reflux, 10 days: 90%); methyl 1-methyl-6-oxo-1,6-dihydro-4-pyridazinecarboxylate (**152**, R = CO_2Me) likewise gave methyl 2,6,7-trimethyl-1-oxo-1,2,4a,5,8,8a-hexahydro-4a-phthalazinecarboxylate (**153**, R = CO_2Me) (70%); such additions by an unsymmetric diene were also investigated.[303]

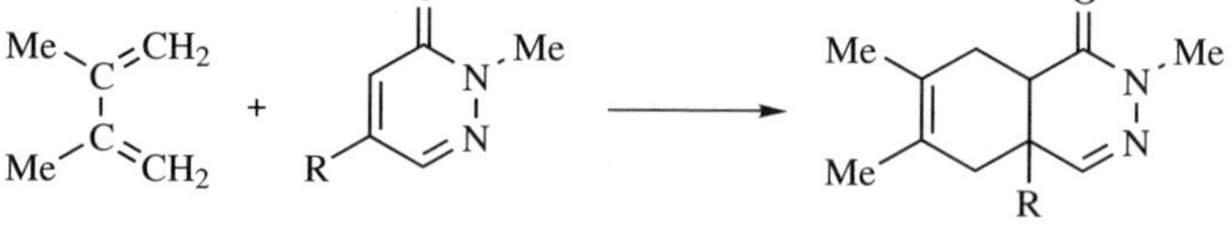

(151) (152) (153)

5-Bromo-2-methoxymethyl-6-phenyl-3(2*H*)-pyridazinone (**154**) with ethyl acrylate gave ethyl 2-methoxymethyl-1-oxo-4-phenyl-1,2-dihydro-6-phthalazinecarboxylate (**154a**) [substrate, synthon (2 mol), $Pd(OAc)_2$, Et_3N, Me_2NCHO, A, 120°C, tlc monitored: 28%, after separation from other products; X-ray structure analysis; mechanism proposed];[1024] also an analogous example.[1036]

2 × $H_2C=CHCO_2Et$

(**154**) (**154a**)

Although not used to make phthalazines, fused phthalazines have been prepared from pyridazine substrates by completing the C6–C7 bond of the phthalazine system therein; for example, 2-benzyl-5-*o*-bromophenyl-6-phenyl-1(2*H*)-pyridazinone (**155**) gave 2-benzyldibenzo[*f,h*]phthalazin-1(2*H*)-one (**155a**) [$AcNMe_2$, AcONa, $Pd(PPh_3)_2Cl_2$, N_2, 130°C, 64 h: 56%].[977] Such a reaction should also be useful to make unfused phthalazines.

Pd catalyst, AcONa (−HBr)

(**155**) (**155a**)

8.5. FROM 1,2,4,5-TETRAZINES AS SUBSTRATES

The only heteromonocyclic system (other than pyridazine) that has been used as a substrate for the primary synthesis of phthalazines is 1,2,4,5-tetrazine. It undergoes Diels–Alder condensation with a variety of appropriate cyclic synthons (with loss of nitrogen) to afford phthalazines or reduced phthalazines. The following examples illustrate such syntheses.

Condensations Affording Aromatic Phthalazines

3,6-Bis(trifluoromethyl)-1,2,4,5-tetrazine (**157**) and anisole (**156**) gave 6-methoxy-1,4-bis(trifluoromethyl)phthalazine (**158**) by a Diels–Alder condensation, loss of N_2, and spontaneous oxidation (neat reactants, Teflon-lined autoclave, 140°C, 12 h: 68% after separation from minor products; thioanisole likewise afforded 6-methylthio-1,4-bis(trifluoromethyl)phthalazine (56%).[491,721]

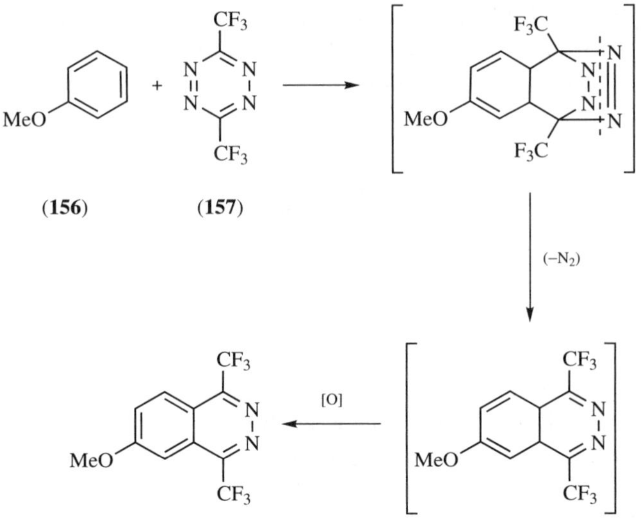

(156) (157)

(158)

Dimethyl 1,2,4,5-tetrazine-3,6-dicarboxylate (**160**) and benzyne (**159**) gave dimethyl 1,4-phthalazinedicarboxylate (**161**) [substrate (**160**), dioxane; *o*-$H_2NC_6H_4CO_2H$/dioxane and PrCHMeONO/dioxane both↓ dropwise, gentle reflux, 3 h: 48%].[116]

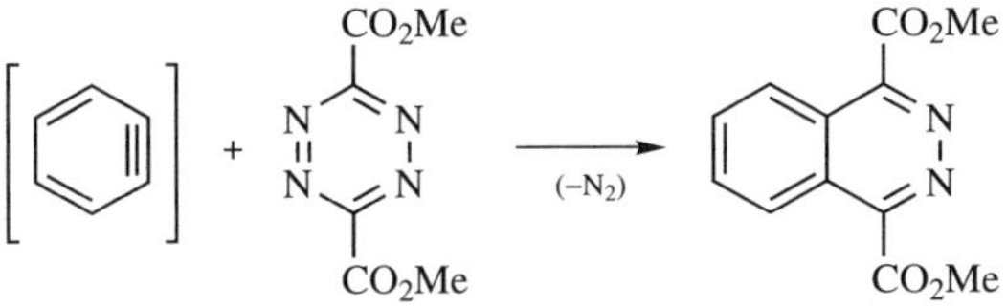

(159) (160) (161)

Also other examples.[131]

Condensations Affording Hydrophthalazines

3,6-Bis(trifluoromethyl)-1,2,4,5-tetrazine (**162**) with cyclohexene (**163**) gave a hexahydro product, formulated as 1,4-bis(trifluoromethyl)-2,4a,5,6,7,8-hexahydrophthalazine (**164**) (neat reactants, ?°C, 1 h: 88%);[699] with cyclohexa-1,4-diene (**165**) the same substrate (**162**) gave 1,4-bis(trifluoromethyl)-2,4a,5,8-tetrahydrophthalazine (**166**) (reactants, CH_2Cl_2, 20°C, 12 h: 95%);[532] or with *N*-(cyclohex-1-enyl)pyrrolidine (**167**) the same substrate gave the isomeric product, 1,4-bis(trifluoromethyl)-5,6,7,8-tetrahydrophthalazine (**168**) [substrate (**159**), CH_2Cl_2; synthon (**163**)/CH_2Cl_2↓ dropwise, 0°C → 20°C, 20 min: 78%].[590]

(162) (163) (164) (165) (166) (167) (168)

3,6-Diphenyl-1,2,4,5-tetrazine (**170**) with *N*-(cyclohex-1-enyl)morpholine (**169**) gave either 4a-morpholino-1,4-diphenyl-4a,5,6,7,8,8a-hexahydrophthalazine (**171**) (reactants, MeCN, reflux, 7 h: 90%) or 1,4-diphenyl-5,6,7,8-tetrahydrophthalazine (**172**) (reactants, PhH, reflux, <12 h: 48%).[41]

(169) (170) (171) (172)

Also other examples.[92,104,111,312,340,923,1041]

8.6. FROM HETEROBICYCLIC DERIVATIVES AS SUBSTRATES

A variety of heterobicyclic derivatives have been used as substrates for the primary synthesis of phthalazines; in this respect, isobenzofurans (including phthalic anhydride and the like) and isoindoles (including phthalimides, etc.) have been particularly heavily employed. The use of such bicyclic substrates is discussed in the following subsections, arranged alphabetically according to the parent system involved.

8.6.1. 6-Azabicyclo[3.1.0]hexanes as Substrates

This rare primary synthesis is represented by the rearrangement of 2-(6-azabicyclo[3.1.0]hexan-6-yl)-1,3-isoindolinedione (**173**) with addition of AcOH to give 2-(1-acetoxy-1,2-dimethylallyl)-1,4(2*H*,3*H*)-phthalazinedione (**174**) (AcOH, reflux, 24 h: >95%).[18]

AcOH

CMe(OAc)CMe=CH_2

(**173**) (**174**)

8.6.2. 2-Benzopyrans as Substrates

These benzopyrans may be converted into phthalazines by several procedures, illustrated in the following examples.

Ethyl 6,8-dimethyl-1,4-dioxo-3,4-dihydro-1*H*-2-benzopyran-7-carboxylate (**175**) with hydrazine hydrate gave ethyl 1-hydroxymethyl-5,7-dimethyl-4-oxo-3,4-dihydro-6-phthalazinecarboxylate (**176**) (EtOH, reflux, 2 h: ~65%).[405]

$H_2NNH_2{\cdot}H_2O$

(**175**) (**176**)

3-Morpholino-1*H*-2-benzopyran-1-one (**177**) with *p*-nitrobenzenediazonium tetrafluoroborate gave a separable mixture of 4-morpholinoformyl-2-*p*-nitrophenyl-1(2*H*)-phthalazinone [**178**, R = $CON(CH_2CH_2)_2O$] and 2-*p*-nitrophenyl-1(2*H*)-phthalazinone (**178**, R = H) (reactants, Et_3N, MeCN, 20°C,

5 min: 53% and ~7%, respectively).[700]

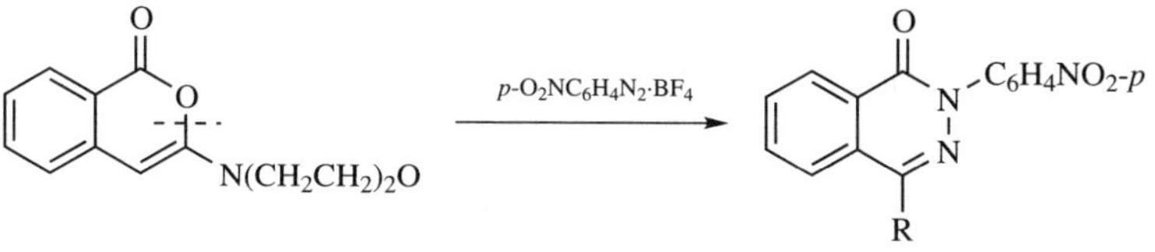

(177) (178)

4-(*p*-Chlorophenylhydrazono)-3,4-dihydro-1*H*-2-benzopyran-1,3-dione (**179**) gave 3-*p*-chlorophenyl-4-oxo-3,4-dihydro-1-phthalazinecarboxylic acid (**180**, R = OH) (1.25M NaOH, reflux, 15 min: 93%),[339] 3-*p*-chlorophenyl-4-oxo-3,4-dihydro-1-phthalazinecarboxamide (**180**, R = NH_2) (substrate, xylene, reflux, $NH_3\downarrow$, 3 h: 75%),[655] methyl 3-*p*-chlorophenyl-4-oxo-3,4-dihydro-1-phthalazinecarboxylate (**180**, R = OMe) (MeONa, MeOH, reflux until clear: >95%),[655] or other such products.[339,655]

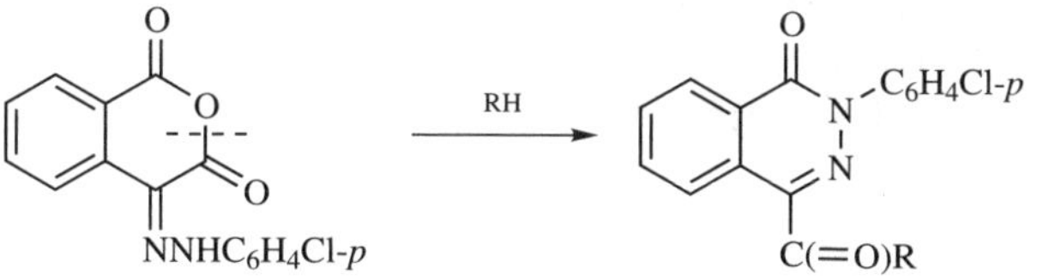

(179) (180)

Ethyl 6,8-dimethyl-1-oxo-4-phenylhydrazono-3,4-dihydro-1*H*-2-benzopyran-7-carboxylate (**181**) gave ethyl 1-hydroxymethyl-5,7-dimethyl-4-oxo-3-phenyl-3,4-dihydro-6-phthalazinecarboxylate (**182**) (NaOH, EtOH, H_2O, reflux, 10 min: 80%).[956]

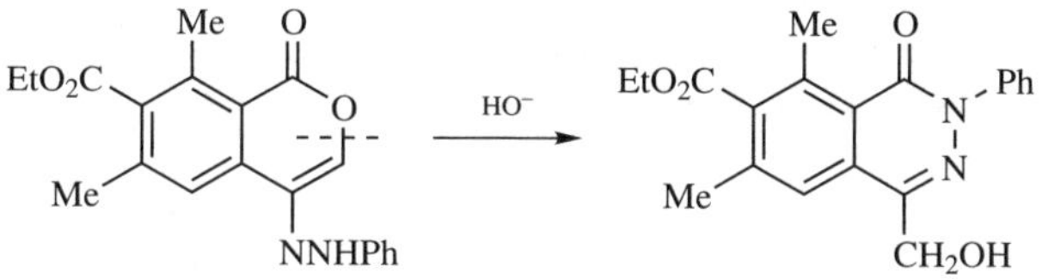

(181) (182)

Also other examples.[240]

8.6.3. 3,2,4-Benzothiadiazepines as Substrates

This interesting transformation is represented by the oxidative desulfurization of 7,8-dimethoxy-1,2,3,4-tetrahydro-3,2,4-benzothiadiazepine 3,3-dioxide (**183**) to give 6,7-dimethoxyphthalazine (**184**) (NaOCl, NaOH, H_2O: ~90%; for conditions, see original).[557]

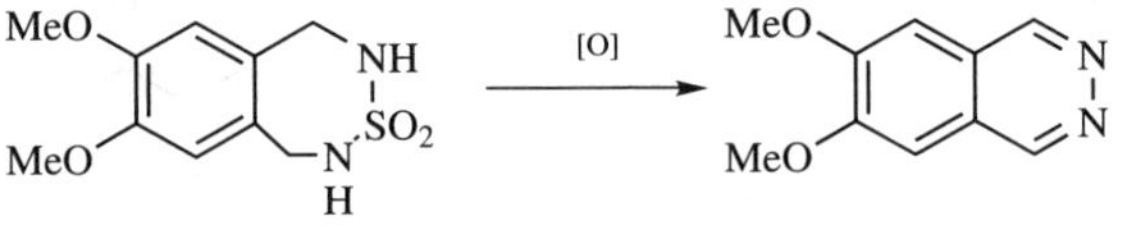

(**183**) (**184**)

8.6.4. 2,3-Benzoxazines as Substrates

This synthetic route to phthalazines has proved reasonably useful, as indicated in the following examples.

4-Phenyl-1*H*-2,3-benzoxazin-1-one (**185**) with hydrazine gave 4-phenyl-1(2*H*)-phthalazinone (**186**, R = H) [substrate, AcOH; excess $H_2NNH_2 \cdot H_2O\downarrow$ dropwise, 20°C, 1 h: 80%; analogs likewise)[760] or with phenylhydrazine gave 2,4-diphenyl-1(2*H*)-phthalazinone (**186**, R = Ph) ($PhNHNH_2$, AcOH, reflux, 2 h: 80%; analogs likewise).[612]

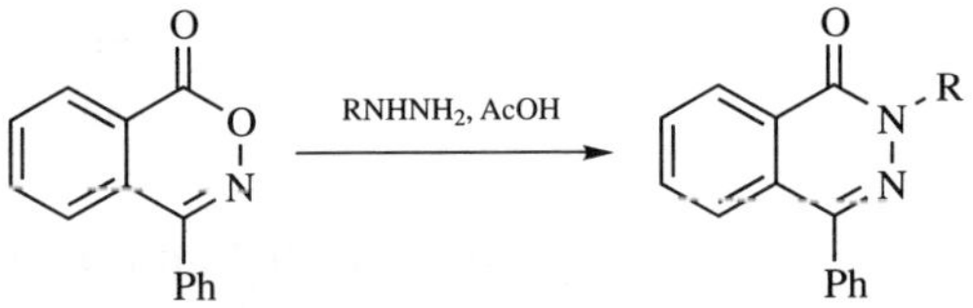

(**185**) (**186**)

In contrast, 5,6,7,8-tetrabromo-4-*p*-tolyl-1*H*-2,3-benzoxazin-1-one (**187**) with hydrazine gave 2-amino-5,6,7,8-tetrabromo-4-*p*-tolyl-1(2*H*)-phthalazinone (**188**) ($H_2NNH_2 \cdot H_2O$, BuOH, reflux, 6 h: 85%; analogs likewise).[666,cf. 367]

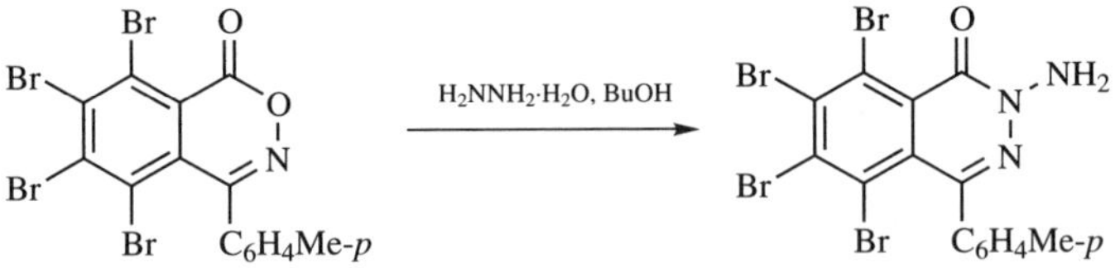

(**187**) (**188**)

Note: The presence or absence of an *N*-amino group in the foregoing products clearly depends on the mechanism involved, but the reason for such a difference is not evident.

8.6.5. Cinnolines as Substrates

This unusual transformation is exemplified in the isomerization of 1-(*o*-methoxycarbonylbenzoyl)-1,4-dihydro-3(2*H*)-cinnolinone (**189**) into 2-[*o*-(methoxycarbonylmethyl)phenyl]-1,4(2*H*,3*H*)-phthalazinedione (**190**) by refluxing in methanolic sulfuric acid or aqueous methanolic potassium hydrogen carbonate for 90 min or 7 h, respectively: yields were unstated.[6]

H^+ or HO^- in MeOH

(**189**) (**190**)

8.6.6. Isobenzofurans as Substrates (*H* 85, 141; *E* 376, 449)

Most of the isobenzofurans that have been used as substrates for phthalazine syntheses are in practice either 1,3-dihydro-1,3-isobenzofurandiones (i.e., phthalic anhydride derivatives) or 1,3-dihydro-1-isobenzofuran-1-ones (i.e., the lactonic phthalides), both of which normally undergo ring fission and reaction with hydrazine or the like to give phthalazinones. Discussion of these syntheses is subdivided according to the type of substrate used.

8.6.6.1. Using 1,3-Dihydro-1,3-isobenzofurandiones (Phthalic Anhydrides)

These substrates usually furnish 1,4(2*H*,3*H*)-phthalazinediones as illustrated in the examples that follow. Please note that the substrates are numbered as heterocycles and not as phthalic anhydride derivatives.

With Hydrazine as Synthon

1,3-Dihydro-1,3-isobenzofurandione (**191**) gave 1,4(2*H*,3*H*)-phthalazinedione (**192**) ($H_2NNH_2 \cdot H_2O$, EtOH, reflux, 30 min: 77%;[875] or neat $H_2NNH_2 \cdot H_2O$, O, $H_3PO_4 + P_2O_5$, ~150°C, 30 min: 80%).[498]

$H_2NNH_2 \cdot H_2O$

(**191**) (**192**)

5,6-Dimethoxy-1,3-dihydro-1,3-isobenzofurandione (**193**, R = OMe) gave 6,7-dimethoxy-1,4(2*H*,3*H*)-phthalazinedione (**194**, R = OMe) ($H_2NNH_2 \cdot H_2O$, EtOH, reflux, 30 min: 93%);[705] 6,7-dimethyl- (**194**, R = Me) (78%) and 6,7-dibromo-1,4(2*H*,3*H*)-phthalazinedione (**194**, R = Br) (94%) were made similarly.[704]

$H_2NNH_2 \cdot H_2O$

(**193**) (**194**)

5-Diethylamino-7-methyl-1,3-dihydro-1,3-isobenzofurandione (**195**) gave 6-diethylamino-8-methyl-1,4(2*H*,3*H*)-phthalazinedione (**196**) ($H_2NNH_2 \cdot H_2O$, AcOH, N_2, reflux, 1 h: 77%).[628]

$H_2NNH_2 \cdot H_2O$, AcOH

(**195**) (**196**)

Also other examples.[222,229,260]

With Alkyl- or Dialkylhydrazines as Synthons

1,3-Dihydro-1,3-isobenzofurandione (**197**) with (2-hydroxyethyl)hydrazine gave 2-(2-hydroxyethyl)-1,4(2*H*,3*H*)-phthalazinedione (**198**, R = CH_2OH) (MeOH, 20°C, 2 h: 40%),[453] with (*o*-nitrobenzyl)hydrazine gave 2-*o*-nitrobenzyl-1,4(2*H*,3*H*)-phthalazinedione (**198**, R = $C_6H_4NO_2$-*o*) (Me_2SO, 125°C, 24 h: 65%),[3] or with *N*-(2-hydroxyethyl)-*N*′-methylhydrazine gave 2-(2-hydroxyethyl)-3-methyl-1,4(2*H*,3*H*)-phthalazinedione (**199**) (synthon · 2HCl, AcONa, AcOH, H_2O, reflux, A, 4 h: 40%).[745]

RH_2CNHNH_2

(**197**) (**198**)

$HOCH_2CH_2NHNHMe$

(**199**)

1,3-Dioxo-1,3-dihydro-5-isobenzofuran-5-carboxylic acid (**200**) gave 2,3-dimethyl-1,4-dioxo-1,2,3,4-tetrahydro-6-phthalazinecarboxylic acid (**201**) (MeNHNHMe·2HCl, Et_3N, Me_2NCHO, reflux, 18 h: 52%).[918]

MeNHNHMe

(**200**) (**201**)

Also other examples.[238]

With Aryl- or Heteroarylhydrazines as Synthons

1,3-Dihydro-1,3-isobenzofurandione (**202**) with phenylhydrazine gave 2-phenyl-1,4(2*H*,3*H*)-phthalazinedione (**203**) (reactants, H_3PO_4, P_2O_5, 150°C, 30 min: 78%;[498] reactants, AcOH, microwave *h*ν, 2 min: 72%)[89] or with 3-hydrazino-6-phenylpyridazine gave 2-(6-phenylpyridazin-3-yl)-1,4(2*H*,3*H*)-phthalazinedione (**204**) (neat reactants, 225°C, 10 h: 64%).[652]

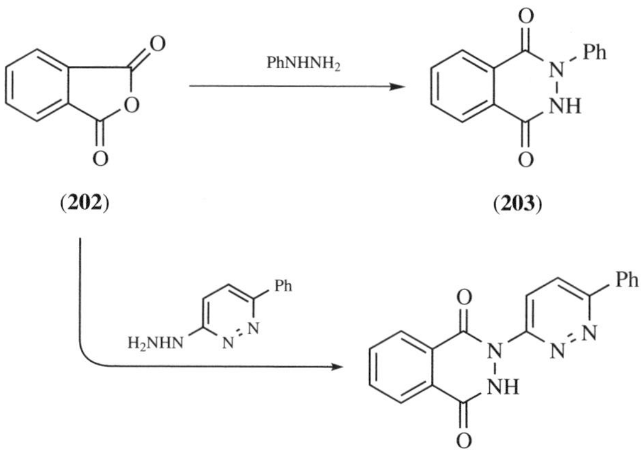

$PhNHNH_2$

(**202**) (**203**)

(**204**)

Also other examples.[261,504]

With Arenecarbohydrazides as Synthons

1,3-Dihydro-1,3-isobenzofurandione (**205**) with 1-amino-2-anthraquinonecarbohydrazide gave 2-(1-amino-2-anthraquinonecarbonyl)-1,4(2*H*,3*H*)-phthalazinedione (**206**) (reactants, $AcNMe_2$, reflux, 12 h: 78%).[679]

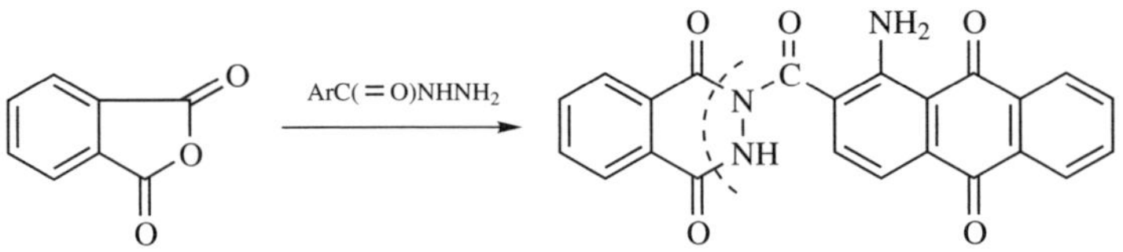

(205) (206)

Also other examples.[992]

With Semicarbazide or Thiosemicarbazide as Synthon

1,3-Dihydro-1,3-isobenzofurandione (**207**) with semicarbazide or thiosemicarbazide gave 1,4-dioxo-1,2,3,4-tetrahydro-2-phthalazinecarboxamide (**208**, X = O) or the corresponding carbothioamide (**208**, X = S), respectively (AcOH, Pr^iOH, reflux, 1 h: 61% or 75%, respectively; analogs similarly);[994] an earlier procedure that employed the same reactants under different conditions (H_3PO_4, P_2O_5, 160°C, 30 min),[499] appears to have given 5,7,12,14-tetrahydrophthalazino[2,3-*b*]phthalazine-5,7,12,14-tetrone (**209**).[994]

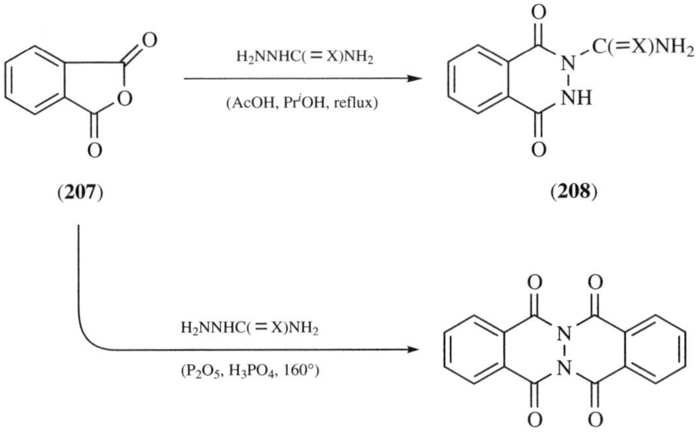

(207) (208)

(209)

8.6.6.2. *Using 1,3-Dihydro-1-isobenzofuranones (Phthalides)*

These substrates have also been used extensively to prepare phthalazines by reaction with hydrazine or its derivatives. The examples that follow are classified according to the 3-substituent present on the substrate.

From 3-Halogeno-1,3-dihydro-1-isobenzofuranones

Note: These halogeno substrates afford 4-unsubstituted or 4-alkyl(or aryl)phthalazinones as final products.

3-Chloro-3-phenyl-1,3-dihydro-1-isobenzofuranone (**210**) and *N*,*N′*-dimethylhydrazine gave (via an unisolated intermediate) 4-hydroxy-2,3-dimethyl-4-phenyl-3,4-dihydro-1(2*H*)-phthalazinone (**211**) (Et_3N, PhH, 20°C, 24 h, aqueous workup: 80%) and thence 2-methyl-4-phenyl-1(2*H*)-phthalazinone (**212**) (pyridine, AcCl↓ dropwise, 20°C, 24 h: 61%); the same substrate (**210**) with methylhydrazine gave the final product (**212**) (47%), apparently without the necessity for an oxidative step.[726]

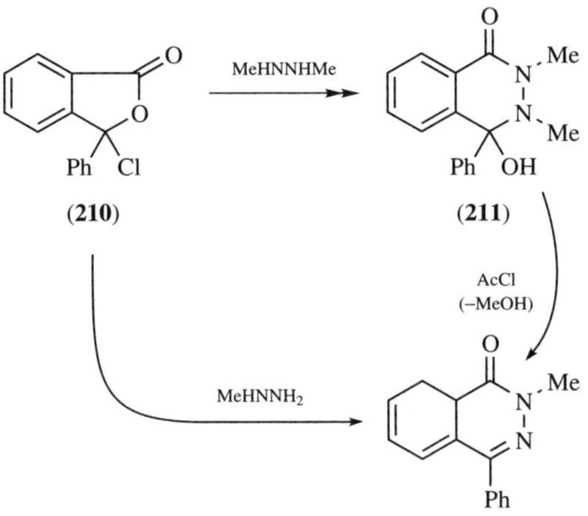

(**210**) (**211**) (**212**)

3-Bromo-6-nitro-1,3-dihydro-1-isobenzofuranone (**213**) gave 7-nitro-2-phenyl-1(2*H*)-phthalazinone (**214**) (3M HCl, reflux, briefly; then $PhNHNH_2$↓, 20°C, ~1 h: 73%); analogs likewise.[411]

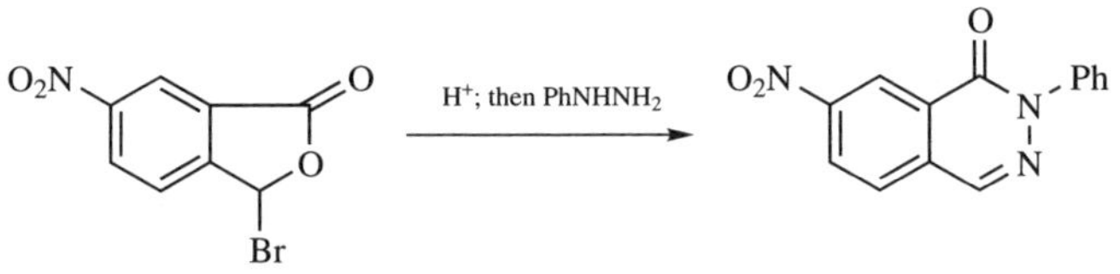

(**213**) (**214**)

Also other examples.[421,622]

From 3-Hydroxy-1,3-dihydro-1-isobenzofuranones

3-Hydroxy-6,7-dimethoxy-1,3-dihydro-1-isobenzofuranone (**215**) gave 7,8-dimethoxy-2-methyl-1(2*H*)-phthalazinone (**216**) ($MeNHNH_2$, EtOH, reflux, 30 min: 83%); analogs likewise.[488]

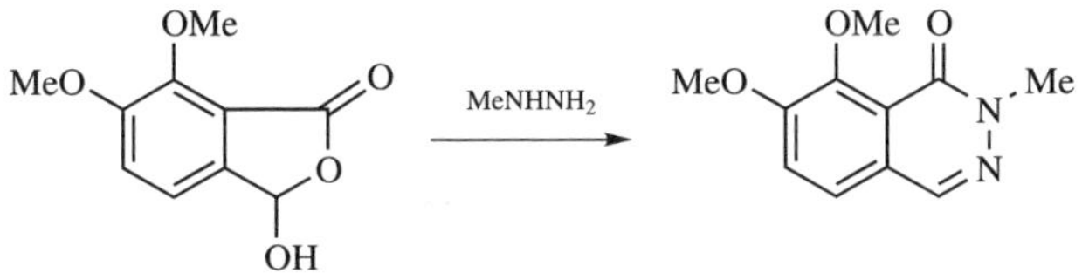

(**215**) (**216**)

1-Hydroxy-3-oxo-1,3-dihydro-4-isobenzofurancarboxylic acid (**217**) gave 4-oxo-3-phenyl-3,4-dihydro-5-phthalazinecarboxylic acid (**218**) ($PhNHNH_2$, AcOH, reflux, 18 h: ~85%); analogs likewise.[623]

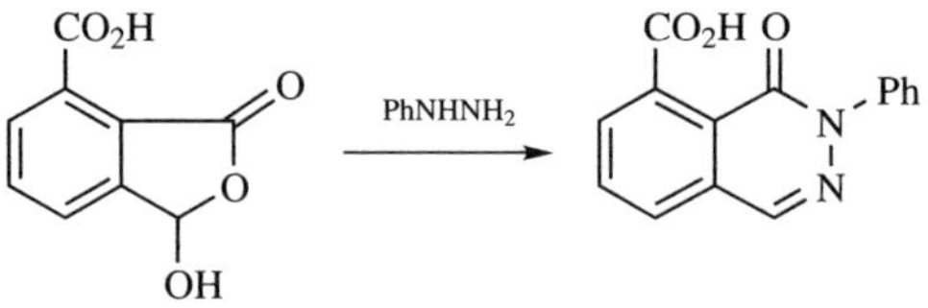

(**217**) (**218**)

1-Hydroxy-3-oxo-1,3-dihydro-4-isobenzofurancarbaldehyde (**219**) with aqueous hydrazine initially gave 8-hydrazonomethyl-1(2*H*)-phthalazinone (**220**), which gradually underwent reduction by the excess of hydrazine to afford 8-methyl-1(2*H*)-phthalazinone (**221**) (reflux, ~5 days: ~40%).[622]

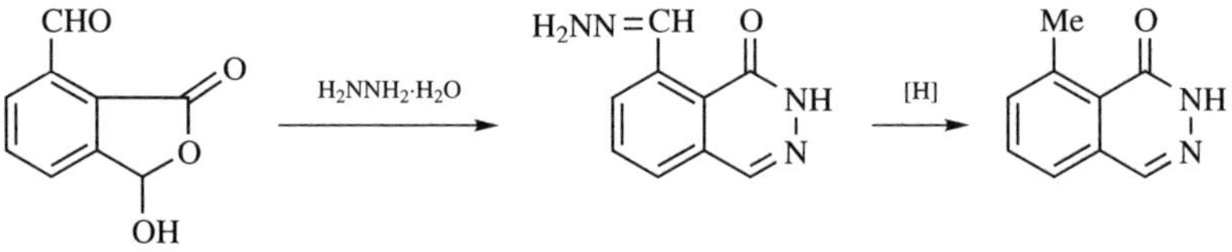

(**219**) (**220**) (**221**)

Also other examples.[270,288,411,861]

From 3-Alkyl(or aryl)-3-hydroxy-1,3-dihydro-1-isobenzofuranones

3-*tert*-Butoxycarbonylmethyl-3-hydroxy-1,3-dihydro-1-isobenzofuranone (**222**) with benzylhydrazine dihydrochloride gave 2-benzyl-4-*tert*-butoxycarbonylmethyl-1(2*H*)-phthalazinone (**223**) (Et_3N, EtOH, reflux, 18 h: 96%).[118]

(**222**) (**223**)

Ethyl 1-hydroxy-1-*o*-methoxyphenyl-4,6-dimethyl-3-oxo-1,3-dihydro-5-isobenzofurancarboxylate (**224**) gave ethyl 1-*o*-methoxyphenyl-5,7-dimethyl-4-oxo-3,4-dihydro-6-phthalazinecarboxylate (**225**) [substrate (made in situ), H_2N $NH_2 \cdot H_2O$, EtOH, reflux, 2 h: >50%].[426]

$H_2NNH_2 \cdot H_2O$

(**224**) (**225**)

Ethyl 1-hydroxy-1,4,6-trimethyl-3-oxo-1,3-dihydro-5-isobenzofurancarboxylate (**226**) gave a separable mixture of 6-ethoxycarbonyl-5,7-dimethyl-4-oxo-3,4-dihydro-1-phthalazinecarboxylic acid (**227**, R = CO_2H) and its decarboxylated derivative, ethyl 5,7-dimethyl-4-oxo-3,4-dihydro-6-phthalazinecarboxylate (**227**, R = H) (substrate, KOH, H_2O, $KMnO_4\downarrow$ slowly, 10°C, 90 min; filtrate, $H_2NNH_2 \cdot HCl\downarrow$, EtOH, 20°C, 48 h: 60% and 16%, respectively).[956,cf. 405]

[O], then $H_2NNH_2 \cdot HCl$

(**226**) (**227**)

Also other examples.[155,262]

From 3,3-Dialkyl(or aryl)-1,3-dihydro-1-isobenzofuranones

Note: These substrates necessarily afford 3,4-dihydro-1(2*H*)-phthalazinones.

3,3-Bis(*p*-dimethylaminophenyl)-1,3-dihydro-1-isobenzofuranone (**228**) gave 4,4-bis(*p*-dimethylaminophenyl)-3,4-dihydro-1(2*H*)-phthalazinone (**229**) ($H_2NNH_2 \cdot H_2O$, EtOH, reflux, 20 h: ~85%);[406] analogs likewise.[406,968]

$H_2NNH_2 \cdot H_2O$

(**228**) (**229**)

From 3-Benzylidene-1,3-dihydro-1-isobenzofuranones or the Like

3-Benzylidene-5-bromo-1,3-dihydro-1-isobenzofuranone (**230**, X = O) with hydrazine hydrate gave 4-benzyl-6-bromo-1(2*H*)-phthalazinone (**231**) (EtOH, reflux, 30 min: 79%);[521,611] similar treatment of 3-benzylidene-5-bromo-1,3-dihydro-1-isobenzofuranthione (**230**, X = S) gave the same product (**231**)

(58%).[642] Analogs were made similarly.[521,522,611,642]

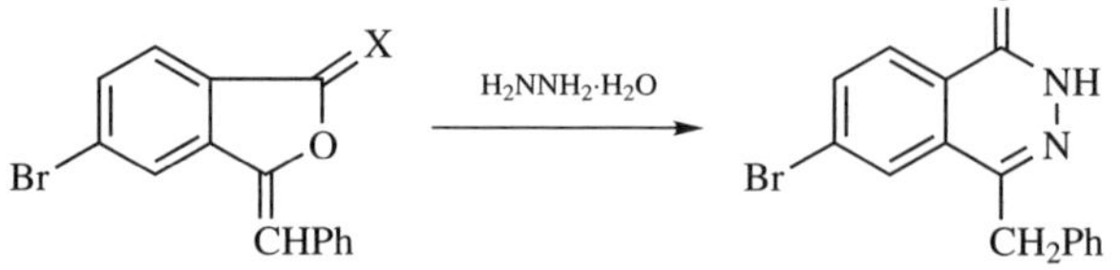

(230) (231)

3-Benzylidene-4,7-diiodo-1,3-dihydro-1-isobenzofuranone (**232**) with phenylhydrazine gave 4-benzyl-5,8-diiodo-2-phenyl-1(2*H*)-phthalazinone (**233**) (EtOH, reflux, 3 h: 41%);[647] analogs likewise.[646]

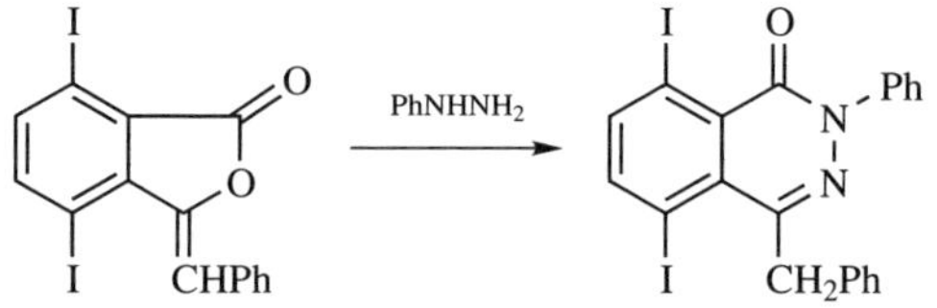

(232) (233)

Also other examples.[68,182,202,207,209,351,643,648,862–864]

From Heterocyclylidene-1,3-dihydro-1-isobenzofuranones

3-(1-Methylpyrrolidin-2-ylidene)-1,3-dihydro-1-isobenzofuranone (**234**) with hydrazine hydrate gave 4-(1-methylpyrrolidin-2-yl)-1(2*H*)-phthalazinone (**235**) (EtOH, H_2O, reflux, 12 h: 85%).[654]

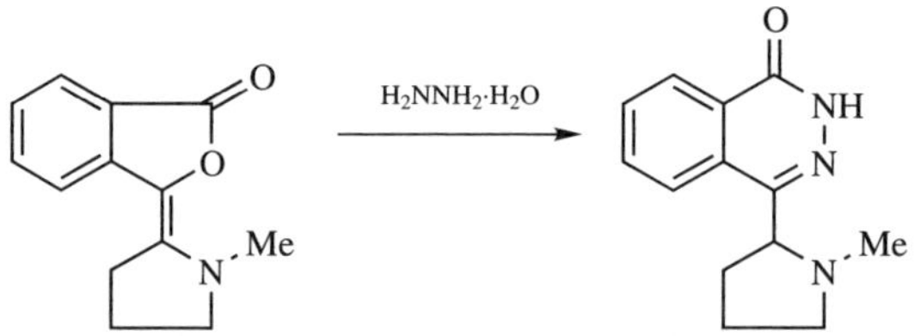

(234) (235)

1,1′-Bi(3-oxo-1,3-dihydroisobenzofuran-1-ylidene) (**236**) with hydrazine (1 equiv) gave 4-(3-oxo-1,3-dihydroisobenzofuran-1-yl)-1(2*H*)-phthalazinone (**237**, R = H) (substrate, EtOH, reflux; $H_2NNH_2 \cdot H_2O$/EtOH↓ dropwise, reflux, 14 h: 65%) or with methylhydrazine gave 1-methyl-4-(3-oxo-1,3-dihydroisobenzofuran-1-yl)-1(2*H*)-phthalazinone (**237**, R = Me) (likewise: 86%).[576,cf. 544]

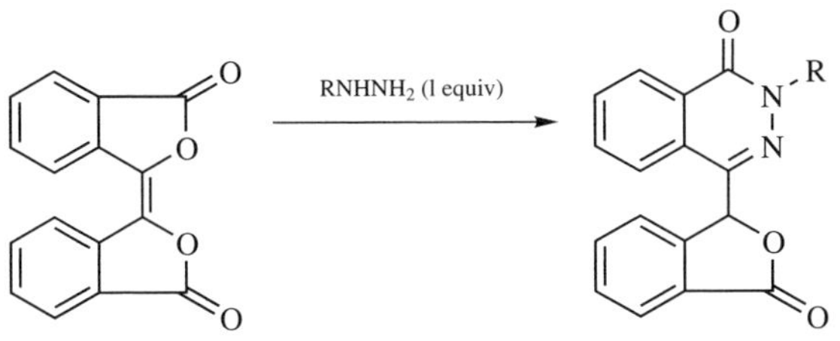

RNHNH$_2$ (1 equiv)

(**236**) (**237**)

Also other examples.[158,159]

From 1-Dimethyliminio-1,3-dihydroisobenzofurans

1-Benzylidene-3-dimethyliminio-1,3-dihydroisobenzofuran perchlorate (**238**) with hydrazine hydrate gave 1-benzyl-4-dimethylaminophthalazine (**239**) (Et_3N, EtOH, 20°C, 1 h: 91%), with methylhydrazine gave 4-benzyl-2-methyl-1(2*H*)-phthalazinone (**240**) (likewise: 83%), or with phenylhydrazine gave 1-benzyl-2-phenyl-4-phenylhydrazono-1,2-dihydrophthalazine (**241**) (likewise but reflux, 20 min: 63%).[689,cf. 70]

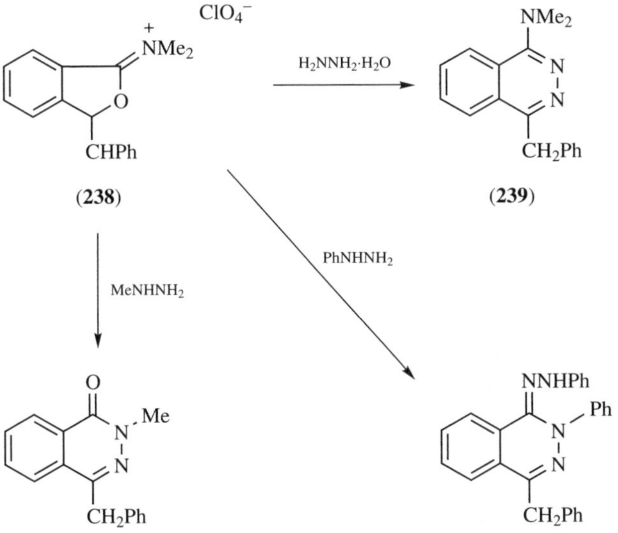

ClO_4^- $\overset{+}{N}Me_2$ CHPh (**238**)

$H_2NNH_2{\cdot}H_2O$

NMe_2 N N CH_2Ph (**239**)

MeNHNH$_2$

PhNHNH$_2$

O Me N N CH_2Ph (**240**)

NNHPh Ph N N CH_2Ph (**241**)

1-Acetoxy-3-dimethyliminio-1-isopropyl-1,3-dihydroisobenzofuran (**242**) gave 1-dimethylamino-4-isopropylphthalazine (**243**) ($H_2NNH_2 \cdot H_2O$, Et_3N, EtOH, 20°C, 20 min: 39%).[689]

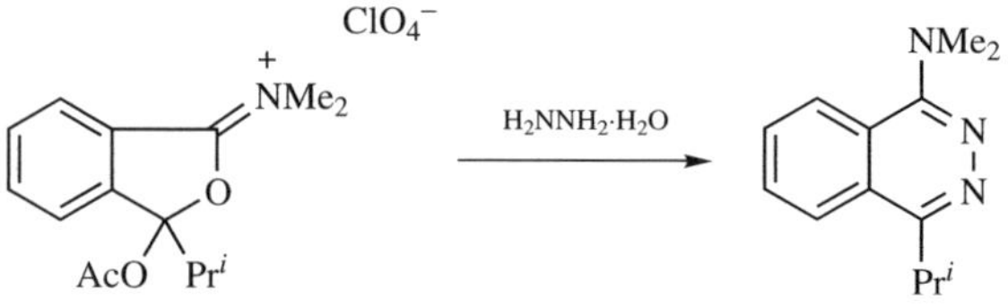

(242) $\xrightarrow{H_2NNH_2 \cdot H_2O}$ (243)

8.6.7. Isoindoles as Substrates (*H* 140, 158; *E* 376, 446)

Most of the isoindole derivatives that have been used as substrates for phthalazine syntheses are in fact 1,3-isoindolinediones [phthalimides (**244**), 1,3-isoindolinediimines (**245**), or 3-hydroxy-1-isoindolinones (**246**)]. According to their substituents, these substrates may or may not require an ancillary synthon to furnish phthalazines. The use of each type of isoindoline derivative is discussed in an appropriate subsection that follows.

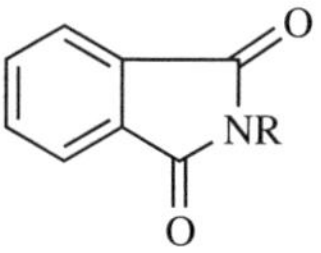

(**244**) 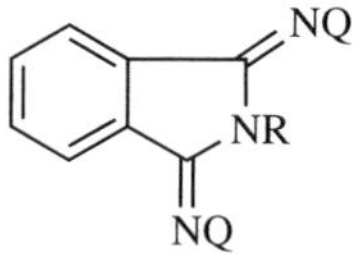 (**245**) 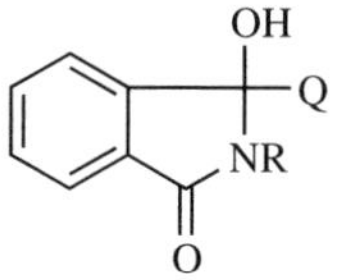 (**246**)

8.6.7.1. Using N-Unsubstituted-1,3-isoindolinediones

These simple substrates react with hydrazines to give 1,4(2*H*,3*H*)-phthalazinediones, as illustrated in the following examples.

1,3-Dioxo-5-isoindolinecarboxamide (**247**) gave 1,4-dioxo-1,2,3,4-tetrahydro-6-phthalazinecarboxamide (**248**) ($H_2NNH_2 \cdot H_2O$, *N*-methyl-2-pyrrolidinone, 20°C, 30 min: 94%).[285]

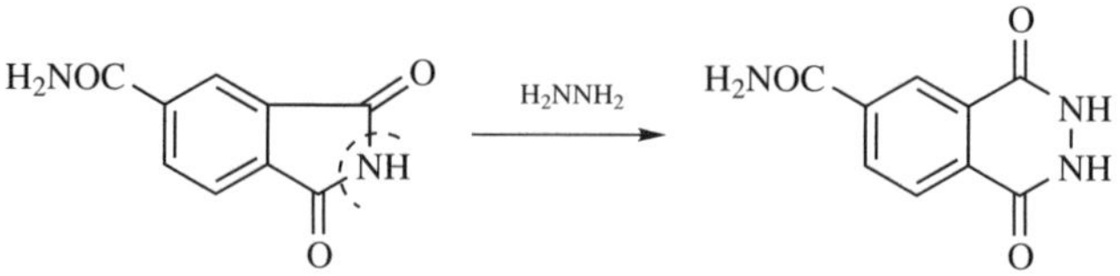

(**247**) $\xrightarrow{H_2NNH_2}$ (**248**)

1,3-Isoindolinedione (**249**) and *threo*-2-hydrazino-1-methylpropanol gave *threo*-2-(2-hydroxy-1,2-dimethylethyl)-1,4(2*H*,3*H*)-phthalazinedione (**250**) (EtOH,

20°C, 18 h: 22% after separation from unchanged substrate); the *erythro*-isomer (25%) was made similarly.[148]

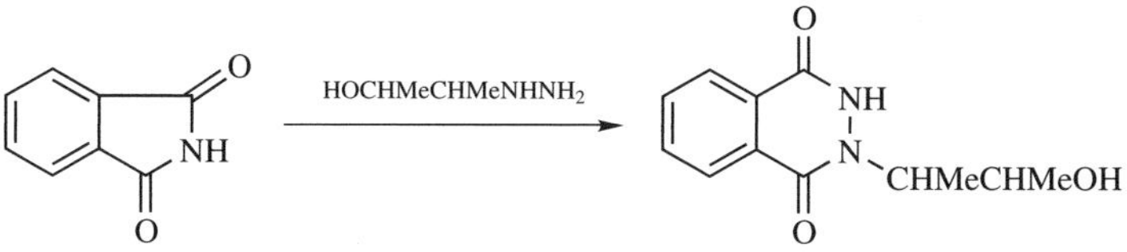

(249) → HOCHMeCHMeNHNH$_2$ → (250)

8.6.7.2. *Using N-Alkyl-1,3-isoindolinediones*

Condensation of these substrates with hydrazine derivatives normally occurs with elimination of the alkylimino moiety of the substrate to afford 1,4(2*H*,3*H*)-phthalazinediones. However, abnormal cases have been reported in which the said moiety is retained to afford 4-alkylamino-1(2*H*)-phthalazinones. The following examples illustrate both outcomes.

Normal Production of 1,4(2*H*, 3*H*)-Phthalazinediones

5-Hydroxy-2-methyl-6-nitro-1,3-isoindolinedione (**251**) with hydrazine hydrate gave 6-hydroxy-7-nitro-1,4(2*H*,3*H*)-phthalazinedione (**252**) (EtOH, reflux, 2 h: 78%); analogs likewise.[48]

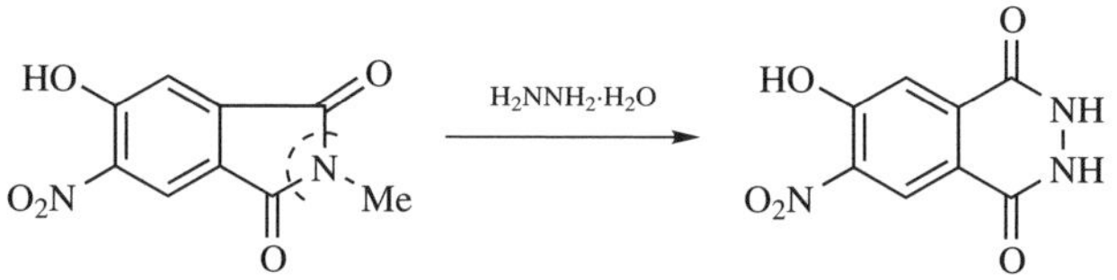

(251) → $H_2NNH_2 \cdot H_2O$ → (252)

5-Butylamino-2-methyl-1,3-isoindolinedione (**253**) gave 6-butylamino-1,4(2*H*, 3*H*)-phthalazinedione (**254**) (H_2NNH_2, EtOH, reflux, 4 h: 60%); analogs likewise.[313]

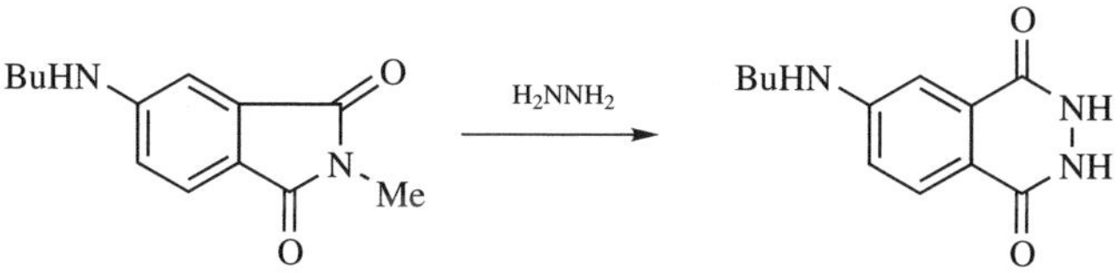

(253) → H_2NNH_2 → (254)

Also other examples.[234,283,316,628]

Abnormal Production of 4-Alkylamino-1(2*H*)-phthalazinone

2-(1-Carboxyethyl)-1,3-isoindolinedione (**255**) with hydrazine hydrate gave 4-(1-carboxyethyl)amino-1(2*H*)-phthalazinone (**256**) (BuOH, reflux, 10 h: 75%).

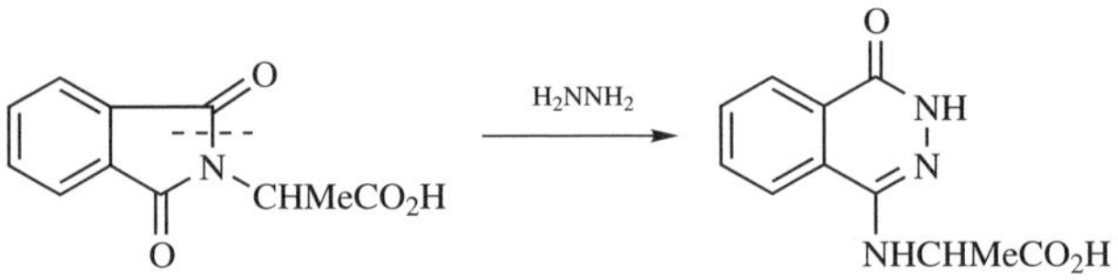

H_2NNH_2

(**255**) (**256**)

Also other examples.[190,197,228,542]

8.6.7.3. *Using N-Hydroxy-1,3-isoindolinediones*

Like their *N*-alkyl analogs, these substrates can eliminate their hydroxyimino moiety during reaction with hydrazine derivatives to afford 1,4(2*H*,3*H*)-phthalazinediones. Thus 2-hydroxy-4,5,6,7-tetramethyl-1,3-isoindolinedione (**257**) and hydrazine hydrate in refluxing ethanol during 3 h gave 5,6,7,8-tetramethyl-1,4(2*H*,3*H*)-phthalazinedione (**258**) in 85% yield; several analogs were made similarly.[114]

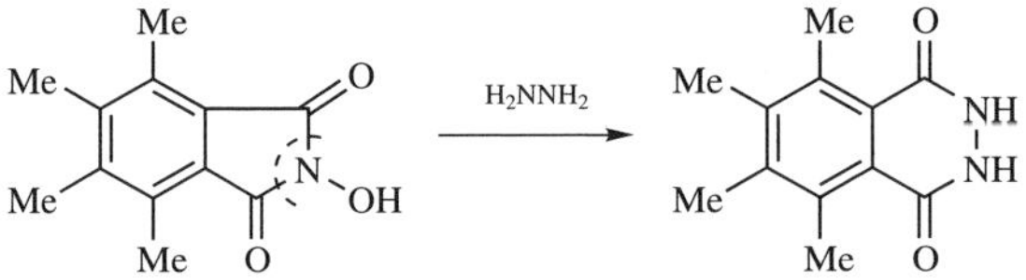

H_2NNH_2

(**257**) (**258**)

8.6.7.4. *Using 3-Hydroxy-1-isoindolinones*

These substrates with hydrazine furnish 1(2*H*)-phthalazinones, as illustrated in the following examples.

5-Amino-6-chloro-3-hydroxy-1-isoindolinone (**259**) with hydrazine gave 6-amino-7-chloro-1(2*H*)-phthalazinone (**260**, R = H) (H_2O, 95°C, N_2, 2 h: 77%) or with methylhydrazine gave 6-amino-7-chloro-2-methyl-1(2*H*)-phthalazinone (**260**, R = Me) (H_2O, 90°C, 3 h: 60%); analogs likewise.[274]

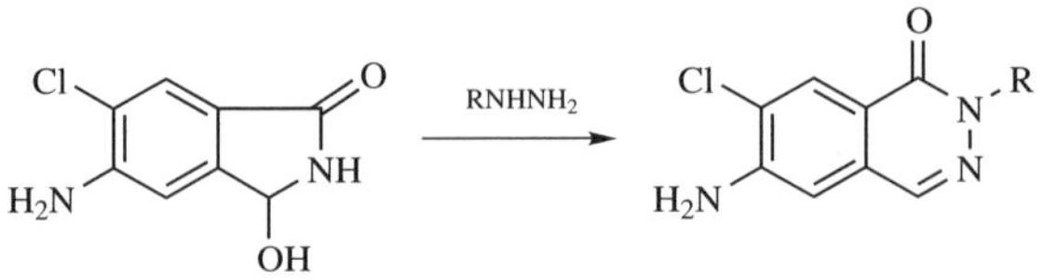

$RNHNH_2$

(**259**) (**260**)

3-Hydroxy-4-methyl-1-isoindolinone (**261**, R = H) gave 5-methyl-1(2*H*)-phthalazinone (**262**) ($H_2NNH_2 \cdot H_2O$, H_2O, reflux, 16 h: 97%); 3-hydroxy-4-methyl-2-phenyl-1-isoindolinone (**261**, R = Ph) gave the same product (**262**) (likewise but ~36 h: 75%).[622]

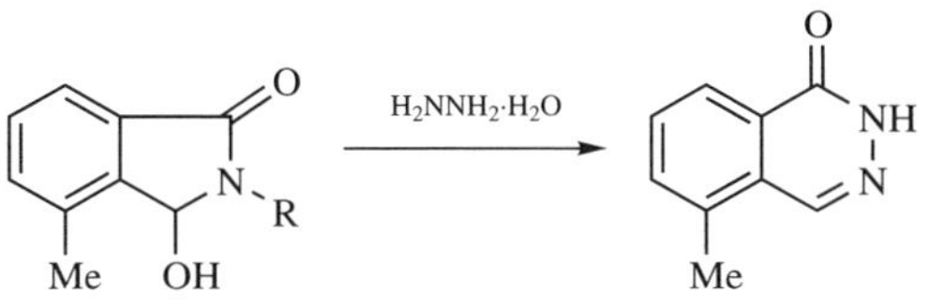

(**261**) (**262**)

3-Hydroxy-3-methyl-1-isoindolinone (**263**) gave 4-methyl-1(2*H*)-phthalazinone (**264**) ($H_2NNH_2 \cdot H_2O$, PrOH, reflux, 5 days: 75%); analogs likewise.[881]

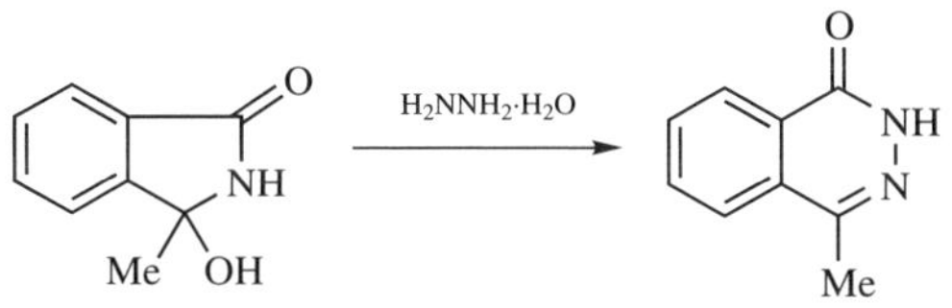

(**263**) (**264**)

3-Hydroxy-3-(pyridin-3-yl)-1-isoindolinone (**265**) gave 4-(pyridin-3-yl)-1(2*H*)-phthalazinone (**266**) (neat $H_2NNH_2 \cdot H_2O$, reflux, 2 h: 90%); analogs likewise, some with added propanol or acetic acid.[391]

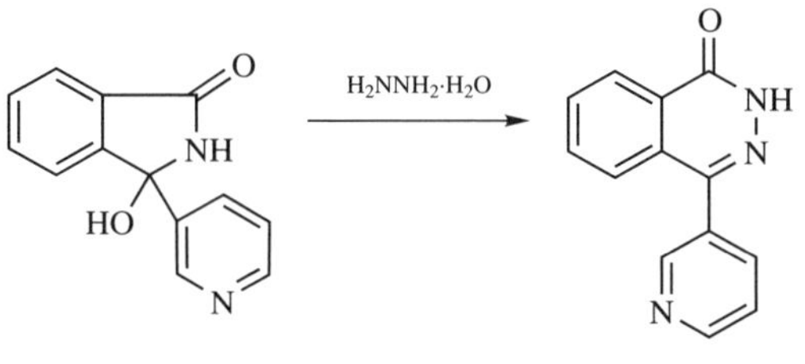

(**265**) (**266**)

Also other examples.[411,519,1015]

8.6.7.5. *Using 1,3-Isoindolinediimines or 3-Imino-1-isoindolinones*

Such substrates give 1,4-phthalazinediamines or 4-amino-1(2*H*)-phthalazinones, respectively, as illustrated in these examples.

1,3-Isoindolinediimine (**267**) gave 1,4-phthalazinediamine (**268**) (substrate, EtOH, warm, then $H_2NNH_2 \cdot H_2O\downarrow$ dropwise: 73%).[685]

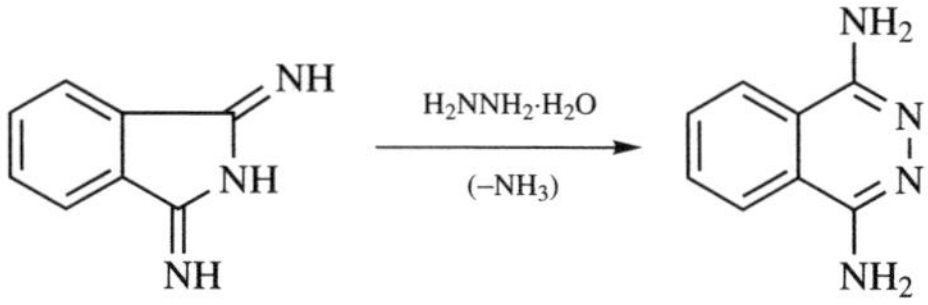

(267) (268)

1,3-Bis[(3-methylpyridin-2-yl)imino]isoindoline (**269**) gave 1,4-bis[(3-methylpyridin-2-yl)amino]phthalazine (**270**) ($H_2NNH_2 \cdot H_2O$, MeOH, reflux, ~4 h: 79%).[29]

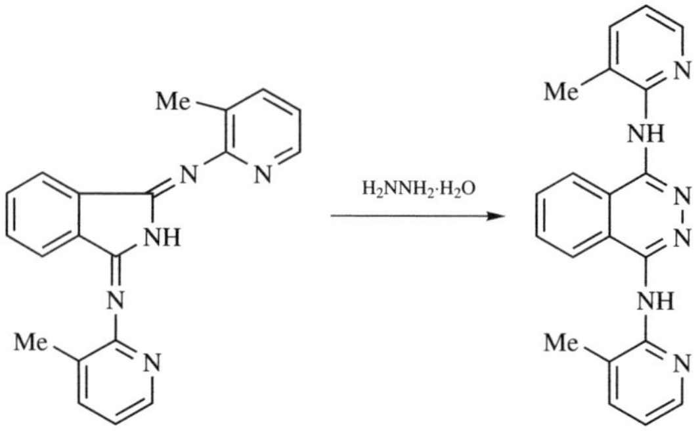

(269) (270)

3-Imino-1-isoindolinone (**271**) gave 4-amino-1(2*H*)-phthalazinone (**272**) (H_2NNH_2, for details, see original).[199]

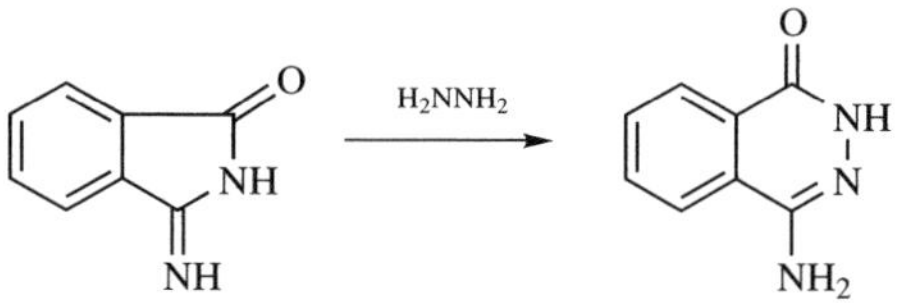

(271) (272)

Also other examples.[624,801]

8.6.7.6. *Using N-Amino-1,3-isoindolinediones*

These or closely related substrates can furnish phthalazines in several ways, some of which require an ancillary synthon. The classified examples that follow illustrate these syntheses.

From 2-Amino-1,3-isoindolinediones or Related Substrates

2-(α,α-Dimethylbenzylamino)-1,3-isoindolinedione (**273**) gave 2-(α,α-dimethylbenzyl)-1,4(2*H*,3*H*)-phthalazinedione (**274**) (EtONa, EtOH, AcOEt, reflux,

20 h: 80%).[103]

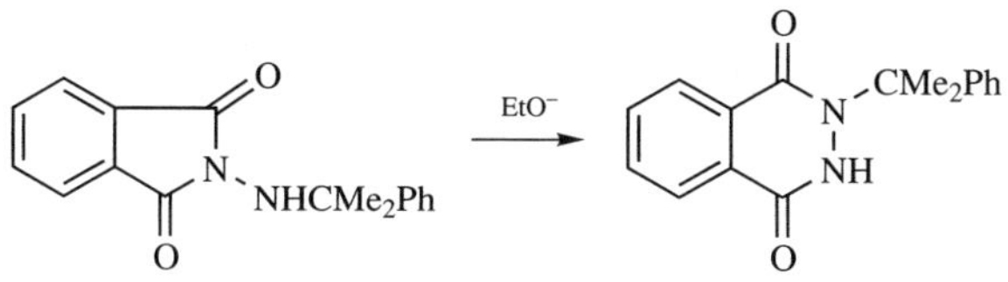

(273) $\xrightarrow{EtO^-}$ (274)

2-Anilino-3-benzyl-3-hydroxy-1-isoindolinone (**275**) gave 1-benzyl-2-phenylphthalazin-2-ium-4-olate (**276**) [AcOH, reflux, 10 min: ~60%; or $OC(CH_2)_5$, reflux, 30 min: ~60%].[698]

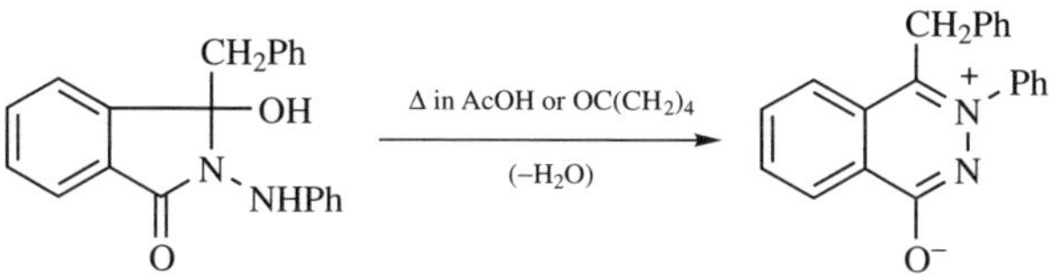

(275) $\xrightarrow[(-H_2O)]{\Delta \text{ in AcOH or } OC(CH_2)_4}$ (276)

2-Anilino-1,3-isoindolinedione (**277**, R = H) gave 2-phenylphthalazin-2-ium-4-olate (**278**, R = H) ($NaBH_4$, THF, H_2O, 20°C, 2 h: 75%);[696] 6,7-dichloro-(**278**, R = Cl) (65%) and 6,7-diiodo-2-phenylphthalazin-2-ium-4-olate (**278**, R = I) (73%) were made similarly.[594]

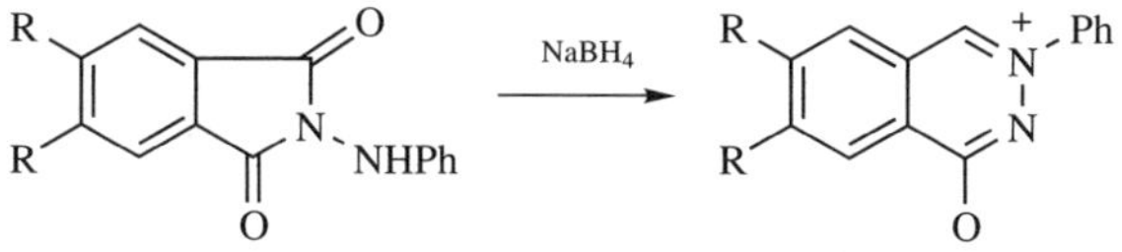

(277) $\xrightarrow{NaBH_4}$ (278)

2-(3,4-Dimethoxybenzylideneamino)-1,3-isoindolinedione (**279**) gave 4-phenyl-1(2*H*)-phthalazinone (**280**) (PhH, $AlCl_3$, 20°C, 16 h: 68%; or PhMgBr, Et_2O, PhH, reflux, 6 h: 57%).[669,cf. 357,537]

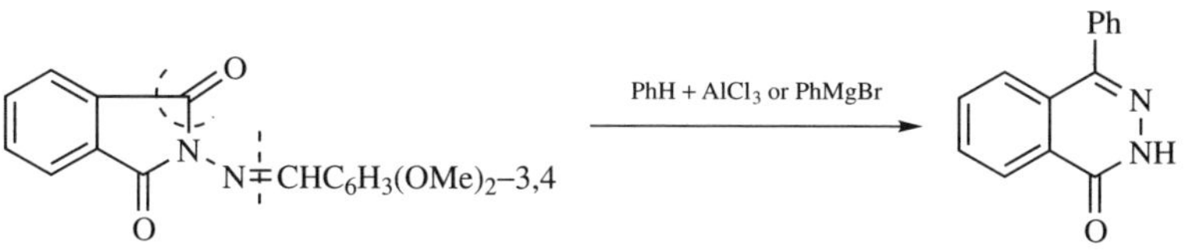

(279) $\xrightarrow{PhH + AlCl_3 \text{ or } PhMgBr}$ (280)

2-Amino-1,3-isoindolinedione (**281**) with phenylhydrazine gave 1,4(2*H*,3*H*)-phthalazinedione (**282**) (EtOH, reflux, 5 min; then 20°C, 12 h: 75%; note absence of the Ph group in the product),[653] with benzene (under

Friedel–Crafts conditions) gave 4-phenyl-1(2*H*)-phthalazinone (**283**) (substrate, PhH, ~5°C, $AlCl_3$↓ portionwise; then 20°C, 1 h: then reflux, 4 h: 77%; analogs likewise),[296] or with benzylmagnesium halide gave 4-benzyl-1(2*H*)-phthalazinone (**284**) (Et_2O, PhH, reflux, 5 h: 52%).[296]

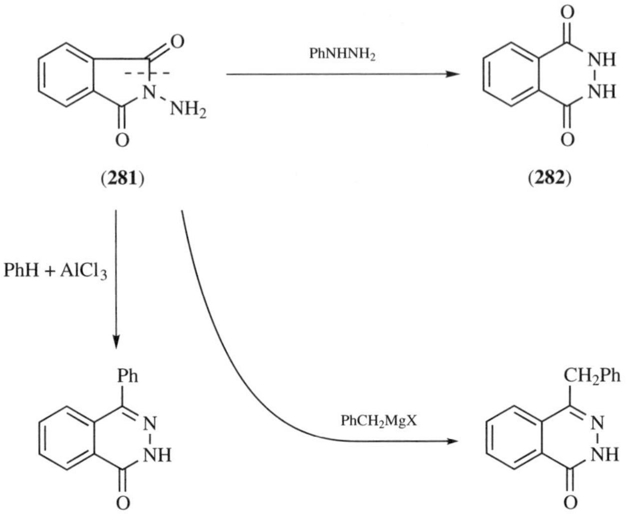

(**281**) —$PhNHNH_2$→ (**282**)

(**281**) —PhH + $AlCl_3$→ (**283**)

(**281**) —$PhCH_2MgX$→ (**284**)

Also other examples.[102,198,267,507,523,549,653,778]

From 2-Azimino-1,3-isoindolinediones (Phthalimidoazimines)

Note: The nomenclature of these substrates is confusing; it seems best to formulate them with sesqui bonds between the three exocyclic nitrogen atoms as in formula 285.

2-(2,3-Dimethylazimino)-1,3-isoindolinedione (**285**, R = Me) gave 2,3-dimethyl-1,4(2*H*,3*H*)-phthalazinedione (**286**, R = Me) (*trans*-substrate, $CHCl_3$, 50°C, 26 days: 45%; *cis*-substrate, likewise but 7 days: 81%).[149]

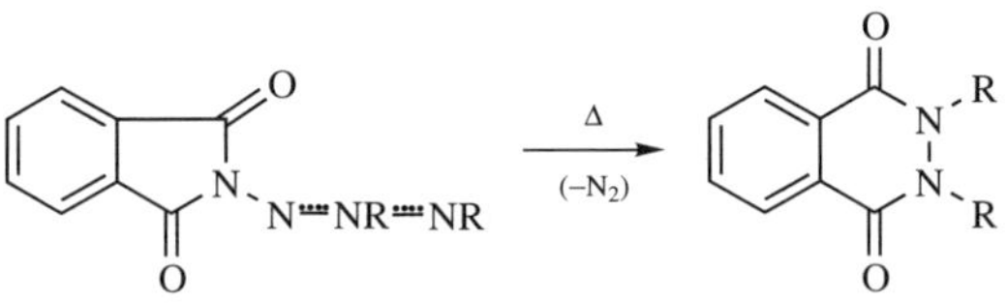

(**285**) —Δ ($-N_2$)→ (**286**)

2-(2,3-Diphenylaziminio)-1,3-isoindolinedione (**285**, R = Ph) gave 2,3-diphenyl-1,4(2*H*,3*H*)-phthalazinedione (**286**, R = Ph) (*cis*-substrate, $CHCl_3$, reflux, 30 h: ~80%).[149,150]

Many analogs were prepared similarly, some from mixtures of *cis*- and *trans*-substrates.[153,154,448,451,457] A mechanism has been proposed.[153]

8.6.8. Isoquinolines as Substrates

The conversion of an isoquinoline to a phthalazine is represented by only one example; 3-oxo-2-phenyl-2,3-dihydro-1,4-isoquinolinequinone (**287**) in ethanolic hydrazine hydrate at 20°C during 2 h afforded 4-oxo-3 4-dihydro-1-phthalazine-carbohydrazide (**288**) but in only ~9% yield after separation from an isoindolinone; the mechanism remains unclear.[484]

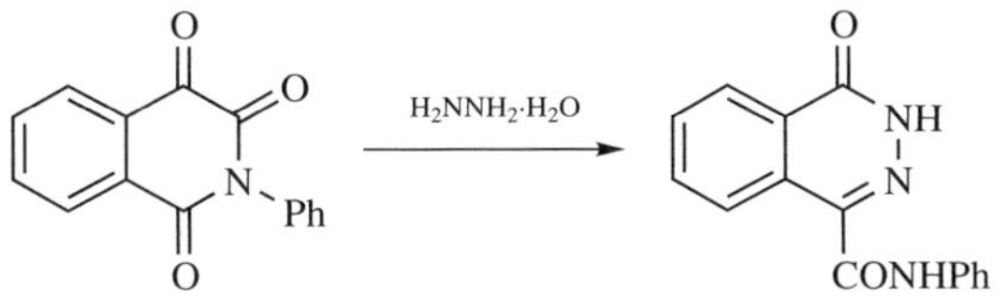

$H_2NNH_2{\cdot}H_2O$

(**287**) (**288**)

8.6.9. Pyridazino[4,5-*d*]pyridazines as Substrates

The conversion of these substrates into phthalazines clearly has synthetic potential. For example, 4-phenylpyridazino[4,5-*d*]pyridazin-1(2*H*)-one (**290**) and *N*-(1-ethylprop-1-enyl)pyrrolidine (**289**) in refluxing dioxane for 1 h gave a mixture of the Diels–Alder adducts (**291**) and (**292**) that lost nitrogen and pyrrolidine on refluxing in dilute propanolic acetic acid during 24 h to afford a 1:2 mixture (61%) of 7-ethyl-6-methyl-4-phenyl-1(2*H*)-phthalazinone (**293**) and 6-ethyl-7-methyl-4-phenyl-1(2*H*)-phthalazinone (**294**), from which only the latter could be obtained in a pure state;[598] analogs similarly.[300]

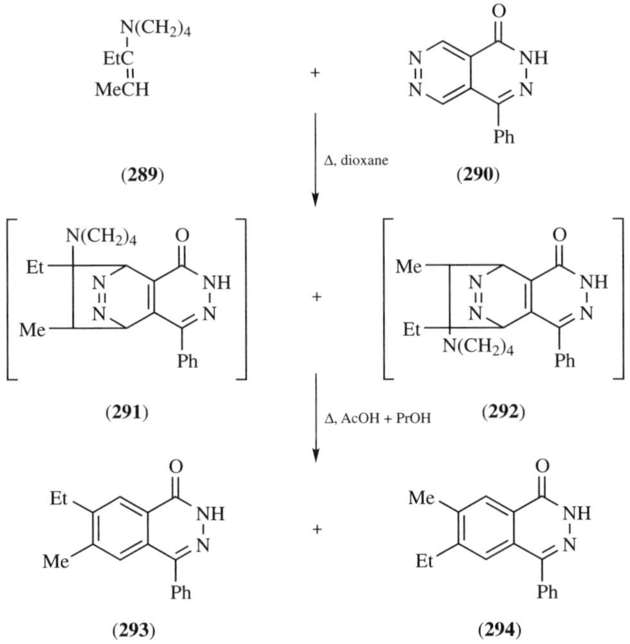

Δ, dioxane

(**289**) (**290**)

Δ, AcOH + PrOH

(**291**) (**292**)

(**293**) (**294**)

8.6.10. Thieno[3,4-*d*]pyridazines as Substrates

These substrates have been used extensively with two-carbon synthons to make phthalazines, apparently via nonisolable cyclic adducts that are degraded spontaneously with or without loss of hydrogen sulfide. Some typical examples follow.

Ethyl 5-amino-4-oxo-3-*p*-tolyl-3,4-dihydrothieno [3,4-*d*] pyridazin-1-carboxylate (**295**) and β-dimethylaminopropiophenone (presumably converted by loss of dimethylamine into phenyl vinyl ketone under the reaction conditions) probably gave the intermediate adduct (**296**) that lost H_2S to afford ethyl 5-amino-7-benzoyl-4-oxo-3-*p*-tolyl-3,4-dihydro-1-phthalazinecarboxylate (**298**) (Me_2NCHO, trace AcOH, reflux, 2 h: 61%); the same substrate (**295**) with *p*-methoxy-β-nitrostyrene likewise gave ethyl 5-amino-7-*p*-methoxyphenyl-6-nitro-4-oxo-3-*p*-tolyl-3,4-dihydro-1-phthalazinecarboxylate (**297**) (80%); and many analogs were made similarly.[553]

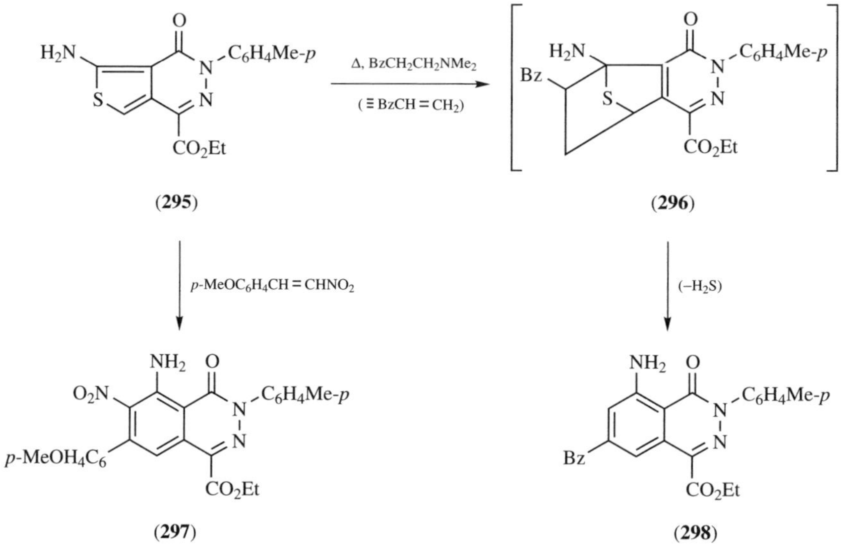

Δ, $BzCH_2CH_2NMe_2$ ($\equiv BzCH=CH_2$)

(**295**) (**296**)

p-$MeOC_6H_4CH=CHNO_2$ ($-H_2S$)

(**297**) (**298**)

In contrast, the closely related substrate, ethyl 5-amino-4-oxo-3-phenyl-3,4-dihydrothieno[3,4-*d*]pyridazine-1-carboxylate (**299**), with acrylonitrile gave ethyl 5-amino-6-cyano-4-oxo-3-phenyl-8-thioxo-2,3,4,8-tetrahydro-1-phtha-

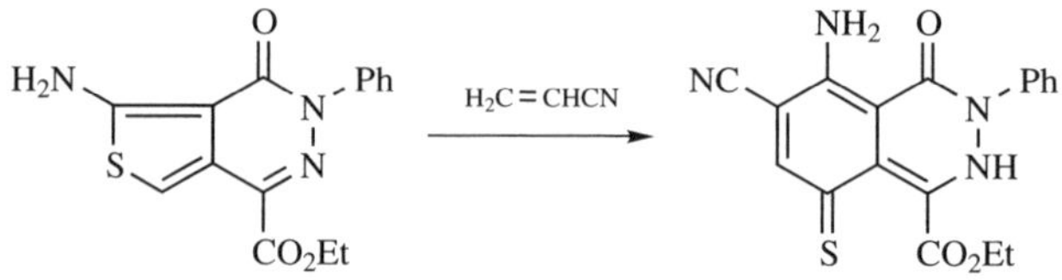

$H_2C=CHCN$

(**299**) (**300**)

lazinecarboxylate (**300**) by an alternative degradation of the intermediate adduct, not involving loss of sulfur (pyridine, reflux, 5 h: 74%).[796]

Also many other examples of both types.[306,379,380,490,520,530,547]

8.6.11. 1,2,3-Triazolo[4,5-*d*]pyridazines as Substrates

The conversion of these substrates into phthalazines is interesting rather than useful in its present state of development. The lead tetracetate oxidation of 4,7-diphenyl-1*H*-1,2,3-triazolo[4,5-*d*]pyridazin-1-amine (**301**) in the presence of 2,3,4,5-tetraphenyl-1-cyclopentadienone (**302**) gave 1,4,5,6,7,8-hexaphenylphthalazine (**303**) [reactants, CaO, CH_2Cl_2, $Pb(OAc)_4$↓ portionwise, 20°C, ? min: ~20%; see original with respect to mechanism].[691]

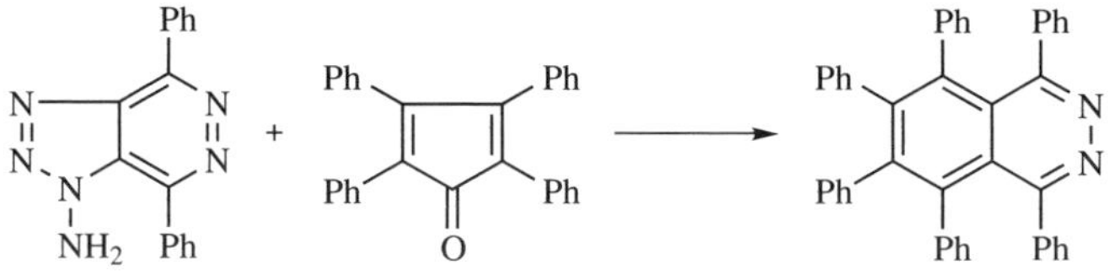

(**301**) (**302**) (**303**)

8.7. FROM HETEROPOLYCYCLIC DERIVATIVES AS SUBSTRATES

Derivatives of 16 heteropolycyclic systems have been used as substrates for the preparation of phthalazines, but none extensively. Accordingly, each such procedure is illustrated by one or more examples that are arranged alphabetically according to parent system used as substrate.

Benzo[1,2-*c*:4,5-*c*′]difurans (1,2,4,5-Benzenetetracarboxylic Anhydrides) as Substrates

5,7-Dihydro-1*H*,3*H*-benzo[1,2-*c*:4,5-*c*′]difuran-1,3,5,7-tetrone (**304**) gave 1,4-dioxo-1,2,3,4-tetrahydrophthalazine-6,7-dicarboxylic acid as its dihydrazinium salt (**305**) (substrate, $H_2NNH_2 \cdot H_2O$, Me_2NCHO, N_2, reflux, 12 h: ?%);[885] an X-ray analysis indicated that this salt existed in the solid state as its dihydrazinium 1-hydroxy-4-oxo-3,4-dihydro-6,7-phthalazinedicarboxylate tautomer.[885,cf. 105]

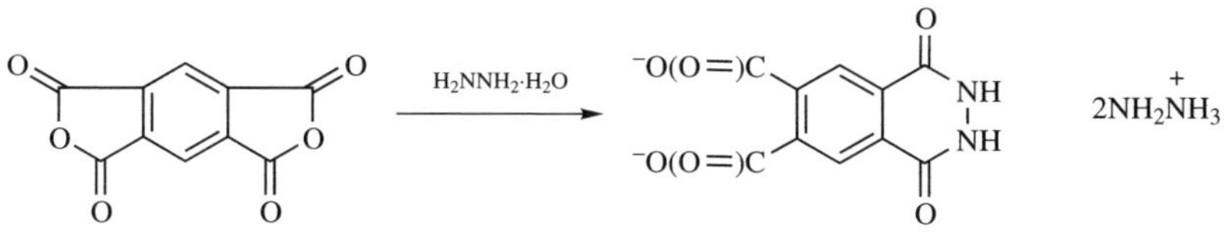

$H_2NNH_2 \cdot H_2O$

$2NH_2NH_3^+$

(**304**) (**305**)

Diazirino[1,2-*b*]phthalazines as Substrates

1,1-Dimethyl-3,8-dihydro-1*H*-diazirino[1,2-*b*]phthalazine-3,8-dione (**306**, Q = Me, R = H) gave 2-isopropenyl-1,4(2*H*,3*H*)-phthalazinedione (**307**, Q = Me, Me, R = H) (PhMe, reflux, 3 h: 76%); 1,1-diethyl-3,8-dihydro-1*H*-diazirino[1,2-*b*]phthalazine-3,8-dione (**306**, Q = Et, R = Me) gave 2-(1-ethylprop-1-enyl)-1,4(2*H*,3*H*)-phthalazinedione (**307**, Q = Et, R = Me) (likewise: 60%); and other homologs behaved similarly.[98]

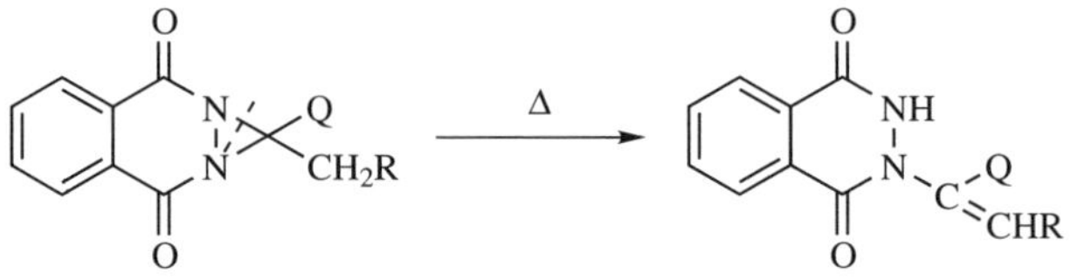

(**306**) (**307**)

5,8-Epidioxyphthalazines as Substrates

Dimethyl 6,7-dihydro-5,8-epidioxyphthalazine-1,4-dicarboxylate (**308**) underwent reductive fission of its 9,10-bond by thiourea to afford dimethyl 5,8-dioxo-2,3,5,6,7,8-hexahydro-1,4-phthalazinedicarboxylate (**309**) (MeOH, 20°C, 2 h: 56%) and thence 1,4-dimethoxycarbonyl-5,8-phthalazinequinone (**310**) (MnO_2, CH_2Cl_2, 20°C, 6 days: 61%).[975]

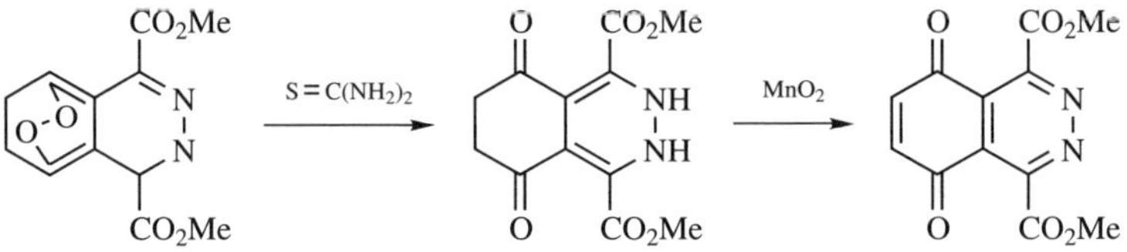

(**308**) (**309**) (**310**)

5,8-Epoxyphthalazines as Substrates

1,4-Diphenyl-5,8-dihydro-5,8-epoxyphthalazine (**311**) underwent vapor-phase pyrolysis to give 1,4-diphenyl-5(3*H*)-phthalazinone (**311a**) (vaporized at 200°C/0.05 mmHg into a tube at 550°C: 25% after purification).[691]

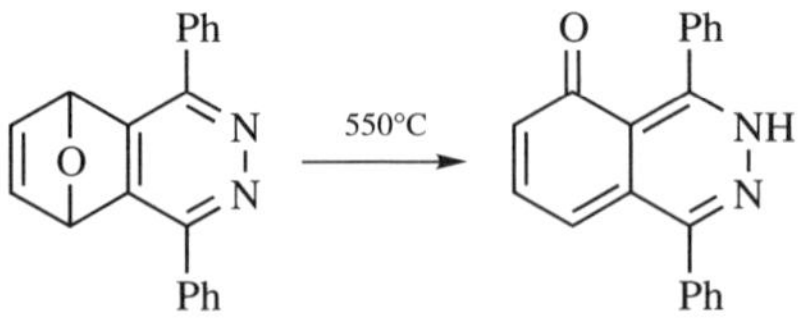

(**311**) (**311a**)

Furo- or Thieno[3,4-*g*]phthalazines as Substrates

1,4,6,8-Tetraphenylfuro[3,4-*g*]phthalazine (**312**, X = O) or 1,4,6,8-tetraphenylthieno[3,4-*g*]phthalazine (**312**, X = S) underwent oxidation to afford 6,7-dibenzoyl-1,4-diphenylphthalazine (**312a**) [HNO_3(*d* 1.38), AcOH, 20°C, 1 min: 80%).[16]

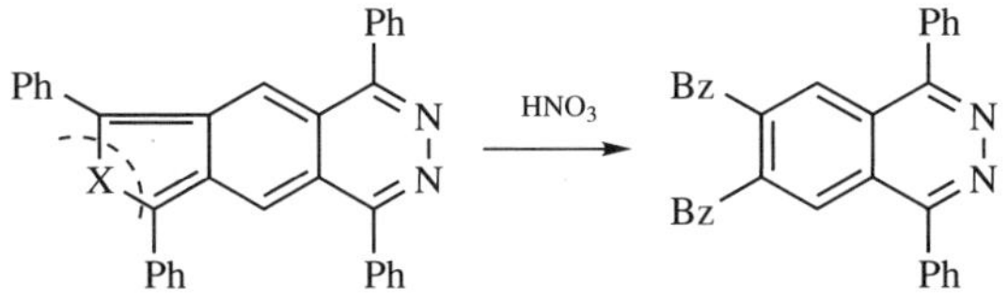

(**312**) (**312a**)

Imidazo[1,2-*b*]isoquinolines as Substrates

5,10-Dihydroimidazo[1,2-*b*]isoquinoline-5,10-dione (**313**) with hydrazine hydrate gave 4-(imidazol-2-yl)-1(2*H*)-phthalazinone (**314**, R = H) (neat reactants, reflux, 3 h: 87%) or with methylhydrazine gave 4-(imidazol-2-yl)-2-methyl-1(2*H*)-phthalazinone (**314**, R = Me) (likewise: 55%); analogs similarly.[99]

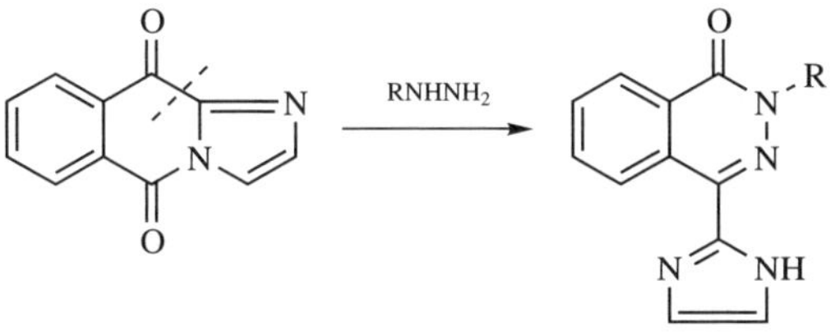

(**313**) (**314**)

Imidazo[2,1-*a*]phthalazines as Substrates

3-Dimethyliminio-2,2-dimethyl-2,3-dihydroimidazo[2,1-*a*]phthalazin-6-olate (**315**) underwent hydrolysis to 4-[1-(dimethylcarbamoyl)-1-methylethyl]-amino-1(2*H*)-phthalazinone (**316**) (Et_3N, H_2O, 20°C, 30 min: 96%).[151]

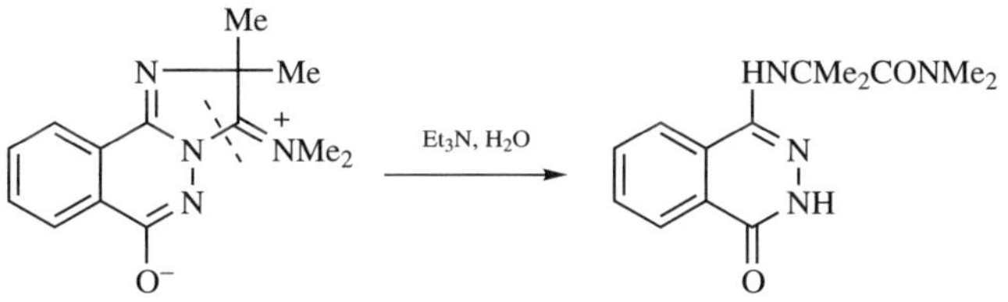

(**315**) (**316**)

2,2-Dimethyl-2,3-dihydroimidazo[2,1-*a*]phthalazine-3,6(5*H*)-dione (**317**) underwent reduction to 4-(2-hydroxy-1,1-dimethylethyl)amino-1 (2*H*)-phthalazinone (**318**) ($NaBH_4$, H_2O, EtOH, 65°C, 16 h: 59%) or alkaline hydrolysis to 4-(1-carboxy-1-methylethyl)amino-1(2*H*)-phthalazinone (**319**) (2M NaOH, 20°C, 1 h: >95%).[151]

Me
Me
N
N
O
NH
O
(**317**)
$NaBH_4$
$NHCMe_2CH_2OH$
N
NH
O
(**318**)
HO^-
$NHCMe_2CO_2H$
N
NH
O
(**319**)

Also other examples.[235,308]

4a,8a-Methanophthalazines as Substrates

1,4-Bis(trifluoromethyl)-4a,8a-methanophthalazine (**320**) underwent thermal carbene elimination to furnish 1,4-bis(trifluoromethyl)phthalazine (**320a**) (PhMe, A, reflux, 3 days: ?%).[536]

CF_3
N
N
CF_3
(**320**)
Δ
CF_3
N
N
CF_3
(**320a**)

Oxazolo[2,3-*a*]isoindoles as Substrates

2,3,5,9b-Tetrahydrooxazolo[2,3-*a*]isoindol-5-one (**321**, R = H) with hydrazine hydrate gave 1(2*H*)-phthalazinone (**321a**, R = H) (EtOH, reflux, 50 h: 48%); 9b-phenyl-2,3,5,9b-tetrahydrooxazolo[2,3-*a*]isoindol-5-one (**321**, R = Ph) with hydrazine hydrate likewise gave 4-phenyl-1(2*H*)-phthalazinone (**321a**, R = Ph) Me_2NCHO, reflux, 20 h: 49%); a rational mechanism was suggested.[707]

$H_2NNH_2 \cdot H_2O$

(**321**) (**321a**)

Oxazolo[2,3-*a*]phthalazin-4-iums as Substrates

2,3-Dihydrooxazolo[2,3-*a*]phthalazin-4-ium perchlorate (**322**) with piperidine gave 3-(2-piperidinoethyl)-1(2*H*)-phthalazinone (**322a**) (EtOH, reflux, 1 h: 81%); several *C*-substituted substrates behaved similarly.[298]

ClO_4^- $HN(CH_2)_5$ $CH_2CH_2N(CH_2)_5$

(**322**) (**322a**)

Phthalazino[2,3-*a*]cinnolines as Substrates

5,6,8,13-Tetrahydrophthalazino[2,3-*a*]cinnoline-6,8,13-trione (**323**) underwent alcoholytic fission of its 6,7-bond to give 2-[*o*-(methoxycarbonylmethyl) phenyl]-1,4(2*H*,3*H*)-phthalazinedione (**323a**) (H_2SO_4, MeOH, reflux, 3 h: 67%).[6]

H^+, MeOH

(**323**) (**323a**)

Phthalazino[2,3-*b*]phthalazines as Substrates

5,7,12,14-Tetrahydrophthalazino[2,3-*b*]phthalazine-5,12-dione (**324**, R = H) underwent oxidative ring fission at either N—CO bond to afford the same product, 2-*o*-carboxybenzyl-1(2*H*)-phthalazinone (**325**, R = H) (substrate, AcOH, H_2SO_4, H_2O; $NaNO_2/H_2O\downarrow$ dropwise, 0°C, 2 h; then 20°C, 12 h: ~85%; or substrate, Br_2, CCl_4, 20°C, 2 days: ~10%).[921]

In contrast, the unsymmetric substrate, 3-chloro-5,7,12,14-tetrahydrophthalazino[2,3-*b*]phthalazine-5,12-dione (**324**, R = Cl), underwent similar oxidative fission of one or other N—CO bond to furnish a separable mixture of 2-*o*-carboxybenzyl-7-chloro-1(2*H*)-phthalazinone (**325**, R = Cl) and 1-(2-carboxy-4-chlorobenzyl)-1(2*H*)-phthalazinone (**326**) (~20% and ~10%, respectively, after separation).[921]

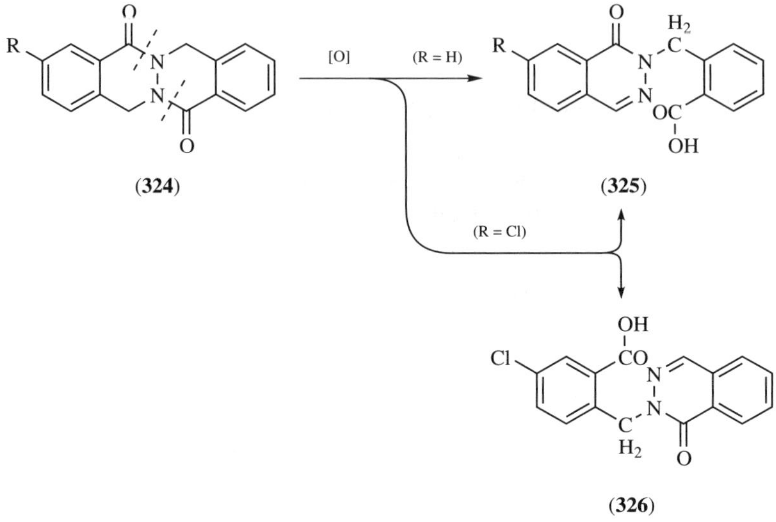

(**324**) (**325**) (**326**)

Also other examples.[254,921]

Pyrazolo[1,2-*b*]phthalazines as Substrates

2,3,5,10-Tetrahydro-1*H*-pyrazolo[1,2-*b*]phthalazine-1,5-dione (**327**) underwent oxidative ring fission to give 2-[α-(carboxymethyl)benzyl]-1(2*H*)-phthalazinone (**328**) (*N*-bromosuccinimide, CCl_4, ?°C, ? h: 42%; a mechanism was postulated).[73]

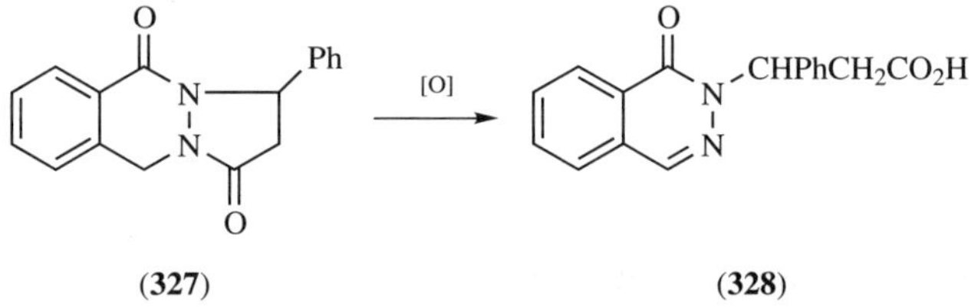

(**327**) (**328**)

Pyrimido[2,1-*a*]phthalazines as Substrates

3,4-Dihydro-2*H*-pyrimido[2,1-*a*]phthalazin-2-one (**329**) underwent alkaline hydrolysis to give 2-(2-carboxyethyl)-1(2*H*)-phthalazinone (**330**), presumably via the corresponding phthalazinimine or carboxamide (NaOH, H_2O, ?°C, ? h: ?%); homologs likewise.[946]

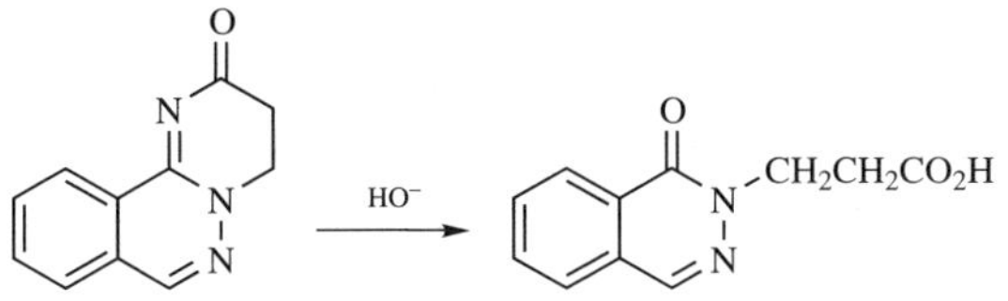

(329) (330)

3,4-Dihydro-2*H*-pyrimido[2,1-*a*]phthalazine-7(6*H*)-one (**331**) suffered acylative ring fission to give 2-(3-acetamidopropyl)-3-acetyl-1,4(2*H*,3*H*)-phthalazinedione (**332**) (Ac_2O, reflux; for further details, see original).[235]

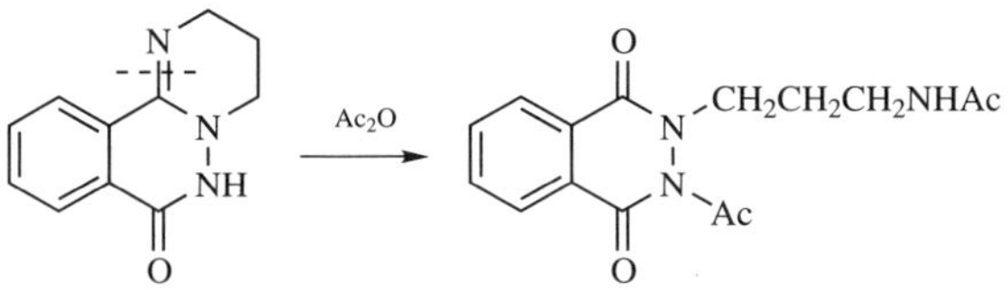

(331) (332)

8.8. FROM SPIRO HETEROCYCLIC SUBSTRATES

This type of synthesis is very poorly represented in the more recent literature by the following examples.

3′,8′-Dihydro{spiro[cyclohexane-1,1′-[1*H*]-diazirino[1,2-*b*]phthalazine]}-3′,8′-dione (**333**) underwent acidic hydrolysis with loss of cyclohexanone to afford 1,4(2*H*,3*H*)-phthalazinedione (**334**) HCl, MeOH, reflux, 90 min: ?%) or thermolytic isomerization to afford 2-(cyclohex-1-enyl)-1,4(2*H*,3*H*)-phthalazinedione (**335**) (PhMe, reflux, 2 h: 70%).[98]

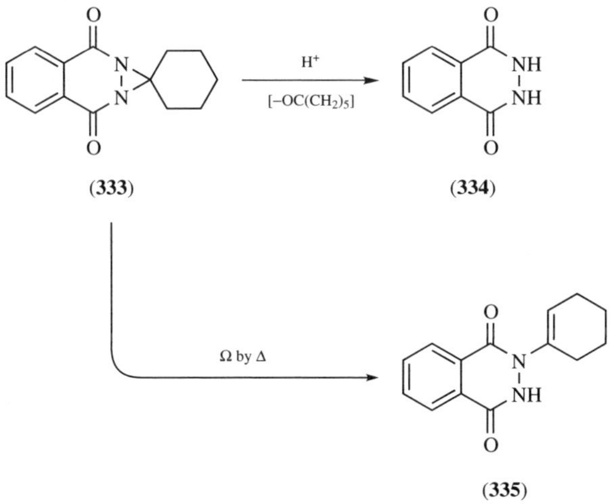

(333) (334)

(335)

In contrast, 2-hydroxy-3′,8′-dihydro{spiro[cyclohexane-1,1′-[1*H*]-diazirino[1,2-*b*]phthalazine]}-3′,8′-dione (**336**) underwent thermal or base-catalyzed isomerization to furnish 4-(2-oxocyclohexyloxy)-1(2*H*)-phthalazinone (**337**) (PhMe, reflux, 2 h: 55%; or Bu^tOK, Me_2SO, 20°C, 12 h: 85%;); a rational mechanism was proposed.[105]

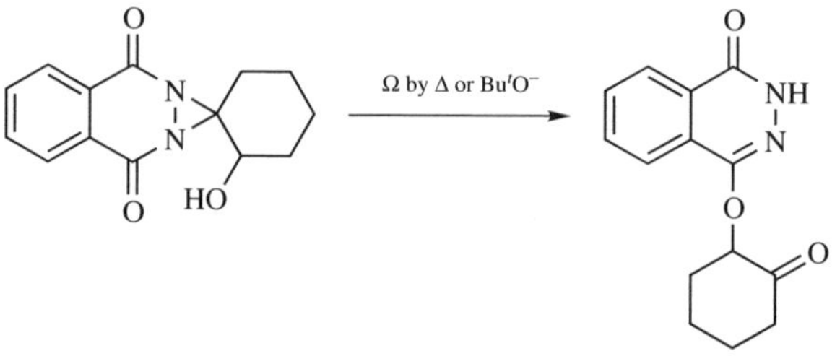

Ω by Δ or Bu^tO^-

(**336**) (**337**)

8.9. GLANCE INDEX TO TYPICAL PHTHALAZINE DERIVATIVES AVAILABLE BY PRIMARY SYNTHESES

This glance index may assist in the choice of a primary synthesis for a required phthalazine derivative; such syntheses are based on aliphatic, carbocyclic, or heterocyclic substrates with or without ancillary synthons. It should be remembered that products analogous to those formulated may often be obtained by quite minor changes to the substrate and/or synthon involved. Products arising from syntheses that appear to be of little preparative value are omitted from this index.

Section	Typical Products
8.1.1 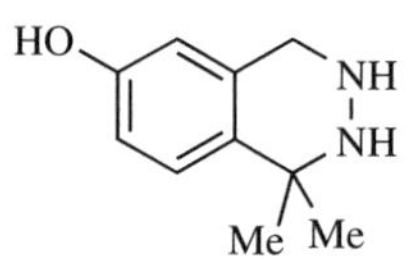 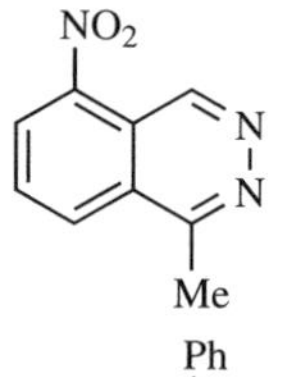	
8.1.2 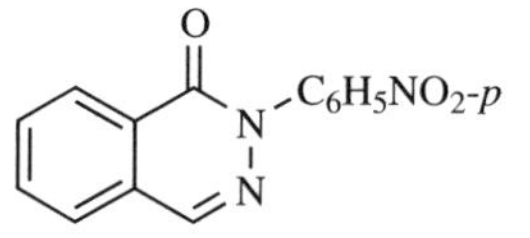 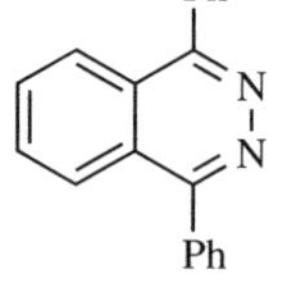	
8.2.1	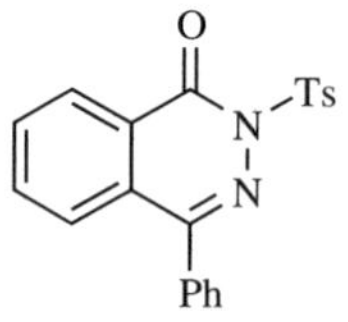

Section		Typical Products
8.2.2.1	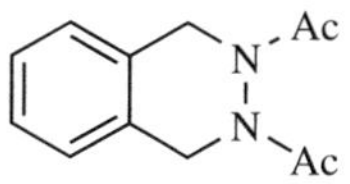 Ac, N, N, Ac	
8.2.2.2	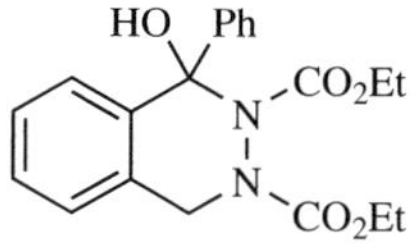 HO, Ph, CO_2Et, N, N, CO_2Et	
8.2.2.4	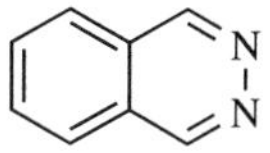 N, N	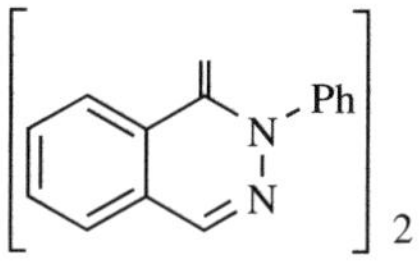 Ph, N, N, 2
8.2.2.5	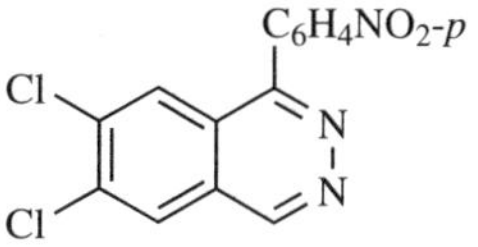 $C_6H_4NO_2$-*p*, Cl, N, N, Cl	
8.2.2.6	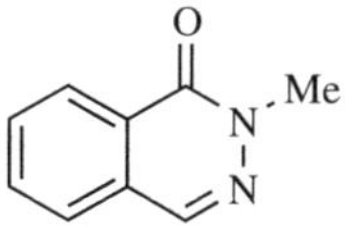 O, Me, N, N	
8.2.2.9	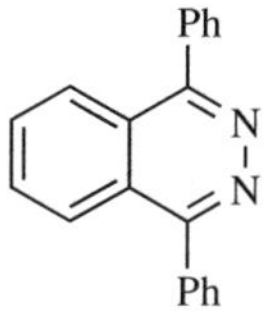 Ph, N, N, Ph	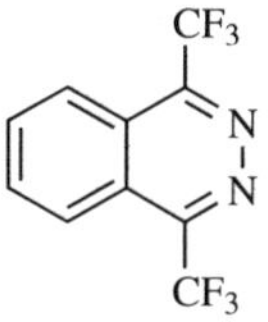 CF_3, N, N, CF_3
8.2.2.10	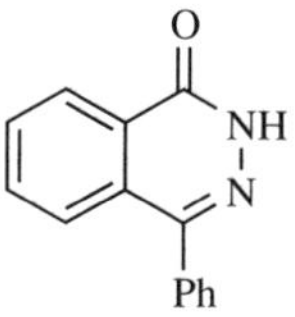 O, NH, N, Ph	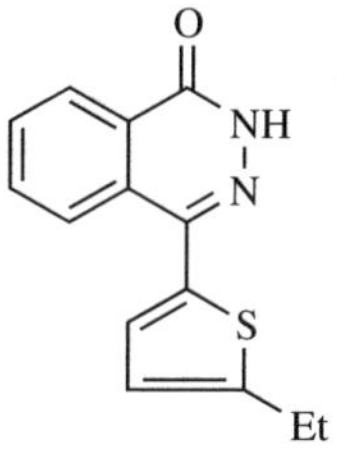 O, NH, N, S, Et
	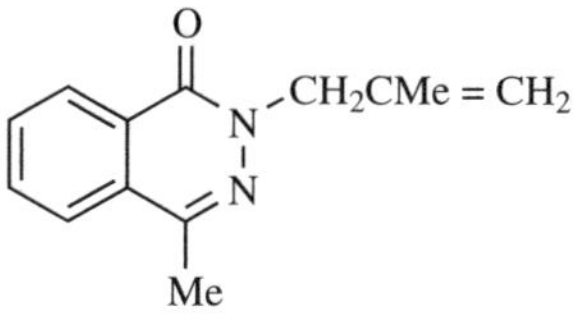 O, $CH_2CMe = CH_2$, N, N, Me	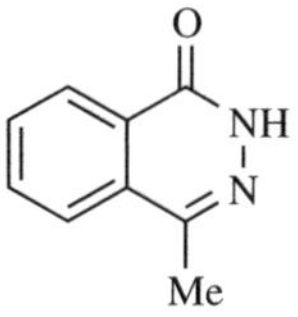 O, NH, N, Me
8.2.2.12	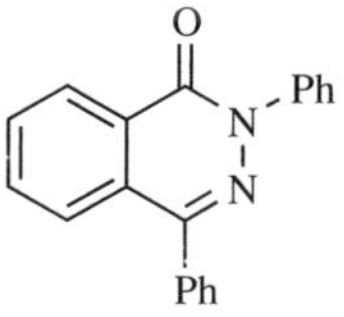 O, Ph, N, N, Ph	

Section	Typical Products
8.2.2.14	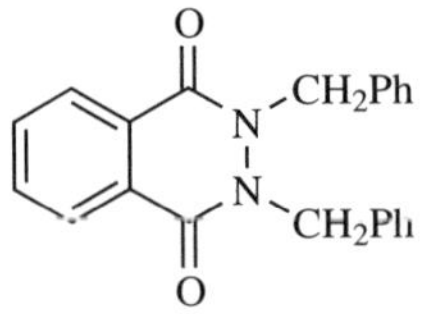
8.2.2.15 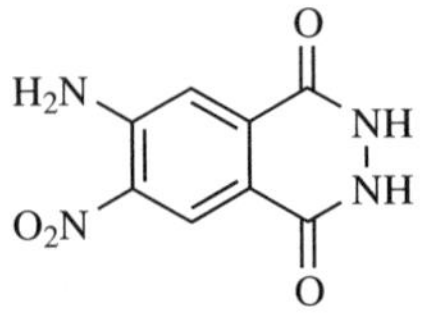 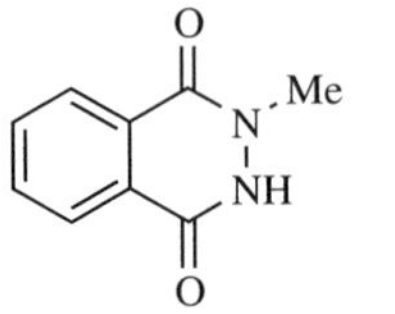	
8.2.2.16	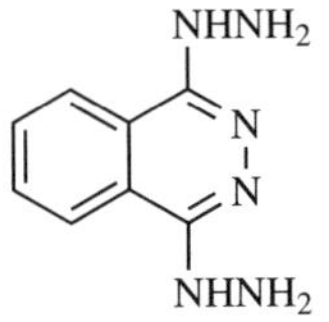
8.2.3	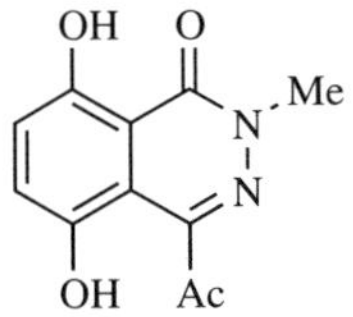
8.3 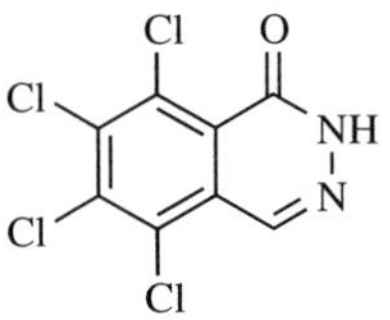 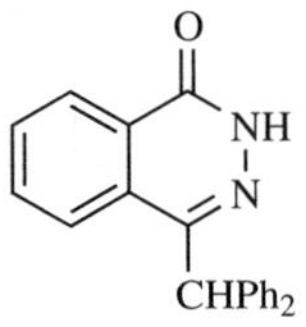	
8.4 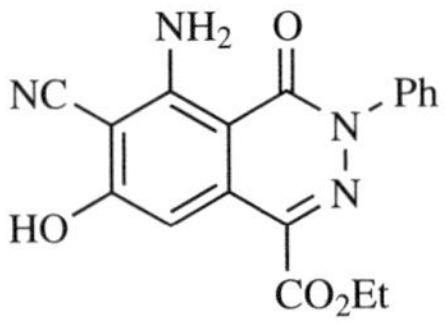 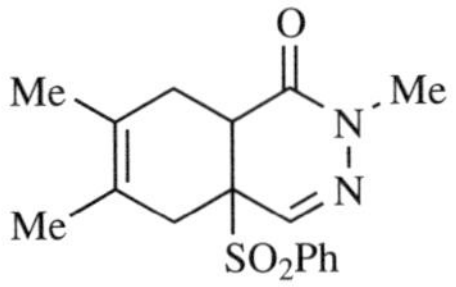	
8.5 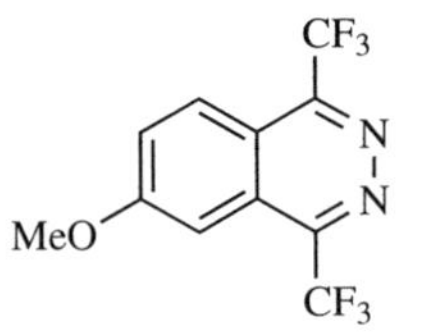 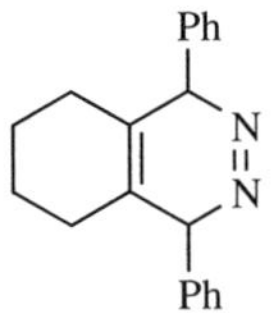	
8.6.2 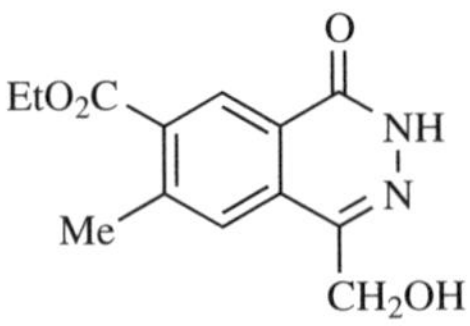 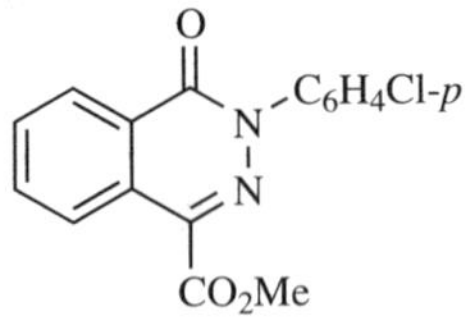	

Section		Typical Products
8.6.4	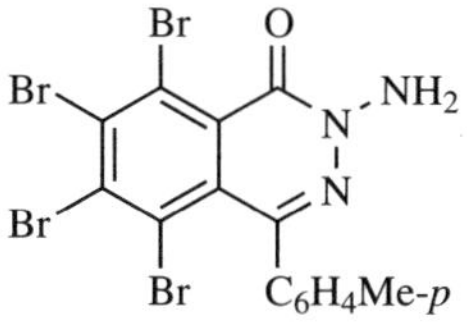	
8.6.6.1	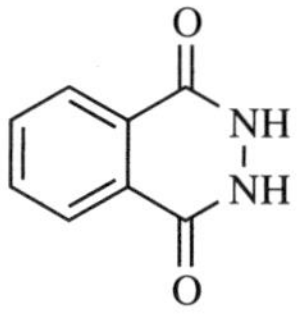	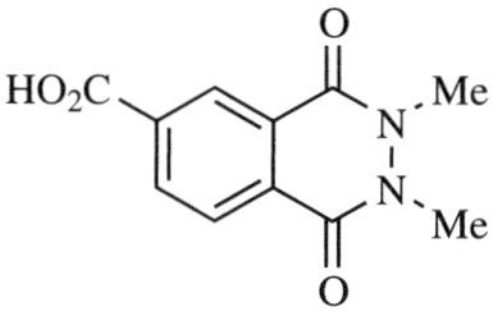
	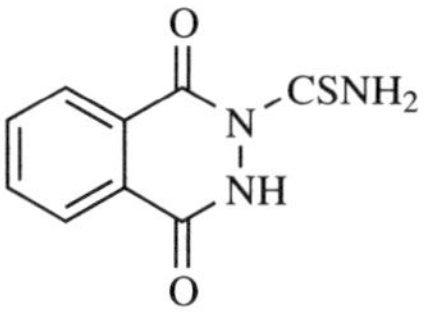	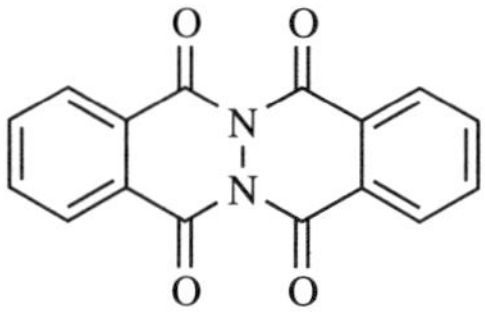
8.6.6.2	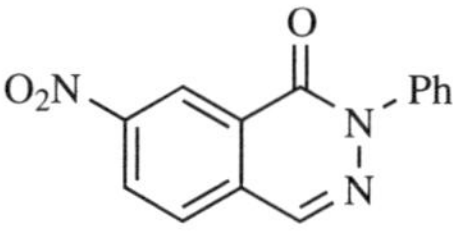	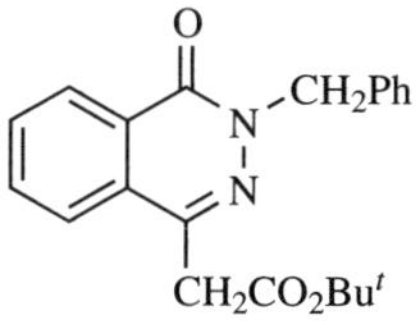
	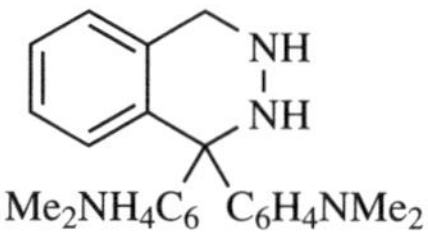	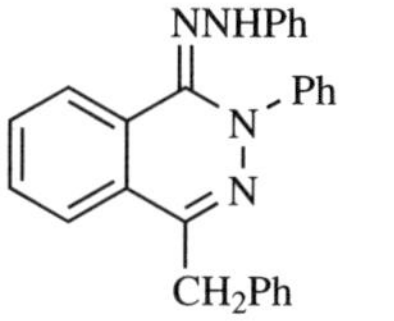
8.6.7.1	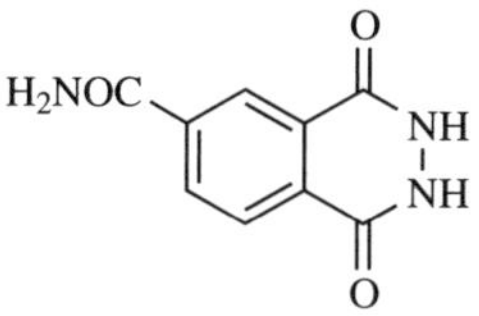	
8.6.7.2	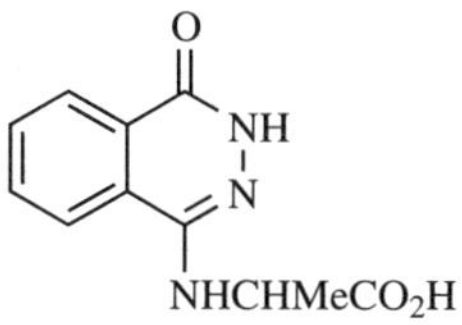	
8.6.7.3	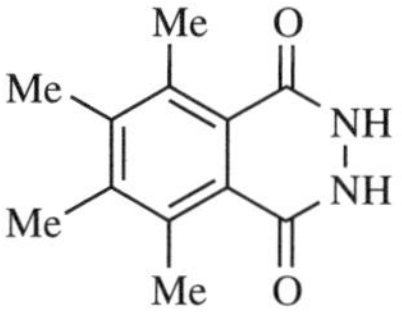	

Section		Typical Products
8.6.7.4	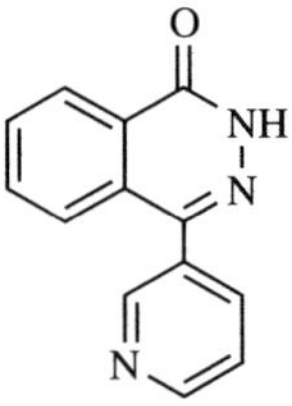	
8.6.7.5	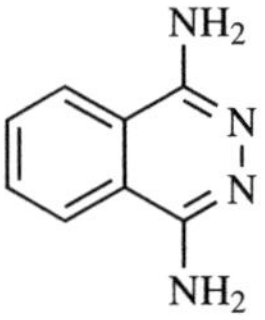	
8.6.7.6	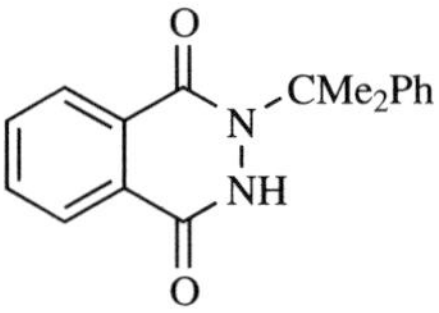	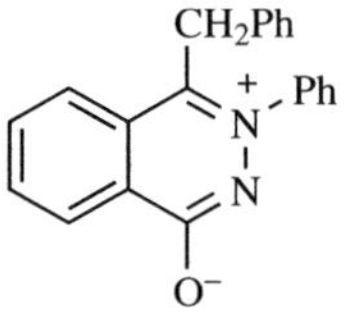
	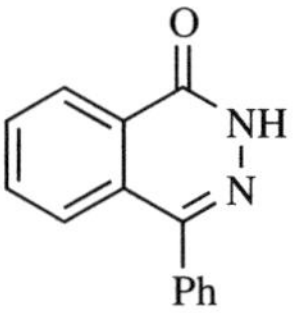	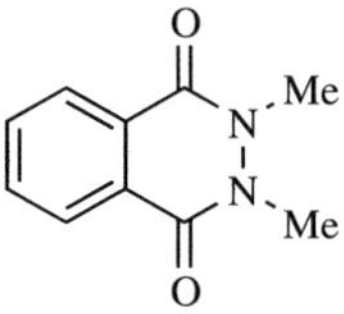
8.6.9	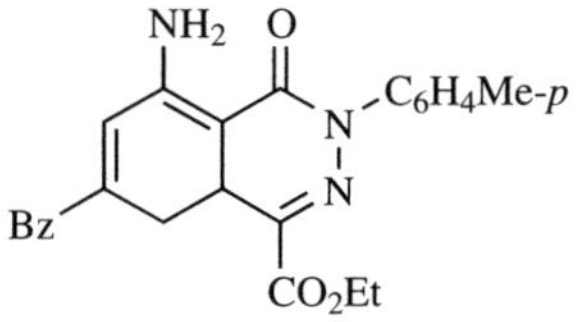	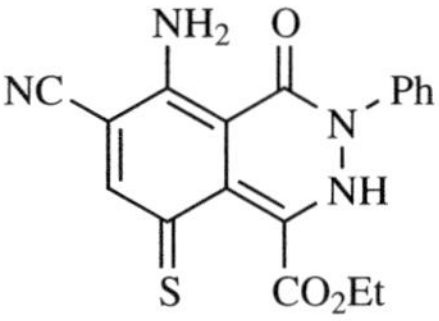
8.7	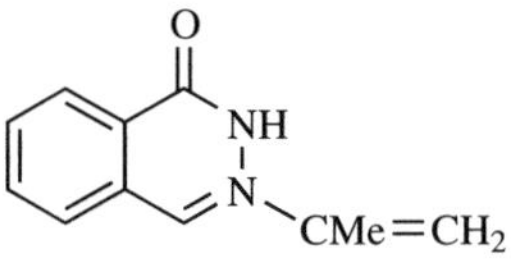	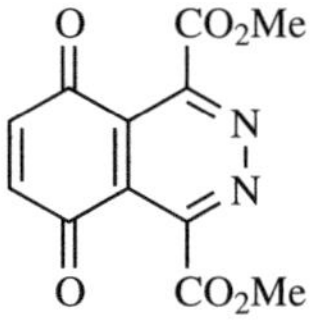
	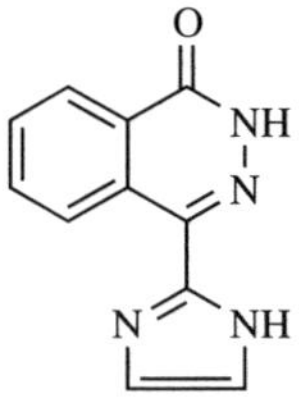	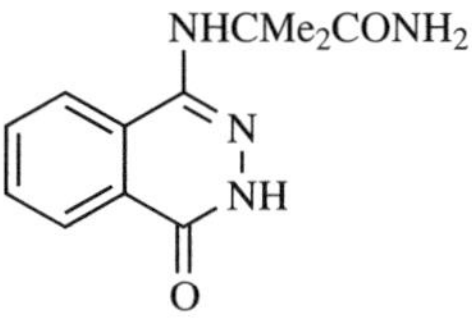
	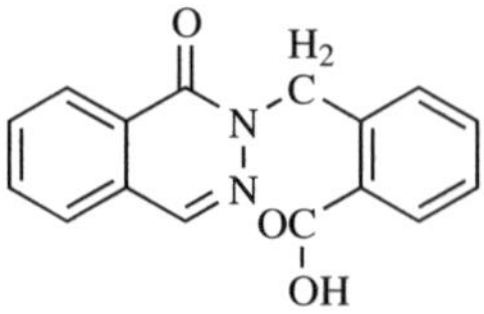	

Section	Typical Products
8.8	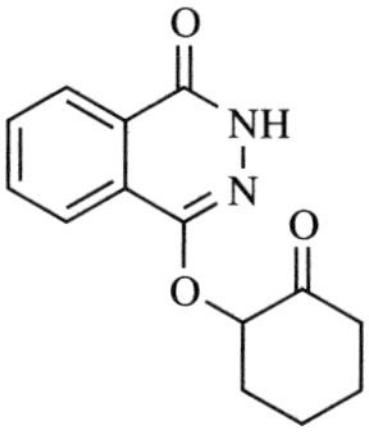 O NH N O O

CHAPTER 9

Phthalazine, Alkylphthalazines, and Arylphthalazines (*H* 69, 72; *E* 324, 338)

This chapter summarizes more recently reported information on the preparation, physical properties, and reactions of phthalazine and its *C*-alkyl, *C*-aryl, *N*-alkyl, and *N*-aryl derivatives as well as those of the nucleus-reduced analogs. It also includes methods for introducing alkyl or aryl groups (substituted or otherwise) into phthalazines already bearing substituents and reactions specific to the alkyl or aryl groups in such compounds. For brevity, the term *alkylphthalazine* includes alkyl, alkenyl, alkynyl, and aralkyl derivatives; likewise, *arylphthalazine* includes both regular and heteroaryl derivatives.

Since the publication of Simpson's volume[906] and Patel's update[907] in this series, several reviews of phthalazine chemistry have appeared,[181,552,903–905,908–916] among which that of Stanovnik[908] is particularly useful.

9.1. PHTHALAZINE (*H* 69; *E* 344)

9.1.1. Preparation of Phthalazine and Hydrophthalazines

Because phthalazine is reasonably inexpensive to purchase, there has been little incentive to develop new or improved synthetic procedures. However, a convenient primary synthesis has been reported (see Section 8.2.2.4); 1,4-dihydrazinophthalazine (**1**) has been oxidized to phthalazine (**2**) (NaOH, EtOH, H_2O, $O_2\downarrow$, 20°C, 3–4 h: ~70%),[275,770] and phthalazine may be recovered from its 2-methiodide by treatment with neat refluxing pyridine hydrochloride.[90]

NHNH$_2$ N N NHNH$_2$ —[O] (−2N$_2$)→ N N

(**1**) (**2**)

Cinnolines and Phthalazines: Supplement II, The Chemistry of Heterocyclic Compounds, Volume 64, by D.J. Brown

Unsubstituted hydrophthalazines have been made more recently by procedures that vary widely in utility, as illustrated by the following examples.

Di-*tert*-Butyl 1,2,3,4-tetrahydro-2,3-phthalazinedicarboxylate (**3**) (made by primary synthesis) gave the hydrochloride of 1,2,3,4-tetrahydrophthalazine ($MeNO_2$, HCl gas↓, 10 min: 98%)[1001] from which the unstable free base (**4**) (42%) was obtained eventually in a pure state.[318]

CO_2Bu^t CO_2Bu^t HCl gas, $MeNO_2$ NH NH

(**3**) (**4**)

Phthalazine (**5**) underwent reduction by the borane–pyridine complex in acetic acid to give 2-acetyl-1,2,3,4-tetrahydrophthalazine (**6**) (reflux, 5 min: 57%),[332] conceivably a potential source of 1,2,3,4-tetrahydrophthalazine.

pyridine·BH_3, AcOH

(**5**) (**6**)

1,4-Dichlorophthalazine (**7**) underwent both hydrogenolysis and hydrogenation to afford 5,6,7,8-tetrahydrophthalazine as its hydrochloride (**8**) [Pd/C, MeOH, H_2 (3 atm), 72 h: "high yield"].[514,cf. 393]

Pd/C, H_2 +HCL

(**7**) (**8**)

Phthalazine (**9**) was reduced by triiron dodecacarbonyl to give 1,2-dihydrophthalazine (**10**) [$Fe_3(CO)_{12}$, MeOH, PhH, reflux, ~12 h: 54%]; the same substrate (**9**) underwent photoreduction to give a separable mixture of 1,2-dihydrophthalazine (**10**) and 1,1′,2,2′-tetrahydro-1,1′-biphthalazine (**11**) (Pr^iOH, *h*ν, 5–8 h: yields ?);[606,614] and the electrochemical reduction of phthalazine has been studied.[725]

$Fe_3(CO)_{12}$

(**9**) (**10**)

Pr^iOH, *h*ν

(**11**)

1,2,3,4-Tetrahydrophthalazine (**12**) has been oxidized to 1,4-dihydrophthalazine (**13**), but the methods appear to be of marginal preparative value.[77,314,318,320]

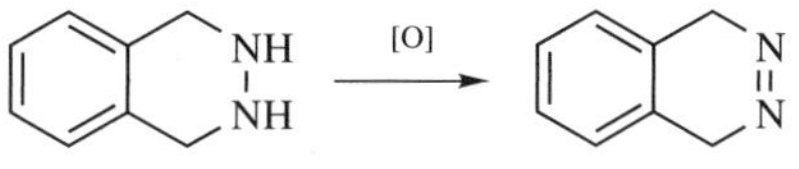

(**12**) (**13**)

9.1.2. Physical Properties of Phthalazine

New or revised physical data for phthalazine and its salts or complexes may be found under "phthalazine" in the Appendix (Table A.2) at the end of this book. More extended studies of such properties are noted in the following list.

Electron Spin Resonance. ESR and CIDEP (chemically induced electron spin polarization) studies have been reported for radicals derived from phthalazine and related diazanaphthalenes.[806]

Energy Determinations. Standard molar enthalpies of formation for phthalazine and related benzodiazines in the gaseous state at 298 K have been derived from experimental data for combustion and sublimation of the solids.[683] The HOMO (highest occupied molecular orbital) values for phthalazine and related heterocycles have been used to calculate their pK_a values.[813]

Nuclear Magnetic Resonance Spectra. In contrast to the behavior of regular aromatic compounds, the ^{1}HNMR solvent shifts for phthalazine and other heteroaromatic molecules showed no correlation with reactivity parameters for simple compounds.[613] The ^{15}NNMR spectrum of phthalazine has been discussed in the context of other diazine spectra.[152]

Ultraviolet Spectra. A comparison of UV spectra for phthalazines and related systems has been prepared and discussed in a comparative way.[84] The electronic spectra for phthalazine and other azanaphthalenes have been calculated by a modified INDO (intermediate neglect of differential overlap) method.[798]

9.1.3. Reactions of Phthalazine

Unsubstituted phthalazine has been used extensively as a substrate for a variety of reactions. Simple *reduction* has been covered in Section 9.1.1, and other reactions are illustrated by the following examples, grouped alphabetically for convenience.

N-Acylation (Reductive)

Phthalazine (**14**) with benzyl chloroformate and sodium cyanoborohydride gave an easily separable mixture of 2-benzyloxycarbonyl- (**15**, R = H) and

2,3-dibenzyloxycarbonyl-1,2,3,4-tetrahydrophthalazine (**15**, R = CO_2CH_2Ph) (MeOH, N_2, 20°C, 16 h: ~55% and ~40%, respectively).[713]

$ClCO_2CH_2Ph$, $NaB(CN)H_3$

(**14**) (**15**)

See also entries on *Reissert* and *Reissert-like reactions* later in this section.

C-Alkylation or Arylation

Note: Such alkylation or arylation occurs in the pyridazine ring by initial 1,2-addition and subsequent oxidation of the adduct; 1,4-dialkylation may occur under appropriate conditions.

Phthalazine (**16**) and phenylmagnesium bromide gave crude 1-phenyl-1,2-dihydrophthalazine [PhMgBr (made in situ), THF, N_2, reflux, 6 h, then 20°C, 12 h: crude (**17**)] that underwent oxidation to give 1-phenylphthalazine (**18**) (MnO_2, PhH, reflux, 8 h: 89% overall); subsequent methylation of this product (**18**) in an analogous way afforded 1-methyl-4-phenylphthalazine (**20**) (MeMgI, Pr^i_2O, reflux, N_2, 17 h; then intermediate, MnO_2, PhH, reflux, 6 h: 50% overall).[290]

Phthalazine (**16**) underwent addition of methanol under irradiation and subsequent dehydration to give 1-methylphthalazine (**19**) (HCl gas in methanol, N_2, reflux, $h\nu$, 3 h: 50%).[610]

PhMgBr MnO_2

(**16**) (**17**) (**18**)

HCl gas, MeOH, $h\nu$

(**19**) (**20**)

Phthalazine (**21**) and perfluorohexyllithium (generated in situ) gave a separable mixture of 1-perfluorohexyl-1,2-dihydrophthalazine (**22**) and 1,4-bis(perfluorohexyl)-1,2,3,4-tetrahydrophthalazine (**23**) [substrate (**21**), $C_6F_{13}I$ (2 mol), $BF_3 \cdot Et_2O$, Et_2O, −78°C; MeLi·LiBr↓ slowly, −78°C, 75 min: 71% and 16%, respectively].[608]

Phthalazine (**21**) and the half adamantyl ester of oxalic acid gave 1-(adamant-1-yl)phthalazine (**24**) [reactants, $(F_3CCO_2)_2IPh$ (as catalyst), PhH, reflux, ? h: 33%; mechanism discussed].[609]

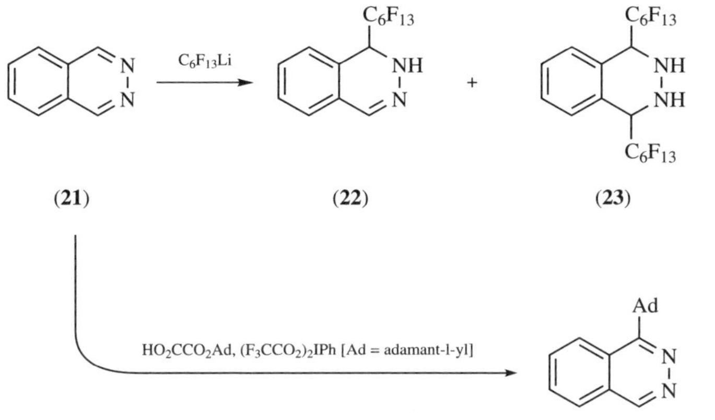

N-Alkylation (Reductive)

Note: *N*-Alkylation of phthalazine can occur only by reduction of at least one C=N bond or by quaternization (exemplified later in this section).

Phthalazine (**25**) underwent reductive formylation to the intermediate (**26**) followed by further reduction to afford 2,3-dimethyl-1,2,3,4-tetrahydrophthalazine (**27**) (substrate, HCO_2H, KBH_4↓ slowly, <15°C; then 0°C, 1 h; then slowly to reflux, 6 h: 97%); use of acetic or propionic acid gave 2,3-diethyl- or 2,3-dipropyl-1,2,3,4-tetrahydrophthalazine, respectively, but in lower yields.[19]

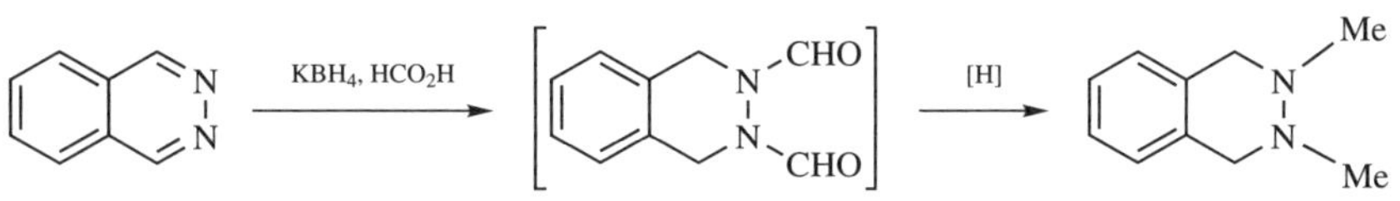

Complex Formation

Phthalazine:phthalic acid cocrystals (2 : 1 and 2 : 3) have been prepared and confirmed in structure by X-ray analyses.[561]

The platinum derivative, *cis*-$[Pt(NH_3)_2Cl(dmf)][NO_3]$, with phthalazine gave the complex (**28**) (Me_2NCHO, 20°C, 12 h: 30%), somewhat akin to cisplatin in structure and anticancer activity.[989]

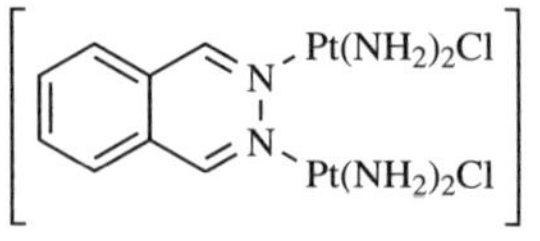

(**28**)

Diiron nonacarbonyl and phthalazine gave a separable mixture of two complexes, formulated as (**29**) and (**30**) (PhH, 20°C, 36 h: 58% and 26%, respectively.[748]

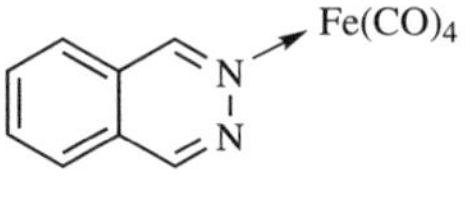

(**29**) 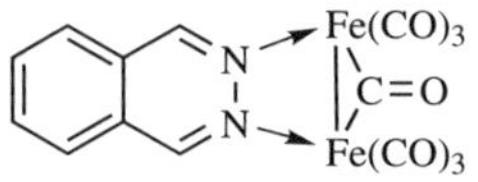 (**30**)

Interesting complexes of phthalazine with copper(II);[867] nickel(II),[1031] tungsten (II),[826,988] ytterbium(III),[866] or zinc(II)[1029] have been studied.

Cyclic Adduct Formation

Note: The formation of such cyclic adducts has not been studied systematically, but diverse examples have been reported.

Phthalazine (**31**) with maleic anhydride (**32**) (2 mol) gave the adduct (**33**), eventually confirmed in structure (reactants, PhMe, reflux, 1 h: 62%).[14,cf. 788 and references cited therein]

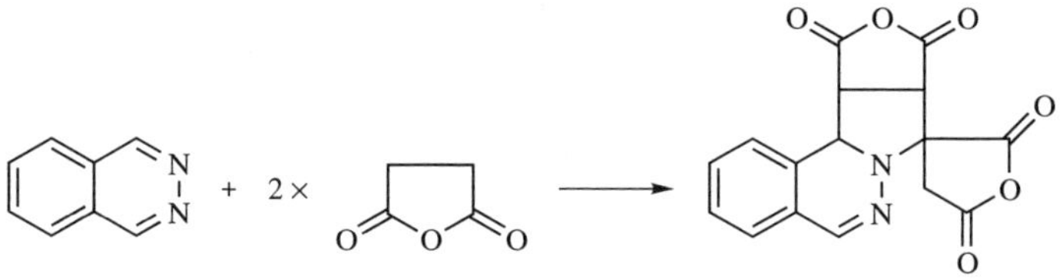

(**31**) (**32**) (**33**)

Phthalazine (**34**) with 2-hydroxy-5-nitrocyclohepta-2,4,6-trien-1-one (**35**) gave the tetracyclic adduct (**36**) (reactants, neat Ac_2O, 20°C, 72 h: 77%).[998]

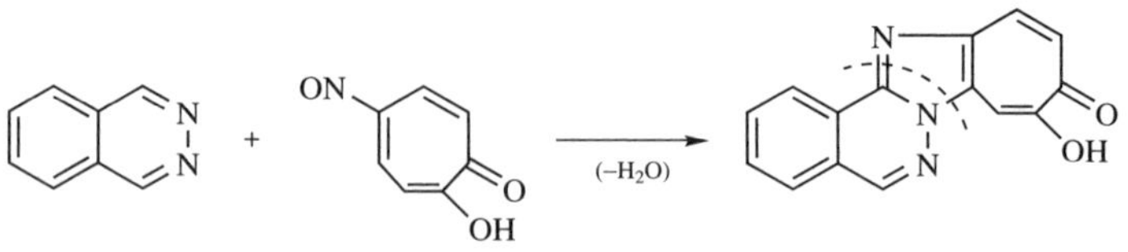

(**34**) (**35**) (**36**)

Phthalazine (**37**) with chloromethyl trimethylsilylmethyl sulfide gave 2-[(trimethylsilylmethylthio)methyl]phthalazinium chloride (**38**) (reactants, MeCN, 60°C, 1 h: 96%) and thence 1,10b-dihydro-3*H*-thiazolo[4,3-*a*]phthalazine (**39**) (CsF, MeCN, 20°C, 48 h: 92%); the reaction may be done as a one-pot procedure without isolation of the intermediate.[115]

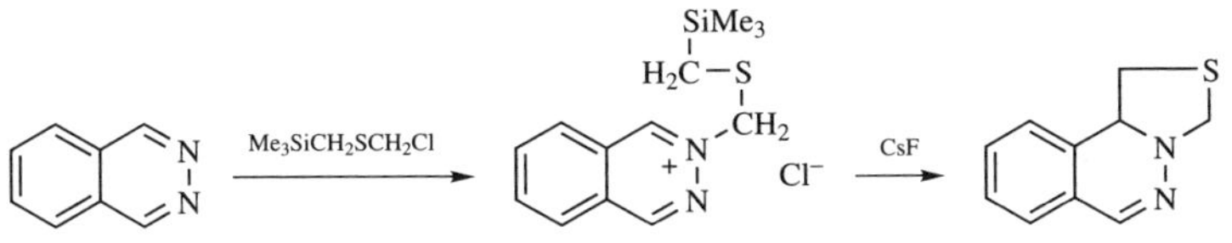

$Me_3SiCH_2SCH_2Cl$; CsF

(**37**) (**38**) (**39**)

Also other examples.[478,1022]

Degradative Reactions

Phthalazine (**40**) underwent hydrogenation to afford *o*-bis(aminomethyl)benzene (**41**) [Pd/C, H_2 (~5 atm), MeOH, 20°C, 18 h; then Raney Ni↓, H_2 (5 atm), 50°C, 5 h: 55%].[640]

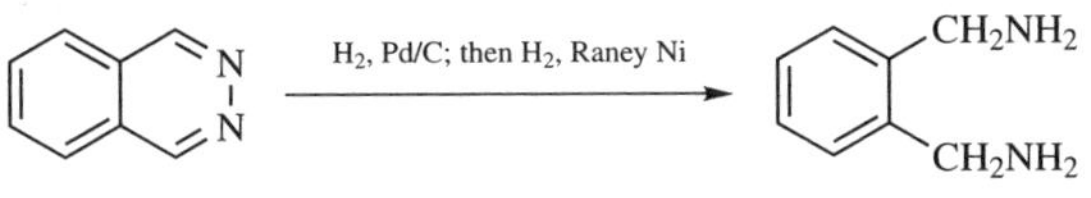

H_2, Pd/C; then H_2, Raney Ni

(**40**) (**41**)

Oxidation of freshly liberated 1,2,3,4-tetrahydrophthalazine (**42**) by mercury(II) oxide in the presence of sulfur dioxide gave an (unseparated) mixture of 1,4-dihydro-2,3-benzoxathiin 3-oxide (**43**) and 1,3-dihydrobenzo[*c*]thiophene 2,2-dioxide (**44**) (substrate, CH_2Cl_2, −20°C, SO_2↓, then HgO↓, 20°C, 12 h: 45% of a 9 : 1 mixture in which the compounds were clearly identified).[77]

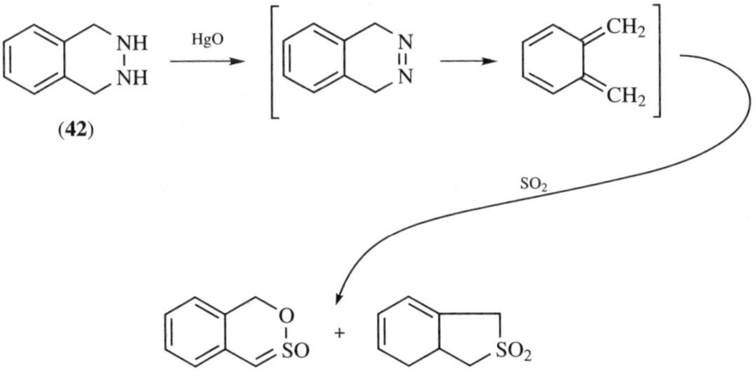

HgO ; SO_2

(**42**) (**43**) (**44**)

The Kjeldahl estimation of nitrogen in phthalazine and related heterocycles has been improved.[481]

Also other examples.[820,843,852]

Deuteration

Phthalazine suffered 80%–complete deuteration in the presence of a special palladium-on-asbestos catalyst (D_2O, Pd/asbestos, N_2, 220°C, sealed, 72 h; preparation of the catalyst and the individual percentage of deuteration at each position of phthalazine are reported).[329]

C-Hydroxylation

Phthalazine (**45**) with benzonitrile oxide gave 1(2*H*)-phthalazinone (**47**) via the unisolated intermediate (**46**) (reactants, PhH, reflux, 3 h: 25%).[305]

Biological oxidation of phthalazine (**45**) also gave 1(2*H*)-phthalazinone (**47**) [*Streptomyces viridosporus* T7a culture, phthalazine, 37°C, 3 days: 46%;[560] or aldehyde oxidase (from human liver): a physicochemical study].[842]

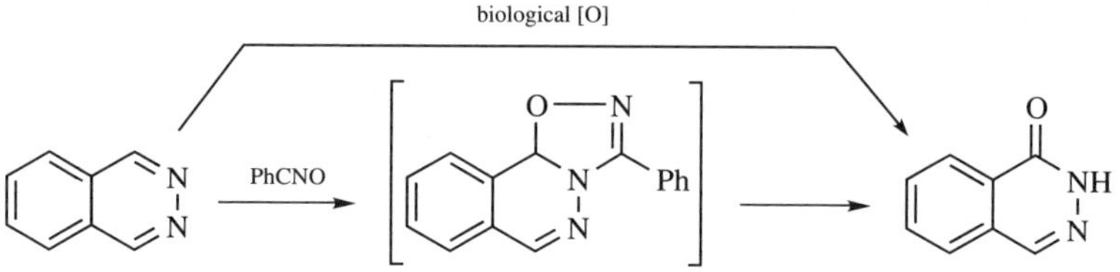

biological [O]

PhCNO

(**45**) (**46**) (**47**)

Nitration

Nitration of phthalazine gave only 5-nitrophthalazine (substrate, 96% H_2SO_4, $KNO_3\downarrow$ slowly, 0°C, then 55°C, 48 h: 79%;[856] or likewise but $KNO_3\downarrow$ at 20°C, then 100°C, 132 h: ~20%).[275]

Quaternization and Betaine or Ylide Formation

Phthalazine with the appropriate alkyl halide afforded 2-methylphthalazinium iodide (**48**, R = H, X = I) (MeI, MeOH, reflux, 3 h: 98%),[117,cf. 47] 2-butylphthalazinium bromide (**48**, R = Pr, X = Br) (BuBr, MeCN, 20°C, 3 days: 97%),[47] 2-phenacylphthalazinium bromide (**48**, R = Bz, X = Br) ($BzCH_2Br$, PhMe, 20°C, 12 h; then 55°C, 5 h: 91%),[847] 2-*p*-fluorophenacylphthalazinium bromide [**48**, R = *p*-$FC_6H_4C(=O)$, X = Br] [*p*-$FC_6H_4C(=O)CH_2Br$, PhH, 20°C, 3 h: 94%],[976] or 2-ethoxycarbonylmethylphthalazinium bromide (**48**, R = CO_2Et, X = Br) (EtO_2CCH_2Br, PhMe, 20°C, 12 h, then 55°C, 5 h: 95%;[847] or EtO_2CH_2Br, CH_2Cl_2, reflux, 1 h: 95%).[997]

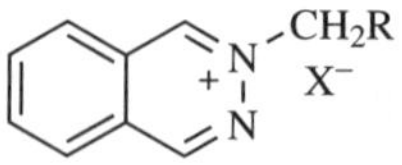

(**48**)

Phthalazine hydrochloride (**49**) and acrylamide gave 2-(2-carbamoylethyl)phthalazinium chloride (**50**) (reactants, EtOH, reflux, 3 h: 86%).[56]

(**49**) + HCl —$H_2C{=}CHCONH_2$→ (**50**) [N^+–$CH_2CH_2CONH_2$ Cl^-]

Phthalazine (**51**) and acrylic acid gave 3-(2-phthalazinio)propionate (**52**) [reactants, AcOEt (or $CHCl_3$, Et_2O, etc.), reflux, ~3 h: 81%, as an acrylic acid solvate] and thence 2-(2-carboxyethyl)phthalazinium chloride (**53**) (HCl, no details).[58]

(**51**) —$H_2C{=}CHCO_2H$→ (**52**) [N^+–$CH_2CH_2CO_2^-$] —HCl→ (**53**) [N^+–$CH_2CH_2CO_2H$ Cl^-]

Phthalazine (**54**) with 3-chloro-2-sulfopropionic acid (**55**, R = H) gave 1-carboxy-2-(2-phthalazinio)ethanesulfonate (**56**, R = H) (synthon, MeOH, substrate↓ slowly, then reflux, 30 min: 40%) or with 3-chloro-2-methoxycarbonylethanesulfonic acid (**55**, R = Me) (made in situ) gave 1-methoxycarbonyl-2-(2-phthalazinio)ethanesulfonate (**56**, R = Me) (MeOH, reflux, ~3 h: 37%).[57]

The same substrate (**54**) with tetracyanoethylene oxide (**57**) gave dicyano(2-phthalazinio)methanide (**58**) (substrate, AcOEt, synthon in AcOEt↓ dropwise, <0°C: 91%).[382]

(**54**) —$ClCH_2CH(CO_2R)SO_3H$ (**55**)→ (**56**) [N^+–$CH_2CH(CO_2R)SO_3^-$]

(**54**) —$(NC)_2C(O)C(CN)_2$ (**57**)→ (**58**) [N^+–$\bar{C}(CN)_2$]

Also other examples.[5,189,232,277,358,362,393,496,631,853,854]

Reissert Reactions

Note: The classical Reissert reaction may be considered as the addition of benzoyl cyanide (supplied as benzoyl chloride and potassium cyanide) across a —HC=N— portion of an appropriate heteroaromatic substrate to afford an *N*-benzoyl-*C*-cyanodihydro derivative. It was applied first to phthalazine circa 1967 (*E* 333) and subsequently improved and extended greatly, as illustrated in the examples that follow.

Phthalazine (**59**) with benzoyl chloride and potassium cyanide in aqueous methylene chloride (classical conditions) gave a mixture of 2-benzoyl-1,2-dihydro-1-phthalazinecarbonitrile (**60**) and 2-benzoyl-1,2-dihydro-1-phthalazinol (**61**, R = H) [the latter isolated as 2-benzoyl-1-ethoxy-1,2-dihydrophthalazine (**61**, R = Et) after recrystallization from ethanol] in variable ratio.[569,1002,cf. 39,568] However, only the nitrile (**60**) resulted when phthalazine and benzoyl chloride were treated in several improved ways [reactants, KCN, $(PhCH_2)Me_3NCl$, H_2O, CH_2Cl_2, 20°C, ~3 h: ~68%;[21,569] reactants, KCN, Bu_4NBr, CH_2Cl_2, reflux, 4 h: 47%;[376] reactants, Me_3SiCN, $AlCl_3$, CH_2Cl_2, 20°C, 12–24 h: ~85%;[21,377,771] or reactants, Bu_3SnCN, $AlCl_3$, CH_2Cl_2, 20°C, 3 h: 95%].[579]

Phthalazine (**59**) underwent related Reissert reactions to furnish, inter alia, 2-isobutyryl-1,2-dihydro-1-phthalazinecarbonitrile (**62**, Q = Pr^i) [Me_3SiCN, $AlCl_3$, CH_2Cl_2, $Pr^iC(=O)Cl$↓ slowly, <33°C; then 20°C, 18 h: 96%],[716] ethyl 1-cyano-1,2-dihydro-2-phthalazinecarboxylate (**62**, Q = OEt) (Me_3SiCN, $AlCl_3$, CH_2Cl_2, $ClCO_2Et$↓ slowly, 20°C, 24 h: 56%),[21] phenyl 1-cyano-1,2-dihydro-2-phthalazinecarboxylate (**62**, Q = OPh) [KCN, $(PhCH_2)Me_3NCl$, H_2O, CH_2Cl_2, $ClCO_2Ph$↓ dropwise, 20°C, 5 h: 41%],[377] and 1-cyano-1,2-dihydro-2-phthalazinecarbonyl chloride (**62**, O = Cl) (Me_3SiCN, CH_2Cl_2, −20°C, A; $BF_3 \cdot Et_2O$↓; $COCl_2$ in PhMe↓ during 1 h; →20°C, 6 h: 52%).[47]

Bz, XCN

(**59**) (**60**) (**61**)

QC(=O)Cl, XCN

(**62**)

Also other examples,[374] some in the presence of passenger groups.[948]

Reissert-Like Reactions

Note: Phthalazine undergoes several two-synthon additions, clearly akin to the Reissert reaction but affording 2-acylphthalazines with a 1-substituent other than cyano.

Phthalazine (**63**) gave 1-dimethoxyphosphinyl-2-methanesulfonyl-1,2-dihydrophthalazine (**64**) [substrate, $MeSO_2Cl$, MeCN, 0°C, 10 min, then $P(OMe)_3\downarrow$, NaI↓, 0°C → 50°C, 10 min: 67%; structure confirmed by X-ray analysis], 1-dimethoxyphosphinyl-*N*,*N*-diphenyl-1,2-dihydro-2-phthalazinecarboxamide (**65**) [substrate, $ClCONPh_2$, MeCN, 80°C, 1 h, then $P(OMe)_3\downarrow$, NaI↓, 0°C → 50°C, 10 min: 38%], or 2-benzoyl-1-dimethoxyphosphinyl-1,2-dihydrophthalazine (**66**) [substrate, BzCl, MeCN, 0°C, 10 min, then $P(OMe)_3\downarrow$, NaI↓, 0°C → 50°C, 10 min: 20%].[419]

$MeSO_2Cl$, $P(OMe)_3$; SO_2Me; $OP(OMe)_2$

(**63**) (**64**)

$ClCONPh_2$, $P(OMe)_3$; BzCl, $P(OMe)_3$; $CONPh_2$; $OP(OMe)_2$; Bz; $OP(OMe)_2$

(**65**) (**66**)

Phthalazine (**67**) with neat acetic anhydride gave 2-acetyl-1-carboxymethyl-1,2-dihydrophthalazine (**68**) (reflux, N_2, 5 h: 45%) or with diethyl malonate and acetic anhydride gave 2-acetyl-1-(diethoxycarbonylmethyl)-1,2-dihydrophthalazine (**69**), contaminated with a little of the product (**68**) (100°C, 15 h: 55% and 14%, respectively, after separation).[408,567]

Ac_2O; Ac; CH_2CO_2H

(**67**) (**68**)

$H_2C(CO_2Et)_2$, Ac_2O; Ac; $CH(CO_2Et)_2$; + (**68**)

(**69**)

Phthalazine with bis(tributylstannyl)acetylene and ethyl chloroformate gave ethyl 1-ethynyl-1,2-dihydro-2-phthalazinecarboxylate (**70**) (substrate, $Bu_3SnC{\equiv}CSnBu_3$, CH_2Cl_2, $ClCO_2Et\downarrow$ dropwise, 0°C → 20°C, <3 days; then $F_3CCO_2H\downarrow$, 20°C, 30 min: 82%).[304,596]

Phthalazine with 6-*p*-fluorophenyl-2,3-dihydroimidazo[2,1-*b*]thiazole and ethyl chloroformate gave ethyl 1-(6-*p*-fluorophenyl-2,3-dihydroimidazo[2.1-*b*]thiazol-5-yl)-1,2-dihydro-2-phthalazinecarboxylate (**71**) (bicyclic reactants, CH_2Cl_2, $ClCO_2Et\downarrow$ dropwise, <10°C → 20°C, 12 h: 62%).[46]

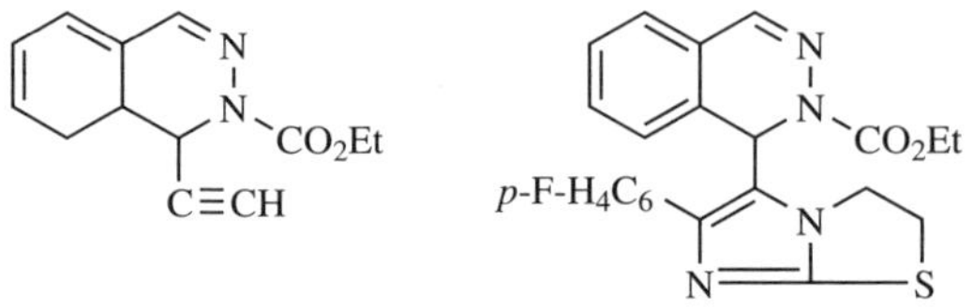

(**70**) (**71**)

Phthalazine (**72**) with diketene (**73**) and formic acid gave 1-acetonyl-1,2-dihydro-2-phthalazinecarbaldehyde (**74**, R = H) (neat reactants, 30°C → 20°C, ~20 h: 62%; rational mechanism suggested); replacement of formic acid by acetic or propionic acid likewise afforded 1-acetonyl-2-acetyl- (**74**, R = Me) (49%) or 1-acetonyl-2-propionyl-1,2-dihydrophthalazine (**74**, R = Et) (69%), respectively.[409,566]

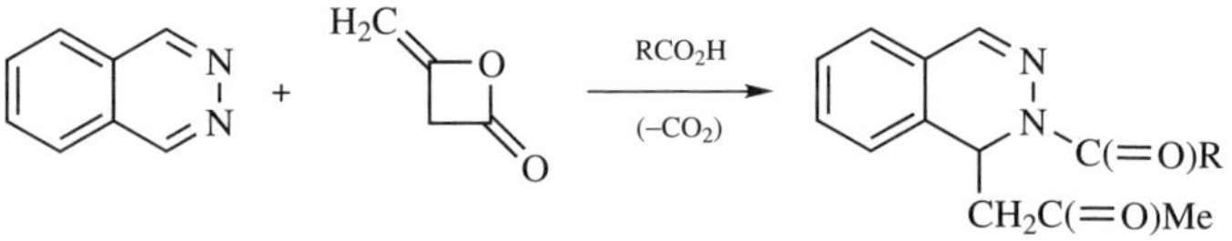

(**72**) (**73**) (**74**)

Also other examples.[588]

9.2. ALKYL- AND ARYLPHTHALAZINES (*H* 72; *E* 137)

This section covers both *C*- and *N*-alkyl/arylphthalazines (including hydrophthalazines) but not alkyl/arylphthalazinium salts (discussed later in Section 9.3) or *N*-alkyl/arylphthalazinones and the like (Section 11.5).

The electron spin resonance and conformations of cation radicals from 1,4-dimethyldecahydrophthalazine[113] and 2,3-diphenyl-1,2,3,4-tetrahydrophthalazine[766] have been studied in detail.

9.2.1. Preparation of Alkyl- and Arylphthalazines

Many alkyl/arylphthalazines have been made by *primary syntheses* (see Chapter 8) and some by alkylation of unsubstituted phthalazine or hydrophthalazines (see Section 9.1.3). Other methods of preparation are illustrated by the following classified examples.

By *C*-Alkylation of Metallated Substrates

6-Chloro-1,4-dimethoxyphthalazine (**75**) was converted into a solution of its 7-lithio derivative (**76**) (substrate, THF, BuLi↓ dropwise, −78°C, A, 30 min); subsequent addition of appropriate electrophiles gave 6-chloro-1,4-dimethoxy-7-methylphthalazine (**77**) [MeI, −78°C, 1 h: ~80% of a 1 : 1 mixture with substrate (**75**)], 6-chloro-7-(1-hydroxyethyl)-1,4-dimethoxyphthalazine (**78**, R = Me) (MeCHO, −78°C, 30 min: 89%), and 6-chloro-7-α-hydroxybenzyl-1,4-dimethoxyphthalazine (**78**, R = Ph) (PhCHO, −78°C, 1 h: 78%).[311]

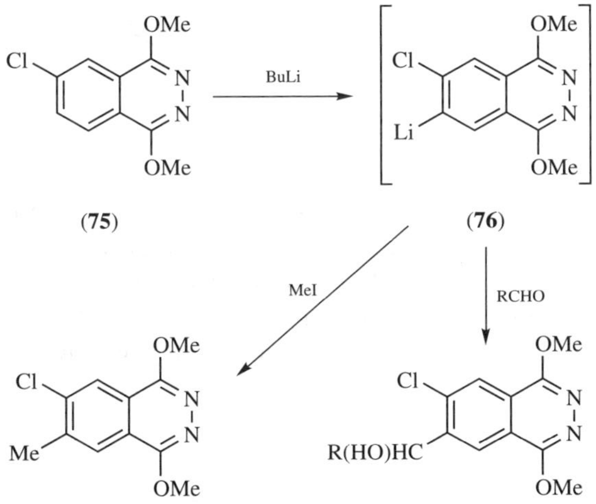

(**75**) (**76**)

(**77**) (**78**)

2-Benzoyl-6,7-dimethoxy-1,2-dihydro-1-phthalazinecarbonitrile (**79**) gave 2-benzoyl-1-(3,4-dimethoxybenzyl)-6,7-dimethoxy-1,2-dihydro-1-phthalazinecarbonitrile (**80**) [substrate, Me_2NCHO, $ClH_2CC_6H_3(OMe)_2$-3,4↓, NaH↓ −15°C → 20°C, ~3 h: 95%].[948]

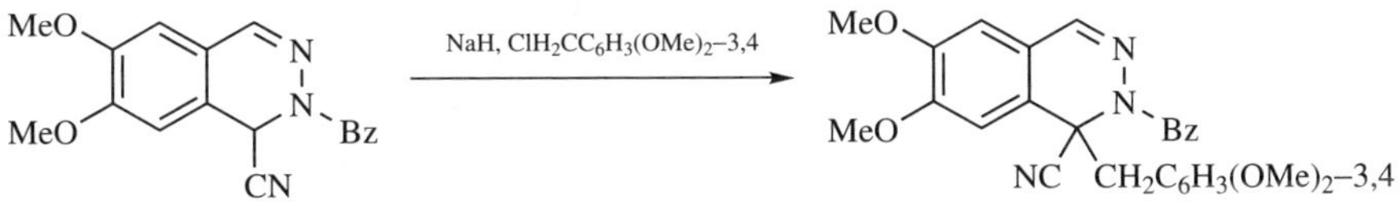

(**79**) (**80**)

In contrast, the related Reissert substrate 2-benzoyl-1,2-dihydro-1-phthalazinecarbonitrile (**81**) gave 1-*o*-chlorobenzylphthalazine (**82**) [substrate,

$ClH_2CC_6H_4Cl$-*o*, $(PhCH_2)Et_3Cl$, CH_2Cl_2, NaOH, H_2O, EtOH, sonication, 1 h: 67%; note loss of CN and Bz with concomitant nuclear oxidation] or 1-α-hydroxybenzylphthalazine (**83**) (likewise but PhCHO: 98%);[39] also analogous alkylations.[35]

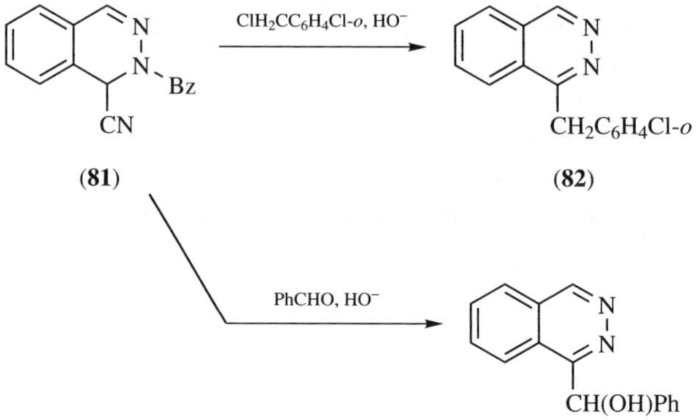

$ClH_2CC_6H_4Cl$-*o*, HO^-

CN

Bz

$CH_2C_6H_4Cl$-*o*

(**81**) (**82**)

PhCHO, HO^-

CH(OH)Ph

(**83**)

Conversion of 1,4-diphenylphthalazine (**84**) into a solution of its highly colored dianion (**85**) (substrate, Na, anhydrous THF, A, 20°C, 8 h) followed by brief treatment with benzyl chloride at −24°C gave a mixture of 1-benzyl- (**86**) and 2-benzyl-1,4-diphenyl-1,2-dihydrophthalazine (**87**), from which only the 1-benzyl isomer (**86**) (43%) was isolated in a pure state.[36]

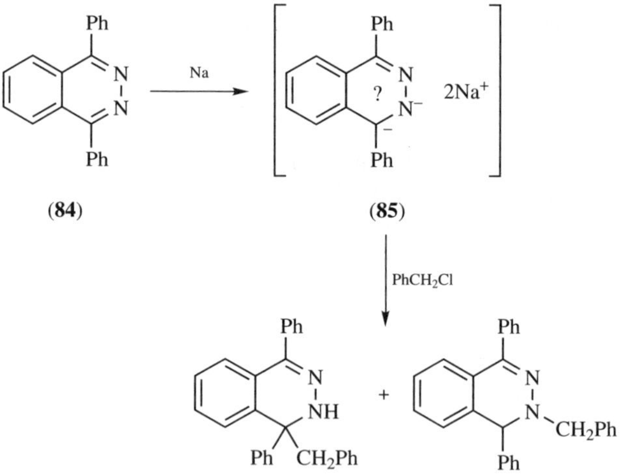

Na

$2Na^+$

(**84**) (**85**)

$PhCH_2Cl$

Ph CH_2Ph

CH_2Ph

(**86**) (**87**)

Also other examples.[50,551]

By *C*-Alkylation with Organometallic Synthons

1-Phenylphthalazine (**88**) gave 1-methyl-4-phenyl-1,2-dihydrophthalazine (**89**) [MeMgI (made in situ), Pr^i_2O, reflux, N_2, 16 h: crude] and thence 1-methyl-4-phenylphthalazine (**90**) (MnO_2, PhH, reflux, 6 h: >95% overall).[290]

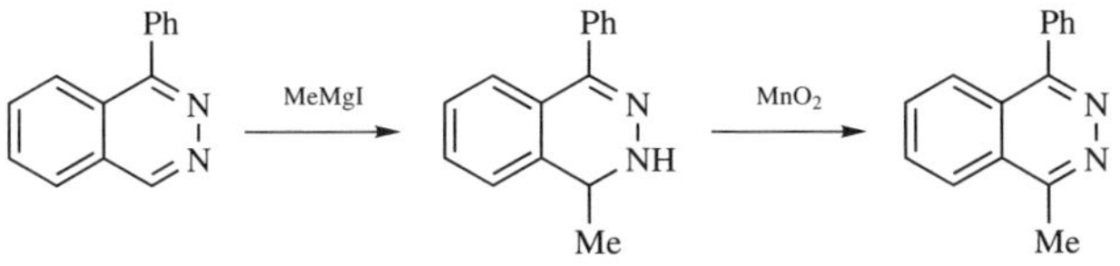

(**88**) (**89**) (**90**)

1(2*H*)-Phthalazinone (**91**) gave 4-phenyl-1(2*H*)-phthalazinone (**92**) [substrate, THF, PhLi (in Et_2O + PhH)↓, N_2, 3°C, ? h: ~50%].[708]

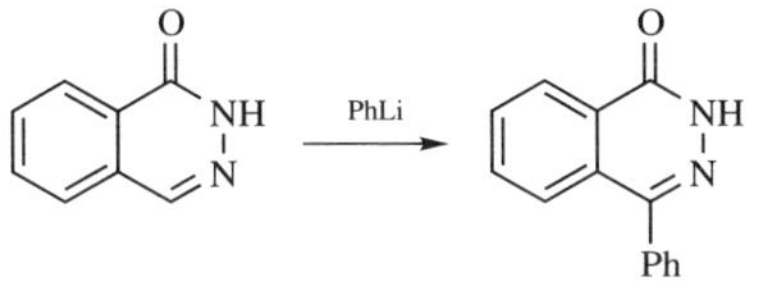

(**91**) (**92**)

2,4-Diphenyl-1(2*H*)-phthalazinone (**93**) gave 1,2,4-triphenyl-1,2-dihydro-1-phthalazinol (**94**) [PhMgX (freshly made in Et_2O), substrate (in warm PhH) ↓, then 20°C, 2 h; NH_4Cl, H_2O, workup: 68%] and thence 1,2,4-triphenylphthalazin-2-ium perchlorate (**96**) [product (**94**), Ac_2O, 100°C, 70% $HClO_4$↓ cautiously: 91%]; analogs likewise.[444]

The same substrate (**93**) gave 1-methyl-2,4-diphenylphthalazin-2-ium perchlorate (**95**) directly [MeMgI (made in Et_2O), substrate (in PhH)↓, 20°C, ~24 h; $HClO_4$, H_2O, workup: 70%];[454] analogs likewise.[454,461]

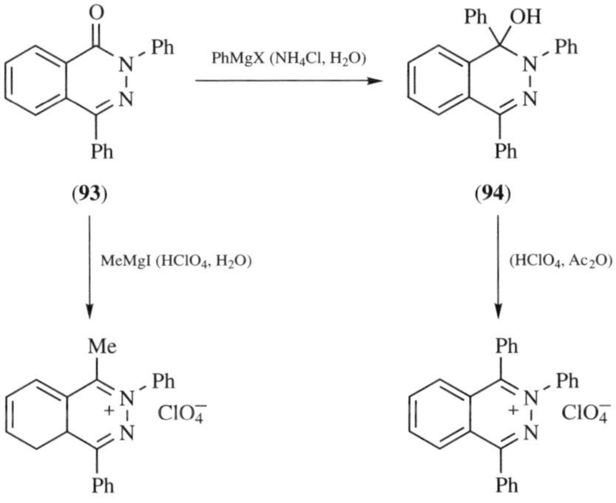

(**93**) (**94**)

(**95**) (**96**)

Also other examples.[183,848,931]

By Alkanelysis of Halogeno- or Alkoxyphthalazines

1-Chloro-4-phenylphthalazine (**97**) with ethyl cyanoacetate gave 1-(α-cyano-α-ethoxtcarbonylmethyl)-4-phenylphthalazine (**98**) (Me_2NCHO, reflux, 30 min, apparently without any base: 75%) or with acetylacetone gave 1-diacetylmethyl-4-phenylphthalazine (likewise: 60%).[664]

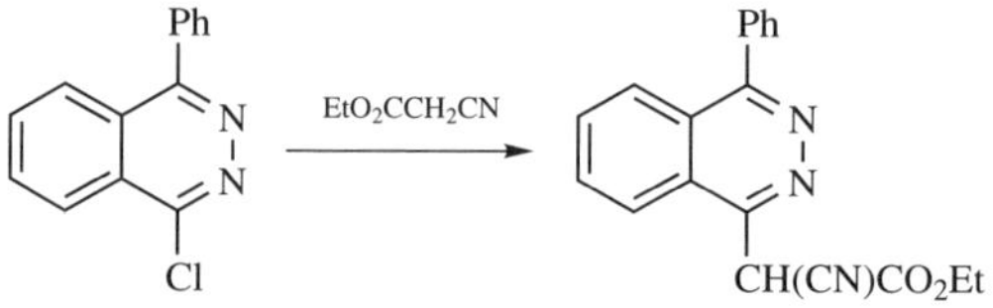

EtO_2CCH_2CN

(**97**) (**98**)

1,4-Dichlorophthalazine with di(pyridin-2-yl)methane gave 1,4-bis[di(pyridin-2-yl)methyl]phthalazine (**99**) (synthon, THF, BuLi↓ slowly, A, −78°C, 10 min; then substrate↓ slowly, −78°C → 20°C: 81%),[344] with 1-methylimidazole gave 1,4-bis(1-methylimidazol-2-yl)phthalazine (**100**) (BuLi etc. likewise: 60%),[626] or with methylenetriphenylphosphorane followed by benzaldehyde gave 1-chloro-4-styrylphthalazine (substrate, $H_2C{=}PPh_3$, PhH, PhCHO↓, reflux, 30 min; 13%).[135]

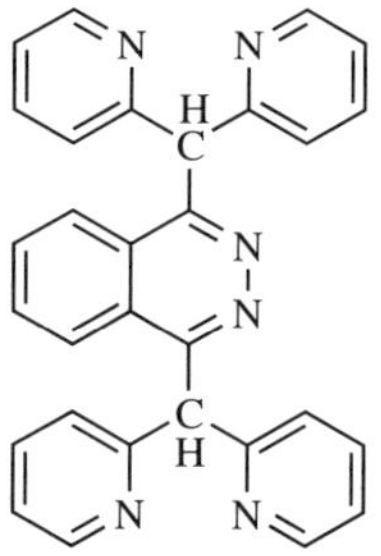

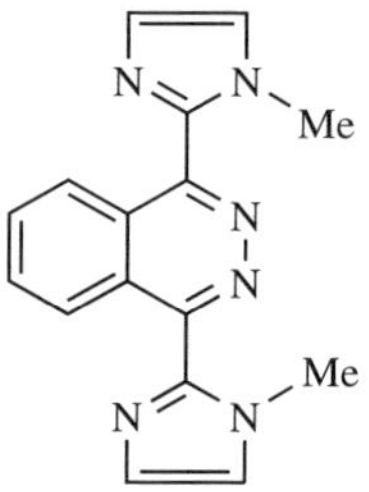

(**99**) (**100**)

1-Chloro-4-(4-methylpiperazin-1-yl)phthalazine gave 1-(4-methylpiperazin-1-yl)-4-phenylphthalazine [substrate, $PhB(OH)_2$, dioxane, Cs_2CO_3↓, A, $Pd_2(PhCH{=}CHCOCH{=}CHPh)_3$↓, PBu^t_3↓, reflux, 24 h: 70%]; substituted-phenyl analogs likewise.[880]

1-Methoxy-gave 1-(α-cyanobenzyl)phthalazine (NaH, $PhCH_2CN$, THF, reflux, 30 min; substrate↓, reflux, N_2, tlc monitored: 46%).[586]

Also other examples.[426,857,861,974]

By Interconversion of Alkyl Groups

Note: There appear to be no simple examples of such interconversions in the 1972–2004 literature.

1,2-Dimethyl-4-phenylphthalazin-2-ium perchlorate (**101**) with benzaldehyde gave 2-methyl-4-phenyl-1-styrylphthalazin-2-ium perchlorate (**102**) (pyridine, reflux, 1 h: 73%) or with 1,1,3,3-tetramethoxypropane gave 1-(4,4-dimethoxybut-1-enyl)-2-methyl-4-phenylphthalazin-2-ium perchlorate (**103**) (pyridine, Ac_2O, 95°C, 4 h: 48%);[461] many other such reactions with quaternary methylphthalazines have been reported.[449,454,455,458]

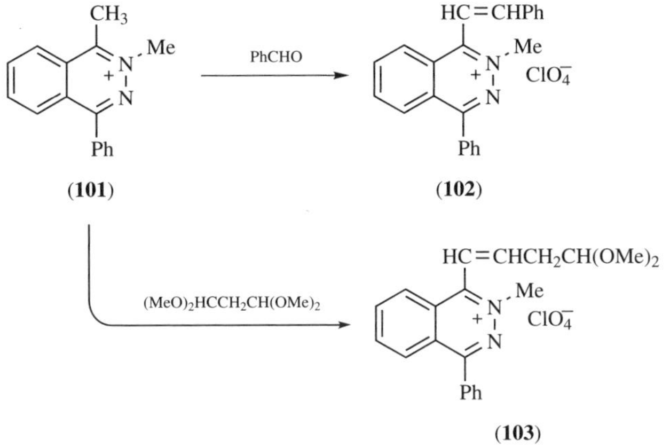

(**101**) —PhCHO→ (**102**)

(**101**) —$(MeO)_2HCCH_2CH(OMe)_2$→ (**103**)

4-(3,4-Dichlorophenyl)-2-phenylacetyl-1(2*H*)-phthalazinone (**104**) with benzaldehyde gave 4-(3,4-dichlorophenyl)-2-(2,3-diphenylacryloyl)-1(2*H*)-phthalazinone (**105**) (NaOH, EtOH, H_2O, 95°C, 30 min: 70%).[621]

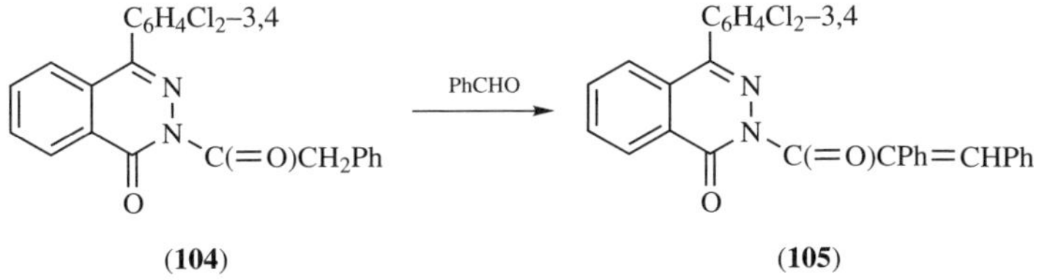

(**104**) —PhCHO→ (**105**)

Also other examples.[595,930,1014]

By Miscellaneous Procedures

1-Methoxy-4-phenylphthalazine (**106**) gave 1,1-dibutyl-4-phenyl-1,2-dihydrophthalazine (**106a**) [substrate, THF, A, BuLi↓ slowly, −78°C, 15 min; then MeCHO↓ (!), 30 min: 45%].[311]

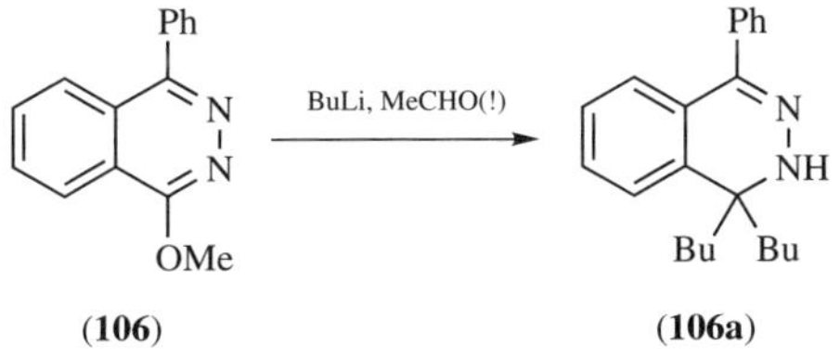

(**106**) —BuLi, MeCHO(!)→ (**106a**)

2-*tert*-Butyl-4-phenyl-1(2*H*)-phthalazinone (**107**) gave 2-*tert*-butyl-4-phenyl-1,2-dihydrophthalazine (**107a**) ($LiAlH_4$, THF, reflux, 5 min: 92%);[119] 2,4-diphenyl-1,2-dihydrophthalazine somewhat similarly.[28,446]

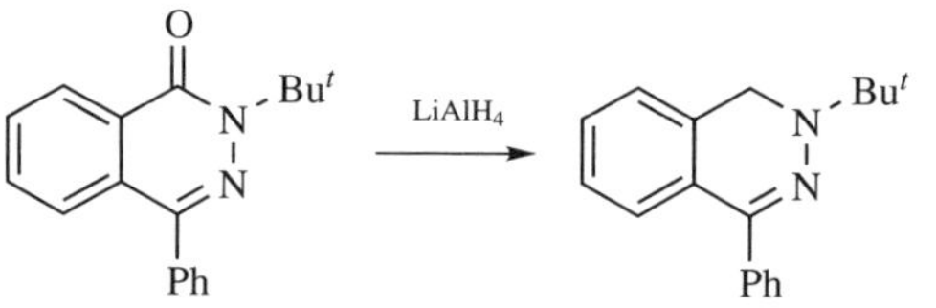

$LiAlH_4$

(**107**) (**107a**)

2-[2-(*p*-Formylphenoxy)ethyl]-1(2*H*)-phthalazinone and 2,4-thiazolidinedione gave 2-{2-[*p*-(2,4-dioxothiazolidin-5-ylidene)phenoxy]ethyl}-1(2*H*)-phthalazinone (**108**) (reactants, BzOH, piperidine, PhMe, reflux, H_2O removal, 1 h: 80%); homologs likewise.[874]

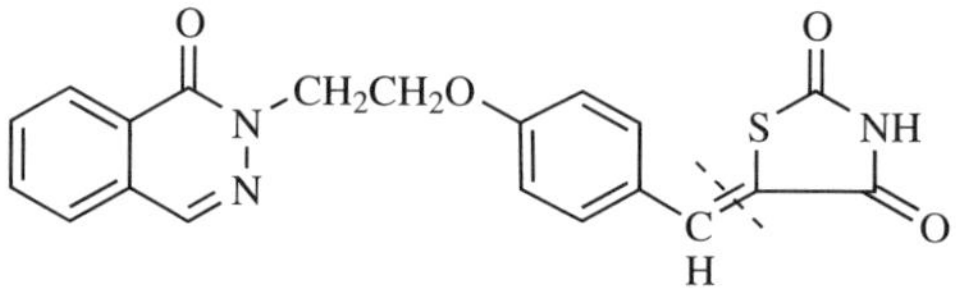

(**108**)

Ethyl 4-*p*-methoxybenzyl-3-methyl-1,2,3,4,5,6,7,8-octahydro-2-phthalazinecarboxylate underwent reduction to give 1-*p*-methoxybenzyl-2,3-dimethyl-1,2,3,4,5,6,7,8-octahydrophthalazine ($LiAlH_4$, THF, reflux, 6.5 h: 89%).[393] Also other examples.[21,399,424,930]

9.2.2. Reactions of Alkyl- and Arylphthalazines

Only reactions specific to the alkyl or aryl substituent(s) are covered in this section. *Interconversion of alkyl groups* has been exemplified in Section 9.2.1, iron complexes of 1,4-dimethylphthalazine have been studied,[990] and other reactions are illustrated by the following classified examples.

Dimerization

1-Chloro-4-styrylphthalazine (**109**) underwent photodimerization to a single dimer, 1,3-bis(4-chlorophthalazin-1-yl)-2,4-diphenylcyclobutane (**109a**) (solid substrate, *hν*, 4 h: 95%).[135]

Halogenation

Note: Alkyl and aryl groups undergo direct halogenation, whereas alkenyl or alkynyl groups undergo addition of halogen. Both types of reaction are represented in these examples.

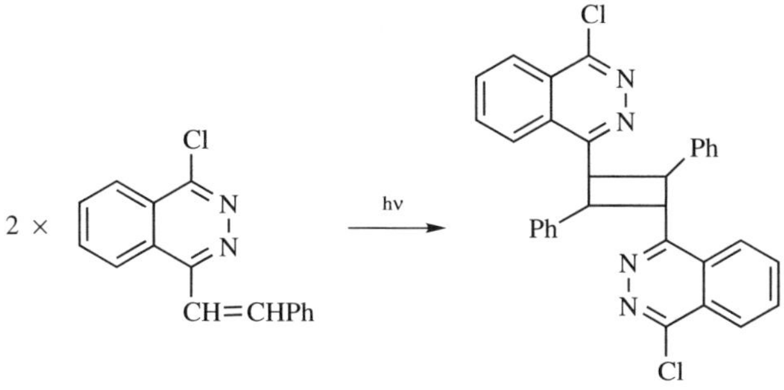

2 × (109) $\xrightarrow{h\nu}$ (109a)

(**109**) (**109a**)

Ethyl 1-*o*-methoxyphenyl-5,7-dimethyl-4-oxo-3,4-dihydro-6-phthalazinecarboxylate (**110**, Q = R = H) gave a separable mixture of ethyl 5-bromomethyl-1-*o*-methoxyphenyl-7-methyl-4-oxo-3,4-dihydro-6-phthalazinecarboxylate (**110**, Q = H, R = Br) and ethyl 1-(3-bromo-6-methoxyphenyl)-5-bromomethyl-7-methyl-4-oxo-3,4-dihydro-6-phthalazinecarboxylate (**110**, Q = R = Br) (substrate, *N*-bromosuccinimide, Bz_2O_2, CCl_4, reflux, 20 h: 38% and 9%, respectively).[426]

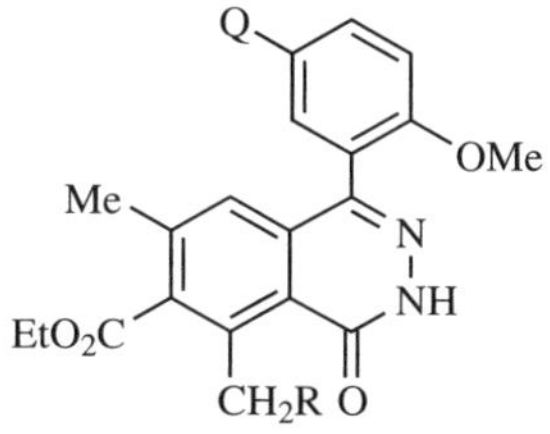

(**110**)

2,8-Dimethyl-1(2*H*)-phthalazinone gave 2-bromomethyl-8-dibromomethyl-1(2*H*)-phthalazinone (substrate, Bz_2O_2, CCl_4, reflux; then Br_2 in CCl_4↓ dropwise, *h*ν, reflux, 3 h: 30% after purification).[622]

5-Phenylethynylphthalazine gave 5-(α,β-dibromostyryl)phthalazine (substrate, CH_2Cl_2, 20°C; Br_2 in CH_2Cl_2↓ dropwise; then 20°C, 3 h: 59%).[857]

2-(3-p-Chlorophenyl-2-phenylacryloyl)- (**111**) gave 2-(2,3-dibromo-3-*p*-chlorophenyl-2-phenylpropionyl)-4-(3,4-dichlorophenyl)-1(2*H*)-phthalazinone (**111a**) (substrate, CCl_4, Br_2↓ dropwise, 20°C, 12 h: 70%).[621]

Also other examples.[250]

Oxidative Reactions

2-(3-*p*-Chlorophenyl-2-phenylacryloyl)- (**111**) gave 2-(3-*p*-chlorophenyl-2-phenyl-2,3-epoxypropionyl)-4-(3,4-dichlorophenyl)-1(2*H*)-phthalazinone (**112**) (H_2O_2, NaOH, H_2O, MeOH, AcMe, 0°C, 2 h: 60%).[621]

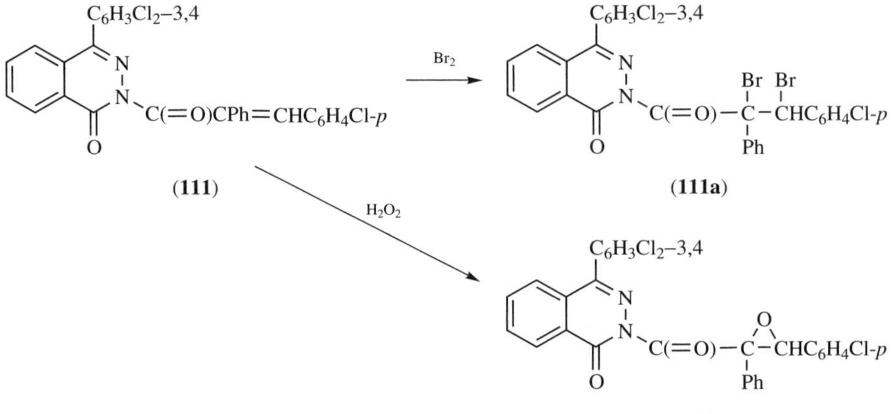

(**111**) (**111a**) (**112**)

1,3-Diphenyl-3,4-dihydrophthalazine (**113**) with chromium trioxide gave either 2,4-diphenyl-1,2-dihydro-1-phthalazinol (**114**, R = H) [CrO_3, H_2O, AcOH, 60°C, 10 min: 47% (from PhH)] or 1-ethoxy-2,4-diphenyl-1,2-dihydrophthalazine (**114**, R = Et) (likewise but recrystallized from EtOH: 33%).[28]

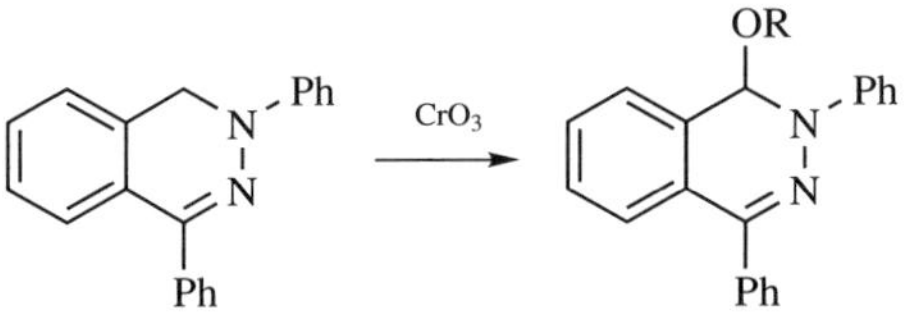

(**113**) (**114**)

1-Phenylphthalazine (**115**) gave 3-phenyl-4,5-pyridazinedicarboxylic acid (**116**) [$KMnO_4$, H_2O, reflux (?), 3 h: 54%; note survival of the Ph substituent].[929]

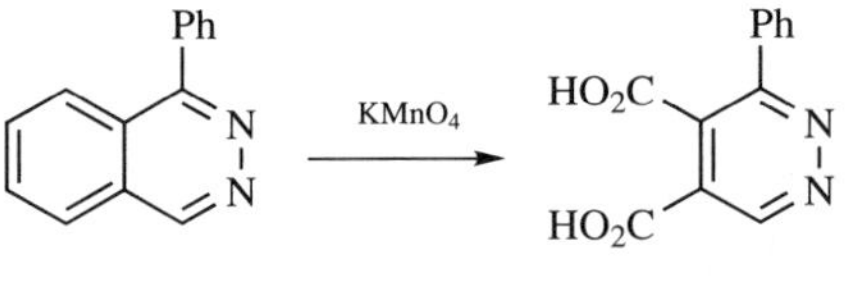

(**115**) (**116**)

4-Benzyl- (**117**) gave 4-benzoyl-1(2*H*)-phthalazinone (**118**) ($Na_2Cr_2O_7$, AcOH: 82%);[545] analogous benzylphthalazines behaved similarly.[541,942]

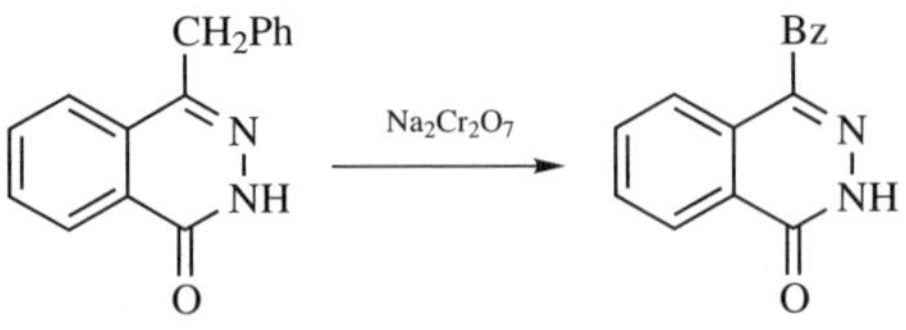

(**117**) (**118**)

6-Methyl-1,4-bis(trifluoromethyl)-3,5,8,8a-tetrahydrophthalazine (**119**) with 4-phenyl-1,2,4-triazoline-3,5-dione (**120**) gave 6-methyl-1,4-bis(trifluoromethyl)-5,8-dihydrophthalazine (**121**) [substrate, synthon (1 mol), PhMe, A, 70°C, 3 h: 85%] or 6-methyl-1,4-bis(trifluoromethyl)phthalazine (**121a**) [substrate, synthon (3 mol), PhMe, A, 70°C, 12 h: 90%]; analogous hydrophthalazines also underwent such clean stepwise nuclear dehydrogenation.[389]

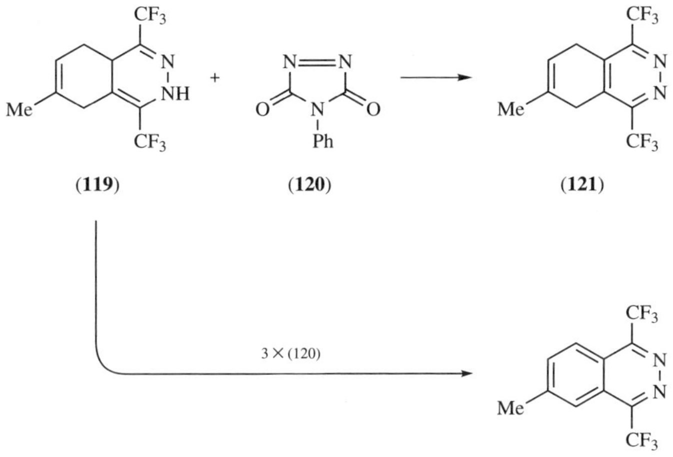

(**119**) (**120**) (**121**)

3 × (120)

(**121a**)

1-Acetoxy-4-benzylidene-3-phenyl-3,4-dihydrophthalazine (**122**) underwent ozonolysis to afford 4-acetoxy-2-phenyl-1(2*H*)-phthalazinone (**122a**) (substrate, MeOH, $O_3 + O_2\downarrow$, −70°C, 3 h: then $Me_2S\downarrow$, 0°C → 20°C, 20 h: 56%).[698]

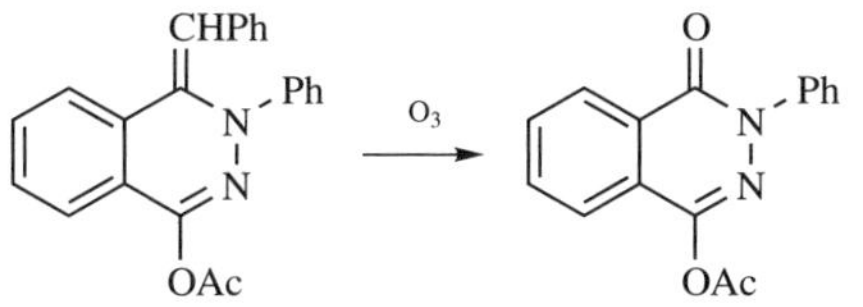

O_3

(**122**) (**122a**)

Rearrangements

1,4,5,6,7,8-Hexaphenylphthalazine (**123**) underwent pyrolytic rearrangement to 2,4,5,6,7,8-hexaphenylquinazoline (**124**) (neat substrate, sealed, 360°C, 30 min: 75%); several heavily substituted analogs behaved similarly.[121]

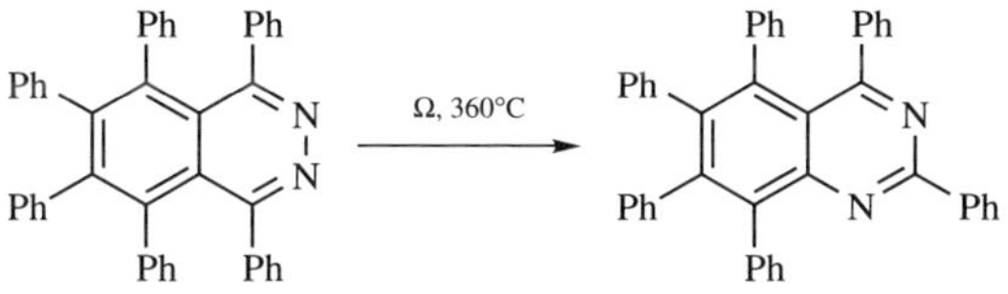

Ω, 360°C

(**123**) (**124**)

Schiff Base Formation

1-Methyl-2,4-diphenylphthalazin-2-ium perchlorate (**125**, R = Ph) and *p*-nitroso-*N*,*N*-dimethylaniline gave 1-[(*p*-dimethylaminophenylimino)methyl]-2,4-diphenylphthalazin-2-ium perchlorate (**126**, R = Ph) (reactants, Ac_2O, reflux, 30 min: 60%); in a somewhat similar way, 1,2-dimethyl-4-phenylphthalazin-2-ium perchlorate (**125**, R = Me) gave 1-[(*p*-dimethylaminophenylimino)methyl]-2-methyl-4-phenylphthalazin-2-ium perchlorate (**126**, R = Me) (reactants, pyridine, 95°C, 2 h: 54%).[462]

The same substrate (**125**, R = Ph) with 1-nitroso-2-naphthol gave either the zwitterionic 1-[*N*-(2-oxidonaphthalen-1-yl)iminomethyl]-2,4-diphenylphthalazin-2-ium (**127**) (reactants, pyridine, reflux, 1 h: 78%) or the isomeric 2′,4′-diphenylspiro{3*H*-naphth[2,1-*b*][1,4]oxazine-3,1′(2′H)phthalazine} (**128**) (pyridine, EtOH, reflux, 90 min: 59%).[462]

CH_3 R ClO_4^- Ph (**125**) —$ONC_6H_4NMe_2$-*p*→ $HC=NC_6H_4NMe_2$-*p* R ClO_4^- Ph (**126**)

(R = Ph) 1-nitroso-2-naphthol, pyridine; plus EtOH

(**127**) (**128**)

Typical Cyclocondensations

Note: One such cyclocondensation (epoxide formation) has been exemplified under *oxidative reactions* earlier in this section.

6,7-Dibenzyl-1,4-diphenylphthalazine (**129**) underwent oxidative cyclization with sulfur to give 1,4,6,8-tetraphenylthieno[3,4-*g*]phthalazine (**130**) (neat reactants, 270°C, 10 min: 56%).[16,942]

PhH_2C PhH_2C Ph Ph (**129**) —S, Δ→ Ph S Ph Ph Ph (**130**)

1,4-Dimethylphthalazine (**131**) with 2-chloroquinoxaline (**132**) gave 5-methylphthalazino[2′,1′:1,5]pyrrolo[2,3-*b*]quinoxaline (**133**) (AcOH, trace HCl, 55°C, 4 days: 35%; presumably involving aerial oxidation).[290]

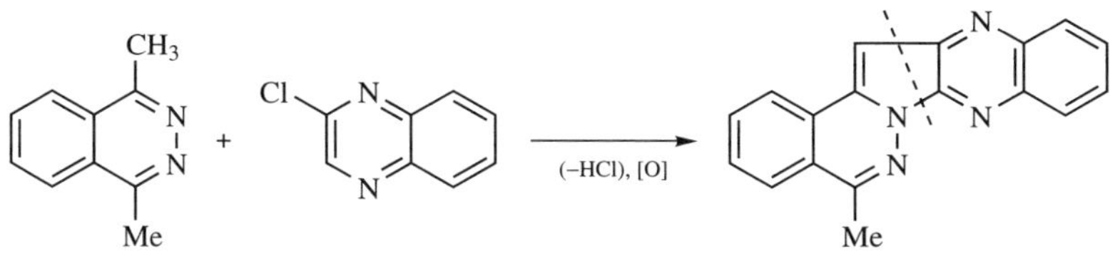

(**131**) (**132**) (**133**)

2-(3-*p*-chlorophenyl-2-phenylacryloyl)-4-(3,4-dichlorophenyl)-1(2*H*)-phthalazinone (**134**) with hydrazine hydrate gave 2-(5-*p*-chlorophenyl-4-phenyl-2-pyrazolin-3-yl)-4-(3,4-dichlorophenyl)-1(2*H*)-phthalazinone (**135**, R = H) (BuOH, reflux, 6 h: 75%) or with phenylhydrazine gave 2-(5-*p*-chlorophenyl-1,4-diphenyl-2-pyrazolin-3-yl)-4-(3,4-dichlorophenyl)-1(2*H*)-phthalazinone (**135**, R = Ph) (likewise: 80%).[621]

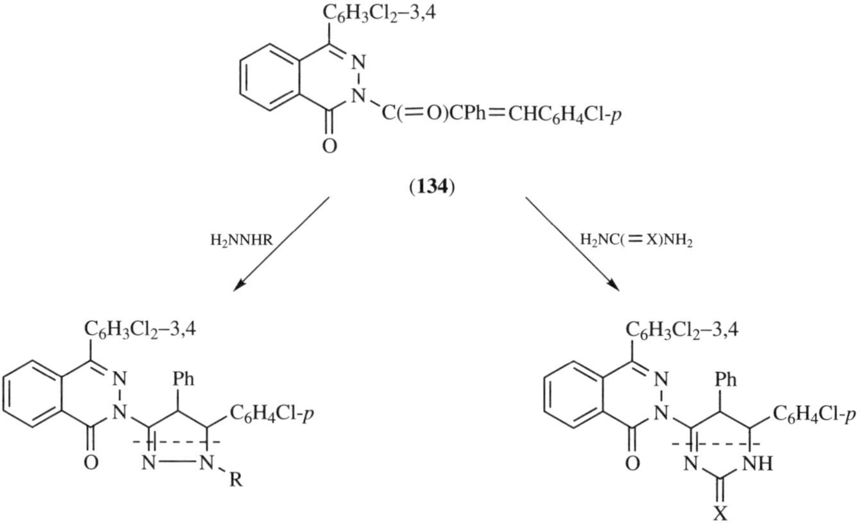

(**134**)

(**135**) (**136**)

The same substrate (**134**) with urea gave 2-(6-*p*-chlorophenyl-2-oxo-5-phenyl-1,2,5,6-tetrahydropyrimidin-4-yl)-4-(3,4-dichlorophenyl)-1(2*H*)-phthalazinone (**136**, X = O) (HCl, EtOH, 95°C, 8 h: 70%) or with thiourea gave 2-(6-*p*-chlorophenyl-5-phenyl-2-thioxo-1,2,5,6-tetrahydropyrimidin-4-yl)-4-(3,4-dichlorophenyl)-1(2*H*)-phthalazinone (**136**, X = S) (KOH, H_2O, EtOH, reflux, 3 h: 85%).[621]

9.3. *N*-ALKYL(OR ARYL)PHTHALAZINIUM SALTS, BETAINES, OR YLIDES (*E* 652)

The formation of such dipolar derivatives of phthalazine is usually quite mundane, but these compounds do have some interesting reactions.

The NMR spectra of 2-carbamoylmethylphthalazinium iodide (**137**, R = H) and its *N*-acyl derivatives (**137**, R = acyl) have been studied in some detail;[853,854] the pK_a values for dibenzoyl(2-phthalazinio)methanide (**138**) and related ylides have been measured,[194] and the interionic association of 2-methylphthalazinium perchlorate (**139**) has been compared with values for other such azinium salts.[466]

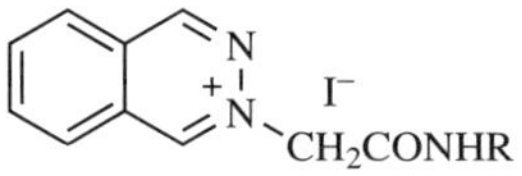

(**137**) 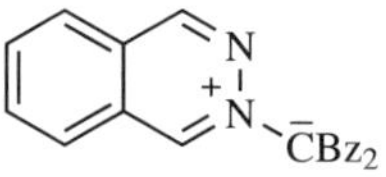 (**138**) 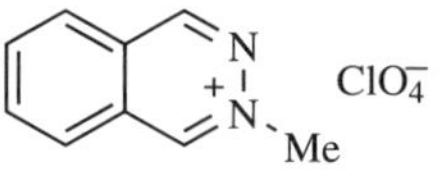 (**139**)

9.3.1. Preparation of *N*-Alkylphthalazinium Salts and the Like

Some of these quaternary entities have been made by *primary synthesis* (see Chapter 8), but most of the more recently described examples have been obtained by *quaternization of phthalazine* (see Section 9.1.3). A few remaining preparative procedures are illustrated in the following examples.

2-Phenyl-1,2-dihydro-1-phthalazinol (**140**) gave 2-phenylphthalazinium perchlorate (**141**) ($HClO_4$, H_2O, MeCN, 20°C, 5 min: 80%);[714] analogs likewise.[444]

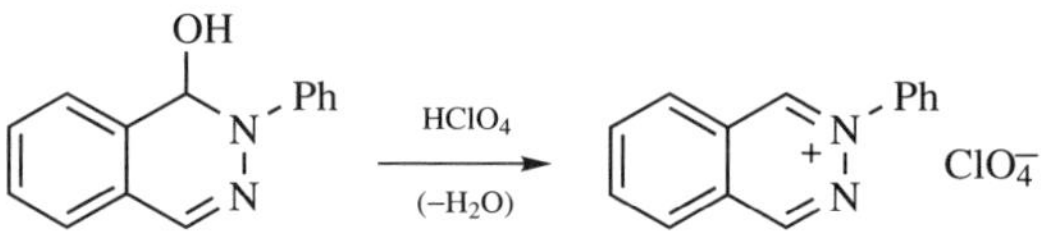

(**140**) (**141**)

1,2,4-Triphenyl-1,2-dihydrophthalazine (**142**) gave 1,2,4-triphenylphthalazinium tetrachloroferrate (**143**, X = $FeCl_4$) ($FeCl_3$, AcOH, 20°C, briefly: ?%) or the corresponding perchlorate (**143**, X = ClO_4) ($HClO_4$: 57%).[445]

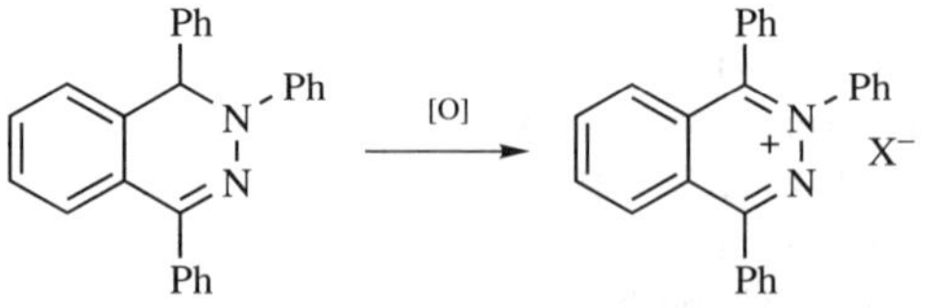

(**142**) (**143**)

1,4-Diphenyl-4a-(pyrrolidin-1-yl)-4a,5,6,7,8,8a-hexahydrophthalazine (**144**) with tetracyanoethylene oxide gave dicyano (1,4-diphenyl-5,6,7,8-tetrahydrophthalazin-2-io)methanide (**145**) [CH_2Cl_2, reflux, 2.5 h: 28%; note partial oxidation by loss of $HN(CH_2)_4$].[312]

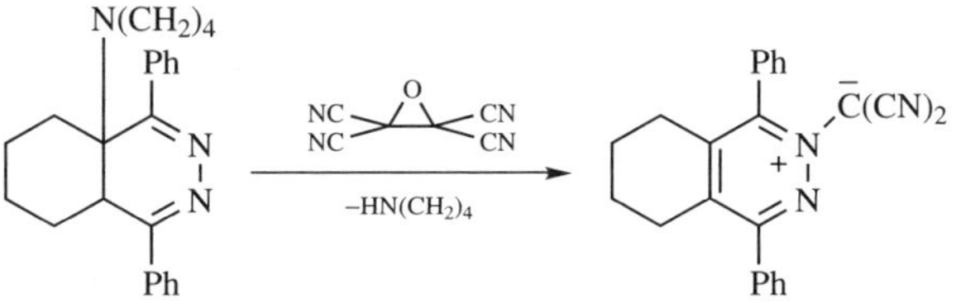

(**144**) (**145**)

1-Methylphthalazine (**146**) with *p*-methoxyphenacyl bromide apparently gave only 2-(*p*-methoxyphenacyl)-4-methylphthalazin-2-ium bromide (**147**) and thence, by treatment with potassium carbonate, *p*-methoxybenzoyl(4-methylphthalazin-2-io)methanide (**148**) (for details, see original).[245]

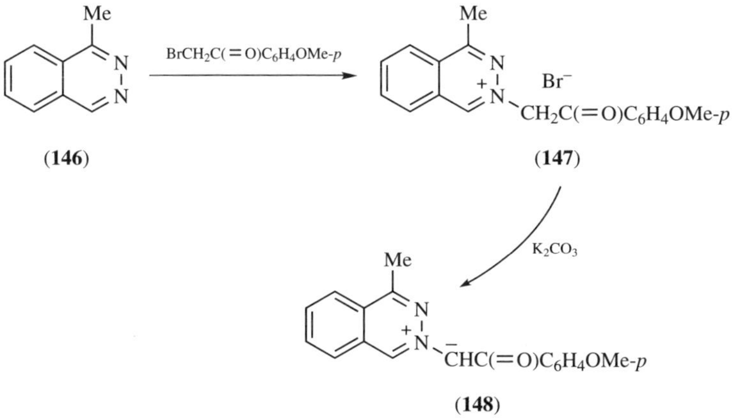

(**146**) (**147**)

(**148**)

9.3.2. Reactions of *N*-Alkylphthalazinium Salts and the Like

The various more recently reported reactions are illustrated in the following classified examples.

C-Alkylation

2-Methylphthalazinium iodide (**149**) gave 1,2-dimethyl-1,2-dihydrophthalazine (**150**) [MeMgI (made in Et_2O), substrate↓ slowly, 20°C, 90 min: 87%] and successively 1,2-dimethylphthalazinium iodide (**151**) (substrate, I_2, CH_2Cl_2, exothermic, 1 h: 65%) and 1,1,2-trimethyl-1,2-dihydrophthalazine (**152**) (MeMgI as in the first step but reflux, 1 h: 50%).[47]

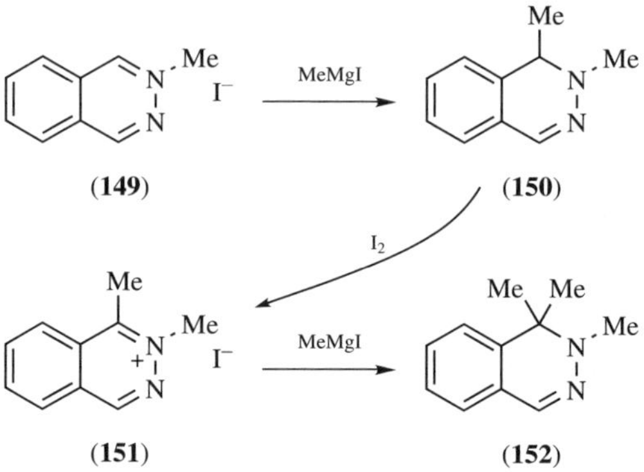

(**149**) (**150**) (**151**) (**152**)

Also other examples.[393]

N-Alkylation

2-Methylphthalazinium iodide (**153**) underwent reduction to afford 2-methyl-1,2-dihydrophthalazine (**154**) ($NaBH_4$, H_2O, exothermic, 3 h: unstable crude) and thence 2,2-dimethyl-1,2-dihydrophthalazinium iodide (**155**) (MeI, MeOH, reflux, 3 h: 61% overall).[117]

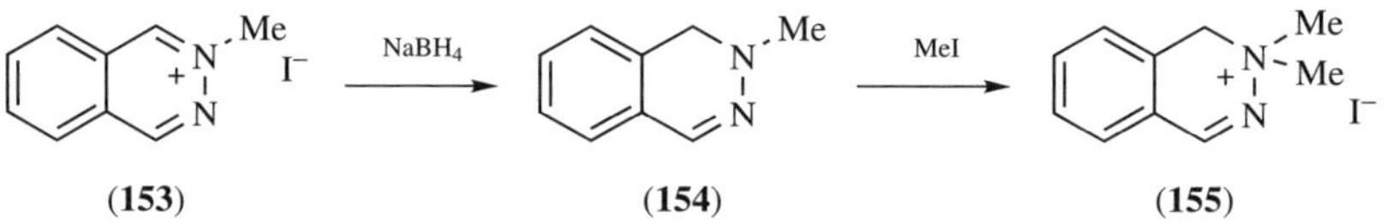

(**153**) (**154**) (**155**)

Bi(dihydrophthalazinylidene) Derivatives Formation

2-Phenylphthalazinium perchlorate (**156**) afforded 2,2′-diphenyl-1,1′-bi(1,2-dihydrophthalazinylidene) (**157**) (AcONa, MeCN, reflux, 3 h: 71%); analogs like 2,2′-bis(*p*-nitrophenyl)-1,1′-bi(1,2-dihydrophthalazinylidene) (60%) were made similarly.[714,cf. 943]

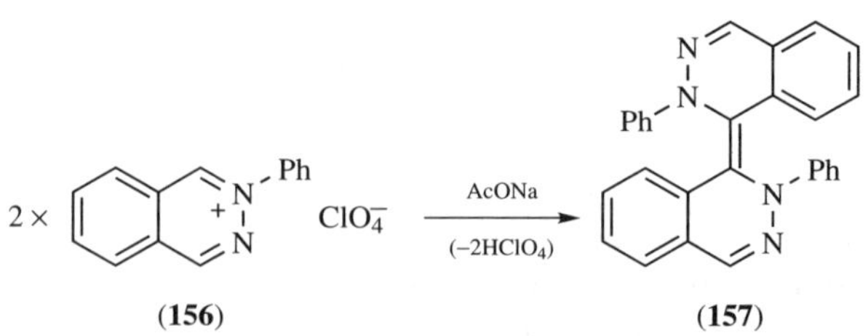

(**156**) (**157**)

Covalent Adduct Formation

Note: Although the products have not been isolated, 2-alkylphthalazinium salts are appreciably converted into the adducts 2-methyl-1,2-dihydro-1-phthalazinamine (**158**) (liquid NH_3),[97] 2-benzyl-1-ethylthio-1,2-dihydrophthalazine (**159**) (EtS^-), and the nitromethanide adduct (**160**) ($MeNO_2$ in liquid NH_3).[100]

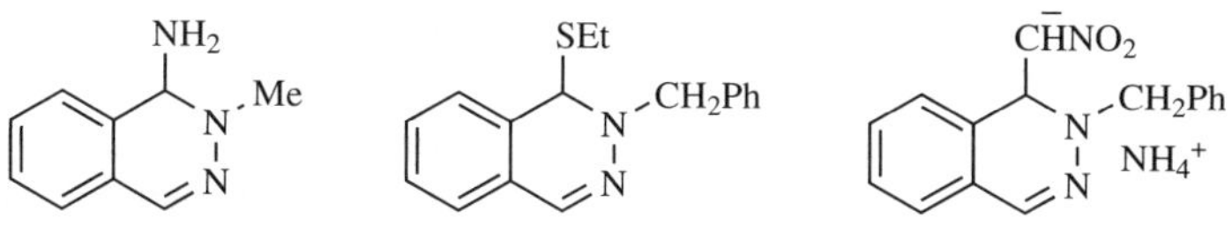

(**158**) (**159**) (**160**)

Cyclocondensations

Note: Most such condensations involve acyclic or cyclic derivatives of ethylene or derivatives of acetylene as synthons.

2-Phenacylphthalazinium bromide (**161**) with acrylonitrile gave 2-benzoyl-1,2,3,10b-tetrahydropyrrolo[2,1-*a*]phthalazine-1-carbonitrile (**162**), which underwent oxidation to afford 3-benzoylpyrrolo[2,1-*a*]phthalazine-1-carbonitrile (**163**) [best done as a one-pot procedure; substrate, $H_2C{=}CHCN$, Et_3N, pyridine$_4$Co$(HCrO_4)_2$ as oxidant, Me_2NCHO, 85°C, 3 h: 67%]; analogs likewise.[847]

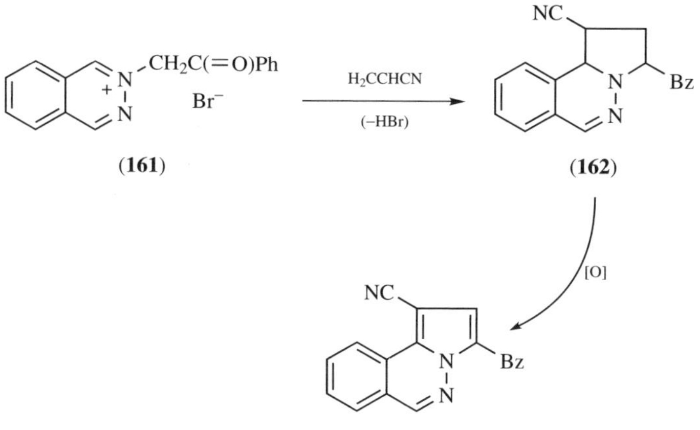

(**161**) $\xrightarrow[(-HBr)]{H_2CCHCN}$ (**162**) $\xrightarrow{[O]}$ (**163**)

2-Ethoxycarbonylmethylphthalazinium bromide (**164**) with 1-nitro-2-phenylthioethylene gave mainly ethyl pyrrolo[2,1-*a*]phthalazine-3-carboxylate (**165**, R = H) accompanied initially by a small amount of its 1-nitro derivative (**165**, R = NO_2) (EtOH, NEt_3, reflux, 6 h: 35% and 7%, respectively; note that the loss of PhSH and HNO_2 precludes any necessity for an oxidant).[582]

CH_2CO_2Et Br$^-$ PhSCH=CHNO$_2$ (–HBr, –PhSH, –HNO$_2$) R CO_2Et

(164) (165)

Dicyano(phthalazin-1-io)methanide (**166**) with acetylethylene gave a separable mixture of 1-*endo*-acetyl- and 1-*exo*-acetyl-1,2,3,10b-tetrahydropyrrolo[2,1-*a*]phthalazine-2,3-dicarbonitrile (**167**) (MeCN, 20°C, 4 h: 72% and 23%, respectively), with stilbene gave 1,2-diphenylpyrrolo[2,1-*a*]phthalazine-3-carbonitrile (**168**) (MeCN, reflux, 24 h: 60%), or with *N*-*p*-methoxyphenylmaleimide gave 9,11-dioxo-8a,9,10,11,11a,11b-hexahydro-8*H*-pyrrolo[3′,4′:3,4]pyrrolo[2, 1-*a*]phthalazine-8,8-dicarbonitrile (**169**) (MeCN, 20°C, 24 h: 91%).[839]

AcCH=CH$_2$ PhC≡CPh (–HCN) C_6H_4OMe-*p*

(166) (167) (168) (169)

The same substrate (**170**) with methyl propiolate gave methyl 3,3-dicyano-3,10b-dihydropyrrolo[2,1-*a*]phthalazine-1-carboxylate (**171**) (MeCN, 20°C, 24 h: 81%) or with phenylacetylene gave 1-phenylpyrrolo[2,1-*a*]phthalazine-3-carbonitrile (**172**) (MeCN, reflux, 24 h: 82%; note loss of HCN).[836]

HC≡CCO$_2$Me PhC≡CPh (–HCN) MeO_2C

(170) (171) (172)

2-Ethoxycarbonylmethylphthalazinium bromide (**173**) with dimethyl acetylenedicarboxylate gave dimethyl 3-ethoxycarbonylpyrrolo[2,1-*a*]phthalazine-1,2-dicarboxylate (**174**) (reactants, MeCN, NEt_3 in MeCN↓ dropwise, reflux, 4 h: 30%).[997]

CH_2CO_2Et Br^- $\xrightarrow{MeO_2CC\equiv CCO_2Me}$ MeO_2C CO_2Me CO_2Et

(**173**) (**174**)

2-Phenacylphthalazinium bromide (**175**) with ammonium acetate and an oxidant gave 2-phenylimidazo[2,1-*a*]phthalazine (**176**) [reactants, pyridine$_4$Co($HCrO_4$)$_2$, AcOH, reflux, 3 h: 55%].[827]

$CH_2C(=O)Ph$ Br^- $\xrightarrow{AcONH_4,\ [O]}$ Ph

(**175**) (**176**)

Also other examples.[165,170,189,215,358,382,578,838,889,976]

Dequaternization

2-Methylphthalazinium iodide (**177**) gave phthalazine (**178**) (neat anhydrous pyridine hydrochloride, reflux (215°C), 10 min; hot mixture↓ NH_4OH, ice: >95%).[90]

Me I^- $\xrightarrow{\text{pyridine hydrochloride}}$

(**177**) (**178**)

Reductive Ring Fission

2-Benzyl-1-phenylphthalazinium chloride (**179**) suffered reductive ring fission to give *o*-aminomethyl-α-benzylamino-α-phenyltoluene (**180**) (substrate, M BH_3/THF (>4 equiv)↓ portionwise during reflux, >3 days: ?%).[119]

Ph CH_2Ph Cl^- $\xrightarrow{BH_3}$ Ph $CHNHCH_2Ph$ CH_2NH_2

(**179**) (**180**)

Self-Condensation

Treatment of 4-dimethylamino-2-phenacylphthalazinium bromide (**181**) with triethylamine gave, not the corresponding ylide, but its dimer 8,16-dibenzoyl-5,13-bis(dimethylamino)-8,8a,16,16a-tetrahydropyrazino[2,1-*a*:5,4-*a*′] diphthalazine (**182**) (Et_3N, $CHCl_3$, reflux: one stereoisomer; or Et_3N, PhMe, reflux: another stereoisomer).[943]

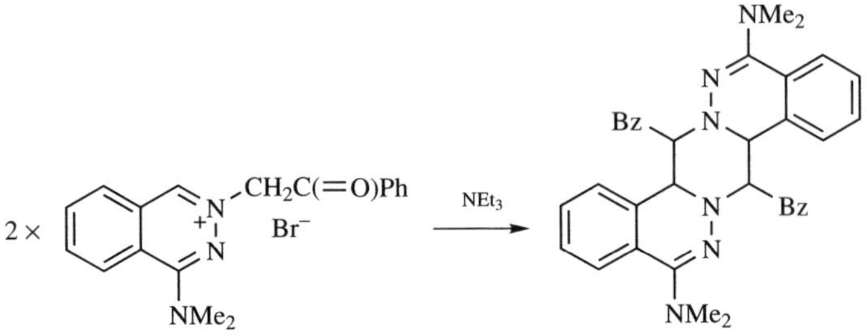

2 × $CH_2C(=O)Ph$ Br^- NMe_2 —NEt_3→ Bz Bz NMe_2 NMe_2

(**181**) (**182**)

CHAPTER 10

Halogenophthalazines (*H* 178; *E* 514)

In the phthalazine system, 1- and 4-halogeno substituents are reasonably activated, each by its adjacent ring nitrogen atom. Halogeno substituents attached to the carbocyclic ring are only marginally more active than those of a halogenonaphthalene. Extranuclear halogeno substituents resemble in activity those of a benzyl halide and are more affected by any adjacent group(s) than by the ring system.

10.1. PREPARATION OF NUCLEAR HALOGENOPHTHALAZINES (*H* 178; *E* 515)

The preparation of such halogenophthalazines by *primary synthesis* has been discussed in Chapter 8. Other methods are outlined in the following subsections.

10.1.1. From Tautomeric Phthalazinones by Halogenolysis

This is the preferred route to 1- and 4-chlorophthalazines. It is illustrated by the following examples, classified according to the reagent(s) used.

Using Neat Phosphoryl Chloride

1(2*H*)-Phthalazinone (**1**, R = H) gave 1-chlorophthalazine (**2**, R = H) ($POCl_3$, 85°C, 4 h: 28%).[282]

O NH N R —POCl3→ Cl N N R

(**1**) (**2**)

4-*p*-Fluorophenyl-1(2*H*)-phthalazinone (**1**, R = C_6H_4F-*p*) gave 1-chloro-4-*p*-fluorophenylphthalazine (**2**, R = C_6H_4F-*p*) ($POCl_3$, 100°C, 90 min: 93%).[958]

Cinnolines and Phthalazines: Supplement II, The Chemistry of Heterocyclic Compounds, Volume 64, by D.J. Brown

4-Phenyl-1(2*H*)-phthalazinone (**3**) gave 1-chloro-4-phenylphthalazine (**4**) ($POCl_3$, 85°C, 4 h: 79%);[282] 4-*p*-dimethylaminophenyl-4-phenyl-3,4-dihydro-1(2*H*)-phthalazinone (**5**) gave 1-chloro-4-*p*-dimethylaminophenyl-4-phenyl-3,4-dihydrophthalazine (**6**) ($POCl_3$, 80°C, 6 h: 58%) and thence 1-chloro-4-phenylphthalazine (**4**) ($POCl_3$, reflux, 6 h: 71%; note nuclear oxidation by loss of $PhNMe_2$);[409] as might be expected, the substrate (**5**) also afforded product (**4**) directly ($POCl_3$, reflux, 6 h: 69%).[409]

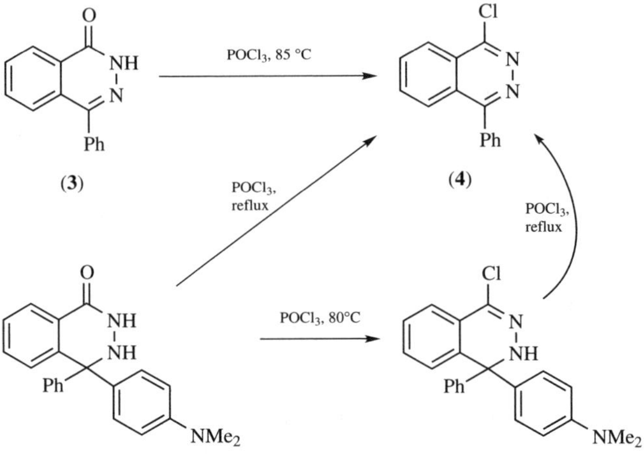

Ethyl 5,8-dimethyl-1-oxo-1,2-dihydro-6-phthalazinecarboxylate gave ethyl 1-chloro-5,8-dimethyl-6-phthalazinecarboxylate (**7**) ($POCl_3$, 90°C, 15 min: 70%);[421] somewhat similarly, ethyl 1-*o*-methoxyphenyl-5,7-dimethyl-4-oxo-3,4-dihydro-6-phthalazinecarboxylate gave 4-chloro-1-*o*-methoxyphenyl-5,7-dimethyl-6-phthalazinecarboxylate (**8**) ($POCl_3$, reflux, 30 min: 86%).[426]

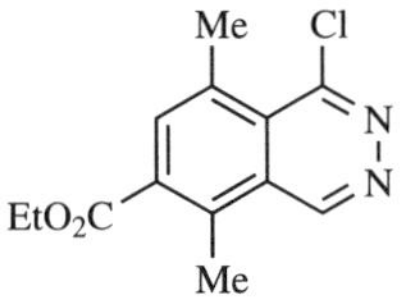

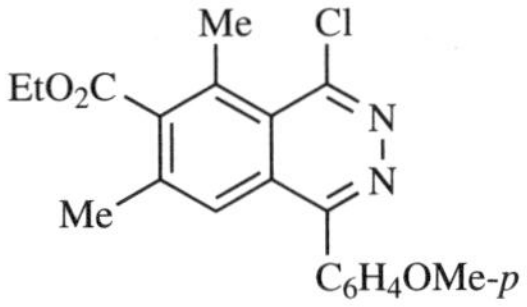

2-Methyl-1,4(2*H*,3*H*)-phthalazinedione (**9**) gave 4-chloro-2-methyl-1(2*H*)-phthalazinone (**10**) ($POCl_3$, reflux, 2 h: ?%; note the immunity of the nontautomeric oxo substituent to halogenolysis).[685]

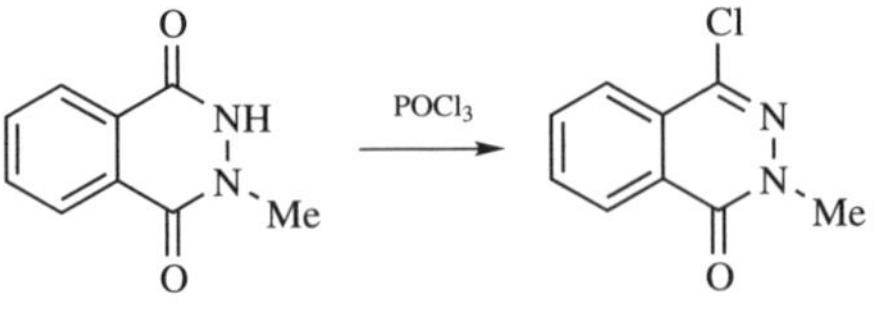

Also many other examples.[184,222,241,249,261,542,650,675,778,783,862,920,936]

Using Phosphoryl Chloride in a Solvent

4-(Pyridin-4-ylmethyl)-1(2*H*)-phthalazinone (**11**, X = CH gave 1-chloro-4-(pyridin-4-ylmethyl)phthalazine (**12**, X = CH) [$POCl_3$, HCl gas in dioxane (2 equiv), MeCN, 50°C, 27 h: 92%]; 4-(pyrimidin-4-ylmethyl)-1(2*H*)-phthalazinone (**11**, X = N) likewise gave 1-chloro-4-(pyrimidin-4-ylmethyl)phthalazine (**12**, X = N) (82%).[870]

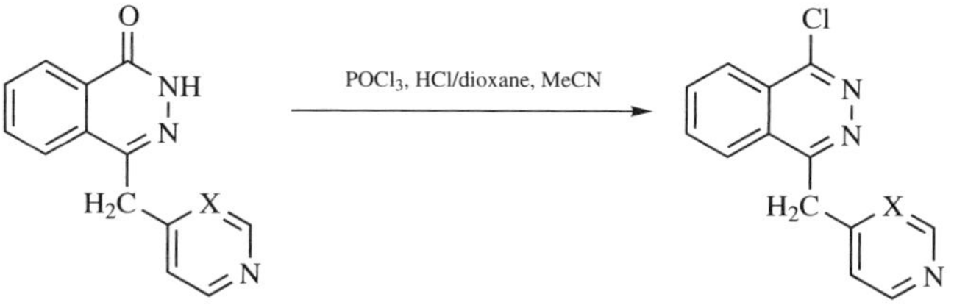

(**11**) (**12**)

Using Phosphorus Pentachloride in Phosphoryl Chloride

1,4(2*H*,3*H*)-Phthalazinedione (**13**) gave 1,4-dichlorophthalazine (**14**) (PCl_5, $POCl_3$, reflux, 5 h: ~55%).[512]

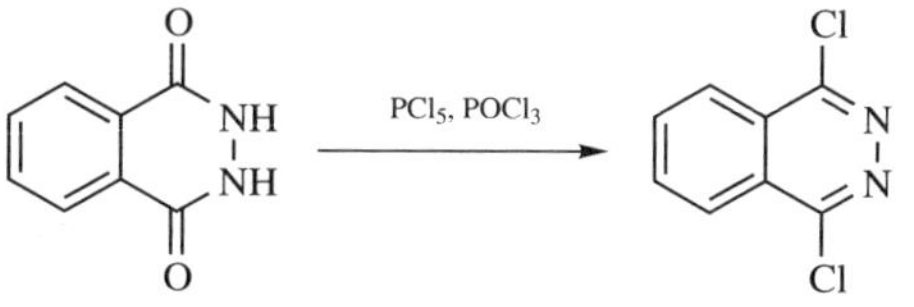

(**13**) (**14**)

4-[*N*-(1-Carboxyethyl)amino]-1(2*H*)-phthalazinone (**15**) gave 1-[*N*-(1-carboxyethyl)amino]-4-chlorophthalazine (**16**) (PCl_5, $POCl_3$, 95°C, 4 h: 60%).[668]

(**15**) (**16**)

1,2-Bis(4-oxo-3,4-dihydrophthalazin-1-yl)ethane (**17**) gave 1,2-bis(4-chlorophthalazin-1-yl)ethane (**18**) (PCl_5, $POCl_3$, 95°C, 3 h: 85%).[665]

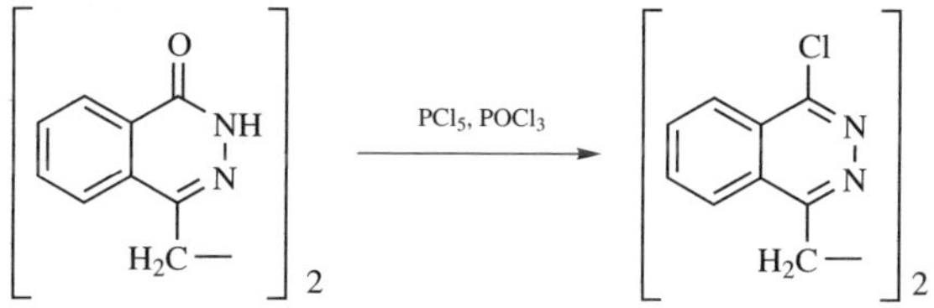

(**17**) (**18**)

Also other examples.[264,268,311,342,996]

Using Phosphoryl Chloride and a Tertiary Base

Note: This excellent procedure has seldom been used in the more recent phthalazine literature.

6-Chloro-1,4(2*H*,3*H*)-phthalazinedione (**19**) gave 1,4,6-trichlorophthalazine (**20**) ($POCl_3$, $EtPr^i_2N$, reflux, 3 h: 82%); analogs likewise.[285]

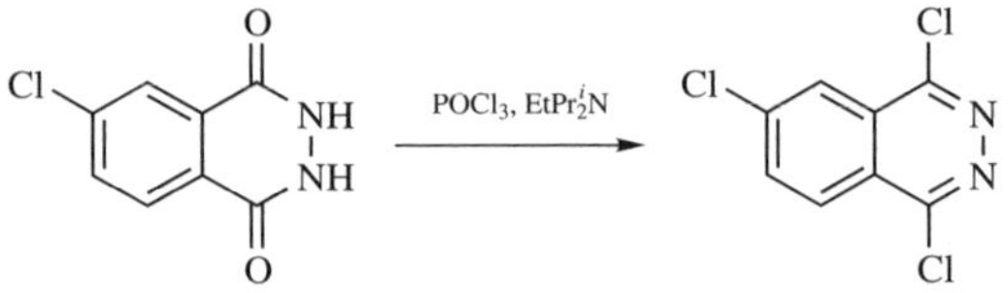

(**19**) (**20**)

Using a Vilsmeier Reagent

Note: This useful procedure (involving dimethylformamide with a phosphorus halide, thionyl halide, or phosgene) has been neglected in the 1972–2004 phthalazine literature.

1,4(2*H*,3*H*)-Phthalazinedione (**21**) gave 1,4-dichlorophthalazine (**22**) (PCl_5, trace Me_2NCHO, 20°C → 145°C during 1 h, then 145°C, 4 h: 78%);[1003] it appears that PCl_5 and/or $POCl_3$ may also be used (no details).[122]

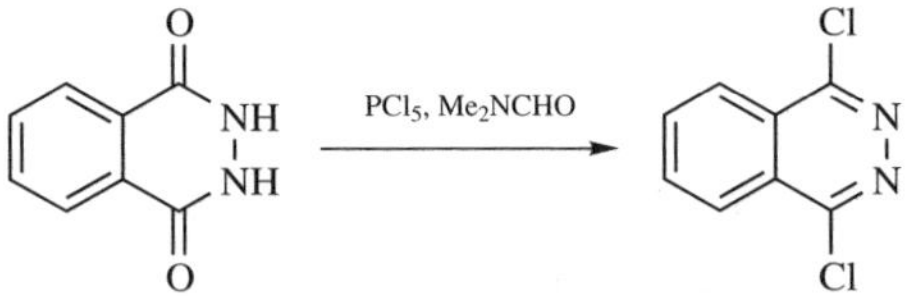

(**21**) (**22**)

10.1.2. From Other Substrates

Other routes to nuclear halogenophthalazines are poorly represented in the 1972–2004 literature, as illustrated by the following examples.

By Direct or Indirect *C*-Halogenation

5,8-Phthalazinequinone (**23**, X = H) gave 6,7-dibromo- (**23**, X = Br) or 6,7-dichloro-5,8-phthalazinequinone (**23**, X = Cl) (Br_2 or Cl_2, CCl_4, $CHCl_3$, 20°C, 1 h: 63% or 87%, respectively).[715]

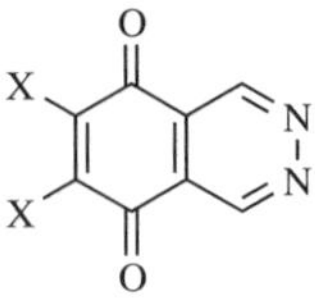

(**23**)

6-Chloro-1,4-dimethoxyphthalazine (**24**, R = H) underwent *C*-lithiation at the 7-position and subsequent halogenolysis to afford 6-chloro-7-iodo-1,4-dimethoxyphthalazine (**24**, R = I) (substrate, THF, −78°C, A; BuLi in C_6H_{14}↓ slowly; −78°C, 30 min; then I_2↓, −78°C, 2 h: 83%).[311]

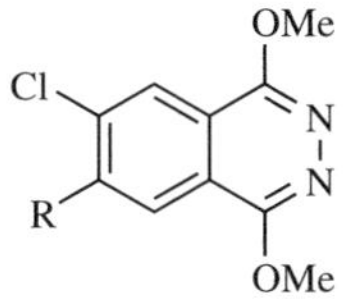

(**24**)

Also other examples.[275,856]

By Transhalogenation

1,4-Dichlorophthalazine (**25**) gave a separable mixture of 1-chloro-4-fluoro- (**25a**) and a little 1,4-difluorophthalazine (**25b**) {substrate, MeCN, [1,8-bis(dimethylamino)naphthalene (3 parts) + $Et_3N \cdot 3HF$ (1 part)]↓ in 3 portions after 0, 24, and 48 h during reflux, 96 h: 37% and ?%, respectively}.[887,891]

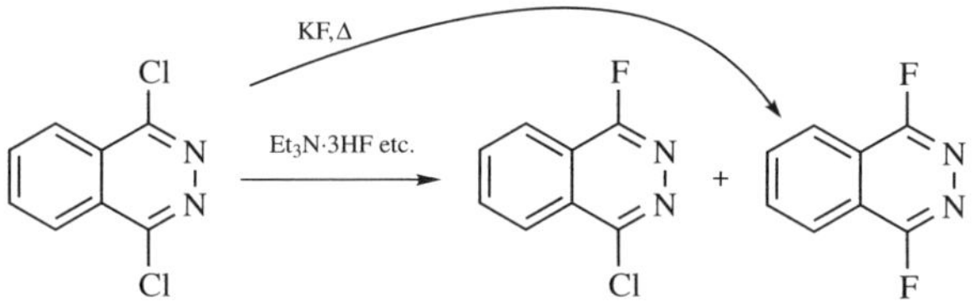

KF,Δ

$Et_3N \cdot 3HF$ etc.

(**25**) (**25a**) (**25b**)

The same substrate (**25**) gave only 1,4-difluorophthalazine (**25b**) (substrate, neat KF, 250°C, 6 h: 73%).[723]

From Phthalazinediazonium Salts

Ethyl 8-amino- (**26**, R = NH_2) underwent diazotization in "fluoroboric acid" to give ethyl 8-fluoro-1-hydroxymethyl-5,7-dimethyl-4-oxo-3,4-dihydro-6-phthalazinecarboxylate (**26**, R = F) [substrate, HBF_4, H_2O, 0°C, $NaNO_2$ in H_2O↓, 30 min; solid, A, Δ, gas↑: 7%, after separation from the product of spontaneous cyclization, ethyl 4,6-dimethyl-3-oxo-2,3-dihydro-8*H*-furo[4,3,2-*de*]phthalazine-5-carboxylate (**26a**) (31%)].[425]

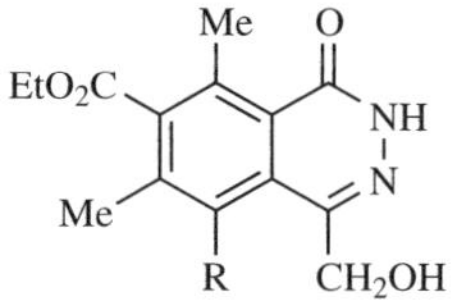

(**26**) 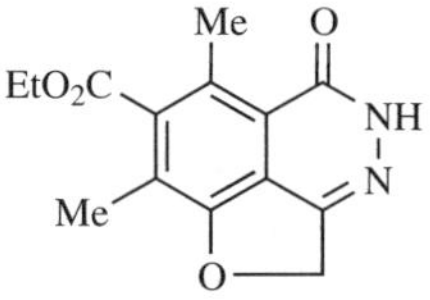 (**26a**)

10.2. PREPARATION OF EXTRANUCLEAR HALOGENOPHTHALAZINES

The formation of extranuclear halogenophthalazines by *halogenation of alkyl- or arylphthalazines* has been covered in Section 9.2.2. Other routes to such halogeno derivatives are illustrated in the following examples.

From Extranuclear Hydroxyphthalazines

Note: This has been done by treatment with a thionyl halide, a phosphorus halide, hydrobromic acid, or a Vilsmeier reagent. The choice is sometimes dependent on the nature of the substrate; for example, the use of thionyl chloride lessens any danger of concomitant halogenolysis of nuclear oxo substituents.

2-(2-Hydroxyethyl)-3-methyl-1,4(2*H*,3*H*)-phthalazinedione (**27**, R = OH) gave 2-(2-chloroethyl)-8-methyl-1,4(2*H*,3*H*)-phthalazinedione (**27**, R = Cl) ($SOCl_2$, $CHCl_3$, reflux, 16 h: 75%);[745] likewise, 7-hydroxymethyl- gave 7-chloromethyl-2-phenyl-1(2*H*)-phthalazinone (neat $SOCl_2$, reflux, 1 h: 78%).[411]

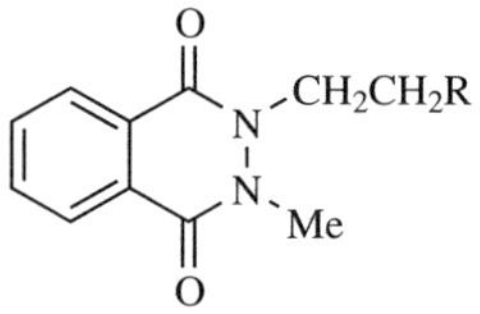

O
N CH2CH2R
N Me
O

(**27**)

4-Ethoxycarbonylmethyl-2-hydroxymethyl-1(2*H*)-phthalazinone (**28**, R = OH) gave 2-bromomethyl-4-ethoxycarbonylmethyl-1(2*H*)-phthalazinone (**28**, R = Br) (PBr_3, Et_2O, 20°C, 12 h: 96%; this reagent posed no threat of halogenolysis toward the nontautomeric oxo substituent).[68]

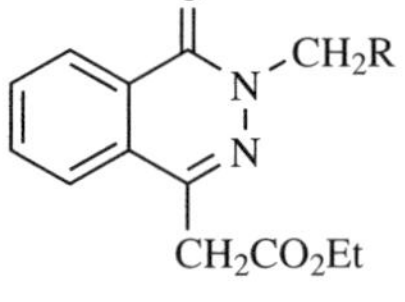

O
N CH2R
N
CH2CO2Et

(**28**)

Ethyl 1-hydroxymethyl- (**29**, R = OH) gave ethyl 1-chloromethyl-5,7-dimethyl-8-nitro-4-oxo-3,4-dihydro-6-phthalazinecarboxylate (**29**, R = Cl) ($POCl_3$, reflux, "short time": 78%);[425] also an analogous transformation.[195]

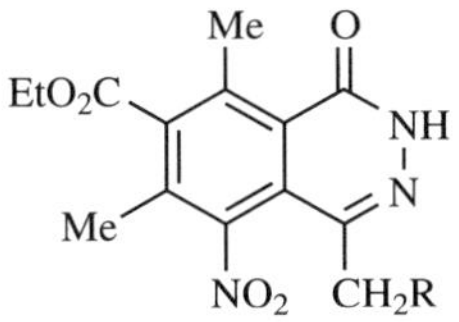

(29)

2-(4-Hydroxybutyl)-4-(2-hydroxyethylamino)-1(2*H*)-phthalazinone (**30**) gave 2-(4-bromobutyl)-4-(2-hydroxyethylamino)-1(2*H*)-phthalazinone (**31**) (48% HBr, reflux, 5 min: 78%; longer reaction times induced a second halogenolysis and subsequent cyclization).[301]

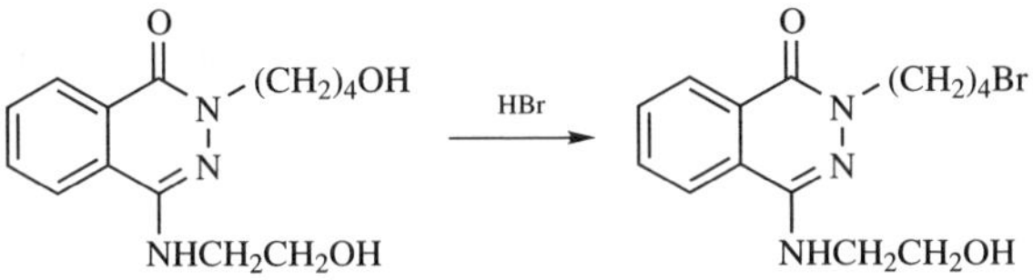

HBr

(30) (31)

2,3-Bis[*m*-(hydroxymethyl)phenyl]-1,4(2*H*,3*H*)-phthalazinedione (**32**, R = OH) gave 2,3-bis[*m*-(bromomethyl)phenyl]-1,4(2*H*,3*H*)-phthalazinedione (**32**, R = Br) [substrate, MeCN, $(Me_2N{=}CHBr)Br$ (preparative details given)↓, reflux, 10 h: 70%; a rare example of the use of an isolated Vilsmeier reagent for halogenolysis].[134]

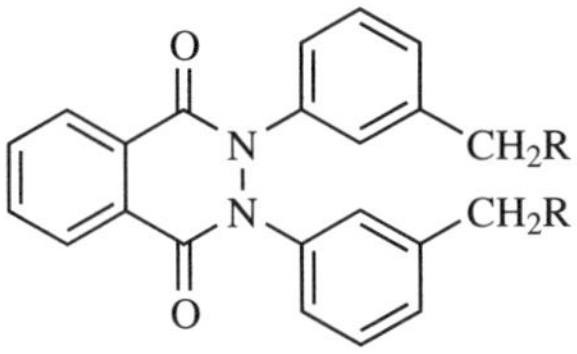

(32)

Also other examples.[405]

By Passenger Introduction

Note: Many examples of such passenger introduction of extranuclear halogeno substituents are scattered through the chapters of this book. A single random example is given here.

1-Chloro- (**33**) gave 1-[(3,5-dichloropyridin-4-yl)methyl]-6-methoxy-5-(5-phenylpent-1-ynyl)phthalazine (**34**) [3,5-Cl_2-4-Me-pyridine, NaH, Me_2NCHO, 20°C, 3 h: 51%].[861]

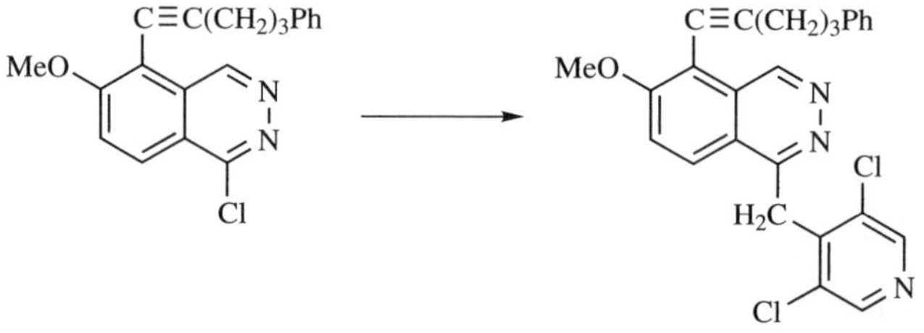

(**33**) (**34**)

10.3. REACTIONS OF HALOGENOPHTHALAZINES (*H* 178, 182; *E* 519)

Both nuclear and extranuclear halogenophthalazines are valuable intermediates for the preparation of other phthalazine derivatives. The *alkanelysis of nuclear halogenophthalazines* has been exemplified in Section 9.2.1, and other reactions of both nuclear and extranuclear halogeno derivatives are covered in the subsections that follow.

The three-dimensional shape of 1-chloro-4-dimethylaminophthalazine has been determined by X-ray analysis,[740] the effect of 1,4-difluoro substituents on the electrochemical reduction of phthalazine has been studied for comparison with similar data for such substitution in other heterocyclic systems,[723] and a calculated structure for 1-chlorophthalazine appears to explain its ready dimerization.[898]

10.3.1. Hydrogenolysis of Halogenophthalazines (*E* 520)

This is done readily by hydrogenation over a palladium catalyst, but electrochemical reduction has also been used. Both processes are illustrated in the following examples.

1-Chloro-4-hydrazinophthalazine (**35**, R = Cl) gave 1-hydrazinophthalazine (**35**, R = H) as its hydrochloride [substrate, H_2 (1 atm), Pd/C, MeOH, HCl, 50°C, 4 h: 91%; likewise but 20°C, 7 h: 84%; likewise but H_2 (15 atm), 80°C, 30 min: 90%];[593] 4-chloro-1(2*H*)-phthalazinone (**36**, R = Cl) somewhat similarly gave 1(2*H*)-phthalazinone (**36**, R = H) [substrate, AcONa, H_2 (1 atm), Pd/C, MeOH, 20°C, 8 h: 89%; substrate, H_2 (6 atm), EtOH, Pd/C, 60°C, 6 h: >95%].[593]

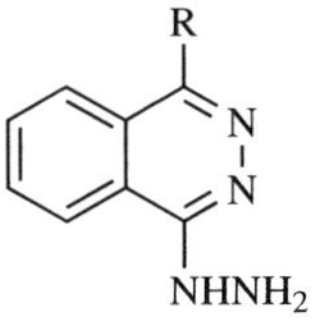

(**35**) 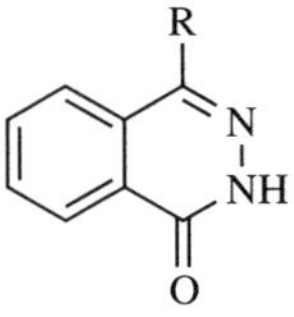 (**36**)

Ethyl 4-chloro-1-*o*-methoxyphenyl-5,7-dimethyl-6-phthalazinecarboxylate (**37**, R = Cl) gave ethyl 1-*o*-methoxyphenyl-5,7-dimethyl-6-phthalazinecarboxylate (**37**, R = H) [substrate, H_2 (1 atm), Pd/C, EtOH, trace NH_4OH, 20°C, 8 h: 44%].[426]

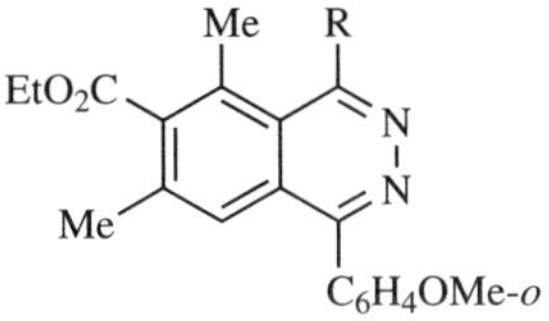

(**37**)

1,4-Dichloro-5,6,7,8-tetrahydrophthalazine (**38**, R = Cl) gave 5,6,7,8-tetrahydrophthalazine (**38**, R = H) [substrate, H_2 (1 atm), Pd/C, KOH, EtOH, until uptake ceased: 95%].[393]

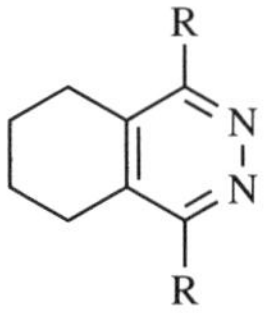

(**38**)

1-Chloro-4-(*N*-methylhydrazino)phthalazine (**39**, R = Cl) underwent electrochemical reductive dechlorination to afford 1-(*N*-methylhydrazino)phthalazine (**39**, R = H) (see original for procedural details: 48%).[734]

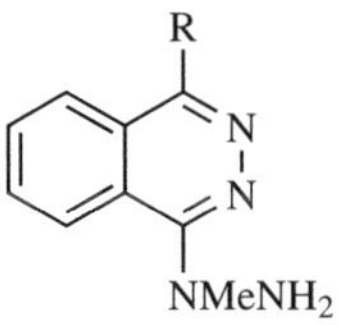

(**39**)

Also other examples.[931]

10.3.2. Aminolysis of Halogenophthalazines (*E* 560)

Aminolysis of both nuclear and extranuclear halogenophthalazines has been employed extensively to prepare a variety of primary, secondary, and tertiary aminophthalazines and related products. The illustrative examples that follow are classified according to the type of product formed.

Formation of (Primary) Phthalazinamines

1,4-Dichlorophthalazine (**40**) gave 4-chloro-1-phthalazinamine (**41**) (substrate, Me_2NCHO, "hot," $NH_4OH\downarrow$ slowly during 2 h: 54%, isolated as sulfate, or 40% as base;[742] NH_4OH, 120°C, sealed, 4 h: ?%).[884,cf. 513]

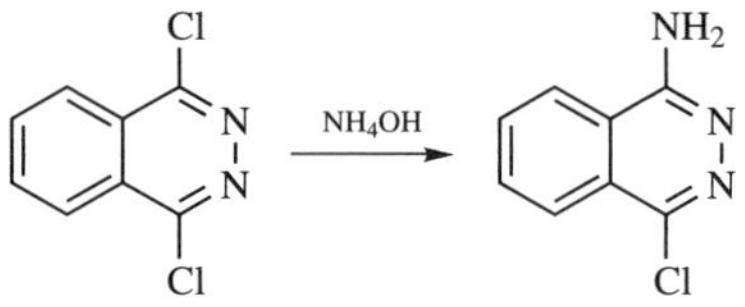

Cl, N, N, Cl; NH_4OH →; NH_2, N, N, Cl

(40) **(41)**

4-Chloro-2-methyl-1(2*H*)-phthalazinone (**42**, R = Cl) gave 4-amino-2-methyl-1(2*H*)-phthalazinone (**42**, R = NH_2) [$(NH_4)_2CO_3$, NH_4OH (*d*.0.88), 213°C, sealed, 20 h: ?%].[685]

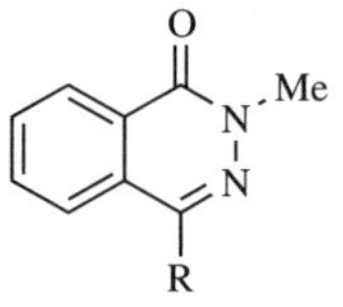

O, N, Me, N, R

(42)

7-Chloromethyl- gave 7-aminomethyl-2-phenyl-1(2*H*)-phthalazinone [neat NH_4OH (*d*.0.88), 20°C, <40 h: 82%].[411]

1-(4-Bromo-3-methylphenyl)-4-chlorophthalazine (**43**, R = Cl) gave 4-(4-bromo-3-methylphenyl)-1-phthalazinamine (**43**, R = NH_2) ($NaNH_2$, see original for detail; note survival of the extranuclear Br substituent).[342]

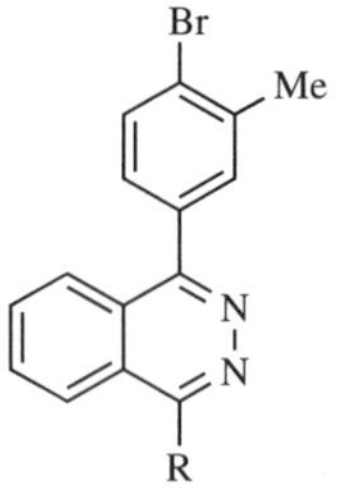

Br, Me, N, N, R

(43)

Also other examples.[62]

Formation of (Secondary) Alkylaminophthalazines

1-Chlorophthalazine (**44**, R = Cl) gave 1-methylaminophthalazine (**44**, R = NHMe) ($MeNH_2$, MeOH, 75°C, sealed, 4 days: 85%) or 1-propylaminophthalazine (**44**, R = NHPr) ($PrNH_2$, likewise: 84%).[62]

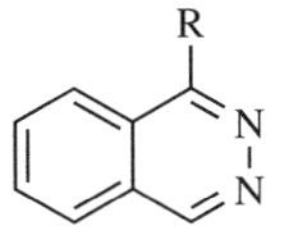

(**44**)

1-Chloro-4-phenylphthalazine (**45**, R = Cl) gave 1-(2-hydroxyethylamino)-4-phenylphthalazine (**45**, R = $NHCH_2CH_2OH$) ($HOCH_2CH_2NH_2$, dioxane, reflux, 19 h: 84%)[692] or 1-[(2-dimethylaminoethyl)amino]-4-phenylphthalazine (**45**, R = $NHCH_2CH_2NMe_2$) (neat $H_2NCH_2CH_2NMe_2$, 130°C, 90 min: 65% as dihydrochloride; analogs likewise).[987]

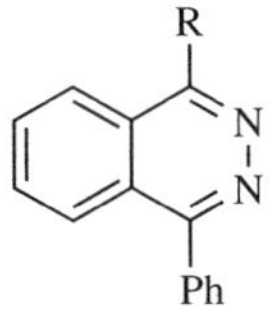

(**45**)

1-[(1-Carboxyethyl)amino]-4-chlorophthalazine (**46**, R = Cl) gave 1-benzylamino-4-[(1-carboxyethyl)amino]phthalazine (**46**, R = $NHCH_2Ph$) ($PhCH_2NH_2$, EtOH, reflux, 3 h: 65%).[668]

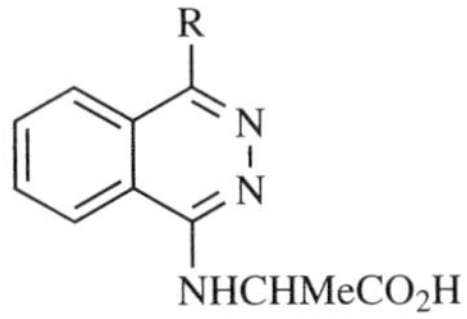

(**46**)

The unsymmetric substrate 1,4-dichloro-6-phthalazinecarbonitrile (**47**) with 3,4-methylenedioxybenzylamine (**48**) gave a mixture of 1-chloro-4-[(3,4-methylenedioxybenzyl)amino]- (**49**), 4-chloro-1-[(3,4-methylenedioxybenzyl)amino]- (**50**), and 1,4-bis[(3,4-methylenedioxybenzyl)amino]-6-phthalazinecarbonitrile [reactants, *N*-methylpyrrolidinone, diazabicycloundecene (catalyst), 20°C, 2 h: 44%, 15%, and a trace (detected but not isolated), respectively];[285] also analogs.[285,871]

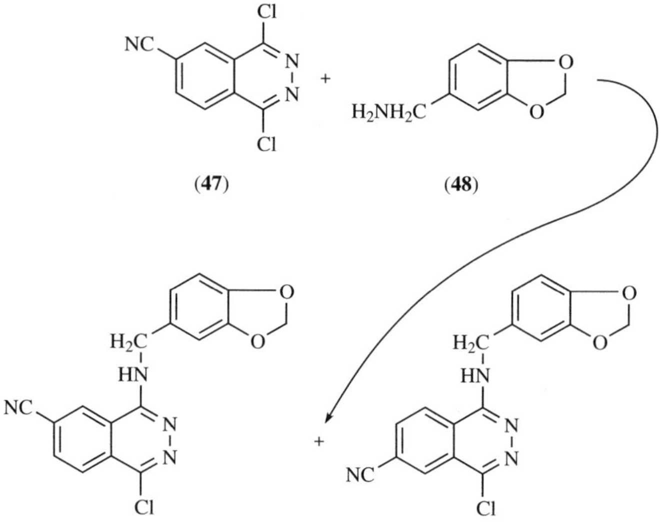

(47) (48)

(49) (50)

Also other examples.[178,196,281,342,420,470,665,742,870]

Formation of (Secondary) Arylaminophthalazines

1-Chlorophthalazine and *o*-aminophenol gave 1-(*o*-hydroxyanilino)phthalazine (**51**) (EtOH, reflux, 2 h: 71%; analogs likewise.[224]

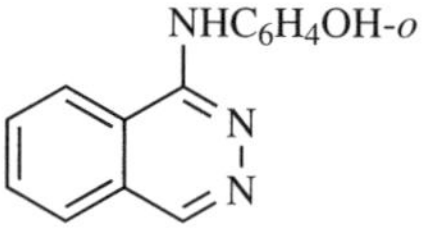

(**51**)

4-Chloro-2-phenyl-1(2*H*)-phthalazinone with *p*-aminoacetophenone gave 4-*p*-acetylanilino-2-phenyl-1(2*H*)-phthalazinone (**52**) [synthon, EtONa in EtOH, reflux, 5 h (presumably to form a sodio derivative?); substrate↓, reflux, 10 h: 75%].[675]

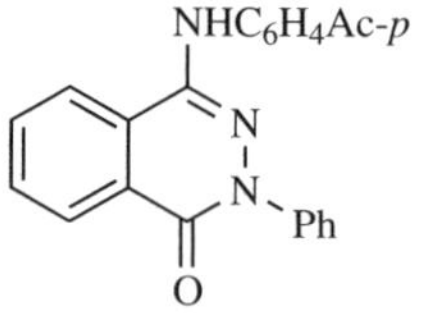

(**52**)

1-Chloro-5-cyclopentyloxy-6-methoxyphthalazine with 3,5-dichloro-4-pyridinamine gave 5-cyclopentyloxy-1-(3,5-dichloropyridin-4-ylamino)-6-methoxyphthalazine (**53**) (reactants, NaH, Me_2NCHO, 20°C, 90 min: >22%).[861]

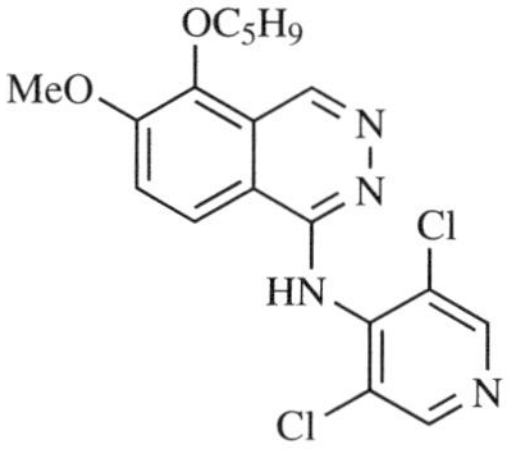

(53)

6,7-Dichloro-5,8-phthalazinequinone (**53a**, R = Cl) gave 6-anilino-7-chloro-5,8-phthalazinequinone (**53a**, R = NHPh) ($PhNH_2$, EtOH, 20°C, 20 h: 96%); analogs likewise.[1017]

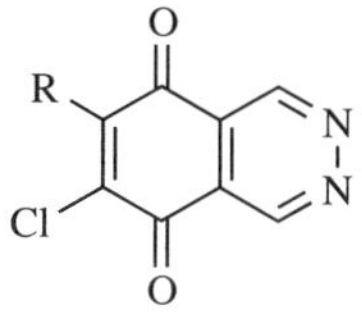

(53a)

Also other examples.[649,680,742,856]

Formation of (Tertiary) Dialkylaminophthalazines or the Like

Note: As might be expected, the rates for aminolysis of 4-chloro-1(2*H*)-phthalazinone by both acyclic and cyclic secondary amines are profoundly affected by the inherent steric hindrance of the attacking amine.[205]

1,4-Dichlorophthalazine with *N*-methylaniline gave 1-chloro-4-(*N*-methylanilino)phthalazine (**54**, R = NMePh) (substrate, EtOH, reflux, synthon↓ slowly, reflux, 35 min: ?%),[742] with morpholine gave 1-chloro-4-morpholinophthalazine [**54**, R = $N(CH_2CH_2)_2O$] [substrate (1 mol), synthon (2 mol), EtOH, reflux, 10 h: 70%],[271] or with 1-methylpiperazine gave 1-chloro-4-(4-methylpiperazin-1-yl)phthalazine [**54**, R = $N(CH_2CH_2)_2NMe$] (reactants, Et_3N, BuOH, 100°C, sealed, 24 h: 70%).[880]

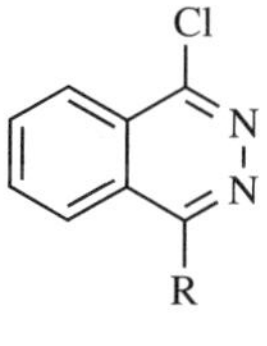

(54)

1,4-Dichloro-6-phenylphthalazine with morpholine gave 1,4-dimorpholino-6-phenylphthalazine (**55**) (excess neat amine, 120°C, 2 h: 60%).[395]

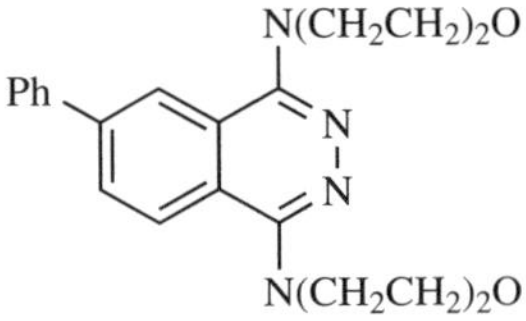

(**55**)

1-Chlorophthalazine with 1-methylpiperazine gave 1-(4-methylpiperazin-1-yl) phthalazine (**56**, R = H) (substrate, excess neat synthon, reflux, 2 h: 73%);[368] 1-chloro-4-phenylphthalazine with 1-methylpiperazine gave 1-(4-methylpiperazin-1-yl)-4-phenylphthalazine (**56**, R = Ph) (reactants, NH_4Cl, BuOH, reflux, 48 h: 61%).[282]

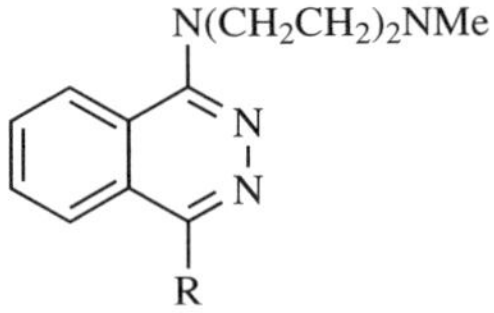

(**56**)

1,4-Dichlorophthalazine (**57**) with imidazole gave (unisolated) 1-chloro-4-(imidazol-1-yl)phthalazine (**58**) and thence 4-(imidazol-1-yl)-1(2*H*)-phthalazinone (**59**) (imidazole, NaH, Me_2NCHO, 20°C, N_2, 30 min; substrate↓, 0°C → 20°C, 2 h; NaOH in H_2O↓, 20°C, 1 h: 34% overall).[279]

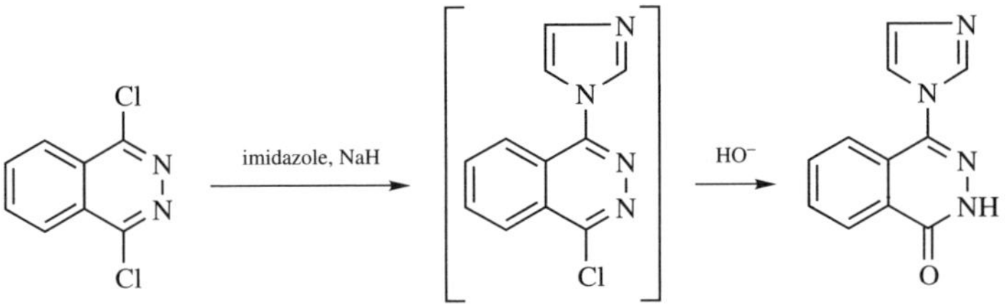

(**57**) (**58**) (**59**)

2-(4-Bromobutyl)-4-methyl-1(2*H*)-phthalazinone with 3-(piperazin-1-yl)-1,2-benzisothiazole gave 2-{4-[4-(1,2-benzisothiazol-3-yl)piperazin-1-yl]butyl}-4-methyl-1(2*H*)-phthalazinone (**60**) (reactants, Et_3N, MeCN, reflux, N_2, 17 h: 68% after conversion to its hydrochloride).[280]

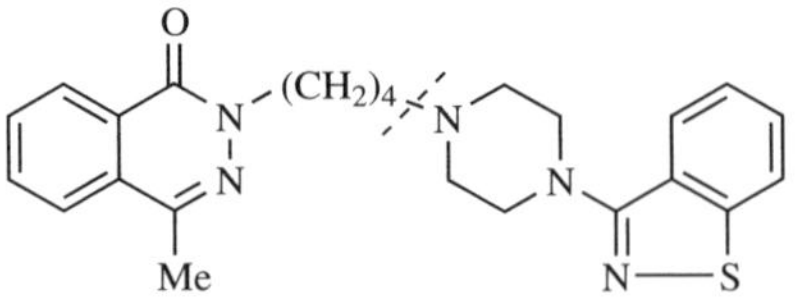

(**60**)

2-(3-Chloro-6-nitrophenyl)-4-methyl-1(2*H*)-phthalazinone with pyrrolidine gave 4-methyl-3-[2-nitro-5-(pyrrolidin-1-yl)phenyl]-1(2*H*)-phthalazinone (**61**) ("easily": 79%; no further details).[972]

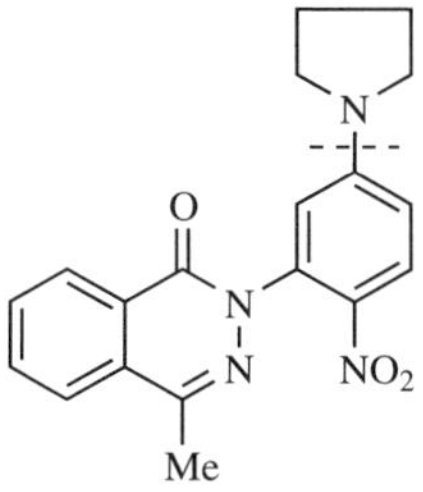

(61)

Also other examples.[164,405,411,421,728,799,871,875,876]

Formation of (Quaternary) Ammoniophthalazine Salts or the Like

1-Chloro-4-phenylphthalazine (**62**) with 3-pyridinol gave a separable mixture of 1-phenyl-4-pyridiniophthalazine chloride (**63**) and 1-butoxy-4-phenylphthalazine hydrochloride (**64**) (reactants, BuOH, reflux, 8 h: 25% and 70%, respectively).[672]

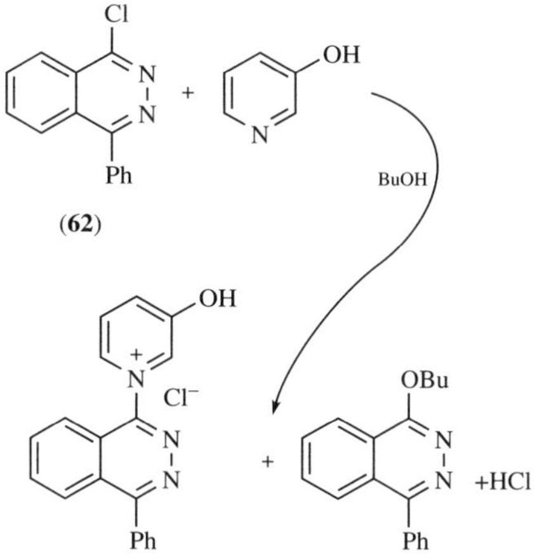

(62)

(63) (64)

6,7-Dichloro-5,8-phthalazinequinone (**65**) with pyridine gave 7-pyridinio-5,8-phthalazinequinon-6-olate (**65a**) (EtOH, 60°C, 3 h: 83%).[856]

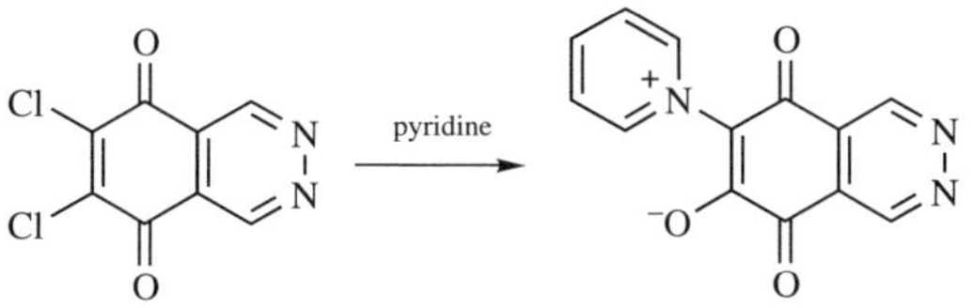

(65) (65a)

Formation of 2-Alkyl-1(2*H*)-phthalazinimines

1,4-Dichloro-2-methylphthalazinium methosulfate (**66**) gave 4-chloro-2-methyl-1(2*H*)-phthalazinimine (**67**, R = H) (liquid NH_3, substrate↓ slowly, A, 30 min: 40%), 1-chloro-3-methyl-4-methylimino-3,4-dihydrophthalazine (**67**, R = Me) (likewise but liquid $MeNH_2$, −30°C: 38%), or 1-chloro-3-methyl-4-phenylimino-3,4-dihydrophthalazine (**67**, R = Ph) (likewise but $PhNH_2$, 20°C: 20%).[742,cf. 728]

(**66**) → (**67**) [RNH_2]

Formation of Hydrazinophthalazines

Note: It should be borne in mind that hydrazine is a diamine and accordingly can react with two molecules of a halogenophthalazine to afford an *N,N′*-diphthalazinylhydrazine under appropriate conditions.

1-Benzyl-4-chlorophthalazine (**68**) gave 1-benzyl-4-hydrazinophthalazine (**69**) [$H_2NNH_2 \cdot H_2O$ (4 mol), H_2O, EtOH, reflux, 1 h: 60%] or *N,N′*-bis(4-benzylphthalazin-1-yl)hydrazine (**70**) [$H_2NNH_2 \cdot H_2O$ (2 mol), EtOH, reflux, 3 h: 90%].[528]

(**68**) → (**69**) [$H_2NNH_2 \cdot H_2O$ (2 mol) (EtOH, H_2)]

(**68**) → (**70**) [$H_2NNH_2 \cdot H_2O$ (4 mol) (EtOH)]

1-Chloro-4-methylphthalazine (**71**) with 1-hydrazino-4-methylphthalazine (**72**) gave *N,N′*-bis(4-methylphthalazin-1-yl)hydrazine (**73**) (EtOH, reflux, ? h: 85%).[33]

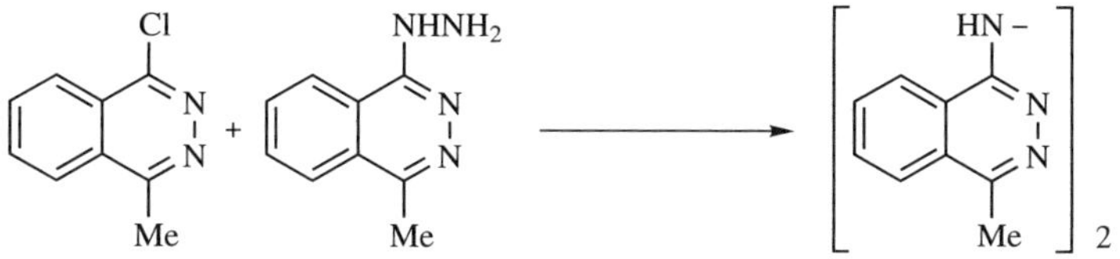

(71) (72) (73)

1,4-Dichlorophthalazine (**74**) gave 1-chloro-4-hydrazinophthalazine (**75**, R = NH_2) (substrate, NH_4OH, EtOH, 60°C; $H_2NNH_2 \cdot H_2O\downarrow$ dropwise; then reflux, 5 min: 94%)[593,cf. 1018] or 1-chloro-4-(*N*′-methyl-*N*′-phenylhydrazino)phthalazine (**75**, R = NMePh) (substrate, PhH, "hot," $H_2NNMePh\downarrow$ dropwise; then reflux, 3.5 h: 60%).[733]

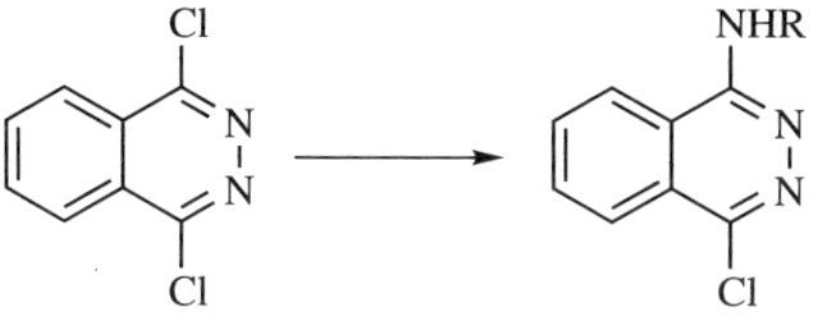

(74) (75)

1-(1-Carboxyethylamino)-4-chlorophthalazine (**76**, R = Cl) gave 1-(1-carboxyethylamino)-4-hydrazinophthalazine (**76**, R = $NHNH_2$) ($H_2NNH_2 \cdot H_2O$, EtOH, reflux, 3 h: 60%).[668]

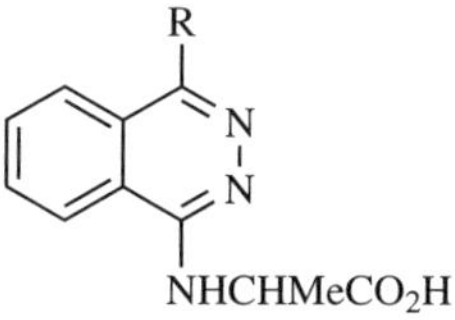

(76)

1-Chloro-4-*p*-tolylphthalazine gave 1-hydrazino-4-*p*-tolylphthalazine (**77**, R = H) ($H_2NNH_2.H_2O$, EtOH, reflux, 1 h: 47%) or 1-(*N*′-benzoylhydrazino)-4-*p*-tolylphthalazine (**77**, R = Bz) (H_2NNHBz, EtOH, reflux, 1 h: 49%); analogs likewise.[650]

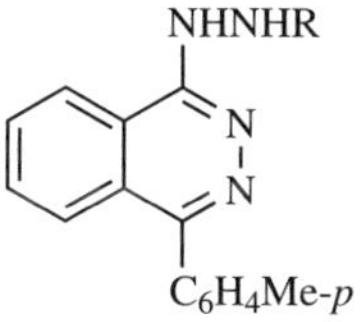

(77)

Also other examples.[32,106,166,192,195,211,223,241,249,392,649,664,675,920,936,996]

Formation of 2-Alkyl-1-hydrazono-1,2-dihydrophthalazines

Note: Like 2-alkyl-1(2*H*)-phthalazinimines, these hydrazono derivatives are made from 2-alkyl-1-chlorophthalazinium salts.

1-Chloro-2-methylphthalazinium chloride (**78**) gave a separable mixture of 1-hydrazono-2-methyl-1,2-dihydrophthalazine (**79**) and *N,N′*-bis(2-methyl-1,2-dihydrophthalazin-1-ylidene)hydrazine (**80**) [15% anhydrous H_2NNH_2 in MeOH, <5°C, substrate (in MeOH?)↓ dropwise; then <5°C, 30 min: 84% and ~5%, respectively]; variations of solvent and conditions produced different ratios of products; analogs likewise.[729]

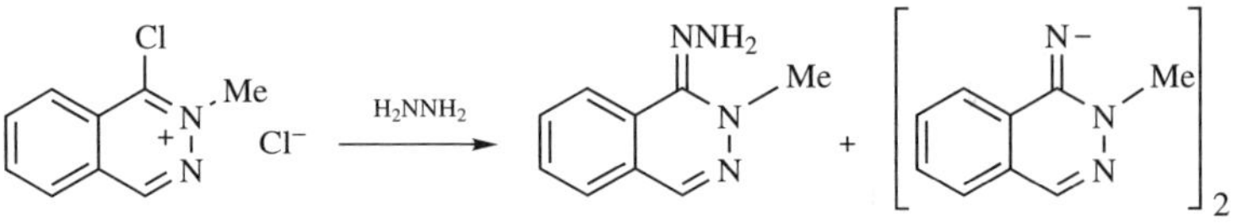

(**78**) (**79**) (**80**)

Formation of Azidophthalazines

6,7-Dichloro- (**81**, R = Cl) gave 6-azido-7-chloro-5,8-phthalazinequinone (**81**, R = N_3) (NaN_3, AcOH, 20°C, 90 min: >95%).[1016]

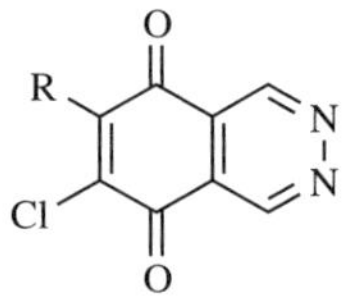

(**81**)

1,2-Bis(4-chlorophthalazin-1-yl)ethane (**82**) gave 1,2-bis(4-azidophthalazin-1-yl)ethane (**82a**), formulated as its cyclic tautomer (**82b**) (NaN_3, Me_2NCHO, reflux, 3 h: 50%).[665]

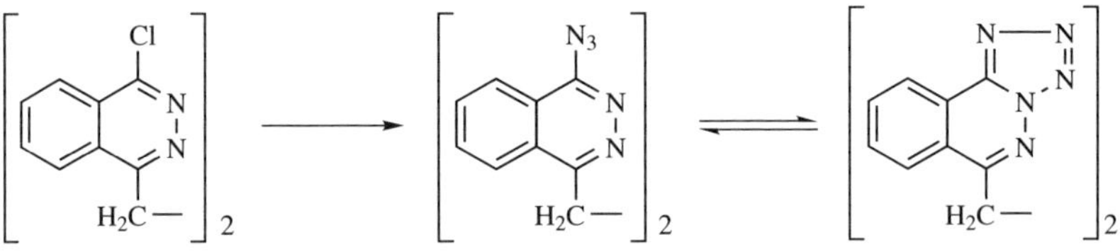

(**82**) (**82a**) (**82b**)

Also other examples.[348]

Formation of Ureidophthalazines and the Like

1-Chlorophthalazine (**84**) gave 1-cyanoaminophthalazine (**83**) (H_2NCN, NaH, Me_2NCHO, 20°C; substrate in Me_2NCHO↓ dropwise; then 100°C, N_2, 2 h:

>95%) or 1-guanidinophthalazine (**85**) [$H_2NC(=NH)NH_2$ (freshly liberated from a salt), MeOH; substrate in dioxane↓; then 60°C, 18 h: 45%]; analogs likewise.[225]

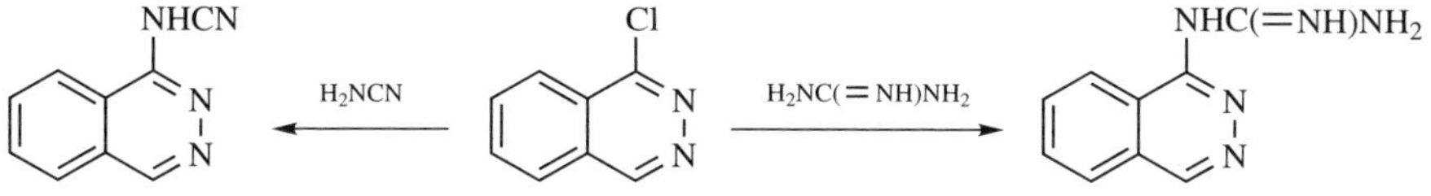

(**83**) ← H_2NCN — (**84**) — $H_2NC(=NH)NH_2$ → (**85**)

1-Chloro-4-phenylphthalazine (**87**) gave 1-phenyl-4-thiosemicarbazidophthalazine (**86**) [$H_2NNHC(=S)NH_2$, Me_2NCHO, H_2O, reflux, 4 h: ?%] or 1-guanidinoamino-4-phenylphthalazine (**88**) [$H_2NNHC(=NH)NH_2 \cdot H_2CO_3$, Me_2NCHO, H_2O, reflux, 4 h: ?%].[658]

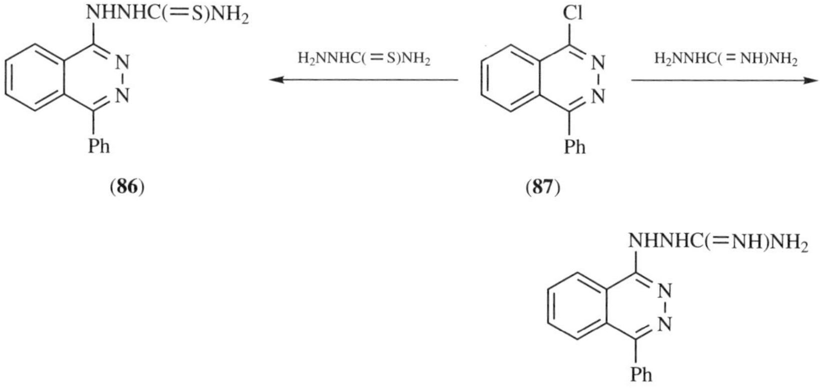

(**86**) ← $H_2NNHC(=S)NH_2$ — (**87**) — $H_2NNHC(=NH)NH_2$ → (**88**)

1,4-Dichlorophthalazine gave 1,4-bis(thiosemicarbazido)phthalazine (**89**) [substrate, $H_2NNHCSNH_2$ (2 mol), Me_2NCHO, reflux, "several" hours, tlc monitored: 58%] that underwent several cyclization reactions.[841]

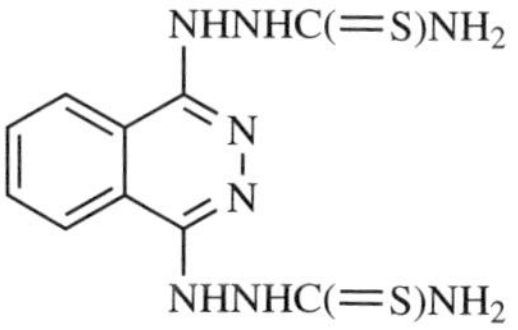

(**89**)

4-Chloro-2-phenyl- gave 2-phenyl-6-{*p*[(2-phenylpyrazol-3-yl)sulfamoyl]phenylsulfamoyl}-4-(*N*′-phenylureido)-1(2*H*)-phthalazinone (**90**) [PhHNC(=O)

NH_2: see original for details].[261]

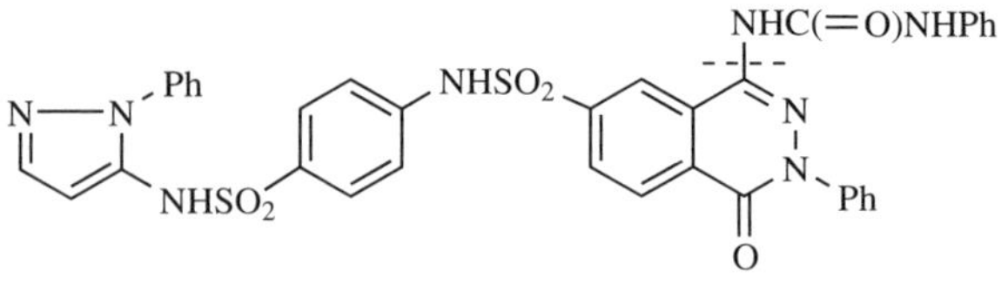

(90)

Also other examples.[664]

10.3.3. Hydrolysis, Alcoholysis, or Phenolysis of Halogenophthalazines (*E* 520)

Of these and related displacement reactions to produce phthalazines with oxygen-joined substituents, only alcoholysis and phenolysis have been used extensively. The following classified examples involve both nuclear and extranuclear halogeno substrates.

Formation of Phthalazinones or Extranuclear Hydroxyphthalazines

Note: Probably because most of the potential halogeno substrates have, in fact, been made from the oxo- or hydroxyphthalazines, the first of these processes is represented only sparingly in the 1972–2004 literature and the second not at all.

1-Chloro-4-(2,4,6-trimethoxyphenyl)phthalazine (**91**) gave 4-(2,4,6-trimethoxyphenyl)-1(2*H*)-phthalazinone (**92**) (AcONa, AcOH, 115°C, 3 h: 89%); several analogs likewise.[974]

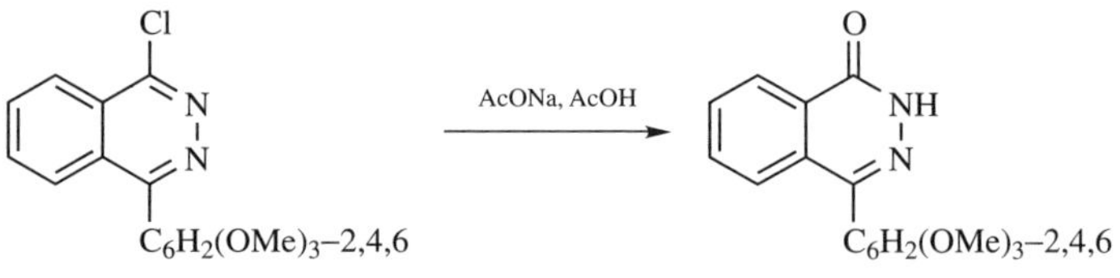

AcONa, AcOH

(91) (92)

Also other examples.[193] including $CHBr_2 \rightarrow CHO$ (Section 14.6).

Formation of (Nuclear) Alkoxyphthalazines

1-Chloro-4-phenylphthalazine (**93**) gave 1-methoxy-4-phenylphthalazine (**94**) (MeONa, MeOH, reflux, >2 h, tlc monitored: >51%).[311]

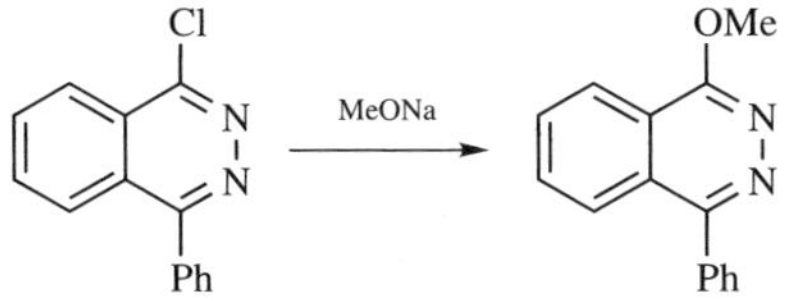

Cl OMe
MeONa
Ph Ph
(**93**) (**94**)

1-Chloro-4-(*N*-methylhydrazino)phthalazine gave 1-ethoxy-4-(*N*-methylhydrazino)phthalazine (**95**) (KOH, EtOH, reflux, 1 h: 85%).[731]

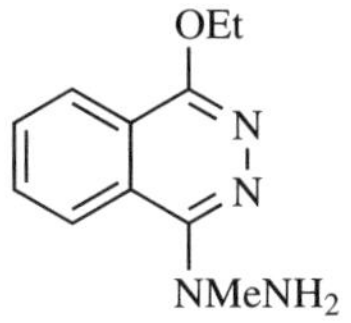

OEt
$NMeNH_2$
(**95**)

Ethyl 4-chloro- gave ethyl 4-methoxy-5,7-dimethyl-6-phthalazinecarboxylate (**96**) (MeONa, MeOH, reflux, 2 h: 67%).[783]

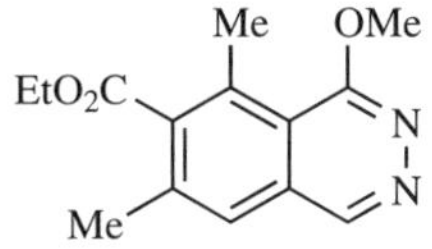

Me OMe
EtO_2C
Me
(**96**)

Also other examples.[88,122,196,690,920,1003,1010]

Formation of (Extranuclear) Alkoxyphthalazines

2-Methyl-1,4(2*H*,3*H*)-phthalazinedione as its anion (**97**) with limited 1-bromo-2-chloroethane gave a mixture of 4-(2-chloroethoxy)-2-methyl-1(2*H*)-phthalazinone (**98**) and its rearrangement product 2-(2-chloroethyl)-3-methyl-1,4(2*H*,3*H*)-phthalazinedione (**100**), each of which reacted further with the anion (**97**) to afford 1,2-bis(3-methyl-4-oxo-3,4-dihydrophthalazin-1-yloxy) ethane (**99**) or 1-(3-methyl-1,4-dioxo-1,2,3,4-tetrahydrophthalazin-2-yl)-2-(3-methyl-4-oxo-3,4-dihydrophthalazin-1-yloxy)ethane (**101**), respectively (as a one-pot procedure: NaH, Me_2NCHO, substrate↓ slowly; then this ↓ slowly to $BrCH_2CH_2Cl$ in Me_2NCHO, A; then 100°C, 1 h: ~4% and ~17%, respectively).[745]

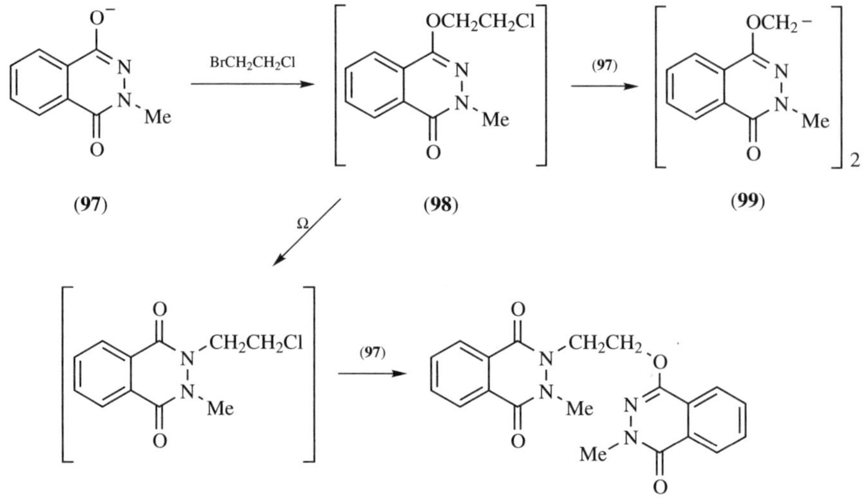

(97) (98) (99)

(100) (101)

Also other examples.[250,411]

Formation of (Nuclear) Aryloxyphthalazines

1-Chloro-4-*p*-tolylphthalazine (**102**) gave 1-phenoxy-4-*p*-tolylphthalazine (**103**) (neat PhOH, K_2CO_3, 110°C, 1 h: 89%);[958] many analogs with different aryl and aryloxy substituents were made similarly.[650,958]

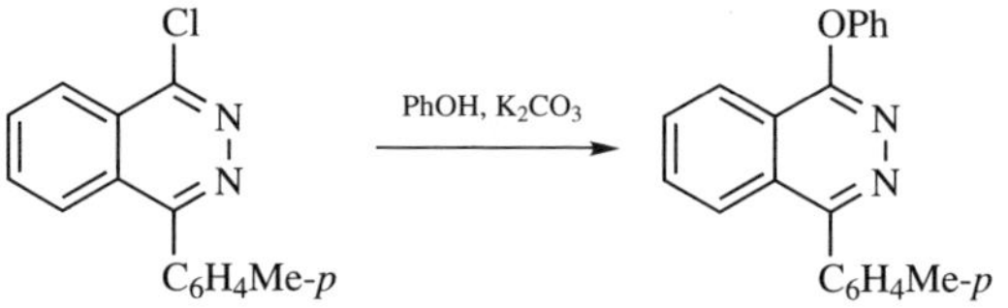

(102) (103)

1-Chloro-4-(pyridin-4-ylmethyl)phthalazine gave 1-(*p*-chlorophenoxy)-4-(pyridin-4-ylmethyl)phthalazine (**104**) (*p*-ClC_6H_4OH, K_2CO_3, Me_2SO, 90°C, 2.5 h: 35%).[870]

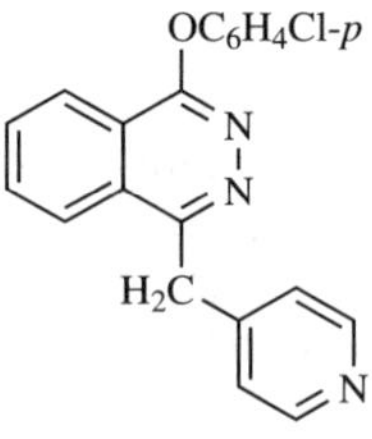

(104)

1-Chloro-4-phenylphthalazine gave 1-[*o*-(2-methylallyl)phenoxy]-4-phenylphthalazine (**105**) (*o*-$HOC_6H_4CH_2CMe{=}CH_2$, Na, PhMe, reflux, 3 h: 96%);[397] analogs somewhat similarly.[394,397,398]

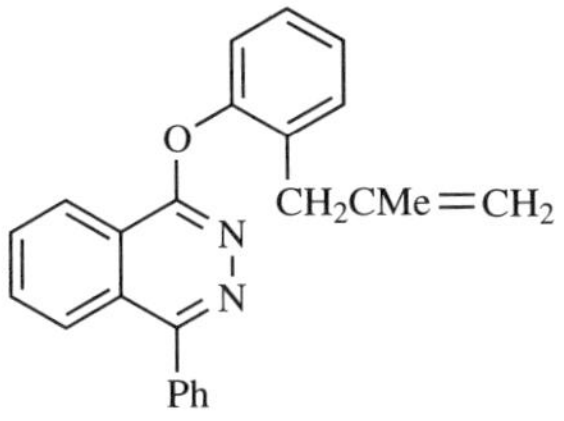

(**105**)

Also other examples.[649,884]

Formation of (Extranuclear) Aryloxyphthalazines

2-(4-Bromobutyl)-4-(3,4-dimethoxyphenyl)-4a,5,8,8a-tetrahydro-1(2*H*)-phthalazine with 6-*p*-hydroxyphenyl-5-methyl-5,6-dihydro-3(2*H*)-pyridazinone gave 4-(3,4-dimethoxyphenyl)-2-{4-[*p*-(4-methyl-6-oxo-1,4,5,6-tetrahydropyridazin-3-yl)phenoxy]butyl}-4a,5,8,8a-tetrahydro-1(2*H*)-phthalazinone (**106**) (reactants, K_2CO_3, trace KI, Me_2NCHO, 60°C, >3 h: 35%); analogs likewise.[985]

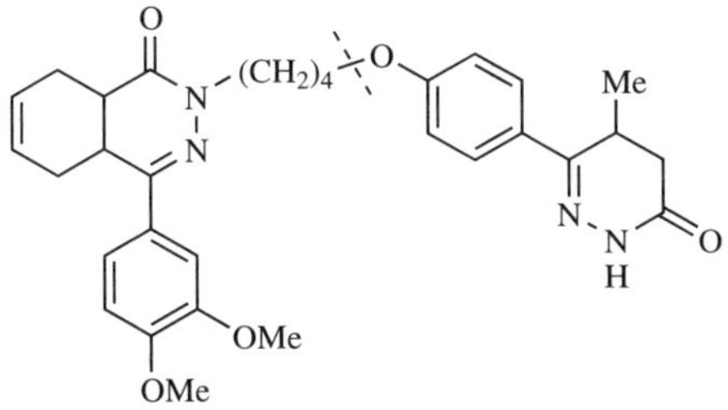

(**106**)

Formation of Acyloxyphthalazines

Note: Although both nuclear and extranuclear acyloxyphthalazines must occur as intermediates during several other processes, according to the more recent literature, they have seldom been isolated.

Ethyl 5-bromomethyl- (**107**) gave ethyl 5-acetoxymethyl-1-*o*-methoxyphenyl-7-methyl-4-oxo-3,4-dihydro-6-phthalazinecarboxylate (**108**) (AcOH, Et_3N, 100°C, 30 min: 53%).[426]

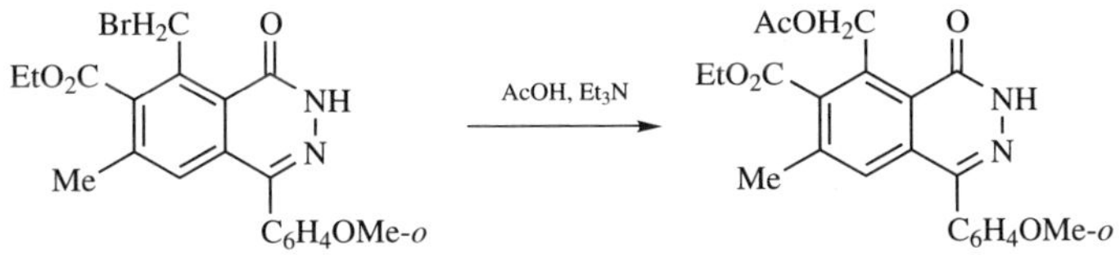

AcOH, Et_3N

(**107**) (**108**)

10.3.4. Thiolysis, Alkanethiolysis, or Arenesulfinolysis of Halogenophthalazines (*E* 536)

These displacement reactions have been used to prepare phthalazinethiones, alkylthiophthalazines, arylthiophthalazines, and arylsulfonylphthalazines.

Formation of Phthalazinethiones

1,4-Dichlorophthalazine (**109**) gave 1,4(2*H*,3*H*)-phthalazinedithione (**110**) (P_2S_5, pyridine, reflux, 2.5 h: ~50%); note treatment of the same substrate with NaHS or of 1,4(2*H*,3*H*)-phthalazinedione with P_2S_5 both proved unsatisfactory.[512]

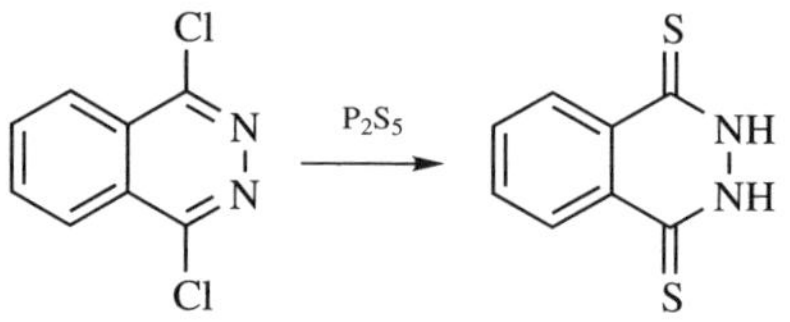

(**109**) (**110**)

1-Chloro-4-methylphthalazine gave 4-methyl-1(2*H*)-phthalazinethione [$H_2NC(=S)NH_2$, EtOH, reflux, 2 h: 70%].[649]

Also other examples.[201,931]

Formation of Alkylthiophthalazines

1-Chloro-4-phenylphthalazine with mercaptoacetic acid gave 1-carboxymethylthio-4-phenylphthalazine (**111**) (reactants, K_2CO_3, AcMe, reflux, 5 h: 81%).[33]

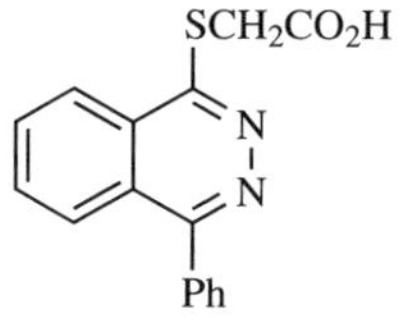

(**111**)

1-Chlorophthalazine with 1-methyl-4-piperidinethiol gave 1-(1-methylpiperidin-4-ylthio)phthalazine (**112**, R = Me) (reactants, K_2CO_3, AcMe, reflux, 10 h: 54%, as its tosylate salt) or with 4-piperidinethiol hydrochloride gave 1-(piperidin-4-ylthio)phthalazine (**112**, R = H) (synthon, NaH, $AcNMe_2$, 0°C, 15 min; then substrate in $AcNMe_2$↓ dropwise, 0°C → 20°C, 3 h: 50%, as its dihydrochloride).[482]

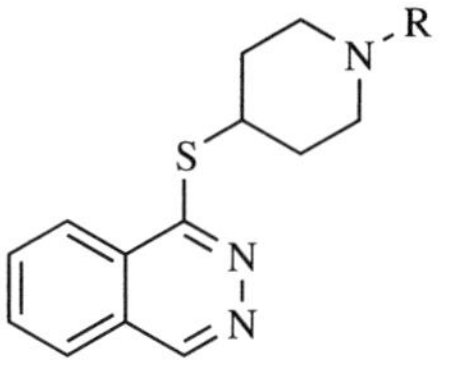

(**112**)

1-Chloro-4-methoxyphthalazine gave 1-*tert*-butylthio-4-methoxyphthalazine [Bu^{t}SNa (made in situ) in THF, 0°C, N_2; substrate in THF↓; then 66°C, 2 h: 95%].[1010]

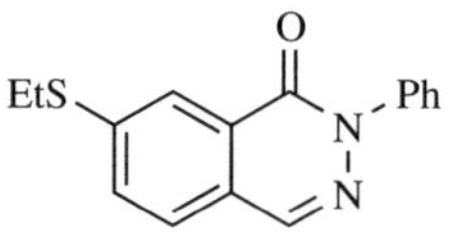

(**113**)

7-Chloro-2-phenyl-1(2*H*)-phthalazinone gave 7-ethylthio-2-phenyl-1(2*H*)-phthalazinone (**113**) (CuSEt, Me_2NCHO, reflux, 15 h: 71%).[411]

Formation of Arylthiophthalazines

1-Chloro-4-methylphthalazine gave 1-methyl-4-phenylthiophthalazine (**114**) (PhSH, K_2CO_3, AcMe, reflux, 5 h: >95%).[33]

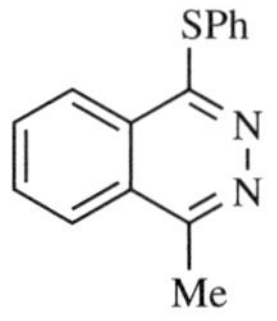

(**114**)

7-Chloro-4-oxo-3,4-dihydro-6-phthalazinesulfonamide gave 4-oxo-7-phenylthio-3,4-dihydro-6-phthalazinesulfonamide (**115**) (PhSH, $NaHCO_3$, H_2O, 100°C, 4 h: 45%).[274]

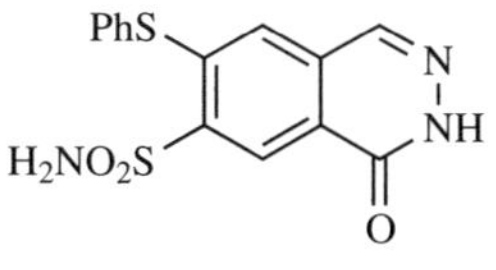

(**115**)

1-Chloro-4-(pyridin-4-ylmethyl)phthalazine gave 1-*p*-chlorophenylthio-4-(pyridin-4-ylmethyl)phthalazine (**116**) (*p*-ClC_5H_4SH, K_2CO_3, Me_2SO, 90°C,

2.5 h: 29%).[876]

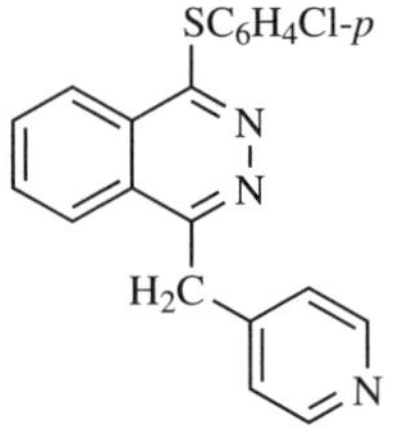

(**116**)

1,4-Dichlorophthalazine with 2(1*H*)-pyridinethione gave 1,4-bis(pyridin-2-ylthio)phthalazine (**117**) (substrate in $ClCH_2CH_2Cl$, 20°C; synthon in $ClCH_2CH_2Cl$↓ dropwise; then 20°C, 30 min: 67%).[682]

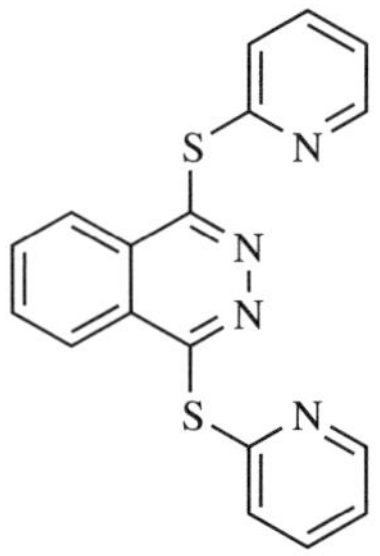

(**117**)

Also other examples.[166,216,650]

Formation of Arylsulfonylphthalazines

Note: This nonoxidative route to heterocyclic sulfones is often overlooked.

1-Chloro-7-methoxyphthalazine (**118**) and sodium *p*-toluenesulfinate gave 6-methoxy-4-(*p*-tolylsulfonyl)phthalazine (**119**) (Me_2NCHO, 100°C, 2 h: 63%).[435]

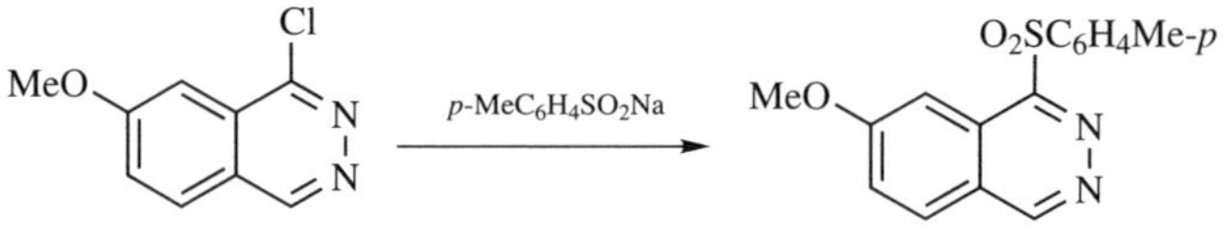

p-$MeC_6H_4SO_2Na$

(**118**) (**119**)

In much the same way, 1-chloro-4-phenylphthalazine gave 1-phenyl-4-(*p*-tolylsulfonyl)phthalazine (**120**) (p-$MeC_6H_4SO_2Na$, Me_2NCHO, 90°C, 1 h: 82%).[591]

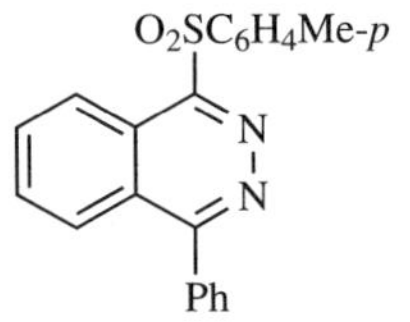

(120)

10.3.5. Cyanolysis or Aroyl Displacement of Halogenophthalazines (*E* 643)

Displacement of nuclear or extranuclear halogeno by the (carbon-joined) cyano or aroyl substituents is not difficult as illustrated in the following examples.

7-Bromo-2-phenyl-1(2*H*)-phthalazinone (**121**) with copper(I) cyanide led to a complex that underwent decomposition on treatment with iron(III) chloride to give 4-oxo-3-phenyl-3,4-dihydro-6-phthalazinecarbonitrile (**122**) (CuCN, Me_2NCHO, reflux, 10 h; residue from evaporation, $FeCl_3$, HCl, 70°C, 20 min: 75%).[411]

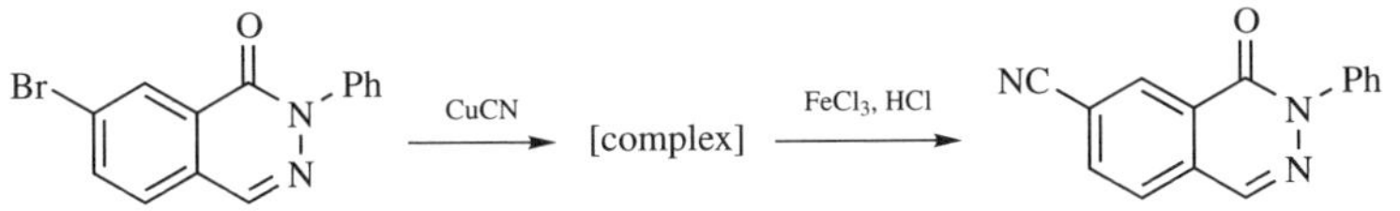

CuCN → [complex] → $FeCl_3$, HCl

(121) (122)

7-Bromo-4-hydroxymethyl-1(2*H*)-phthalazinone (**123**, R = Br) gave 1-hydroxymethyl-4-oxo-3,4-dihydro-6-phthalazinecarbonitrile (**123**, R = CN) (CuCN, Me_2NCHO, reflux, 15 h: 60%).[574]

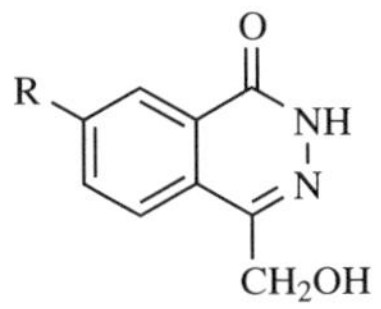

(123)

2-Bromomethyl- (**124**, R = Br) gave 2-cyanomethyl-4-ethoxycarbonylmethyl-1(2*H*)-phthalazinone (**124**, R = CN) [substrate, AcMe, 0°C; KCN + trace KI in H_2O↓ dropwise; then 0°C (?), 2 h: 63%].[68,784]

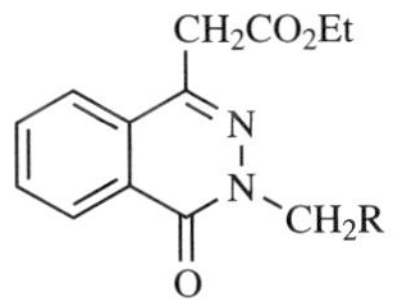

(124)

1-Chloro-4-phenylphthalazine (**125**) gave 4-phenyl-1-phthalazinecarbonitrile (**126**) (KCN, Me_2NCHO, 90°C, 1 h: 44% with 39% recovery of substrate; KCN, *p*-$MeC_6H_4SO_2Na$, Me_2NCHO, 90°C, 1 h: 90%; note the marked catalytic effect of the sulfinate).[591]

1-Chloro-4-phenylphthalazine (**125**) gave 1-benzoyl-4-phenylphthalazine (**127**) [substrate, PhCHO, 1,3-dimethylimidazolium iodide (catalyst), *p*-MeC_6H_4-SO_2Na or $MeSO_2Na$ (catalyst), Me_2NCHO, NaH↓, 80°C, 20 min: 63% or 79%, respectively; similarly but without a sulfinate: 0%; substituted-benzoyl and other such analogs likewise].[601]

KCN ± *p*-$MeC_6H_4SO_2Na$

(**125**) (**126**)

PhCHO, 1,3-dimethylimidazolium iodide, *p*-$MeC_6H_4SO_2Na$, NaH

(**127**)

Also other examples.[405,930]

10.3.6. Ring Fission and Cyclization Reactions of Halogenophthalazines

A variety of transformations to the nucleus of halogenophthalazines have been reported. The following examples illustrate some typical processes.

Ring Fission Reactions

The electrochemical degradation of 1,4-dichlorophthalazine (**128**) to phthalonitrile (**129**) and analogous reductions of other chlorophthalazines have been studied in detail.[850,991]

electroreduction

(**128**) (**129**)

Hexachlorophthalazine (**130**) underwent thermolysis to afford tetrachlorophthalonitrile (**131**) (vapor over SiO_2, 775°C, vacuum: 92%).[688]

775°C

(**130**) (**131**)

1-Chlorophthalazine (**132**) with 1-diethylaminoprop-1-yne (**133**, R = Me) gave 2-(*o*-cyanophenyl)-4,6-bis(diethylamino)-3,5-dimethylpyridine (**134**, R = Me) [substrate, synthon (2.1 mol), dioxane, 80°C, 10 min: 68%] or with 1-diethylaminobut-1-yne (**133**, R = Et) gave 2-(*o*-cyanophenyl)-4,6-bis(diethylamino)-3,5-diethylpyridine (**134**, R = Et) (likewise: 35%).[429,435] 1-Bromophthalazine behaves similarly to give analogous products.[599]

$2 \times RC{\equiv}CNEt_2$

(**132**) (**133**) (**134**)

Cyclocondensations Involving Two Nuclear Halogeno Substituents

1,4-Dichlorophthalazine (**135**) with disodium 1,3-propanedithiolate gave the macrocyclic product (**136**) (THF, reflux, N_2; substrate in THF and synthon in EtOH both ↓ dropwise simultaneously during 6 h; then reflux, 12 h: 48%) or with disodium 3,7-dithianonane-1,9-dithiolate gave the product (**137**) (similar procedure: 44%).[800]

$NaSCH_2CH_2CH_2SNa$

(**135**) (**136**)

$NaCH_2CH_2SCH_2CH_2CH_2SCH_2CH_2SNa$

(**137**)

Cyclocondensations Involving a Nuclear Halogeno Substituent and a Ring Nitrogen

1-Chloro-4-methylphthalazine (**138**) with anthranilic acid (**139**) gave 5-methyl-8*H*-phthalazino[1,2-*b*]quinazolin-8-one (**140**) (neat reactants, 155°C, 4 h: 78%; analogs likewise).[162]

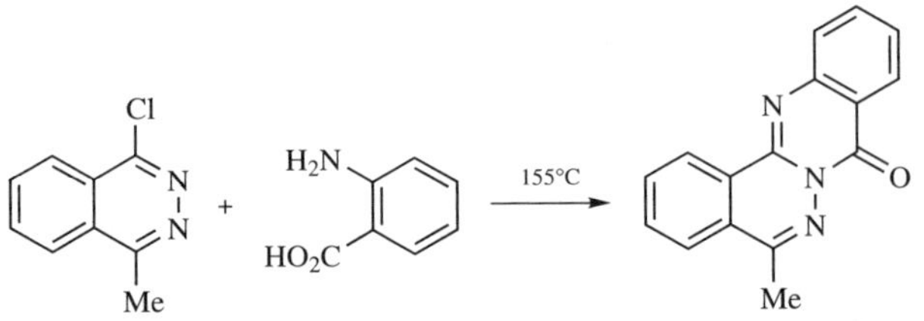

(**138**) (**139**) (**140**)

1,4-Dichlorophthalazine (**141**) with ethyl 2-amino-5-phenyl-3-thiophenecarboxylate (**142**) gave 5-chloro-10-phenyl-8*H*-thieno[2′,3′:4,5]pyrimido[2,1-*a*] phthalazin-8-one (**143**) [neat reactants, Δ (oil bath, ?°C) until HCl↑ ceased: 20%];[44] analogs similarly.[44,349]

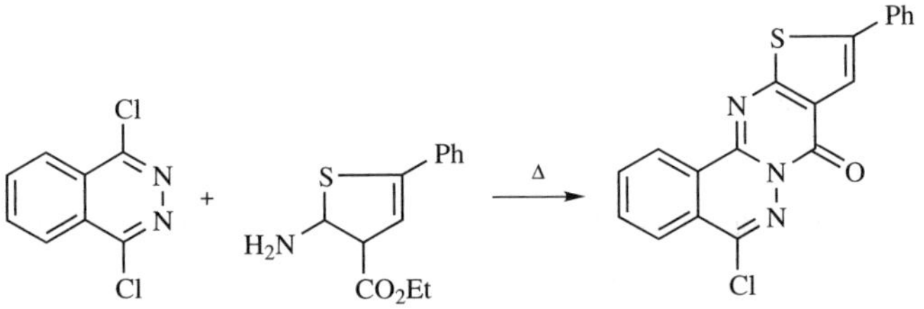

(**141**) (**142**) (**143**)

Also other examples.[1020]

Cyclizations Involving an Extranuclear Halogeno Substituent and a Ring Nitrogen

2-(*o*-Fluorobenzoyl)-1,2,3,4-tetrahydrophthalazine (**144**) underwent intramolecular cyclization to 6,11-dihydro-13*H*-indazolo[1,2-*b*]phthalazin-13-one (**145**) (K_2CO_3, $C_5H_{11}OH$, reflux, 14 days: 70%); analogs somewhat similarly.[493]

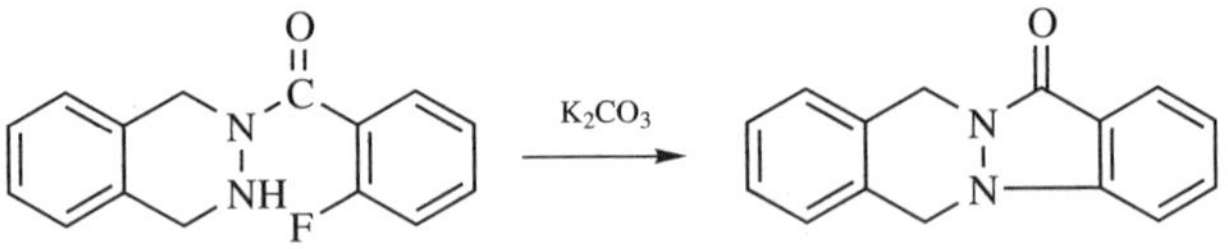

(**144**) (**145**)

Also other examples.[301,692]

Cyclocondensations Involving Two Extranuclear Halogeno Substituents

1,3-Bis[*m*-(Bromomethyl)phenyl]-1,4(2*H*,3*H*)-phthalazinedione (**146**) gave 5,21-dihydro-12*H*,14*H*-7,11:15,19-dimetheno[2,8,9]thiadiazacyclopentadecino[8,9-*b*] phthalazine-5,21-dione (**147**) (PhH, EtOH, vigorous stirring, reflux; substrate + $MeCSNH_2$ in PhH + EtOH↓ and KOH in EtOH↓ dropwise simultaneously during 40 h; then reflux 90 min: 36%).[134]

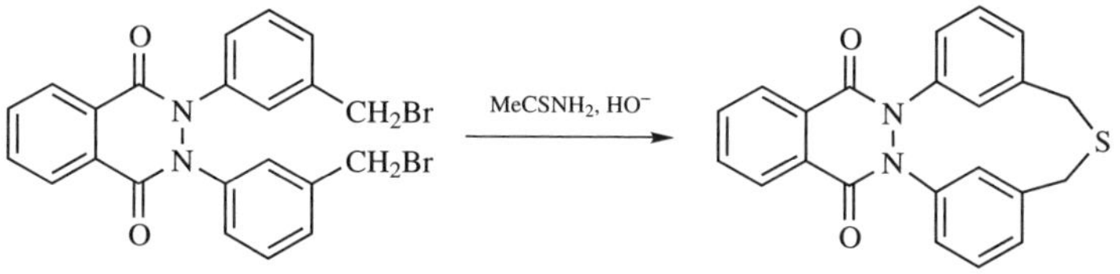

CH2Br
CH2Br
MeCSNH2, HO−
S

(**146**) (**147**)

2,3-Bis(α-bromopropionyl)-1,2,3,4-tetrahydrophthalazine (**148**) with 4-(2-aminoethyl)morpholine gave 2,4-dimethyl-3-(2-morpholinoethyl)-2,3,4,5,7,12-hexahydro-1*H*-[1,2,5]triazepino[1,2-*b*]phthalazine-1,5-dione (**149**) (substrate Na_2CO_3, EtOH, reflux; synthon in EtOH↓ dropwise during 1 h; then reflux, 50 h: 64%);[927] analogs likewise.[925,927]

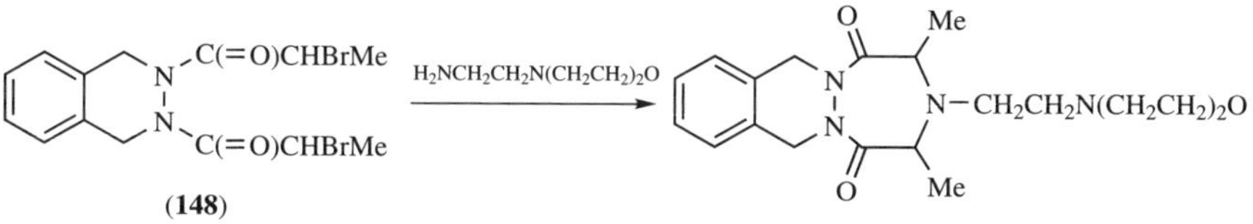

C(=O)CHBrMe
C(=O)CHBrMe
H2NCH2CH2N(CH2CH2)2O
Me
N−CH2CH2N(CH2CH2)2O
Me

(**148**) (**149**)

Di-*tert*-butyl 6,7-bis(bromomethyl)-1,2,3,4-tetrahydro-2,3-phthalazinedicarboxylate (**151**) and the "gylcoluril" diester [diisopentyl 2,5-dioxooctahydroimidazo[4,5-*d*]imidazole-3a,6a-dicarboxylate (**150**)] gave the product (**152**) (synthon, Bu^{t}OK, Me_2SO, 50°C, 20 min; then substrate in Me_2SO↓ dropwise, 50°C → 20°C, 25 min: 56%);[325] analogs likewise.[326,327]

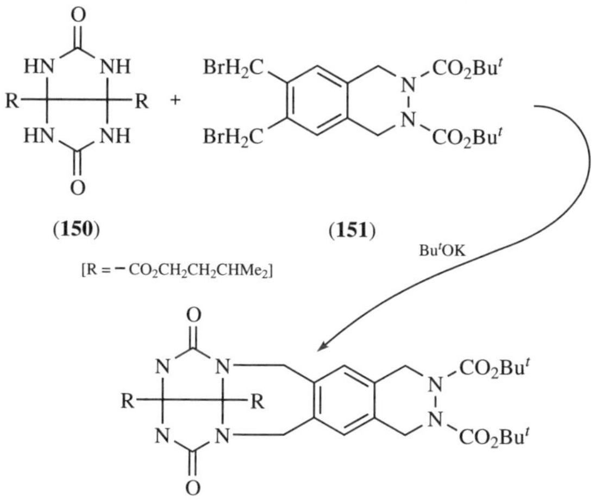

BrH2C
BrH2C
CO2Bu^t
CO2Bu^t
(150)
(151)
[R = −CO2CH2CH2CHMe2]
Bu^tOK

(**152**)

5-(α,β-Dibromostyryl)phthalazine (**153**) and sodium trithiocarbonate gave 5-phenylthiino[2,3,4-*de*]phthalazine (**154**) (substrate, MeOH, THF, 62°C; Na_2CS_3 in H_2O↓ dropwise; then reflux, 18 h: 50%; a rational mechanism was suggested).[857]

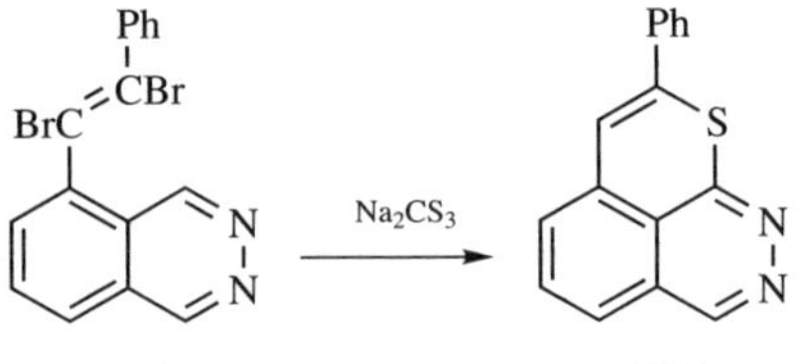

(**153**) (**154**)

Miscellaneous Cyclocondensations

6,7-Dichloro-5,8-phthalazinequinone (**156**) with 2-pyridinamine (**155**) gave 6,6-diethoxy-5,6-dihydropyrido[1′,2′:1,2]imidazo[4,5-*f*]phthalazin-5-one (**157**) (EtOH, 60°C, 6 h: 48%; structure confirmed by X-ray analysis).[856]

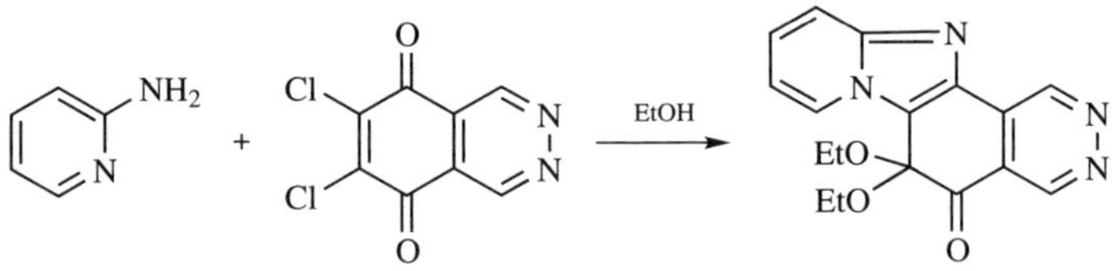

(**155**) (**156**) (**157**)

6-Acetamido-7-chloro-5,8-phthalazinequinone (**158**) gave 1,2-dimethyl-1*H*-imidazo[4,5-*g*]phthalazine-4,9-quinone (**159**) ($MeNH_2$, THF, EtOH, reflux, 30 min: 30%); analogs likewise.[1016]

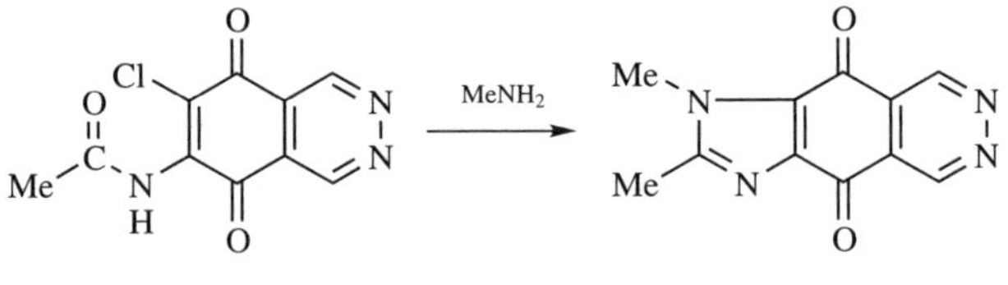

(**158**) (**159**)

Also other examples.[1017]

CHAPTER 11

Oxyphthalazines (*H* 78, 177; *E* 364, 375, 445)

The term *oxyphthalazine* is used here to cover compounds such as the tautomeric phthalazinone (**1**), its nontautomeric *O*-methyl (**2**), and *N*-methyl derivatives (**3** and **4**), the extranuclear hydroxymethylphthalazine (**5**), the phthalazinequinone (**6**), the phthalazine *N*-oxide (**7**), and related derivatives.

(**1**)

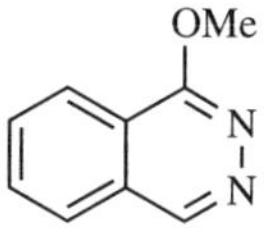

(**2**) 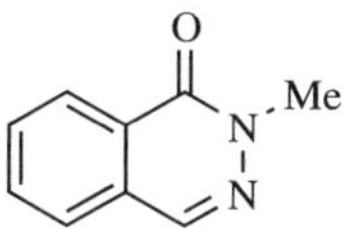 (**3**) 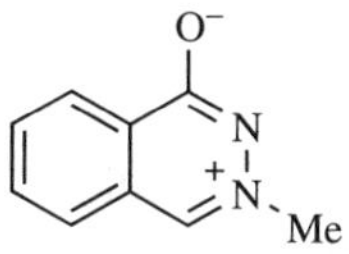 (**4**)

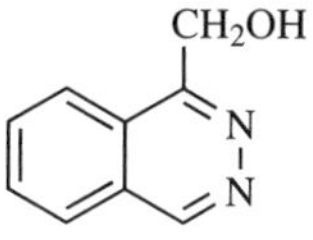

(**5**) 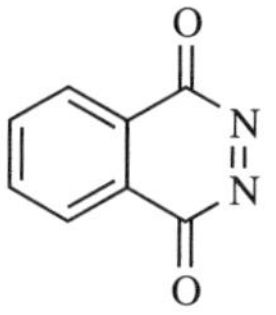 (**6**) 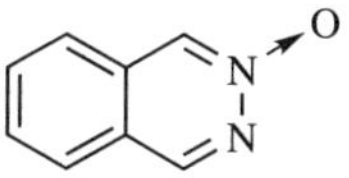 (**7**)

The important phenomenon of *chemiluminescence* clearly falls within the scope of this chapter because the oxidation of luminol [5-amino-1,4(2*H*,3*H*)-phthalazinedione (**8**)] or related phthalazinones by hydrogen peroxide is commonly used to produce light in this way. However, the whole specialized subject, including the use of chemiluminescence in analysis and of biochemiluminescence in diagnostic

Cinnolines and Phthalazines: Supplement II, The Chemistry of Heterocyclic Compounds, Volume 64, by D.J. Brown

medicine, has been well reviewed,[966] and it seems appropriate simply to draw attention here to several papers that warrant special attention.[142,319,323,418,849,883]

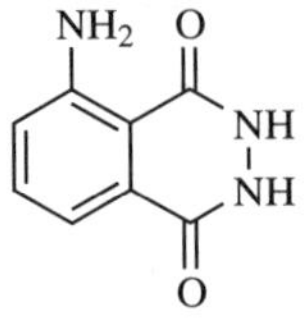

(8)

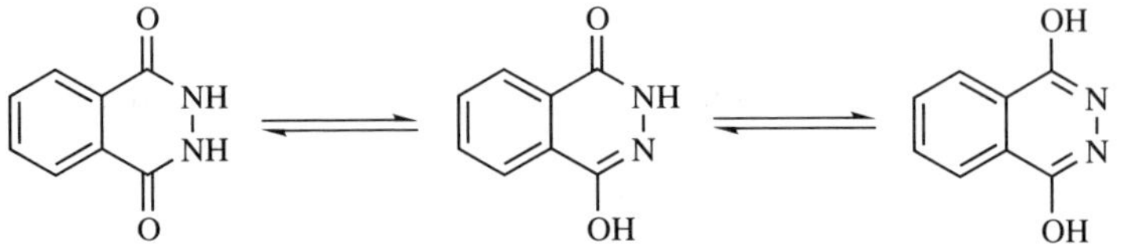

(9)

11.1. TAUTOMERIC PHTHALAZINONES (*H* 78, 148; *E* 375, 445)

The tautomerism of these phthalazinones has been studied experimentally in some detail,[681,690] and some theoretical calculations seem to uphold experimental findings that 1(2*H*)-phthalazinone (**1**) exists as its oxo tautomer and that 1,4(2*H*,3*H*)-phthalazinedione (**9**) exists predominantly as its hydroxyphthalazinone tautomer.[761] The X-ray analysis of 4-*p*-hydroxyphenyl-1(2*H*)-phthalazinone indicates that it exists as such, at least in the solid state.[558]

The ionization constants for several 5- or 6-substituted 1,4(2*H*,3*H*)-phthalazinediones (**10**) have been measured for comparison with those of the unsubstituted molecule (**10**, R = H);[805] the ^{1}H and ^{13}C NMR spectra for some 4-substituted 1(2*H*)-phthalazinones (**11**) have been reported;[773] the predominant conformation of 4-phenyl-5,6,7,8-tetrahydro-1(2*H*)-phthalazinone (**12**) has been determined by ^{1}H and ^{13}C NMR;[792] the MS fragmentation patterns of simple 4-(ω-hydroxyalkylamino)-1(2*H*)-phthalazinones (**13**) have been studied;[755] 4-benzoyloxy-1(2*H*)-phthalazinone (**14**) has been found to catalyze trimethylsilylation of otherwise resistant substrates by hexamethyldisilazane;[110] and a series of substituted 1,4(2*H*,3*H*)-phthalazinediones (**15**) include some with potent hypolipidemic activity in rodents.[237]

11.1.1. Preparation of Tautomeric Phthalazinones

Most of these phthalazinones have been made by *primary synthesis* (see Chapter 8). A few have been made also by *direct C-hydroxylation* (see Sections 9.1.3 and 9.2.2), by *hydrolysis of halogenophthalazines* (see Section 10.3.3), or by miscellaneous routes illustrated in the following classified examples.

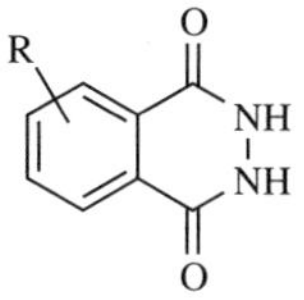

(10) 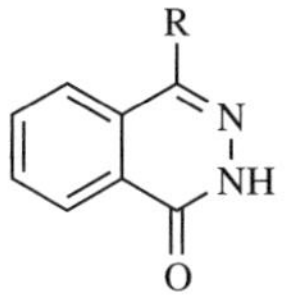 (11) 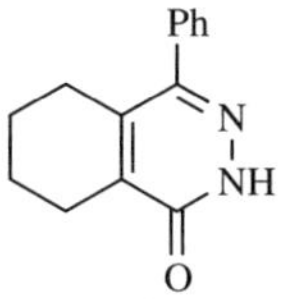 (12)

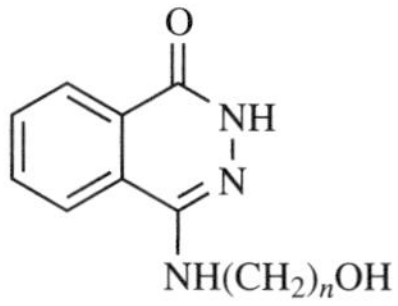

(13) 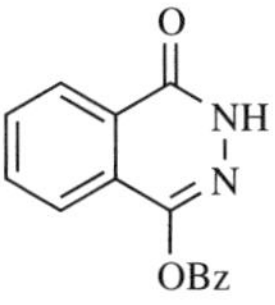 (14) 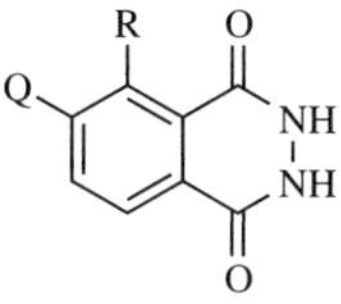 (15)

From Phthalazinamines

Note: This can be done by direct hydrolysis or via diazotization.

1,4-Phthalazinediamine (**17**) gave 4-amino-1(2*H*)-phthalazinone (**16**) (H_2O, 190°C, sealed, 20 h: ~75%) or 1,4(2*H*,3*H*)-phthalazinedione (**18**) (substrate, HCl, H_2O, 0°C; $NaNO_2$ in H_2O↓ dropwise: 78%).[685]

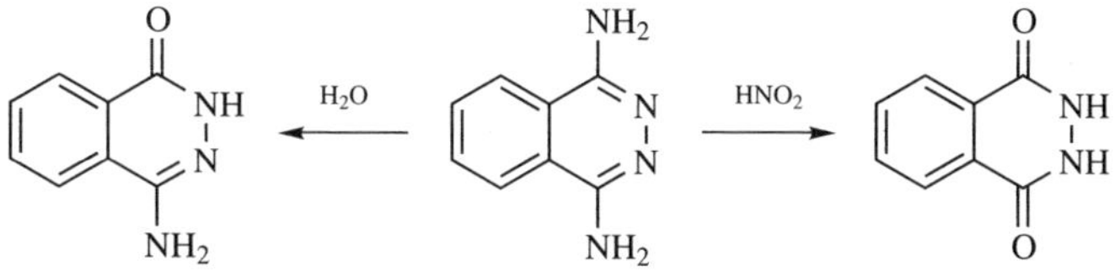

(16) (17) (18)

Also other examples.[856]

From Alkoxyphthalazines

4-(2-Oxocyclohexyloxy)-1(2*H*)-phthalazinone (**19**, R = H) gave 1,4(2*H*,3*H*)-phthalazinedione (**20**, R = H) (5M NaOH, reflux, 30 min: 95%);[105] 2-methyl-4-(2-oxocyclohexyloxy)-1(2*H*)-phthalazinone (**19**, R = Me) gave 2-methyl-1,4(2*H*,3*H*)-phthalazinedione (**20**, R = Me) (10M HCl, reflux, 1 h: >95%).[105]

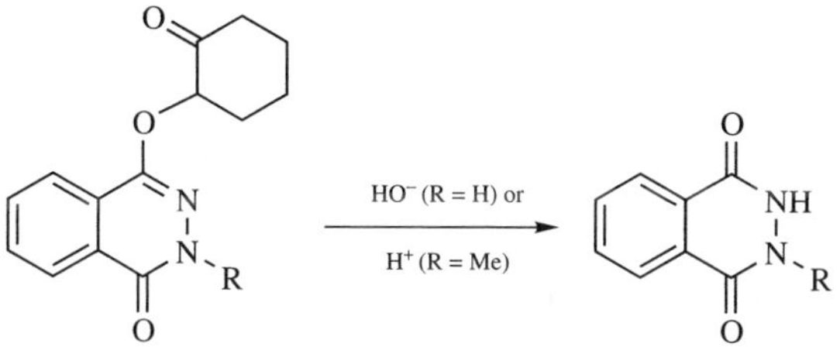

(19) (20)

4-Hydroxymethyl-7-methoxy-1(2*H*)-phthalazinone gave 4-hydroxymethyl-1,7 (2*H*,3*H*)-phthalazinedione ($AlCl_3$, PhH, reflux; see original for further details).[195]

By Direct *C*-Hydroxylation

Note: Some examples of this process have been given in Sections 9.1.3 and 9.2.2.

1-Benzoylphthalazine (**21**) gave 4-benzoyl-1(2*H*)-phthalazinone (**22**) [substrate, H_2O, Me_2SO, 20°C, 10 min: $K_4Fe(CN)_6$ in 5M KOH↓, 20°C, 30 min: 23% with 20% recovery of substrate).[412]

HO^-, $K_4Fe(CN)_6$

(**21**) (**22**)

From Phthalazine *N*-Oxides

Note: Only a few examples of this process appear to have been reported in the more recent literature.

1-Phenylphthalazine 3-oxide (**23**) gave 4-phenyl-1(2*H*)-phthalazinone (**24**) (BzCl, $CHCl_3$, 95°C, 15 min: 2% after separation from other products).[938]

BzCl

(**23**) (**24**)

From Alkylsulfonylphthalazines

1-Methyl-4-methylsulfonylphthalazine gave 4-methyl-1(2*H*)-phthalazinone (~4M NaOH: 93%).[931]

11.1.2. Reactions of Tautomeric Phthalazinones

The halogenolysis of tautomeric phthalazinones has been covered in Section 10.1.1. Other recently used reactions are discussed in the following subsections.

11.1.2.1. Alkylation or Arylation of Tautomeric Phthalazinones

All such alkylations probably afford mixtures of *O*- and *N*-alkylated isomers in which the alkoxyphthalazine may sometimes predominate but usually the *N*-alkylphthalazinone and occasionally (under abnormal conditions) the *N*-alkylphthalaziniumolate predominate; the minor product(s) is (are) normally lost during workup or purification. The predominant site of alkylation clearly depends on the type of reagent, steric hindrance inherent in both substrate and reagent, conditions, and other factors difficult to assess. Accordingly, the best hope of obtaining a given product by alkylation (despite a creditable theoretical approach[452] to this problem) is to follow the procedure for the nearest analogous product in the following classified examples.

Formation of Alkoxyphthalazines

Note: An example of *O*-trimethylsilylation is also included in this subsection. Traditionally diazoalkanes are the reagents of choice for *O*-alkylation, but they have not been used extensively in this series.

4-Benzyl-1(2*H*)-phthalazinone (**25**) gave 1-benzyl-4-methoxyphthalazine (**26**) (CH_2N_2, Et_2O, 20°C, ? h: 82%).[525]

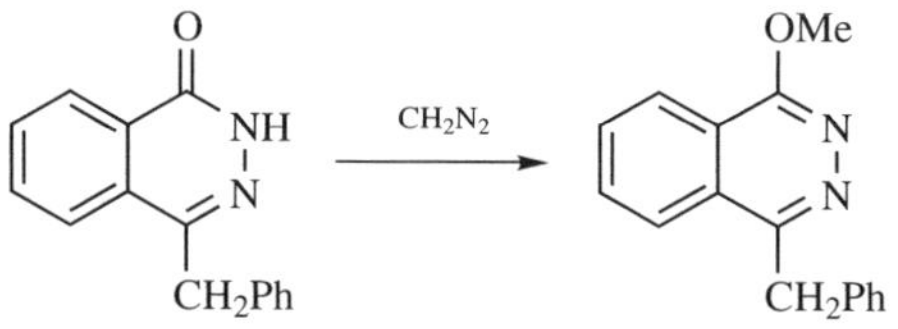

O NH N CH_2Ph —CH_2N_2→ OMe N N CH_2Ph

(**25**) (**26**)

2-(5-Methyl-2-oxotetrahydrofuran-3-yl)-1,4(2*H*,3*H*)-phthalazinedione gave 4-methoxy-2-(5-methyl-2-oxotetrahydrofuran-3-yl)-1(2*H*)-phthalazinone (**27**) (Me_2SO_4, K_2CO_3, AcMe, reflux, 10 h: 78%).[126]

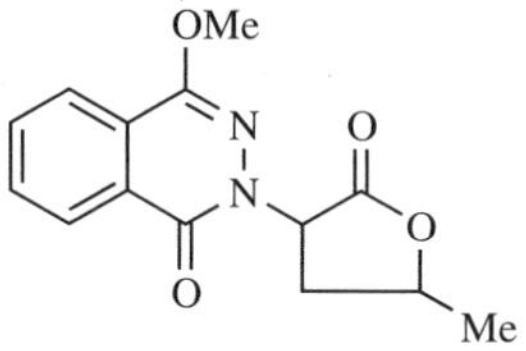

OMe N N O O O Me

(**27**)

2-Methyl-1,4(2*H*,3*H*)-phthalazinedione (**28**) reacted with 2-chloroethanol to give 4-(2-hydroxyethoxy)-2-methyl-1(2*H*)-phthalazinone (**29**, R = OH) (NaH, Me_2NCHO, A; substrate↓ slowly; stirred until gas↑ ceased; then synthon in Me_2NCHO↓ during 30 min, 20°C; then reflux, 30 min: 70%) or with 1-bromo-2-chloroethane to give 4-(2-chloroethoxy)-2-methyl-1(2*H*)-phthalazinone

(**29**, R = Cl) (similarly but synthon in Me_2NCHO↓, 0°C; then 20°C, 72 h: 71%; note preferential reactivity of the bromo substituent in the synthon).[745]

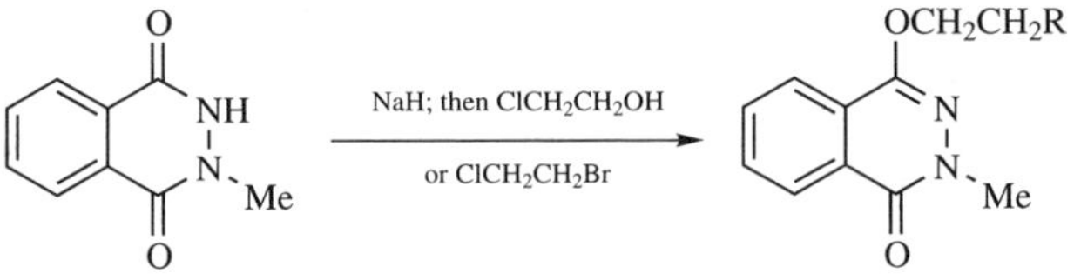

NaH; then $ClCH_2CH_2OH$ or $ClCH_2CH_2Br$

(**28**) (**29**)

1,4(2*H*,3*H*)-Phthalazinedione (**30**) gave 4-(*o*-nitrobenzyloxy)-1(2*H*)-phthalazinone (**31**) (NaH, Me_2SO; substrate↓; $ClCH_2C_6H_4NO_2$↓; 20°C, 3 h: 53%; see original for essential details) and thence 1,4-bis(*o*-nitrobenzyloxy)phthalazine (**32**) (similarly but in Me_2NCHO; finally 100°C, 48 h: <72%; see original for details).[3]

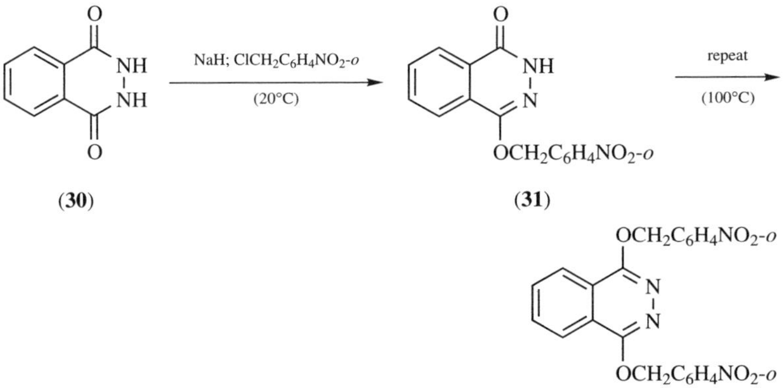

NaH; $ClCH_2C_6H_4NO_2$-*o* (20°C) repeat (100°C)

(**30**) (**31**) (**32**)

2-Methyl-1,4(2*H*,3*H*)-phthalazinedione (**33**) with the epoxide, 2,2-dimethyl-1a,7b-dihydro-2*H*-oxireno[*c*][1]benzopyran-6-carbonitrile (**33a**), gave 4-(6-cyano-3-hydroxy-2,2-dimethyl-3,4-dihydro-2*H*-[1]benzopyran-4-yloxy)-2-methyl-1(2*H*)-phthalazinone (**34**) (pyridine, EtOH, reflux, 14 h: 20%).[439,752]

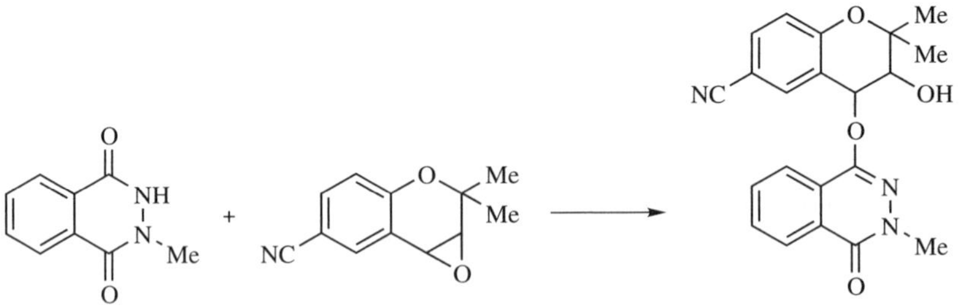

(**33**) (**33a**) (**34**)

1(2*H*)-Phthalazinone (**35**) gave exclusively 1-(trimethylsiloxy)phthalazine (**35a**) [neat $Me_3SiNHSiMe_3$, trace $(NH_4)_2SO_4$, reflux, 5 h: crude].[986]

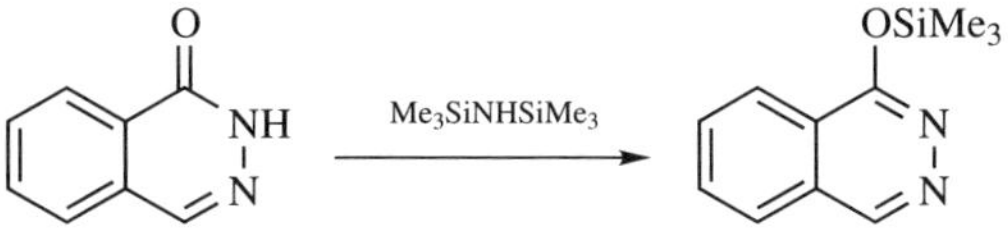

(**35**) (**35a**)

Also other examples.[351,363,418,469,471,811,875,949,1014,1040]

Formation of *N*-Alkylphthalazinones

Note: This is usually done by treatment with an alkyl halide in a basic medium, but the Mannich procedure or an unsaturated reagent (e.g., acrylonitrile) may be used.

8-Methyl-1(2*H*)-phthalazinone (**36**, R = H) gave 2,8-dimethyl-1(2*H*)-phthalazinone (**36**, R = Me) (MeI, NaOH, H_2O, MeOH, reflux, 1 h: 95%).[622]

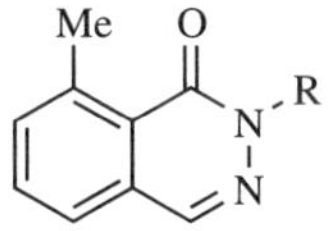

(**36**)

4-(2-Methoxynaphthalen-1-yl)-1(2*H*)-phthalazinone (**37**, R = H) gave 4-(2-methoxynaphthalen-1-yl)-2-propyl-1(2*H*)-phthalazinone (**37**, R = Pr) (PrBr, NaOH, H_2O, MeOH, reflux, 2 h: 72%).[371]

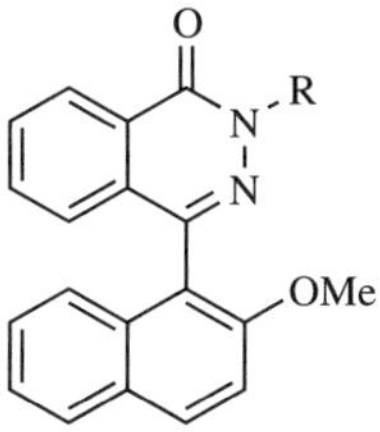

(**37**)

5,8-Dimethoxy-1,4(2*H*,3*H*)-phthalazinedione gave 5,8-dimethoxy-2,3-dimethyl-1,4(2*H*,3*H*)-phthalazinedione (MeI, K_2CO_3, AcMe, reflux, 3 h: 67%).[715]

Methyl 4-oxo-3,4-dihydro-5-phthalazinecarboxylate (**38**, R = H) gave methyl 3-methyl-4-oxo-3,4-dihydro-5-phthalazinecarboxylate (**38**, R = Me) (MeI, NaOH, H_2O, MeOH, reflux, 1 h: ~88%).[623]

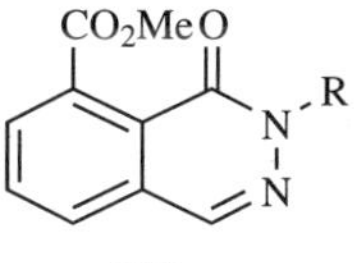

(**38**)

4-Methyl-1(2*H*)-phthalazinone (**39**, R = H) gave 2-(4-bromobutyl)-4-methyl-1(2*H*)-phthalazinone (**39**, R = $CH_2CH_2CH_2Br$) (NaH, Me_2NCHO; substrate in Me_2NCHO↓, 20°C, 30 min; this solution↓ to $BrCH_2CH_2CH_2CH_2Br$ in Me_2NCHO, 20°C, 4 h: 34%).[280]

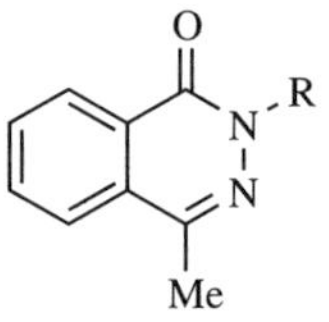

(**39**)

4-Ethoxycarbonylmethyl-1(2*H*)-phthalazinone (**40**, R = H) gave 2-cyanomethyl-4-ethoxycarbonylmethyl-1(2*H*)-phthalazinone (**40**, R = CH_2CN) ($ClCH_2CN$, Bu^tOK, Me_2NCHO, 20°C, 30 min: 91%).[68]

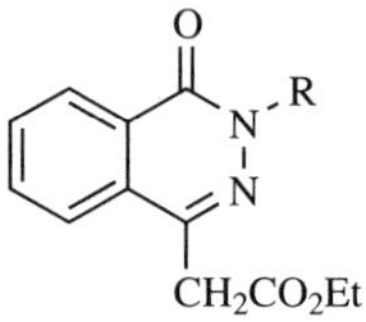

(**40**)

4-Benzyl-2(1*H*)-phthalazinone (**41**, R = H) gave 4-benzyl-2-ethoxycarbonylmethyl-1(2*H*)-phthalazinone (**41**, R = CH_2CO_2Et) ($ClCH_2CO_2Et$, NaOH, EtOH, 95°C, 1 h: 87%);[676] 2-ethoxycarbonylmethyl-4-phenyl-1(2*H*)-phthalazinone (80%) was made somewhat similarly.[659]

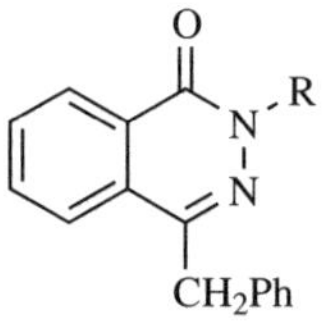

(**41**)

4-Phenyl-2(1*H*)-phthalazinone (**42**, R = Ph) with acrylonitrile gave 2-(2-cyanoethyl)-4-phenyl-1(2*H*)-phthalazinone (**43**, R = Ph) (reactants, trace Triton B, MeOH, dioxane, reflux, 3 h: ~65%);[650] somewhat similarly, 4-(*p*,*p*′-dibromobenzhydryl)-1(2*H*)-phthalazinone [**42**, R = $CH(C_6H_4Br\text{-}p)_2$] gave 2-(2-cyanoethyl)-4-(*p*,*p*′-dibromobenzhydryl)-1(2*H*)-phthalazinone [**43**, R = CH $(C_6H_4Br\text{-}p)_2$] ($H_2C=CHCN$, pyridine, reflux, 2 h: ~90%; structure confirmed by X-ray analysis; analogs likewise).[995]

(42) (43)

4-Benzyl-1(2*H*)-phthalazinone (**44**) underwent Mannich aminoalkylation by *p*-aminophenylacetic acid and formaldehyde to give 4-benzyl-2-[*p*-(carboxymethyl)anilinomethyl]-1(2*H*)-phthalazinone (**45**) (reactants, H_2O, MeOH, reflux, 30 min; then 20°C, 12 h: 70%; analogs likewise).[525]

(44) (45)

4-Ethoxycarbonylmethyl-1(2*H*)-phthalazinone (**46**) with formaldehyde gave 4-ethoxycarbonylmethyl-2-hydroxymethyl-1(2*H*)-phthalazinone (**47**) (reactants, EtOH, H_2O, reflux, 40 h: 65%).[68]

(46) (47)

Also other examples.[136,155,159,179,199,214,218,241,257,279,348,351,365,369,405,469,483,495,573,753,855,869,949,961,985,1040]

Formation of *N*-Aralkylphthalazines or Related Products

1(2*H*)-Phthalazinone (**48**, R = H) gave 2-*o*-methylbenzyl-1(2*H*)-phthalazinone (**48**, R = $CH_2C_6H_4Me$-*o*) ($ClCH_2C_6H_4Me$-*o*, K_2CO_3, Me_2NCHO, intermittent microwave irradiation, 30 min: 63%); analogs likewise but no evidence as to the efficacy of such irradiation.[52]

(48)

4-(3,4-Dimethoxyphenyl)-1(2*H*)-phthalazinone (**49**, R = H) gave 2-benzyl-4-(3,4-dimethoxyphenyl)-1(2*H*)-phthalazinone (**49**, R = CH_2Ph) (substrate, NaH, Me_2NCHO, 20°C, 3 h; then $PhCH_2Cl$↓, 20°C, 4 h: 61%); analogs like 2-benzyl-4-(3,4-dimethoxyphenyl)-5,6,7,8-tetrahydro-1(2*H*)-phthalazinone (57%) were made likewise.[872]

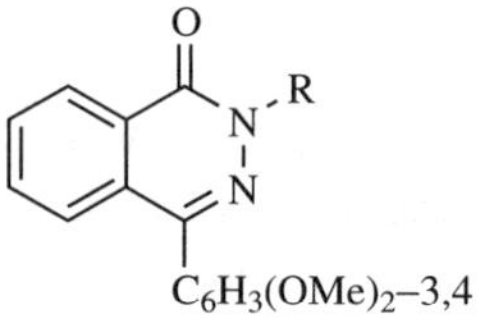

(**49**)

4-Methyl-1(2*H*)-phthalazinone with *p*-(2-bromoethoxy)benzaldehyde gave 2-[2-(*p*-formylphenoxy)ethyl]-4-methyl-1(2*H*)-phthalazinone (**50**) (substrate, K_2CO_3, Me_2NCHO, 25°C, 30 min; $BrCH_2CH_2OC_6H_4CHO$-*p*↓, 65%, 24 h: 77%); analogs likewise.[874]

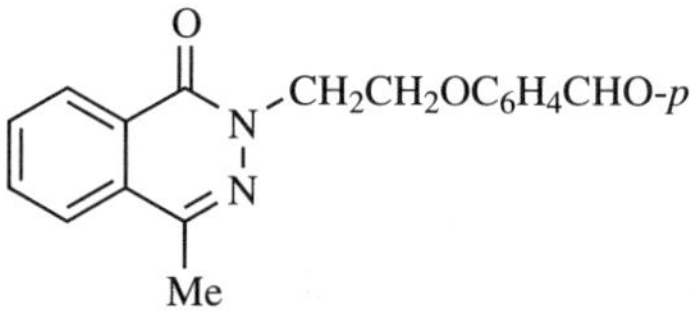

(**50**)

4-Phenyl-1(2*H*)-phthalazinone with 1-(2-bromoethyl)imidazole hydrochloride gave 2-[2-(imidazol-1-yl)ethyl]-4-phenyl-1(2*H*)-phthalazinone (**51**) (reactants, K_2CO_3, Me_2NCHO, 80°C, 6 h: 82%); analogs likewise.[278]

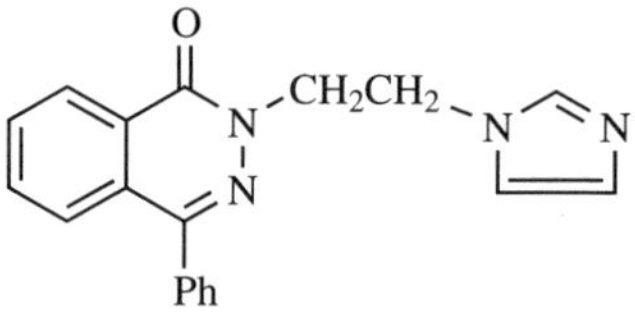

(**51**)

Methyl 4-oxo-3,4-dihydro-5-phthalazinecarboxylate with 2-chloromethylpyridine hydrochloride gave methyl 4-oxo-3-(pyridin-2-ylmethyl)-3,4-dihydro-5-phthalazinecarboxylate (**52**) (substrate, $LiNH_2$, Me_2SO, N_2, 20°C, 2 h; then synthon↓ portionwise, 20°C, 3 h: 60%, characterized as its methanesulfonate salt).[623]

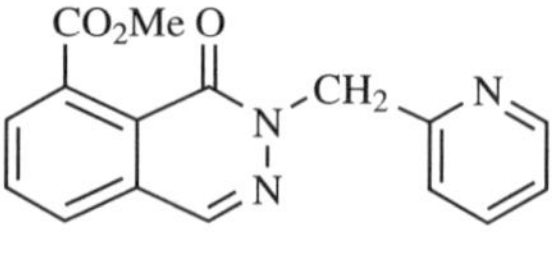

(**52**)

7-Hydroxy-6(2*H*)-phthalazinone (**53**) with three different (3-iodopropenyl) cephalosporins (**54**) gave appropriate 2-(cephalosporinpropenyl)-7-hydroxy-6(2*H*)-phthalazinones (**55**) (reactants, Et_3N, Me_2NCHO, <5°C, 1 h: ~20% in each case; note the interannular *N*-alkylation).[790,791]

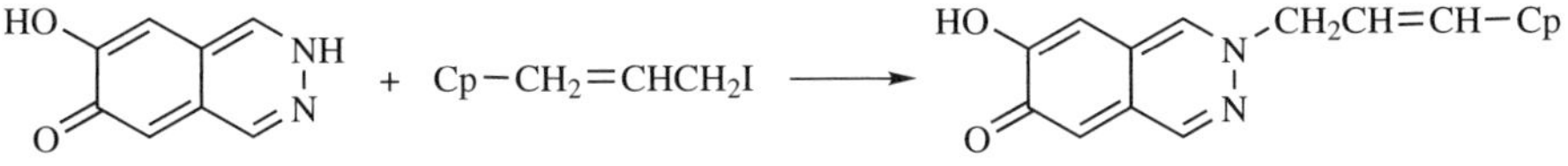

HO, O, NH, N + Cp−CH_2=$CHCH_2I$ → HO, O, N, N, CH_2CH=CH−Cp

(**53**) (**54**) (**55**)

Also other examples.[33,108,173,432,665,873,955,1007]

Formation of *N*-Arylphthalazinones

Note: Such arylation or heteroarylation has been reported occasionally, usually under Ullmann–Goldberg[1006] conditions employing a copper catalyst.

1(2*H*)-Phthalazinone (**56**, R = H) with iodobenzene gave 2-phenyl-1(2*H*)-phthalazinone (**56**, R = Ph) (reactants, K_2CO_3, CuI, Me_2NCHO, 150°C, 6 h: 78%).[437]

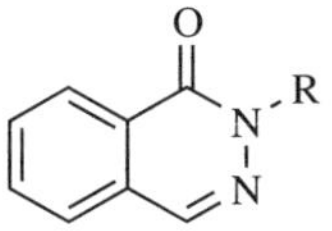

(**56**)

4-(Pyridin-3-yl)-1(2*H*)-phthalazinone with dimethyl 1-(2-bromopyridin-4-yl)-6,7-dimethoxy-2,3-naphthalenedicarboxylate gave 2-[4-(6,7-dimethoxy-2,3-dimethoxycarbonylnaphthalen-1-yl)pyridin-2-yl]-4-(pyridin-3-yl)-1(2*H*)-phthalazinone (**57**) (reactants, K_2CO_3, CuI, Me_2NCHO, 120°C, 5 h: 17%).[286]

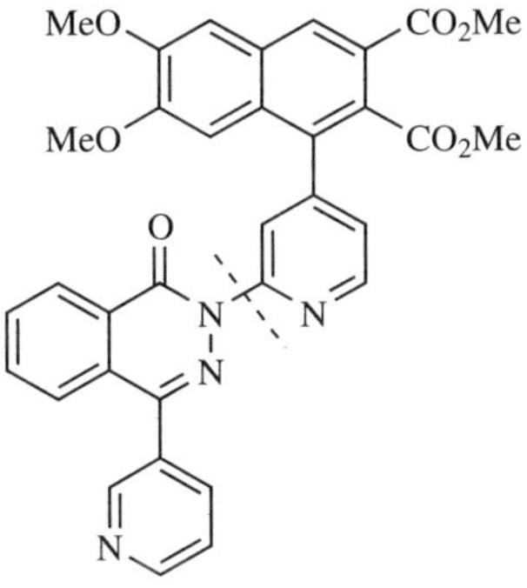

MeO, MeO, CO_2Me, CO_2Me, O, N, N, N, N

(**57**)

Also other examples.[893]

Formation of 2-Alkylphthalazin-2-ium-4-olates

1(2*H*)-Phthalazinone (**58**) with methyl *p*-toluenesulfonate gave 4-hydroxy-2-methylphthalazin-2-ium tosylate (**59**) (reactants, kerosene, 20°C, 90 min; then 140°C, 6 h: >95%)[681,690] and thence 2-methylphthalazin-2-ium-4-olate (**60**) (the tosylate, H_2O, 2M NaOH↓ to pH 6: 35%).[690]

TsOMe; HO^-

(**58**) (**59**) (**60**)

11.1.2.2. Acylation of Tautomeric Phthalazinones

From the limited number of available examples, it appears that tautomeric phthalazinones undergo *O*-acylation under mild conditions but *N*-acylation under more severe conditions, possibly involving an $O \rightarrow N$ rearrangement.[253] Analogous *N*-arenesulfonations have been reported.[33]

4-[(1-Carboxyethyl)amino]-1(2*H*)-phthalazinone (**62**) gave 1-acetoxy-4-[(1-carboxyethyl)amino]phthalazine (**61**) (AcCl, AcOH, reflux, 2 h: 70%) or 2-acetyl-4-[(1-carboxyethyl)amino]-1(2*H*)-phthalazinone (**63**) (neat Ac_2O, AcONa, reflux, 6 h: 70%).[668]

AcOH, AcCl; AcONa, Ac_2O

(**61**) (**62**) (**63**)

5-Amino-1(2*H*)-phthalazinone (**64**) gave 5-acetamido-1-acetoxyphthalazine (**65**) (Ac_2O, 140°C, ? h: 93%).[936]

Ac_2O

(**64**) (**65**)

4-(Biphenyl-4-yl)-1(2*H*)-phthalazinone (**66**) gave a mixture of 1-acetoxy-4-(biphenyl-4-yl)phthalazine (**67**) and 2-acetyl-4-(biphenyl-4-yl)-1(2*H*)-phthalazinone (**68**) (Ac_2O, AcOH),[528,543]

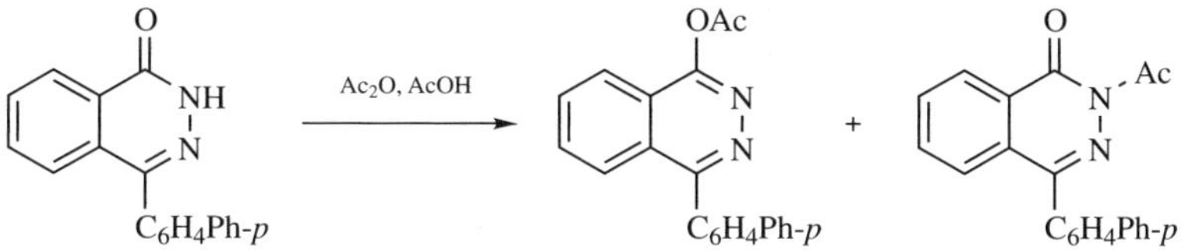

(**66**) (**67**) (**68**)

Also other examples.[76,195,253,351,354,542]

11.1.2.3. Reductive and Oxidative Reactions of Tautomeric Phthalazinones

Tautomeric phthalazinones undergo several reductive or oxidative reactions, the most important of which is the production of phthalazinequinones or semiquinones. The following examples illustrate such reactions.

1(2*H*)-Phthalazinone (**69**) gave 3,4-dihydro-1(2*H*)-phthalazinone (**70**) (substrate, Zn powder, H_2O; 5M HCl↓ during 10 h, 0°C; 89% as hydrochloride)[386] or a mixture of 5,6,7,8-tetrahydro-1(2*H*)-phthalazinone (**70a**) and 3,4,4a,5,6,7,8,8a-octahydro-1(2*H*)-phthalazinone (**70b**) [H_2 (70 atm), PtO_2, AcOH, 60°C, 2 h: 16% and 25%, respectively, after chromatographic separation).[920]

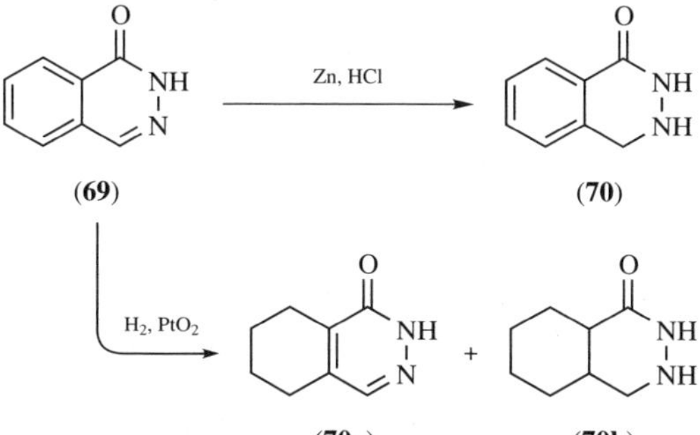

(**69**) (**70**)

(**70a**) (**70b**)

4-Phenyl-4a,5,6,7,8,8a-hexahydro-1(2*H*)-phthalazinone (**71**) gave 4-phenyl-5,6,7,8-tetrahydro-1(2*H*)-phthalazinone (**72**) ($CuCl_2$, MeCN, reflux, 1 h: 89%; analogs likewise).[337]

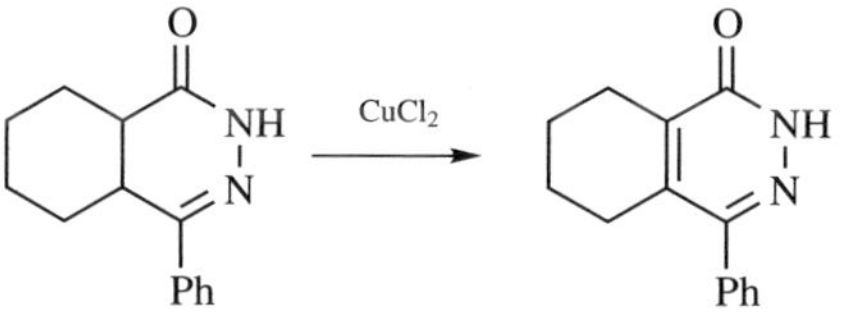

(**71**) (**72**)

1,4(2*H*,3*H*)-Phthalazinedione (**73**) gave 2,2′-biphthalazine-1,1′,4,4′(2*H*,2′*H*,3*H*, 3′*H*)-tetrone (**74**) (Ph_2SeO, AcOH, ?°C, 2 h: 86%).[156]

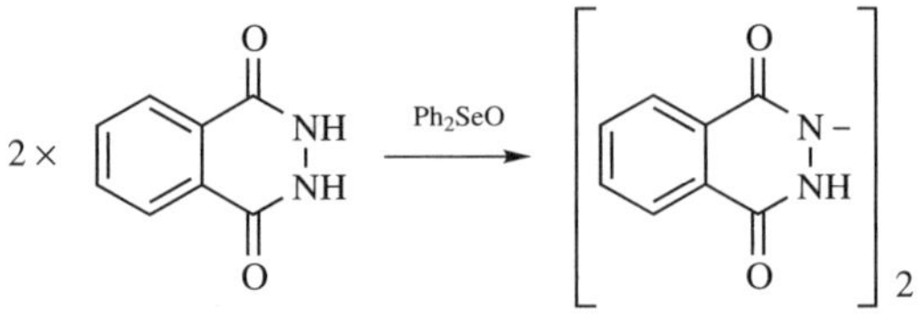

(**73**) (**74**)

4-Acetyl-8-hydroxy-2-methyl-1,5(2*H*,3*H*)-phthalazinedione (**75**) (or tautomer) underwent oxidation to 1-acetyl-3-methyl-4-oxo-3,4-dihydro-5,8-phthalazinequinone (**76**) (Ag_2O, Na_2SO_4, AcMe, A, 20°C, 35 min: 80%).[890]

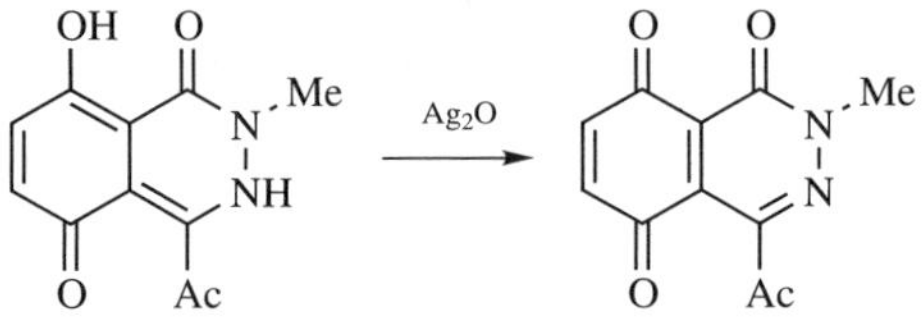

(**75**) (**76**)

5,8-Dihydroxy-5,6,7,8-tetrahydro-1,4-phthalazinedicarboxylic acid (**77**) (or an oxo tautomer) underwent oxidation to 1,4-dicarboxy-5,8-phthalazinequinone (**78**) (MnO_2, CH_2Cl_2, 20°C, 6 days: 61%).[975]

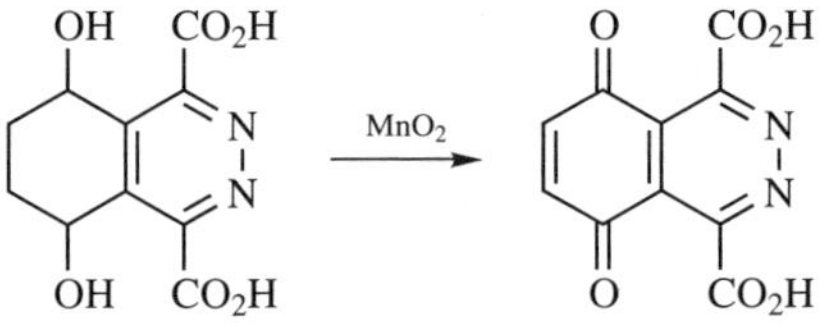

(**77**) (**78**)

6-Dimethylamino-1,4(2*H*,3*H*)-phthalazinedione (**79**) underwent oxidation to 6-dimethylamino-1,4-phthalazinequinone (**80**), too unstable for isolation as such [$Pb(OAc)_4$, AcOH, dioxane, 20°C, ~2 min; solution decomposed ~1% per minute and was therefore converted into a stable adduct (see Section 11.3.2) immediately];[775,1004] other 1,4-phthalazinequinones were made and converted into adducts and the like somewhat similarly.[787,860,892]

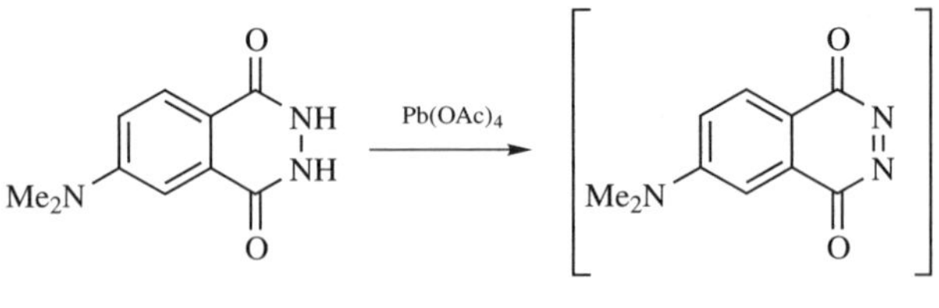

(**79**) (**80**)

In much the same way, 4,4-bis(*p*-dimethylaminophenyl)-3,4-dihydro-1(2*H*)-phthalazinone (**81**) gave the unstable semiquinone (?), 4,4-bis(*p*-dimethylaminophenyl)-1,4-dihydro-1-phthalazinone (**82**) [substrate, Et_3N, CH_2Cl_2, $-80°C$; $Pb(OAc)_4$ in $CH_2Cl_2\downarrow$ slowly; $-80°C$, 3 h) (see also Section 11.5.2).[718,719]

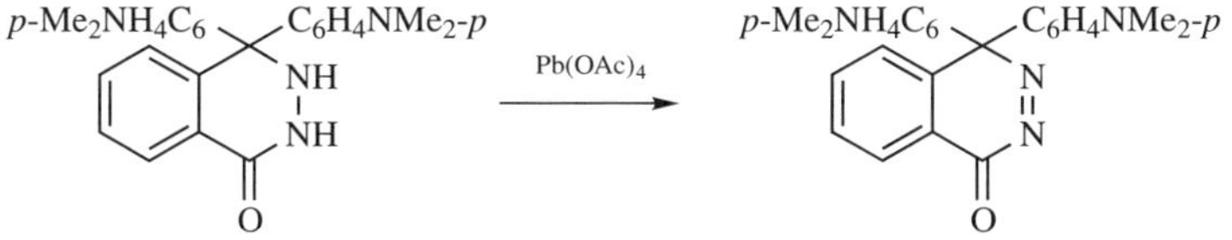

(**81**) → $Pb(OAc)_4$ → (**82**)

11.1.2.4. Other Reactions of Tautomeric Phthalazinones

Several other useful reactions of tautomeric phthalazinones that have been employed in the more recent literature are illustrated by the following classified examples.

Thiation

Note: The usual reagent is phosphorus pentasulfide; Lawesson's reagent (**83**), may also be used,[592] but there appear to be no well-described examples in the more recent literature.

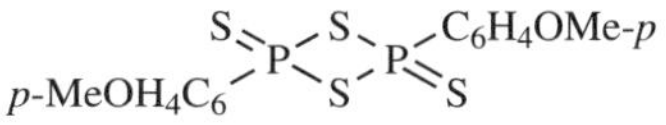

(**83**)

4-Methyl-1(2*H*)-phthalazinone (**84**, R = Me) gave 4-methyl-1(2*H*)-phthalazinethione (**85**, R = Me) (P_2S_5, pyridine, reflux, 4.5 h: 75%);[33] 4-*p*-nitrobenzyl-1(2*H*)-phthalazinone (**84**, R = $CH_2C_6H_4NO_2$-*p*) likewise gave 4-*p*-nitrobenzyl-1(2*H*)-phthalazinethione (**85**, R = $CH_2C_6H_4NO_2$-*p*) (3 h: 81%).[641]

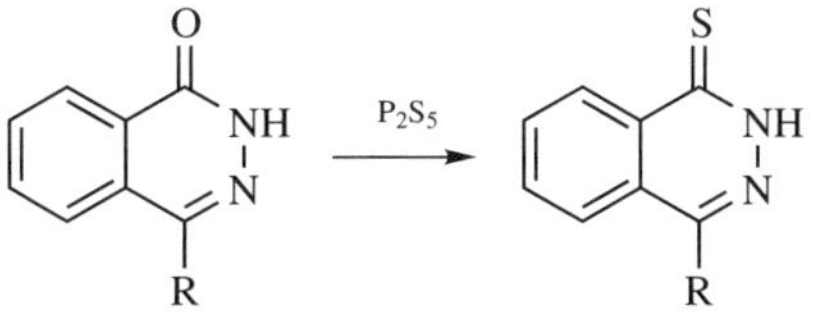

(**84**) → P_2S_5 → (**85**)

4-Benzyl-8-chloro-1(2*H*)-phthalazinone (**86**, X = O) gave 4-benzyl-8-chloro-1(2*H*)-phthalazinethione (**86**, X = S) (P_2S_5, xylene, reflux, 5 h: 50%).[646]

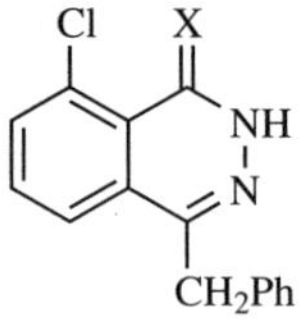

(**86**)

Also other examples.[172,241,344,470,665,668,859]

Aminolysis (Indirect)

Note: The indirect aminolysis of tautomeric phthalazinones may be done via halogeno, alkoxy, thioxo, alkylthio, or other such derivatives. Convenient one-pot syntheses are exemplified here.

1,4(2*H*,3*H*)-Phthalazinedione (**87**) gave 1,4-bis(benzylamino)phthalazine (**87a**) [$Me_3SiNHSiMe_3$, $PhCH_2NH_2$, $(NH_4)_2SO_4$, ~160°C, 24 h: 87%].[133]

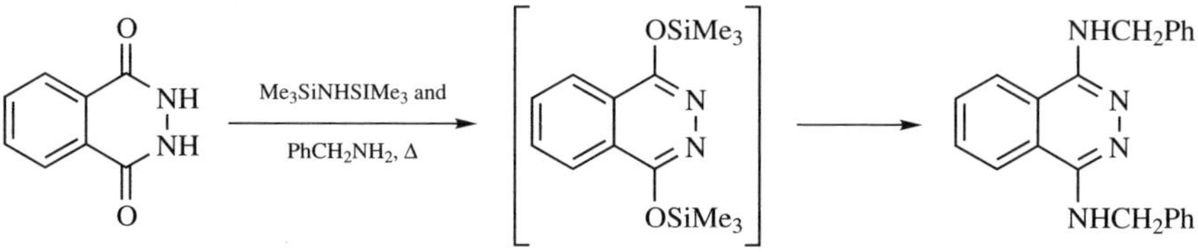

(**87**) (**87a**)

4-(Pyridin-4-ylmethyl)-1(2*H*)-phthalazinone (**88**) gave 1-*p*-chloroanilino-4-(pyridin-4-ylmethyl)phthalazine (**88a**) ($H_2NC_6H_4Cl$-*p*, P_2O_6, $Et_3N{\cdot}HCl$, 170°C, 40 min; substrate↓, 170°C, 2 h: 82%).[870]

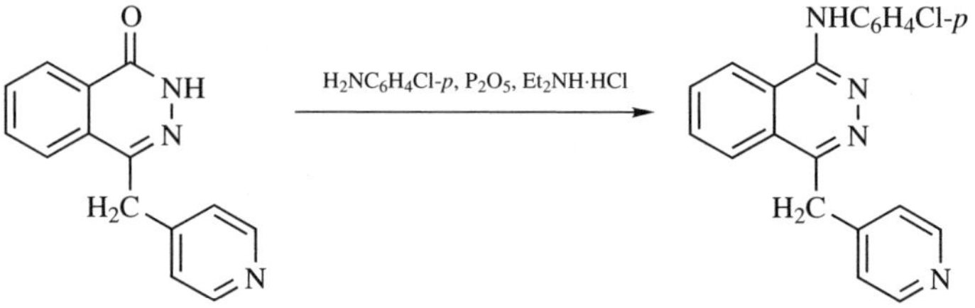

(**88**) (**88a**)

Cyclocondensations

1,4(2*H*,3*H*)-Phthalazinedione (**89**) with 2-dimethylamino-3,3-dimethyl-3*H*-azirine (**90**) gave 3-dimethyliminio-2,2-dimethyl-2,3-dihydroimidazo[2,1-*a*]phthalazin-6-olate (**91**) (substrate, Me_2NCHO, 60°C → 20°C, 30 min; synthon in Me_2NCHO↓ dropwise; 20°C, 18 h: 84%).[151]

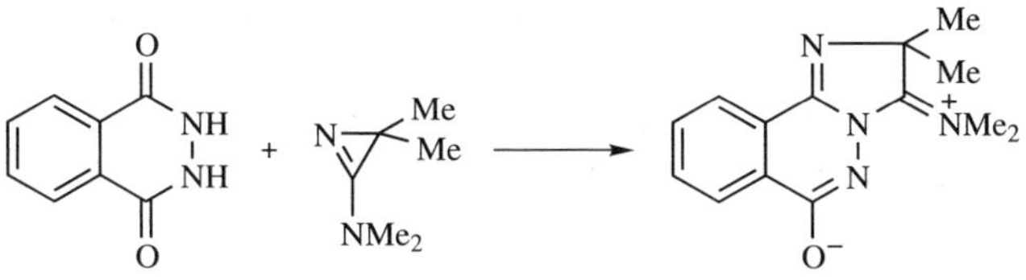

(89) (90) (91)

3,4,4a,5,6,7,8,8a-Octahydro-1(2*H*)-phthalazinone (**92**) with *o*-(bromomethyl)-benzoyl chloride (**93**) gave a single product, 1,2,3,4,4a,5,7,12,14,14a-decahydrophthalazino[2,3-*b*]phthalazine-5,12-dione (**94**) (substrate, dioxane, 15°C; synthon in dioxane↓ dropwise; then Et_3N in dioxane↓; 60°C, 2 h: 65%);[920] analogs likewise.[919]

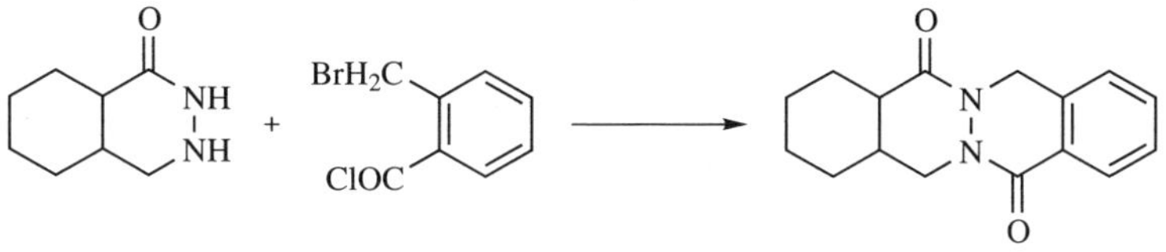

(92) (93) (94)

For another example, see Section 14.1.2.

Ring Fission

The conversion of 7-dimethylamino-4,4-bis(*p*-dimethylaminophenyl)-3,4-dihydro-1(2*H*)-phthalazinone (**95**) into 2-[α,α-bis(*p*-dimethylaminophenyl)-α-ethoxymethyl]-5-dimethylaminobenzohydrazide (**96**) (and/or related compounds) by irradiation in ethanol has been studied spectrally, apparently without isolation of the product(s).[410,968]

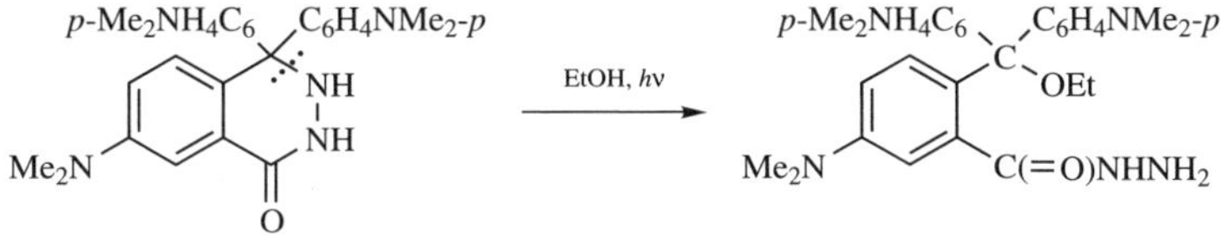

EtOH, *hν*

(95) (96)

Ring Contraction

3,4,4a,5,6,7,8,8a-Octahydro-1(2*H*)-phthalazinone (**97**) underwent a de facto isomerization to afford 2-aminoperhydro-1-isoindolinone (**97a**) (neat H_2N $NH_2 \cdot H_2O$, reflux, 6 h: 70%).[920]

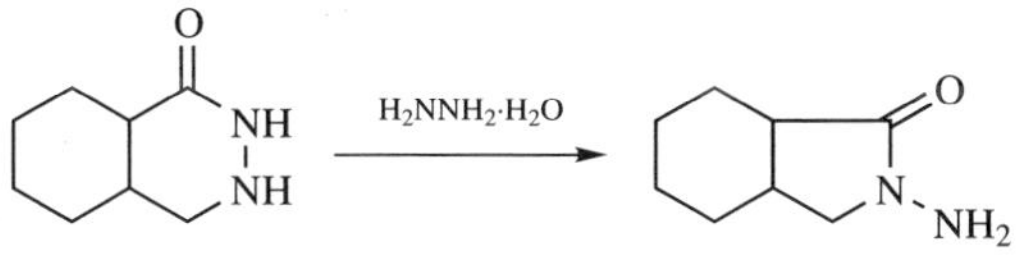

$H_2NNH_2 \cdot H_2O$

(97) (97a)

5,8-Dimethoxy-1,4(2*H*,3*H*)-phthalazinedione (**98**) with ammonium cerium(IV) nitrate unexpectedly gave 4,7-dimethoxy-1,3-isoindolinedione (3,6-dimethoxyphthalimide) (**98a**) [substrate, MeCN, H_2O, 20°C; $(NH_4)_2Ce(NO_3)_6$ in H_2O↓ dropwise during 10 min; then 20°C, 4 h: 68%]; analogs likewise.[715]

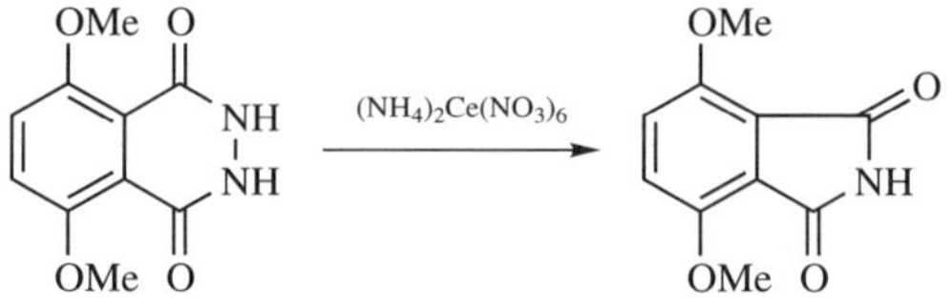

(**98**) (**98a**)

11.2. EXTRANUCLEAR HYDROXYPHTHALAZINES

Such hydroxyphthalazines naturally resemble benzyl alcohol or an hydroxybiphenyl in both preparative procedures and reactions.

11.2.1. Preparation of Extranuclear Hydroxyphthalazines

The preparation of these hydroxy compounds by *primary synthesis* has been covered in Chapter 1 and by *hydrolysis of extranuclear halogenophthalazines* in Section 10.3 (no other examples). The remaining routes are illustrated by the following examples.

By Reduction of Phthalazinecarboxylic Esters or Amides

Note: Sodium borohydride is usually the reagent of choice, but lithium aluminum hydride may also be used.

Diethyl 4-oxo-3,4-dihydro-1,7-phthalazinedicarboxylate (**100**) gave ethyl 4-hydroxymethyl-1-oxo-1,2-dihydro-6-phthalazinecarboxylate (**99**) [$NaBH_4$ (0.2 part), EtOH, 0°C; substrate (1 part)↓ portionwise; $CaCl_2$ in EtOH↓ slowly, −5°C, 2 h; the 20°C, 1 h: 56%] or 4, 6-bis(hydroxymethyl)-1(2*H*)-phthalazinone (**101**) [substrate (1 part), EtOH, 20°C; $NaBH_4$ (0.4 part)↓ portionwise; 20°C, 12 h: 90%];[404] analogs likewise.[405]

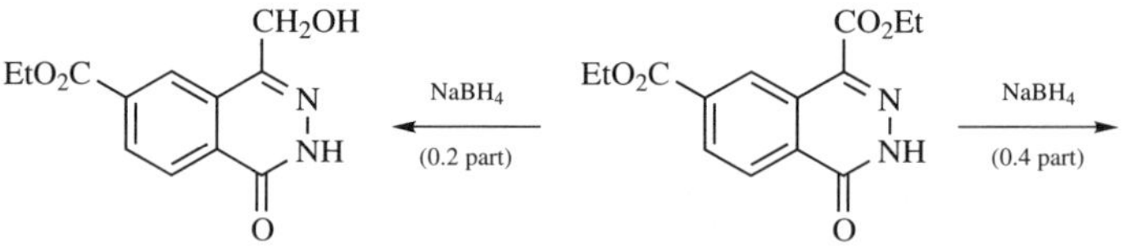

(**99**) (**100**)

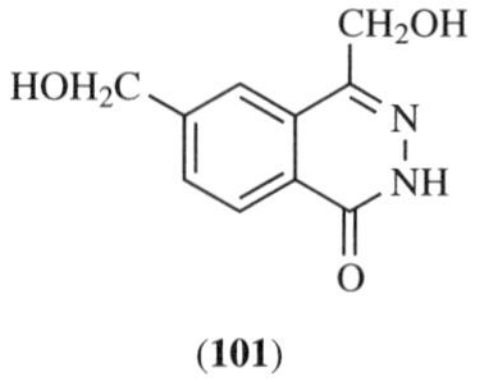

(**101**)

Ethyl 4-oxo-3-phenyl-3,4-dihydro-6-phthalazinecarboxylate (**102**, R = CO_2Et) gave 7-hydroxymethyl-2-phenyl-1(2*H*)-phthalazinone (**102**, R = CH_2OH) ($NaBH_4$, EtOH, 20°C → reflux, 2 h: 76%); analogs likewise.[411]

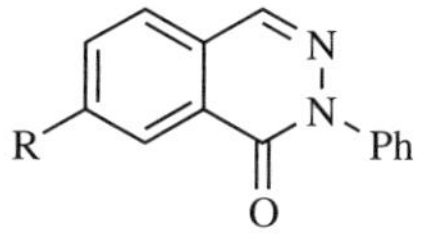

(**102**)

Diethyl 5,7-dimethyl-4-oxo-3-phenyl-3,4-dihydro-1,6-phthalazinedicarboxylate (**103**, R = CO_2Et) selectively gave ethyl 1-hydroxymethyl-5,7-dimethyl-4-oxo-3-phenyl-3,4-dihydro-6-phthalazinecarboxylate (**103**, R = CH_2OH) (substrate, EtOH, −5°C; excess $NaBH_4$↓ portionwise; 20°C, 12 h: 89%); a score of substituted-phenyl analogs were made similarly.[956]

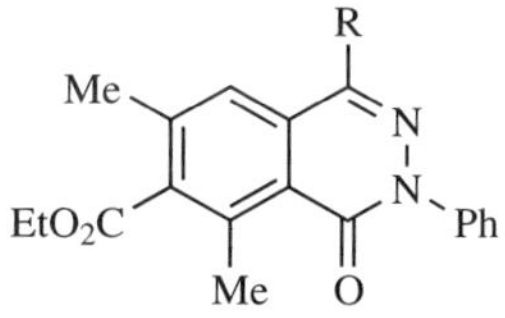

(**103**)

Diethyl 6,7-phthalazinedicarboxylate (**104**) gave 6,7-bis(hydroxymethyl)phthalazine (**104a**) ($LiAlH_4$, $AlCl_3$, THF, 60°C, 24 h: 58%).[584]

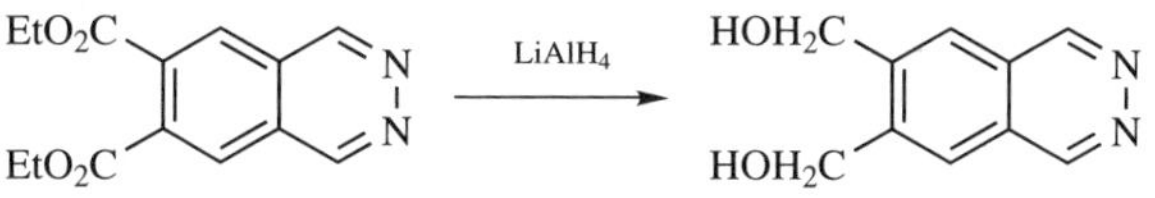

$LiAlH_4$

(**104**) (**104a**)

4-[1-(Dimethylcarbamoyl)-1-methylethyl]amino-1(2*H*)-phthalazinone (**105**) gave 4-(2-hydroxy-1,1-dimethylethyl)amino-2(1*H*)-phthalazinone (**105a**) ($NaBH_4$, EtOH, H_2O, 65°C, 20 h: 59%).[151]

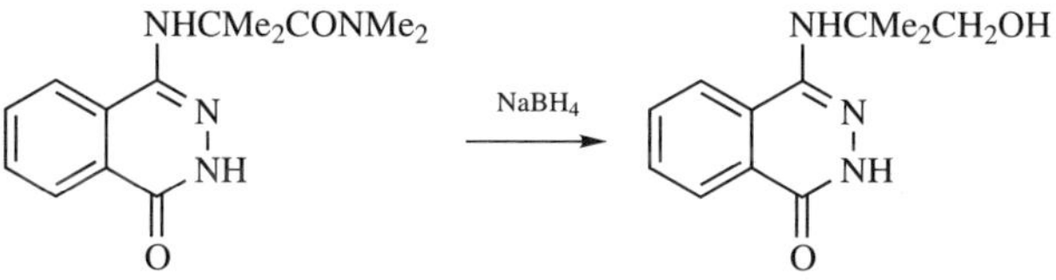

$NaBH_4$

(**105**) (**105a**)

Also other examples.[573,574,783]

From Phthalazinecarboxylic Esters with a Grignard Reagent

Ethyl 4-oxo-3,4-dihydro-1-phthalazinecarboxylate (**106**) with methylmagnesium bromide gave 4-(1-hydroxy-1-methylethyl)-1(2*H*)-phthalazinone (**107**) (substrate, THF, <5°C; MeMgBr in $Et_2O\downarrow$ dropwise; reflux, 2 h: >80%).[404,574]

CO_2Et MeMgBr Me_2COH

(**106**) (**107**)

Also other examples.[405]

From Phthalazinecarbaldehydes with Ethyl Diazoacetate or a Grignard Reagent

Ethyl 1-formyl-5,7-dimethyl-4-oxo-3,4-dihydro-6-phthalazinecarboxylate (**108**) with ethyl diazoacetate gave ethyl 1-(2-ethoxycarbonyl-2-diazo-1-hydroxyethyl)-5,7-dimethyl-4-oxo-3,4-dihydro-6-phthalazinecarboxylate (**109**) (reactants, EtOH, 0°C; 1.2M NaOH↓ dropwise; <5°C, 5 h: 58%)[424] or with methylmagnesium bromide gave ethyl 1-(1-hydroxyethyl)-5,7-dimethyl-4-oxo-3,4-dihydro-6-phthalazinecarboxylate (**109a**) (substrate, Et_2O, MeMgBr in $Et_2O\downarrow$ dropwise; then reflux, 3 h: ~60%).[405]

EtO_2CCHN_2 $HOCHC(N_2)CO_2Et$ MeMgBr HOCHMe

(**108**) (**109**) (**109a**)

By Reduction of *C*-Acylphthalazines

1-(3,4-Dimethoxybenzoyl)phthalazine (**110**) underwent reduction by sodium borohydride to give 1-(α-hydroxy-3,4-dimethoxybenzyl)phthalazine (**111**) (substrate, MeOH, $NaBH_4\downarrow$ slowly, 0°C, until clear; then 20°C, 30 min: 72%).[21]

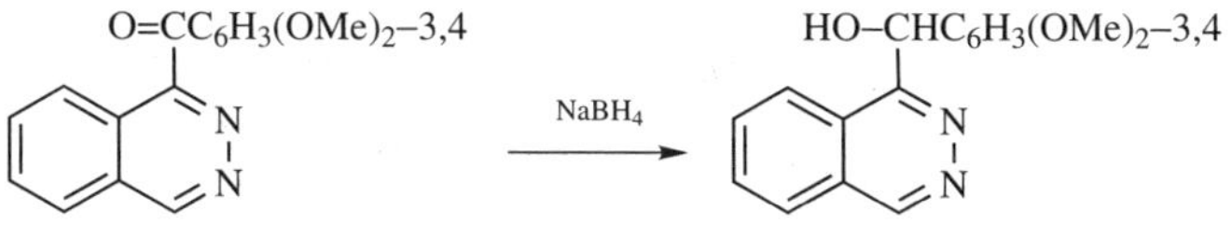

(110) (111)

4-(4-Dimethylaminobutyryl)-1(2*H*)-phthalazinone likewise gave 4-(4-dimethylamino-1-hydroxybutyl)-1(2*H*)-phthalazinone ($NaBH_4$, EtOH, 20°C: no other details).[574]

Also other examples.[743]

By Hydrolysis of Extranuclear Acyloxyphthalazines

2-(2-Acetoxyethyl)-4-(pyridin-3-yl)-1(2*H*)-phthalazinone (**112**, R = Ac) gave 2-(2-hydroxyethyl)-4-(pyridin-3-yl)-1(2*H*)-phthalazinone (**112**, R = H) (1M NaOH, THF, 40°C, 3 h: 92%).[279]

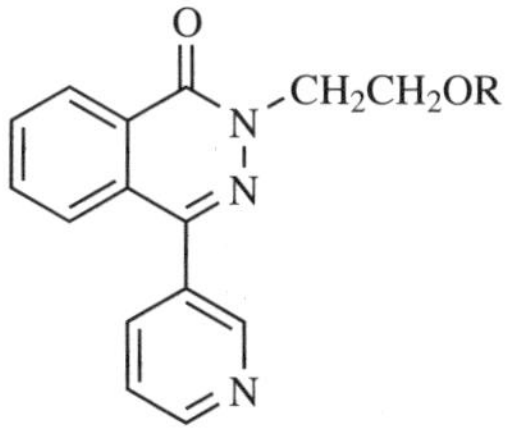

(112)

1-(α-Benzoyloxybenzyl)phthalazine (**113**, R = Bz) gave 1-(α-hydroxybenzyl)phthalazine (**113**, R = H) (KOH, EtOH, H_2O, reflux, N_2, 8 h: 57%); analogs likewise.[377]

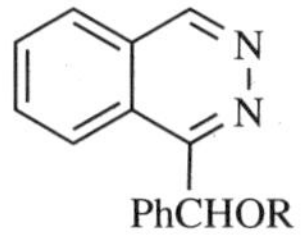

(113)

Also other examples.[986]

From Extranuclear Alkoxy- or Epoxyphthalazines

Note: The formation of extranuclear hydroxyphthalazines from such ethers is often possible by hydrolysis, but reductive cleavage is a good procedure for benzyloxy substrates.

1-*p*-Methoxybenzyl- (**114**, R = Me) gave 1-*p*-hydroxybenzylphthalazine (**114**, R = H) (10M HCl, reflux, 66 h: 74%).[393]

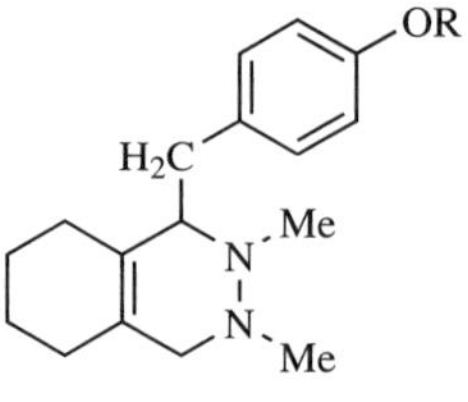

(**114**)

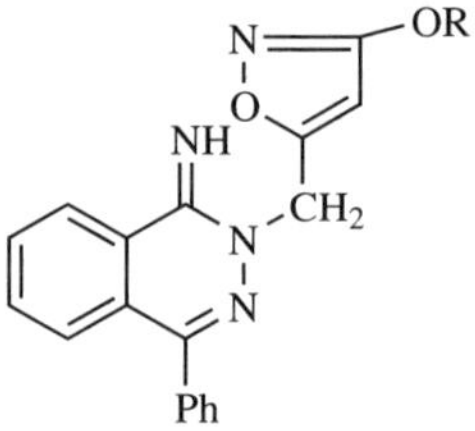

(**115**)

2-[(3-Methoxyisoxazol-5-yl)methyl]- (**115**, R = Me) gave 2-[(3-hydroxyisoxazol-5-yl)methyl]-4-phenyl-1(2*H*)-phthalazinimine (**115**, R = H) (HBr, AcOH, 100°C, 12 h: 60%).[109]

2-Cycloheptyl-4-(3-cyclopentyloxy-4-methoxyphenyl)- (**116**, R = OC_5H_9) underwent selective hydrolysis to 2-cycloheptyl-4-(3-hydroxy-4-methoxyphenyl)-4a,5,8,8a-tetrahydro-1(2*H*)-phthalazinone (**116**, R = H) (TsOH · H_2O, PhMe, reflux, 3 h: 71%).[873]

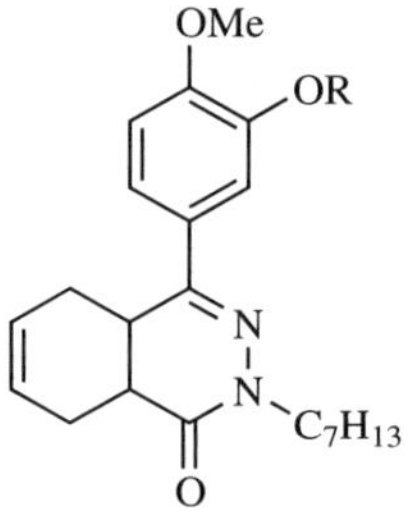

(**116**)

The isopropylidene ether, 4-[(2,2-dimethyl-1,3-dioxolan-4-yl)methyl]-2-phenyl-1(2*H*)-phthalazinone (**117**) gave 4-(2,3-dihydroxypropyl-2-phenyl-1(2*H*)-phthalazinone (**118**) (70% AcOH, reflux, 2 h: 30%).[483]

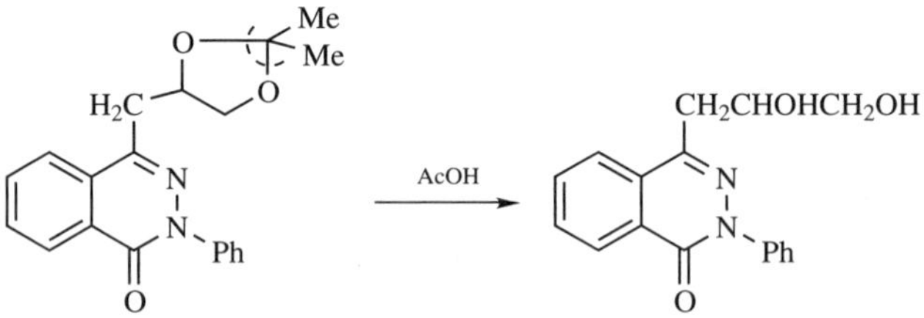

AcOH

(**117**) (**118**)

Ethyl 1-*o*-benzyloxyphenyl- (**119**, R = CH_2Ph) underwent reductive cleavage to ethyl 1-*o*-hydroxyphenyl-5,7-dimethyl-4-oxo-3,4-dihydro-6-phthalazinecarboxylate (**119**, R = H) (H_2, Pd/C, EtOH, 2 h: 91%).[426]

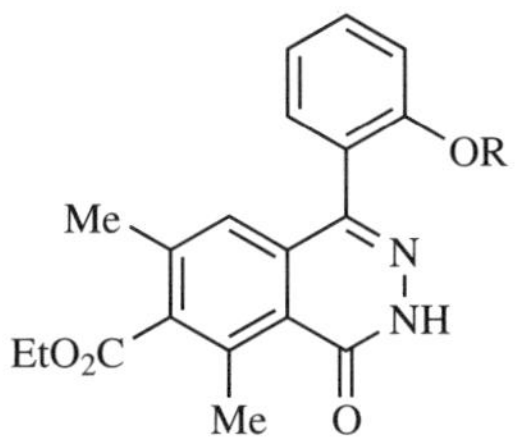

(**119**)

4-(2,3-Epoxypropoxy)-1(2*H*)-phthalazinone (**120**) with diethylamine gave 4-(3-diethylamino-2-hydroxypropoxy)-1(2*H*)-phthalazinone (**121**) (reactants, KOH, MeOH, reflux, 5 h: 82%); many analogs likewise.[875]

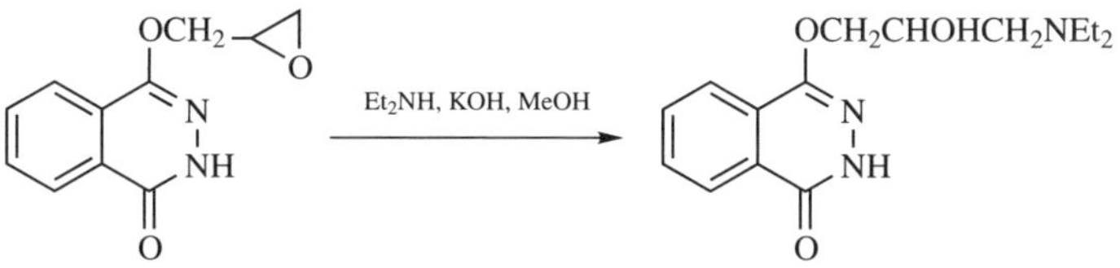

(**120**) (**121**)

Ethyl 3-(2,3-epoxypropyl)- (**122**) gave ethyl 3-(2-hydroxy-3-isopropylamino-propyl)-1,5,7-trimethyl-4-oxo-3,4-dihydro-6-phthalazinecarboxylate (**123**) (Pr^iNH_2, Me_3Al, CH_2Cl_2, N_2, 20°C, 30 min; substrate↓, 20°C, 10 h: 56%).[424]

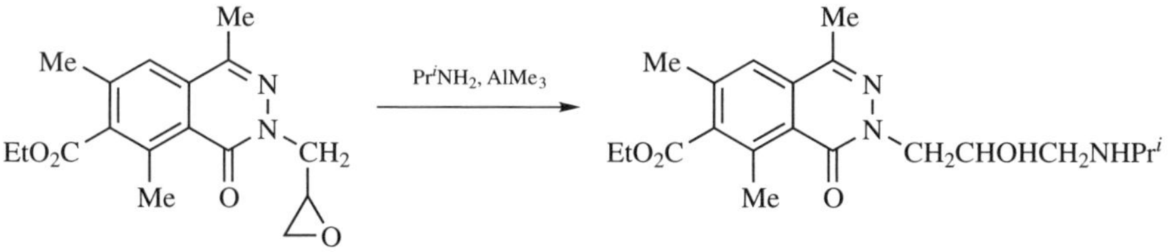

(**122**) (**123**)

From Extranuclear Primary Aminophthalazines

Ethyl 1-(2-aminoethyl)- (**124**, R = NH_2) gave ethyl 1-(2-hydroxyethyl)-5,7-dimethyl-4-oxo-3,4-dihydro-6-phthalazinecarboxylate (**124**, R = OH) (substrate, AcOH, ~5°C; $NaNO_2$ in H_2O↓ dropwise; 5°C until N_2↑ ceased: 39%).[405]

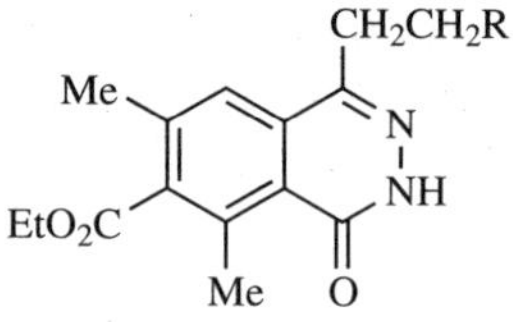

(**124**)

Ethyl 3-*p*-aminophenyl- (**125**, R = NH_2) gave ethyl 3-*p*-hydroxyphenyl-5,7-dimethyl-4-oxo-3,4-dihydro-6-phthalazinecarboxylate (**125**, R = OH) (substrate, THF, 1M HCl; $NaNO_2$ in $H_2O\downarrow$ dropwise, <5°C; then 5°C → 20°C, 60 min: 16%).[270]

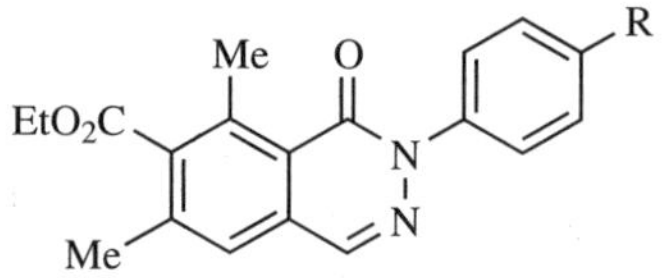

(**125**)

From Reissert Derivatives of Phthalazine

The Reissert compound, 1-cyano-*N,N*-diethyl-1,2-dihydro-2-phthalazinecarboxamide (**126**) with benzaldehyde in an alkaline medium gave 1-(α-hydroxybenzyl)phthalazine (**128**), allegedly via the intermediate (**127**) (PhCHO, NaOH, H_2O, MeCN, $PhCH_2Et_3NCl$, 20°C, 2 h: 15%).[35]

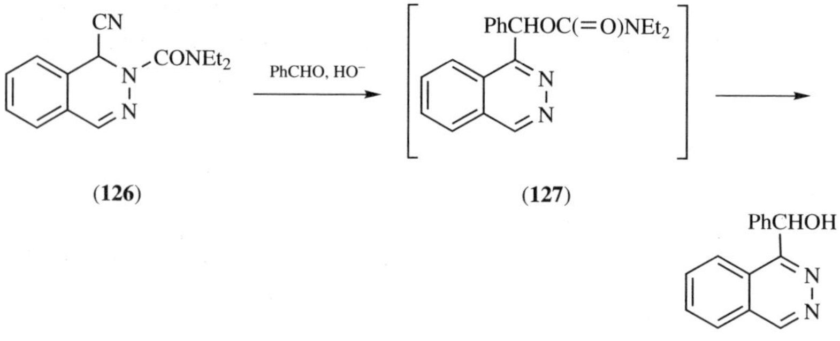

PhCHO, HO^-

(**126**) (**127**)

(**128**)

By Passenger Introduction

Note: Examples of this type of synthesis occur in most chapters of this book. A single typical example is given here.

1-Chloro-4-phenylphthalazine (**129**) gave 1-(2-hydroxyethylamino)-4-phenylphthalazine (**130**) ($H_2NCH_2CH_2OH$, dioxane, reflux, 19 h: 84%).[692]

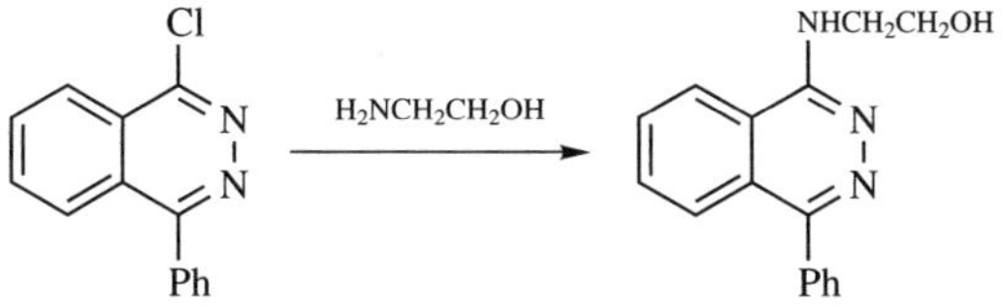

Cl NHCH$_2$CH$_2$OH

$H_2NCH_2CH_2OH$

Ph Ph

(**129**) (**130**)

11.2.2. Reactions of Extranuclear Hydroxyphthalazines

Recently used reactions of these alcohols are typified by the classified examples that follow.

Oxidative Reactions

Ethyl 1-hydroxymethyl- (**131**) was oxidized to ethyl 1-formyl-5,7-dimethyl-4-oxo-3,4-dihydro-6-phthalazinecarboxylate (**132**) (substrate, *N*-bromosuccinimide, Bz_2O_2, CCl_4, reflux, 2 h: 80%).[405]

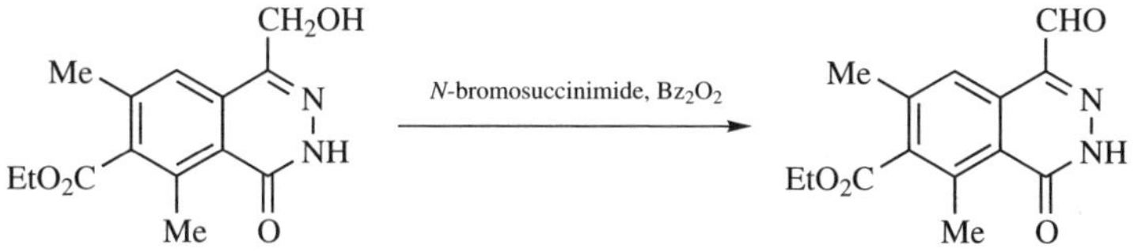

N-bromosuccinimide, Bz_2O_2

(**131**) (**132**)

1-(α-Hydroxy-3,4-dimethoxybenzyl)phthalazine (**133**) gave 1-(3,4-dimethoxybenzoyl)phthalazine (**134**) (CrO_3, pyridine, 20°C, 12 h: 77%).[21]

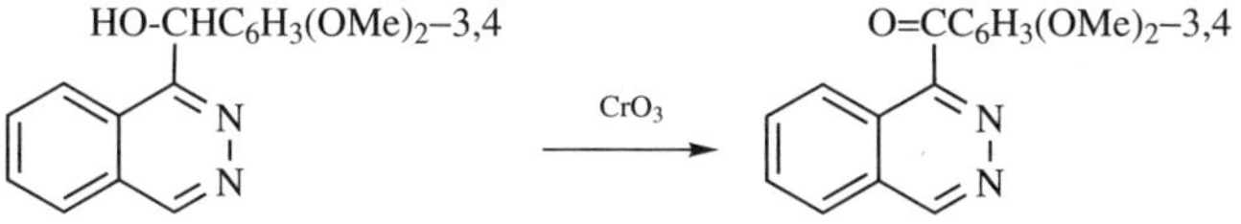

HO-CHC$_6$H$_3$(OMe)$_2$–3,4 O=CC$_6$H$_3$(OMe)$_2$–3,4

CrO_3

(**133**) (**134**)

O-Alkylations

2-Cycloheptyl-4-(3-hydroxy-4-methoxybenzyl)- (**135**, R = H) gave 2-cycloheptyl-4-[4-methoxy-3-(5-phenylpentyloxy)benzyl]-4a,5,8,8a-tetrahydro-1(2*H*)-phthalazinone [**135**, R = $(CH_2)_5$Ph] [Br$(CH_2)_5$Ph, K_2CO_3, *N*-methylpyrrolidine, 60°C, 5 h: 72%]; several analogs likewise.[873]

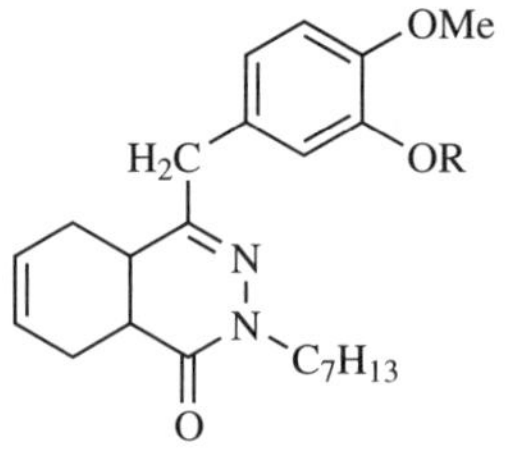

(**135**)

Ethyl 1-hydroxymethyl-5,7-dimethyl-4-oxo-3,4-dihydro-6-phthalazinecarboxylate (**136**, Q = R = H) gave a separable mixture of ethyl 1-hydroxymethyl-3,5,7-trimethyl- (**136**, Q = Me, R = H) and ethyl 1-methoxymethyl-3,5,7-trimethyl-4-oxo-3,4-dihydro-6-phthalazinecarboxylate (**136**, Q = R = Me) (MeI, NaOH, EtOH, reflux, 90 min: ~60% and ~15%, respectively).[405]

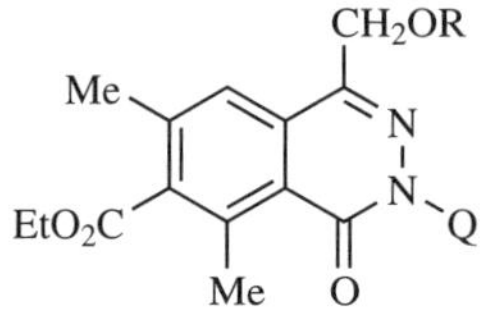

(**136**)

Also other examples.[893]

O-Acylation

Note: This subsection also includes the analogous formation of nitroxyalkyl derivatives.

4-Hydroxymethyl-1,7(2*H*,3*H*)-phthalazinedione (**137**) apparently underwent acetylation to afford 7-acetoxy-4-acetoxymethyl-1(2*H*)-phthalazinone (**138**) (for details, see original).[195]

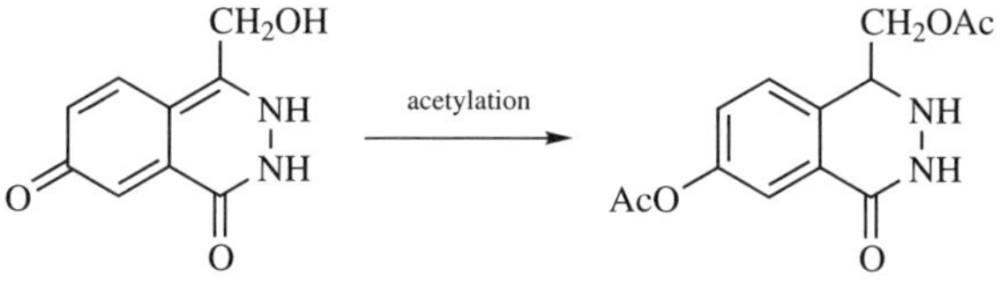

acetylation

(**137**) (**138**)

Ethyl 1-hydroxymethyl-5,7-dimethyl-4-oxo-3,4-dihydro-6-phthalazinecarboxylate (**139**) with fuming nitric acid and acetic anhydride gave a separable mixture of ethyl 5,7-dimethyl-1-nitroxymethyl-4-oxo-3,4-dihydro-6-phthalazinecarboxylate (**140**) and ethyl 5,7-dimethyl-3-nitro-1-nitroxymethyl-4-oxo-3,4-dihydro-6-phthalazinecarboxylate (**142**) [initially described as the dinitroxy isomer (**141**) but later corrected on X-ray evidence] [HNO_3 (*d* 1.5), Ac_2O, <0°C, 30 min; then substrate↓ slowly, <0°C, 2 h: 79% and 11%, respectively];[415,428] repetition at 20°C gave only product (**142**) in 93% yield.[415]

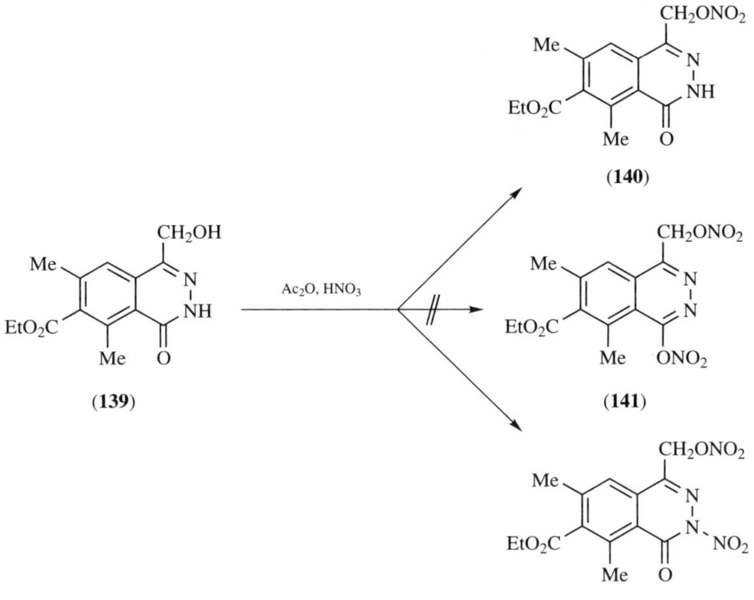

(140)

(139)

Ac_2O, HNO_3

(141)

(142)

Also other examples.[168,425]

Retro Benzoin Condensations

1-(α-Hydroxy-α-*p*-methoxyphenylphenethyl)-4-phenylphthalazine (**143**) appears to have given a separable mixture of 1-phenylphthalazine (**144**), 1-benzyl-4-phenylphthalazine (**145**), *p*-methoxybenzaldehyde (**146**), and benzyl *p*-methoxyphenyl ketone (**147**) [substrate, KCN (1 mol), Me_2NCHO, reflux, 1 h: 35%, 31%, 3%, and 40%, respectively).[438]

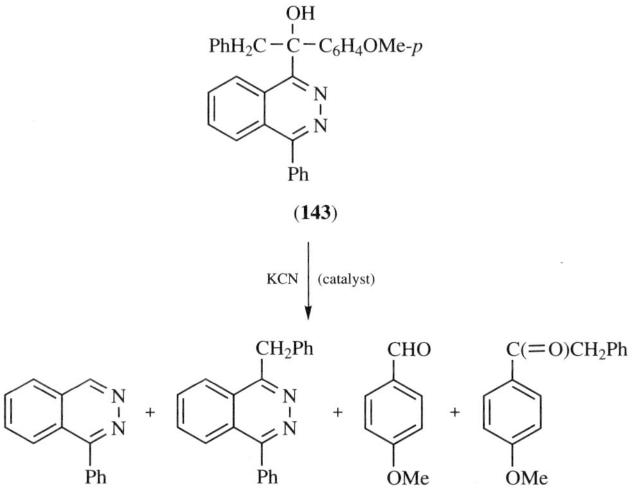

(143)

KCN (catalyst)

(144) (145) (146) (147)

Cyclization Reactions

2-(3-Hydroxypropyl)-4-(3-hydroxypropylamino)-1(2*H*)-phthalazinone (**148**) with 70% perchloric acid gave 2,3,7,8,9,10-hexahydro[1,3]oxazino[2,3-*d*]pyrimido[1,2-*c*]phthalazine-5,8-diium bisperchlorate (**149**) (~95°C, 6 h: 94%); many analogous cyclizations similarly.[299]

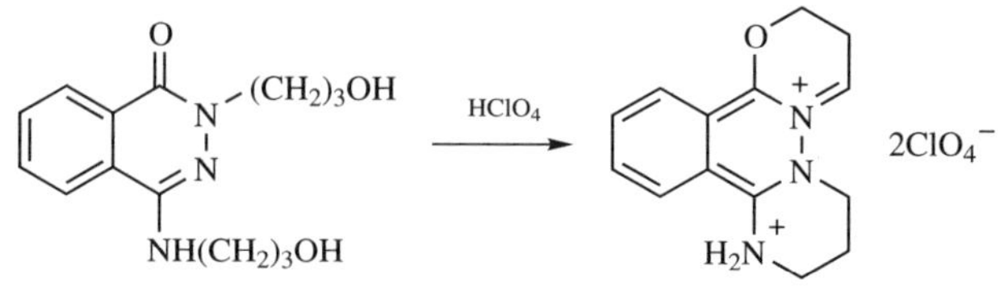

(**148**) (**149**)

4-(3-Hydroxypropylamino)-1(2*H*)-phthalazinone (**150**) in sulfuric acid gave 3,4-dihydro-2*H*-pyrimido[2,1-*a*]phthalazin-7(6*H*)-one (**152**) via the sulfate ester (**151**) (for details, see original).[221]

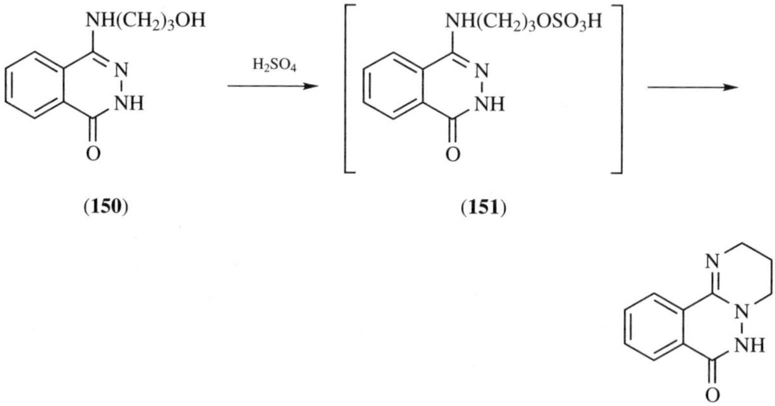

(**150**) (**151**)

(**152**)

1-*o*-Hydroxyanilino-4-phenylphthalazine (**153**) underwent dehydrative cyclization to give 5-phenylbenzimidazo[2,1-*a*]phthalazine (**154**) (polyphosphoric acid; for details, see original); analogs likewise.[680]

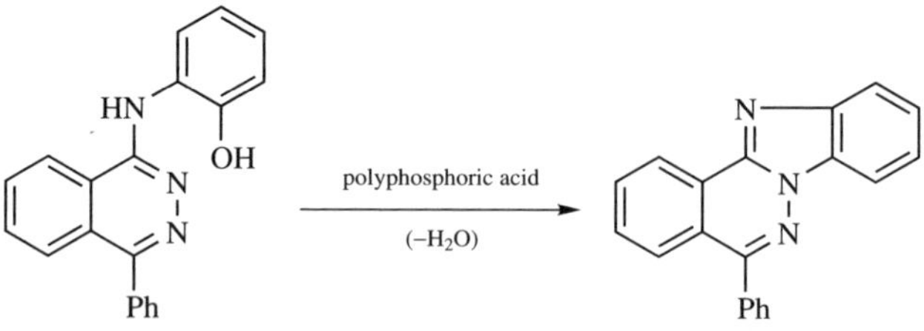

(**153**) (**154**)

For another example, see Section 10.1.2.

11.3. PHTHALAZINEQUINONES

Of the four possible intraannular phthalazinequinones, 5,8-phthalazinequinone (**155**) and its derivatives are stable and isolable; 1,4-phthalazinequinones (e.g., **156**) are inisolable but are sufficiently stable in acetonic or like solution, especially at subzero temperatures, for examination of their spectral and chemical properties; and the 5,6- and 6,7-phthalazinequinones, (e.g., **157** and **158**), appear to be unknown. Interannular phthalazinequinones are also unknown.

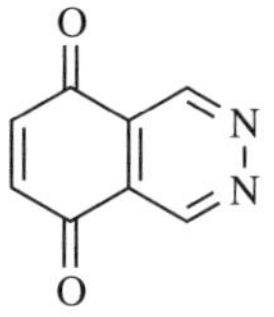

(**155**) 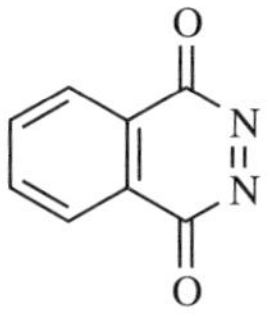 (**156**) 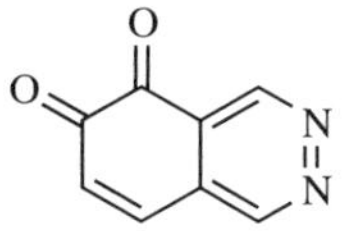 (**157**) 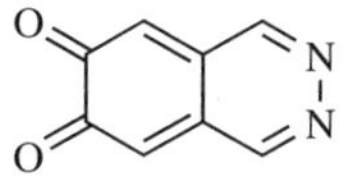 (**158**)

11.3.1. Preparation of Phthalazinequinones (*E* 465)

The available information on the preparation of phthalazinequinones is summarized in the following classified examples.

5,8-Phthalazinequinones from 5,8(2*H*,3*H*)-Phthalazinediones

Note: This route to such quinones, using silver oxide or manganese dioxide as oxidant, has been illustrated in Section 11.1.2.3.

5,8-Phthalazinequinones from Phthalazinamines

5-Phthalazinamine (**159**) underwent oxidative hydrolysis to afford 5,8-phthalazinequinone (**160**) (substrate, H_2SO_4, H_2O, 5°C; $K_2Cr_2O_7$ in dilute H_2SO_4↓, ice↓; then 5°C, 2 h: 58%).[715]

The same substrate (**159**) underwent a similar oxidative hydrolysis and an additional chlorination to give 6,7-dichloro-5,8-phthalazinequinone (**161**) (substrate, 10M HCl, 20°C, $NaClO_3$↓ slowly; then 20°C, 1 h: 13%) but somewhat similar treatment of 5,8-phthalazinediamine (**162**) gave a better yield of the same product (**161**) (substrate, 10M HCl, $NaClO_3$↓ portionwise, 0°C; then 20°C, 1 h: 94%).[856]

1,4-Phthalazinequinones

Note: The early work on formation of these phthalazinequinones in solution has been summarized.[571] Subsequent oxidative routes from 1,4(2*H*,3*H*)-phthalazinediones has been exemplified in Section 11.1.2.3.

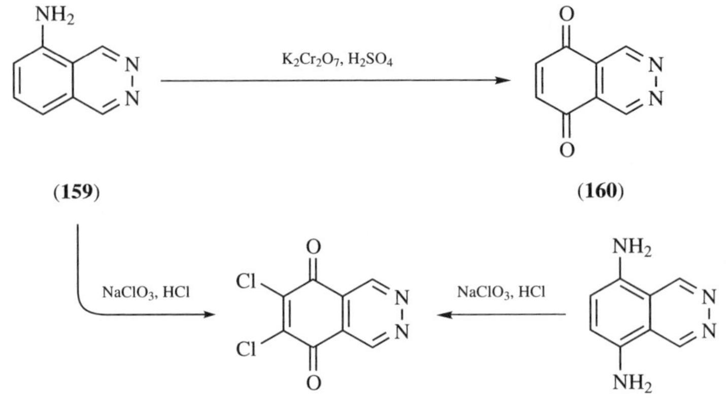

(159) (160) (161) (162)

11.3.2. Reactions of Phthalazinequinones (*E* 461)

Some early reports on the reactions of 1,4-phthalazinediones have been reviewed briefly.[571] The *halogenation* of 5,8-phthalazinequinones has been covered in Section 10.1.2, and other reactions of both 5,8- and 1,4-phthalazinequinones are illustrated by the following classified examples.

Acylation and Acyloxylation of 5,8-Phthalazinequinones

5,8-Phthalazinequinone (**163**) with acetic anhydride underwent a Thiele–Winter reaction[1006] to afford 5,6,8-triacetoxyphthalazine (**164**) (Ac_2O, $BF_3 \cdot Et_2O$, 20°C, 36 h: 12%).[715]

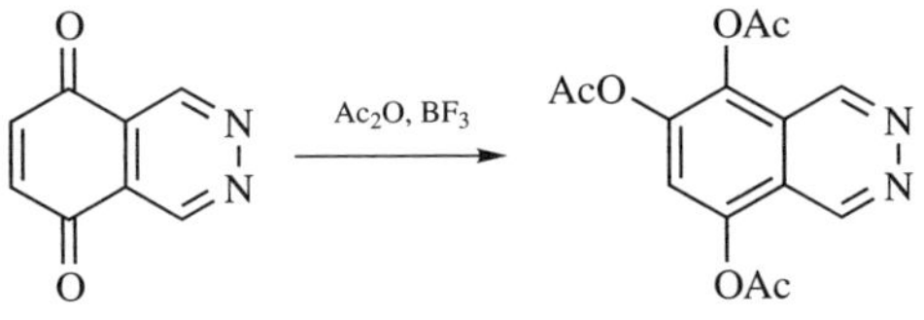

(163) (164)

Degradation and Aggregation of 1,4-Phthalazinequinones

An acetonic solution of 1,4-phthalazinequinone (**168**) gave a separable mixture of phthalic acid (**165**), 1,4(2*H*,3*H*)-phthalazinedione (**166**), and 5,7,12,14-tetrahydrophthalazino[2,3-*b*]-phthalazine-5,7,12,14-tetrone (**167**) (KOH, H_2O, 20°C, 12 h: 70%, 18%, and 5% respectively); a separable mixture of 2,2′-biphthalazine-1,1′,4,4′(2*H*,2′*H*,3*H*,3′*H*)-tetrone (**169**) and phthalic acid (**170**) (H_2O, 20°C, 48 h: 48% and 45%, respectively); or a separable mixture of the same products (**169** and **170**) with some polymeric material (acetonic solution open to air, 20°C, 25 h: ~15%, ~14%, and ~52%, respectively).[511] The mechanisms of the foregoing and related changes associated with chemiluminescense have been studied in some detail.[775–777]

(165) + (166) + (167)

(AcMe solution)

(168)

H_2O → (169) + (170)

air → (169) + (170) + polymers

Formation of Adducts from 1,4-Phthalazinequinones

Note: Both acyclic and cyclic adducts can be formed, but the latter are prepared more frequently. In some cases, the 1,4(2*H*,3*H*)-phthalazinedione is oxidized in the presence of a synthon so as to trap the quinone as it is formed.

A freshly prepared acetonic solution of 1,4-phthalazinequinone (**171**) with ethyl 2-amino-5-methyl-4,5-dihydro-3-furancarboxylate (**172**) gave 2-(3-ethoxy-carbonyl-2-imino-5-methyltetrahydrofuran-3-yl)-1,4(2*H*,3*H*)-phthalazined-ione (**173**) (AcOMe, −30°C → 20°C, 30 min: 76%).[126]

(171) + (172) → (173)

1,4(2*H*,3*H*)-Phthalazinedione (**174**) was oxidized to 1,4-phthalazinequinone (**175**) and condensed with 2-furancarbaldehyde (**176**) in a one-pot reaction to afford 5,10-dioxo-5,10-dihydro-1*H*-pyrazolo[1,2-*b*]phthalazine-1-carboxylic acid (**178**) via the unisolated glyoxyloyl intermediate (**177**) [substrate, synthon, CH_2Cl_2, $Pb(OAc)_4\downarrow$ during 10 min, 20°C: 64%).[892]

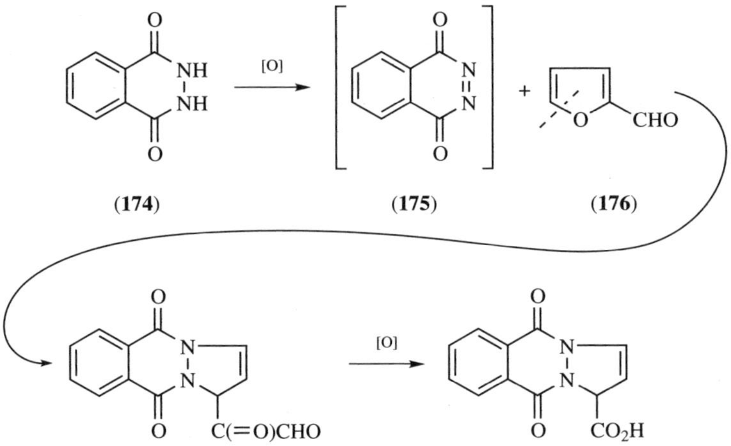

(**174**) (**175**) (**176**)

(**177**) (**178**)

A fresh acetonic solution of 1,4-phthalazinequinone (**179**) with 2-acetoxy-6-isopropenyl-3-methyl-1,3-cyclohexadiene (**180**) gave 2-acetoxy-14-isopropenyl-3-methyl-1,4,6,11-tetrahydro-1,4-ethanopyridazino[1,2-*b*]phthalazine-6,11- dione (**181**, R = CHMe:CH_2) (substrate, AcMe, ~−70°C; synthon in AcMe↓ dropwise, −70°C → 20°C, ~4 h: 20%).[583]

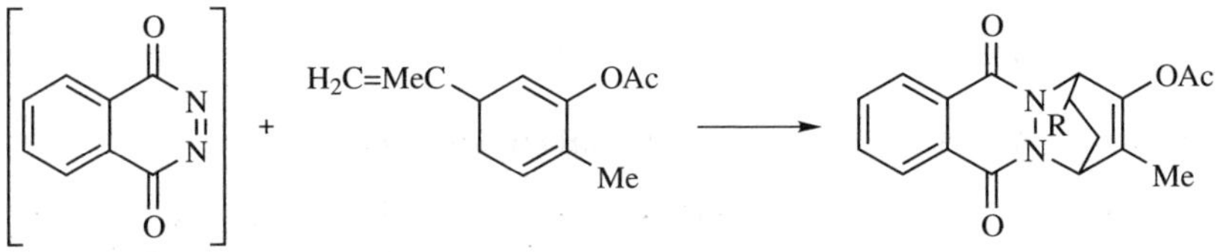

(**179**) (**180**) (**181**)

A fresh aqueous solution of 6-dimethylamino-1,4-phthalazinequinone (**182**) with 2,3-dimethylbutadiene (**183**) gave 8-dimethylamino-2,3-dimethyl-1,4,6,11-tetrahydropyridazino[1,2-*b*]phthalazine-6,11-dione (**184**) (synthon, AcOMe; substrate in aqueous pH 7 buffer↓: 30%).[1004]

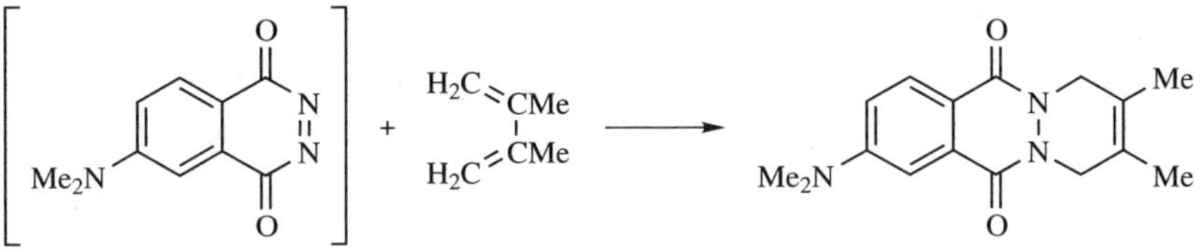

(**182**) (**183**) (**184**)

Also other examples.[51,427,564,705,879]

11.4. ALKOXY- AND ARYLOXYPHTHALAZINES
(*H* 84, 157, 168; *E* 392, 469)

These phthalazine ethers have received less attention as useful intermediates than they deserve. Accordingly, their literature is limited relative to those of related systems.

11.4.1. Preparation of Alkoxy- and Aryloxyphthalazines

Such compounds are usually made by *primary synthesis* (see Chapter 8), by *epoxidation of alkenylphthalazines* (see Section 9.2.2), by *alcoholysis or phenolysis of halogenophthalazines* (see Section 10.3.3), by *alkylation of tautomeric phthalazinones or extranuclear hydroxyphthalazines* (see Sections 11.1.2 and 11.2.2), or by other routes that are illustrated in the following classified examples.

From Alkylsulfonylphthalazines

1-Ethylsulfonyl-4-methylphthalazine (**185**) gave 1-methoxy-4-methylphthalazine (**186**) (MeONa, MeOH, reflux, 2 h: ?%).[649]

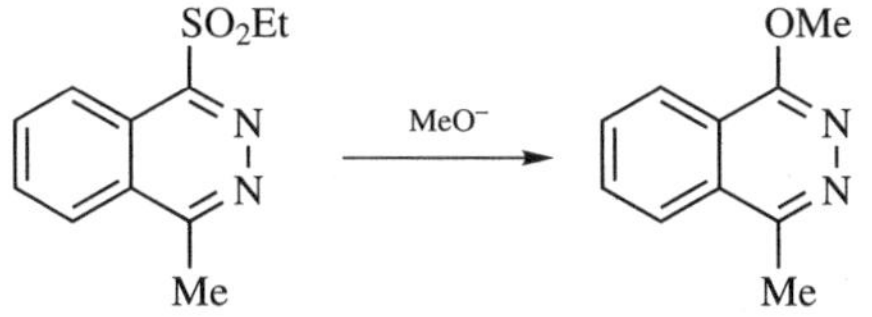

SO_2Et ... Me → (MeO^-) → OMe ... Me

(**185**) (**186**)

From Trimethylsiloxyphthalazines

1-Trimethylsiloxyphthalazine (**187**) with α-D-galactopyranose pentaacetate (**188**) gave 1-(2,3,4,6-tetra-*O*-acetyl-β-D-galactopyranosyloxy)phthalazine (**189**) ($SnCl_4$, $ClCH_2CH_2Cl$, reflux, 2 h: 85%); analogs likewise.[986]

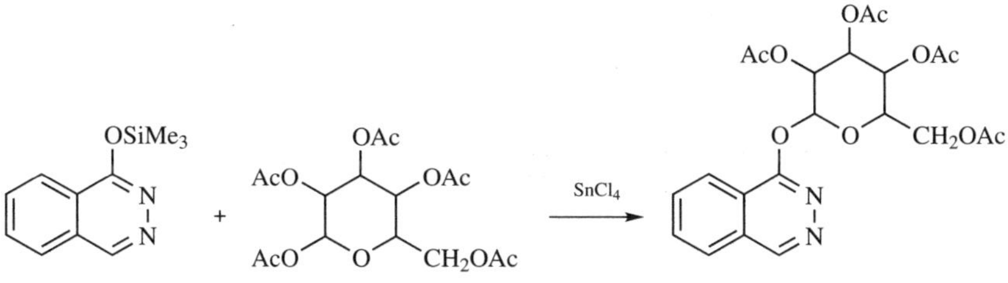

$OSiMe_3$ + AcO, OAc, OAc, AcO, O, CH_2OAc → ($SnCl_4$) → OAc, AcO, OAc, O, O, CH_2OAc

(**187**) (**188**) (**189**)

From Hydrazinophthalazines

1-Chloro-4-hydrazinophthalazine (**190**) with methanolic thalium(III) nitrate gave 1-chloro-4-methoxyphthalazine (**191**) [substrate, MeOH, $Tl(NO_3)_3 \cdot 2H_2O$, in MeOH↓ dropwise; then 20°C, 90 min: 65%; note survival of the Cl substituent under these mild conditions].[120]

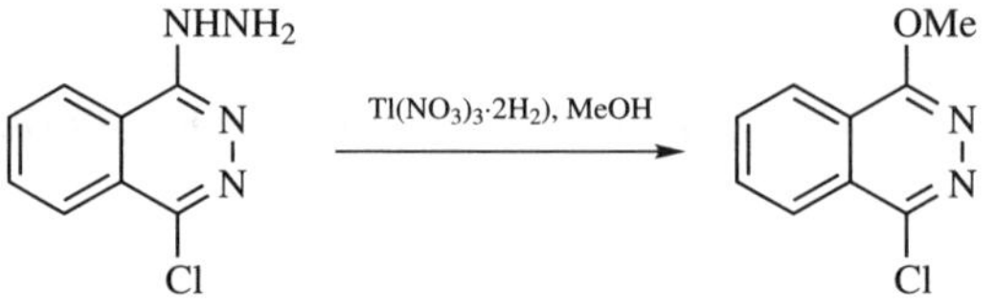

$Tl(NO_3)_3 \cdot 2H_2O$, MeOH

(**190**) (**191**)

From Phthalazinecarbaldehydes

Ethyl 1-formyl- (**192**) and dimethylsulfonium methylide (prepared in situ) gave ethyl 1-epoxyethyl-5,7-dimethyl-4-oxo-3,4-dihydro-6-phthalazinecarboxylate (**193**) (Me_2SO, NaH, 65°C, N_2, THF; Me_3SI in THF + Me_2SO↓ dropwise, −10°C; then substrate in THF + Me_2SO↓, −10°C, 30 min: ~50%).[424]

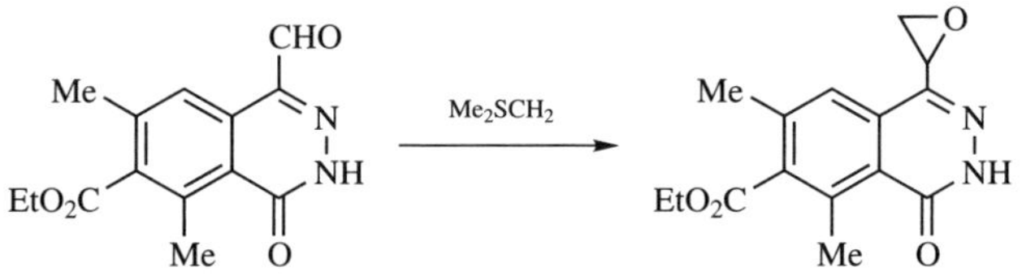

Me_2SCH_2

(**192**) (**193**)

From Ammoniophthalazine Salts

1-(3-Hydroxypyridinio)-4-phenylphthalazine chloride (**194**) underwent thermal rearrangement to 1-phenyl-4-(pyridin-3-yloxy)phthalazine (**195**) (CH_2Cl_2, reflux, 2 h: 40%).[672]

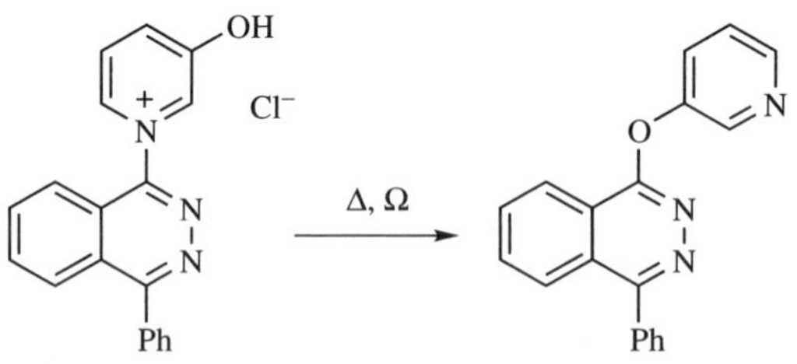

Δ, Ω

(**194**) (**195**)

11.4.2. Reactions of Alkoxy- and Aryloxyphthalazines

The *hydrolysis of such ethers to tautomeric phthalazinones or extranuclear hydroxyphthalazines* has been covered in Sections 11.1.1 and 11.2, respectively. Other reactions are illustrated by the following classified examples.

Aminolysis

1-Methyl-4-phenoxyphthalazine (**196**) gave 4-methyl-1-phthalazinamine (**197**) [excess neat $AcONH_4$, 155°C, 1 h: 55%];[508] analogs likewise.[685,958]

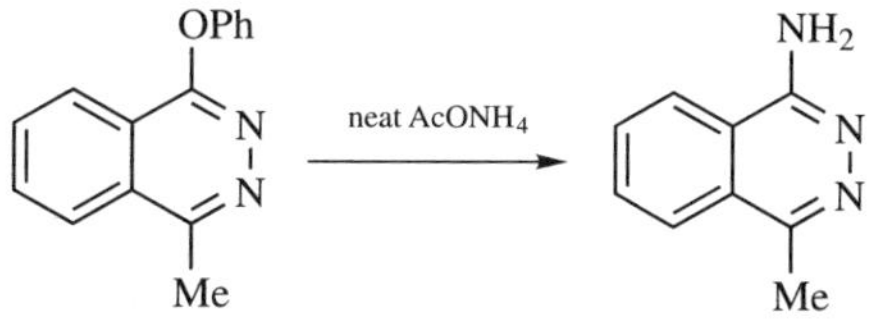

1,4-Dimethoxyphthalazine (**198**, R = OMe) gave 4-methoxy-1-phthalazinamine (**199**, R = NH_2) ($NaNH_2$, PhH, AcEt, reflux, 30 min: 16% with 77% recovery of substrate); similar treatment of 1-methoxy-4-phenylphthalazine (**198**, R = Ph) gave 4-phenyl-1-phthalazinamine (**199**, R = Ph) (likewise: 6% with 85% recovery of substrate).[935]

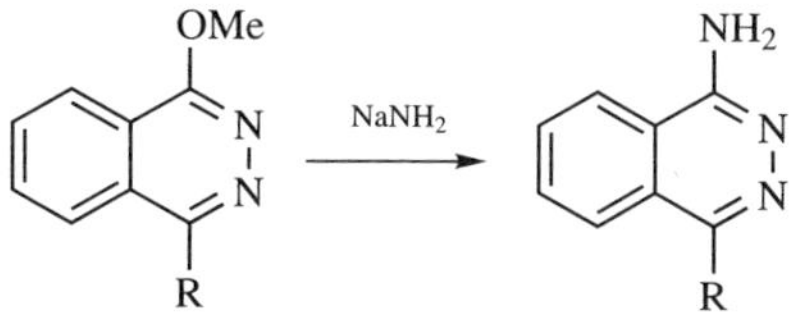

Rearrangement to *N*-Alkylphthalazinones

Note: This thermal rearrangement of alkoxyphthalazines appears to occur under conditions that are relatively mild in comparison with those required in analogous systems.

1-Butoxy-4-phenylphthalazine (**200**) gave 2-butyl-4-phenyl-1(2*H*)-phthalazinone (**201**) (BuOH, reflux, 2 h: ?%).[672]

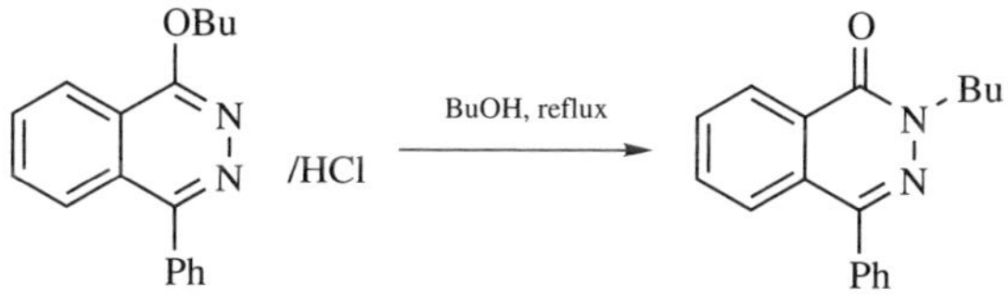

4-(2-Chloroethoxy)-2-methyl-1(2*H*)-phthalazinone (**202**) underwent rearrangement (and slower subsequent hydrolysis?) to afford a separable mixture of 2-(2-chloroethyl)-3-methyl-1,4(2*H*,3*H*)-phthalazinedione (**203**) and 2-(2-hydroxyethyl)-3-methyl-1,4(2*H*,3*H*)-phthalazinedione (**204**) (Me_2NCHO, reflux, 6 h: 38% and ~8%, respectively, after separation).[745]

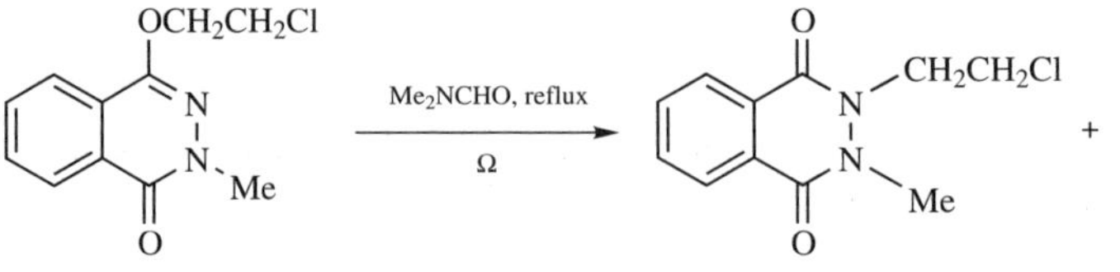

(**202**) (**203**) 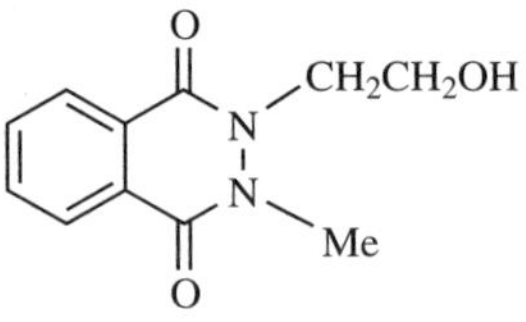 (**204**)

4-Methoxy-2-methylphthalazin-2-ium *p*-toluenesulfonate (**205**) gave 2,3-dimethyl-4-oxo-3,4-dihydrophthalazin-2-ium *p*-toluenesulfonate (**206**) (neat substrate, 160°C, 2.5 h: 94%).[681,690]

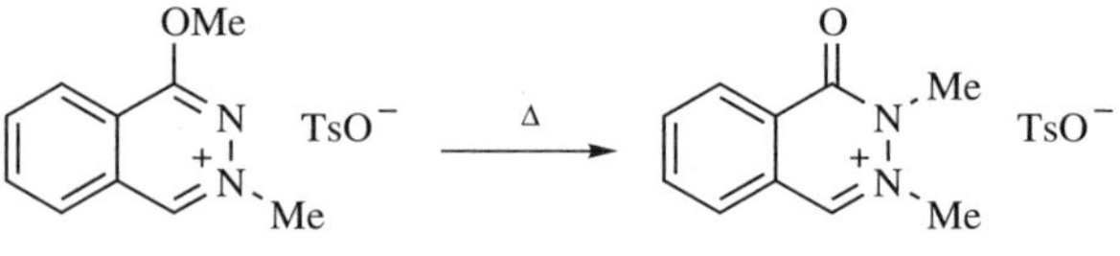

(**205**) (**206**)

Cyclization Reactions

1-[*o*-(2-Methylallyl)phenoxy]-4-phenylphthalazine (**207**) underwent intramolecular cycloaddition with loss of nitrogen to afford 6a-methyl-5-phenyl-6a,12a-dihydrobenzo[*c*]xanthene (**208**) ($PhNEt_2$, reflux, 3 h: 60%);[397] analogs likewise.[294,298]

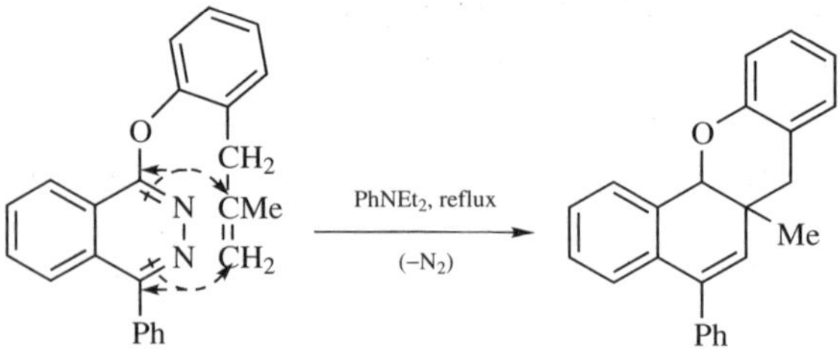

(**207**) (**208**)

11.5. NONTAUTOMERIC PHTHALAZINONES AND PHTHALAZINIUMOLATES

The chemistry of these phthalazines has been better explored than might be expected, probably because of the biological activities possessed by some. For example, the marked bronchodilatory effects of 2-ethyl-4-(pyridin-3-yl)-1(2*H*)-phthalazinone (**209**)[279] and 4-(*p*-chlorobenzyl)-2-(1-methylhexahydro-1*H*-azepin-4-yl)-1(2*H*)-phthalazinone hydrochloride (Azelastin) (**210**)[951,960] led to the synthesis and investigation of a variety of phthalazinone, fused phthalazine, and nonphthalazine analogs as antiasthmatics.[278,434,866,877,878,1007]

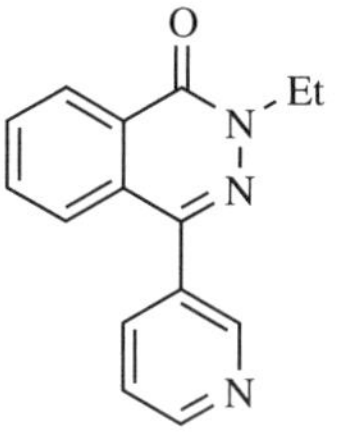

(**209**) 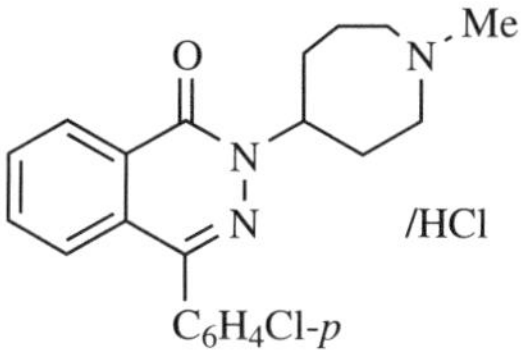 (**210**)

11.5.1. Preparation of Nontautomeric Phthalazinones and Phthalaziniumolates

Most such compounds have been made by *primary synthesis* (see Chapter 8), by *ozonolysis of alkenylphthalazines* (see Section 9.2.2), or by *alkylation or acylation of tautomeric phthalazinones* (see Section 11.1.2.1). Minor preparative routes are exemplified here.

4-Phenylphthalazin-2-ium-2-benzimidate (**211**) underwent ring fission, aerial oxidation, and recyclization in alkali to give 2-benzamido-4-phenyl-1(2*H*)-phthalazinone (**212**) (KOH, EtOH, H_2O, reflux, 3 h: 49%).[9]

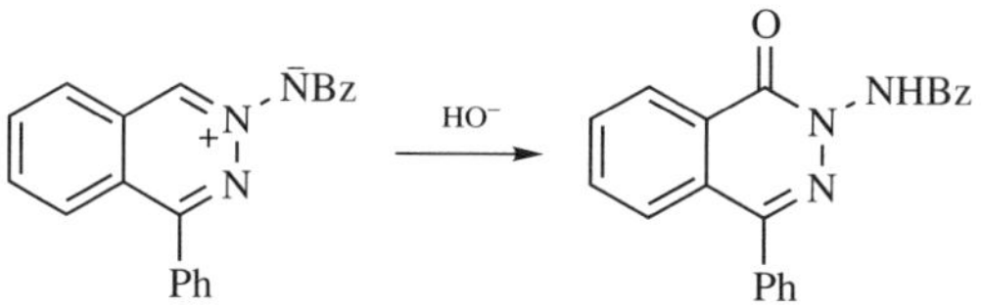

(**211**) (**212**)

2-Methylphthalazin-2-ium-4-olate (**213**) underwent photolytic rearrangement in water, probably via the bridged intermediate (**214**), to give 2-methyl-1(2*H*)-phthalazinone (**215**) (H_2O, 20°C, *hν*, 10 h: 90%).[603,698]

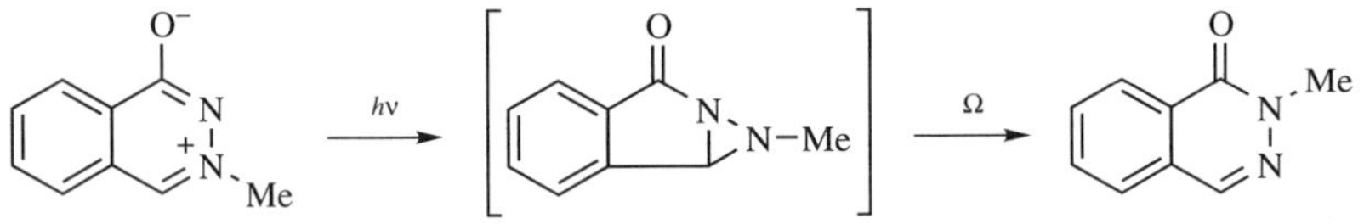

(213) (214) (215)

2,4-Diphenyl-1,2-dihydro-1-phthalazinol (**216**) gave 2,4-diphenyl-1(2*H*)-phthalazinone (**216a**) (1% $KMnO_4$ in Et_2O, reflux, 2 h: 49%).[446]

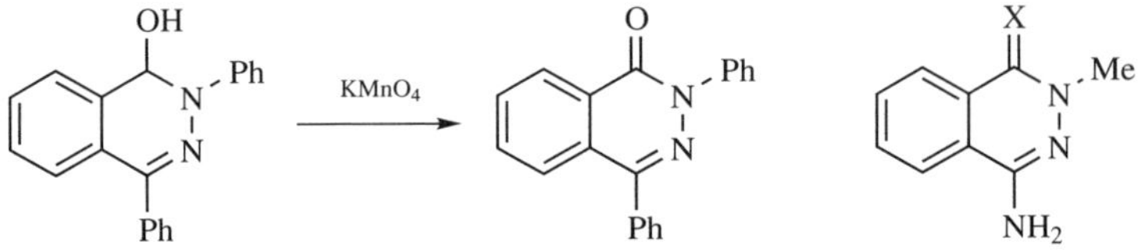

(216) (216a) (217)

4-Imino-3-methyl-3,4-dihydro-1-phthalazinamine (**217**, X = NH) underwent hydrolysis to 4-amino-2-methyl-1(2*H*)-phthalazinone (**217**, X = O) (hot H_2O: ?%).[685]

11.5.2. Reactions of Nontautomeric Phthalazinones and Phthalaziniumolates

The *reductive deoxygenation of nontautomeric phthalazones* has been exemplified toward the end of Section 9.2.1. Other reactions of both entities are illustrated in the following classified examples.

Thiation

Note: Thiation may be effected by phosphorus pentasulfide, Lawesson's reagent (**83**), or occasionally elemental sulfur.

2-Phenyl-4-*p*-tolyl-1(2*H*)-phthalazinone (**218**, X = O) gave 2-phenyl-4-*p*-tolyl-1(2*H*)-phthalazinethione (**218**, X = S) (P_2S_5, xylene, reflux, 30 min: 82%).[463]

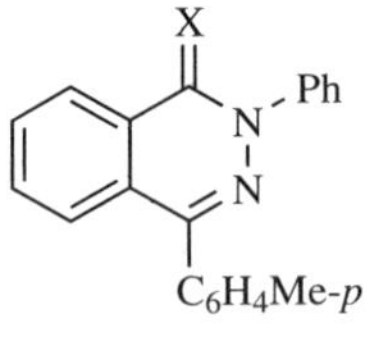

(218)

2,3-Diphenyl-1,4(2*H*,3*H*)-phthalazinedione (**219**, X = Y = O) gave 2,3-diphenyl-4-thioxo-3,4-dihydro-1(2*H*)-phthalazinone (**219**, X = O, Y = S)

(Lawesson's reagent, PhMe, 80°C, 2 h: 85%) or 2,3-diphenyl-1,4(2*H*,3*H*)-phthalazinedithione (**219**, X = Y = S) (Lawesson's reagent, PhMe, 100°C, 4 h: 91%).[295]

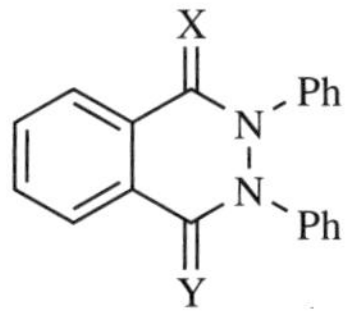

(**219**)

2-Phenyl-1,2-dihydro-1-phthalazinol (**220**) gave 2-phenyl-1(2*H*)-phthalazinethione (**221**) (S, MeCN, reflux, "prolonged": ~35%; clearly involved more than one step);[712] the same substrate (**220**) gave 2,2′-diphenyl-1,1′,2,2′-tetrahydro-1,1′-biphthalazinylidene (**222**) (AcOH, MeCN, reflux, 2 h: 48% after purification from other products) and thence the product (**221**) (neat substrate, S, 270°C, N_2, 20 min: 70%).[712]

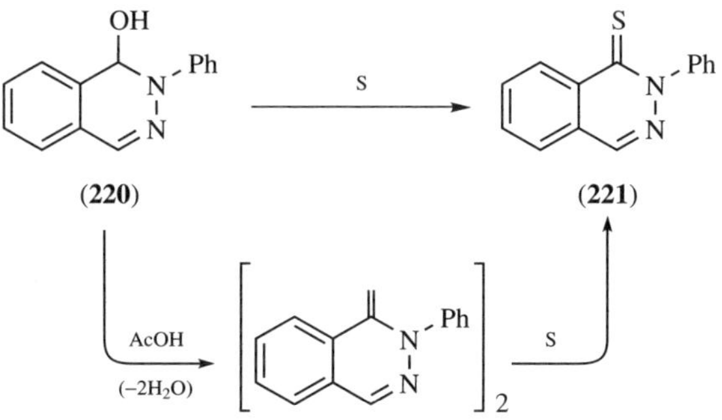

(**220**) (**221**)

(**222**)

Also other examples.[460,464,646]

Cyclization Reactions

Methyl 3-methyl-4-oxo-3,4-dihydro-5-phthalazinecarboxylate (**223**) gave 9-methyl- 3*H*-2,9-dihydrophthalazino[3,4,5-*de*]phthalazin-3-one (**224**) ($H_2NNH_2 \cdot H_2O$, H_2O, reflux, 66 h: 96%).[623]

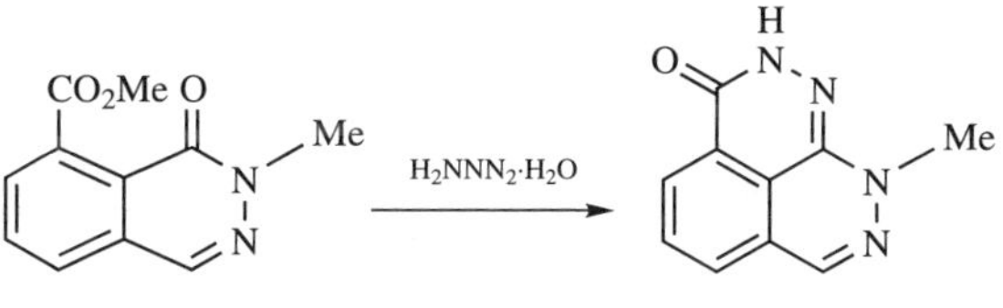

(**223**) (**224**)

2-Methylphthalazin-2-ium-4-olate (**225**, R = Me) with diphenylacetylene gave 10-methyl-3,4-diphenyl-1,5-dihydro-2,5-imino-2*H*-2-benzazepin-1-one (**226**, R = Me) (ClC_6H_4Cl-*o*, reflux, 24 h: 80%);[690] 2-phenylphthalazin-2-ium-4-olate (**225**, R = Ph) likewise gave 3,4,10-triphenyl-1,5-dihydro-2,5-imino-2*H*-2-benzazepin-1-one (**226**, R = Ph) (reflux, 18 h: 86%).[696]

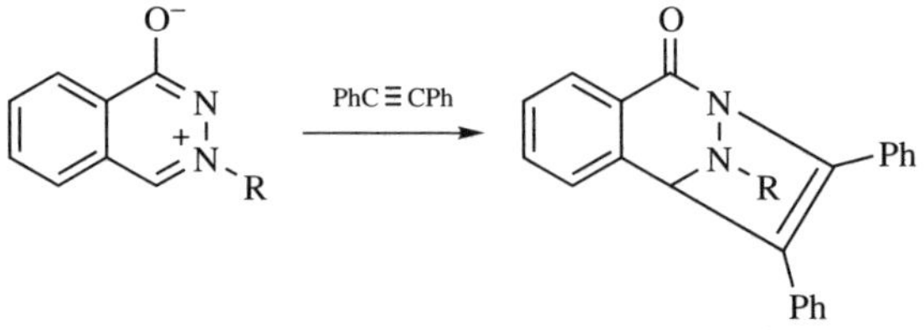

PhC≡CPh

(**225**) (**226**)

In contrast, 2-phenylphthalazin-2-ium-4-olate (**227**) with dimethyl acetylenedicarboxylate in refluxing xylene gave dimethyl 1-oxo-10-phenyl-1,5-dihydro-2,5-imino-2*H*-benzazepine-3,4-dicarboxylate (**228**) (xylene, reflux, 12 h: 80%) but in refluxing chloroform gave the isomeric dimethyl 1-oxo-2-phenyl-1,2-dihydro-2,3-benzodiazocine-4,5-dicarboxylate (**229**); thermolysis of either product gave a third isomer, dimethyl 4-oxo-1,3a,4,8b-tetrahydroindeno[1,2-*b*]pyrazol-3,3a-dicarboxylate (**230**) (neat substrate, 180–190°C, 3–5 h: >95%).[74,696]

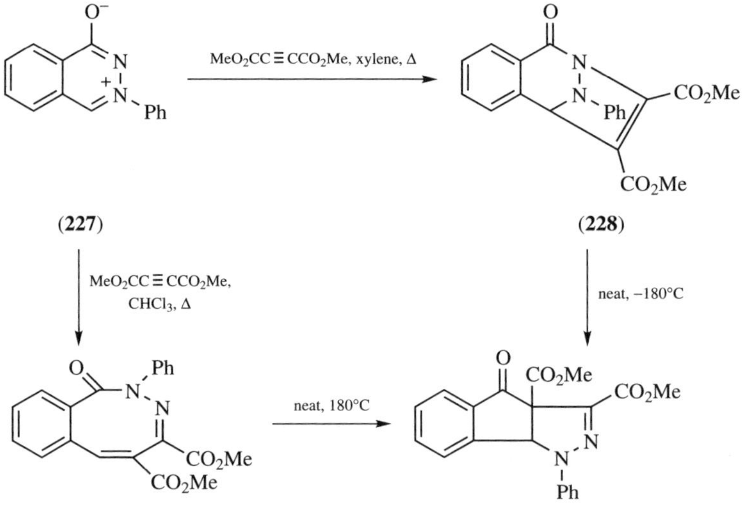

$MeO_2CC{\equiv}CCO_2Me$, xylene, Δ

(**227**) (**228**)

$MeO_2CC{\equiv}CCO_2Me$, $CHCl_3$, Δ

neat, ~180°C

neat, 180°C

(**229**) (**230**)

Also other examples.[76,594,698,897]

Miscellaneous Reactions

2-Methylphthalazin-2-ium-4-olate (**231**) underwent photolysis in methanol to give 2-(*o*-methoxycarbonylphenyl)-1-methyldiaziridine (**232**, R = OMe)

(MeOH, $h\nu$, 2 h: 70%), in aqueous ethylamine to give 2-[*o*-(ethylcarbamoyl)-phenyl]-1-methyldiaziridine (**232**, R = (NHEt) ($EtNH_2$, H_2O, $h\nu$, 5 h: 80%), or in acetonitrile to give 3-oxo-1-methyl-3,7a-dihydro-1*H*-diazirino[3,1-*a*]isoindol-3-one (**233**) (MeCN, $h\nu$, 3 h: ?%); the last product (**233**), "set aside" in water, methanol, or aqueous ethylamine, gave in "high yield" 2-methyl-1(2*H*)-phthalazinone (**234**), the ester (**232**, R = OMe), or the amide (**232**, R = NHEt), respectively.[703]

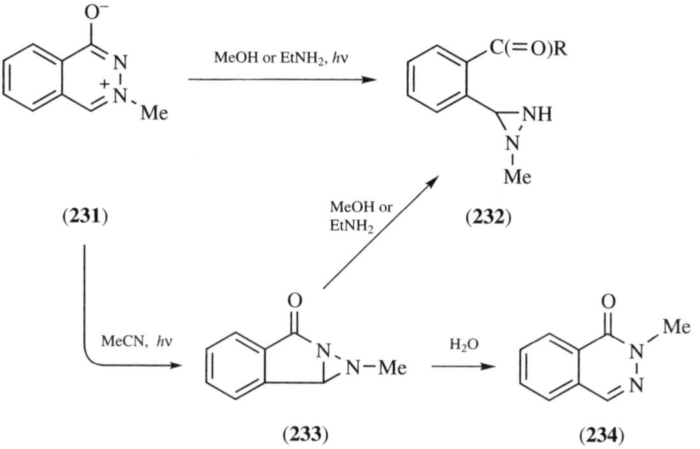

MeOH or $EtNH_2$, $h\nu$

(**231**) (**232**)

MeOH or $EtNH_2$

MeCN, $h\nu$

H_2O

(**233**) (**234**)

2,4-Diphenyl-1(2*H*)-phthalazinone (**235**, R = Ph) with triethyloxonium tetrafluoroborate gave 2-ethyl-4-oxo-1,3-diphenyl-3,4-dihydrophthalazin-2-ium tetrafluoroborate (**236**) (reactants, CH_2Cl_2, 20°C, 1 week: 86%), but similar treatment of the analogous substrate, 2-methyl-4-phenyl-1(2*H*)-phthalazinone (**235**, R = Me), gave only 1-ethoxy-2-methyl-4-phenylphthalazin-2-ium tetrafluoroborate (**237**) (20%); analogs behaved similarly.[447]

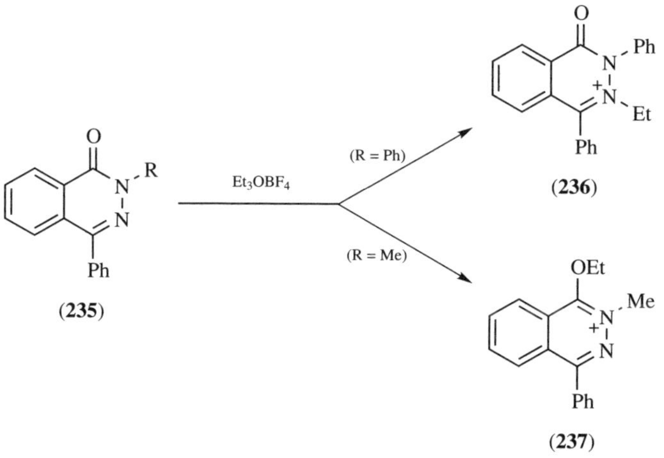

Et_3OBF_4

(R = Ph)

(R = Me)

(**235**) (**236**) (**237**)

The unstable substrate, 4,4-bis(*p*-dimethylaminophenyl)-1,4-dihydro-1-phthalazinone (**240**) [prepared in solution at −80°C by oxidation of 4,4-bis(*p*-dimethylaminophenyl)-3,4-dihydro-1(2*H*)-phthalazinone (**238**): see Section 11.1.2.3], appears on spectral evidence to have undergone isomerization in acidic media to give 1,2-bis(*p*-dimethylaminophenyl)phthalazin-2-ium-4-olate (**239**) (F_3CCO_2H, CH_2Cl_2, −80°C → 20°C)[79,718] or a more profound change with loss of N_2 (in nonacidic media) to give 7-dimethylamino-4b-(*p*-dimethylaminophenyl)-4b,10-dihydrobenzo[*a*]azulin-1-one (**241**) (CH_2Cl_2, −80°C → 20°C).[719]

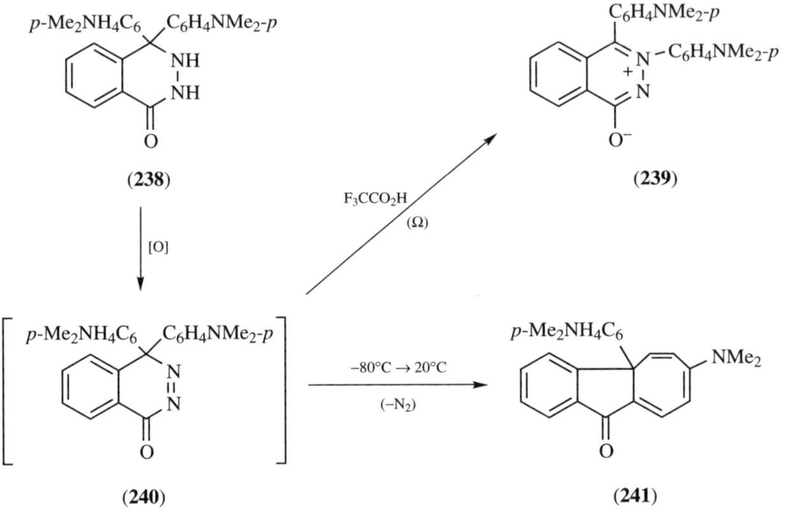

p-Me2NH4C6 C6H4NMe2-p NH NH O (**238**)

[O]

F3CCO2H (Ω)

C6H4NMe2-p N C6H4NMe2-p + N O− (**239**)

p-Me2NH4C6 C6H4NMe2-p N N O (**240**)

−80°C → 20°C (−N2)

p-Me2NH4C6 NMe2 O (**241**)

11.6. PHTHALAZINE *N*-OXIDES (*E* 364)

The 1972–2004 literature on phthalazine *N*-oxides is so sparse that it is best summarized as a series of classified examples.

Preparation

Note: Some *N*-oxides have been made by *primary synthesis* (see Chapter 8) and others by *direct oxidation*, as illustrated here.

1,4-Diphenylphthalazine gave 1,4-diphenylphthalazine 2-oxide (**242**) (*m*-$ClC_6H_4CO_3H$, $CHCl_3$, 20°C, 4 days: >80%).[315]

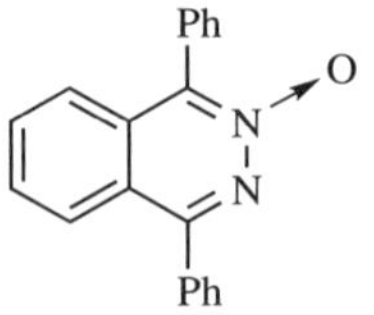

Ph O N N Ph

(**242**)

4-Phenyl-1-phthalazinamine gave only 4-phenyl-1-phthalazinamine 2-oxide (**243**) (m-$ClC_6H_4CO_3H$, AcMe, 20°C, 1 h: 84%).[508]

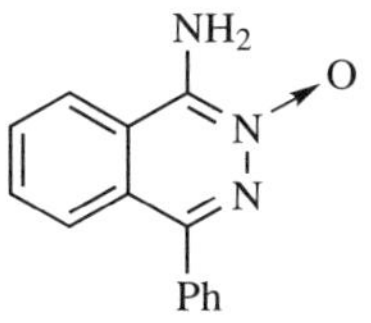

(**243**)

5-Cyclopentyloxy-1-(3,5-dichloropyridin-4-ylmethyl)-6-methoxyphthalazine gave only its 3-oxide (**244**) (m-$ClC_6H_4CO_3H$, CH_2Cl_2, 20°C, 1 h: 64%).[861]

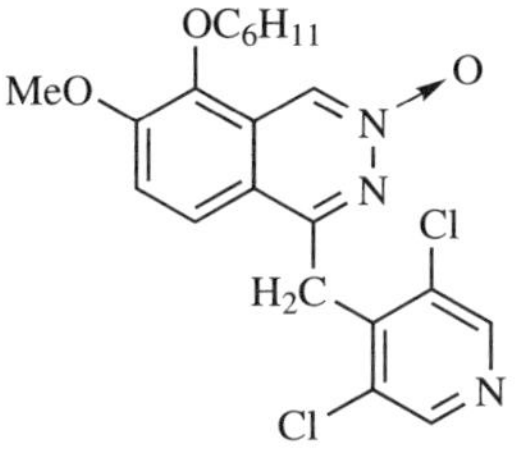

(**244**)

In contrast, 1-phenylphthalazine gave a separable mixture of its 2- and 3-oxide (see original for details)[938] and 7-dimethylamino-2-phenyl-1(2*H*)-phthalazinone apparently gave its ω-*N*-oxide (**245**) (m-$ClC_6H_4CO_3H$, $CHCl_3$, 20°C, 21 h: 54%).[411]

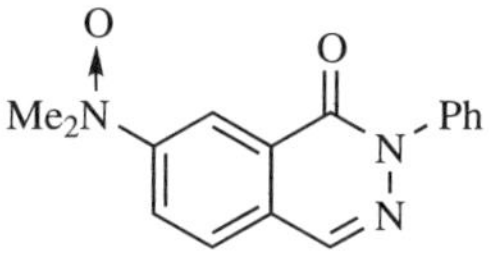

(**245**)

Also other examples.[426,783]

Properties

The ^{14}N NMR chemical shifts for the oxide bearing nitrogen atom in phthalazine 1-oxide and related azine and diazine *N*-oxides have been shown to reflect the calculated π-charges thereat.[760]

Reactions

Note: The *formation of phthalazinones from phthalazine N-oxides* has been exemplified in Section 11.1.1.

1-Phenylphthalazine 3-oxide (**246**, R = Ph) underwent a Reissert–Henze type of reaction to give 4-phenyl-1-phthalazinecarbonitrile (**247**, R = Ph) [substrate, Me_3SiCN, THF; diazabicycloundecene (DBU)↓ slowly, 20°C; then reflux, 40 min: 76%].[587] Using a less effective procedure, 1-methylphthalazine 3-oxide (**246**, R = Me) gave 4-methyl-1-phthalazinecarbonitrile (**247**, R = Me) (substrate, KCN, MeOH, H_2O; BzCl↓ slowly; then 50°C, 30 min: 31%);[930] also similar examples.[783]

The same substrate (**246**, R = Ph) with methanesulfonyl chloride gave 1-chloro-4-phenylphthalazine (**247a**) ($CHCl_3$, reflux, 10 min: 39%); other such acylation reagents gave a variety of products (see original).[938]

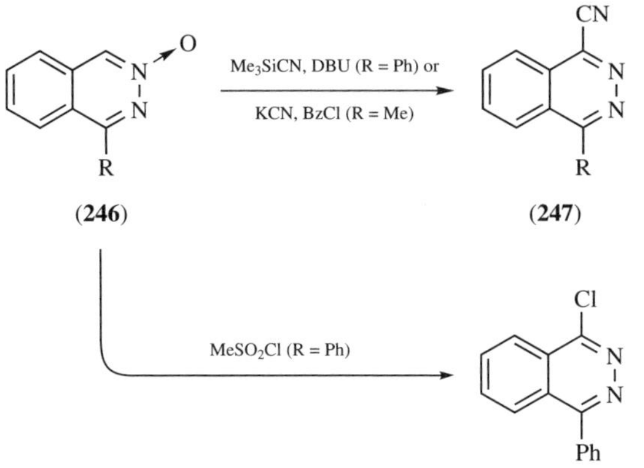

Me_3SiCN, DBU (R = Ph) or KCN, BzCl (R = Me)

$MeSO_2Cl$ (R = Ph)

(**246**) (**247**) (**247a**)

1-Methoxyphthalazine 3-oxide (**248**) with acetic anhydride underwent fission of the pyridazine ring with loss of nitrogen to afford a separable mixture of methyl *o*-(acetoxymethyl)benzoate (**249**), methyl *o*-(1-acetoxyacetonyl)-benzoate (**250**), and dimethyl 2,2′-stilbenedicarboxylate (**251**) (neat Ac_2O, reflux, 5 h: 60%, 33%, and 10%, respectively).[565]

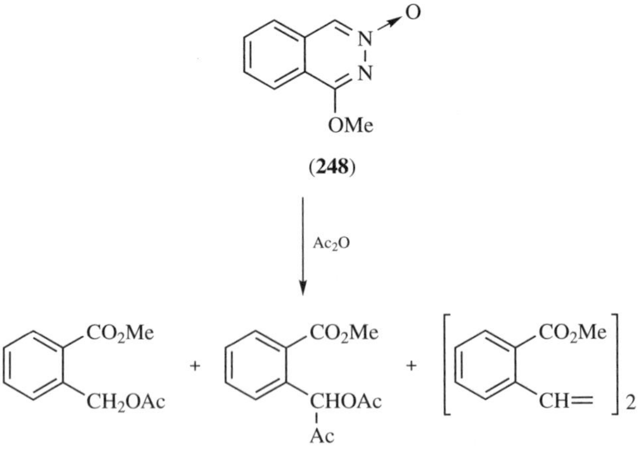

(**248**)

Ac_2O

(**249**) (**250**) (**251**)

1,4-Diphenylphthalazine 2-oxide (**252**) suffered photolysis in the absence of air to give 1,3-diphenylisobenzofuran (**253**) (AcMe, *h*ν, N_2, 2 h) that was rapidly transformed in air to *o*-dibenzoylbenzene (**254**); a similar photolysis in the presence of oxygen gave only product (**254**).[315]

hν, AcMe
(N_2)
(**252**)
(**253**)
air
hν, AcMe
(O_2)
(**254**)

Irradiation of 1-methoxyphthalazine 3-oxide (**255**) in cyclohexane gave only 1,3-isoindolinedione (**256**) (λ 3500 Å, 15 h: 55%); with maleic anhydride (**257**) likewise gave 4-methyl-1,3-dihydronaphtho[2,3-*c*]furan-1,3-dione (**258**) (45%); or with benzoquinone (**259**) gave 9-methoxy-1,4-anthraquinone (**260**) (35%).[10]

hν (in C_6H_{12})
(**256**)
(**255**)
hν
(**257**)
(**258**)
hν
(**259**)
(**260**)

The reactions of simple phthalazine *N*-oxides with dialkyl acetylenedicarboxylates have been studied in detail.[963,965] For example, phthalazine 2-oxide (**261**) with dimethyl acetylenedicarboxylate (**262**) gave a separable mixture of 1,2-dimethoxycarbonyl-2-phthalazinioethenolate (**263**) (or equivalent formulation) (4%), trimethyl pyrrolo[2,1-*a*]phthalazine-1,2,3-tricarboxylate (**264**) (11%), dimethyl 2-hydroxy-1,3-naphthalenedicarboxylate (**265**) (24%), dimethyl 1,2-dihydro-1,2-epoxynaphthalene-2,3-dicarboxylate (**266**) (7%), and other products.[965]

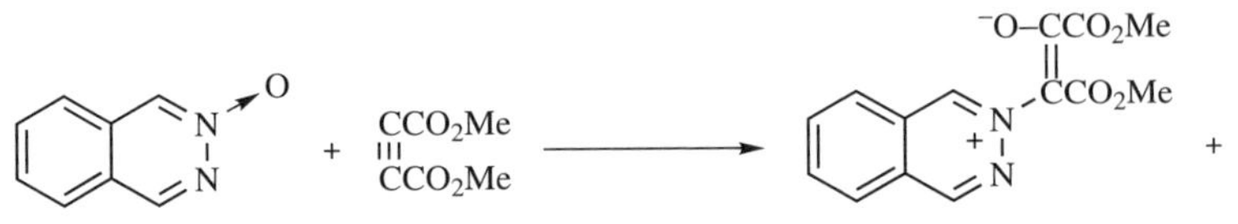

(**261**) (**262**) (**263**)

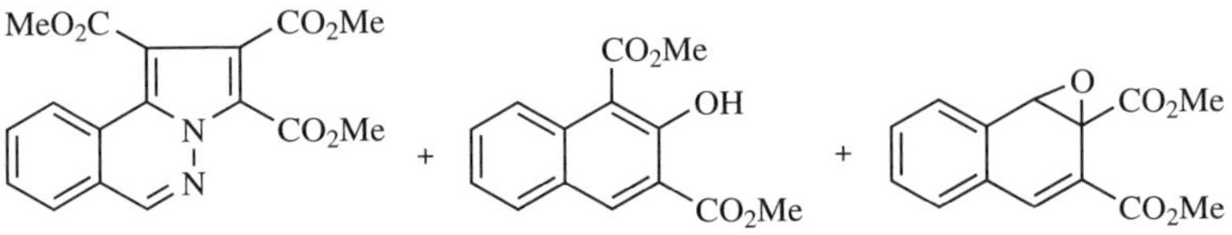

(**264**) (**265**) (**266**)

4-Methyl-1-phthalazinamine 2-oxide (**267**) with α-bromoacetophenone gave 6-methyl-2-phenylimidazo[2,1-*a*]phthalazin-3(5*H*)-one (**268**) (EtOH, reflux, 1 h: 49% as hydrobromide).[508]

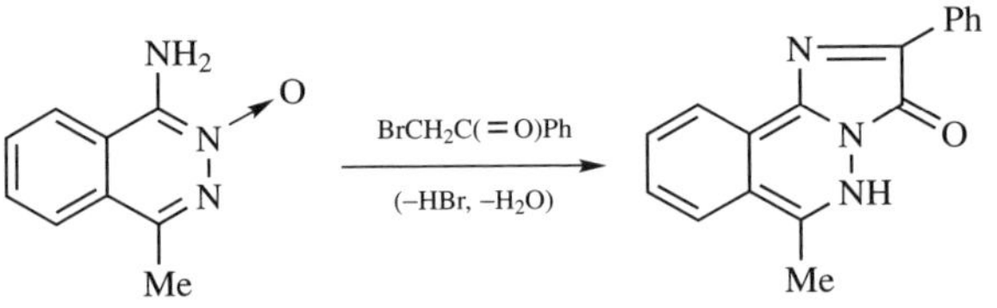

$BrCH_2C(=O)Ph$ ($-HBr$, $-H_2O$)

(**267**) (**268**)

CHAPTER 12

Thiophthalazines (*E* 535)

The term *thiophthalazine* includes all phthalazines that have a sulfur-containing substituent joined directly or indirectly to the nucleus through its sulfur atom: phthalazinethiones (both tautomeric and nontautomeric), extranuclear phthalazinethiols, alkylthiophthalazines, alkylsulfinyl- or alkylsulfonylphthalazines, phthalazinesulfonic acids and derivatives, diphthalazine sulfides, and so on. With the exception of phthalazinethiones, very little work has been reported on any thiophthalazines during the period under review (1972–2004).

12.1. PHTHALAZINETHIONES AND PHTHALAZINETHIOLS

Phthalazinethiones have considerable potential as intermediates for the preparation of other phthalazines, but it remains largely unused. A spectral study has confirmed that the tautomeric system, 1-phthalazinethiol ↔ 1(2*H*)-phthalazinethione, strongly favors the thione form in solution.[171] Several sulfur-containing derivatives of phthalazine inhibit corrosion of copper in dilute sulfuric acid.[1027]

12.1.1. Preparation of Phthalazinethiones and Phthalazinethiols

These phthalazines have been made by *primary synthesis* (see Chapter 8), by *thiolysis of halogenophthalazines* (see Section 10.3.4), and by *thiation of both tautomeric and nontautomeric phthalazinones* (see Sections 11.1.2.4 and 11.5.2). No minor preparative routes appear to have been used.

12.1.2. Reactions of Phthalazinethiones and Phthalazinethiols

The more recently employed reactions of these thiophthalazines are illustrated by the classified examples that follow.

Cinnolines and Phthalazines: Supplement II, The Chemistry of Heterocyclic Compounds, Volume 64, by D.J. Brown

Alkylation

Note: These alkylations usually afford only alkylthiophthalazines, but occasionally an *N*-alkylated phthalazinethione has been reported.[171,173]

4-Methyl-1(2*H*)-phthalazinethione (**1**) gave 1-ethylthio-4-methylphthalazine (**2**, R = Et) (EtI, NaOH, EtOH, reflux, 2 h: 50%)[649] or 1-methyl-4-methylthiophthalazine (**2**, R = Me) (MeI, 2M NaOH: 69%).[931]

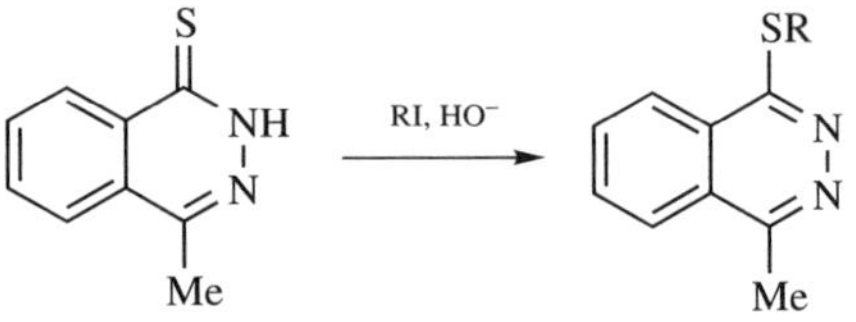

1(2*H*)-Phthalazinethione (**3**) gave 1-(α-carboxybenzylthio)phthalazine (**4**) (substrate, THF; $BrCHPhCO_2H$ in THF↓; then Et_3N in THF↓; 20°C, 1 h: 40%).[112]

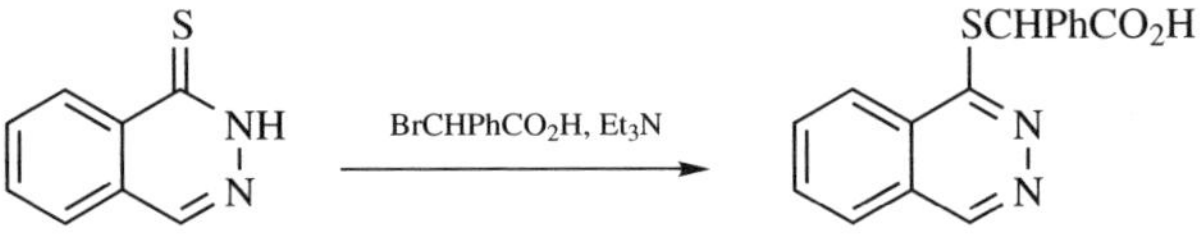

2-Phenyl-4-*p*-tolyl-1(2*H*)-phthalazinethione (**5**) with triethyloxonium tetrafluoroborate gave 1-ethylthio-2-phenyl-4-*p*-tolylphthalazin-2-ium tetrafluoroborate (**6**) (reactants, $CHCl_3$, reflux, briefly: 52%).[464]

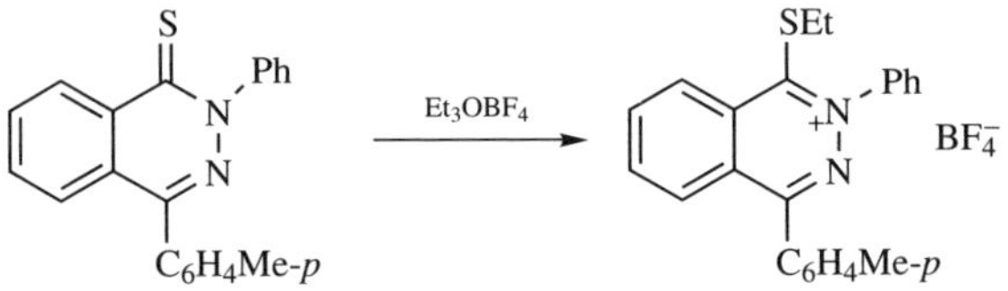

The extranuclear phthalazinethione, 4-benzyl-2-(5-thioxo-4,5-dihydro-1,3,4-oxadiazol-2-yl)-1(2*H*)-phthalazinone (**7**, R = H) underwent regular *S*-alkylation to give, for example, 4-benzyl-2-(5-methylthio-1,3,4-oxadiazol-2-yl)-1(2*H*)-phthalazinone (**8**) but Mannich *N*-(aminoalkylation) to give, for example, 4-benzyl-2-(4-anilinomethyl-5-thioxo-4,5-dihydro-1,3,4-oxadiazol-2-yl)-1(2*H*)-phthalazinone (**7**, R = $PhNHCH_2$) (for details, see original).[256]

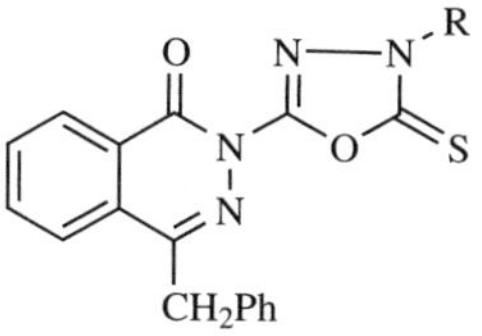

(7) 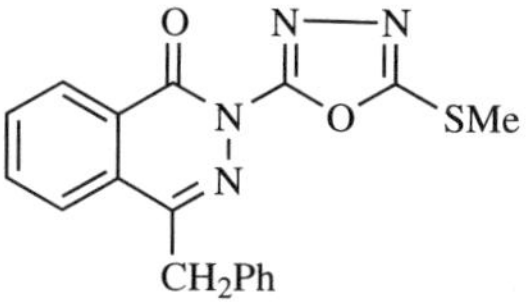 (8)

Also other examples.[33,171,216,256,516]

Aminolysis

1(2*H*)-Phthalazinethione (**9**) with azidotrimethylsilane gave 1-azidophthalazine (**10**) (substrate, Me_3SiN_3, CH_2Cl_2, A, 20°C until clear; $SnCl_4$ in Ch_2Cl_2↓ dropwise, 20°C, during 30 min; then 20°C, 24 h: 95%).[592]

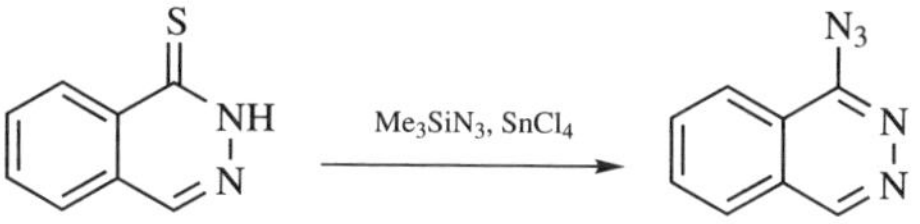

Me_3SiN_3, $SnCl_4$

(9) (10)

Also other examples.[344]

Photolysis with Alkenes

1(2*H*)-Phthalazinethione (**11**) with ethoxyethylene gave 1-[1-ethoxy-2-(2-ethoxyethylthio)ethyl]phthalazine (**12**) (substrate, excess $EtOCH{=}CH_2$, MeOH, $h\nu$, 3 h: 23%).[413]

The same substrate (**11**) with 2,3-dimethylbut-2-ene gave 1-(2-mercapto-1,1,2-trimethylpropyl)phthalazine (**13**) (MeOH, $h\nu$, 2 h: 71%) and thence 1-isopropylphthalazine (**14**) (vacuum distillation: 31%); analogs likewise.[417,607]

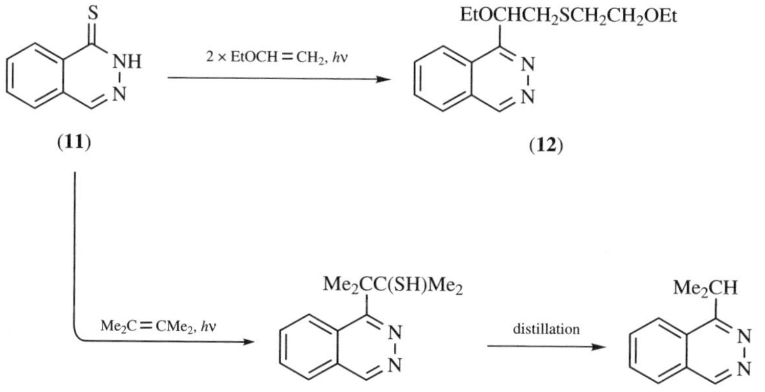

$EtOCHCH_2SCH_2CH_2OEt$

2 × $EtOCH{=}CH_2$, $h\nu$

(11) (12)

$Me_2C{=}CMe_2$, $h\nu$

$Me_2CC(SH)Me_2$

distillation

Me_2CH

(13) (14)

Cyclization Reactions

1(2*H*)-Phthalazinethione (**15**) with 2-bromo-2-phenylacetyl chloride or 2-phenyl-2-(tosylhydrazono)acetyl chloride gave 2-phenylthiazolo[2,3-*a*]phthalazin-4-ium-3-olate (**16**) (substrate, Et_2O; PhCHBrCOCl in Et_2O↓ dropwise during 15 min; then Et_3N↓ slowly: 91%)[112] or [substrate, PhC(=NNHTs)COCl, Et_2O; Et_3N↓, 20°C, 12 h: 93%].[143]

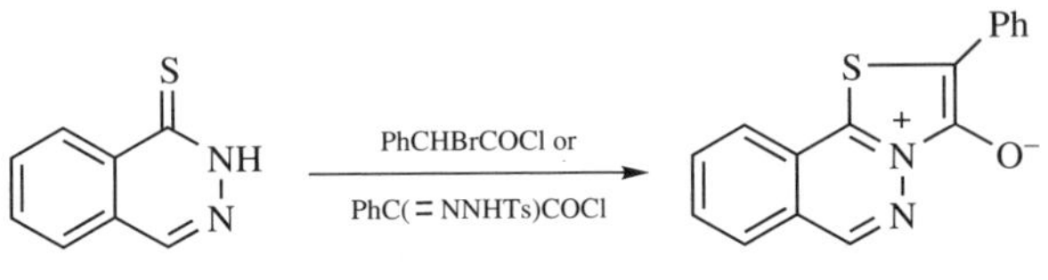

PhCHBrCOCl or PhC(=NNHTs)COCl

(**15**) (**16**)

2-Phenyl-4-*p*-tolyl-1(2*H*)-phthalazinethione (**17**) with α-(phenylhydrazono)benzyl chloride gave 2,3′,5′-triphenyl-4-*p*-tolylspiro{phthalazine-1(2*H*),2′(3′*H*)–[1,3,4]thiadiazole} (**18**) (reactants, Et_3N, PhH, reflux, 30 min: 80%); also an analog likewise.[463]

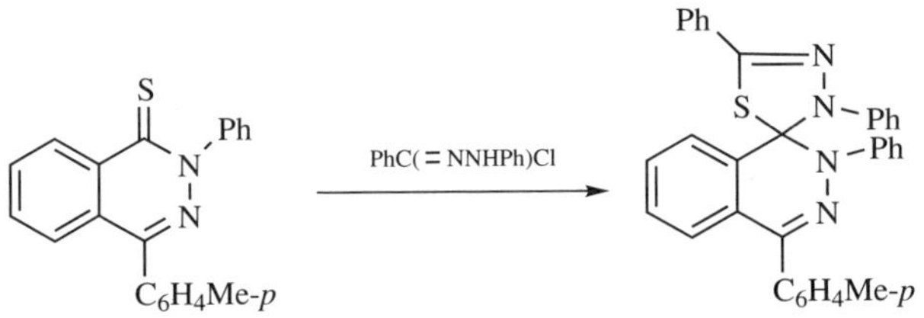

PhC(=NNHPh)Cl

(**17**) (**18**)

12.2. ALKYLTHIOPHTHALAZINES

The preparation and reactions of these useful thioethers are illustrated by the following examples.

Preparation

Note: Alkyl- and arylthiophthalazines have been made by *primary synthesis* (see Chapter 8), by *alkane- or arenethiolysis of halogenophthalazines* (see Section 10.3.4), by *alkylation of phthalazinethiones* (see Section 12.1.2), or by *passenger introduction* (for extranuclear thioethers) as exemplified here.

1,4(2*H*,3*H*)-Phthalazinedione (**19**) gave 2-(2-methylthioethyl)-1,4(2*H*,3*H*)-phthalazinedione (**20**) (substrate, NaH, Me_2NCHO; $ClCH_2CH_2SMe$↓, reflux, 12 h: 43%).[949]

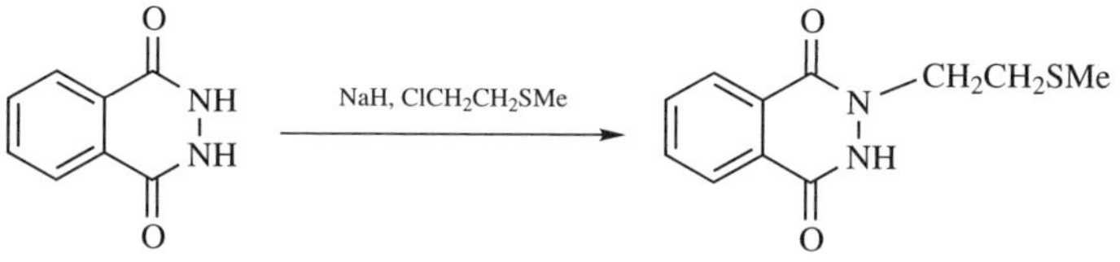

NaH, $ClCH_2CH_2SMe$
CH_2CH_2SMe

(19) (20)

Reactions

Note: Several reactions of these thioethers have been reported, as exemplified here.

1-Ethylthio-4-methylphthalazine (**21**, R = Et) underwent *oxidation* to 1-ethylsulfonyl-4-methylphthalazine (**22**, R = Et) (substrate, $CHCl_3$, H_2SO_4, H_2O, ice; $KMnO_4\downarrow$ slowly, 0°C: 50%);[649] somewhat similarly, 1-methyl-4-methylthiophthalazine (**21**, R = Me) gave 1-methyl-4-methylsulfonylphthalazine (**22**, R = Me) ($KMnO_4$, $MgSO_4$, H_2O, dioxane: 64%).[931]

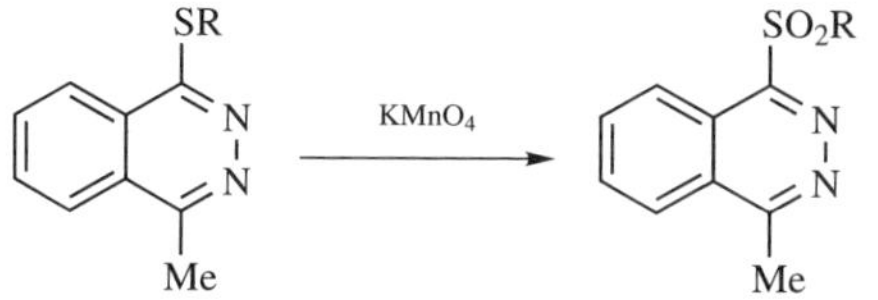

SR
$KMnO_4$
SO_2R
Me

(21) (22)

2-(2-Methylthioethyl)-1,4(2*H*,3*H*)-phthalazinedione (**23**) underwent stepwise *oxidation* to afford 2-(2-methylsulfinylethyl)- (**23a**, $n = 1$) [H_2O_2 (1 mol), AcOH, H_2SO_4, 4°C; substrate↓ slowly, 4°C→ 20°C, 12 h: 97%] or 2-(2-methylsulfonylethyl)-1,4(2*H*,3*H*)-phthalazinedione (**23a**, $n = 2$) [H_2O_2 (2 mol), likewise: 64%];[949] also other such examples.[263]

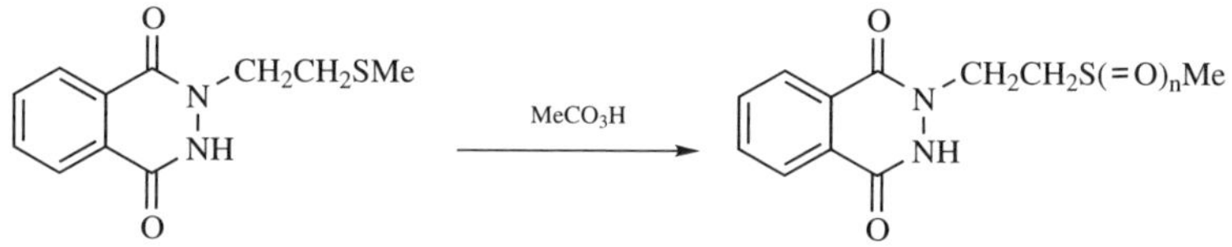

CH_2CH_2SMe
$MeCO_3H$
$CH_2CH_2S(=O)_nMe$

(23) (23a)

1-*tert*-Butylthio-4-methoxyphthalazine (**24**) likewise gave 1-*tert*-butylsulfinyl-4-methoxyphthalazine (**24a**) [substrate, CH_2Cl_2, −10°C; m-$ClC_6H_4CO_3H$ (1 mol) in $CH_2Cl_2\downarrow$ dropwise during 1 h: 83%];[1010] also other examples of the use of this oxidant,[411,426]

SBu^t
m-$ClC_6H_4CO_3H$
$S(=O)Bu^t$
OMe

(24) (24a)

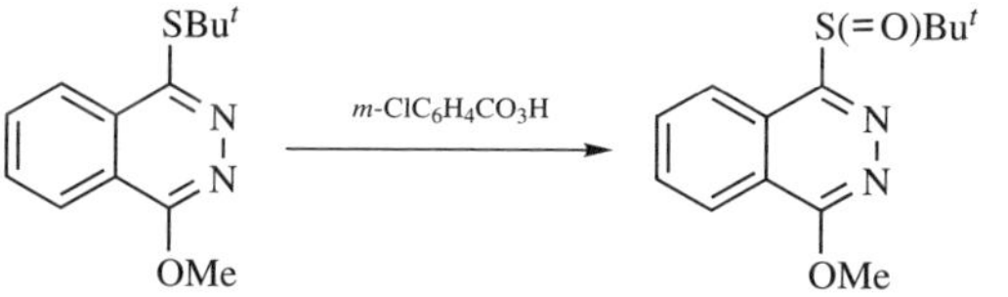

1-Ethylthio-2,4-diphenylphthalazin-2-ium tetrafluoroborate (**25**) underwent *aminolysis* by 2,4-dinitrophenylhydrazine to give 1-(2,4-dinitrophenylhydrazono)-2,4-diphenyl-1,2-dihydrophthalazine (**26**) (reactants, AcOH, trace Et_3N, hot, briefly: ?%);[464] more conventional examples include the conversion of 1-(ethoxycarbonylmethylthio)- (**27**) into 1-hydrazino-4-methylphthalazine (**27a**) (excess $H_2NNH_2 \cdot H_2O$, EtOH, reflux, ? h: 65%, as hydrochloride)[33] and other such aminolyses.[216]

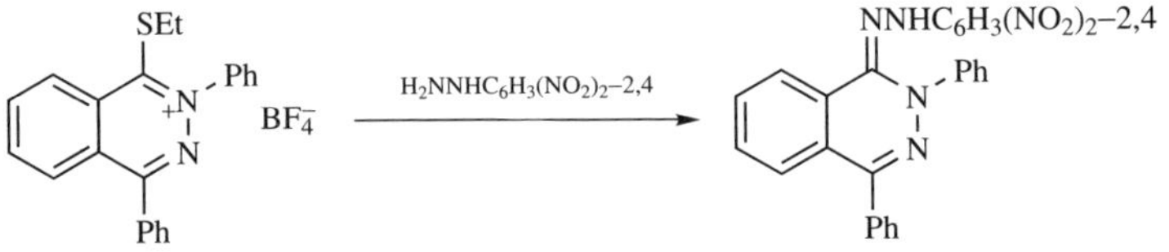

(**25**) (**26**)

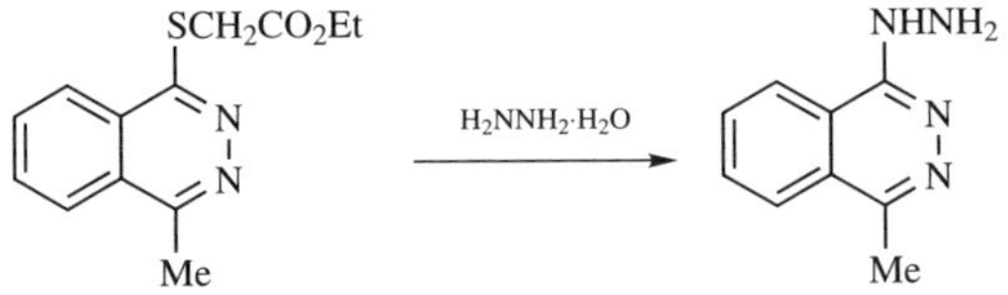

(**27**) (**27a**)

1-Methylthiophthalazine (**28**) underwent *adduct formation* with 1-(cyclobut-1-enyl)piperidine and subsequent loss of nitrogen to give 4-methylthio-2,3-dihydro-1*H*-benzo[*f*]indene (**29**) (neat reactants, 160°C, 3 h: 33%; note that the MeS group was not involved directly).[422]

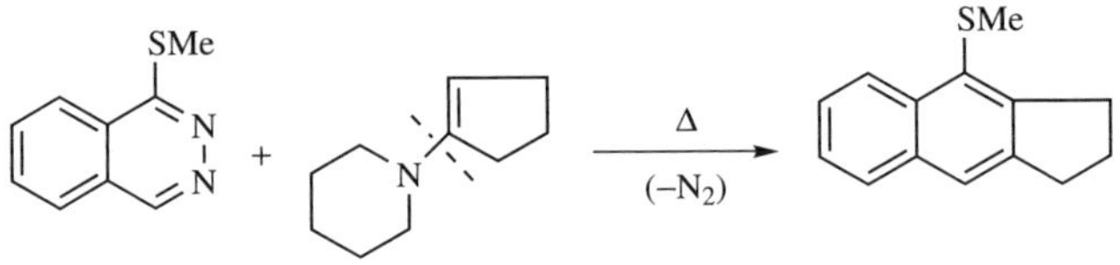

(**28**) (**29**)

1,4-Bis(pyridin-2-ylthio)phthalazine underwent *chelation* with copper(II) chloride hydrate to give the complex (**30**) (H_2O, MeOH, reflux, 30 min).[682]

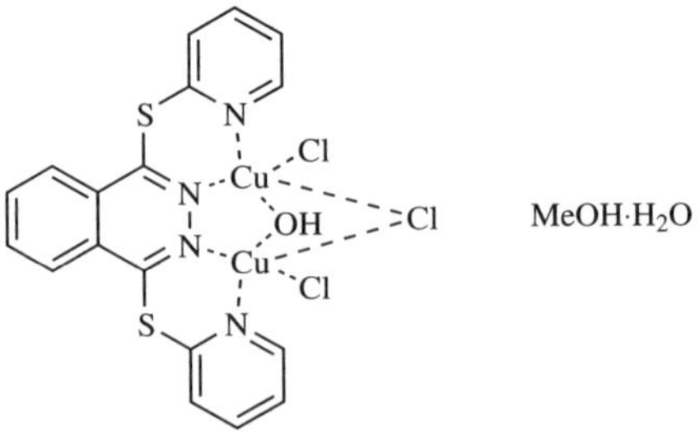

(**30**)

12.3. ALKYLSULFINYL- AND ALKYLSULFONYLPHTHALAZINES

The preparation and more recently reported reactions of these potentially useful intermediates are illustrated in the following examples.

Preparation

Note: Examples have been given already for the formation of sulfones by *primary synthesis* (Chapter 8), of the sulfones by *arenesulfinolysis of halogenophthalazines* (Section 10.3.4), and of both sulfoxides and sulfones by *oxidation of alkylthiophthalazines* (Section 12.1). A minor route is exemplified here.

2-Phenyl-1,4(2*H*,3*H*)-phthalazinedione (**31**) with chlorosulfonic acid at 0°C gave bis[*p*-(1,4-dioxo-1,2,3,4-tetrahydrophthalazin-2-yl)phenyl] sulfone (**32**) (for details, see original).[184]

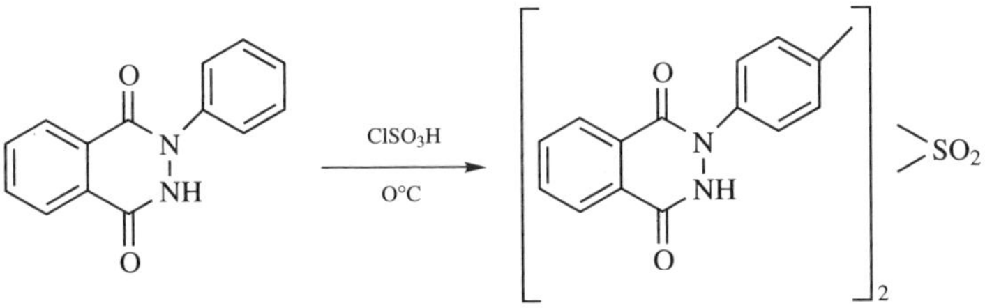

(**31**) (**32**)

Reactions

Note: Examples have been given already for *hydrolysis of the sulfones to phthalazinones* (Section 11.1.1) and for their *alcoholysis to alkoxyphthalazines* (Section 11.4.1). Other reactions are illustrated here.

1-Phenyl-4-*p*-tolylsulfonylphthalazine (**33**) underwent *cyanolysis* to give 4-phenyl-1-phthalazinecarbonitrile (**34**) (KCN, Me_2NCHO, 140°C, 2 h: 96%).[591] Such cyanolyses were also performed in dimethyl sulfoxide.[931,932]

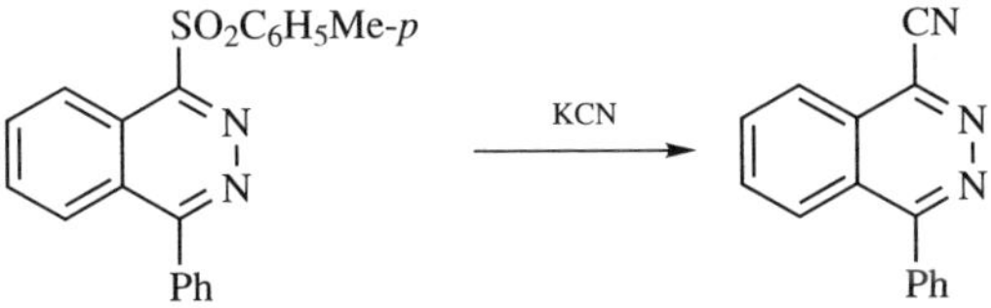

(**33**) (**34**)

1-Methylsulfonylphthalazine (**35**) underwent *adduct formation* with 1-styrylpiperidine (**36**) followed by loss of nitrogen to afford 1-methylsulfonyl-3-

phenylnaphthalene (**37**) (neat reactants, 120°C, 10 min: 59%; analogs likewise);[422] in a subtly different way, 6-methoxy-4-*p*-tolylsulfonylphthalazine (**38**) with 1-diethylaminopropyne gave 2-diethylamino-7-methoxy-3-methyl-1-*p*-tolylsulfonylnaphthalene (**39**) (reactants, dioxane, 80°C, 10 min: 88%).[435]

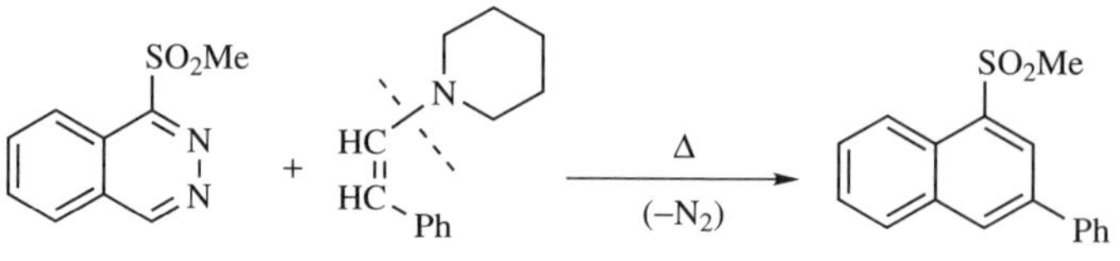

(**35**) (**36**) (**37**)

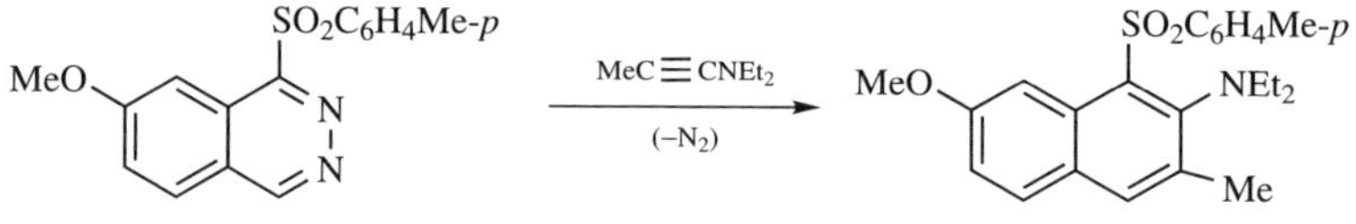

(**38**) (**39**)

2,3-Dibenzenesulfonyl-1,4(2*H*,3*H*)-phthalazinedione (**40**) suffered mono*desulfonylation* to give 2-benzenesulfonyl-1,4(2*H*,3*H*)-phthalazinedione (**41**) [excess Bu_3SnH, $(=NCMe_2CN)_2$ (catalyst), PhMe, reflux, N_2; more catalyst ↓ after 30 min if incomplete (tlc): 59%].[87]

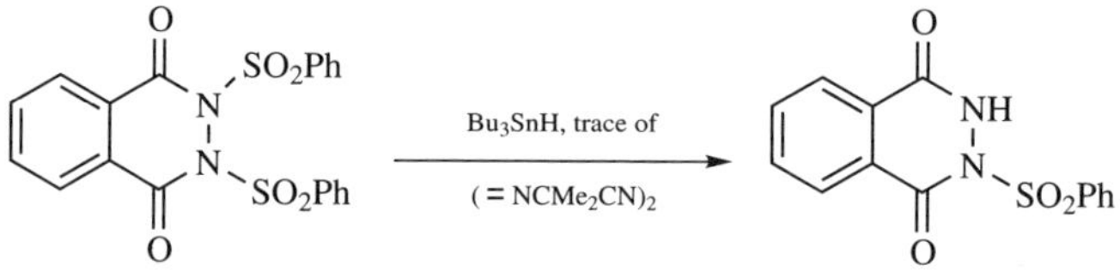

(**40**) (**41**)

12.4. PHTHALAZINESULFONIC ACID DERIVATIVES

Little work has been reported in this whole area of late. It is summarized briefly in the examples that follow.

6-Amino-7-chloro-1(2*H*)-phthalazinone (**42**, R = H) gave 7-chloro-1-oxo-1,2-dihydro-6-phthalazinesulfonyl chloride (**43**, R = H) (substrate, 10M HCl, 0°C; $NaNO_2$ in H_2O↓ dropwise, 5–10°C, 20 min; this solution↓ to AcOH saturated with SO_2, CuCl: 68%); 2-benzyl-7-chloro- (**43**, R = CH_2Ph) (61%) and 7-chloro-2-methyl-1-oxo-1,2-dihydro-6-phthalazinesulfonyl chloride (**43**, R = Me) (52%) were made similarly.[274]

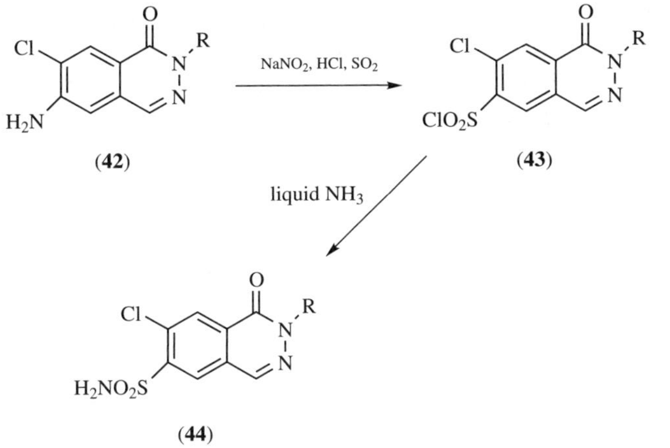

NaNO2, HCl, SO2

(42) (43)

liquid NH3

(44)

Each of the foregoing products (**43**) underwent *aminolysis* to afford 7-chloro-1-oxo- (**44**, R = H) (substrate, liquid NH_3, ~−60°C, 6 h: 55%; note survival of the 7-Cl substituent under these conditions), 2-benzyl-7-chloro-1-oxo- (**44**, R = CH_2Ph) (likewise: 66%), and 7-chloro-2-methyl-1-oxo-1,2-dihydro-6-phthalazinesulfonamide (**44**, R = Me) (likewise: 80%), respectively.[274]

2-Phenyl-1,4(2*H*,3*H*)-phthalazinedione (**45**) underwent ω-*chlorosulfonation* to give 2-*p*-chlorosulfonylphenyl-1,4(2*H*,3*H*)-phthalazinedione (**46**) ($ClSO_3H$, ~115°C) and thence *aminolysis* to afford 2-*p*-sulfamoylphenyl-1,4(2*H*,3*H*)-phthalazinedione (**47**, R = NH_2) or analogs (appropriate amine: for details, see original);[184] hydrazinolysis of 4-chloro-2-*p*-chlorosulfonylphenyl- likewise gave 2-*p*-(aminosulfamoyl)phenyl-4-chloro-2(1*H*)-phthalazinone (**48**).[192]

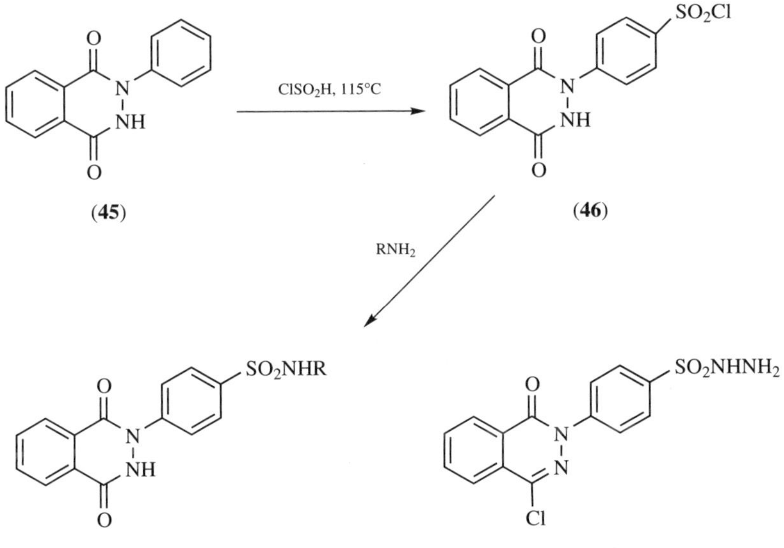

ClSO2H, 115°C

(45) (46)

RNH2

(47) (48)

4-*p*-Tolyl-1-phthalazinamine (**49**) with methyl 3-chloropropanesulfonate gave a separable mixture of the ω-sulfonic esters, 1-amino-2-(3-methoxysulfonylpropyl)-4-*p*-tolylphthalazin-2-ium chloride (**50**) and 4-amino-2-(3-methoxysulfonylpropyl)-1-*p*-tolylphthalazin-2-ium chloride (**51**) (neat reactants, 120°C, 30 min: 8% of each after separation).[958]

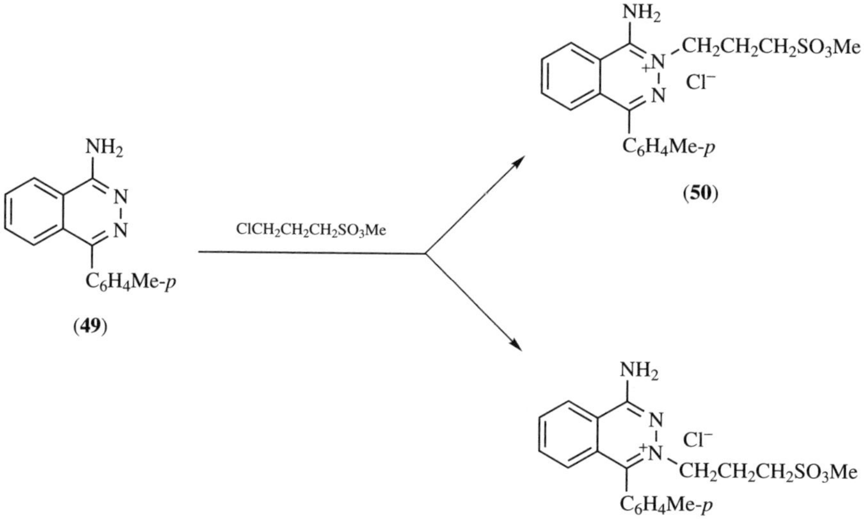

NH_2 $CH_2CH_2CH_2SO_3Me$ Cl^- C_6H_4Me-*p*

(**50**)

NH_2 C_6H_4Me-*p*

(**49**)

$ClCH_2CH_2CH_2SO_3Me$

NH_2 Cl^- $CH_2CH_2CH_2SO_3Me$ C_6H_4Me-*p*

(**51**)

CHAPTER 13

Nitro-, Amino-, and Related Phthalazines (*H* 183; *E* 560,596)

This chapter embraces the preparation and reactions of phthalazines with nitrogenous substituents joined directly or indirectly to the nucleus through their nitrogen atom: nitro, nitroso, amino, hydrazino, and related derivatives.

13.1. NITRO- AND NITROSOPHTHALAZINES

Much of the available information on nitrophthalazines refers to extranuclear nitro derivatives. The related *nitroxyphthalazines* (nitrate esters derived from extranuclear hydroxyphthalazines) have been mentioned in Section 11.2.2.

13.1.1. Preparation of Nitrophthalazines

Many nuclear and extranuclear nitrophthalazines have been made by *primary synthesis* (see Chapter 8). The remainder have been made by *passenger introduction* (examples in most chapters) or by *direct nitration* of phthalazine substrates, as illustrated in the following examples.

Note: The nitration of unsubstituted phthalazine to give 5-nitrophthalazine has been described in Section 9.1.3.

Ethyl 1-hydroxymethyl-7-methyl-4-oxo-3,4-dihydro-6-phthalazinecarboxylate (**1**, R = CH_2OH) gave ethyl 1-hydroxymethyl-7-methyl-8-nitro-4-oxo-3,4-dihydro-6-phthalazinecarboxylate (**2**, R = CH_2OH) (substrate, 96% H_2SO_4, 20°C; KNO_3↓ portionwise; then 20°C, 7 h: 56%); appropriate substrates (**1**) likewise gave ethyl 1,7-dimethyl-8-nitro-4-oxo-3,4-dihydro-6-phthalazinecarboxylate (**2**, R = Me) and diethyl 7-methyl-5-nitro-4-oxo-3,4-dihydro-1,6-phthalazinedicarboxylate (**2**, R = CO_2Et).[425]

Cinnolines and Phthalazines: Supplement II, The Chemistry of Heterocyclic Compounds, Volume 64, by D.J. Brown

(1) $\xrightarrow{KNO_3, H_2SO_4}$ (2)

Ethyl 5,7-dimethyl-4-oxo-3-phenyl-3,4-dihydro-6-phthalazinecarboxylate (**3**) gave either ethyl 5,7-dimethyl-3-*p*-nitrophenyl-4-oxo-3,4-dihydro-6-phthalazinecarboxylate (**4**) [substrate, 95% H_2SO_4; KNO_3 (1 mol)↓ portionwise, ~0°C, 5 h: 50%] or ethyl 3-(2,4-dinitrophenyl)-5,7-dimethyl-8-nitro-4-oxo-3,4-dihydro-6-phthalazinecarboxylate (**5**) [substrate, 95% H_2SO_4, KNO_3 (3 mol)↓ portionwise, 0°C; then 20°C, 12 h: 80%].[270]

(3) → H_2SO_4, KNO_3 (1 mol), 0°C → (4)

(3) → H_2SO_4, KNO_3 (3 mol), 20°C → (5)

Also other examples.[209,573]

13.1.2. Reactions of Nitrophthalazines

The only widely used reaction of nitrophthalazines is their reduction to the corresponding phthalazinamines. The various reduction procedures employed recently (as of early 2005) are illustrated in the following examples.

Hydrogenation

Ethyl 1-hydroxymethyl-5,7-dimethyl-8-nitro-4-oxo-3,4-dihydro-6-phthalazinecarboxylate (**6**, R = NO_2) gave 8-amino-1-hydroxymethyl-5,7-dimethyl-4-oxo-3,4-dihydro-6-phthalazinecarboxylate (**6**, R = NH_2) (H_2, Pd/C, MeOH, AcOEt, 20°C: 85%); analogs likewise.[425]

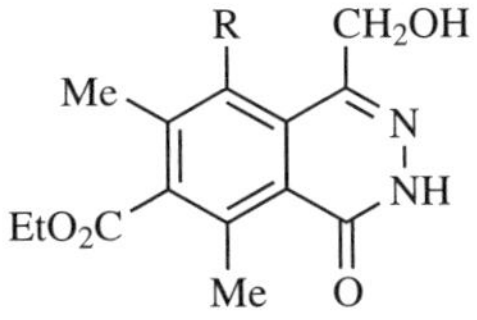

(6)

2-*p*-Nitrobenzyl- (**7**, R = NO_2) gave 2-*p*-aminobenzyl-1(2*H*)-phthalazinone (**7**, R = NH_2) [H_2 (3 atm), Pd/C, EtOH, AcOEt, 3 h: 69%].[52]

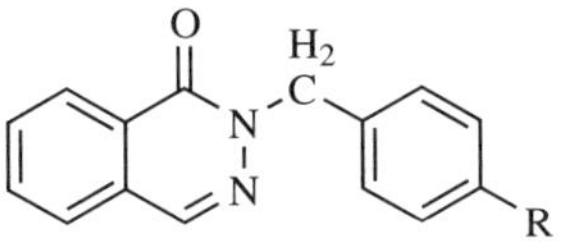

(7)

Ethyl 5,7-dimethyl-3-*p*-nitrophenyl- (**8**, R = NO_2) gave ethyl 3-*p*-aminophenyl-5,7-dimethyl-4-oxo-3,4-dihydro-6-phthalazinecarboxylate (**8**, R = NH_2) (H_2, Pd/C, dioxane, 20°C, until complete: 87%).[270]

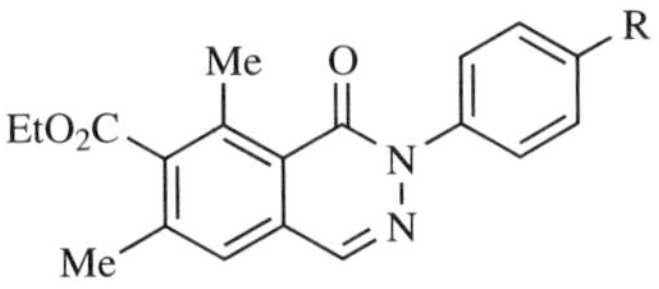

(8)

Ethyl 8-nitro- (**9**, R = NO_2) gave ethyl 8-amino-4-oxo-3,4-dihydro-1-phthalazinecarboxylate (**9**, R = NH_2) (H_2, Raney Ni, no details).[573]

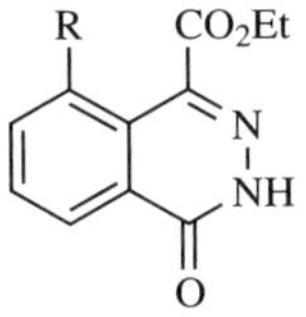

(9)

Also other examples.[203,921,972]

Reduction by Catalyzed Hydrazine

8-Nitro-5-phthalazinamine (**10**, R = NO_2) gave 5,8-phthalazinediamine (**10**, R = NH_2) (substrate, $H_2NNH_2 \cdot H_2O$, charcoal, MeOH; $FeCl_3\downarrow$, reflux, 5 h: >95%).[856]

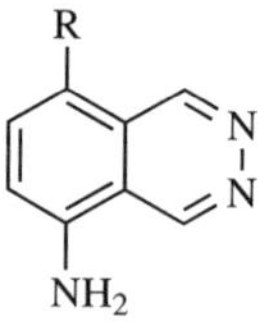

(**10**)

4-[*p*-(*p*-Nitrophenoxy)phenyl]-2-*p*-nitrophenyl-1(2*H*)-phthalazinone (**11**, R = NO_2) gave 4-[*p*-(*p*-aminophenoxy)phenyl]-2-*p*-aminophenyl-1(2*H*)-phthalazinone (**11**, R = NH_2) ($H_2NNH_2 \cdot H_2O$, Pd/C: no details).[893]

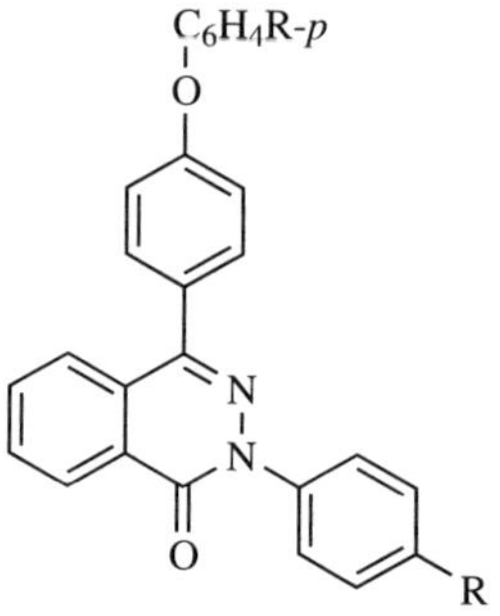

(**11**)

Other Reductive Procedures

5-Nitro- (**12**, R = NO_2) gave 5-amino-1,4(2*H*,3*H*)-phthalazinedione (**12**, R = NH_2) ($Na_2S_2O_4$: for details, see original).[546]

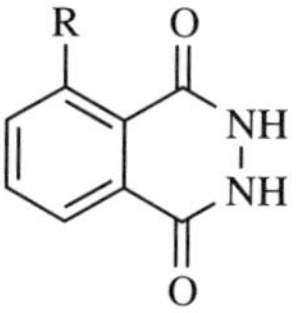

(**12**)

6,7-Dichloro-1-*p*-nitrophenylphthalazine (**13**, R = NO_2) gave 1-*p*-aminophenyl-6,7-dichlorophthalazine (**13**, R = NH_2) (Fe powder, HCl, H_2O, EtOH, reflux, 1 h: 85%).[848]

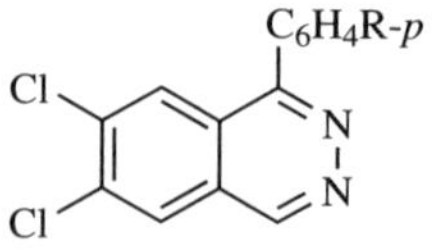

(**13**)

4-Methyl-2-[2-nitro-5-(pyrrolidin-1-yl)phenyl]-1(2*H*)-phthalazinone (**14**, R = NO_2) underwent a one-pot reduction to the corresponding amino analog (**14**, R = NH_2) and dehydrative cyclization to give 5-methyl-9-(pyrrolidin-1-yl)benzimidazo[2,1-*a*]phthalazine (**15**) (substrate, polyphosphoric acid, 100°C; Fe powder↓ portionwise; then 140°C briefly: ~25%).[972]

Me N N N(CH$_2$)$_4$ O R → Me N N N(CH$_2$)$_4$ N

(**14**) (**15**)

13.1.3. Nitrosophthalazines: Preparation and Reactions

This class appears to be represented by only one *N*-nitrosodihydrophthalazine. Treatment of 3,4-dihydro-1(2*H*)-phthalazinone (**16**) with nitrous acid at 0°C gave 3-nitroso-3,4-dihydro-2(1*H*)-phthalazinone (**17**) (~85%), with analysis and spectra consistent with such formulation. Subsequent treatment with nitrous acid at 20°C for 12 h gave 1(2*H*)-phthalazinone (**18**).[921]

O NH NH —HNO_2, 0°C→ O N N NO —HNO_2, 20°C→ O NH N

(**16**) (**17**) (**18**)

13.2. AMINOPHTHALAZINES
(*E* 560)

This section covers only primary, secondary, and tertiary aminophthalazines; hydrazinophthalazines and their derivatives are discussed separately in Section 13.3.

13.2.1. Preparation of Aminophthalazines

Most aminophthalazines have been made by *primary synthesis* (see Chapter 8), by the *addition of amines to alkylphthalazines* (Section 9.3.2), by *aminolysis of halogenophthalazines* (Section 10.3.2), by *aminolysis of phthalazinones* (Section 11.1.2.4), by *aminolysis of alkoxyphthalazines* (Section 11.4.2), by *aminolysis of alkylthiophthalazines* (Section 12.2), or by *reduction of nitrophthalazines*

(Section 13.1.2). Some minor routes are illustrated by the following classified examples.

By Amination

5-Nitrophthalazine (**19**, R = H) underwent semidirect amination to 8-nitro-5-phthalazinamine (**19**, R = NH_2) (substrate, EtOH, KOH, ~50°C; $H_2NOH \cdot HCl$ in MeOH↓ dropwise during 3 h: 66%).[856]

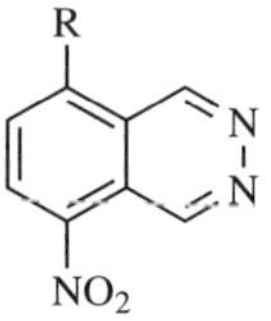

R

NO2

(**19**)

Phthalazine (**20**, R = H) and *O*-mesitylenesulfonyl hydroxylamine gave 2-aminophthalazin-2-ium mesitylenesulfonate (**21**, R = H) (substrate, CH_2Cl_2; synthon in CH_2Cl_2↓ dropwise; 20°C, 10 min: 53%); 1-phenylphthalazine (**20**, R = Ph) likewise gave 2-amino-4-phenylphthalazin-2-ium mesitylenesulfonate (**21**, R = Ph) (60%, including a little of the 1-phenyl isomer).[5]

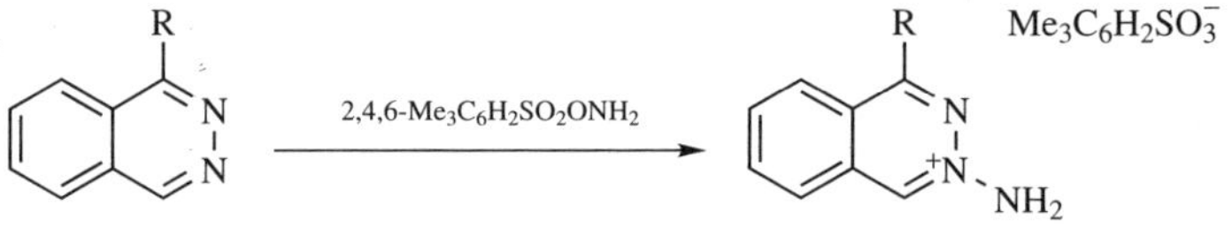

$2,4,6\text{-}Me_3C_6H_2SO_2ONH_2$; $Me_3C_6H_2SO_3^-$

(**20**) (**21**)

From Acylaminophthalazines

Note: Available examples appear to be confined largely to the deacylation of extranuclear acetamido- or phthalimidophthalazines.

Ethyl 5,7-dimethyl-4-oxo-1-phthalimidomethyl-3,4-dihydro-6-phthalazinecarboxylate (**22**) gave ethyl 1-aminomethyl-5,7-dimethyl-4-oxo-3,4-dihydro-6-phthalazinecarboxylate (**23**) (HCl, H_2O, EtOH, reflux, 2 h: ~70% as hydrochloride).[405]

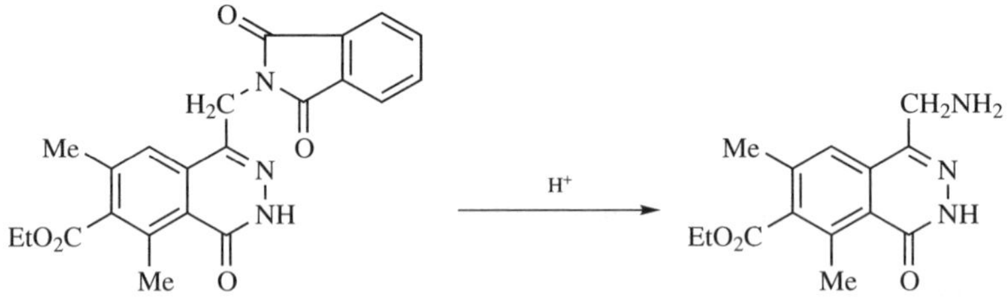

H^+

(**22**) (**23**)

2-(2-Phthalimidoethyl)-4-(pyridin-3-yl)-1(2*H*)-phthalazinone (**24**) gave 2-(2-aminoethyl)-4-(pyridin-3-yl)-1(2*H*)-phthalazinone (**25**) ($H_2NNH_2 \cdot H_2O$, EtOH, reflux, N_2, 3 h: 87%).[279]

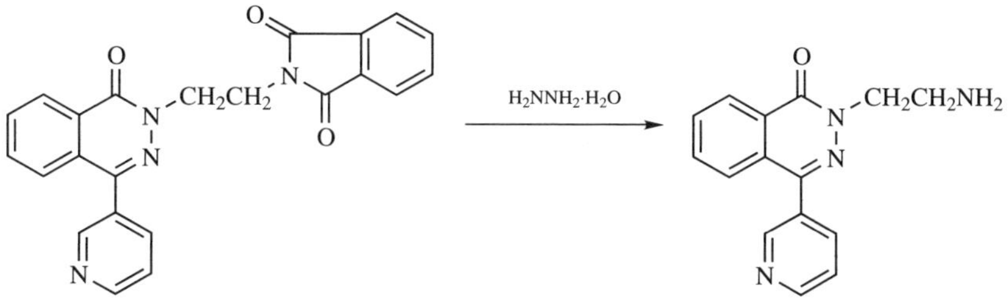

(**24**) (**25**)

4-*p*-Acetamidophenyl- (**26**, R = Ac) gave 4-*p*-aminophenyl-*N*-butyl-6,7-dichloro-1-methyl-1,2-dihydro-2-phthalazinecarboxamide (**26**, R = H) (NaOH, H_2O, MeOH, reflux, 3 h: 93%); analogs likewise.[848]

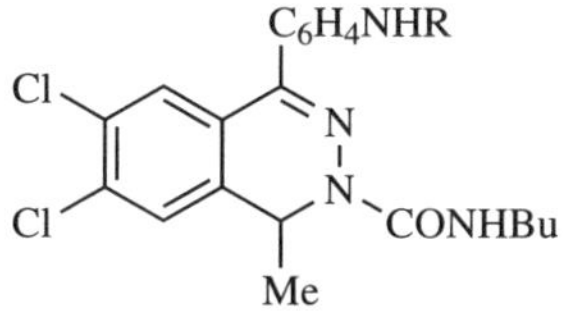

(**26**)

Also other examples.[158]

By Interconversion of Aminophthalazines

Note: The conversion of one aminophthalazine into another can be done in several ways. The best of these, transamination, is represented in the 1972–2004 literature by only a $RNHNH_2 \rightarrow RNH_2$ transformation (see Section 13.2.2), but other processes are illustrated here.

1-Hydrazinophthalazine (**27**, R = NH_2) gave 1-phthalazinamine (**27**, R = H) (H_2, Raney -Ni, H_2O, EtOH, 20°C, 12 h: 50%).[32,cf. 1007]

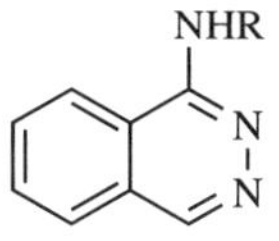

(**27**)

1-Dimethylaminophthalazine (**28**, R = Me) gave 1-methylaminophthalazine (**28**, R = H) [10M HBr, reflux, 3 h: >95%(?)].[59]

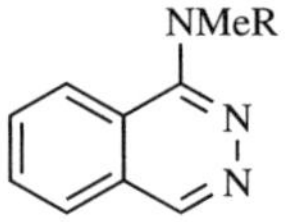

(**28**)

4-[(2-Hydroxy-1,1-dimethylethyl)amino]-1(2*H*)-phthalazinone (**29**, R = CMe_2CH_2OH) gave 4-amino-1(2*H*)-phthalazinone (**29**, R = H) (6M HCl, 125°C, sealed, 15 h: 40%).[151]

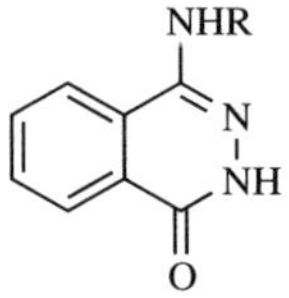

(**29**)

7-Amino- (**30**, R = H) gave 7-dimethylamino-2-phenyl-1(2*H*)-phthalazinone (**30**, R = Me) (substrate, NaH, THF, reflux; Me_2SO_4↓ dropwise; reflux, 4 h: 45%).[411]

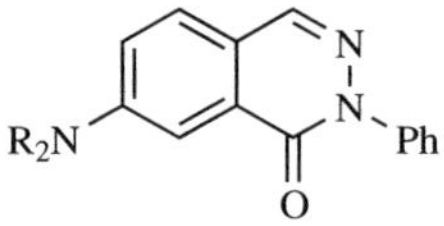

(**30**)

2-(2-Aminoethyl)-1(2*H*)-phthalazinone with 6-chloropurine gave 2-[2-(purin-6-ylamino)ethyl]-1(2*H*)-phthalazinone (**31**) (Et_3N, EtOH, reflux, 20 h: 60%).[369]

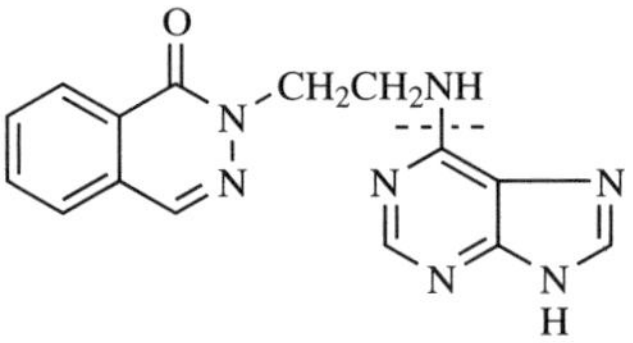

(**31**)

6-Azido-7-chloro-5,8-phthalazinequinone gave 6-amino-7-chloro-5,8-phthalazinequinone (substrate, EtOH, 0°C; $NaBH_4$↓, 0°C → 20°C, 2 h: 45%).[1016]

By Covalent Addition of Ammonia

Note: The formation of covalent amino adducts from *N*-benzylphthalazinium salts in liquid ammonia has been studied.[317]

By the Curtius Reaction

4-Oxo-3-phenyl-3,4-dihydro-6-phthalazinecarbonyl azide (**32**) gave 7-(benzyloxycarbonylamino)-2-phenyl-1(2*H*)-phthalazinone (**33**) ($PhCH_2OH$, PhH, gradually → reflux, 6 h: 72%) and thence 7-amino-2-phenyl-1(2*H*)-phthalazinone (**34**) (H_2, Pd/C, 20°C: 86%).[411]

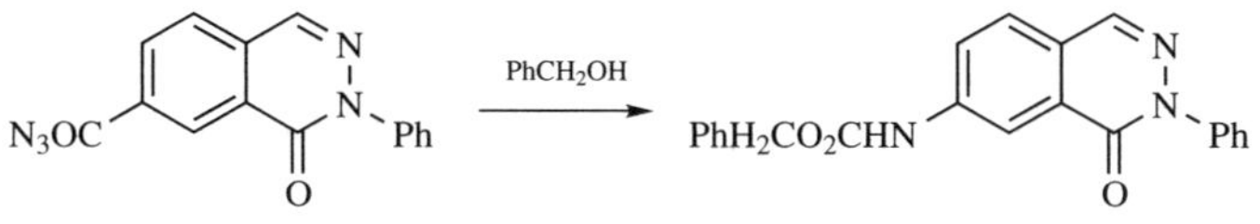

(**32**) (**33**)

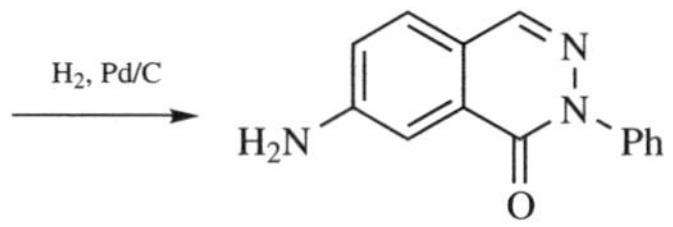

(**34**)

Also a related example, described with little detail.[573]

13.2.2. Reactions of Aminophthalazines

The *conversion of primary aminophthalazines into halogenophthalazines* (Section 10.1.2), of *aminophthalazines into phthalazinones or extranuclear hydroxyphthalazines* (Section 11.1.1 and 11.2.1), of *primary phthalazinamines into phthalazinequinones* (Section 11.3.1), and of *one amino- into another aminophthalazine* (Section 13.2.1) have been covered already. Other reactions, including some of the ring NH groups in reduced phthalazines, are illustrated in the following classified examples.

N-Acylation

7-Amino-4-(pyridin-4-yl)methyl-1(2*H*)-phthalazinone (**35**, R = H) gave 4-(pyridin-4-yl)methyl-7-trifluoroacetamido-1(2*H*)-phthalazinone (**35**, R = $COCF_3$) (neat F_3CCO_2H, 20°C, 48 h: 80%, as its trifluoroacetate salt).[870]

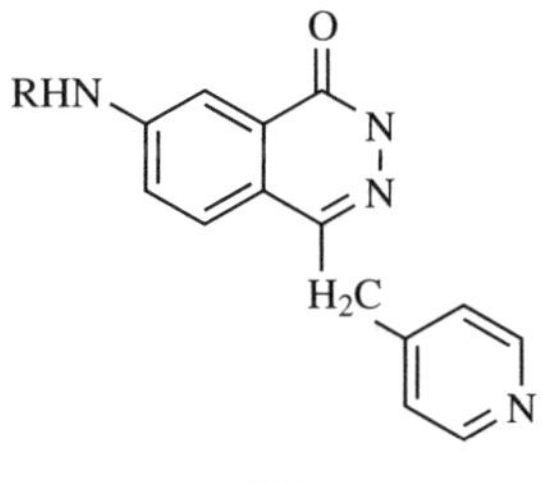

(**35**)

Ethyl 3-*p*-aminophenyl- (**36**, R = H) gave ethyl 3-*p*-acetamidophenyl-5,7-dimethyl-4-oxo-3,4-dihydro-6-phthalazinecarboxylate (**36**, R = Ac) (Ac_2O, pyridine, 20°C, ? h: >95%).[270]

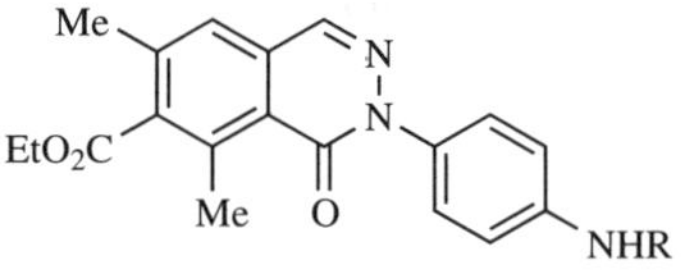

(**36**)

1,4-Phthalazinediamine (**37**, R = H) gave 1,4-dibenzamidophthalazine (**37**, R = Bz) (BzCl, pyridine, 55°C, 5 h: 48%).[685]

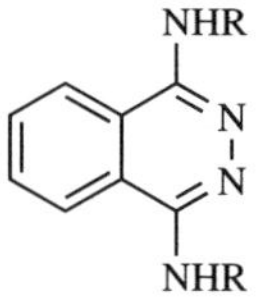

(**37**)

4-Amino-2-methyl-1(2*H*)-phthalazinone (**38**) gave 4-dibenzoylamino-2-methyl-1(2*H*)-phthalazinone (**39**) (BzCl, pyridine, 75°C, 6 h: 42%) and thence 4-benzamido-2-methyl-1(2*H*)-phthalazinone (**40**) NH_4OH, EtOH, 20°C, 3 days: 75%, as hemihydrate).[685]

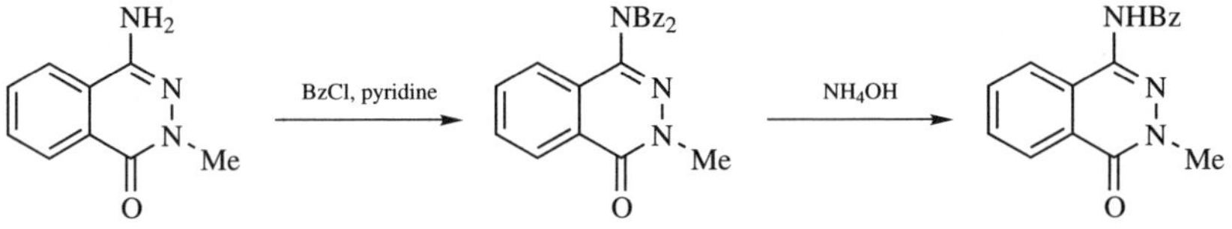

BzCl, pyridine; NH_4OH

(**38**) (**39**) (**40**)

2-Aminophthalazin-2-ium mesitylenesulfonate (**41**) gave phthalazin-2-ium-2-benzamidate (**42**) (neat BzCl, 90°C, 3 h; residue from evaporation, NaOH, H_2O: 86%);[5] analogs likewise.[5,177]

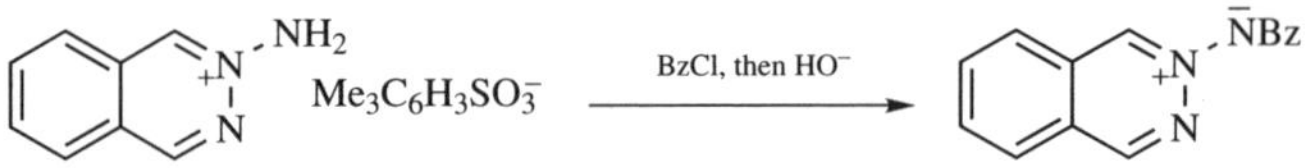

$Me_3C_6H_3SO_3^-$; BzCl, then HO^-

(**41**) (**42**)

1,2,3,4-Tetrahydrophthalazine (**43**) with appropriate acyl chlorides gave 2,3-bis(chloroacetyl)- (**44**, R = CH_2Cl) (substrate · HCl, H_2O, $CHCl_3$; simultaneously $ClCH_2COCl$ (4 mol)↓ and K_2CO_3 in H_2O↓ dropwise, <5°C, during 1 h; then 20°C, 13 h: 91%), 2,3-bis(bromoacetyl)- (**44**, R = CH_2Br)

(likewise: 89%), and related 1,2,3,4-tetrahydrophthalazines;[924] the same substrate (**43**) with *o*-fluorobenzoyl chloride gave 2-*o*-fluorobenzoyl-1,2,3,4-tetrahydrophthalazine (**45**) [substrate · HCl, $NaHCO_3$, H_2O; synthon (1 mol) in Et_2O↓ slowly; 20°C, 1 h: 70%] and analogs similarly.[493]

NH NH —RCOCl (excess)→ N–C(=O)R, N–C(=O)R

(**43**) (**44**)

o-FC_6H_4COCl (1 mol) → N–$C(=O)C_6H_4F$-*o*, NH

(**45**)

Also other examples.[243,393,573,666,848,865,936,1016]

Alkoxy- and Aryloxycarbonylation

6,7-Dichloro-4-*p*-nitrophenyl-1,2-dihydrophthalazine (**46**) with ethyl chloroformate gave ethyl 6,7-dichloro-4-*p*-nitrophenyl-1,2-dihydro-2-phthalazinecarboxylate (**47**, R = Et) (neat reactants, 140°C, 2 h: 85%) or with phenyl chloroformate to give phenyl 6,7-dichloro-4-*p*-nitrophenyl-1,2-dihydro-2-phthalazinecarboxylate (**47**, R = Ph) (likewise: 89%).[848]

$C_6H_4NO_2$-*p*, Cl, Cl, N, NH —$ClCO_2R$→ $C_6H_4NO_2$-*p*, Cl, Cl, N, N–CO_2R

(**46**) (**47**)

Arene- or Alkanesulfonylation

2-(2-Aminoethyl)-4-(pyridin-3-yl)-1(2*H*)-phthalazinone (**48**) gave 2-(2-benzenesulfonamidoethyl)-4-(pyridin-3-yl)-1(2*H*)-phthalazinone (**49**) (substrate, NaH, Me_2NCHO, 20°C, N_2, 30 min; then $PhSO_2Cl$↓, 0°C, 1 h: 45%).[279]

O, N–$CH_2CH_2NH_2$, N, N —NaH; then $PhSO_2Cl$→ O, N–$CH_2CH_2NHSO_2Ph$, N, N

(**48**) (**49**)

5-Cyclopentyloxy-1-(3,5-dichloropyridin-4-ylmethyl)-6-methoxy-3,4-dihydrophthalazine (**50**, R = H) with methanesulfonyl chloride gave 5-cyclopentyloxy-1-(3,5-dichloropyridin-4-ylmethyl)-3-methanesulfonyl-6-methoxy-3,4-dihydrophthalazine (**50**, R = SO_2Me) (Et_3N, CH_2Cl_2, 20°C, 1 h: >60%).[865]

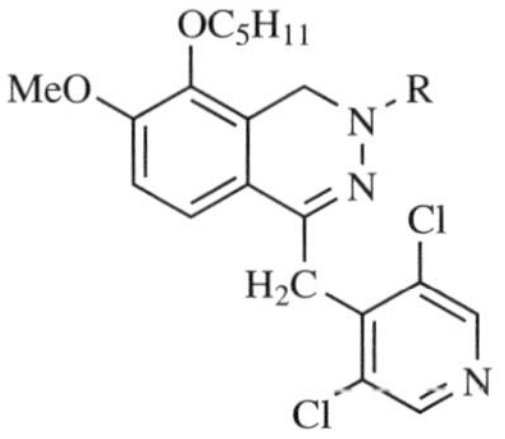

(**50**)

Alkylidenation

2-Amino-1,4(2*H*,3*H*)-phthalazinedione (**51**) with acetylacetone gave 2-(1-acetonylethylideneamino)-1,4(2*H*,3*H*)-phthalazinedione (**52**) (Ac_2CH_2, AcOH: for details, see original).[255]

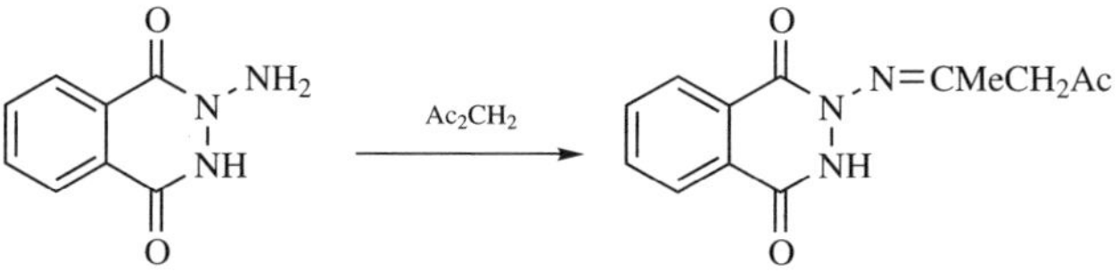

(**51**) → (**52**)

Hydrazinolysis (a Transamination)

1-Phthalazinamine with hydrazine hydrate gave 1-hydrazinophthalazine (EtOH, reflux, 20 h: ?%).[32]

Conversion into Guanidino-, Ureido-, or Thioureidophthalazines

4-Aminomethyl-1(2*H*)-phthalazinone (**53**, R = H) gave 4-guanidinomethyl-1(2*H*)-phthalazinone [**53**, R = C(=NH)NH_2] [MeSC(=NH)$NH_2 \cdot H_2SO_4$: for details, see original).[157]

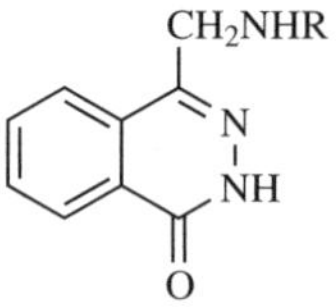

(**53**)

4-Amino-1(2*H*)-phthalazinone (**54**, R = H) gave 4-(*N*′-phenylureido)-1(2*H*)-phthalazinone [**54**, R = C(=O)NHPh] (PhNCO: for details, see original); analogs likewise.[199,cf. 1037]

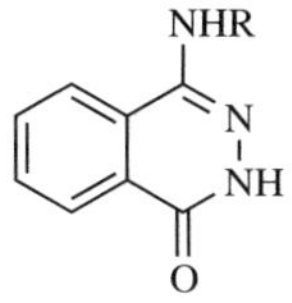

(54)

2-Amino-5,6,7,8-tetrabromo-4-tolyl-1(2*H*)-phthalazinone (**55**, R = H) gave 5,6,7,8-tetrabromo-2-[*N*′-phenyl(thioureido)]-4-tolyl-1(2*H*)-phthalazinone [**55**, R = C(=S)NHPh] (PhNCS, EtOH, reflux, 6 h: 80%; whether *o*-, *m*-, or *p*-tolyl is unspecified).[666]

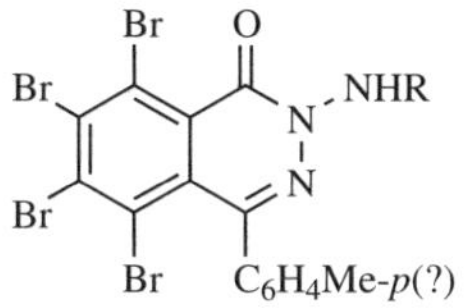

(55)

Conversion into Azo Derivatives

5-Amino-1,4(2*H*,3*H*)-phthalazinedione (**56**) gave a solution of the corresponding diazonium chloride (**57**) (substrate, 6M HCl, <5°C; $NaNO_2$ in H_2O↓ dropwise; <5°C, 1 h; urea↓ to remove any HNO_2) that underwent coupling to afford 5-(2-hydroxynaphthalen-1-ylazo)-1,4(2*H*,3*H*)-phthalazinedione (**58**) [2-naphthol, NaOH, H_2O, EtOH, 20°C; solution (**57**)↓, 20°C, 1 h: 70%],[657] 5-(diacetylmethylazo)-1,4(2*H*,3*H*)-phthalazinedione (**59**) (likewise but Ac_2CH_2, AcONa: 90%),[500] or analogs.[500]

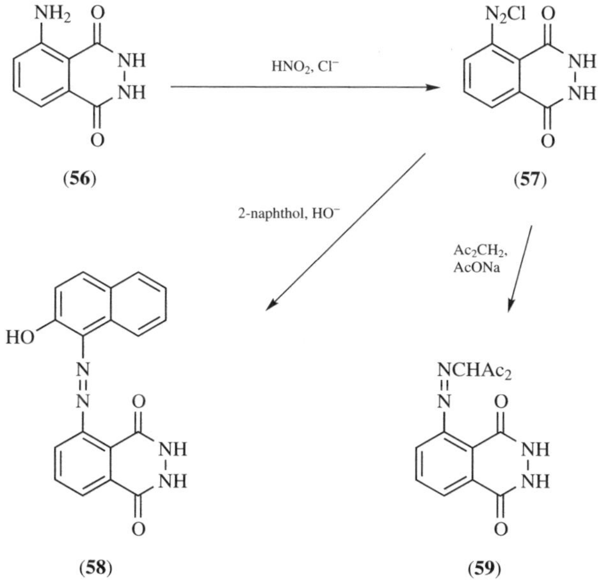

(56) (57) (58) (59)

Conversion into Quaternary Salts or Nontautomeric Iminophthalazines

Note: Treatment of primary or secondary aminophthalazines with alkyl halides in the presence of a base may afford a secondary or tertiary aminophthalazine, respectively (see Section 13.2.1); similar treatment in the absence of a base usually produces quaternization at a ring nitrogen and, when that quaternary nitrogen is adjacent to the amino substituent, basification will produce a nontautomeric phthalazinimine.

1-Phthalazinamine (**60**) gave a separable mixture of 1-amino- (**61**) and 4-amino-2-methylphthalazin-2-ium iodide (**62**) (MeI, MeOH, 50°C, sealed, 24 h: 90% before separation).[62]

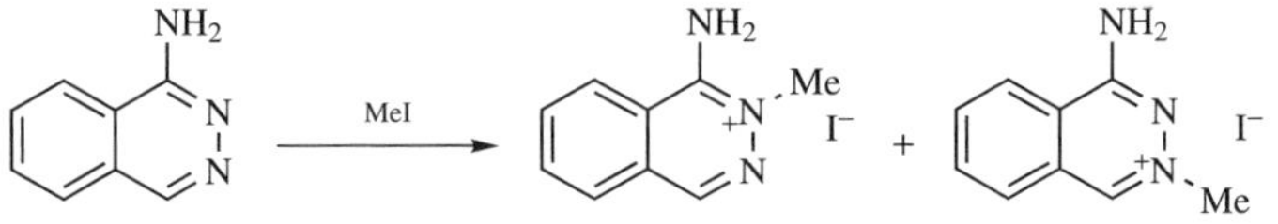

(**60**) —MeI→ (**61**) + (**62**)

In contrast, 1-dimethylaminophthalazine (**63**) with methyl iodide gave only 4-dimethylamino-2-methylphthalazin-2-ium iodide (**64**, R = Me) (likewise but 75°C: >95%) or with propyl iodide gave 4-dimethylamino-2-propylphthalazin-2-ium iodide (**64**, R = Pr) (likewise, 75°C: 95%);[62] analogs likewise.[62,939]

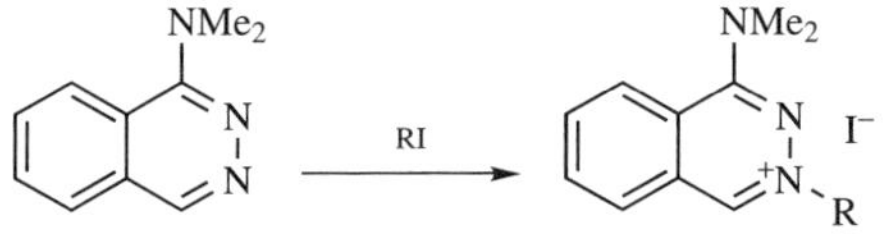

(**63**) —RI→ (**64**)

4-Benzyl-1-phthalazinamine (**65**) with ethyl 4-bromobutyrate gave a separable mixture of 1-amino-4-benzyl- (**66**) and 4-amino-1-benzyl-2-(3-ethoxycarbonylpropyl)phthalazin-2-ium bromide (**67**) (Me_2NCHO, 80°C, 2 h: 46% and 7%, respectively, after chromatographic separation); analogs likewise.[958]

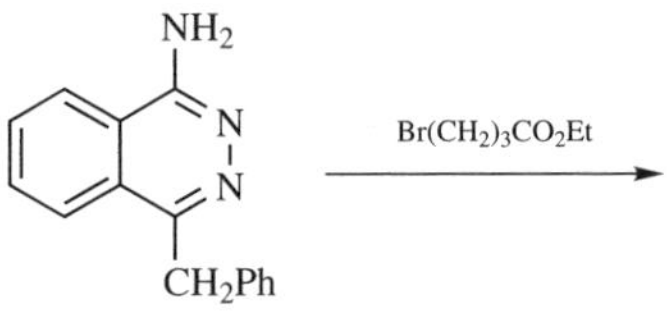

(**65**) —$Br(CH_2)_3CO_2Et$→ (**66**) + (**67**)

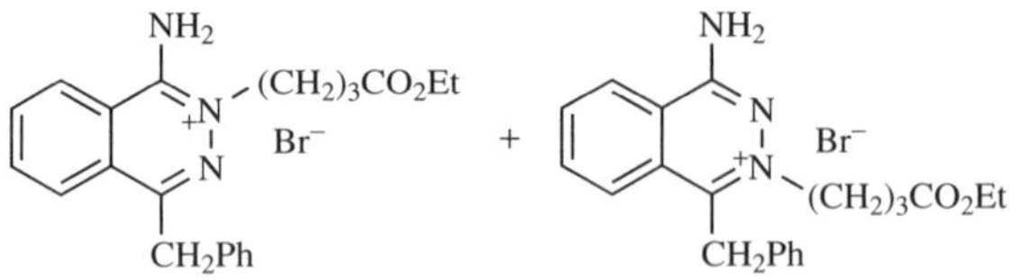

Because of symmetry, a similar monoquaternization of 1,4-phthalazinediamine (**68**) gave only one product, 1,4-diamino-2-(3-ethoxycarbonylpropyl)phthalazin-2-ium bromide (**69**) (Me_2NCHO, 80°C, 2 h: 70%), which furnished its nontautomeric free base, 3-(3-ethoxycarbonylpropyl)-4-imino-3,4-dihydro-1-phthalazinamine (**70**) (KOH, H_2O, 20°C: 20%).[947]

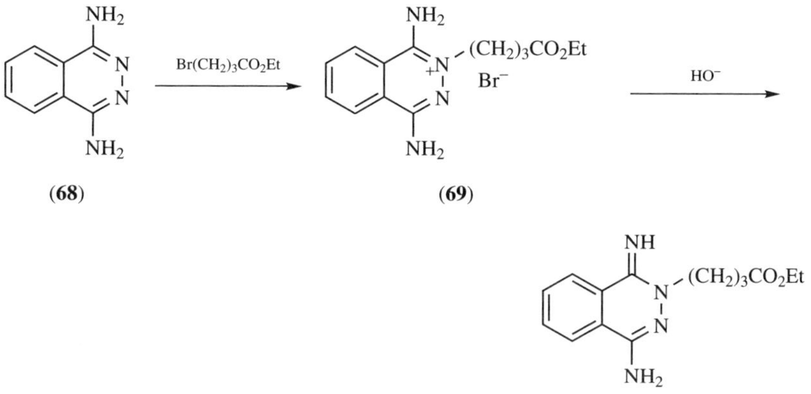

$Br(CH_2)_3CO_2Et$ → ; HO^- →

(**68**) (**69**) (**70**)

Metal Complex Formation

Note: Phthalazines bearing two tertiary amino groups attached indirectly to the nucleus at the 1- and 4-positions have proved particularly interesting as ligands for heavy metals. Some such extranuclear diaminophthalazines are listed here with the metals involved.

1,4-Bis(pyridin-2-ylamino)phthalazine (**71**): Cu;[802,803] Co, Ni.[1023]

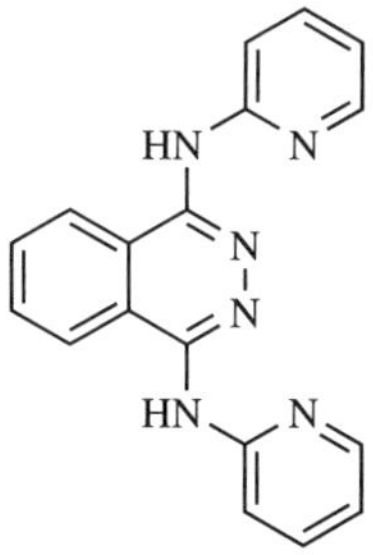

(**71**)

1,4-Bis(3-, 5-, or 6-methylpyridin-2-ylamino)phthalazine: Cu.[625,802]

1,4-Bis(4,6-dimethylpyridin-2-ylamino)phthalazine: Cu.[625]

1,4-Bis(1-methylimidazol-2-yl)phthalazine (**72**): Cu.[626]

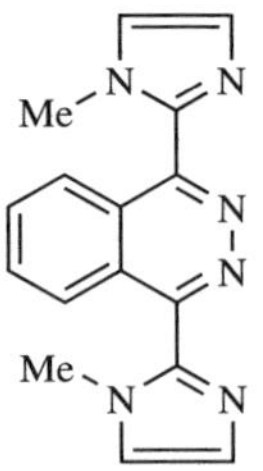

(72)

1,4-Bis[di(pyridin-2-yl)methyl]phthalazine (**73**): Cu;[882,1021] Mn, Fe, Zn;[882] Ni.[845]

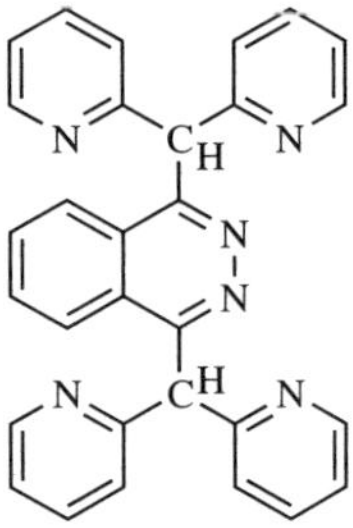

(73)

1,4-Bis[di(6-phenylpyridin-2-yl)methyl]phthalazine: Fe.[846]
Also other examples.[1025]

Cyclization to Imidazo[2,1-*a*]phthalazines

4-Chloro-1-phthalazinamine (**74**) with phenylglyoxal gave 6-chloro-2-phenylimidazo[2,1-*a*]phthalazin-3(5*H*)-one (**75**) (EtOH, trace HCl, 20°C, 3.5 days: 78%)[513] or with phenacyl bromide gave 6-chloro-2-phenylimidazo[2,1-*a*] phthalazine (**76**) (EtOH, reflux, 4 h: 81%).[884]

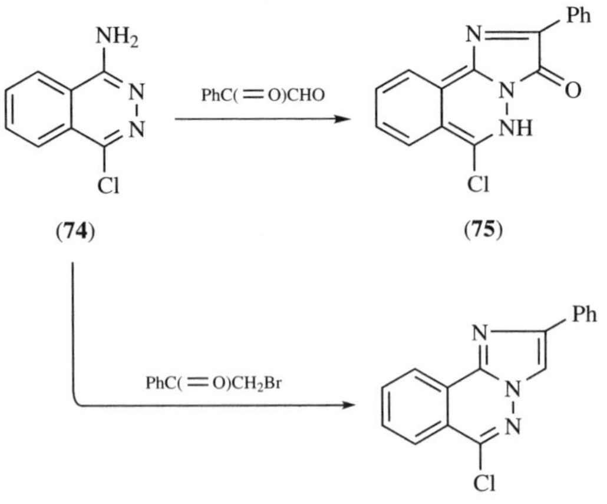

PhC(=O)CHO

(74) (75)

PhC(=O)CH$_2$Br

(76)

4-Phenyl-1-phthalazinamine (**77**) with ethyl 4-chloroacetoacetate gave 2-ethoxycarbonylmethyl-6-phenylimidazo[2,1-*a*]phthalazine (**78**) (Et_3N, Me_2NCHO, 80°C, tlc monitored: 55%).[958]

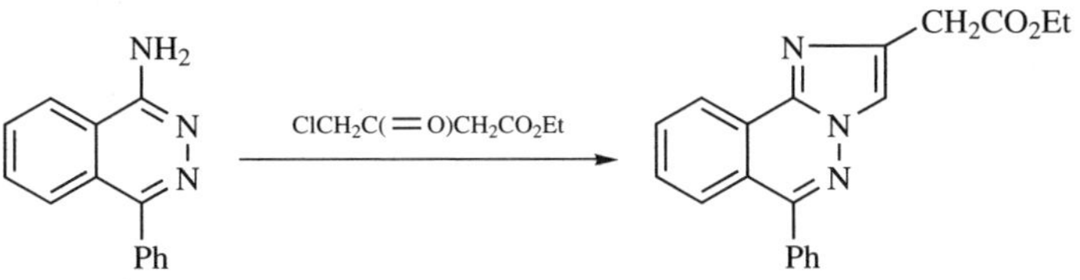

$ClCH_2C(=O)CH_2CO_2Et$

(**77**) (**78**)

4-[1-(Dimethylcarbamoyl)-1-methylethyl]amino-1(2*H*)-phthalazinone (**79**) underwent thermal cyclization to give 2,2-dimethyl-2,3-dihydroimidazo[2,1-*a*] phthalazine-3,6(5*H*)-dione (**80**) (kerosene, reflux, 2 h: >95%).[151]

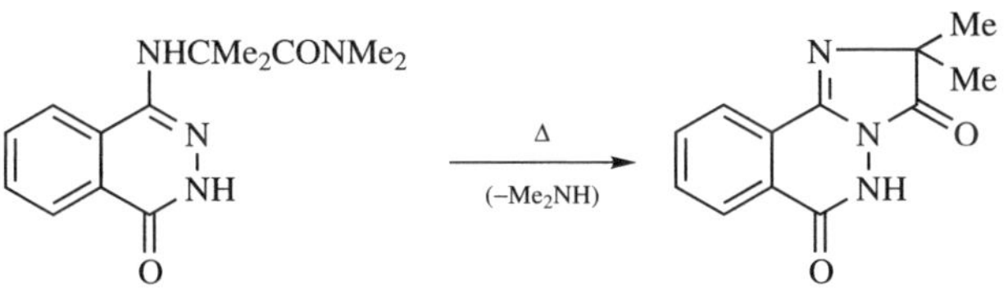

Δ ($-Me_2NH$)

(**79**) (**80**)

Also other examples.[163,597]

Cyclization to Pyrazolo[1,2-*b*]phthalazines

3,4-Dihydro-1(2*H*)-phthalazinone (**81**) with 3-chloro-3-diethylaminomethacryloyl chloride (**82**) gave 3-diethylamino-2-methyl-5,10-dihydro-1*H*-pyrazolo[1,2-*b*]phthalazine-1,5-dione (**83**) (substrate, Et_3N, PhMe, 0°C; synthon in PhMe↓ during 30 min; then 20°C, 2 h: 81%); analogs likewise.[922]

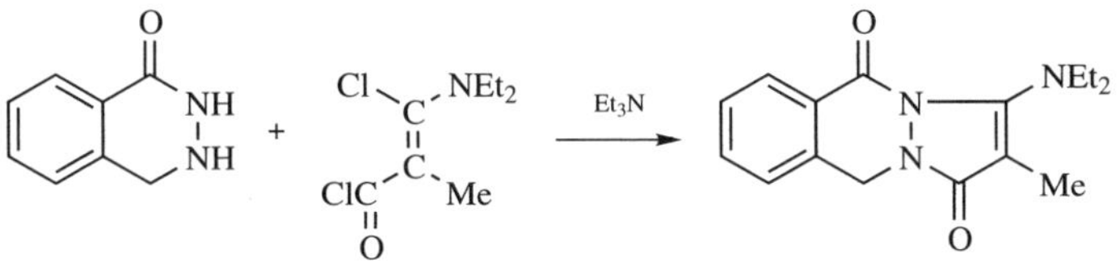

Et_3N

(**81**) (**82**) (**83**)

Also other examples.[210]

Cyclization to Pyrazolo[5,1-*a*]phthalazines

2-Amino-4-phenylphthalazin-2-ium mesitylenesulfonate (**84**) with dimethyl acetylenedicarboxylate gave dimethyl 6-phenylpyrazolo[5,1-*a*]phthalazine-1,2-dicarboxylate (**85**) (substrate, K_2CO_3, Me_2NCHO, 20°C, 10 min; synthon in $Me_2NCHO\downarrow$, 20°C, 2 days: 12%).[8]

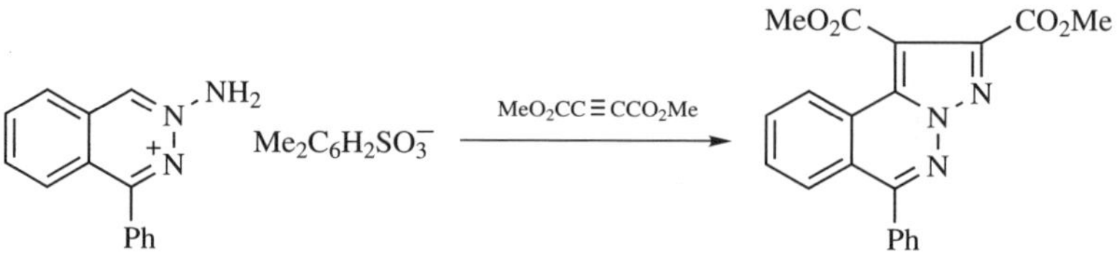

(**84**) $Me_2C_6H_2SO_3^-$ —$MeO_2CC\equiv CCO_2Me$→ (**85**)

Cyclization to Pyridazino[1,2-*b*]phthalazines

1,4(2*H*,3*H*)-phthalazinedione (**86**) with penta-1,4-diene gave 1,4,6,11-tetrahydropyridazino[1,2-*b*]phthalazine-6,11-dione (**87**) (for details, see original);[210] analogs likewise.[167,210,1033]

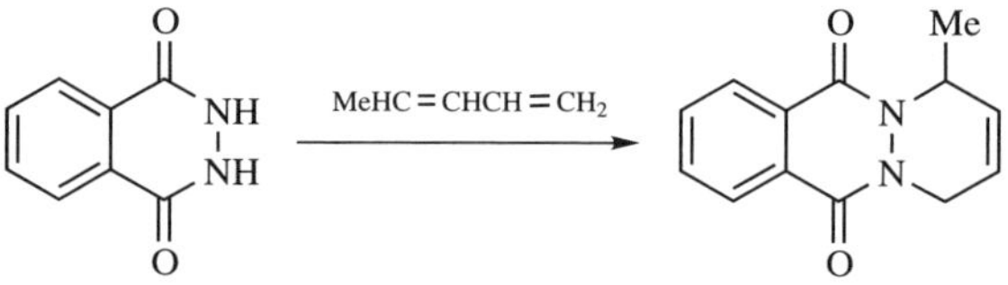

(**86**) —MeHC=CHCH=CH_2→ (**87**)

Cyclization to Pyridazino[4,5-*g*]quinoxalines

6,7-Diamino-1,4(2*H*,3*H*)-phthalazinedione (**88**) with propylglyoxylic acid gave 3-propylpyridazino[4,5-*g*]quinoxaline-2,6,9(1*H*,7*H*,8*H*)-trione (**89**) (reactants, $HSCH_2CH_2OH$, HCl, H_2O, EtOH, reflux, 3 h: ~50%); analogs likewise.[423]

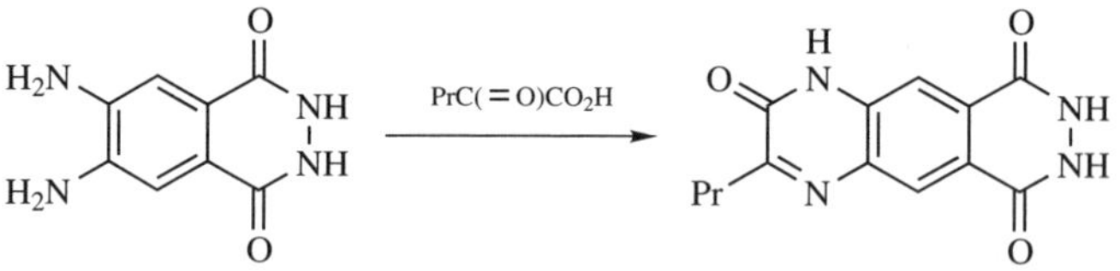

(**88**) —PrC(=O)CO_2H→ (**89**)

Cyclization to Pyrimido[2,1-*a*]phthalazin-5-ium Salts

1,4-Phthalazinediamine (**90**) (as perchlorate?) with 3-chloro-2-methylcrotonaldehyde gave 7-amino-3,4-dimethylpyrimido[2,1-*a*]phthalazin-5-ium perchlorate (**91**) (87%; for details, see original); analogs likewise.[163]

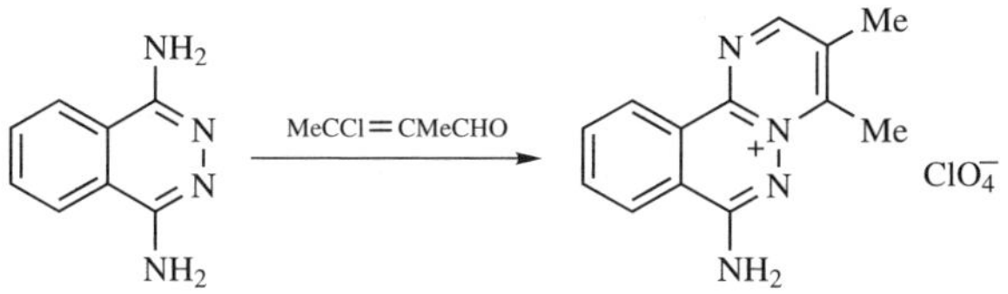

(90) (91)

Cyclization to Pyrrolo(4,3,2-*de*]phthalazines

6-Ethoxycarbonyl-5,7-dimethyl-8-nitro-4-oxo-3,4-dihydro-1-phthalazinecarboxylic acid (**92**) underwent reduction to its 8-amino analog followed by spontaneous cyclization to afford ethyl 4,6-dimethyl-3,8-dioxo-2,3,7,8-tetrahydropyrrolo[4,3,2-*de*]phthalazine-5-carboxylate (**93**) (H_2, Pd/C, EtOH, AcOEt, 20°C, 65%).[425]

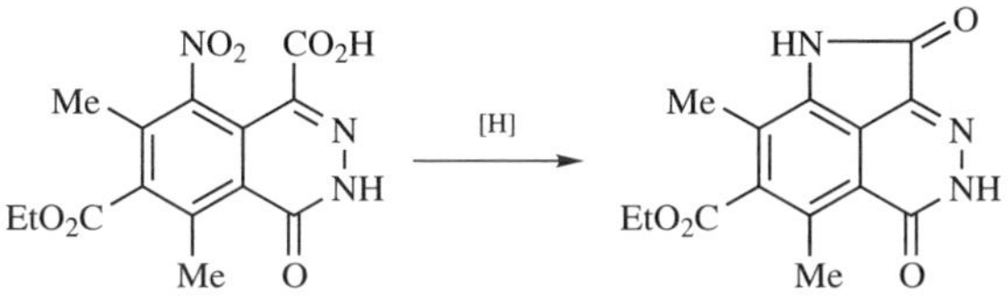

(92) (93)

Cyclization to [1,2,3]Triazino[1,2-*b*]phthalazines

2-Amino-1,4(2*H*,3*H*)-phthalazinedione (**94**) with isocrotonaldehyde gave 4-methyl-3,4,6,11-tetrahydro[1,2,3]triazino[1,2-*b*]phthalazine-6,11-dione (**95**) (see original for details)[266] or with acetylacetone gave initially 2-(1-acetonylethylideneamino)-1,4(2*H*,3*H*)-phthalazinedione (**96**) (Ac_2CH_2, AcOH) and thence 2,4-dimethyl-6,11-dihydro[1,2,3]triazino[1,2-*b*]phthalazine-6,11-dione (**97**) (polyphosphoric acid: for details, see original).[255]

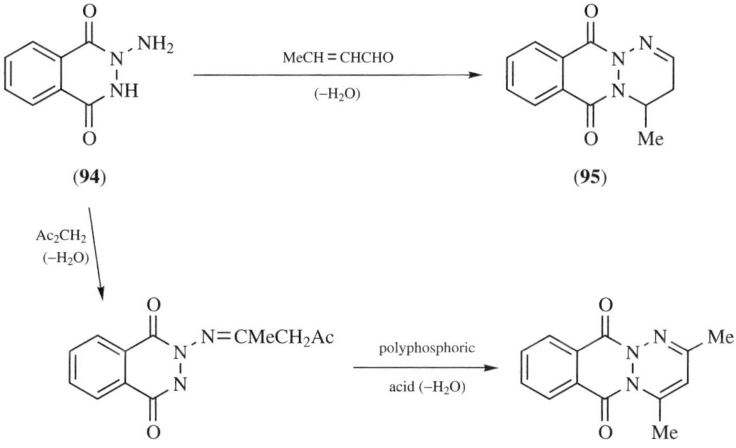

(94) (95)

(96) (97)

Also other examples.[239]

Extranuclear Cyclizations

1-[α-Ethoxycarbonyl-α-(hydroxyamino)methyl]-4-phenylphthalazine (**98**) with hydrazine hydrate gave 1-(5-oxo-1,2,3-triazolidin-4-yl)-4-phenylphthalazine (**99**) (neat reactants, fused, 1 h: 60%).[664]

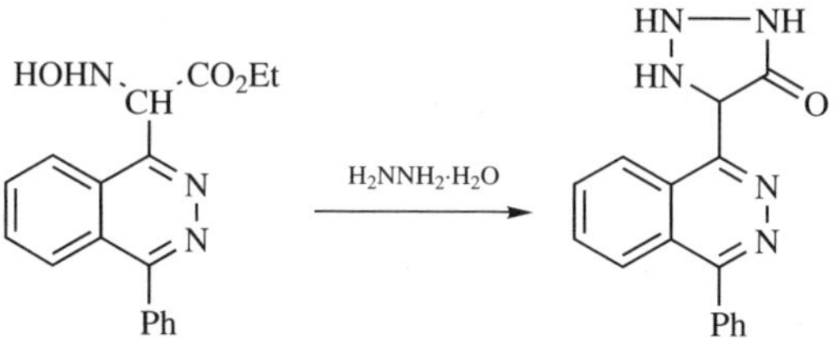

HOHN, CH, CO_2Et, N, N, Ph — $H_2NNH_2 \cdot H_2O$ → HN—NH, HN, O, N, N, Ph

(**98**) (**99**)

Also other examples.[1039]

13.3. HYDRAZINOPHTHALAZINES (*E* 596)

Probably because of their marked antihypertensive,[191,242,249,661,1008] antimicrobial,[259,501,675] and other bioactivities,[957] hydrazinophthalazines and their derivatives have a considerable literature.

The X-ray crystal structure of *N,N′*-di(phthalazin-1-yl)hydrazine (**100**) has been reported,[900] and the fine structures of 1-chloro-4-hydrazinophthalazine (**101**, R = Cl),[78] 1,4-dihydrazinophthalazine (**101**, R = $NHNH_2$),[895] and some derivatives thereof have been studied.

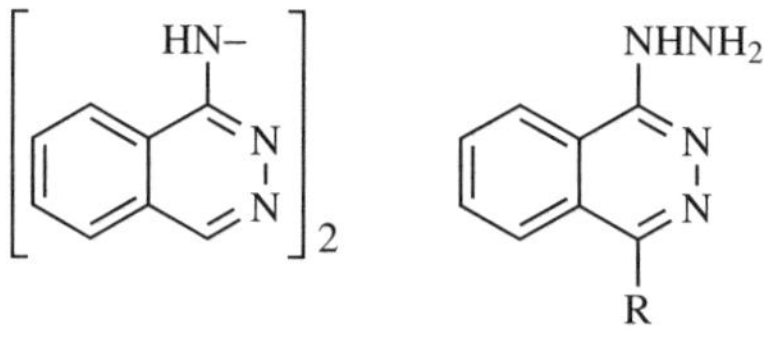

[HN–, N, N]$_2$ — $NHNH_2$, N, N, R

(**100**) (**101**)

13.3.1. Preparation of Hydrazinophthalazines

Various preparative routes to hydrazinophthalazines have been covered already: by *primary synthesis* (Chapter 8), by *hydrazinolysis of halogenophthalazines* (Section 10.3.2), by *hydrazinolysis of alkylthiophthalazines* (Section 12.2), and by *hydrazinolysis of phthalazinamines* (Section 13.2.2).

No other methods appear to have been used in the more recent literature.

13.3.2. Reactions of Hydrazinophthalazines

Some reactions of hydrazinophthalazines have been covered already: *oxidative dehydrazination* (Section 9.1.1), *conversion into alkoxyphthalazines* (Section 11.4), and *reductive conversion into phthalazinamines* (Section 13.2.1). The remaining reactions are summarized in the following subsections.

13.3.2.1. Conversion into Arylazo- or Azidophthalazines

Only *N*′-arylhydrazinophthalazines may be oxidized to the corresponding arylazophthalazines, and only (unsubstituted-hydrazino)phthalazines react with nitrous acid to give azidophthalazines. These reactions are illustrated by the following examples.

1-[*N*′-(2,4-Dinitrophenyl)hydrazino]phthalazine (**102**) underwent *oxidation* to 1-(2,4-dinitrophenylazo)phthalazine (**103**) (yellow HgO, THF, 20°C, 8 h: 70%; or 14% HNO_3, trituration, 20°C, 25 min: 75%); several analogs likewise.[733] It is interesting that the foregoing reaction was reversed by reduction (not displacement) on treatment of the product (**103**) with either 2,4-dinitro- or *p*-nitrophenylhydrazine (EtOH, reflux, ~1 h: ~75%).[733]

$NHNHC_6H_3(NO)_2$–2,4   HgO or HNO_3 → ; ← $H_2NNHC_6H_3(NO_2)_2$–2,4 or $H_2NNHC_6H_4NO_2$-*p* $N{=}NC_6H_3(NO_2)_2$–2,4

(**102**) (**103**)

1-Hydrazino-4-methylphthalazine (**104**) gave 1-azido-4-methylphthalazine (**105**) (substrate, H_3PO_4; $NaNO_2$ in $H_2O\downarrow$, 20°C, ? min: 85%).[33] A similar reaction of 1-hydrazinophthalazine forms the basis for a fluorometric assay of NO_2 in air.[1028]

$NHNH_2$, Me HNO_2 → N_3, Me

(**104**) (**105**)

13.3.2.2. N′-Acylation, Arylation, or Carbamoylation and Subsequent Reactions

These seldom used reactions are illustrated by the following classified examples.

N′-Acylation

1-Benzyl-2-hydrazinophthalazine (**106**) in acetic acid or acetic anhydride gave 1-(*N*′-acetylhydrazino)-4-benzylphthalazine (**107**) (neat reactants, reflux, 5 h: 80%), which underwent dehydrative cyclization to 6-benzyl-3-methyl-1,2,4-triazolo[3,4-*a*]phthalazine (**108**) ($POCl_3$, 90°C, 2 h: 35%); use of triethyl orthoacetate as the acylating agent gave the tricyclic product (**108**) directly (neat reactants, reflux, 8 h: 90%); homologs likewise.[528]

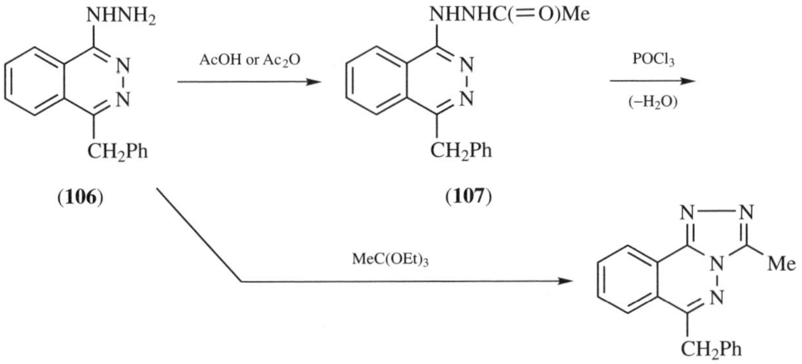

NHNH$_2$ / CH$_2$Ph (**106**) —AcOH or Ac$_2$O→ NHNHC(=O)Me / CH$_2$Ph (**107**) —$POCl_3$ (−H_2O)→ (**108**)

(**106**) —MeC(OEt)$_3$→ N—N, Me, CH$_2$Ph (**108**)

4-Hydrazino-2-phenyl-1(2*H*)-phthalazinone (**109**, R = H) gave 4-[*N*′-(chloroacetyl)hydrazino]-2-phenyl-1(2*H*)-phthalazinone [**109**, R = C(=O)CH$_2$Cl] (ClCH$_2$COCl, PhH: 70%, no other details).[675]

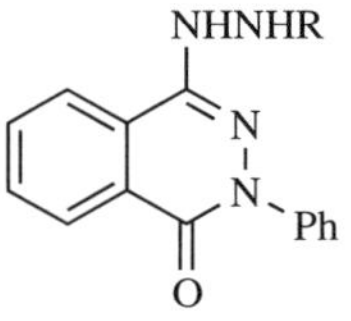

NHNHR / Ph / O

(**109**)

1-Hydrazino-4-phenylphthalazine (**110**, R = H) gave 1-(*N*′-benzenesulfonylhydrazino)-4-phenylphthalazine (**110**, R = SO$_2$Ph) (for details, see original).[259]

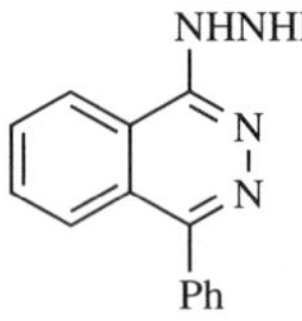

NHNHR / Ph

(**110**)

Also other examples.[276,472,645,996,1019]

N′-Arylation

1-Chloro-4-(*N*-methylhydrazino)phthalazine (**111**) with picryl chloride gave 1-chloro-4-(*N*-methyl-*N′*-picrylhydrazino)phthalazine (**112**) (reactants, dioxane, 20°C, 30 min: 49%).[733]

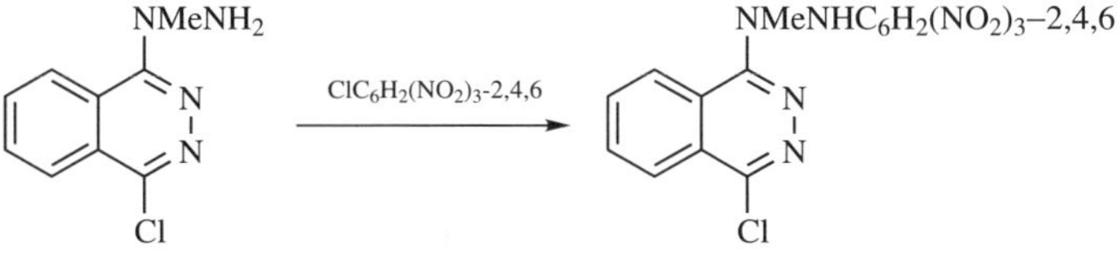

NMeNH$_2$ → $ClC_6H_2(NO_2)_3$-2,4,6 → NMeNHC$_6$H$_2$(NO$_2$)$_3$–2,4,6

(**111**) (**112**)

N′-Carbamolyation to Semicarbazidophthalazines

4-Hydrazino-1(2*H*)-phthalazinone gave 4-(4-methylsemicarbazido)-1(2*H*)-phthalazinone (**113**) (MeNCO: for details, see original).[342]

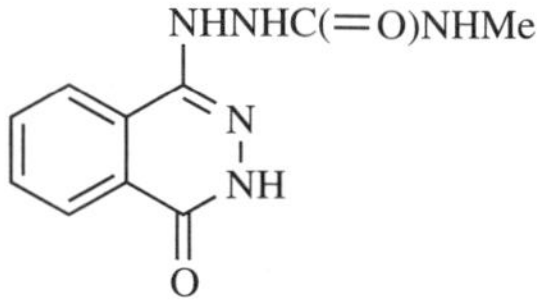

NHNHC(=O)NHMe

(**113**)

1,4-Dihydrazinophthalazine likewise gave 1,4-bis[4-methyl(thiosemicarbazido)]phthalazine (**114**) (MeNCS).[160]

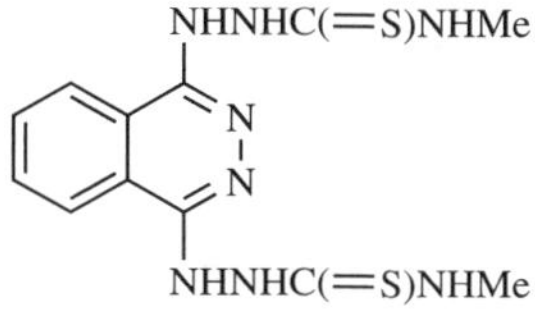

NHNHC(=S)NHMe

NHNHC(=S)NHMe

(**114**)

Also other examples.[180]

13.3.2.3. Alkylidenation and Subsequent Reactions

Many alkylidenehydrazinophthalazines have been isolated, often as intermediates for subsequent cyclization or further elaboration. Although such Schiff bases or hydrazones are usually formulated as such, there is now compelling evidence that at least some exist as alkylidenehydrazonodihydrophthalazines. Thus X-ray analysis and NMR spectra confirm that 1-[*N′*-(1-methoxycarbonylethylidene)hydrazino] phthalazine (**115**) exists entirely as the tautomeric 1-(1-methoxycarbonylethylidene) hydrazono-1,2-dihydrophthalazine (**116**), both as a solid and in solution;[837] nevertheless, to avoid confusion, all such Schiff bases mentioned here are named traditionally.

$NHN{=}CMeCO_2Me$ ⇌ $NN{=}CMeCO_2Me$

(**115**) (**116**)

These alkylidenations and some subsequent reactions are illustrated in the following examples.

1-Hydrazino-4-phenylphthalazine (**117**) with benzaldehyde gave 1-benzylidenehydrazino-4-phenylphthalazine (**118**) (EtOH, reflux, 2 h: 80%), which underwent oxidative cyclization to give 3,6-diphenyl-1,2,4-triazolo[3,4-*a*]phthalazine (**119**) ($PhNO_2$, reflux, 4 h: 71%).[661] Analogs of the Schiff base (**118**) were made similarly[661,678] and underwent other subsequent reactions; for example, reduction of 1-*m*-nitrobenzylidenehydrazino-4-phenylphthalazine (**120**, R = NO_2-*m*) gave 1-[*N*′-(*m*-acetamidobenzyl)hydrazino]-4-phenylphthalazine (**121**) (Zn, AcOH, reflux, 2 h: ?%) and 1-*p*-hydroxybenzylidenehydrazino-4-phenylphthalazine (**120**, R = OH-*p*) gave 1-[*N*-acetyl-*N*′-(*p*-hydroxybenzylidene)hydrazino]-4-phenylphthalazine (**122**) (AcCl, Me_2NCHO, reflux, 8 h: ?%).[678]

$NHNH_2$ —PhCHO→ $NHN{=}CHPh$ —[O]→ (triazolo, Ph)

(**117**) (**118**) (**119**)

$NHN{=}CHC_6H_4R$ —Zn, AcOH (R = NO_2-*m*)→ $NHNHCH_2C_6H_4NHAc$-*m* (**121**)

—AcCl (R = OH-*p*)→ $NAcN{=}CHC_6H_4OH$-*p* (**122**)

(**120**)

1-Hydrazino-4-methylphthalazine (**123**) with *p*-nitrobenzaldehyde gave either 1-(*p*-nitrobenzylidenehydrazino)-4-methylphthalazine (**124**) (EtOH, reflux, ? h: 85%) or 6-methyl-3-*p*-nitrophenyl-1,2,4-triazolo[3,4-*a*]phthalazine (**125**) [AcOH, reflux, ? h: 75%; presumably by aerial oxidative cyclization of the intermediate (**124**)].[33]

$NHNH_2$ (**123**) —p-$O_2NC_6H_4CHO$ in EtOH→ $NHN{=}CHC_6H_4NO_2$-p, Me (**124**) —?→ (**125**); (**123**) —AcOH [O]→ N—N, $C_6H_4NO_2$-p, Me (**125**)

1-Chloro-4-hydrazinophthalazine (**126**) with benzaldehyde gave 1-benzylidene-hydrazino-4-chlorophthalazine (**127**, R = H) (substrate, 0.1M HCl, 20°C; PhCHO↓ dropwise; 60°C, 10 min: 92%) and thence 6-chloro-3-phenyl-1,2,4-triazolo[3,4-*a*]phthalazine (**128**, R = H) (intermediate, AcONa, AcOH; Br_2 in AcOH↓ dropwise, 20°C, 1 h: 74%); analogs likewise, but in some cases oxidative cyclization was done differently; for example, 1-chloro-4-(*p*-dimethylaminobenzylidenehydrazino)phthalazine (**127**, R = NMe_2) gave 6-chloro-3-*p*-dimethylaminophenyl-1,2,4-triazolo[3,4-*a*]phthalazine (**128**, R = NMe_2) [$Pb(OAc)_4$, AcOH, 45°C → 20°C, 1 h: 49%].[276]

$NHNH_2$, Cl (**126**) —p-RC_6H_4CHO, HCl→ $NHN{=}CHC_6H_4R$-p, Cl (**127**) —[O]→ N—N, C_6H_4R-p, Cl (**128**)

1,4-Dihydrazinophthalazine with salicylaldehyde gave 1,4-bis(salicylidenehydrazino)phthalazine (**129**) (EtOH, H_2O, reflux, 1 h: 90%) that was used (with ancillary ligands) to make a variety of molybdenum complexes.[804]

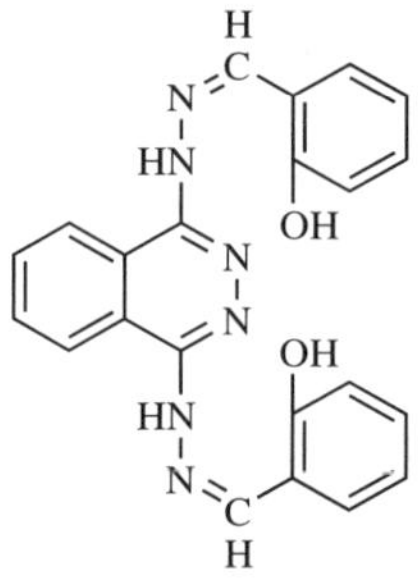

(**129**)

1-Chloro-4-(*N*-methylhydrazino)phthalazine (**130**) with formaldehyde gave 1-chloro-4-(*N*-methyl-*N*′-methylenehydrazino)phthalazine (**131**, Q = R = H) (substrate, EtOH, trace H_2SO_4; CH_2O gas↓, 20°C, 30 min: 82%), with acetaldehyde gave 1-chloro-4-(*N*′-ethylidene-*N*-methylhydrazino)phthalazine (**131**, Q = H, R = Me) (reactants, EtOH, trace H_2SO_4, 20°C, 5 min: 90%), with acetone gave 1-chloro-4-(*N*′-isopropylidene-*N*-methylhydrazino)phthalazine (**131**, Q = R = Me) (likewise, 20 min: 90%), and so on.[731,738]

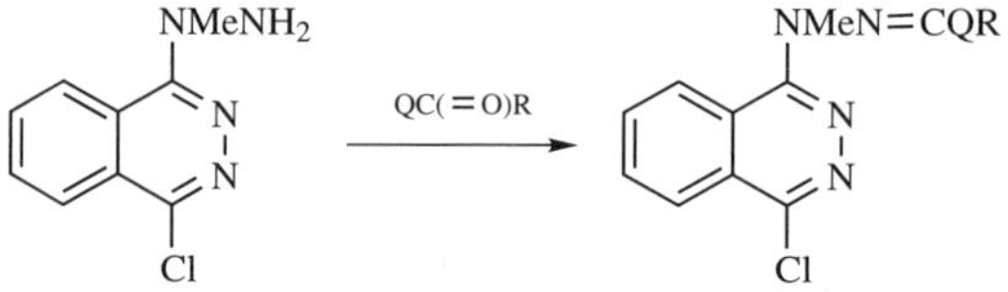

NMeNH2 → QC(=O)R → NMeN=CQR

(**130**) (**131**)

1-Benzylidenehydrazinophthalazine (**132**)[1008] with benzenediazonium chloride gave 1,3-diphenyl-5-(phthalazin-1-yl)formazan (**133**) (reactants, Me_2NCHO, 6M HCl, <0°C; 2M NaOH↓ dropwise, 2 h: 85%; analogs likewise).[732,cf. 440]

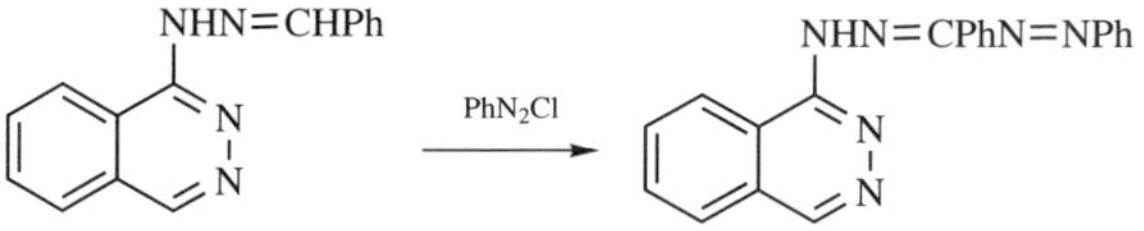

NHN=CHPh → PhN_2Cl → NHN=CPhN=NPh

(**132**) (**133**)

Also other examples.[246,249,259,501,675,840,901,957,1026]

13.3.2.4. Cyclization Reactions

Hydrazinophthalazines with dicarbonyl synthones usually give (unfused) heterocyclylphthalazines but with monocarbonyl synthons usually give (fused) heterocyclophthalazines.[cf. 580] Examples of the latter type have been given in Sections 13.3.2.2 and 13.3.2.3 when intermediates were isolated and subsequently cyclized. One-pot cyclizations of both types are illustrated by the following typical examples.

Formation of Pyrazolylphthalazines

1-Hydrazinophthalazine (**135**, R = H) with acetylacetone gave 1-(3,5-dimethylpyrazol-1-yl)phthalazine (**134**) (reactants, EtOH, reflux, 2 h: ~8%);[396,200] 1,4-dihydrazinophthalazine (**135**, R = $NHNH_2$) gave 1,4-bis(3,5-dimethylpyrazol-1-yl)phthalazine (**136**) (neat reactants, 95°C, 1 h: 80%; note improved yield);[581] and analogs were made similarly.[31,206,396,400]

(**134**) (**135**) (**136**)

Formation of Pyrrolylphthalazines

1-Hydrazinophthalazine (**137**) with acetonylacetone gave 1-[(2,5-dimethylpyrrol-1-yl)amino]phthalazine (**138**) (substrate, AcOH; synthon↓ slowly; 65°C; 3 h: 15%); analogs likewise.[271]

(**137**) (**138**)

Formation of 1,2,4-Triazolo[3,4-*a*]phthalazines

1-Chloro-4-hydrazino-5,6,7,8-tetrahydrophthalazine (**139**) and triethyl orthoformate gave 6-chloro-7,8,9,10-tetrahydro-1,2,4-triazolo[3,4-*a*]phthalazine (**140**) (neat reactants, reflux, 6 h: 72%).[106]

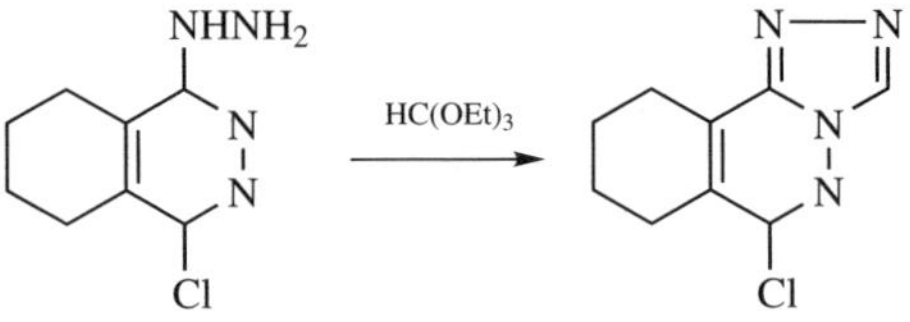

(139) (140)

1-Hydrazinophthalazine (**141**) with *o*-formylbenzoic acid (**142**) gave an unseparated mixture of two intermediates (substrate·HCl, synthon, H_2O, reflux, 30 min: 93%) that gave 3-(*o*-carboxyphenyl)-1,2,4-triazolo[3,4-*a*]phthalazine (**143**) (mixture, EtOH, 20°C, 2 weeks: ?%); mechanistic details are discussed.[25]

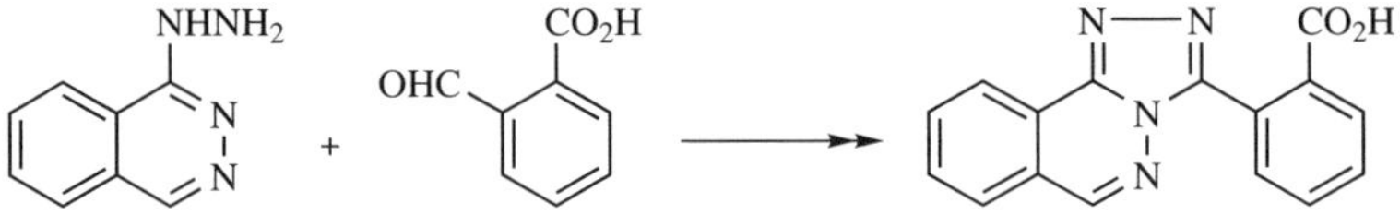

(141) (142) (143)

Also other examples.[350,392,580,1018,1019]

Formation of [1,2,4]Triazino[3,4-*a*]phthalazines

1-Hydrazinophthalazine (**144**) with ethyl 3-benzoylpyruvate gave 3-phenyl-4*H*-[1,2,4]triazino[3,4-*a*]phthalazin-4-one (**145**) (EtOH, reflux, 30 min: 32%); analogs likewise.[31,580]

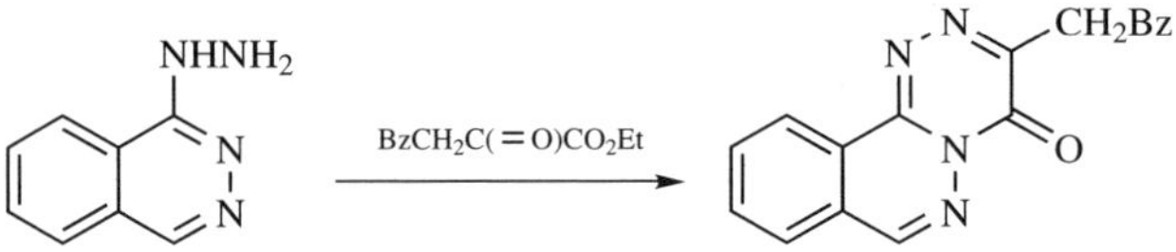

$BzCH_2C(=O)CO_2Et$

(144) (145)

CHAPTER 14

Phthalazinecarboxylic Acids and Related Derivatives (*E* 638)

This chapter covers nuclear and extranuclear phthalazinecarboxylic acids, their derivatifes, phthalazinecarbaldehydes, and the ketonic *C*-acylthalazines. To avoid repetition, the interconversions of these entities are discussed only at the first opportunity; for example, the conversion of esters into amides appears as a reaction of esters rather than as a preparative route to amides, simply because treatment of esters precedes that of amides.

14.1. PHTHALAZINECARBOXYLIC ACIDS

14.1.1. Preparation of Phthalazinecarboxylic Acids

The formation of nuclear or extranuclear phthalazinecarboxylic acids by *primary synthesis* (Chapter 8) or by *Reissert-like reactions* (Section 9.1.3) has been covered. Classical *oxidative approaches* from alkyl- or hydroxyalkylphthalazines do not appear to have been used in the 1972–2004 period. Other routes are illustrated by the following classified examples.

By Hydrolysis of Phthalazinecarboxylic Esters

Note: This hydrolysis can be done directly in alkaline or acidic media or indirectly according to the sensitivity of passenger groups attached to the substrate.

Methyl 4-oxo-3-(3-phenylpropyl)-3,4-dihydro-5-phthalazinecarboxylate (**1**, R = Me) gave 4-oxo-3-(3-phenylpropyl)-3,4-dihydro-5-phthalazinecarboxylic acid (**1**, R = H) (2M NaOH, reflux, 1 h: ~60%).[623]

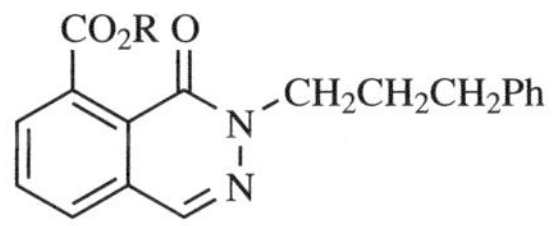

CO_2R O

N $CH_2CH_2CH_2Ph$

N

(**1**)

Cinnolines and Phthalazines: Supplement II, The Chemistry of Heterocyclic Compounds, Volume 64, by D.J. Brown

Ethyl 5-amino-6-cyano-4-oxo-3,7-diphenyl-3,4-dihydro-1-phthalazinecarboxylate (**2**, R = Et) gave 5-amino-6-cyano-4-oxo-3,7-diphenyl-3,4-dihydro-1-phthalazinecarboxylic acid (**2**, R = H) (NaOH, EtOH, reflux, 30 min: 75%).[489]

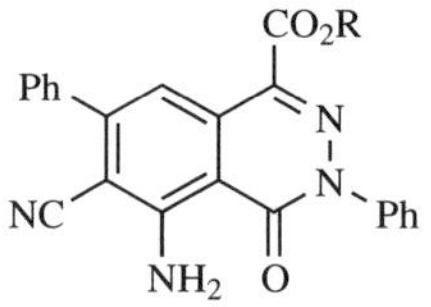

(**2**)

2-*o*-(Methoxycarbonylmethyl)phenyl-1,4(2*H*, 3*H*)-phthalazinedione (**3**, R = Me) gave 2-*o*-(carboxymethyl)phenyl-1,4(2*H*, 3*H*)-phthalazinedione (**3**, R = H) (KOH, MeOH, H_2O, reflux, 1 h: 90%).[6]

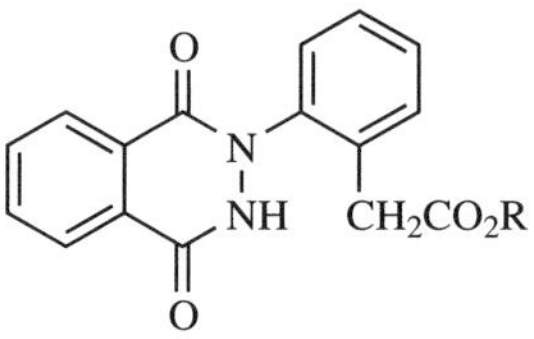

(**3**)

2-Ethoxycarbonylmethyl-4-methyl-1(2*H*)-phthalazinone (**4**, R = Et) gave 2-carboxymethyl-4-methyl-1(2*H*)-phthalazinone (**4**, R = H) 2M HCl, reflux, 2 h: 80%).[649]

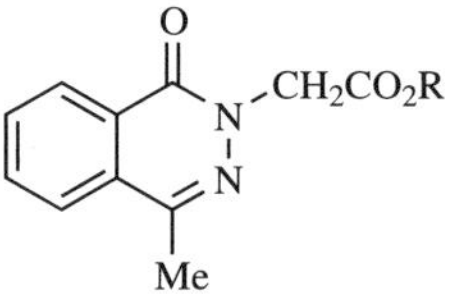

(**4**)

1-(4-*tert*-Butoxycarbonylpiperidino)- (**5**, R = Bu^t) gave 1-(4-carboxypiperidino)-4-[(3-chloro-4-methoxybenzyl)amino]-6-phthalazinecarboitrile (**5**, R = H) (neat HCO_2H, 20°C, 20 h: >95% or 82% as hydrochloride).[871]

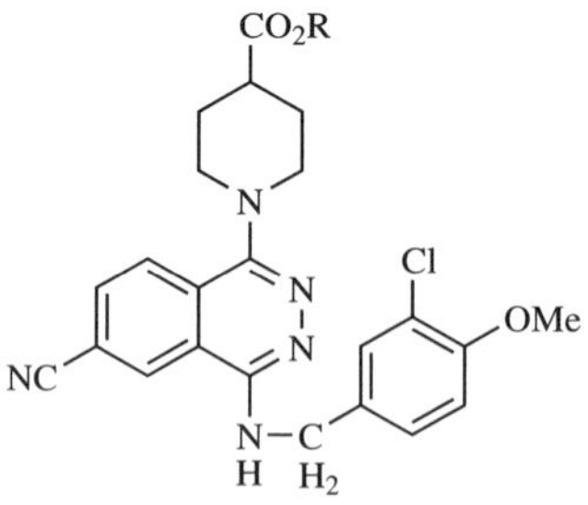

(**5**)

4-(p-Methoxybenzyloxycarbonyl)methyl- (**6**, R = $CH_2C_6H_4OMe$-p) gave 4-carboxymethyl-2-(5-trifluoromethylbenzoxazol-2-yl)methyl-1(2H)-phthalazinone (**6**, R = H) (substrate, BBr_3, CH_2Cl_2, ~−70°C, 30 min; then →20°C, ice/water↓: 75%).[108]

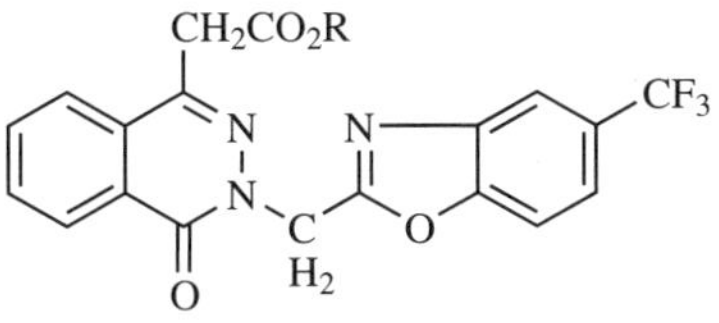

(**6**)

Also other examples.[243,279,584,784,947,958,985]

By Hydrolysis of Phthalazinecarboxamides

4-[α-Benzamido-α-(hydrazinocarbonyl)methyl]-1(2H)-phthalazinone (**7**, R = $NHNH_2$) gave 4-(α-benzamido-α-carboxymethyl)-1(2H)-phthalazinone (**7**, R = OH) (1M KOH: for details, see original).[159]

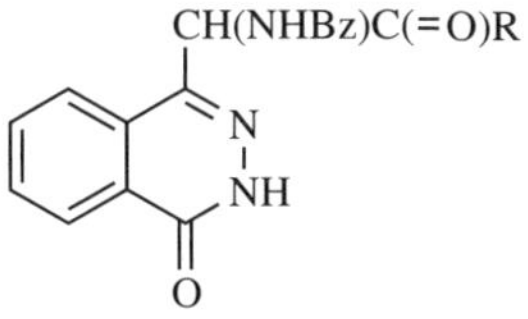

(**7**)

By Hydrolysis of Phthalazinecarbonitriles

4-Oxo-3-phenyl-3,4-dihydro-6-phthalazinecarbonitrile (**8**) gave 4-oxo-3-phenyl-3,4-dihydro-6-phthalazinecarboxylic acid (**9**) (NaOH, H_2O, EtOH, reflux, 3 h: 91%).[411]

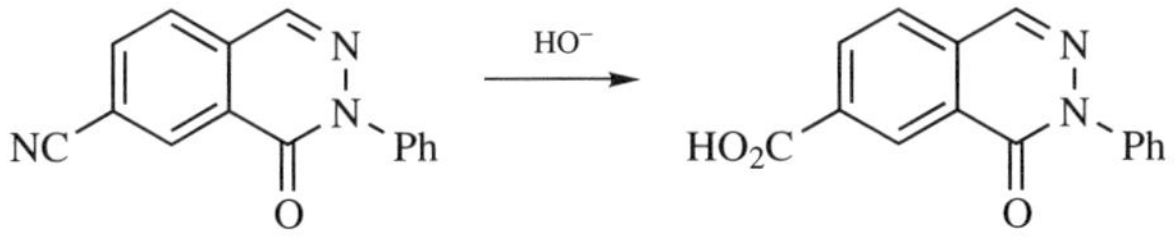

(**8**) (**9**)

Ethyl 1-cyano-4-methoxy-5,7-dimethyl-6-phthalazinecarboxylate (**10**) underwent selective hydrolysis to 6-ethoxycarbonyl-4-methoxy-5,7-dimethyl-1-phthalazinecarboxylic acid (**11**) [$Ba(OH)_2$, EtOH, H_2O, reflux, 6 h: 78%].[783]

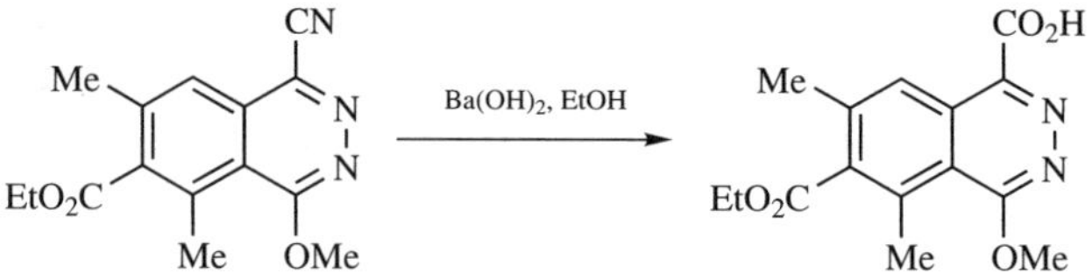

(10) $\xrightarrow{Ba(OH)_2,\ EtOH}$ (11)

Also other examples.[871]

From Diazoacetylphthalazines by the Arndt–Eistert Reaction

Ethyl 1-diazoacetyl-5,7-dimethyl-4-oxo-3-phenyl-3,4-dihydro-6-phthalazinecarboxylate (**12**) underwent an Arndt–Eistert reaction[1006] to give ethyl 1-carboxymethyl-5,7-dimethyl-4-oxo-3-phenyl-3,4-dihydro-6-phthalazinecarboxylate (**13**) [colloidal Ag, H_2O, 65°C; substrate in dioxane↓ dropwise; then 65°C (?), 2 h: 66%].[956]

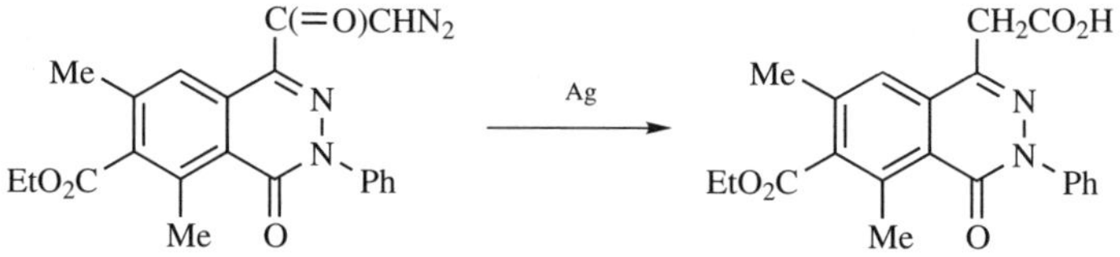

(12) $\xrightarrow{Ag}$ (13)

14.1.2. Reactions of Phthalazinecarboxylic Acids

The various reactions of phthalazinecarboxylic acids are illustrated by the following classified examples.

Decarboxylation

6-Ethoxycarbonyl-5,7-dimethyl-4-oxo-3,4-dihydro-1-phthalazinecarboxylic acid (**14**) underwent thermal decarboxylation to give ethyl 5,7-dimethyl-4-oxo-3,4-dihydro-6-phthalazinecarboxylate (**15**) (neat substrate, 190°C, 15 min: ~60%).[405]

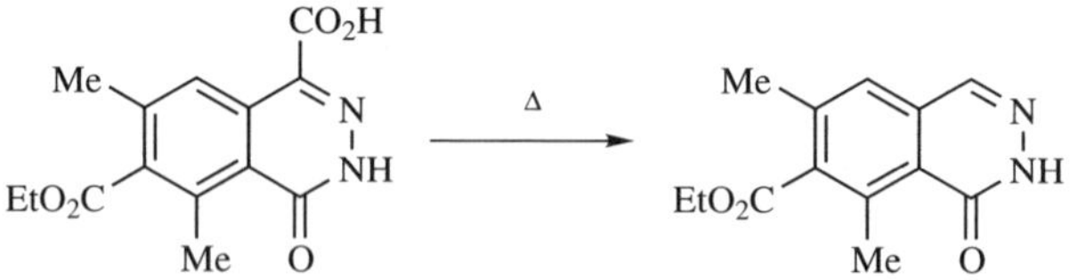

(14) $\xrightarrow{\Delta}$ (15)

Also other examples.[159,926]

Formation of Anhydrides

Note: Although linear phthalazinecarboxylic anhydrides appear to be unrepresented, phthalazinedicarboxylic anhydrides have been made.

6,7-Phthalazinedicarboxylic acid (**16**) with acetic anhydride gave 6,7-phthalazinedicarboxylic anhydride (**17**) (neat reactants, reflux, 2 h: 78%).[584]

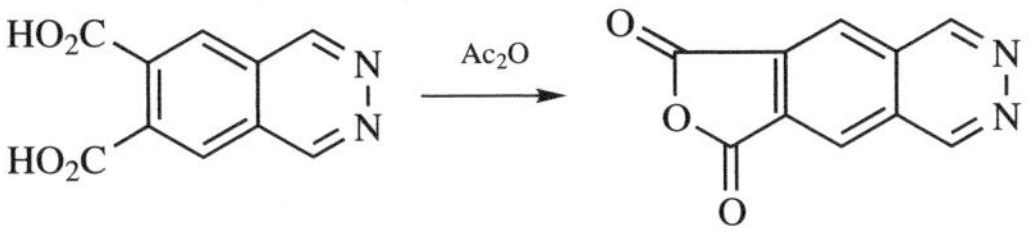

HO_2C HO_2C N N $\xrightarrow{Ac_2O}$ O O O N N

(**16**) (**17**)

Formation of Phthalazinecarbonyl Halides

3-*p*-Chlorophenyl-4-oxo-3,4-dihydro-1-phthalazinecarboxylic acid (**18**, R = OH) gave 3-*p*-chlorophenyl-4-oxo-3,4-dihydro-1-phthalazinecarbonyl chloride (**18**, R = Cl) (neat $SOCl_2$, reflux, 30 min: 92%).[339]

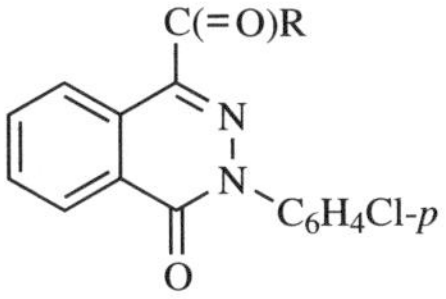

C(=O)R N N C_6H_4Cl-*p* O

(**18**)

4-Oxo-3,4-dihydro-5-phthalazinecarboxylic acid (**19**, R = OH) gave 4-oxo-3,4-dihydro-5-phthalazinecarbonyl chloride (**19**, R = Cl) (neat $SOCl_2$, reflux, 2.5 h: 88%;[622] or $SOCl_2$, PhCl, reflux, 4 h: >75%).[623]

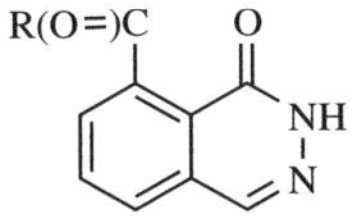

R(O=)C O NH N

(**19**)

2-(3-Carboxypropyl)-4-(pyridin-3-yl)-1(2*H*)-phthalazinone (**20**, R = OH) gave 2-[3-(chloroformyl)propyl]-4-(pyridin-3-yl)-l(2*H*)-phthalazinone (**20**, R = Cl) (substrate, Et_3N, CH_2Cl_2, 0°C; $PCl_5\downarrow$; 0°C, 15 min: >12%; note survival of the oxo substituent under these gentle conditions).[279]

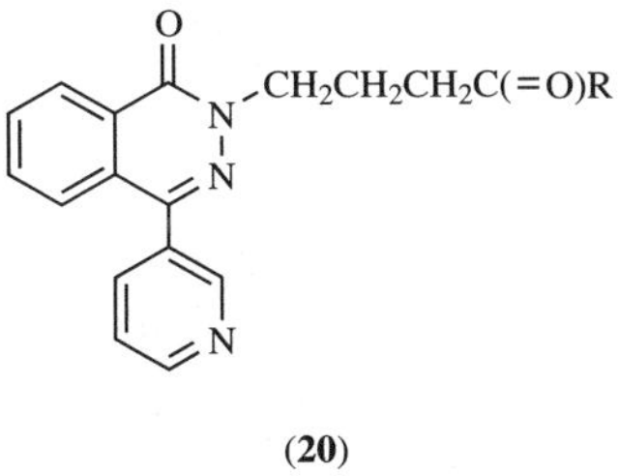

O N $CH_2CH_2CH_2C(=O)R$ N N

(**20**)

Also other examples.[411,956]

Esterification

Note: Esterification may be done indirectly via phthalazinecarbonyl halides or directly (as illustrated here) by treatment with alcoholic sulfuric acid, a diazoalkane, or some other alkylating agent.

1-(5-Ethylthien-2-yl)-4-oxo-3,4-dihydro-6-phthalazinecarboxylic acid (**21**, R = H) gave ethyl 1-(5-ethylthien-2-yl)-4-oxo-3,4-dihydro-6-phthalazinecarboxylate (**21**, R = Et) (EtOH, H_2SO_4, reflux, 12 h: 88%).[430]

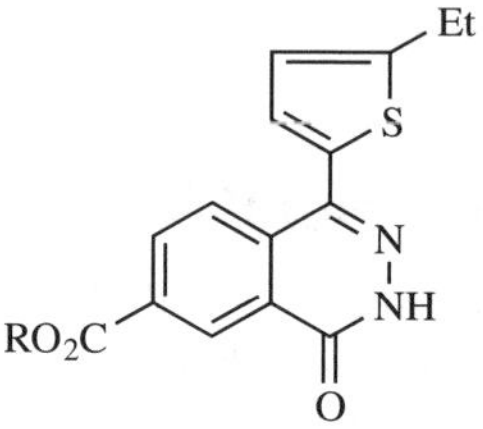

(**21**)

4-Oxo-3,4-dihydro-1,7-phthalazinedicarboxylic acid (**22**, R = H) gave diethyl 4-oxo-3,4-dihydro-1,7-phthalazinedicarboxylate (**22**, R = Et) (EtOH, H_2SO_4, reflux, 20 h: ~80%).[404]

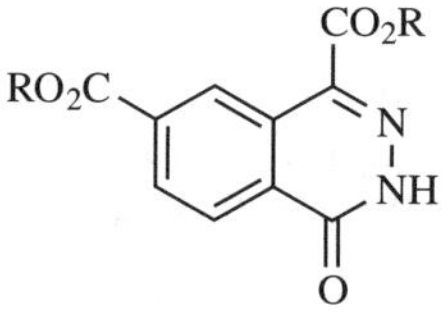

(**22**)

4-Oxo-3,4-dihydro-5-phthalazinecarboxylic acid (**23**, R = H) gave methyl 4-oxo-3,4-dihydro-5-phthalazinecarboxylate (**23**, R = Me) (CH_2N_2, Et_2O, MeOH, ?°C, ? h: 82%).[622]

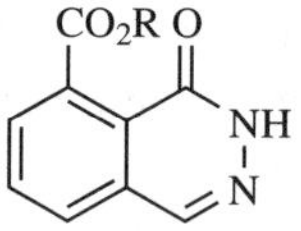

(**23**)

1-Hydroxymethyl-5,7-dimethyl-4-oxo-3,4-dihydro-6-phthalazinecarboxylic acid (**24**, R = H) gave methyl 1-hydroxymethyl-5,7-dimethyl-4-oxo-3,4-dihydro-6-phthalazinecarboxylate (**24**, R = Me) (CH_2N_2, Et_2O, 20°C: ~45%) or the corresponding isopropyl ester (**23**, R = Pr^i) (substrate, Et_2O; Me_2CN_2 in $Et_2O\downarrow$ dropwise, 20°C; 2 h: ~50%).[405]

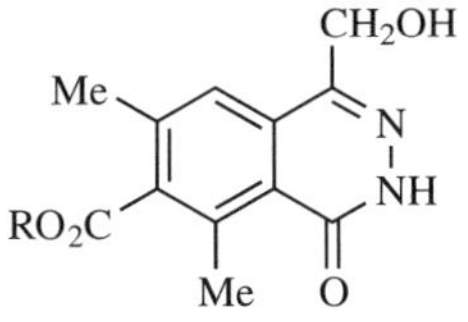

(**24**)

4-Carboxymethyl-1(2*H*)-phthalazinone (**25**) with *p*-methoxybenzyl chloride gave 4-*p*-methoxybenzyloxycarbonylmethyl-1(2*H*)-phthalazinone (**25a**) (substrate, NaI, Et_3N, Me_2NCHO; synthon↓, 20°C, 3 h: 42%).[108]

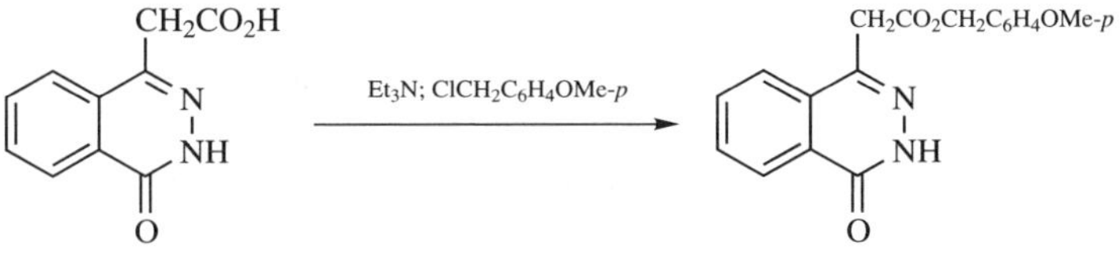

Et_3N; $ClCH_2C_6H_4OMe$-*p*

(**25**) (**25a**)

Also other examples.[574,622,783,921,926,956]

Conversion into Amides

Note: This is usually done indirectly via acyl halides or esters, but direct conversion is possible as exemplified here.

4-(3-Chloro-4-methoxybenzyl)amino-1-(4-hydroxypiperidino)-6-phthalazinecarboxylic acid (**26**, R = OH) gave 4-(3-chloro-4-methoxybenzyl) amino-1-(4-hydroxypiperidino)-*N*,*N*-dimethyl-6-phthalazinecarboxamide (**26**, R = NMe_2) (substrate, $Me_2NH \cdot HCl$, *N*,*N*′-dicyclohexylcarbodiimide, 1-hydroxybenzotriazole, Et_3N, AcMe, H_2O, 60°C, 8 h: 82%).[871]

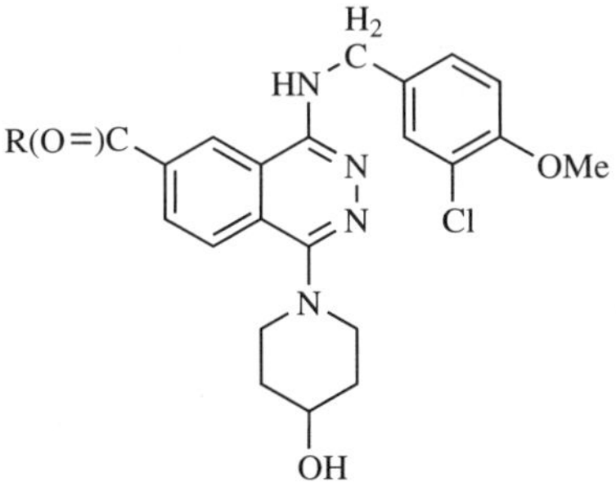

(**26**)

2-Carboxymethyl-1(2*H*)-phthalazinone (**26a**, R = OH) with ethyl chloroformate gave the unisolated intermediate (**26a**, R = OCO_2Et) and thence with

1-phenylpiperazine, the amide, 2-(4-phenylpiperazin-1-yl-carbonylmethyl)-1(2*H*)-phthalazinone [**26a**, R = $N(CH_2CH_2)_2NPh$] (substrate, Et_3N, $ClCO_2Et$, CH_2Cl_2, 0°C, 15 min; then amine↓, 0°C→25°C, 24 h: 60%); analogs likewise.[1032]

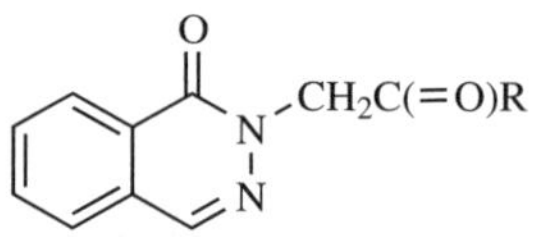

(**26a**)

Cyclizations

Note: Most types of cyclization involving carboxy and other substituents have been exemplified in previous chapters.

1-[1-Carboxy-2-(*p*-methylbenzoyl)ethyl]thio-5,6,7,8-tetrachloro-4-(3,4-dimethylphenyl)phthalazine (**27**) with hydrazine hydrate gave 5,6,7,8-tetrachloro-1-(3,4-dimethylphenyl)-4-(3-oxo-6-*p*-tolyl-2,3,4,5-tetrahydropyridazin-4-ylthio) phthalazine (**28**) (substrate, Et_2), hot; synthon↓ portionwise; reflux, 6 h: 70%).[671]

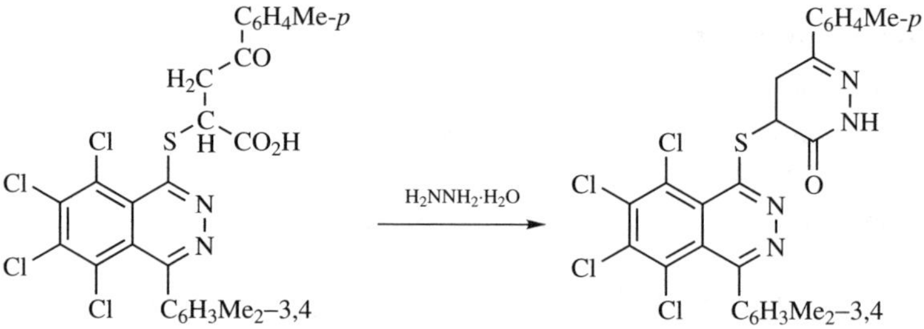

(**27**) $\xrightarrow{H_2NNH_2 \cdot H_2O}$ (**28**)

4-Oxo-3,4-dihydro-5-phthalazinecarboxylic acid (**29**) with hydrazine hydrate gave 1*H*-pyridazino[3,4,5-*de*]phthalazin-3(2*H*)-one (**30**) (reactants, H_2O, reflux, 84 h: 65%).[622]

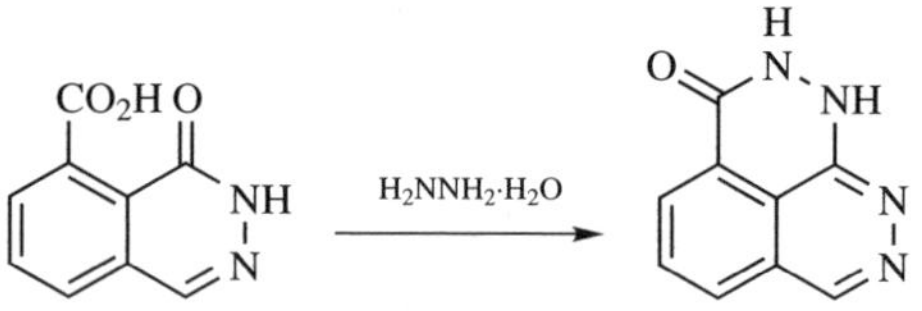

(**29**) $\xrightarrow{H_2NNH_2 \cdot H_2O}$ (**30**)

4-*o*-Carboxybenzyl-1(2*H*)-phthalazinone (**31**) gave 8*H*-isoquino[3,2-*a*]phthalazine-5,8(6*H*)-dione (**32**) (neat Ac_2O, reflux, 5 min: 95%).[926]

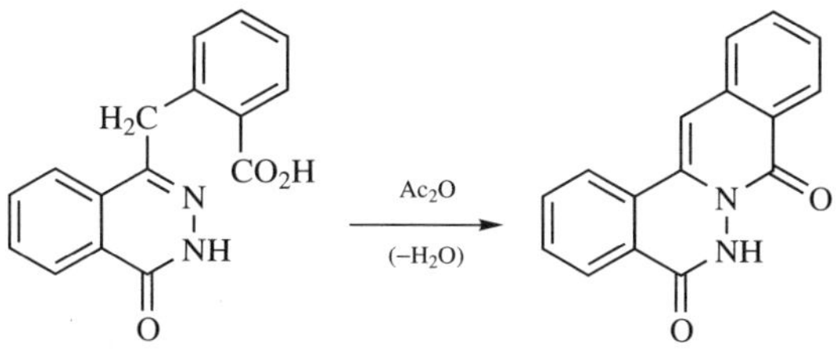

(31) (32)

14.2. PHTHALAZINECARBONYL HALIDES

Like most acyl halides, these useful intermediates are moisture-sensitive and unsuitable for storage. Accordingly, they are usually prepared and then used immediately as crude intermediates. However, it is quite possible to purify and characterize such halides with the obvious precautions. Their chemistry is illustrated by the following classified examples.

Preparation

Note: Phthalazinecarbonyl halides may be made by *Reissert reactions* (see Section 9.1.3) but usually from *phthalazinecarboxylic acids with thionyl chloride or the like* (see Section 14.1.2).

Reactions

Note: The conversions of phthalazinecarbonyl halides into the corresponding *esters*, *amides*, or *diazoacetates* are illustrated here; also their *cyclizations*.

4-Oxo-3,4-dihydro-5-phthalazinecarbonyl chloride (**33**, R = Cl) [crude, from the acid (**31**, R = OH) with thionyl chloride] gave ethyl 4-oxo-3,4-dihydro-5-phthalazinecarboxylate (**33**, R = OEt) (EtOH, reflux, 18 h: >85%)[623] or methyl 4-oxo-3,4-dihydro-5-phthalazinecarboxylate (**33**, R = OMe) (MeOH, likewise: 97%).[622]

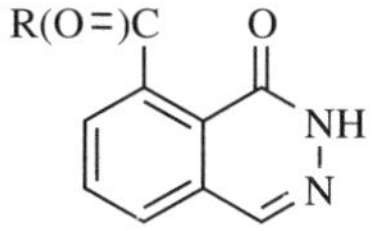

(33)

4-Oxo-3-phenyl-3,4-dihydro-6-phthalazinecarbonyl chloride (**34**, R = Cl) gave 4-oxo-3-phenyl-3,4-dihydro-6-phthalazinecarboxamide (**34**, R = NH_2) [crude substrate, CH_2Cl_2, NH_3, solvent (?), 20°C, ~4 h: 60%], *N*,*N*-dimethyl-4-oxo-3-phenyl-3,4-dihydro-6-phthalazinecarboxamide (**34**, R = NMe_2) (Me_2NH, likewise: 52%), or analogs.[411]

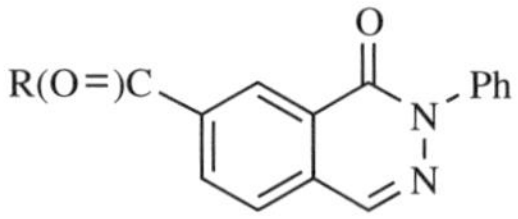

(**34**)

3-*p*-Chlorophenyl-4-oxo-3,4-dihydro-1-phthalazinecarbonyl chloride (**35**, R = Cl) gave 3-*p*-chlorophenyl-4-oxo-3,4-dihydro-1-phthalazinecarboxanilide (**35**, R = NHPh) ($PhNH_2$, EtOH, reflux, 2 h: 66%); analogs similarly using appropriate amines.[339]

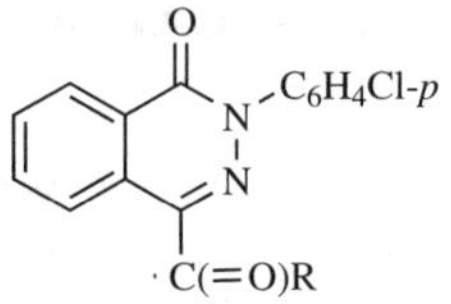

(**35**)

Ethyl 1-chloroformyl- (**36**, R = Cl) with ammonia gave ethyl 1-carbamoyl- (**36**, R = NH_2) (substrate, PhH; NH_3 in EtOH↓; 20°C, 5 days: 90%) or with diazomethane gave ethyl 1-diazoacetyl-5,7-dimethyl-4-oxo-3-phenyl-3,4-dihydro-6-phthalazinecarboxylate (**36**, R = CHN_2) (substrate, PhH, $<5°C$; CH_2N_2 in Et_2O↓ slowly; 20°C, 2 h: 91%).[956] The latter product was converted successively into ethyl 1-hydroxyacetyl- (**36**, R = CH_2OH) (H_2SO_4, dioxane, H_2O, 50°C, 90 min: 79%) and ethyl 1-(acetoxyacetyl)-5,7-dimethyl-4-oxo-3-phenyl-3,4-dihydro-6-phthalazinecarboxylate (**36**, R = CH_2OAc) (Ac_2O, PhH, pyridine, 20°C, 2 days: 65%).[956]

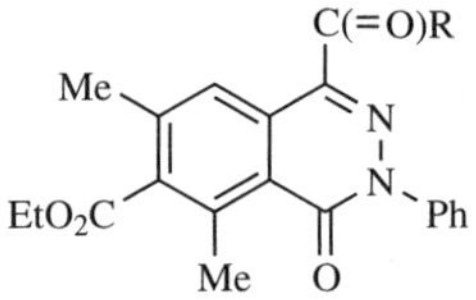

(**36**)

1-Cyano-1,2-dihydro-2-phthalazinecarbonyl chloride (**37**) underwent nitrosation (substrate, MeOH, −20°C; $NaNO_2$↓; then H_2O↓; 20°C, 12 h: crude solid) and subsequent cyclization (MeOH, 0°C; HCl/MeOH↓; 30 min) to afford 1-amino[1,2,3]oxadiazolo[4,3-*a*]phthalazin-4-ium chloride (**38**) (12%).[47]

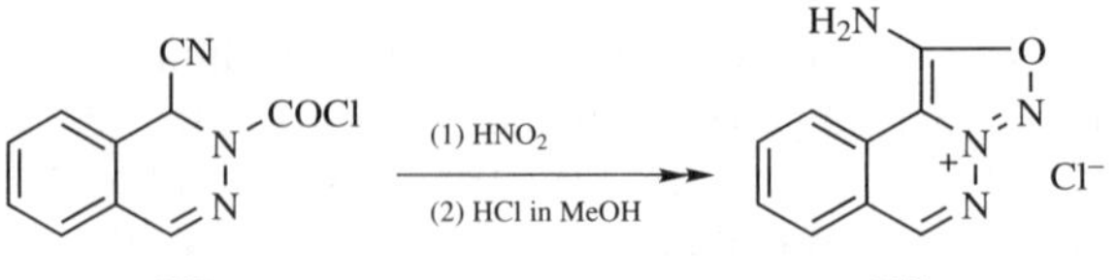

(1) HNO_2
(2) HCl in MeOH

(**37**) (**38**)

14.3. PHTHALAZINECARBOXYLIC ESTERS

14.3.1. Preparation of Phthalazinecarboxylic Esters

Several routes to such esters have been covered already: by *primary synthesis* (Chapter 8), by *Reissert or Reissert-like reactions* (Section 9.1.3), by *esterification of phthalazinecarboxylic acids* (Section 14.1.2), by *alcoholysis of phthalazinecarbonyl halides* (Section 14.2), or by a variety of *passenger introductions* (several chapters). Other minor preparative methods are illustrated in the following examples.

By Transesterification

Ethyl (**39**, R = H) gave 2-dimethylaminoethyl 1-hydroxymethyl-4-oxo-3,4-dihydro-6-phthalazinecarboxylate (**39**, R = NMe_2) (neat $Me_2NCH_2CH_2OH$, trace solid KOH, 115°C, 6 h: ~25%); analogs likewise.[404]

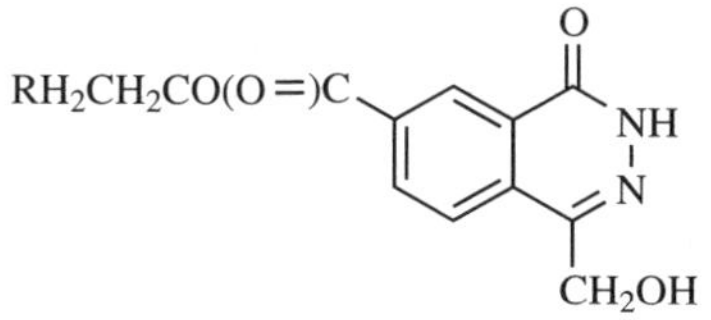

$RH_2CH_2CO(O{=})C$–[phthalazinone ring]–NH, N, CH_2OH

(**39**)

From Acylphthalazines

Note: This route may well be confined to the use of *N*-acylated hydrophthalazines as substrates; it is akin to transacylation of an amine.

2-Acetyl-4-*p*-methoxybenzyl-3-methyl-1,2,3,4,5,6,7,8-octahydrophthalazine (**40**, R = Me) gave ethyl 4-*p*-methoxybenzyl-3-methyl-1,2,3,4,5,6,7,8-octahydro-2-phthalazinecarboxylate (**40**, R = OEt) (substrate·HCl, $ClCO_2Et$, CH_2Cl_2, 20°C; $NaHCO_3$↓ portionwise during 1 h; then 30 min: 60%).[393]

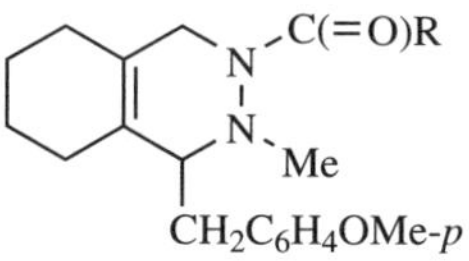

N–C(=O)R, N–Me, $CH_2C_6H_4OMe$-*p*

(**40**)

By Alkoxycarbonylation or the Like

Note: This procedure has been applied to the ring nitrogen of reduced phthalazines, to anionized phthalazines, or to Reissert derivatives of phthalazine.

tert-Butyl 1,2,3,4-tetrahydro-2-phthalazinecarboxylate (**41**) with *p*-nitrophenyl chloroformate gave *tert*-butyl *p*-nitrophenyl 1,2,3,4-tetrahydro-2,3-phthalazinedicarboxylate (**42**) (substrate, Et_3N, AcOEt, 0°C; synthon in AcOEt ↓ slowly; 45°C, 4 h: 69%).[310]

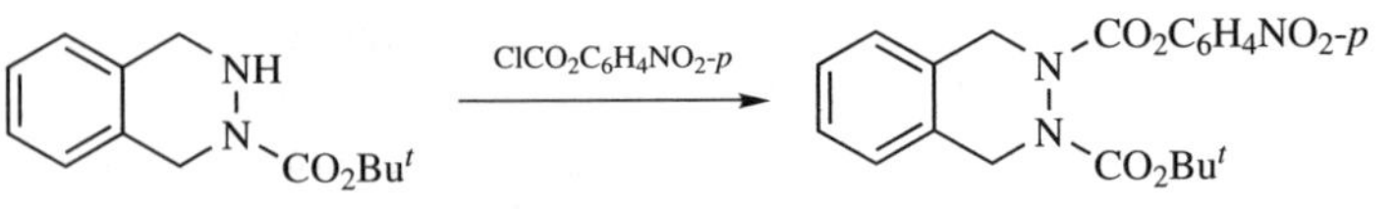

$ClCO_2C_6H_4NO_2$-*p*

(**41**) (**42**)

Conversion of 1,4-diphenylphthalazine into a solution of its dianion (**43**) (substrate, anhydrous THF, Na, A, 20°C, 8 h) and subsequent treatment with methyl chloroformate gave dimethyl 1,4-diphenyl-1,2-dihydro-1,2-phthalazinedicarboxylate (**44**) (dianion solution, −78°C; synthon↓; 20°C, ? min: 62%).[36]

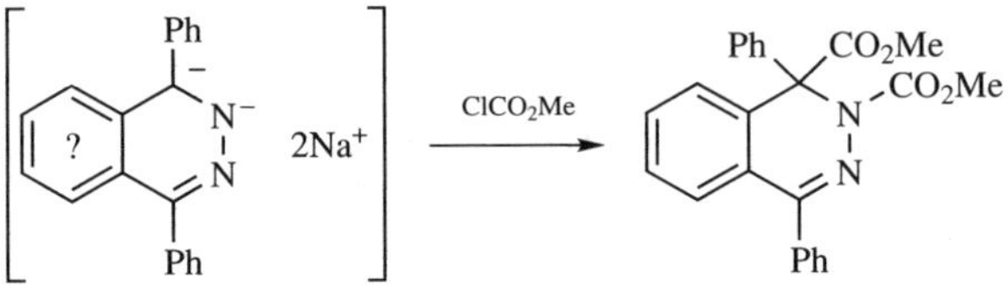

$ClCO_2Me$

(**43**) (**44**)

Phenyl 1-cyano-1,2-dihydro-2-phthalazinecarboxylate (**45**) gave phenyl 1-cyano-1-methylthio(thiocarbonyl)-1,2-dihydro-2-phthalazinecarboxylate (**46**) (substrate, Me_2NCHO; NaH↓, N_2; CS_2↓, 10 min; MeI↓, 20°C, 6 h: 30%).[374]

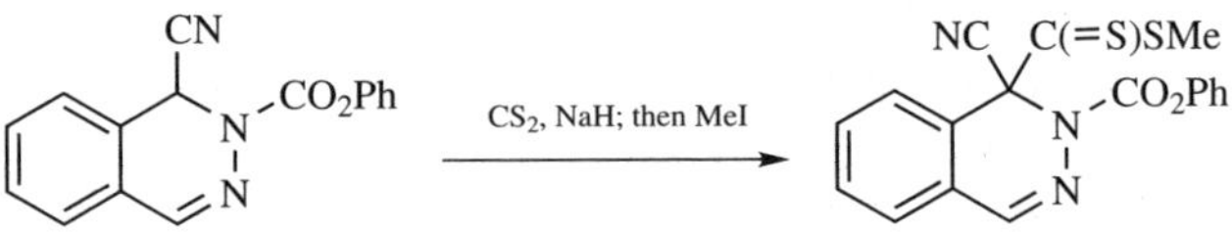

CS_2, NaH; then MeI

(**45**) (**46**)

14.3.2. Reactions of Phthalazinecarboxylic Esters

Several reactions of these esters have been covered already: *reduction to alkylphthalazines* (Section 9.2.1), *reduction to extranuclear hydroxyphthalazines* (Section 11.2.1), *conversion into extranuclear hydroxyphthalazines with Grignard reagents* (Section 11.2.1), *conversion into aminophthalazines by the Curtius reaction reaction* (Section 13.2.1), *hydrolysis to phthalazinecarboxylic acids* (Section 14.1.1), and *transesterification* (Section 14.3.1). Other reactions are illustrated in the following classified examples.

Removal of the Ester Grouping

Note: This is usually done by hydrolysis and decarboxylation of the product, often in one pot, but hydrogenation has been used to remove *N*-alkoxycarbonyl groups directly.

2-(3-Ethoxycarbonyl-2-imino-5-methyltetrahydrofuran-3-yl)-1,4(2*H*,3*H*)-phthalazinedione (**47**) gave 2-(5-methyl-2-oxotetrahydrofuran-3-yl)-1,4(2*H*,3*H*)-phthalazinedione (**48**) (3M NaOH, 20°C: 93%; note hydrolysis of the ester and imino groupings and decarboxylation).[126]

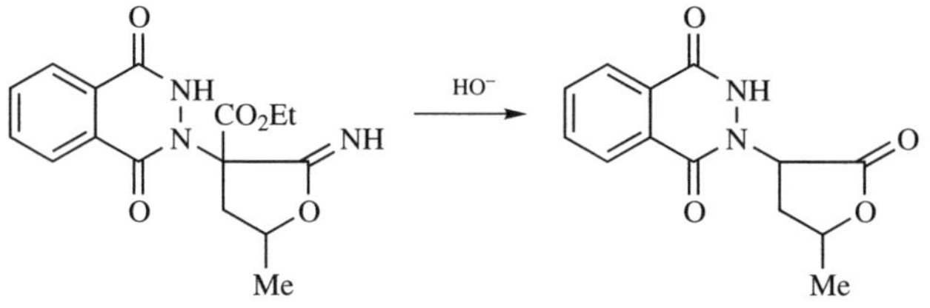

(**47**) (**48**)

Benzyl 3-carbamoyl-1,2,3,4-tetrahydro-2-phthalazinecarboxylate (**49**) gave 1,2,3,4-tetrahydro-2-phthalazinecarboxamide (**50**) (H_2, Pd/C, MeOH, 20°C, 1 h: 97%; conformation by X-ray analysis).[309]

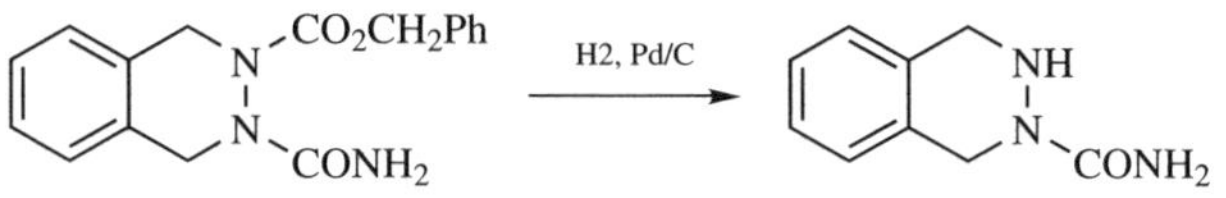

(**49**) (**50**)

Aminolysis

2-Ethoxycarbonylmethyl-4-methyl-1(2*H*)-phthalazinone (**51**, R = OEt) gave 2-carbamoylmethyl-4-methyl-1(2*H*)-phthalazinone (**51**, R = NH_2) (NH_4OH, reflux, 2 h: 50%) or 4-methyl-2-(phenylcarbamoyl)methyl-1(2*H*)-phthalazinone (**51**, R = NHPh) (neat $PhNH_2$, reflux, 2 h: 60%).[649]

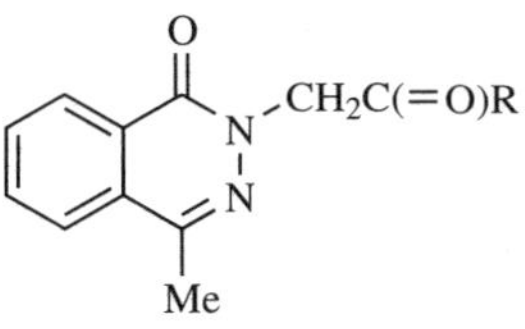

(**51**)

Ethyl 1-hydroxymethyl-4-oxo-3,4-dihydro-6-phthalazinecarboxylate (**52**, R = OEt) gave *N*-(2-dimethylaminoethyl)-1-hydroxymethyl-4-oxo-3,4-dihydro-6-phthalazinecarboxamide (**52**, R = $NHCH_2CH_2NMe_2$) (neat $H_2NCH_2CH_2NMe_2$, 90°C, 4 h: ~45%); analogs likewise.[404]

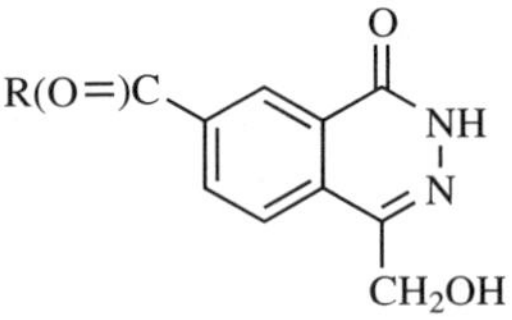

(**52**)

Phenyl 4-*p*-aminophenyl-6,7-dichloro-1,2-dihydro-2-phthalazinecarboxylate (**53**, R = OPh) gave 4-*p*-aminophenyl-6,7-dichloro-*N*-propyl-1,2-dihydro-2-phthalazinecarboxamide (**53**, R = NHPr) ($PrNH_2$, Me_2NCHO, 60°C, 3 h: 91%); analogs likewise.[848]

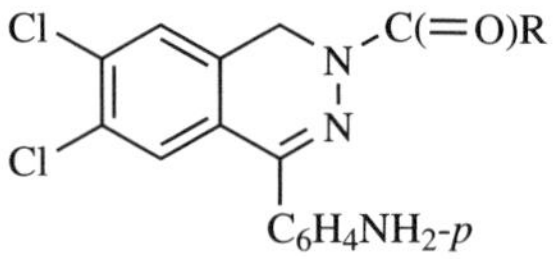

(**53**)

tert-Butyl *p*-nitrophenyl 1,2,3,4-tetrahydro-2,3-phthalazinedicarboxylate (**54**, R = $OC_6H_4NO_2$-*p*) gave *tert*-butyl 3-methylcarbamoyl-1,2,3,4-tetrahydro-2-phthalazinecarboxylate (**54**, R = NHMe) (substrate, 4-Me_2N-pyridine, Me_2NCHO; $MeNH_2$ in H_2O↓ dropwise, 20°C; addition repeated after 7 h and 28 h; then 20°C, 27 h: 79%; note selective aminolysis of the nitrophenyl ester grouping).[310]

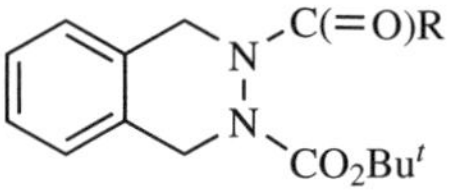

(**54**)

Hydrazinolysis

2-Ethoxycarbonylmethyl-4-methyl-1(2*H*)-phthalazinone (**55**, R = OEt) gave 2-hydrazinocarbonylmethyl-4-methyl-1(2*H*)-phthalazinone (**55**, R = $NHNH_2$) ($H_2NNH_2 \cdot H_2O$, EtOH, reflux, 3 h: 80%);[649] analogs likewise.[659,676]

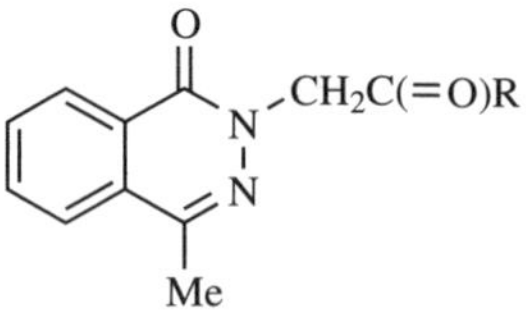

(**55**)

Ethyl 5-amino-6-cyano-4-oxo-7-phenyl-3-*p*-tolyl-3,4-dihydro-1-phthalazinecarboxylate (**56**, R = OEt) gave 5-amino-6-cyano-4-oxo-7-phenyl-3-*p*-tolyl-3,4-

dihydro-1-phthalazinecarbohydrazide (**56**, R = $NHNH_2$) ($H_2NNH_2 \cdot H_2O$, EtOH, reflux, 1 h: ?%).[530]

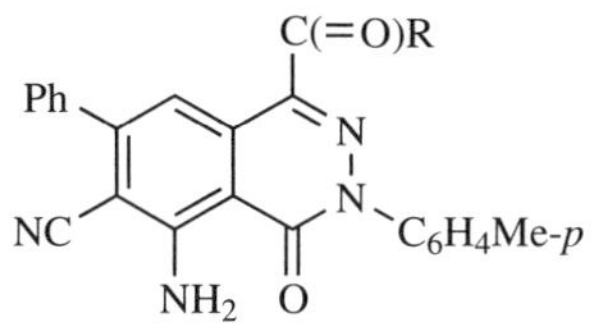

(**56**)

Cyclization Reactions

1-(α-Cyano-α-ethoxycarbonylmethyl)-4-phenylphthalazine (**57**) with guanidine hydrochloride gave 1-(2,4-diamino-6-oxo-1,6-dihydropyrimidin-5-yl)-4-phenylphthalazine (**58**) (reactants, EtOH, trace H_2 trace piperidine, reflux, 10 h; 70%); also analogous cyclizations.[664]

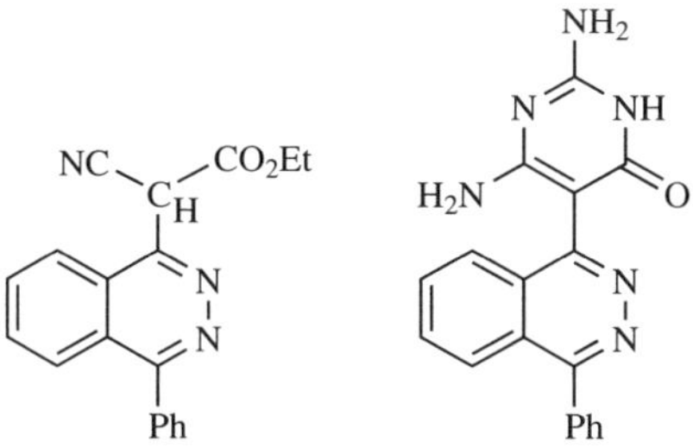

(**57**) (**58**)

Ethyl 5-bromomethyl-1-*o*-methoxyphenyl-7-methyl-4-oxo-3,4-dihydro-6-phthalazinecarboxylate (**59**) with ethanolic alkali gave 4-*o*-methoxyphenyl-6-methyl-7,9-dihydrofuro[3,4-*f*]phthalazine-1,7(2*H*)-dione (**60**, X = O) (KOH, EtOH, H_2O, 60°C, 2 h: ~75%) or with ammonia gave 4-*o*-methoxyphenyl-6-methyl-7,9-dihydro-8*H*-pyrrolo[3,4-*f*]phthalazine-1,7(2*H*)-dione (**60**, X = NH) (NH_3 in EtOH, 60°C, 2 h: ~65%).[426]

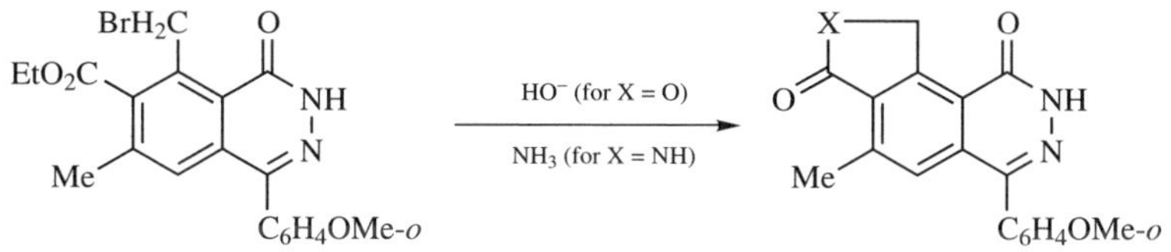

(**59**) (**60**)

Cyano ethoxycarbonyl (phthalazin-2-io)methanide (**61**) gave dimethyl 3-ethoxycarbonylpyrrolo[2,1-*a*]phthalazine-1,2-dicarboxylate (**62**) (for details and possible mechanism, see original).[161]

(61) $\xrightarrow[(-HCN)]{MeO_2CC\equiv CCO_2Me}$ (62)

14.4. PHTHALAZINECARBOXAMIDES AND PHTHALAZINECARBOHYDRAZIDES

In this section, the general term *phthalazine amides* is used to cover phthalazinecarboxamides, phthalazinecarbohydrazides, and their thio analogs.

14.4.1. Preparation of Phthalazine Amides

Most of the routes to such amides have been covered already: by *primary synthesis* (Chapter 8), by *aminolysis of phthalazinecarboxylic acids* (Section 14.1.2), by *aminolysis of phthalazinecarbonyl halides* (Section 14.2), by *aminolysis of phthalazinecarboxylic esters* (Section 14.3.2), and by *passenger introduction* (most chapters). One major and several minor preparative methods are illustrated in the following examples.

By Hydrolysis or Thiolysis of Phthalazinecarbonitriles

Note: These processes may be used to afford amides or thioamides but not the corresponding hydrazides. The hydrolysis of phthalazinecarbonitriles can give phthalazinecarboxylic acids unless conditions are controlled carefully or the Radziszewski procedure (H_2O_2, HO^-)[1009] is adopted.

1-Phthalazinecarbonitrile (**63**, R = H) gave 1-phthalazinecarboxamide (**64**, R = H) (K_2CO_3, H_2O_2, AcMe, 20°C, 12 h: 24%);[425] 4-methyl-1-phthalazinecarbonitrile (**63**, R = Me) gave 4-methyl-1-phthalazinecarboxamide (**64**, R = Me) (likewise but 2 h: 54%).[930]

(63) $\xrightarrow{K_2CO_3,\ H_2O_2}$ (64)

4-(3-Chloro-4-methoxybenzylamino)-1-(4-hydroxypiperidino)-6-phthalazinecarbonitrile (**65**, R = CN) gave the corresponding 6-phthalazinecarboxamide (**65**, R = $CONH_2$) (NaOH, EtOH, THF, H_2O, 20°C, 15 h: 44%).[871]

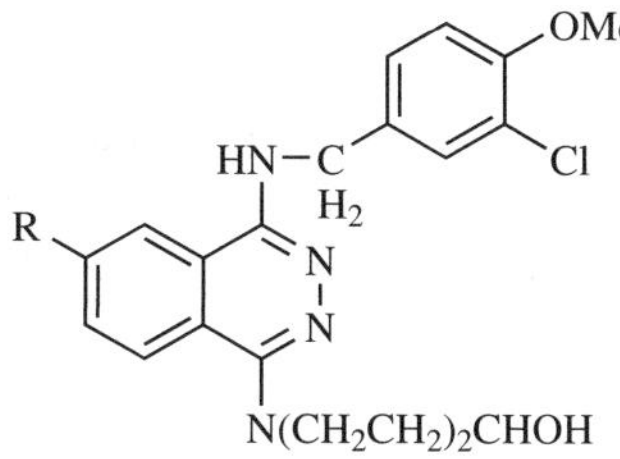

OMe
HN–C
H_2
Cl
R
N
N
$N(CH_2CH_2)_2CHOH$

(**65**)

2-(2-Cyanoethyl)-4-phenyl-1(2*H*)-phthalazinone (**66**) gave 4-phenyl-2-(2-thiocarbamoylethyl)-1(2*H*)-phthalazinone (**67**) (substrate, pyridine, Et_2N, <5°C, 2 h; H_2S↓, 0°C, 15 min; sealed, 20°C, 2 days: 89%); homologs likewise.[264]

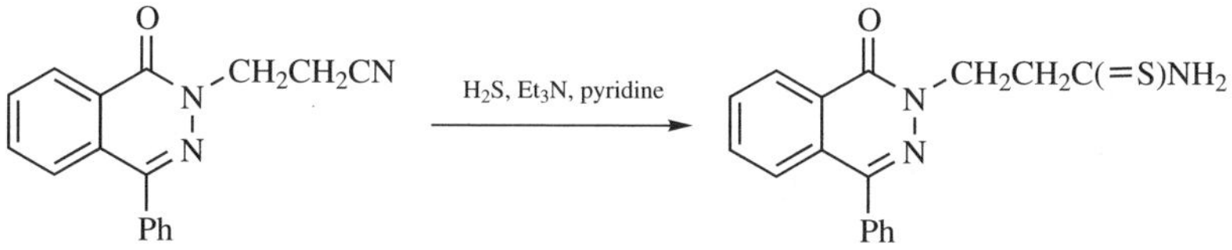

O
CH_2CH_2CN
N
N
Ph
H_2S, Et_3N, pyridine
O
$CH_2CH_2C(=S)NH_2$
N
N
Ph

(**66**) (**67**)

By Miscellaneous Routes

tert-Butyl 1,2,3,4-tetrahydro-2-phthalazinecarboxylate (**68**) underwent *N*-carbamoylation by methyl 2-isocyanato-4-methylvalerate to afford *tert*-butyl 3-[(1-methoxycarbonyl-3-methylbutyl)carbamoyl]-1,2,3,4-tetrahydro-2-phthalazinecarboxylate (**69**) [synthon (made in situ), CH_2Cl_2, 0°C; substrate in CH_2Cl_2↓; 20°C, 20 h: 83%]; conformation by X-ray analysis.[310]

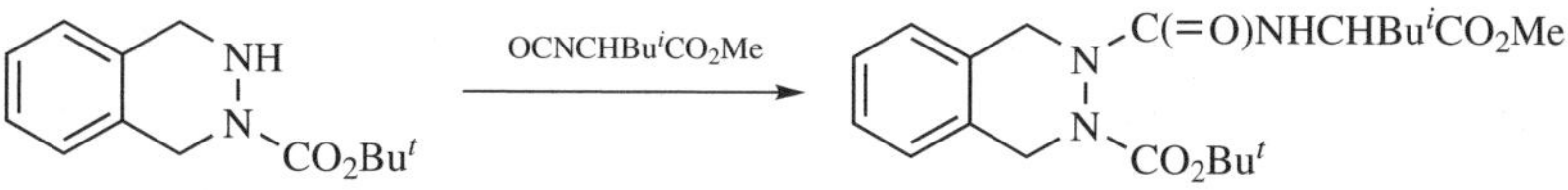

NH
N
CO_2Bu^t
$OCNCHBu^iCO_2Me$
$C(=O)NHCHBu^iCO_2Me$
N
N
CO_2Bu^t

(**68**) (**69**)

Methyl 1-cyano-1,2-dihydro-2-phthalazinecarboxylate (**70**) with phenyl isothiocyanate gave methyl 1-cyano-1-[*N*-phenyl(thiocarbamoyl)]-1,2-dihydro-2-phthalazinecarboxylate (**71**) (NaH, Me_2NCHO, 0°C, N_2; substrate in Me_2NCHO↓ dropwise during 10 min; 0°C, 15 min; synthon in Me_2NCHO↓ dropwise; 0°C → 20°C, 3 h: 67%).[374]

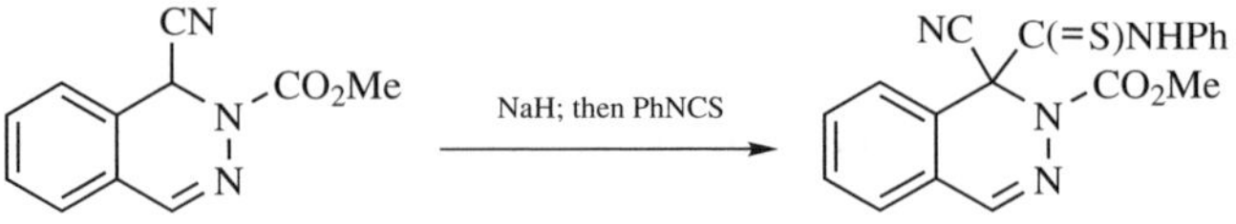

(70) NaH; then PhNCS → (71)

2-Amino-5,6,7,8-tetrabromo-4-tolyl-1(2*H*)-phthalazinone (**72**) underwent substituted-alkylation by 2-chloroacetanilide to give 5,6,7,8-tetrabromo-2-[(*N*-phenylcarbamoylmethyl)amino]-4-tolyl-1(2*H*)-phthalazinone (**72a**) (reactants, EtOH, reflux, 4 h: 73%; *o*, *m*, or *p*-tolyl unspecified).[666]

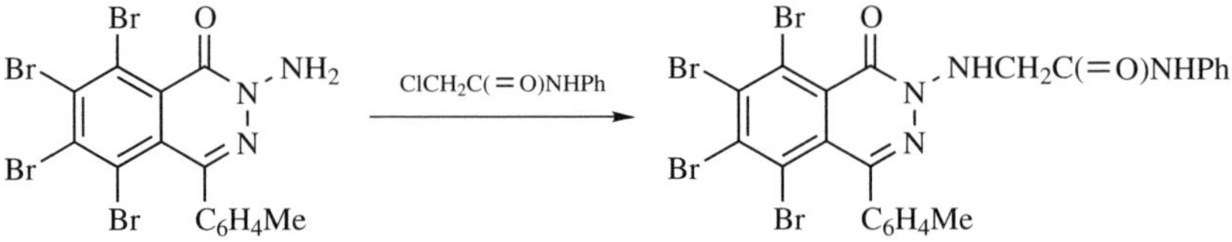

(72) $ClCH_2C(=O)NHPh$ → (72a)

2-[*N*-(2-Bromo-3,5-bistrifluoromethylphenyl)carbamoylmethyl]- (**73**, X = O) underwent thiation to the corresponding thioamide, 2-[*N*-(2-bromo-3,5-bis trifluoromethylphenyl)thiocarbamoylmethyl]-4-ethoxycarbonylmethyl-1(2*H*)-phthalazinone (**73**, X = S) (P_2S_5, PhH, 70°C, 4 h: 59%).[68]

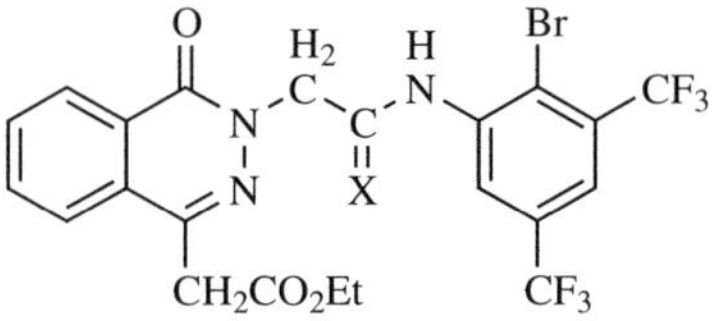

(73)

14.4.2. Reactions of Phthalazine Amides

The *hydrolysis of phthalazinecarboxamides to phthalazinecarboxylic acids* has been exemplified in Section 14.1.1. Some other reactions are illustrated in the following classified examples.

Removal of Amidic Groups

Note: This has been done indirectly as a one-pot procedure by hydrolysis and decarboxylation.

4-(α-Benzamido-α-hydrazinocarbonylmethyl)-1(2*H*)-phthalazinone (**74**) gave 4-aminomethyl-1(2*H*)-phthalazinone (**75**) (HCl; then NaOH: for details, see original; note additional deacylation).[158,cf. 159]

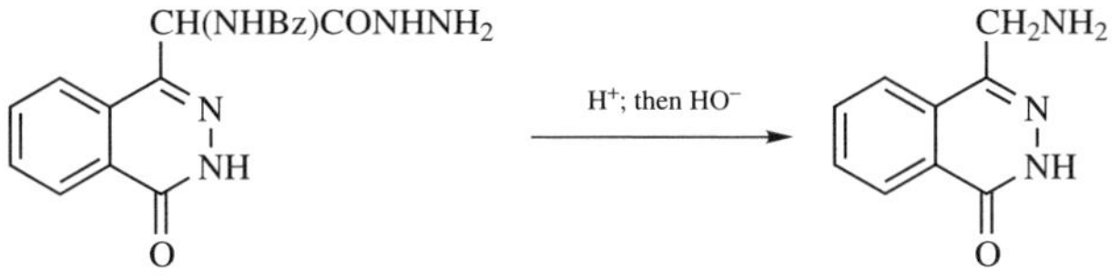

Dehydration to Phthalazinecarbonitrile

1,4-Dioxo-1,2,3,4-tetrahydro-6-phthalazinecarboxamide (**76**) gave 1,4-dichloro-6-phthalazinecarbonitrile (**77**) (neat $POCl_3 + SOCl_2$, reflux, 12 h: 70%; note concomitant chlorolysis of the oxo substituents).[285]

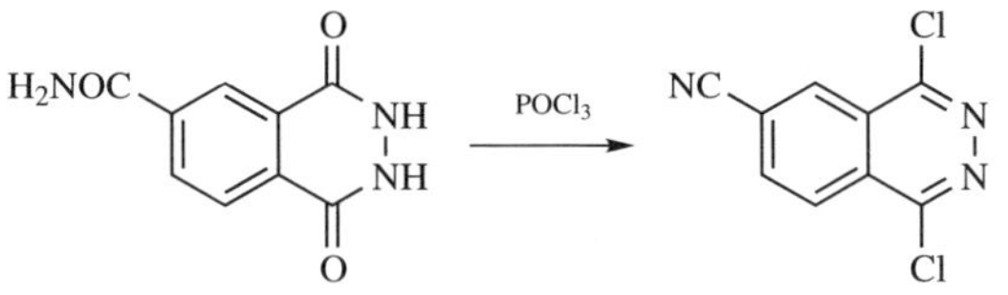

Formation of Linear Derivatives

Note: Carbohydrazides, rather than carboxamides, are prone to the formation of such derivatives.

2-Hydrazinocarbonylmethyl-4-phenyl-1(2*H*)-phthalazinone (**78**) with *p*-bromobenzaldehyde gave 2-[(*p*-bromobenzylidenehydrazinocarbonyl)methyl]-4-phenyl-1(2*H*)-phthalazinone (**79**) (reactants, trace AcOH, EtOH, reflux, 2 h: 70%) or with *o*-hydroxyacetophenone gave 2-[(*o*-hydroxy-α-methylbenzylidenehydrazinocarbonyl)methyl]-4-phenyl-1(2*H*)-phthalazinone (**80**) (likewise: 80%); other aldehydes or ketones produced analogous products.[659,676]

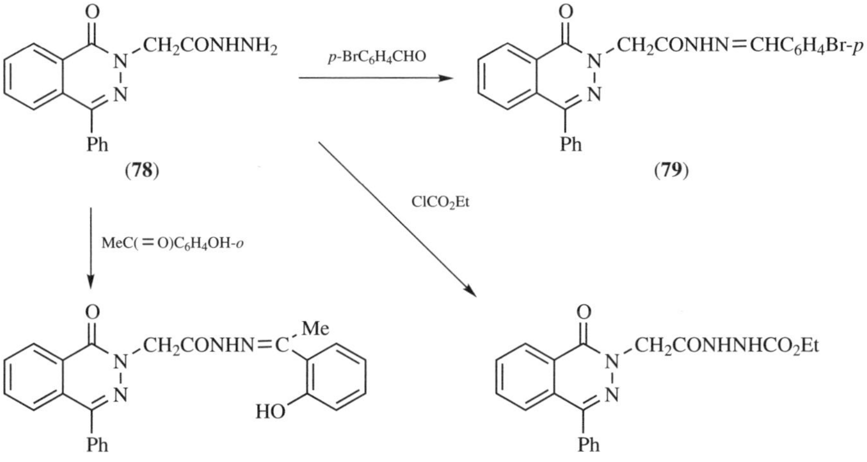

The same substrate (**78**) with ethyl chloroformate gave 2-[(*N′*-ethoxycarbonylhydrazinocarbonyl)methyl]-4-phenyl-1(2*H*)-phthalazinone (**81**) (substrate, Me_2NCHO, reflux, 15 min: 80%).[660]

4-Benzyl-2-(hydrazinocarbonylmethyl)-1(2*H*)-phthalazinone (**82**) with phenyl isocyanate gave 4-benzyl-2-[(4-phenylsemicarbazido)carbonylmethyl]- (**83**, X = O) (reactants, PhH, reflux, 5 h: 95%) or with phenyl isothiocyanate gave 4-benzyl-2-{[4-phenyl(thiosemicarbazido)]carbonylmethyl}-1(2*H*)-phthalazinone (**83**, X = S) (likewise: 93%).[676]

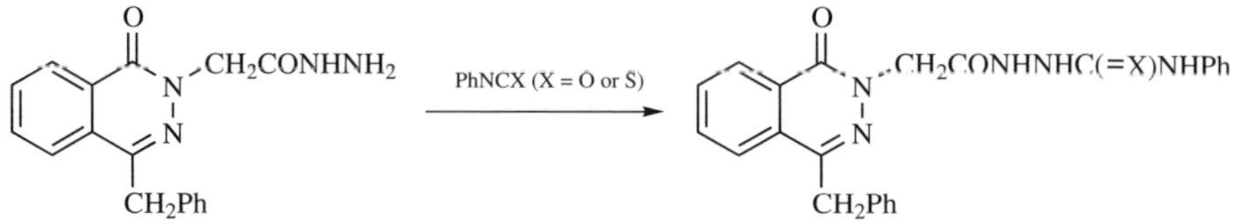

$CH_2CONHNH_2$; CH_2Ph ; PhNCX (X = O or S) ; $CH_2CONHNHC(=X)NHPh$; CH_2Ph

(**82**) (**83**)

Also other examples.[180,204,247,347]

Formation of Cyclized Derivatives

2-Hydrazinocarbonylmethyl-4-phenyl-1(2*H*)-phthalazinone (**84**) with triethyl orthoformate gave 2-(5-oxo-2-pyrazolin-4-yl)-4-phenyl-1(2*H*)-phthalazinone (**85**) (neat reactants, reflux, 4 h: 80%), with carbon disulfide/potassium hydroxide gave 2-(3-oxo-5-thioxopyrazolin-4-yl)-4-phenyl-1(2*H*)-phthalazinone (**86**) (reactants, EtOH, reflux, 4 h: 75%), or simply on fusion gave 3,4-dihydro-2*H*-[1,2,4]triazino[3,4-*a*]phthalazin-3-one (**87**) (200°C, 1 h: 75%);[659] also analogous cyclizations.[659,660,676]

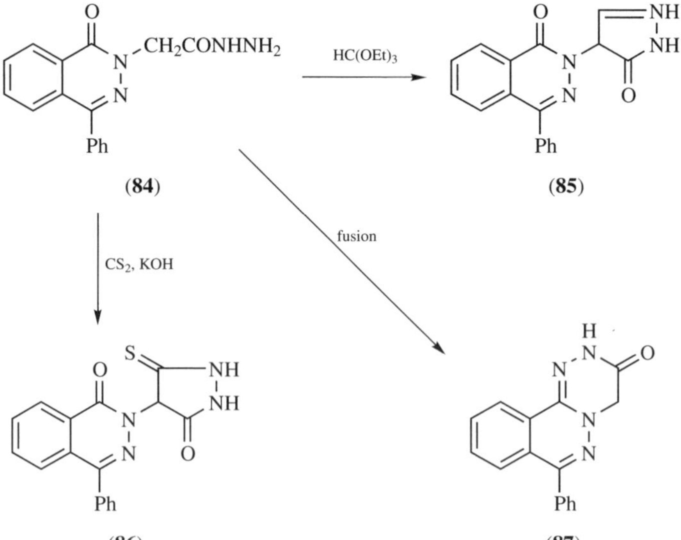

$HC(OEt)_3$; CS_2, KOH ; fusion

(**84**) (**85**)

(**86**) (**87**)

N-Methyl-1,2,3,4-tetrahydro-2-phthalazinecarbothioamide (**88**) with thiophosgene gave 3-methylimino-1*H*,3*H*-[1,3,4]thiadiazolo[3,4-*b*]phthalazine-1-thione (**89**) (mild conditions; see original for further details).[479]

S=CCl$_2$ (−2HCl)

(**88**) (**89**)

Also many other examples.[68,200,204,247,248,269,347,355,356,467,666]

14.5. PHTHALAZINECARBONITRILES

Perhaps the most interesting phthalazinecarbonitriles are the Reissert derivatives (see Section 9.1.3); thus compounds such as 2-benzoyl-1,2-dihydro-1-phthalazinecarbonitrile (**90**) exist as such but their salts have been shown by careful NMR and MS studies to exist predominantly as tricyclic tautomers like the 1-aminooxazolo[4,3-*a*]phthalazin-4-ium cation (**91**).[321,cf. 22] The MS results of simple phthalazinecarbonitriles have also been studied.[188]

HX

(**90**) (**91**)

14.5.1. Preparation of Phthalazinecarbonitriles

The main routes to such nitriles have been covered already: by *primary synthesis* (Chapter 8), by the *Reissert reaction* (Section 9.1.3), by *cyanolysis of halogenophthalazines* (Section 10.3.5), from *phthalazine N-oxides* (Section 11.6), by *cyanolysis of alkylsulfonylphthalazines* (Section 12.3), and by *dehydration of phthalazinecarboxamides* (Section 14.4.2). Other preparative methods are exemplified here.

4-Phenyl-2-phthalazin-2-ium-2-benzimidate (**92**) with cyanide ion gave 4-phenyl-1-phthalazinecarbonitrile (**94**), probably by loss of benzamide from the intermediate adduct (**93**) (substrate, MeOH; KCN in H_2O↓ dropwise, trace KOH; 20°C, 3 h: 75%).[9]

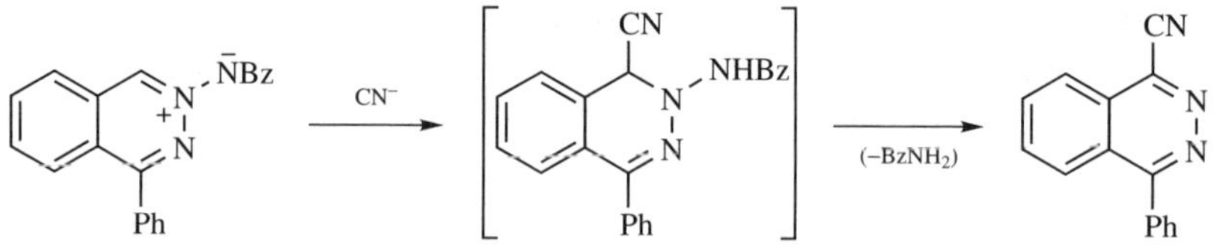

(92) (93) (94)

1-Phthalazinecarbonitrile (**95**) underwent passenger introduction of a cyanoalkyl group to give 4-(α-cyano-α-ethoxycarbonylmethyl)-1-phthalazinecarbonitrile (**96**) (substrate, $NaNH_2$, $NCCH_2CO_2Et$, 7 h: 17%; for details, see original).[930]

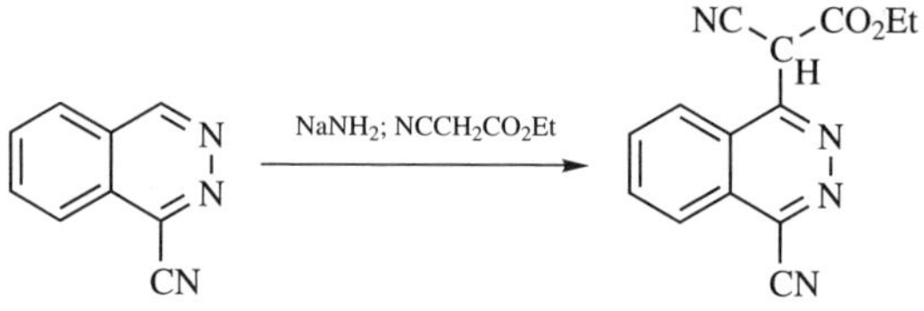

(95) (96)

14.5.2. Reactions of Phthalazinecarbonitriles

Reactions of phthalazinecarbonitriles already covered include *conversion of Reissert nitriles into extracyclic hydroxyphthalazines* (Section 11.2.1), *hydrolysis to phthalazinecarboxylic acids* (Section 14.1.1), *controlled hydrolysis to phthalazinecarboxamides* (Section 14.4.1), and *thiolysis to phthalazinecarbothioamides* (Section 14.4.1). A variety of other reactions are illustrated in the following examples.

Alcoholysis to Phthalazinecarboximidic Esters

Ethyl 1-cyanomethyl-5,7-dimethyl-4-oxo-3,4-dihydro-6-phthalazinecarboxylate (**97**) gave ethyl 1-(2-imino-2-methoxyethyl)-5,7-dimethyl-4-oxo-3,4-dihydro-6-phthalazinecarboxylate (**98**) (HCl gas, MeOH, reflux, 2 h: ~80%).[243]

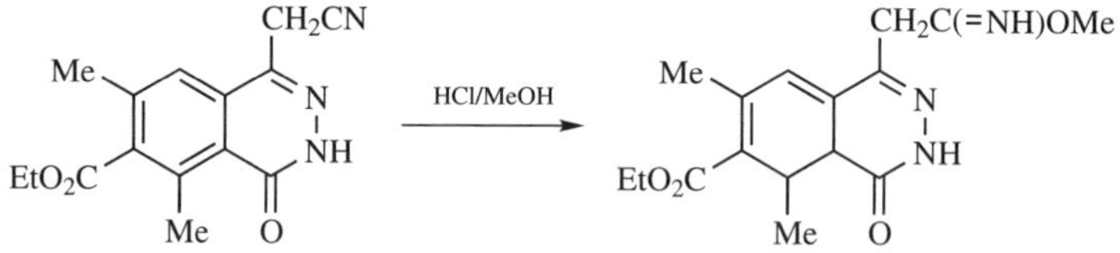

(97) (98)

Reissert Nitriles to α-(Acyloxy)benzylphthalazines

2-Benzoyl-1,2-dihydro-1-phthalazinecarbonitrile (**99**) gave 1-(α-benzoyloxybenzyl)phthalazine (**100**) (NaH, Me_2NCHO, <5°C; substrate + PhCHO +

Me_2NCHO↓ dropwise during 15 min; <5°C → 20°C, 12 h: 72%); analogs likewise.[377]

(**99**) (**100**)

Oxidation of α-Cyanobenzyl- to Benzoylphthalazines

1-(α-Cyanobenzyl)phthalazine (**101**) gave 1-benzoylphthalazine (**102**) (substrate, NaH, THF, 5 min; then O_2↓ until colorless: 92%).[586]

(**101**) (**102**)

Conversion into Amidines

Note: There appear to be no simple examples in the 1972–2004 period but the conversion of cyanoamino- to guanidinophthalazines is virtually the same process.

1-Cyanoaminophthalazine (**103**) with butylamine gave 1-(N'-butylguanidino) phthalazine (**104**, R = Bu) (neat reactants, 70°C, 8 h: 20%) or with *O*-methylhydroxylamine gave 1-(N'-methoxyguanidino)phthalazine (**104**, R = OMe) ($MeONH_2 \cdot HCl$, NaOH, MeOH, 20°C, 30 min; substrate↓, reflux, 15 h: 7%).[225]

(**103**) (**104**)

Reaction with Grignard Reagents

Note: Treatment of 1-phthalazinecarbonitrile (**105**) with a Grignard reagent leads to several products in each case; however, such products from different Grignards are not necessarily analogs. Since such a procedure is of little practical utility, one example will suffice. Extranuclear cyanophthalazines appear to react more predictably with Grignards to afford ketones.

1-Phthalazinecarbonitrile (**105**) with benzylmagnesium chloride gave a separable mixture of 1-benzylphthalazine (**106**), 4-benzyl-3,4-dihydro-1-phthalazinecarbonitrile (**107**), 4-benzyl-1-phthalazinecarbonitrile (**108**), and 4-benzyl-3-(phthalazin-1-yl)-3,4-dihydro-1-phthalazinecarbonitrile (**109**) (substrate, THF; $PhCH_2MgCl$ in Et_2O↓, reflux, 3 h: 5%, 17%, 3%, and 3%, respectively, after chromatographic separation)[399]

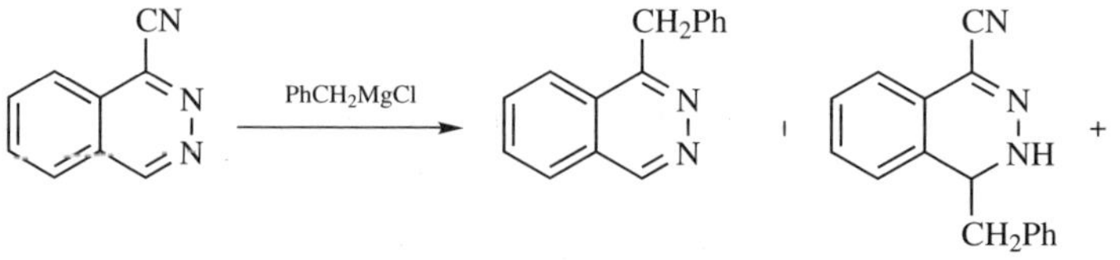

CN N N PhCH₂MgCl CH₂Ph N N CN N NH CH₂Ph +

(**105**) (**106**) (**107**)

CN N N CH₂Ph + CN N N PhH₂C N N

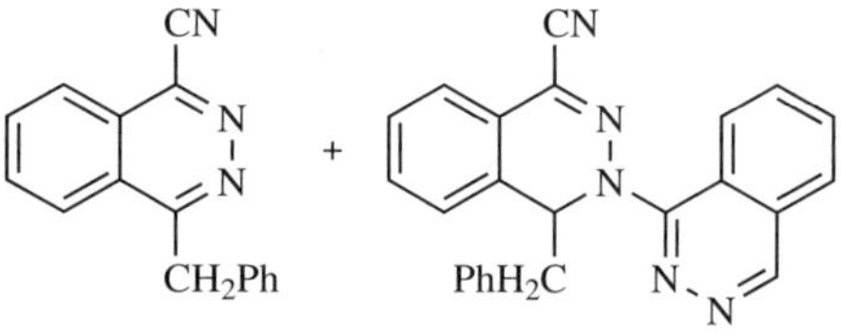

(**108**) (**109**)

2-(4-*p*-Chlorophenyl-5-cyano-6-oxo-3-phenyl-1,6-dihydropyridin-2-yl)- (**110**, R = CN) with methylmagnesium iodide gave 2-(5-acetyl-4-*p*-chlorophenyl-6-oxo-3-phenyl-1,6-dihydropyridin-2-yl)-4-(3,4-dichlorophenyl)-1(2*H*)-phthalazinone (**110**, R = Ac) (MeMgI in Et_2O; substrate in PhH↓ dropwise, reflux, 2 h; 20°C, 24 h: 75% ?).[621]

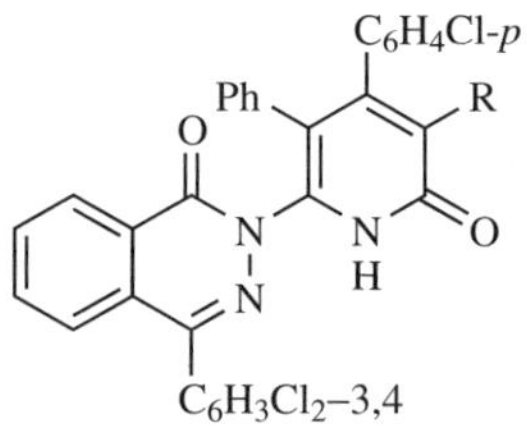

C_6H_4Cl-*p* Ph R O N N H O N $C_6H_3Cl_2$–3,4

(**110**)

Extranuclear Cyclizations

Ethyl 1-cyanomethyl-5,7-dimethyl-4-oxo-3,4-dihydro-6-phthalazinecarboxylate (**111**) with sodium azide gave ethyl 5,7-dimethyl-4-oxo-1-(tetrazol-5-ylmethyl)-3,4-dihydro-6-phthalazinecarboxylate (**112**) (NaN_3, NH_4Cl, LiCl, Me_2NCHO, 125°C, 1 h: 46%).[243]

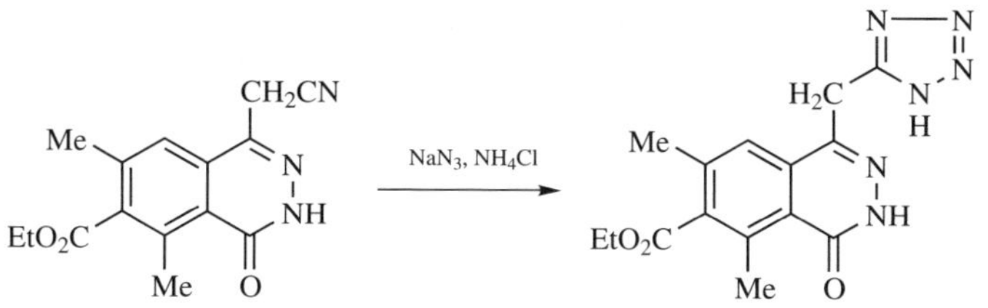

(**111**) $\xrightarrow{NaN_3,\ NH_4Cl}$ (**112**)

2-Cyanomethyl-4-ethoxycarbonylmethyl-1(2*H*)-phthalazinone (**113**) with 2-amino-5-fluoro-1-benzenethiol hydrochloride (**114**) gave 4-ethoxycarbonylmethyl-2-(6-fluorobenzothiazol-2-ylmethyl)-1(2*H*)-phthalazinone (**115**) (EtOH, reflux, 48 h: 53%).[784]

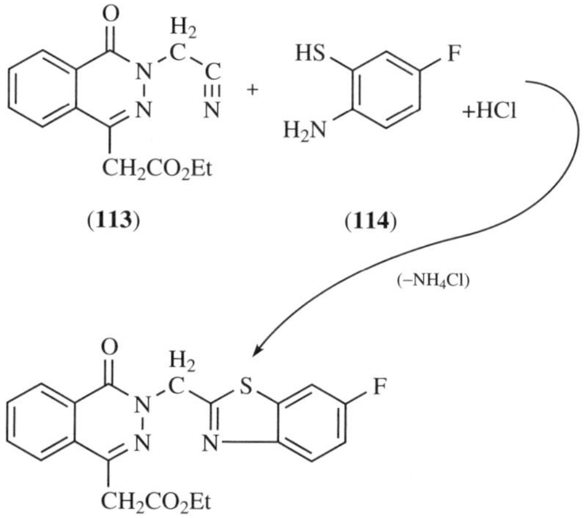

(**113**) + (**114**) + HCl $\xrightarrow{(-NH_4Cl)}$ (**115**)

Also other examples.[621]

Nuclear Cyclizations

1-Cyanoaminophthalazine (**116**) with methylhydrazine gave 3-methylamino-1,2,4-triazolo[3,4-*a*]phthalazine (**117**) of confirmed structure (neat reactants, 70°C, 3 h: 17%; an explanation, involving Dimroth rearrangement, is offered for the formation of this unlikely product).[225]

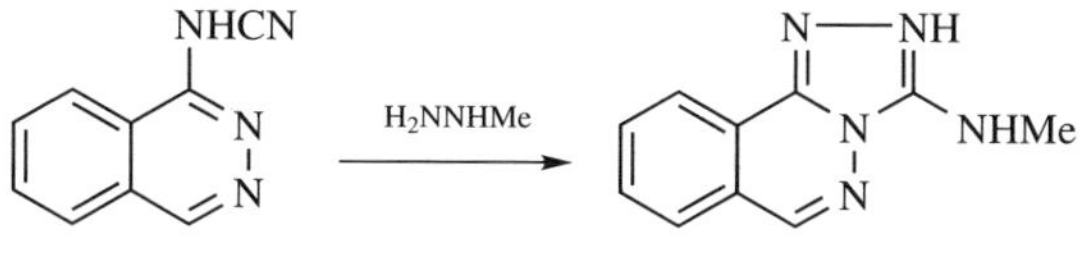

(**116**) $\xrightarrow{H_2NNHMe}$ (**117**)

2-Benzoyl-1,2-dihydro-1-phthalazinecarbonitrile hydrotrifluoroborate (**118**) and dimethyl acetylenedicarboxylate gave dimethyl 3-phenylpyrrolo[2,1-*a*]phthalazine-1,2-dicarboxylate (**119**) (Me_2NCHO, 20°C → 100°C slowly; then 100°C, 24 h: 82%);[22] analogs somewhat similarly.[22,107,716]

The same substrate, converted into its carbanion, reacted with acrylonitrile to give eventually 3-phenylpyrrolo[2,1-*a*]phthalazine-2-carboxamide (**120**) (substrate, Bu^tOK, Me_2SO, N_2, 20°C, 15 min; $H_2C{=}CHCN\downarrow$, 20°C, 30 min: 65%); analogs likewise.[331]

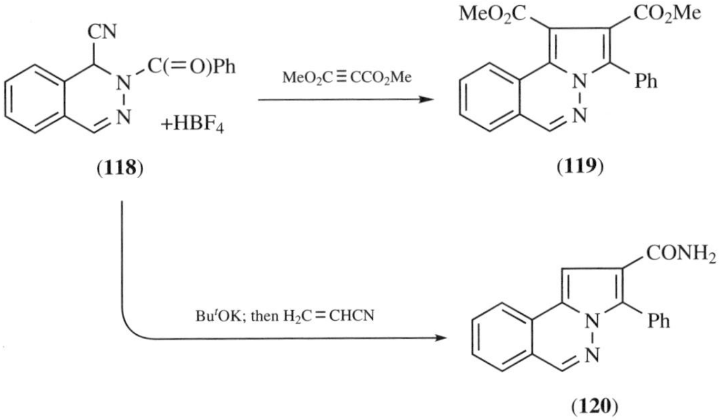

(**118**) (**119**) (**120**)

14.6. PHTHALAZINE ALDEHYDES AND KETONES

Most 1972–2004 information on these aldehydes and ketones has been covered already, as indicated in the following lists.

Preparation

Note: Acylphthalazines have been made by *primary synthesis* (see Chapter 8), *Reissert-type additions to simple phthalazines* (Section 9.1.3), *oxidation of alkylphthalazines* (Section 9.2.2), *displacement of halogeno substituents* (Section 10.3.4), *acylation of tautomeric phthalazinones* (Section 11.1.2.2), *oxidation of extranuclear hydroxyphthalazines* (Section 11.2.2), or as illustrated in these examples.

2-Bromomethyl-8-dibromomethyl-1(2*H*)-phthalazinone (**121**) gave 4-oxo-3,4-dihydro-5-phthalazinecarbaldehyde (**122**) (Na_2CO_3, H_2O, 95°C, 1 h: 39%; note removal of the bromomethyl substituent).[622]

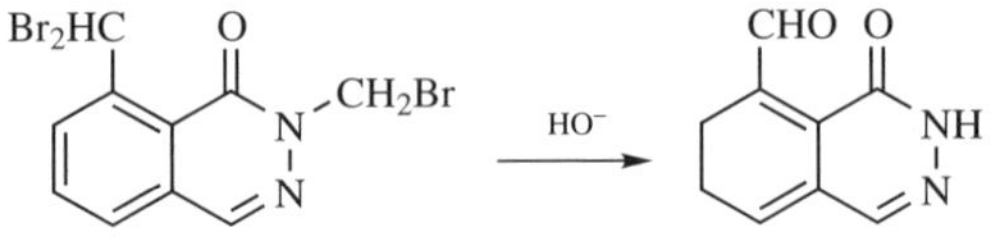

(**121**) (**122**)

4-[β-(2-Hydrazonocyclohexyl)phenethyl]-1(2*H*)-phthalazinone (**123**, X = NNH_2) gave 4-[β-(2-oxocyclohexyl)phenethyl]-1(2*H*)-phthalazinone (**123**, X = O) (10M HCl, 95°C, 4 h: >95%).[743]

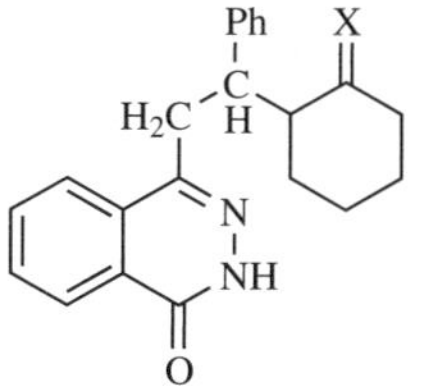

(**123**)

2-(3-Ethoxycarbonyl-2-imino-5-methyltetrahydrofuran-3-yl)-1,4(2*H*,3*H*)-phthalazinedione (**124**, X = NH) gave 2-(3-ethoxycarbonyl-5-methyl-2-oxotetrahydrofuran-3-yl)-1,4(2*H*,3*H*)-phthalazinedione (**124**, X = O) (HCl, H_2O, EtOH, reflux, briefly: 92%).[126]

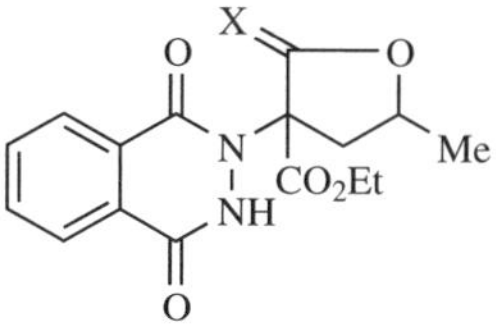

(**124**)

1-Phthalazinamine (**125**) gave 4-amino-2-phenacylphthalazin-2-ium bromide (**126**) ($BzCH_2Br$, EtOH: no details).[944]

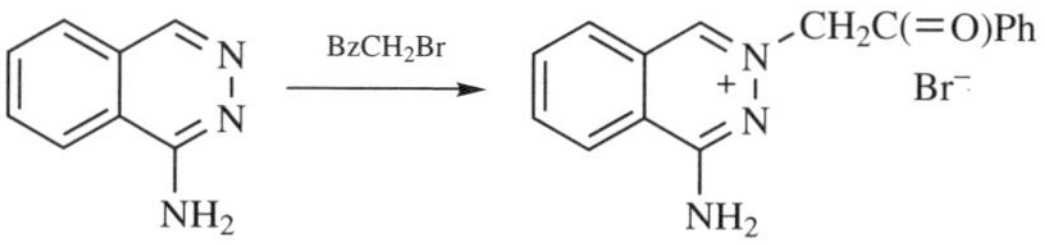

(**125**) (**126**)

Reactions

Note: Reactions already covered include *conversion into extranuclear hydroxyphthalazines by reductive or other means* (Section 11.2.1) and *conversion into epoxyphthalazines* (Section 11.4.1); their *oxidation to phthalazinecarboxylic acids* appears to be unrepresented, and some other reactions are illustrated in the following examples.

4-*p*-Acetylanilino-2-phenyl-1(2*H*)-phthalazinone (**127**) with 4-butylsemicarbazide gave 4-*p*-[1-(4-butylsemicarbazono)ethyl]anilino-2-phenyl-1(2*H*)-phthalazinone (**128**, R = Bu, X = O) (EtOH, reflux, 8 h: 81%) or with 4-phenyl (thiosemicarbazide) gave 2-phenyl-4-*p*-{1-[4-phenyl (thiosemicarbazono)] ethyl}anilino-1(2*H*)-phthalazinone (**128**, R = Ph, X = S) (likewise: 80%); analogs similarly.[675]

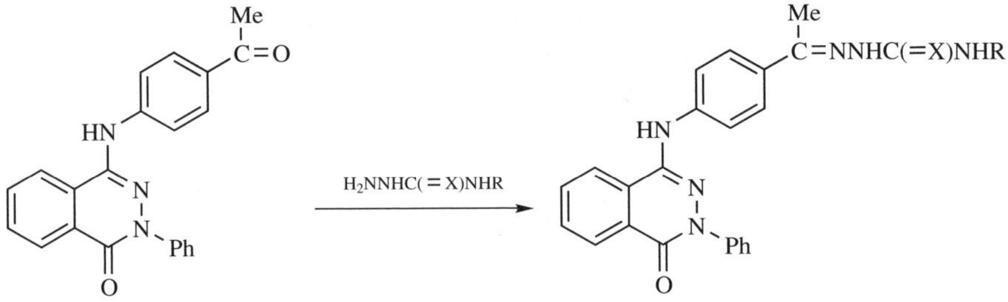

Me
C=O
HN
N
N
Ph
O
(127)

$H_2NNHC(=X)NHR$

Me
C=NNHC(=X)NHR
HN
N
N
Ph
O
(128)

1-(Diacetylmethyl)-4-phenylphthalazine (**130**) with hydrazine hydrate gave 1-(3,5-dimethylpyrazol-4-yl)-4-phenylphthalazine (**129**) (trace piperidine, H_2O, EtOH, reflux, 1 h: 75%) or with guanidine hydrochloride gave 1-(2-amino-4,6-dimethylpyrimidin-5-yl)-4-phenylphthazazine (**131**) (likewise: 79%).[664]

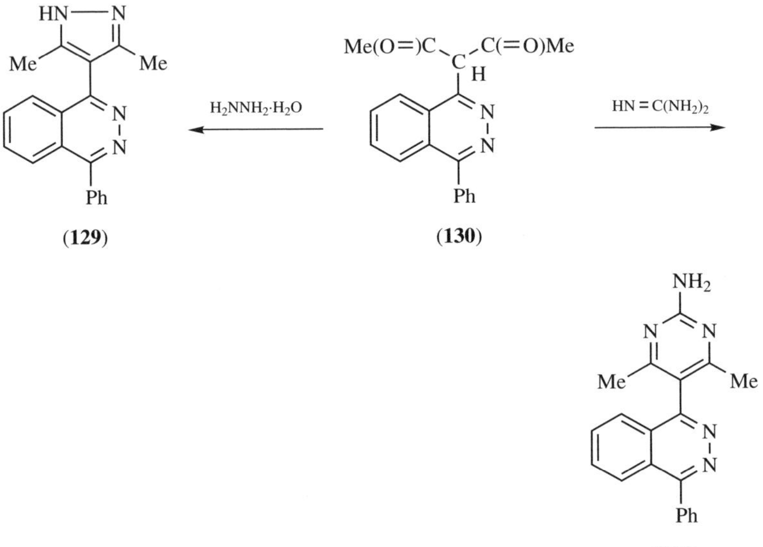

HN—N
Me Me
N
N
Ph
(129)

$H_2NNH_2{\cdot}H_2O$

Me(O=)C C(=O)Me
C
H
N
N
Ph
(130)

$HN=C(NH_2)_2$

NH_2
N N
Me Me
N
N
Ph
(131)

6,7-Dibenzoyl-1,4-diphenylphthalazine (**132**) with sulfur gave 1,4,6,8-tetraphenylthieno[3,4-*g*]phthalazine (**133**) (neat reactants, 270°C, 10 min: 56%).[16]

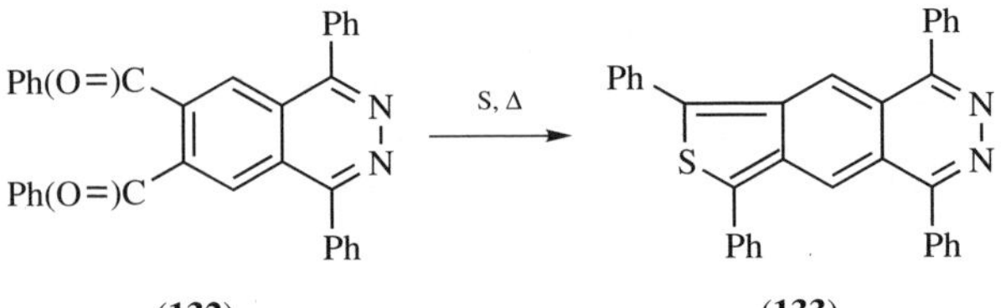

Ph
Ph(O=)C
N
N
Ph(O=)C
Ph
(132)

S, Δ

Ph Ph
N
S
N
Ph Ph
(133)

4-Dimethylamino-2-phenacylphthalazin-2-ium bromide (**134**) with hydrazine hydrate gave 7-dimethylamino-3-phenyl-1,11b-dihydro-4*H*-[1,2,4]triazino[3,4-*a*]phthalazine (**135**) (EtOH, 20°C, 24 h: 95%);[63] also analogous cyclizations.[944,945]

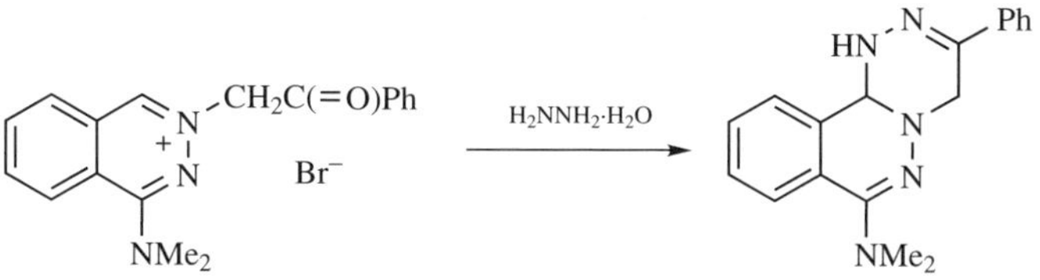

(**134**) (**135**)

In contrast, 2-*p*-fluorophenacylphthalazin-2-ium bromide (**136**) underwent cyclization with alkenes or alkynes without involving the acyl group; for example, with ethyl propiolate it gave ethyl 3-*p*-fluorobenzoylpyrrolo[2,1-*a*]phthalazine-1-carboxylate (**137**) (Et_3N, $CHCl_3$, reflux, 2 h: 79%; analogs similarly).[976]

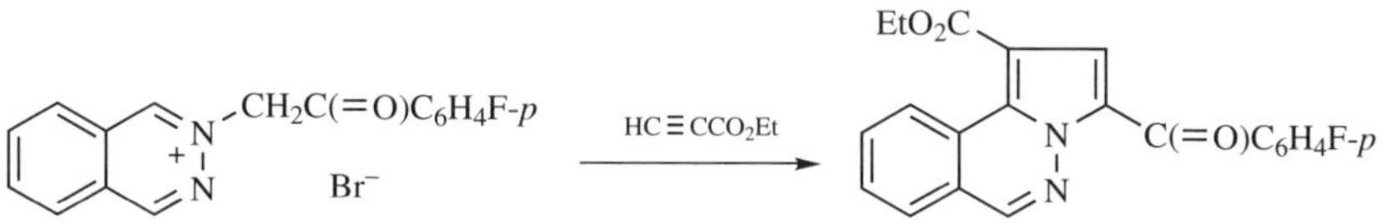

(**136**) (**137**)

4-[(3-Dimethylaminoacryloyl)methoxy]-1(2*H*)-phthalazinone (**138**) condensed with ethyl acetoacetate and ammonia (supplied as $AcONH_4$) to give 4-[(5-ethoxycarbonyl-6-methylpyridin-2-yl)methoxy]-1(2*H*)-phthalazinone (**139**) (reactants, AcOH, reflux, <2 h: 60%).[1014]

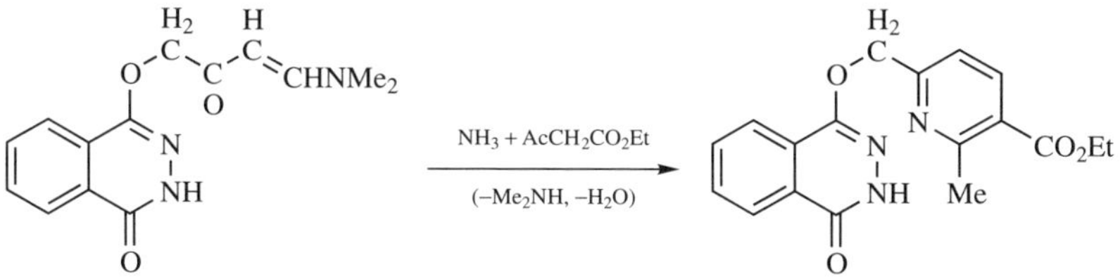

(**138**) (**139**)

Also other examples.[251]

APPENDIX

Tables of Simple Cinnolines and Simple Phthalazines

These two tables are intended as reasonably comprehensive lists of simple cinnolines (Table A.1) and simple phthalazines (Table A.2) described before 2005. For each compound are recorded (1) melting and/or boiling point(s); (2) an indication of reported spectra or other physical properties; (3) any reported salts or simple derivatives, especially when the parent compound was un- or ill-characterized; (4) an indication of an any simple complexes reported; and (5) direct references to the original literature from 1972 onward, preceded by page references (in parentheses) to any earlier data reported in Simpson's *Hauptwerk*[906] (e.g., *H* 70) or in Singerman and Patel's *Ergänzungswerk*[907] (e.g., *E* 324).

To keep the tables within manageable proportions, the following categories of cinnolines and phthalazines have been *excluded* on the grounds that they are not simple:

- All fused or nucleus-reduced derivatives
- Those with a cyclic substituent other than an unsubstituted cycloalkyl, morpholino, phenyl, or piperidino group
- Those bearing a substituent with more than six carbon atoms except for an unsubstituted benzoyl, benzyl, benzylidene, phenethyl, phenylethynyl, or styryl group
- Those with two or more independent functional groups on any one substituent

The following conventions and abbreviations have been used in the tables.

Melting Point. This term covers not only a regular melting point or melting range but also such variations as "decomposing at" or "melting with decomposition at." The symbol > before the melting point indicates that the substance melts or decomposed above that temperature or that it does not melt or decompose below that temperature. Where two differing melting points/ranges are reported in the

Cinnolines and Phthalazines: Supplement II, The Chemistry of Heterocyclic Compounds, Volume 64,
by D.J. Brown

literature, they appear in the tables as, for example, "93–94 or 97–99"; when more than two melting points/ranges are reported, they appear in the tables as, for example, "207 to 219."

Boiling Point. Boiling points/ranges are distinguished from melting points/ranges by the presence of a pressure in millimeters of mercury (mmHg) after the temperature(s): for example. 71–73/1.3.

Abbreviations for Physical Data

anal	Analytical data (usually assumed)
biol	Bioactivity only reported
crude	Compound not purified
dip	Dipole moment
fl sp	Fluorescent spectral data
IR	Infrared spectral data
liq	Liquid at room temperature (no boiling point reported)
MS	Mass spectral data
NMR	Nuclear magnetic resonance data (any nucleus)
solid	Solid at room temperature (no melting point reported)
st	Fine structure, for instance, tautomerism, discussed
th	Theoretical calculations reported
UV	Ultraviolet/visible spectral data
xl st	Crystal structure (X-ray data)

Abbreviations for Salts, Associated Anions, or Solvates

AcOH	Acetate salt
EtOH	Ethanolate
HBr, etc.	Appropriate hydrohalide salt
H_2O	Hydrate
HSO_4^-	Sulfate anion
H_2SO_4	Sulfate salt
I, etc.	Appropriate halide anion
MeI	Quaternary methiodide
NH_4	Ammonium salt
Na, etc.	Appropriate alkali metal salt
pic	Picrate salt of anion
TsOH	*p*-Toluenesulfonate salt

Abbreviations for Derivatives

dnp	2,4-Dinitrophenylhydrazone
Et_2 acetal, etc.	Appropriate dialkyl acetal
$H_2NN=$	Hydrazone
MeCH= etc.	Appropriate alkylidene derivative

PhNHN=	Phenylhydrazone
PhN=	Anil (Schiff base)
sc	Semicarbazone
tsc	Thiosemicarbazone

Other Notes. The use of "cf." before a reference usually indicates some inconsistent or doubtfully relevant information therein. A query mark (?) indicates some doubt associated with a datum or reference. A dash (—) in the data column indicates that no new physical data were gleaned from original literature of the period 1972–2004 (references, if any, in the next column).

TABLE A.1. ALPHABETICAL LIST OF SIMPLE CINNOLINES REPORTED BEFORE 2005

Cinnoline	Melting Point (°C) etc.	Reference(s)
7-Acetamido-4-acetoxycinnoline	—	(*E* 234)
8-Acetamido-4-acetoxycinnoline	—	(*E* 234)
6-Acetamido-4-anilinocinnoline	—	(*E* 242)
4-Acetamido-3-benzoyl-6,8-dimethylcinnoline	172, IR, NMR	830
6-Acetamido-4-cinnolinamine	—	(*E* 236)
7-Acetamido-4-cinnolinamine	—	(*H* 35)
3-Acetamidocinnoline	225–226	(*E* 238) 908
4-Acetamidocinnoline	—	(*E* 228, 239, 240)
8-Acetamidocinnoline	NMR	(*E* 240) 289
7-Acetamido-4(1*H*)-cinnolinone	—	(*H* 18; *E* 234)
7-Acetamido-4-hydroxyaminocinnoline	—	(*E* 234)
8-Acetamido-4-methylcinnoline	—	(*E* 240)
6-Acetamido-1-methyl-4-phenylimino-1,4-dihydrocinnoline	—	(*H* 40)
4-Acetamido-6-nitrocinnoline	—	(*H* 35; *E* 228)
4-Acetamido-7-nitrocinnoline	—	(*E* 234)
4-Acetamido-8-nitrocinnoline	—	(*E* 240)
4-Acetamido-6-phenylazocinnoline	—	(*E* 236)
4-Acetonylcinnoline	92–94	933
4-Acetoxy-6-chloro-3-cinnolinecarbonitrile	164	505
4-Acetoxy-7-chloro-3-cinnolinecarbonitrile	83	505
4-Acetoxy-8-chloro-3-cinnolinecarbonitrile	68	505
8-Acetoxycinnoline	NMR	289
4-Acetoxy-3-cinnolinecarbonitrile	147, IR, NMR	505
4-Acetoxy-5,6-dichlorocinnoline	—	(*E* 143)
4-Acetoxy-6-methyl-3-cinnolinecarbonitrile	110	505
4-Acetoxy-7-methyl-3-cinnolinecarbonitrile	80	505
4-Acetoxy-8-methyl-3-cinnolinecarbonitrile	128	505
3-Acetyl-6-bromo-4(1*H*)-cinnolinone	220, IR, NMR; tsc: 235, IR	619
3-Acetyl-6-chlorocinnoline	—	(*E* 144, 267)
3-Acetyl-7-chlorocinnoline	—	(*E* 144, 267)
3-Acetylcinnoline	155–156	(*E* 467) 908
4-Acetylcinnoline	100–101	(*E* 269) 908
3-Acetyl-4(1*H*)-cinnolinone	155–156, IR, NMR; tsc: 256–257, IR, NMR	619

TABLE A.1. (*Continued*)

Cinnoline	Melting Point (°C) etc.	Reference(s)
7-Acetyl-4(1*H*)-cinnolinone	—	(*H* 18)
4-(1-Acetylethyl)cinnoline	—	(*E* 56)
7-Acetyl-4-hydroxyaminocinnoline	—	(*E* 234)
3-Acetyl-6-methoxy-4(1*H*)-cinnolinone	174–175, IR, NMR; tsc: 225, IR	619
3-Acetyl-6-methyl-4(1*H*)-cinnolinone	195, IR; tsc: 262, IR	619
3-Acetyl-4-oxo-1, 4-dihydro-6-cinnolinesulfonamide	193–195, IR	663
3-Acetyl-4-oxo-1, 4-dihydro-6-cinnolinesulfonanilide	196–198, IR, NMR	663
3-Acetyl-4-oxo-1,4-dihydro-6-cinnolinesulfonic acid	>360, IR	663
3-Acetyl-4-oxo-1-phenylazo-1,4-dihydro-6-cinnolinesulfonanilide	102–104, IR	663
7-Acetyl-4-phenoxycinnoline	—	(*H* 33)
3-Acetyl-1-phenylazo-4(1*H*)-cinnolinone(?)	125–127, NMR, UV	620, cf. 982
3-Acetyl-6-phenylazo-4(1*H*)-cinnolinone	220, NMR, UV	620
3-Acetyl-1-phenyl-4(1*H*)-cinnolinone	165	140
1-Allyl-*N*-cyclopentyl-6-nitro-4-oxo-1, 4-dihydro-3-cinnolinecarboxamide	NMR	1034
1-Allyl-3-(2-ethoxycarbonylethyl)-6-ethyl-4 (1*H*)-cinnolinone	80–82	213
4-Allylthio-3-bromo-6,7-dimethoxycinnoline	165	635
3-Amino-7-chloro-4-cinnolinecarboxylic acid 1-oxide	254	147
3-Amino-6-chloro-4(1*H*)-cinnolinone	—	(*E* 113, 244)
6-Amino-8-chloro-4(1*H*)-cinnolinone	—	(*E* 114, 244)
8-Amino-6-chloro-4(1*H*)-cinnolinone	—	(*E* 113)
4-Amino-7-chloro-6-fluoro-3-cinnolinecarboxamide	350; PhCH=: 310	556, 670
4-Amino-7-chloro-6-fluoro-3-cinnolinecarboxylic acid	305; PhCH=: 290	556, 670
4-Amino-3-cinnolinecarbonitrile	318–320 or >360, IR, NMR, UV	291, 487, 677, 1011
4-Amino-3-cinnolinecarbothioamide	315–317, UV	487
4-Amino-3-cinnolinecarboxamide	286–287 or 287–289, IR, UV	54, 487, 502, 556, 667
4-Amino-3-cinnolinecarboxylic acid	Na($4H_2O$): xl st	537, 556
3-Amino-4-cinnolinecarboxylic acid 1-oxide	215	147
2-Amino-3(2*H*)-cinnolinone	130–131, IR, MS, NMR	687
4-Amino-3(2*H*)-cinnolinone	—	772
6-Amino-4(1*H*)-cinnolinone	—	(*H* 19; *E* 242)
7-Amino-4(1*H*)-cinnolinone	—	(*H* 18; *E* 234)
8-Amino-4(1*H*)-cinnolinone	—	(*H* 18; *E* 234)
8-Amino-7-cyano-3-oxo-2,6-diphenyl-2, 3-dihydro-4-cinnolinecarboxylic acid	>270, IR	535
4-Amino-5,7-dimethyl-3-cinnolinecarbonitrile	270–275, IR, NMR	1011
4-Amino-5,8-dimethyl-3-cinnolinecarbonitrile	305–306, IR, NMR	830
4-Amino-6,8-dimethyl-3-cinnolinecarbonitrile	189, IR, NMR	830

TABLE A.1. (*Continued*)

Cinnoline	Melting Point (°C) etc.	Reference(s)
4-Amino-*N*-isopropyl-5,7-dimethyl-3-cinnolinecarboxamidine	170–173, IR, NMR	1011
4-Amino-7-methoxy-3-cinnolinecarboxamide	326–329, NMR, UV	487
4-Amino-7-methoxy-3-cinnolinecarboxylic acid	252–254, NMR	487
3-Amino-7-methoxy-4-cinnolinecarboxylic acid 1-oxide	215	147
4-Amino-6-methyl-3-cinnolinecarboxamide	323–324	502, 667
4-Amino-7-methyl-3-cinnolinecarboxamide	331–333, NMR, UV	487, 502, 667
4-Amino-8-methyl-3-cinnolinecarboxamide	318–319	502, 667
4-Amino-6-methyl-3-cinnolinecarboxylic acid	complexes	562
4-Amino-7-methyl-3-cinnolinecarboxylic acid	complexes	562
4-Amino-8-methyl-3-cinnolinecarboxylic acid	complexes	562
3-Amino-7-methyl-4-cinnolinecarboxylic acid 1-oxide	264	147
4-Amino-7-oxo-1,7-dihydro-3-cinnolinecarboxamide	HCl: >360, IR, NMR	487
6-Amino-4-oxo-1,4-dihydro-3-cinnolinecarboxylic acid	—	(*E* 242)
8-Amino-3-oxo-2,6-diphenyl-2,3-dihydro-4,7-cinnolinedicarbonitrile	>270, IR, NMR	535
5-Amino-3-oxo-7-phenyl-2,3-dihydro-4,6-cinnolinedicarbonitrile	275, IR, NMR	67
1-Amino-3-phenylcinnolin-1-ium mesylate	186–188	5
3-Amino-5,7,8-tribromo-6-oxo-2,6-dihydro-4-cinnolinecarbonitrile	>360, IR, MS, NMR	973
3-Amino-5,7,8-trichloro-6-oxo-2,6-dihydro-4-cinnolinecarbonitrile	>360, IR, MS, NMR	973
4-Anilino-3-bromocinnoline	194–196	293
4-Anilino-6-cinnolinamine	—	(*E* 242)
4-Anilinocinnoline	—	(*H* 37; *E* 228, 229, 244)
4-Anilino-6,7-dimethoxycinnoline	—	(*E* 164, 244)
4-Anilino-6-methoxycinnoline	—	(*H* 37)
4-Anilino-3-methoxycinnoline 1-oxide	193–195	72, 709
4-Anilino-3-methylcinnoline	—	(*H* 37)
4-Anilino-8-methylcinnoline	—	(*H* 37; *E* 242)
4-Anilino-7-methyl-8-nitrocinnoline	—	(*H* 37; *E* 202, 242)
4-Amino-8-methyl-5-nitrocinnoline	—	(*E* 202)
4-Anilino-8-methyl-5/7-nitrocinnoline	—	(*H* 37)
4-Anilino-3-nitrocinnoline	—	(*E* 201, 240)
4-Anilino-6-nitrocinnoline	—	(*H* 37; *E* 200, 228)
4-Azidocinnoline	245–246, IR, NMR	402
4-Azidocinnoline 1-oxide	129, IR, NMR	402
4-Azidocinnoline 2-oxide	184–185, IR, NMR	402
3-Benzoyl-7-chlorocinnoline	181–183, IR, NMR	824
3-Benzoyl-4-cinnolinamine	150–153, IR, MS, NMR; oxime: 210, IR, NMR	1011 824
3-Benzoylcinnoline	138–139, IR, NMR	824
4-Benzoylcinnoline	103–104, IR, NMR	586, 601
3-Benzoyl-5,7-dimethyl-4-cinnolinamine	168, IR, NMR	1011

TABLE A.1. *(Continued)*

Cinnoline	Melting Point (°C) etc.	Reference(s)
3-Benzoyl-5,8-dimethyl-4-cinnolinamine	160, IR, NMR	830
3-Benzoyl-6,8-dimethyl-4-cinnolinamine	330, IR, NMR	830
3-Benzoyl-6-methoxycinnoline	195–197, IR, MS	824
3-Benzoyl-6-nitrocinnoline	188–190, IR	824
3-Benzoyloxy-6-phenyl-4(1*H*)-cinnolinone	165	140
3-Benzoyl-1-phenyl-4(1*H*)-cinnolinone	183	140
4-Benzylaminocinnoline	193–195, IR, NMR, UV	993
4-Benzylamino-3-cinnolinecarbonitrile	231–232, IR, NMR	291
4-Benzylamino-3-phenylcinnoline	—	(*E* 230) 980
1-Benzyl-6-chloro-4(1*H*)-cinnolinone	173–174, NMR	334
3-Benzylcinnoline	—	(*E* 54)
4-Benzylcinnoline	—	(*E* 46)
2-Benzylcinnolin-2-ium-4-olate	NMR	693
1-Benzyl-4(1*H*)-cinnolinone	125–126, NMR	334
2-Benzyl-3(2*H*)-cinnolinone	146–149, IR	42
4-Benzyl-3(2*H*)-cinnolinone	—	(*E* 96, 117)
1-Benzyl-3-(2-ethoxycarbonylethyl)-6-ethyl-4(1*H*)-cinnolinone	110–112	213
3-Benzyl-2-methylcinnolin-2-ium-4-olate	NMR	693
1-Benzyl-8-methyl-4(1*H*)-cinnolinone	120–121, NMR	334
6-Benzyloxy-3-(2-carboxyethyl)-4(1*H*)-cinnolinone	245–247	213
4-Benzyloxycinnoline	—	(*E* 158)
7-Benzyloxy-6-methoxycinnoline	—	(*E* 163)
3-Benzyl-4-phenylcinnoline	—	(*H* 8)
3-Benzyl-4-phenylcinnoline 1-oxide	—	(*H* 9)
4-Benzylsulfonyl-6-chlorocinnoline	—	(*E* 186)
4-Benzylthiocinnoline	—	(*E* 183)
2-Benzylthio-6-ethylthiocinnoline	—	(*E* 183)
4-Benzylthio-6-fluorocinnoline	—	(*E* 184)
4,5-Bismethylamino-6-nitrocinnoline	—	(*E* 204, 244)
4,6-Bismethylthiocinnoline	—	(*E* 183)
3-Bromo-4-chlorocinnoline	—	(*E* 142)
6-Bromo-4-chlorocinnoline	—	(*H* 30)
3-Bromo-6-chloro-4(1*H*)-cinnolinone	—	(*H* 20)
6-Bromo-3-chloro-4(1*H*)-cinnolinone	—	(*E* 113)
3-Bromo-4-chloro-6,7-dimethoxycinnoline	—	(*E* 145, 163)
3-Bromo-4-cinnolinamine	—	(*E* 240)
3-Bromocinnoline	93–94	(*E* 142, 144) 908
6-Bromocinnoline	127–128, IR, NMR	(*E* 145) 816, 817, 819
8-Bromocinnoline	NMR	289
6-Bromo-4(1*H*)-cinnolinethione	—	(*E* 180, 183)
8-Bromo-4(1*H*)-cinnolinethione	—	(*E* 180, 183)
3-Bromo-4(1*H*)-cinnolinone	—	(*H* 19; *E* 112)
5-Bromo-4(1*H*)-cinnolinone	—	(*E* 99, 112)
6-Bromo-4(1*H*)-cinnolinone	—	(*H* 18; *E* 99)
7-Bromo-4(1*H*)-cinnolinone	—	(*E* 100, 112)
8-Bromo-4(1*H*)-cinnolinone	—	(*H* 18; *E* 101, 112)
8-Bromo-3-cyclohexyl-6,7-dimethoxy-4(1*H*)-cinnolinone	153–155, NMR	20

TABLE A.1. (*Continued*)

Cinnoline	Melting Point (°C) etc.	Reference(s)
4-Bromo-6,8-difluoro-3-pentylcinnoline	—	94
3-Bromo-6,7-dimethoxy-4(1*H*)-cinnolinethione	219	635
3-Bromo-6,7-dimethoxy-4(1*H*)-cinnolinone	—	(*E* 113)
3-Bromo-6,7-dimethoxy-1-methyl-4(1*H*)-cinnolinethione	208	635
3-Bromo-6,7-dimethoxy-4-methylthiocinnoline	164	635
6-Bromo-2-ethylcinnolin-2-ium-4-olate	168–169, NMR	12
6-Bromo-2-ethyl-4-oxo-1,4-dihydrocinnolin-2-ium-3-carboxylate	212–213, NMR	12
6-Bromo-4-hydrazinocinnoline	—	(*E* 238)
6-Bromo-4-methoxycinnoline	—	(*E* 168)
3-Bromo-4-methylcinnoline	—	(*E* 144)
6-Bromo-4-methylcinnoline	—	(*E* 40)
3-Bromo-2-methylcinnolin-2-ium-4-olate	—	(*E* 95)
6-Bromo-2-methylcinnolin-2-ium-4-olate	229–230, NMR	12
3-Bromo-1-methyl-4(1*H*)-cinnolinone	—	(*E* 95)
3-Bromo-8-methyl-4(1*H*)-cinnolinone	285–287	(*E* 102, 112) 908
5-Bromo-6-methyl-4(1*H*)-cinnolinone	—	(*E* 94)
5/7-Bromo-6-methyl-4(1*H*)-cinnolinone	—	(*E* 111)
6-Bromo-3-methyl-4(1*H*)-cinnolinone	—	(*H* 20)
6-Bromo-7-methyl-4(1*H*)-cinnolinone	—	(*H* 20)
5-Bromo-6-methyl-4-oxo-1,4-dihydro-3-cinnolinecarboxylic acid	—	(*E* 94)
5/7-Bromo-6-methyl-4-oxo-1,4-dihydro-3-cinnolinecarboxylic acid	—	(*E* 110)
6-Bromo-1-methyl-4-oxo-1,4-dihydro-3-cinnolinecarboxylic acid	290–291, NMR	12
6-Bromo-2-methyl-4-oxo-1,4-dihydrocinnolin-2-ium-3-carboxylate	crude, NMR	12
3-Bromo-6-nitro-4(1*H*)-cinnolinone	—	(*E* 103, 112, 202)
3-Bromo-8-nitro-4(1*H*)-cinnolinone	—	(*E* 112, 202)
4-Bromo-6-nitro-3-phenylcinnoline	184–185, IR, NMR	492
6-Bromo-4-oxo-1,4-dihydro-3-cinnolinecarboxylic acid	—	(*H* 21; *E* 110)
8-Bromo-4-oxo-1,4-dihydro-3-cinnolinecarboxylic acid	—	(*E* 111)
6-Bromo-4-oxo-1-propyl-1,4-dihydro-3-cinnolinecarboxylic acid	210–212, NMR	12
6-Bromo-4-phenoxycinnoline	—	(*H* 33)
4-Bromo-3-phenylcinnoline	150–151, IR, NMR	94, 388, 492
6-Bromo-4-phenylcinnoline	—	(*H* 8)
6-Bromo-2-propylcinnolin-2-ium-4-olate	151–153, NMR	12
6-Bromo-1-propyl-4(1*H*)-cinnolinone	136–138, NMR	12
4-Bromo-3-trimethylsilyl-6-trimethylsilylethynylcinnoline	—	94
4-Butylaminocinnoline	—	(*E* 238)
3-*tert*-Butyl-4-*tert*-butylsulfinylcinnoline	liq, IR, NMR	1010
3-*tert*-Butyl-4-*tert*-butylthiocinnoline	86, IR, NMR	1010
3-*tert*-Butyl-4-chlorocinnoline	113, IR, NMR	1010
4-Butylcinnoline	—	(*E* 58)

TABLE A.1. (*Continued*)

Cinnoline	Melting Point (°C) etc.	Reference(s)
4-*tert*-Butylcinnoline	—	(*E* 40, 54)
6-*tert*-Butylcinnoline	liq, IR, NMR	817, 819
3-Butyl-4-cinnolinecarbonitrile	MS	401
3-*tert*-Butyl-4(1*H*)-cinnolinone	244, IR, NMR	1010
3-Butyl-2-methylcinnolin-2-ium-4-olate	NMR	693
4-*tert*-Butyl-8-nitrocinnoline	—	(*E* 40, 204)
3-*tert*-Butylsulfinylcinnoline	129, IR, NMR	1010
4-*tert*-Butylsulfinylcinnoline	117, IR, NMR	1010
4-*tert*-Butylsulfinyl-6,7-dimethoxycinnoline	172, IR, NMR	1010
4-Butylsulfonylcinnoline	—	(*E* 186)
4-*tert*-Butylsulfonylcinnoline	140, IR, NMR	1010
3-*tert*-Butylthiocinnoline	69, IR, NMR	1010
4-*tert*-Butylthiocinnoline	70, IR, NMR	1010
4-*tert*-Butylthio-6,7-dimethoxycinnoline	180, IR, NMR	1010
4-Butyrylmethylcinnoline	—	(*E* 56)
3-(2-Carboxyethyl)-4(1*H*)-cinnolinone	226–228	213
3-(2-Carboxyethyl)-6-ethyl-4(1*H*)-cinnolinone	240–241	213
3-(2-Carboxyethyl)-6-isopropoxy-4(1*H*)-cinnolinone	210–213	213
3-(2-Carboxyethyl)-6-isopropyl-4(1*H*)-cinnolinone	228–230	213
3-(2-Carboxyethyl)-5-methoxy-4(1*H*)-cinnolinone	257–259	213
3-(2-Carboxyethyl)-6-methoxy-4(1*H*)-cinnolinone	256–258	213
3-(2-Carboxyethyl)-5-phenyl-4(1*H*)-cinnolinone	262–263	213
3-(2-Carboxyethyl)-6-phenyl-4(1*H*)-cinnolinone	306–310	213
3-(2-Carboxyethyl)-6-propyl-4(1*H*)-cinnolinone	230–231	213
3-Carboxymethyl-6,7-dimethoxy-4(1*H*)-cinnolinone	—	(*H* 21; *E* 103)
4-Carboxymethyl-3,6,7-trimethyl-5-cinnolinamine	>300, IR, NMR, UV	656
4-(5-Carboxypentyl)cinnoline	—	(*E* 56)
3-(3-Carboxypropyl)-4(1*H*)-cinnolinone	—	(*H* 19)
3-Chloro-4-cinnolinamine	—	(*E* 240)
6-Chloro-3-cinnolinamine	—	(*E* 227, 229, 244)
6-Chloro-4-cinnolinamine	—	(*H* 35; *E* 229, 244)
7-Chloro-3-cinnolinamine	—	(*E* 227, 229, 244)
7-Chloro-4-cinnolinamine	—	(*H* 35)
3-Chlorocinnoline	90 or 91–92	(*E* 143) 307, 908, 1010
4-Chlorocinnoline	76–77 or 77–78, NMR	(*H* 30) 94, 307, 822, 908, 1012
6-Chlorocinnoline	128–130 or 131–132, IR, NMR	(*E* 145) 816, 817, 819, 908
7-Chlorocinnoline	—	(*E* 145)
8-Chlorocinnoline	NMR	(*E* 145) 289
4-Chloro-3-cinnolinecarbonitrile	179–181	695
4-Chloro-3-cinnolinecarboxylic acid	>260, IR, NMR	307
7-Chloro-4-cinnolinecarboxylic acid 1-oxide	252	146
6-Chloro-3,4-cinnolinediamine	—	(*E* 244)

TABLE A.1. (*Continued*)

Cinnoline	Melting Point (°C) etc.	Reference(s)
6-Chloro-3,4(1*H*,2*H*)-cinnolinedione	180–181, IR, MS, NMR	809
3-Chlorocinnoline 1-oxide	—	(*E* 295)
4-Chlorocinnoline 1-oxide	—	(*E* 295)
4-Chlorocinnoline 2-oxide	—	(*E* 297)
5-Chloro-4(1*H*)-cinnolinethione	—	(*E* 179, 183)
6-Chloro-4(1*H*)-cinnolinethione	—	(*E* 179, 183)
7-Chloro-4(1*H*)-cinnolinethione	—	(*E* 180, 183)
8-Chloro-4(1*H*)-cinnolinethione	—	(*E* 180, 183)
3-Chloro-4(1*H*)-cinnolinone	278–279	(*H* 19) 908
4-Chloro-3(2*H*)-cinnolinone	220–223	(*E* 117) 687
5-Chloro-3(2*H*)-cinnolinone	—	(*E* 117)
5-Chloro-4(1*H*)-cinnolinone	—	(*E* 92, 98, 111)
6-Chloro-3(2*H*)-cinnolinone	—	(*E* 97, 108, 116)
6-Chloro-4(1*H*)-cinnolinone	296–297, NMR, xl st; 2-MeOTs: 165, IR, UV	(*H* 18; *E* 92, 99, 113) 335, 634, 690
7-Chloro-3(2*H*)-cinnolinone	—	(*E* 97, 108, 117)
7-Chloro-4(1*H*)-cinnolinone	—	(*H* 18; *E* 92, 100)
8-Chloro-4(1*H*)-cinnolinone	—	(*H* 18; *E* 93, 101)
4-Chloro-3-cyclohexyl-6,7-dimethoxycinnoline	184–185, NMR	20
4-Chloro-6,8-difluoro-3-pentylcinnoline	—	94
4-Chloro-3,8-diiodocinnoline	242, IR, NMR	307
4-Chloro-6,7-dimethoxycinnoline	—	(*E* 143, 161)
4-Chloro-6,7-dimethoxy-3-methoxycarbonylmethylcinnoline	—	(*E* 41, 142, 158, 168)
4-Chloro-6,7-dimethoxy-3-methylcinnoline	—	(*E* 41, 142, 158, 168)
3-Chloro-6,7-dimethyl-4(1*H*)-cinnolinone	—	(*H* 21)
6-Chloro-1-ethyl-4(1*H*)-cinnolinone	134–135, NMR	334
7-Chloro-1-ethyl-3-methylsulfonyl-4(1*H*)-cinnolinone	153–156	485
4-Chloro-6-fluorocinnoline	—	(*E* 142)
7-Chloro-6-fluoro-1-methyl-4-oxo-1,4-dihydro-3-cinnolinecarboxylic acid	214–215, IR, MS, NMR	416
4-Chloro-3-hexylcinnoline	71–72	388
3-Chloro-4-hydrazinocinnoline	—	(*E* 238)
6-Chloro-4-hydrazinocinnoline	—	(*E* 238)
3-Chloro-4-(1-hydroxyethyl)cinnoline	144, IR, NMR	307
4-Chloro-3-(1-hydroxyethyl)cinnoline	90, NMR	307
4-Chloro-3-(3-hydroxy-3-methylbut-1-ynyl)cinnoline	90–91, IR, NMR	492
4-Chloro-3-(1-hydroxy-1-methylethyl)cinnoline	90–91	388
4-Chloro-3-iodocinnoline	165–166, IR, NMR	293, 307
4-Chloro-8-iodo-3-methoxycinnoline	182, IR, NMR	307
4-Chloro-3-methoxycinnoline	110, IR, NMR	307
4-Chloro-5-methoxycinnoline	—	(*E* 142, 162)
4-Chloro-6-methoxycinnoline	—	(*H* 30)
4-Chloro-7-methoxycinnoline	—	(*E* 142, 162)
4-Chloro-8-methoxycinnoline	—	(*E* 142, 162)
6-Chloro-4-methoxycinnoline	—	(*H* 31)

TABLE A.1. (*Continued*)

Cinnoline	Melting Point (°C) etc.	Reference(s)
4-Chloro-3-methoxycinnoline 1-oxide	165–167	(*E* 295) 709
4-Chloro-6-methoxy-4(1*H*)-cinnolinone	154–155, IR, NMR	27
6-Chloro-4-methylaminocinnoline	—	(*E* 244)
3-Chloro-4-methylcinnoline	—	(*E* 144)
4-Chloro-3-methylcinnoline	88, IR, NMR	(*H* 30) 307
4-Chloro-7-methylcinnoline	—	(*H* 30)
4-Chloro-8-methylcinnoline	—	(*H* 30)
6-Chloro-4-methylcinnoline	—	(*H* 14, 40)
7-Chloro-4-methylcinnoline	—	(*H* 14, 40)
8-Chloro-4-methylcinnoline	—	(*H* 14; *E* 144)
6-Chloro-1-methyl-4(1*H*)-cinnolinimine	—	(*H* 40)
7-Chloro-1-methyl-4(1*H*)-cinnolinimine	—	(*H* 40)
6-Chloro-2-methylcinnolin-2-ium-4-olate	218 or 221–223, IR, UV	690, 917
8-Chloro-2-methylcinnolin-2-ium-4-olate	NMR	693
3-Chloro-6-methyl-4(1*H*)-cinnolinone	—	(*H* 20)
6-Chloro-1-methyl-4(1*H*)-cinnolinone	—	(*H* 49)
6-Chloro-3-methyl-4(1*H*)-cinnolinone	—	(*H* 20)
6-Chloro-7-methyl-4(1*H*)-cinnolinone	—	(*H* 20)
8-Chloro-7-methyl-4(1*H*)-cinnolinone	—	(*H* 20)
4-Chloromethyl-3-methylcinnoline	139–140, NMR	291
4-Chloro-3-methyl-6-nitrocinnoline	—	(*H* 30)
4-Chloro-3-methyl-8-nitrocinnoline	—	(*H* 30)
4-Chloro-7-methyl-8-nitrocinnoline	—	(*H* 30; *E* 202)
4-Chloro-8-methyl-5-nitrocinnoline	—	(*E* 202)
4-Chloro-8-methyl-5/7-nitrocinnoline	—	(*H* 30)
6-Chloro-7-methyl-4-phenoxycinnoline	—	(*H* 33)
6-Chloro-3-methyl-4-phenylcinnoline	165–166, MS, NMR, UV	290
6-Chloro-8-methyl-4-phenylcinnoline	—	(*E* 56)
6-Chloro-4-methylsulfonylcinnoline	—	(*E* 186)
8-Chloro-4-methylsulfonylcinnoline	—	(*E* 186)
7-Chloro-3-methylsulfonyl-4(1*H*)-cinnolinone	325–330	485
6-Chloro-4-methylthiocinnoline	—	(*E* 185)
6-Chloro-3-nitro-4-cinnolinamine	—	(*E* 204, 244)
4-Chloro-3-nitrocinnoline	—	(*E* 142, 201)
4-Chloro-5-nitrocinnoline	—	(*H* 30; *E* 203)
4-Chloro-6-nitrocinnoline	228–229	(*H* 30; *E* 143, 200) 908
4-Chloro-7-nitrocinnoline	—	(*H* 30; *E* 203)
4-Chloro-8-nitrocinnoline	—	(*H* 30; *E* 203)
6-Chloro-3-nitrocinnoline	—	(*E* 145, 203)
7-Chloro-3-nitrocinnoline	—	(*E* 144, 203)
4-Chloro-5-nitrocinnoline 1-oxide	—	(*E* 296)
5-Chloro-6-nitro-4(1*H*)-cinnolinone	—	(*E* 94, 114, 204)
5-Chloro-8-nitro-4(1*H*)-cinnolinone	—	(*E* 94, 114, 204)
6-Chloro-3-nitro-4(1*H*)-cinnolinone	—	(*E* 103, 113, 204)
6-Chloro-5-nitro-4(1*H*)-cinnolinone	—	(*E* 94, 113, 204)
6-Chloro-8-nitro-4(1*H*)-cinnolinone	—	(*E* 94, 113, 204)
7-Chloro-6-nitro-4(1*H*)-cinnolinone	—	(*H* 20)
7-Chloro-8-nitro-4(1*H*)-cinnolinone	—	(*H* 20)
8-Chloro-6-nitro-4(1*H*)-cinnolinone	—	(*E* 114, 204)

TABLE A.1. (*Continued*)

Cinnoline	Melting Point (°C) etc.	Reference(s)
6-Chloro-8-nitro-4-oxo-1,4-dihydro-3-cinnolinecarboxylic acid	—	(*E* 94, 113, 204)
8-Chloro-6-nitro-4-oxo-1,4-dihydro-3-cinnolinecarboxylic acid	—	(*E* 94, 114, 204)
6-Chloro-3-nitro-4-phenoxycinnoline	—	(*E* 146, 163, 204)
4-Chloro-6-nitro-3-phenylcinnoline	178–179, IR, NMR	492
6-Chloro-4-oxo-1,4-dihydro-3-cinnolinecarbohydrazide	—	(*E* 114)
5-Chloro-4-oxo-1,4-dihydro-3-cinnolinecarboxylic acid	—	(*E* 92, 110)
6-Chloro-4-oxo-1,4-dihydro-3-cinnolinecarboxylic acid	—	(*H* 21; *E* 110)
8-Chloro-4-oxo-1,4-dihydro-3-cinnolinecarboxylic acid	247–248	(*E* 93, 110) 908
5-Chloro-4-oxo-1,4-dihydro-8-cinnolinesulfonic acid	—	(*E* 94, 114)
8-Chloro-4-oxo-1,4-dihydro-5-cinnolinesulfonic acid	—	(*E* 94, 114)
5-Chloro-3,4,6,7,8-pentafluorocinnoline	—	(*E* 146)
3-Chloro-4-phenoxycinnoline	—	(*E* 142, 160)
5-Chloro-4-phenoxycinnoline	—	(*E* 143, 163)
6-Chloro-4-phenoxycinnoline	—	(*H* 33; *E* 145, 163)
7-Chloro-4-phenoxycinnoline	—	(*H* 33)
8-Chloro-4-phenoxycinnoline	—	(*H* 33; *E* 145)
4-Chloro-6-phenylazocinnoline	—	(*E* 143)
6-Chloro-3-phenyl-4-cinnolinamine	280, IR	940
4-Chloro-3-phenylcinnoline	118–119 or 120–121	(*H* 30; *E* 143) 94, 388, 492, 908
4-Chloro-6-phenylcinnoline	—	(*E* 142)
4-Chloro-7-phenylcinnoline	—	(*E* 142)
6-Chloro-4-phenylcinnoline	141–142	(*E* 56) 637
6-Chloro-4-phenylcinnoline 1-oxide	165–167, NMR, UV	11
6-Chloro-4-phenylcinnoline 2-oxide	195–197, NMR, UV	11
5-Chloro-3-phenyl-4(1*H*)-cinnolinone	—	(*H* 21)
4-Chloro-3-phenylethynylcinnoline	135–136	293
3-Chloro-4-(*N*′-phenylhydrazino)cinnoline	—	(*E* 238)
6-Chloro-4-phenylsulfonylcinnoline	—	(*E* 186)
6-Chloro-3-propionyloxy-4(1*H*)-cinnolinone	—	(*E* 113)
6-Chloro-4-propylaminocinnoline	—	(*E* 244)
6-Chloro-1-propyl-4(1*H*)-cinnolinone	141–142, NMR	334
3-Cinnolinamine	158 or 165–166; 1-MeI: 174–176; 2-MeI: 246–248	(*E* 226, 229, 238, 248) 147, 908
4-Cinnolinamine	205 to 214, IR, NMR, UV	(*H* 35; *E* 228, 229, 238, 248) 307, 908, 993
5-Cinnolinamine	156–159; HCl: 192–196, IR	(*E* 228, 229, 240, 248) 217
6-Cinnolinamine	—	(*E* 227, 229, 240)
7-Cinnolinamine	—	(*E* 227, 229, 240)

TABLE A.1. (*Continued*)

Cinnoline	Melting Point (°C) etc.	Reference(s)
8-Cinnolinamine	89–92 or 95–96, NMR; HCl: 171–174 or 177–179	(*E* 227, 229, 240, 248) 217, 289, 403, 908
6-Cinnolinamine 2-oxide	—	(*E* 297)
8-Cinnolinamine 2-oxide	—	(*E* 297)
Cinnoline	38, IR, MS, NMR, pK_a, UV; pic: 191	(*H* 4, 40; *E* 1, 12–15) 152, 289, 341, 527, 605, 757, 763, 813, 817, 819, 908, 967
3-Cinnolinecarbaldehyde	pK_a	(*E* 266) 101
4-Cinnolinecarbaldehyde	147–149; H_2NN=: 292 294; sc: 234–235	(*E* 56, 265) 908
4-Cinnolinecarbohydrazide	177–178, IR, NMR, UV	616
4-Cinnolinecarbonitrile	139–140 or 140–141, IR, MS	(*E* 265) 401, 908, 937
5-Cinnolinecarbonitrile	178, IR, MS, NMR	217
6-Cinnolinecarbonitrile	129–131, IR, NMR	817, 819
4-Cinnolinecarbonyl chloride	—	(*E* 266)
3-Cinnolinecarboxylic acid	—	(*E* 266)
4-Cinnolinecarboxylic acid	—	(*E* 266) 291, 908
4-Cinnolinecarboxylic acid 1-oxide	222	146
4-Cinnolinecarboxylic acid 2-oxide	—	(*E* 293, 298)
3,4-Cinnolinediamine	220 or 228–230	(*E* 240) 370
4,6-Cinnolinediamine	—	(*E* 229, 236)
4,7-Cinnolinediamine	—	(*E* 232)
4,8-Cinnolinediamine	—	(*H* 35; *E* 234)
3,4-Cinnolinedicarbaldehyde	144–145, IR, MS, NMR	711
3,4-Cinnolinedicarbonitrile	185–186, Raman	291
4,6-Cinnolinedicarbonitrile	—	(*E* 265)
3,4-Cinnolinedicarboxamide	340	291
3,4-Cinnolinedicarboxylic acid	178, IR, UV	291
3,4(1*H*,2*H*)-Cinnolinedione	—	772
Cinnoline 1,2-dioxide	235	(*E* 291, 293) 908
Cinnoline 1-oxide	110–111 or 111–112, NMR	(*E* 291, 292, 295, 296) 760, 908
Cinnoline 2-oxide	122–123, NMR	(*E* 291, 293, 297) 760, 908
4(1*H*)-Cinnolinethione	204–205	(*E* 179, 182) 908
3(2*H*)-Cinnolinone	198 to 204, st	(*E* 96, 108, 116) 2, 38, 54, 187, 373, 687, 760, 908
4(1*H*)-Cinnolinone	225 to 235, NMR, st, xl st	(*H* 18; *E* 92, 97, 108) 94, 335, 509, 760, 807, 822
5(1*H*)-Cinnolinone	—	(*E* 104, 108, 118)
6(2*H*)-Cinnolinone	—	(*E* 105, 108, 118)
7(1*H*)-Cinnolinone	—	(*E* 106, 108, 118)
8(2*H*)-Cinnolinone	.NMR	(*E* 107, 108, 118) 289

TABLE A.1. (*Continued*)

Cinnoline	Melting Point (°C) etc.	Reference(s)
4(1*H*)-Cinnolinone 2-oxide	—	(*E* 297)
4-(2-Cyanoethyl)cinnoline	—	605
4-Cyclohexylaminocinnoline	160–162, IR, NMR, UV	993
3-Cyclohexyl-6,7-dimethoxy-4(1*H*)-cinnolinone	222–223, IR, NMR	20
N-Cyclopentyl-6-nitro-4-oxo-1,4-dihydro-3-cinnolinecarboxamide	NMR	1034
4,6-Diacetamidocinnoline	—	(*H* 40)
4,7-Diacetamidocinnoline	—	(*H* 35; *E* 234)
4,8-Diacetamidocinnoline	—	(*H* 35; *E* 234)
4,6-Diacetamido-3-methylcinnoline	—	(*H* 42)
3,4-Dibenzoylcinnoline	—	(*E* 265)
3,4-Dibenzylcinnoline	—	(*E* 54)
6,7-Dibenzyloxycinnoline	—	(*E* 163)
3,6-Dibromo-4(1*H*)-cinnolinone	—	(*E* 113)
6,8-Dibromo-4(1*H*)-cinnolinone	—	(*H* 20)
3,6-Dichloro-4-cinnolinamine	—	(*E* 244)
3,4-Dichlorocinnoline	—	(*E* 142)
4,6-Dichlorocinnoline	—	(*H* 30; *E* 145)
4,7-Dichlorocinnoline	—	(*H* 30)
4,8-Dichlorocinnoline	—	(*H* 30; *E* 145)
3,6-Dichloro-4(1*H*)-cinnolinone	—	(*H* 20; *E* 93, 114)
5,6-Dichloro-4(1*H*)-cinnolinone	—	(*E* 93, 111)
5,8-Dichloro-4(1*H*)-cinnolinone	—	(*E* 93)
6,7-Dichloro-4(1*H*)-cinnolinone	—	(*H* 20; *E* 93)
6,8-Dichloro-4(1*H*)-cinnolinone	—	(*E* 93, 111)
7,8-Dichloro-4(1*H*)-cinnolinone	—	(*H* 20; *E* 93, 111)
3,4-Dichloro-6,7-dimethoxycinnoline	—	(*E* 145, 164)
4,6-Dichloro-7-methylcinnoline	—	(*H* 30)
7,8-Dichloro-4-methylsulfonylcinnoline	—	(*E* 186)
4,5-Dichloro-6-nitrocinnoline	—	(*E* 146)
4,6-Dichloro-3-nitrocinnoline	—	(*E* 146, 204)
6,7-Dichloro-3-nitro-4(1*H*)-cinnolinone	—	(*E* 93, 114)
7,8-Dichloro-6-nitro-4(1*H*)-cinnolinone	—	(*E* 94, 114, 205)
7,8-Dichloro-6-nitro-4-oxo-1,4-dihydro-3-cinnolinecarboxylic acid	—	(*E* 94, 114, 204)
5,6-Dichloro-4-oxo-1,4-dihydro-3-cinnolinecarboxylic acid	—	(*E* 110)
5,8-Dichloro-4-oxo-1,4-dihydro-3-cinnolinecarboxylic acid	—	(*E* 110)
6,8-Dichloro-4-oxo-1,4-dihydro-3-cinnolinecarboxylic acid	—	(*E* 93, 110)
7,8-Dichloro-4-oxo-1,4-dihydro-3-cinnolinecarboxylic acid	—	(*E* 93, 110)
3,6-Dichloro-4-phenoxycinnoline	—	(*E* 146)
5,6-Dichloro-4-phenoxycinnoline	—	(*E* 143, 163)
6,7-Dichloro-4-phenoxycinnoline	—	(*H* 33; *E* 145)
7,8-Dichloro-4-phenoxycinnoline	—	(*H* 33; *E* 145)
5,7-Dichloro-3-phenyl-4-cinnolinamine	>280, IR	940
6,7-Dichloro-3-phenyl-4-cinnolinamine	>280, IR	940

TABLE A.1. (*Continued*)

Cinnoline	Melting Point (°C) etc.	Reference(s)
6,8-Dichloro-3-phenyl-4-cinnolinamine	>280, IR	940
6-Diethylaminocinnoline	liq, IR, NMR	816
4-Diethylamino-3-phenylethynylcinnoline	116–117	293
3,4-Diethylcinnoline	2-PhBF_4: 216–217, NMR	769
6,7-Difluoro-2-methylcinnolin-2-ium-4-olate	218–219, IR, NMR, UV	416
6,7-Difluoro1-methyl-4-oxo-1,4-dihydro-3-cinnolinecarboxylic acid	228–229, MS, NMR	416
4,8-Diiodo-3-methoxycinnoline	162, IR, NMR	307
6,7-Dimethoxy-4-cinnolinamine	—	236
6,7-Dimethoxycinnoline	—	(*E* 163, 168)
3,4-Dimethoxycinnoline 1-oxide	—	(*E* 295)
6,7-Dimethoxy-4(1*H*)-cinnolinethione	—	(*E* 182)
6,7-Dimethoxy-4(1*H*)-cinnolinone	—	(*E* 92, 111)
6,7-Dimethoxy-3-methoxycarbonylmethyl-4(1*H*)-cinnolinone	—	(*E* 104, 112)
6,7-Dimethoxy-4-methylcinnoline	—	(*E* 163, 168)
6,7-Dimethoxy-2-methylcinnolin-2-ium-4-olate	NMR	(*E* 95) 693
6,7-Dimethoxy-1-methyl-4(1*H*)-cinnolinone	205–206	(*E* 95) 509
6,7-Dimethoxy-3-methyl-4(1*H*)-cinnolinone	—	(*E* 104, 112)
6,7-Dimethoxy-4-methylsulfonylcinnoline	—	(*E* 186)
6,7-Dimethoxy-4-methylthiocinnoline	—	(*E* 182)
6,7-Dimethoxy-4-phenoxycinnoline	—	162
6,7-Dimethoxy-4-phenylcinnoline 2-oxide	—	(*E* 198)
6,7-Dimethoxy-4-phenylthiocinnoline	—	(*E* 184)
3-Dimethylaminocinnoline	—	(*E* 240)
4-Dimethylaminocinnoline	—	(*E* 240)
3-Dimethylamino-4-cinnolinecarbonitrile	138–139, NMR	291
4-Dimethylamino-3-cinnolinecarbonitrile	180–181	291
Dimethyl 6-bromo-3,4-cinnolinedicarboxylate	147–149, IR, NMR	43
3,4-Dimethylcinnoline	117–120, MS, NMR	(*E* 40, 54) 40, 908
4,6-Dimethylcinnoline	—	(*E* 40, 54)
4,8-Dimethylcinnoline	—	(*E* 40, 54)
5,7-Dimethyl-3,4(1*H*,2*H*)-cinnolinedione	213–214, IR, MS, NMR	809
3,4-Dimethylcinnoline 1,2-dioxide	—	(*E* 294)
3,4-Dimethylcinnoline 2-oxide	—	(*E* 293, 298)
4,6-Dimethylcinnoline 1-oxide	—	(*E* 292, 296)
4,6-Dimethylcinnoline 2-oxide	—	(*E* 293, 298)
2,3-Dimethylcinnolin-2-ium-4-olate	—	(*E* 95)
1,3-Dimethyl-4(1*H*)-cinnolinone	—	(*E* 95)
6,7-Dimethyl-4(1*H*)-cinnolinone	—	(*H* 20)
3,4-Dimethyl-8-nitrocinnoline	—	(*E* 40, 204)
1,3-Dimethyl-6-nitro-4(1*H*)-cinnolinimine	—	(*H* 42)
1,3-Dimethyl-8-nitro-4(1*H*)-cinnolinimine	—	(*H* 42)
1,3-Dimethyl-6-nitro-4(1*H*)-cinnolinone	—	(*H* 49)
1,8-Dimethyl-5/7-nitro-4(1*H*)-cinnolinone	—	(*H* 49)
6,7-Dimethyl-3-phenyl-5-cinnolinamine	209–211, IR, NMR	139, 656
6,7-Dimethyl-4-phenylcinnoline	—	(*E* 56)

TABLE A.1. (*Continued*)

Cinnoline	Melting Point (°C) etc.	Reference(s)
6,8-Dimethyl-4-phenylcinnoline	—	(*E* 56)
N′,*N*′-Dimethyl-3-phenyl-4-cinnolinecarbohydrazide	—	(*E* 269)
4,5-Dinitrocinnoline 1-oxide	191–192	(*E* 296) 908
3,4-Diphenylcinnoline	151–152; 2-PhBF_4: 277–278, NMR	(*H* 8; *E* 41, 44) 768, 769, 908
3,5-Diphenylcinnoline	—	(*E* 44)
3,4-Diphenylcinnoline 1-oxide	—	(*H* 9)
3-Ethoxycarbonyl-6,7-difluoro-2-methylcinnolin-2-ium-4-olate	117–118, IR, NMR, UV	416
3-Ethoxycarbonyl-2-ethyl-6,7-difluorocinnolin-2-ium-4-olate	117–118, IR, MS, NMR, UV	416
3-(2-Ethoxycarbonylethyl)-6-ethyl-4(1*H*)-cinnolinone	179–180	213
3-(2-Ethoxycarbonylethyl)-6-ethyl-1-methyl-4(1*H*)-cinnolinone	92–94	213
3-(2-Ethoxycarbonylethyl)-1-ethyl-5-phenyl-4(1*H*)-cinnolinone	128–131	213
3-(2-Ethoxycarbonylethyl)-6-isopropoxy-4(1*H*)-cinnolinone	184–186	213
3-(2-Ethoxycarbonylethyl)-6-isopropyl-4(1*H*)-cinnolinone	155–160	213
3-(2-Ethoxycarbonylethyl)-5-phenyl-4(1*H*)-cinnolinone	161–163	213
3-(2-Ethoxycarbonylethyl)-6-phenyl-4(1*H*)-cinnolinone	201–202	213
3-(2-Ethoxycarbonylethyl)-6-propyl-4(1*H*)-cinnolinone	175–177	213
4-Ethoxycarbonylmethylcinnoline	—	(*E* 54)
2-Ethoxycarbonylmethylcinnolin-2-ium-4-olate	xl st	480
1-Ethoxycarbonylmethyl-4(1*H*)-cinnolinone	xl st	480
3-Ethoxycarbonylmethyl-6,7-dimethoxy-4(1*H*)-cinnolinone	—	(*H* 21)
4-Ethoxycinnoline	—	(*H* 31; *E* 158)
4-Ethoxycinnoline 2-oxide	—	(*E* 297)
4-Ethoxy-6,7-dimethoxycinnoline	—	(*E* 158)
2-(1-Ethoxyethyl)-3(2*H*)-cinnolinone	94–97, IR, NMR	687
4-Ethoxy-6-phenylcinnoline	—	(*E* 161)
4-Ethylamino-3-phenylethynylcinnoline	168–169	293
Ethyl 4-bromo-3-cinnolinecarboxylate	138–139	54
Ethyl 6-chloro-3-cinnolinecarboxylate	—	(*E* 144, 267)
Ethyl 7-chloro-3-cinnolinecarboxylate	—	(*E* 144, 267)
Ethyl 7-chloro-6-fluoro-1-methyl-4-oxo-1,4-dihydro-3-cinnolinecarboxylate	178–179, IR, NMR, UV	414
Ethyl 6-chloro-4-oxo-1,4-dihydro-3-cinnolinecarboxylate	—	(*E* 114)
4-Ethylcinnoline	pic: 143	(*E* 58) 933
3-Ethyl-4-cinnolinecarbonitrile	MS	(*E* 58) 401
Ethyl 3-cinnolinecarboxylate	89–91 or 97, NMR	(*E* 267) 292, 570, 702, 908

TABLE A.1. *(Continued)*

Cinnoline	Melting Point (°C) etc.	Reference(s)
Ethyl 4-cinnolinecarboxylate	—	(*E* 267)
1-Ethyl-4(1*H*)-cinnolinone	87–88, NMR	334
3-Ethyl-4(1*H*)-cinnolinone	—	(*H* 19)
2-Ethyl-6,7-difluorocinnolin-2-ium-4-olate	186–187, IR, NMR, UV	416
Ethyl 6,7-difluoro-1-methyl-4-oxo-1,4-dihydro-3-cinnolinecarboxylate	180–181, IR, MS, NMR, UV	416
Ethyl 6,7-difluoro-4-oxo-1,4-dihydro-3-cinnolinecarboxylate	270–272, IR, NMR, UV	414
1-Ethyl-6,7-difluoro-4-oxo-1,4-dihydro-3-cinnolinecarboxylic acid	193–194, MS, NMR	416
Ethyl 2-ethoxycarbonylmethyl-4-oxo 1,4-dihydro-7-cinnolinecarboxylate	—	(*H* 21)
Ethyl 1-ethyl-6,7-difluoro-4-oxo-1,4-dihydro-3-cinnolinecarboxylate	146–147, IR, NMR, UV	416
Ethyl 1-ethyl-7-methoxy-4-oxo-1,4-dihydro-3-cinnolinecarboxylate	132–133, IR, NMR; HI: 168–172 then 227–245	387
Ethyl 6-ethyl-4-oxo-1,4-dihydro-3-cinnolinecarboxylate	—	782
Ethyl 1-ethyl-6,7,8-trifluoro-4-oxo-1,4-dihydro-3-cinnolinecarboxylate	95–96, IR, NMR	416
6-Ethyl-3-(2-isopropoxycarbonylethyl)-4(1*H*)-cinnolinone	182–183	213
6-Ethyl-3-(2-methoxycarbonylethyl)-4(1*H*)-cinnolinone	173–174	213
3-Ethyl-4-methoxycinnoline	228, IR, NMR	307
Ethyl 7-methoxy-3-cinnolinecarboxylate	121–126, IR, NMR	292, 570
Ethyl 7-methoxy-4-methyl-3-cinnolinecarboxylate	156–159, IR, NMR	292, 570
1-Ethyl-7-methoxy-3-methylsulfonyl-4(1*H*)-cinnolinone	193–196	485
1-Ethyl-7-methoxy-4-oxo-1,4-dihydro-3-cinnolinecarbonitrile	248–258	485
Ethyl 6-methoxy-4-oxo-1,4-dihydro-3-cinnolinecarboxylate	—	(*H* 21)
Ethyl 7-methoxy-4-oxo-1,4-dihydro-3-cinnolinecarboxylate	258–259, IR, MS, NMR	387
1-Ethyl-7-methoxy-4-oxo-1,4-dihydro-3-cinnolinecarboxylic acid	274–275 or 277–282	387, 485
3-Ethyl-4-methylcinnoline	—	(*E* 56)
4-Ethyl-3-methylcinnoline	—	(*E* 40, 54)
Ethyl 4-methyl-7-cinnolinecarboxylate	—	(*H* 14, 40)
4-Ethyl-3-methylcinnoline 1,2-dioxide	—	(*E* 294)
4-Ethyl-3-methylcinnoline 2-oxide	—	(*E* 293, 298)
1-Ethyl-8-methyl-4(1*H*)-cinnolinone	100–101, NMR	334
1-Ethyl-4-methylimino-6-nitro-1,4-dihydrocinnoline	—	(*E* 201)
1-Ethyl-6-methyl-3-methylsulfonyl-4(1*H*)-cinnolinone	182	485

TABLE A.1. (*Continued*)

Cinnoline	Melting Point (°C) etc.	Reference(s)
4-Ethyl-3-methyl-8-nitrocinnoline	—	(*E* 40, 204)
1-Ethyl-6-methyl-4-oxo-1,4-dihydro-3-cinnolinecarbonitrile	201	485
Ethyl 1-methyl-4-oxo-1,4-dihydro-3-cinnolinecarboxylate	135–136, NMR	54
1-Ethyl-6-methyl-4-oxo-1,4-dihydro-3-cinnolinecarboxylic acid	228	485
1-Ethyl-3-methylsulfonyl-4(1*H*)-cinnolinone	161–163	485
Ethyl 4-methylthio-3-cinnolinecarboxylate	115–116	54
Ethyl 1-methyl-4-thioxo-1,4-dihydro-3-cinnolinecarboxylate	104–105	54
1-Ethyl-6-nitro-4(1*H*)-cinnolinimine	—	(*E* 201)
1-Ethyl-4-oxo-1,4-dihydro-3-cinnolinecarbonitrile	220–225	485
Ethyl 4-oxo-1,4-dihydro-3-cinnolinecarboxylate	187–188	(*H* 19; *E* 113) 54
1-Ethyl-4-oxo-1,4-dihydro-3-cinnolinecarboxylic acid	215–217	485
6-Ethyl-4-oxo-1,4-dihydro-3-cinnolinecarboxylic acid	—	782
1-Ethyl-4-oxo-1,4-dihydro-3,7-cinnolinedicarbonitrile	—	485
1-Ethyl-4-oxo-1,4-dihydro-2,7-cinnolinedicarboxylic acid	305–325	485
Ethyl 4-oxo-1-phenyl-1,4-dihydro-3-cinnolinecarboxylate	152 or 154–156	140, 333
Ethyl 3-phenyl-4-cinnolinecarboxylate	—	(*E* 41, 266)
Ethyl 3-phenyl-4-cinnolinecarboxylate 1-oxide	—	(*E* 293, 296)
4-Ethylsulfonylcinnoline	—	(*E* 186)
4-Ethylthiocinnoline	—	(*E* 183)
Ethyl 4-thioxo-1,4-dihydro-3-cinnolinecarboxylate	141–142	54
2-Ethyl-6,7,8-trifluorocinnolin-2-ium-4-olate	185–186, IR, NMR	416
Ethyl 6,7,8-trifluoro-1-methyl-4-oxo-1,4-dihydro-3-cinnolinecarboxylate	101–102, IR, NMR	416
Ethyl 6,7,8-trifluoro-4-oxo-1,4-dihydro-3-cinnolinecarboxylate	228–229, IR, NMR, UV	414
1-Ethyl-6,7,8-trifluoro-4-oxo-1,4-dihydro-3-cinnolinecarboxylic acid	140–141, IR, NMR	416
3-Ethynylcinnoline	120–121	338
6-Ethynylcinnoline	145–146, IR, NMR	817, 819
3-Ethynyl-4(1*H*)-cinnolinone	192–193	293
3-Ethynyl-4-phenoxycinnoline	109–110	338
6-Fluorocinnoline	liq, IR, NMR	816, 817, 819
8-Fluorocinnoline	NMR	289
6-Fluoro-4(1*H*)-cinnolinethione	—	(*E* 179, 183)
7-Fluoro-4(1*H*)-cinnolinethione	—	(*E* 180, 183)
8-Fluoro-4(1*H*)-cinnolinethione	—	(*E* 180, 183)
6-Fluoro-4(1*H*)-cinnolinone	—	(*E* 99, 112)
7-Fluoro-4(1*H*)-cinnolinone	—	(*E* 100, 112)

TABLE A.1. (*Continued*)

Cinnoline	Melting Point (°C) etc.	Reference(s)
8-Fluoro-4(1*H*)-cinnolinone	—	(*E* 100, 112)
6-Fluoro-4-oxo-1,4-dihydro-3-cinnolinecarboxylic acid	xl st	(*E* 112) 981
7-Fluoro-4-oxo-1,4-dihydro-3-cinnolinecarboxylic acid	—	(*E* 112)
8-Fluoro-4-oxo-1,4-dihydro-3-cinnolinecarboxylic acid	—	(*E* 111)
3,4,5,6,7,8-Hexachlorocinnoline	—	(*E* 145)
3,4,5,6,7,8-Hexafluorocinnoline	—	(*E* 145)
1-(2-Hydrazinocarbonylethyl)-4(1*H*)-cinnolinone	xl st	550
4-Hydrazinocinnoline	285 or 293–294	(*E* 238) 908
4-Hydrazino-3-cinnolinecarbohydrazide	>300	54
4-Hydrazino-6-nitrocinnoline	—	(*E* 200, 240)
4-Hydrazino-3-phenylcinnoline	—	(*E* 230)
4-Hydroxyaminocinnoline	—	(*E* 244)
1-Hydroxy-4(1*H*)-cinnolinone	153	(*E* 295) 908
5-Hydroxy-4(1*H*)-cinnolinone	—	(*E* 93, 114)
6-Hydroxy-3(2*H*)-cinnolinone	—	(*E* 116, 118)
8-Hydroxy-4(1*H*)-cinnolinone	—	(*E* 93, 114)
5-Hydroxy-6,8-diiodo-4(1*H*)-cinnolinone	—	(*E* 114)
8-Hydroxy-5,7-diiodo-4(1*H*)-cinnolinone	—	(*E* 94)
3-(1-Hydroxyethyl)-4-methoxycinnoline	96, IR, NMR	307
4-(1-Hydroxyethyl)-3-methoxycinnoline	125, IR, NMR	307
5-Hydroxy-6-iodo-4-oxo-1,4-dihydro-8-cinnolinesulfonic acid	—	(*E* 94, 114)
8-Hydroxy-7-iodo-4-oxo-1,4-dihydro-5-cinnolinesulfonic acid	—	(*E* 94, 114)
3-(3-Hydroxy-3-methylbut-1-ynyl)cinnoline	119–120	338
3-(3-Hydroxy-3-methylbut-1-ynyl)-4-phenoxycinnoline	129–130	338
4-Hydroxymethyl-3-cinnolinecarbaldehyde	153–154, IR, NMR	711
5-Hydroxy-6-nitro-4(1*H*)-cinnolinone	—	(*E* 114, 204)
5-Hydroxy-4-oxo-1,4-dihydro-8-cinnolinesulfonic acid	—	(*E* 94, 114)
8-Hydroxy-4-oxo-1,4-dihydro-5-cinnolinesulfonic acid	—	(*E* 94, 114)
1-Hydroxy-3-phenyl-4(1*H*)-cinnolinone	—	(*E* 291, 295)
4-Imino-1,3-dimethyl-1,4-dihydro-6-cinnolinamine	—	(*H* 42)
4-Imino-1,7-dimethyl-1,4-dihydro-8-cinnolinamine	—	(*H* 40)
4-Imino-1-methyl-1,4-dihydro-6-cinnolinamine	—	(*H* 40; *E* 237)
4-Imino-1-methyl-1,4-dihydro-3-cinnolinecarboxamide	173–175, xl st	54, 555
3-Iodocinnoline	102–103	293
8-Iodocinnoline	NMR	289
3-Iodo-4(1*H*)-cinnolinone	294–296	293
6-Iodo-4(1*H*)-cinnolinone	—	(*H* 18)

TABLE A.1. (*Continued*)

Cinnoline	Melting Point (°C) etc.	Reference(s)
8-Iodo-4(1*H*)-cinnolinone	—	(*E* 109)
3-Iodo-4-methoxycinnoline	124, IR, NMR	307
4-Iodo-3-methoxycinnoline	155, IR, NMR	307, 1010
8-Iodo-3-methoxycinnoline	124, IR, NMR	307
4-Isobutylcinnoline	—	(*E* 58)
4-Isobutyrylmethylcinnoline	—	(*E* 56)
4-Isopropoxycinnoline	—	(*E* 158)
3-(2-Isopropoxyethyl)-5-phenyl-4(1*H*)-cinnolinone	141–142	213
6-Isopropoxy-3-(2-methoxycarbonylethyl)-4(1*H*)-cinnolinone	205–207	213
6-Isopropoxy-3-(2-methoxycarbonylethyl)-1-methyl-4(1*H*)-cinnolinone	124–126	213
5-Isopropoxy-4-oxo-1-phenyl-1,4-dihydro-3-cinnolinecarboxylic acid	248–250	765
3-Isopropylaminocinnoline	208–210, IR, NMR, UV	993
3-Isopropyl-4-cinnolinecarbonitrile	MS	401
2-Isopropyl-3(2*H*)-cinnolinone	78–79, IR	42
4-Isopropylidenehydrazino-6-nitrocinnoline	—	(*E* 200, 240)
N′-Isopropylidene-3-phenyl-4-cinnolinecarbohydrazide	—	(*E* 269)
3-Isopropyl-4-methylcinnoline	—	(*E* 43, 56)
4-Isovalerylmethylcinnoline	—	(*E* 56)
6-Mercapto-4(1*H*)-cinnolinethione	245	(*E* 184) 908
7-Mercapto-4(1*H*)-cinnolinethione	—	(*E* 184)
3-Methoxy-4,8-bis(trimethylsilyl)cinnoline	70, IR, NMR	307
4-(2-Methoxycarbonylethyl)cinnoline	—	605
3-(2-Methoxycarbonylethyl)-1-methyl-6-phenyl-4(1*H*)-cinnolinone	130–132	213
3-(2-Methoxycarbonylethyl)-6-phenyl-4(1*H*)-cinnolinone	255–257	213
6-Methoxy-8-cinnolinamine	185–186, IR, NMR	27
3-Methoxycinnoline	41, NMR; pic: 155–157	(*E* 161) 307, 908
4-Methoxycinnoline	126 or 128–129	(*H* 31; *E* 158, 165) 908
5-Methoxycinnoline	—	(*E* 162, 166)
6-Methoxycinnoline	—	(*E* 162, 166)
7-Methoxycinnoline	—	(*E* 162, 166)
8-Methoxycinnoline	NMR	(*E* 163, 165) 289
4-Methoxy-5-cinnolinecarbaldehyde	198, IR, NMR	307
7-Methoxy-4-cinnolinecarboxylic acid 1-oxide	320	146
6-Methoxy-3,4(1*H*,2*H*)-cinnolinedione	281–282, IR, MS, NMR	809
3-Methoxycinnoline 1-oxide	92–93	(*E* 292, 295) 908
4-Methoxycinnoline 1-oxide	—	(*E* 295)
4-Methoxycinnoline 2-oxide	—	(*E* 297)
6-Methoxy-4(1*H*)-cinnolinethione	—	(*E* 185)
5-Methoxy-4(1*H*)-cinnolinone	—	(*E* 109)
6-Methoxy-4(1*H*)-cinnolinone	—	(*H* 19)

TABLE A.1. (*Continued*)

Cinnoline	Melting Point (°C) etc.	Reference(s)
7-Methoxy-4(1*H*)-cinnolinone	—	(*E* 109)
8-Methoxy-4(1*H*)-cinnolinone	—	(*E* 109)
7-Methoxy-*N*,*N*-dimethyl-4-oxo-1,4-dihydro-3-cinnolinesulfonamide	260–322 (?)	485
6-Methoxy-4-methylaminocinnoline	—	(*E* 244)
4-Methoxy-3-methylcinnoline	77, IR, NMR	(*E* 158, 166) 307
4-Methoxy-8-methylcinnoline	—	(*E* 158, 167)
8-Methoxy-4-methylcinnoline	130–132	(*E* 163) 908
4-Methoxy-3-methyl-6-nitrocinnoline	—	(*H* 31)
8-Methoxy-4-methylsulfonylcinnoline	—	(*E* 186)
7-Methoxy-3-methylsulfonyl-4(1*H*)-cinnolinone	275–285	485
4-Methoxy-5-nitrocinnoline	—	(*E* 203)
4-Methoxy-6-nitrocinnoline	—	(*H* 31; *E* 167, 200)
4-Methoxy-7-nitrocinnoline	200	(*E* 158, 204) 908
4-Methoxy-8-nitrocinnoline	—	(*E* 167)
3-Methoxy-4-nitrocinnoline 1-oxide	—	(*E* 295)
4-Methoxy-5-nitrocinnoline 1-oxide	—	(*E* 296)
6-Methoxy-8-nitro-4(1*H*)-cinnolinone	227, IR, NMR	27
6-Methoxy-8-nitro-4-oxo-1,4-dihydro-3-cinnolinecarboxylic acid	229–231, IR, NMR	27
6-Methoxy-4-oxo-1,4-dihydro-3-cinnolinecarboxylic acid	—	(*H* 21; *E* 110)
8-Methoxy-4-oxo-1,4-dihydro-3-cinnolinecarboxylic acid	—	(*E* 110)
7-Methoxy-4-oxo-1,4-dihydro-3-cinnolinesulfonamide	310–332	485
7-Methoxy-4-oxo-1,4-dihydro-3-cinnolinesulfonanilide	303–310	485
5-Methoxy-4-oxo-1-phenyl-1,4-dihydro-3-cinnolinecarboxylic acid	289–290 (?)	765
6-Methoxy-4-phenoxycinnoline	—	(*H* 33)
8-Methoxy-4-phenoxycinnoline	—	(*E* 162)
4-Methoxy-3-phenylcinnoline	—	(*E* 158, 166)
8-Methoxy-4-phenylcinnoline	—	(*E* 163)
6-Methoxy-3-phenyl-4-cinnolinecarboxylic acid	—	(*E* 163, 265)
6-Methoxy-3-phenyl-4(1*H*)-cinnolinone	—	(*H* 21)
3-Methoxy-4-phenylsulfinylcinnoline	146, IR, NMR	1010
3-Methoxy-4-phenylthiocinnoline	87 or 98, IR, NMR	821, 1010
3-(3-Methoxypropyl)-4(1*H*)-cinnolinone	148–149, IR	693
3-Methoxy-4-trimethylsilylcinnoline	52, IR, NMR	307
Methyl 3-acetyl-6-bromo-4-cinnolinecarboxylate	176, IR, NMR	43
4-Methylamino-6-cinnolinamine	—	(*E* 236)
4-Methylaminicinnoline	238–239, IR, NMR	(*E* 238) 627
3-Methylamino-4-cinnolinecarboxylic acid 1-oxide	182	147
4-Methylamino-6(2*H*)-cinnolinone	—	(*E* 118, 244)
4-Methylamino-6-nitrocinnoline	—	(*E* 201, 236)
4-Methylamino-3-phenylethynylcinnoline	165–166, UV	293

TABLE A.1. (*Continued*)

Cinnoline	Melting Point (°C) etc.	Reference(s)
Methyl 6-chloro-4-oxo-1,4-dihydro-3-cinnolinecarboxylate	—	(*E* 113)
4-Methyl-3-cinnolinamine	—	(*E* 242)
4-Methyl-5-cinnolinamine	—	(*E* 240)
4-Methyl-8-cinnolinamine	—	(*H* 14; *E* 240)
8-Methyl-4-cinnolinamine	—	(*H* 35; *E* 242)
3-Methylcinnoline	47–48, MS, NMR; 2-EtI: 196; 2-MeI: 200–202; pic: 175–177	(*E* 44) 40, 604, 605, 636
4-Methylcinnoline	72–73, xl st	(*H* 14, 40; *E* 40, 42) 137, 604, 605, 630
6-Methylcinnoline	liq, IR, NMR	817, 819
8-Methylcinnoline	NMR	(*E* 42) 289
3-Methyl-4-cinnolinecarbaldehyde	143–145 or 151–153, IR, MS, NMR	291, 711
3-Methyl-4-cinnolinecarbonitrile	MS	(*E* 58) 401
Methyl 6-cinnolinecarboxylate	130–131, IR, NMR	817, 819
4-Methyl-7-cinnolinecarboxylic acid	—	(*H* 14)
7-Methyl-4-cinnolinecarboxylic acid 1-oxide	250	146
3-Methyl-4,6-cinnolinediamine	—	(*E* 236)
6-Methyl-3,4(1*H*,2*H*)-cinnolinedione	292–293, IR, MS, NMR	809
4-Methylcinnoline 1,2-dioxide	—	(*E* 291, 294)
4-Methylcinnoline 1-oxide	—	(*E* 291, 292, 295, 296)
4-Methylcinnoline 2-oxide	—	(*E* 291, 293, 297, 298)
1-Methyl-4(1*H*)-cinnolinethione	—	(*E* 180)
2-Methyl-3(2*H*)-cinnolinethione	—	(*E* 180)
1-Methyl-4(1*H*)-cinnolinimine	—	(*H* 40; *E* 239)
2-Methyl-3(2*H*)-cinnolinimine	HI: 246–248	908
2-Methylcinnolin-2-ium-4-olate	NMR, xl st	(*E* 95) 693, 807
2-Methylcinnolin-2-ium-4-thiolate	—	(*E* 180)
1-Methyl-4(1*H*)-cinnolinone	113 to 117, NMR, xl st	(*H* 49; *E* 95) 334, 399, 509, 807
2-Methyl-3(2*H*)-cinnolinone	135–137, xl st	37, 38, 373, 808
3-Methyl-4(1*H*)-cinnolinone	—	(*H* 19; *E* 101)
4-Methyl-3(2*H*)-cinnolinone	235–236	(*E* 96, 117) 54
4-Methyl-8(2*H*)-cinnolinone	—	(*E* 107, 108, 118)
6-Methyl-4(1*H*)-cinnolinone	—	(*E* 111)
7-Methyl-4(1*H*)-cinnolinone	—	(*H* 18)
8-Methyl-4(1*H*)-cinnolinone	219–221, NMR	(*H* 18; *E* 102) 335
6-Methyl-3,5-diphenylcinnoline	—	(*E* 44)
7-Methyl-3,4-diphenylcinnoline	—	(*E* 44)
7-Methyl-3,5-diphenylcinnoline	—	(*E* 44)
8-Methyl-3,4-diphenylcinnoline	—	(*E* 44)
1-Methyl-4-methylimino-1,4-dihydro-6-cinnolinamine	—	(*E* 237)
1-Methyl-4-methylimino-6-nitro-1,4-dihydrocinnoline	—	(*E* 201)

TABLE A.1. *(Continued)*

Cinnoline	Melting Point (°C) etc.	Reference(s)
6-Methyl-4-methylsulfonylcinnoline	—	(*E* 186)
1-Methyl-3-methylsulfonyl-4(1*H*)-cinnolinone	204–207	485
3-Methyl-6-nitro-4-cinnolinamine	—	(*H* 35; *E* 204, 242)
3-Methyl-8-nitro-4-cinnolinamine	—	(*H* 35)
7-Methyl-8-nitro-4-cinnolinamine	—	(*H* 35; *E* 242)
8-Methyl-5-nitro-4-cinnolinamine	—	(*E* 202)
8-Methyl-5/7-nitro-4-cinnolinamine	—	(*H* 35)
4-Methyl-3-nitrocinnoline	—	(*E* 40, 203)
4-Methyl-8-nitrocinnoline	138–139	(*H* 14; *E* 40, 200) 908
1-Methyl-6-nitro-3,4(1*H*,2*H*)-cinnolinedione	263–266, IR, NMR	717
4-Methyl-3-nitrocinnoline 1-oxide	—	(*E* 292, 296)
1-Methyl-6-nitro-4(1*H*)-cinnolinimine	—	(*H* 40; *E* 201, 235)
1-Methyl-7-nitro-4(1*H*)-cinnolinimine	—	(*E* 235)
1-Methyl-8-nitro-4(1*H*)-cinnolinimine	—	(*H* 40)
2-Methyl-6-nitrocinnolin-2-ium-4-olate	—	(*E* 95)
1-Methyl-5-nitro-4(1*H*)-cinnolinone	—	(*H* 49)
1-Methyl-6-nitro-4(1*H*)-cinnolinone	—	(*H* 49; *E* 95)
1-Methyl-7-nitro-4(1*H*)-cinnolinone	—	(*H* 42)
1-Methyl-8-nitro-4(1*H*)-cinnolinone	—	(*H* 49)
3-Methyl-6-nitro-4(1*H*)-cinnolinone	—	(*H* 21)
3-Methyl-8-nitro-4(1*H*)-cinnolinone	—	(*H* 20; *E* 113)
7-Methyl-6-nitro-4(1*H*)-cinnolinone	—	(*H* 20; *E* 202)
7-Methyl-8-nitro-4(1*H*)-cinnolinone	—	(*H* 20; *E* 202)
8-Methyl-5-nitro-4(1*H*)-cinnolinone	—	(*E* 103, 112, 202)
8-Methyl-6-nitro-4(1*H*)-cinnolinone	—	(*E* 103, 112, 202)
8-Methyl-5-nitro-4-oxo-1,4-dihydro-3-cinnolinecarboxylic acid	—	(*E* 112, 202)
8-Methyl-6-nitro-4-oxo-1,4-dihydro-3-cinnolinecarboxylic acid	—	(*E* 112, 202)
3-Methyl-6-nitro-4-phenoxycinnoline	—	(*H* 33)
3-Methyl-8-nitro-4-phenoxycinnoline	—	(*H* 33)
7-Methyl-8-nitro-4-phenoxycinnoline	—	(*H* 33; *E* 202)
8-Methyl-5/7-nitro-4-phenoxycinnoline	—	(*H* 33)
8-Methyl-5-nitro-4-phenylcinnoline	—	(*E* 202)
1-Methyl-6-nitro-4-phenylimino-1,4-dihydrocinnoline	—	(*H* 40)
1-Methyl-6-nitro-4-propylimino-1,4-dihydrocinnoline	—	(*E* 201)
Methyl 4-oxo-1,4-dihydro-3-cinnolinecarboxylate	—	(*E* 113)
6-Methyl-4-oxo-1,4-dihydro-3-cinnolinecarboxylic acid	—	(*E* 110)
7-Methyl-4-oxo-1,4-dihydro-3-cinnolinecarboxylic acid	xl st	981
8-Methyl-4-oxo-1,4-dihydro-3-cinnolinecarboxylic acid	—	(*E* 110)
1-Methyl-3-phenethyl-4(1*H*)-cinnolinone	111–112, MS	293
3-Methyl-4-phenoxycinnoline	2-EtI: 188; 2-MeI: 199–200	(*H* 33)
7-Methyl-4-phenoxycinnoline	—	(*H* 33)

TABLE A.1. (*Continued*)

Cinnoline	Melting Point (°C) etc.	Reference(s)
8-Methyl-4-phenoxycinnoline	—	(*H* 33)
1-Methyl-6-phenylazo-4(1*H*)-cinnolinone	—	(*E* 237)
6-Methyl-3-phenyl-4-cinnolinamine	250, IR	940
3-Methyl-4-phenylcinnoline	—	(*H* 8, 42; *E* 56)
6-Methyl-3-phenylcinnoline	—	(*E* 41, 46)
6-Methyl-4-phenylcinnoline	124–125	(*E* 56) 637
N-Methyl-3-phenyl-4-cinnolinecarboxamide	—	(*E* 266)
Methyl 4-phenyl-3-cinnolinecarboxylate	2-$PhBF_4$: 211–212, NMR	769
6-Methyl-3-phenyl-4-cinnolinecarboxylic acid	—	(*E* 267)
3-Methyl-4-phenylcinnoline 1-oxide	—	(*H* 9)
2-Methyl-7-phenyl-3(2*H*)-cinnolinone	151–153, IR, NMR, UV	517
6-Methyl-3-phenyl-4(1*H*)-cinnolinone	—	(*E* 109)
1-Methyl-3-phenylethynyl-4(1*H*)-cinnolinone	156–157	293
1-Methyl-4-phenylimino-1,4-dihydro-6-cinnolinamine	—	(*H* 40)
4-Methyl-3-phenylthiocinnoline	95, NMR	518
1-Methyl-3-styryl-4(1*H*)-cinnolinone	142–144, NMR	293
4-Methylsulfinylcinnoline	157	908
4-Methylsulfonylcinnoline	183–184	(*E* 186) 908
4-Methylsulfonyl-6-cinnolinecarbonitrile	—	(*E* 186)
3-Methylsulfonyl-4(1*H*)-cinnolinone	265–302 (!)	485
4-Methylsulfonyl-8-nitrocinnoline	—	(*E* 186, 204)
4-Methylthiocinnoline	96, NMR; pic: 168–169; 1-MeI: 191–193; 2-MeI: 214–216	(*E* 181, 182) 908, 1010
4-Methylthio-3-cinnolinecarbonitrile	133–134, IR	695
3-Methylthiocinnoline 1-oxide	—	(*E* 295)
4-Methylthio-6-nitrocinnoline	—	(*E* 185)
4-Morpholinocinnoline	140–141, IR, NMR, UV	993
3-Nitro-4-cinnolinamine	305–306 or 307–308	(*E* 201, 240) 370, 908
6-Nitro-4-cinnolinamine	289–291	(*H* 35; *E* 201, 228, 229, 234) 908
7-Nitro-4-cinnolinamine	—	(*H* 35; *E* 203, 234)
8-Nitro-4-cinnolinamine	—	(*H* 33; *E* 202, 234)
3-Nitrocinnoline	205–206 or 209–210	(*E* 203, 206) 370, 908
5-Nitrocinnoline	150–152	(*E* 200, 206) 217, 908
6-Nitrocinnoline	194–195, IR, NMR	(*E* 200) 817, 819
7-Nitrocinnoline	—	(*E* 203)
8-Nitrocinnoline	133–134 or 135–136, NMR	(*E* 200, 206) 217, 289, 908
3-Nitrocinnoline 1-oxide	214–215	(*E* 296) 908
3-Nitrocinnoline 2-oxide	215–217	(*E* 297) 908
4-Nitrocinnoline 1-oxide	161–162	(*E* 295) 908
5-Nitrocinnoline 1-oxide	182–183	(*E* 295) 908
5-Nitrocinnoline 2-oxide	219–220	(*E* 297) 908
6-Nitrocinnoline 2-oxide	214–215	(*E* 297) 908

TABLE A.1. (*Continued*)

Cinnoline	Melting Point (°C) etc.	Reference(s)
8-Nitrocinnoline 2-oxide	228 or 232–234	(*E* 297) 908
6-Nitro-4(1*H*)-cinnolonethione	—	(*E* 185)
3-Nitro-4(1*H*)-cinnolinone	284–285	(*H* 19; *E* 109, 201) 908
5-Nitro-4(1*H*)-cinnolinone	184–186	(*H* 19; *E* 203) 908
5/7-Nitro-4(1*H*)-cinnolinone	—	(*H* 20)
6-Nitro-4(1*H*)-cinnolinone	330–331	(*H* 19; *E* 102, 200) 497, 908
7-Nitro-4(1*H*)-cinnolinone	—	(*H* 18; *E* 203)
8-Nitro-4(1*H*)-cinnolinone	181–183 or 274–275	(*H* 18; *E* 102, 203) 497, 908
4-Nitromethylcinnoline	152, IR, NMR	95
6-Nitro-4-oxo-1,4-dihydro-3-cinnolinecarbonyl chloride	crude	1034
6-Nitro-4-oxo-1,4-dihydro-3-cinnolinecarboxylic acid	NMR	(*E* 111, 201) 1034
8-Nitro-4-oxo-1,4-dihydro-3-cinnolinecarboxylic acid	—	(*E* 93, 111, 203)
3-Nitro-4-phenoxycinnoline	—	(*E* 160, 201)
5-Nitro-4-phenoxycinnoline	—	(*E* 160, 204)
6-Nitro-4-phenoxycinnoline	190–191	(*H* 33; *E* 201) 908
7-Nitro-4-phenoxycinnoline	—	(*H* 33; *E* 203)
8-Nitro-4-phenoxycinnoline	—	(*H* 33; *E* 202)
3-Nitro-1-phenyl-4(1*H*)-cinnolinone	190	140
6-Nitro-3-phenyl-4(1*H*)-cinnolinone	325, IR, NMR	(*E* 109) 492
6-Nitro-4-propylaminocinnoline	—	(*E* 201, 242)
4-(2-Nitrovinyl)cinnoline	—	(*E* 48)
4-Oxo-1,4-dihydro-3-cinnolinecarbohydrazide	—	(*E* 114)
4-Oxo-1,4-dihydro-3-cinnolinecarbonitrile	260–261, IR; H_2O: 166–169, IR, NMR	487, 695
4-Oxo-1,4-dihydro-6-cinnolinecarbonitrile	—	(*H* 19)
4-Oxo-1,4-dihydro-3-cinnolinecarboxamide	—	(*E* 114)
4-Oxo-1,4-dihydro-3-cinnolinecarboxylic acid	268–269, NMR	(*E* 92) 908, 1034
8-Oxo-2,8-dihydro-3-cinnolinecarboxylic acid	—	(*E* 118)
4-Oxo-1,4-dihydro-3-cinnolinesulfonamide	275–286	485
4-Oxo-1,4-dihydro-3-cinnolinesulfonic acid	Na: >340	485
4-Oxo-1-phenyl-1,4-dihydro-3-cinnolinecarbonitrile	224	140
4-Oxo-1-phenyl-1,4-dihydro-3-cinnolinecarboxamide	294	140
4-Oxo-1-phenyl-1,4-dihydro-3-cinnolinecarboxylic acid	275	140, 917
3,5,6,7,8-Pentafluoro-4-cinnolinamine	—	(*E* 244)
3,5,6,7,8-Pentafluoro-4(1*H*)-cinnolinone	—	(*E* 113)
4-Phenoxy-5-cinnolinamine	—	(*E* 160, 234)
4-Phenoxy-7-cinnolinamine	—	(*E* 160, 232)
4-Phenoxy-8-cinnolinamine	—	(*E* 160, 224)
4-Phenoxycinnoline	—	(*H* 33)
4-Phenoxy-6-phenylazocinnoline	—	(*E* 160)
4-Phenoxy-3-phenylethynylcinnoline	152–153	293
4-Phenoxy-3-styrylcinnoline	185–186	293

TABLE A.1. *(Continued)*

Cinnoline	Melting Point (°C) etc.	Reference(s)
6-Phenylazo-4-cinnolinamine	—	(*E* 236)
3-Phenylazo-4(1*H*)-cinnolinone (?)	UV	982, cf. 620
6-Phenylazo-4(1*H*)-cinnolinone	—	(*E* 112)
3-Phenyl-4-cinnolinamine	156 or 238, IR, NMR	85, 93, 940
3-Phenyl-5-cinnolinamine	219–221, NMR	139, 656
3-Phenylcinnoline	117–119, MS, NMR	(*E* 41, 44) 40, 139, 605, 747, 908
4-Phenylcinnoline	64 to 67, MS	(*H* 8, 42; *E* 40, 43, 44) 486, 637, 908
6-Phenylcinnoline	—	(*E* 44)
7-Phenylcinnoline	—	(*E* 44)
5-Phenyl-4-cinnolinecarbohydrazide	—	(*E* 269)
3-Phenyl-4-cinnolinecarbonitrile	169–170, IR, NMR	91
3-Phenyl-4-cinnolinecarbonyl chloride	—	(*E* 266)
3-Phenyl-4-cinnolinecarboxamide	—	(*E* 266)
3-Phenyl-4-cinnolinecarboxylic acid	224–225	(*E* 467) 908
1-Phenyl-3,4(1*H*,2*H*)-cinnolinedione	227	140
3-Phenylcinnoline 1-oxide	134–136, MS	(*E* 292, 296) 747
3-Phenylcinnoline 2-oxide	—	(*E* 293, 297)
3-Phenylcinnolin-1-ium-1-benzamidate	226–228, UV	5
1-Phenyl-4(1*H*)-cinnolinone	133–134	399, 917
3-Phenyl-4(1*H*)-cinnolinone	259 to 262	(*H* 19; *E* 101, 109) 94, 492, 585, 908
4-Phenyl-3(2*H*)-cinnolinone	300–302	(*E* 97, 116) 908
4-Phenyl-8(2*H*)-cinnolinone	—	(*E* 118)
6-Phenyl-4(1*H*)-cinnolinone	—	(*E* 109)
7-Phenyl-4(1*H*)-cinnolinone	—	(*E* 109)
3-Phenylethynylcinnoline	127–128, IR, UV	293
4-Phenylethynyl-4(1*H*)-cinnolinone	180–181	293
3-Phenylethynyl-4-piperidinocinnoline	165–166	293
1-Phenyl-3-phenylsulfonyl-4(1*H*)-cinnolinone	276	140
4-Phenyl-3-phenylthiocinnoline	131, NMR	518
4-Phenylsulfinylcinnoline	157, IR, NMR	1010
3-Phenylsulfonyl-4(1*H*)-cinnolinone	339–343	485
3-Phenylthiocinnoline	85, NMR	518
4-Phenylthiocinnoline	135, IR, NMR	1010
3-Piperidinocinnoline	125–127, IR, NMR, UV	993
4-Piperidinocinnoline	134–136, IR, NMR, UV	993
4-Pivalamidocinnoline	150, NMR	307
1-Pivaloyl-4-pivaloylimino-1,4-dihydrocinnoline	100, NMR	307
4-(Propionylmethyl)cinnoline	112–113	933
4-Propoxycinnoline	—	(*E* 158)
3-Propylaminocinnoline	182–184, IR, NMR, UV	993
4-Propylaminocinnoline	158–160, IR, NMR, UV	(*E* 238) 993
4-Propylcinnoline	—	(*E* 58)
3-Propyl-4-cinnolinecarbonitrile	MS	401

TABLE A.1. (*Continued*)

Cinnoline	Melting Point (°C) etc.	Reference(s)
1-Propyl-4(1*H*)-cinnolinone	81–83	334
4-Styrylcinnoline	crude: 113–118	(*E* 41, 48) 908
4-Styrylcinnoline 2-oxide	—	(*E* 297)
3-Styryl-4(1*H*)-cinnolinone	295, NMR	293
4-Styryl-8(2*H*)-cinnolinone	—	(*E* 118)
3,4,7,8-Tetrachlorocinnoline	—	(*E* 145)
5,6,7,8-Tetrafluorocinnoline	—	(*E* 145)
3,4,6,7-Tetramethyl-5-cinnolinamine	206–207, NMR	656
4-Thioxo-1,4-dihydro-3-cinnolinecarbonitrile	292–293, IR, UV; Al(OH)$_3$: 300	695
4-Thioxo-1,4-dihydro-3-cinnolinecarbonyl chloride	crude	54
4-Thioxo-1,4-dihydro-3-cinnolinecarboxamide	205–206	54
4-Thioxo-1,4-dihydro-3-cinnolinecarboxylic acid	238–240	54
3,4,6-Trichlorocinnoline	—	(*E* 143) 908
4,5,6-Trichlorocinnoline	—	(*E* 143)
4,6,7-Trichlorocinnoline	—	(*H* 30; *E* 145)
4,7,8-Trichlorocinnoline	—	(*H* 30; *E* 145)
6,7,8-Trifluoro-2-methylcinnolin-2-ium-4-olate	188–189, NMR, UV	416
6,7,8-Trifluoro-1-methyl-4-oxo-1,4-dihydro-3-cinnolinecarboxylic acid	178–179, NMR	416
4,6,7-Trimethoxycinnoline	206–209, NMR	(*E* 158, 168) 509
4,6,7-Trimethoxy-3-methylcinnoline	—	(*E* 41, 158, 165, 168)
3,6,7-Trimethyl-5-cinnolinamine	235–236, NMR, UV; HCl: 268–271	139, 656
5,7,8-Trimethyl-3-phenyl-6(2*H*)-cinnolinone	266, IR, NMR, UV	706

TABLE A.2. ALPHABETICAL LIST OF SIMPLE PHTHALAZINES REPORTED BEFORE 2005

Phthalazine	Melting Point (°C) etc.	Reference(s)
	156–158, IR, NMR	936
5-Acetamido-1-acetoxyphthalazine		
8-Acetamido-4-acetoxy-1(2*H*)-phthalazinone	—	(*E* 489)
5-Acetamido-2-acetyl-1,4(2*H*,3*H*)-phthalazinedione	—	(*E* 485)
6-Acetamido-7-chloro-5,8-phthalazinequinone	200–201, IR, NMR	1016
2-(2-Acetamidoethyl)-3-methyl-1,4(2*H*,3*H*)-phthalazinedione	—	235
5-Acetamido-4-hydroxymethyl-1(2*H*)-phthalazinone	206–207	573
5-Acetamido-2-methyl-1,4(2*H*,3*H*)-phthalazinedione	—	(*H* 164)
4-Acetamidomethyl-1(2*H*)-phthalazinone	—	(*E* 411)
1-Acetamido-7-nitrophthalazine	—	(*E* 580)
1-Acetamidophthalazine	—	(*E* 578)
5-Acetamido-1,4(2*H*,3*H*)-phthalazinedione	—	(*H* 147; *E* 473)
6-Acetamido-1,4(2*H*,3*H*)-phthalazinedione	—	(*H* 149)
5-Acetamido-1(2*H*)-phthalazinone	307–308, IR, NMR	936

TABLE A.2. (*Continued*)

Phthalazine	Melting Point (°C) etc.	Reference(s)
2-(3-Acetamidopropyl)-3-acetyl-1,4(2*H*,3*H*)-phthalazinedione	—	235
1-Acetonyl-4-phenylphthalazine	—	(*E* 359)
4-Acetoxy-2-(2-acetoxyethyl)-1(2*H*)-phthalazinone	—	(*E* 491)
6-Acetoxy-1-acetoxymethyl-4-chlorophthalazine	—	195
7-Acetoxy-4-acetoxymethyl-1(2*H*)-phthalazinone	—	185
4-Acetoxy-2-acetyl-1(2*H*)-phthalazinone	—	(*E* 491)
1-Acetoxy-4-benzylidene-3-phenyl-3,4-dihydrophthalazine	183–184, IR, NMR	698
1-Acetoxy-4-benzylphthalazine	—	351
1-Acetoxy-4-isopropylphthalazine	—	(*E* 439)
4-Acetoxymethyl-2-methyl-1(2*H*)-phthalazinone	—	(*H* 170)
4-Acetoxymethyl-1(2*H*)-phthalazinone	—	(*E* 410)
1-Acetoxy-4-phenylphthalazine	—	253
4-Acetoxy-2-phenyl-1(2*H*)-phthalazinone	132–134, IR	698
4-Acetoxy-1(2*H*)-phthalazinone	—	(*E* 486)
4-Acetyl-8-amino-7-benzoyl-1(2*H*)-phthalazinone	>300, IR, NMR	553
4-Acetyl-8-amino-7-nitro-2,6-diphenyl-1(2*H*)-phthalazinone	210, IR, NMR	553
2-Acetyl-4-*sec*-butyl-1(2*H*)-phthalazinone	—	(*E* 422)
1-(1-Acetylethyl)-4-phenylphthalazine	—	(*E* 359)
2-(2-Acetylethyl)-1,4(2*H*,3*H*)-phthalazinedione	—	(*E* 480)
1-(*N*′-Acetylhydrazino)-4-benzylphthalazine	180–182, IR, NMR	528
4-Acetyl-8-hydroxy-2-methyl-1,5(2*H*,3*H*)-phthalazinedione	190, IR, MS, NMR, UV	890
1-Acetyl-3-methyl-4-oxo-3,4-dihydro-5,8-phthalazinequinone	146, IR, MS, NMR, UV	890
2-Acetyl-3-methyl-1,4(2*H*,3*H*)-phthalazinedione	—	(*E* 499)
2-Acetyl-4-phenyl-1(2*H*)-phthalazinone	—	253
2-Acetyl-1,4(2*H*,3*H*)-phthalazinedione	—	(*E* 479)
1-(1-Acetylpropyl)-4-phenylphthalazine	—	(*E* 359)
5-Acryloyloxy-2,3-diphenyl-1,4(2*H*,3*H*)-phthalazinedione	—	(*E* 500)
2-Allyl-4-allyloxy-1(2*H*)-phthalazinone	—	(*E* 491)
2-Allyl-4-hydrazino-1(2*H*)-phthalazinone	—	(*E* 618)
2-Allyl-4-methoxycarbonylmethyl-1(2*H*)-phthalazinone	—	218
N-Allyl-4-oxo-3,4-dihydro-1-phthalazinecarboxamide	—	(*E* 647)
1-Allyloxy-4-chlorophthalazine	—	(*E* 525)
4-Allyloxy-2-phenyl-1(2*H*)-phthalazinone	—	(*E* 493)
5-Amino-5-benzoyl-4-oxo-3-phenyl-3,4-dihydro-1-phthalazinecarbothioamide	173 or 200, IR, NMR	380, 553
6-Amino-2-benzyl-7-chloro-1(2*H*)-phthalazinone	190–193, IR, MS, NMR	274
8-Amino-2-benzyl-4-methyl-1(2*H*)-phthalazinone	—	(*E* 434)

TABLE A.2. (*Continued*)

Phthalazine	Melting Point (°C) etc.	Reference(s)
6-Amino-2/3-benzyl-1,4(2*H*,3*H*)-phthalazinedione	—	(*H* 165)
5-Amino-8-bromo-6-cyano-1-ethoxycarbonyl-2-phenylphthalazin-2-ium-4-olate	100, IR, NMR	520
4-Amino-2-(2-bromoethyl)-1(2*H*)-phthalazinone	164–167, IR, NMR	298
5-Amino-7-*tert*-butyl-1,4(2*H*,3*H*)-phthalazinedione	solid, IR	316
4-Amino-2-(4-carboxybutyl)-1(2*H*)-phthalazinone	189–190, IR, NMR	947
4-Amino-2-(3-carboxypropyl)-1(2*H*)-phthalazinone	193–194, IR, NMR	947
6-Amino-7-chloro-2-methyl-1(2*H*)-phthalazinone	>300, IR, MS, NMR	274
6-Amino-7-chloro-5,8-phthalazinequinone	>300, IR, NMR	1016
6-Amino-7-chloro-1(2*H*)-phthalazinone	>310, IR, MS, NMR	274
5-Amino-6-cyano-1-ethoxycarbonyl-2,7-diphenylphthalazin-2-ium-4-olate	260–261, IR, NMR	520
5-Amino-6-cyano-1-ethoxycarbonyl-2-phenylphthalazin-2-ium-4-olate	—	520
5-Amino-6-cyano-4-oxo-3,7-diphenyl-3,4-dihydro-1-phthalazinecarboxylic acid	240 or 252, IR, NMR	489, 535
5-Amino-1,6-diethoxycarbonyl-2,7-diphenylphthalazin-2-ium-4-olate	180–182, IR, NMR	520
5-Amino-7,8-dimethoxy-1,4(2*H*,3*H*)-phthalazinedione	290–292	(*E* 474) 908
5-Amino-2,4-dimethyl-1(2*H*)-phthalazinone	—	(*E* 434)
8-Amino-2,4-dimethyl-1(2*H*)-phthalazinone	—	(*E* 434)
5-Amino-2,3-diphenyl-1,4(2*H*,3*H*)-phthalazinedione	—	(*E* 500)
6-Amino-2,3-diphenyl-1,4(2*H*,3*H*)-phthalazinedione	—	(*E* 500)
6-Amino-2/3-ethyl-1,4(2*H*,3*H*)-phthalazinedione	—	(*H* 165)
2-(2-Aminoethyl)-1(2*H*)-phthalazinone	crude: 84–88, NMR	369
4-Amino-2-(4-hydroxybutyl)-1(2*H*)-phthalazinone	135–137, IR	301
4-Amino-2-(2-hydroxyethyl)-1(2*H*)-phthalazinone	—	199
4-Amino-2-(3-hydroxypropyl)-1(2*H*)-phthalazinone	164–165, IR	301
5-Amino-8-isopropyl-1,4(2*H*,3*H*)-phthalazinedione	solid, IR, UV	316
5-Amino-8-methoxymethyl-1,4(2*H*,3*H*)-phthalazinedione	solid, IR	316
5-Amino-4-methoxy-2-methyl-1(2*H*)-phthalazinone	—	(*H* 173)
8-Amino-4-methoxy-2-methyl-1(2*H*)-phthalazinone	—	(*H* 173)
5-Amino-8-methoxy-1,4(2*H*,3*H*)-phthalazinedione	243–244, IR, NMR, UV	(*E* 474) 48

TABLE A.2. *(Continued)*

Phthalazine	Melting Point (°C) etc.	Reference(s)
6-Amino-7-methoxy-1,4(2*H*,3*H*)-phthalazinedione	321–322, IR, NMR, UV	48
4-Aminomethyl-2-benzyl-1(2*H*)-phthalazinone	—	159
4-Aminomethyl-2-ethyl-1(2*H*)-phthalazinone	—	159
4-Aminomethyl-2-methyl-1(2*H*)-phthalazinone	—	159
7-Aminomethyl-2-phenyl-1(2*H*)-phthalazinone	70–72, IR, NMR	411
8-Amino-4-methyl-2-phenyl-1(2*H*)-phthalazinone	—	(*E* 433)
5-Amino-2-methyl-1,4(2*H*,3*H*)-phthalazinedione	—	(*H* 164)
5-Amino-3-methyl-1,4(2*H*,3*H*)-phthalazinedione	—	(*H* 164)
5-Amino-8-methyl-1,4(2*H*,3*H*)-phthalazinedione	solid, anal, IR	313
6-Amino-2/3-methyl-1,4(2*H*,3*H*)-phthalazinedione	—	(*H* 165)
6-Amino-5-methyl-1,4(2*H*,3*H*)-phthalazinedione	solid, IR	316
6-Amino-7-methyl-1,4(2*H*,3*H*)-phthalazinedione	solid, IR, NMR	316
4-Aminomethyl-1(2*H*)-phthalazinone	—	(*E* 411) 158
4-Amino-2-methyl-1(2*H*)-phthalazinone	160 or 164–165, IR, NMR	(*E* 586) 34, 199, 685
4-Amino-8-methyl-1(2*H*)-phthalazinone	281, IR, NMR	615
5-Amino-4-methyl-1(2*H*)-phthalazinone	—	(*E* 414)
8-Amino-4-methyl-1(2*H*)-phthalazinone	—	(*E* 414)
5-Amino-6-nitro-1,4(2*H*,3*H*)-phthalazinedione	342, IR, NMR	4
6-Amino-5-nitro-1,4(2*H*,3*H*)-phthalazinedione	348, IR, NMR	4
6-Amino-7-nitro-1,4(2*H*,3*H*)-phthalazinedione	382, IR, NMR	4
6-Amino-2/8-phenyl-1,4(2*H*,3*H*)-phthalazinedione	—	(*H* 165)
2-Amino-4-phenylphthalazin-2-ium mesylate	144–145	5
7-Amino-2-phenyl-1(2*H*)-phthalazinone	216–217, IR, NMR	411
5-Amino-1,4(2*H*,3*H*)-phthalazinedione (Luminol)	335–336, pK_a, UV	(*H* 147; *E* 472) 48, 203, 252, 546, 805
6-Amino-1,4(2*H*,3*H*)-phthalazinedione (Isoluminol)	>350, IR, NMR, pK_a, UV	(*H* 149) 48, 252, 805
2-Aminophthalazin-2-ium mesylate	157–158	5
4-Amino-1(2*H*)-phthalazinone	260 to 274	(*H* 79, 185; *E* 586) 151, 199, 685, 908
5-Amino-1(2*H*)-phthalazinone	288–289, IR, NMR	936
5-Amino-8-propyl-1,4(2*H*,3*H*)-phthalazinedione	solid, IR, UV	316
5-Amino-6,7,8-trimethoxy-1,4(2*H*,3*H*)-phthalazinedione	—	(*E* 475)
1-Anilino-4-benzylphthalazine	—	(*H* 185)
1-Anilino-4-butylphthalazine	—	(*E* 583)
1-Anilino-4-chlorophthalazine	205–206, dip, IR, NMR, UV; H_2SO_4: 204–206, IR, NMR	(*H* 180; *E* 583) 742
6-Anilino-7-chloro-5,8-phthalazinequinone	226–227 or 245–246, IR, NMR	856, 1017
1-Anilino-4-hydroxyaminophthalazine	—	(*E* 592)
1-Anilino-4-phenylphthalazine	—	(*H* 185; *E* 583)
1-Anilinophthalazine	—	(*E* 577)
5-Anilino-1,4(2*H*,3*H*)-phthalazinedione	—	(*H* 148)
4-Anilino-1(2*H*)-phthalazinone	—	(*E* 587)

TABLE A.2. (*Continued*)

Phthalazine	Melting Point (°C) etc.	Reference(s)
1-Azido-4-methylphthalazine	222	33
6-Azido-7-methyl-5,8-phthalazinequinone	135–136, IR, NMR	1016
1-Azidophthalazine	208, MS, NMR	592
4-Benzamido-2-methyl-1(2*H*)-phthalazinone	153–154, IR, NMR	185
2-Benzamido-4-phenyl-1(2*H*)-phthalazinone	258–260, IR, NMR	9
1-Benzamidophthalazine	—	(*E* 578)
5-Benzamido-1,4(2*H*,3*H*)-phthalazinedione	—	(*H* 147)
6-Benzamido-1,4(2*H*,3*H*)-phthalazinedione	329	499
4-Benzamido-2(1*H*)-phthalazinone	273, IR, NMR	685
2-Benzenesulfonyl-4-methyl-1(2*H*)-phthalazinone	109	33
4-Benzenesulfonyloxy-1(2*H*)-phthalazinone	—	(*E* 488)
2-Benzenesulfonyl-1,4(2*H*,3*H*)-phthalazinedione	—	87
4-Benzoyl-2-(2-carbamoylethyl)-1(2*H*)-phthalazinone	—	541
4-Benzyl-2-(2-cyanoethyl)-1(2*H*)-phthalazinone	—	541
1-(*N*′-Benzoylhydrazino)-4-phenylphthalazine	210–212	650
4-Benzoyloxy-2-phenyl-1(2*H*)-phthalazinone	—	(*E* 494)
4-Benzoyloxy-1(2*H*)-phthalazinone	—	(*E* 486)
1-Benzoyl-4-phenylphthalazine	134–136, IR, NMR	601, 938
1-Benzoylphthalazine	122–123, IR	586
2-Benzoyl-1,4(2*H*,3*H*)-phthalazinedione	—	(*E* 479)
4-Benzoyl-2(1*H*)-phthalazinone	202	412, 541, 545
1-Benzylamino-4-phenylphthalazine	—	(*E* 581)
1-Benzylamino-5-phthalazinamine	249–251	936
1-Benzylaminophthalazine	—	(*E* 577)
4-Benzylamino-1(2*H*)-phthalazinethione	—	(*E* 548)
4-Benzylamino-1(2*H*)-phthalazinone	—	196
2-Benzyl-4-benzyloxy-1(2*H*)-phthalazinone	—	(*H* 170; *E* 492)
4-Benzyl-6-bromo-1(2*H*)-phthalazinone	196–198	521, 611, 642
4-Benzyl-7-bromo-1(2*H*)-phthalazinone	175–177	521, 611, 642
2-Benzyl-4-*tert*-butoxycarbonylmethyl-1(2*H*)-phthalazinone	103, NMR	118
2-Benzyl-4-carbamoylmethyl-1(2*H*)-phthalazinone	—	(*E* 431)
2-Benzyl-4-carboxymethyl-1(2*H*)-phthalazinone	—	(*E* 431)
4-Benzyl-2-(3-carboxypropyl)-1(2*H*)-phthalazinimine	HBr: 236–238, NMR	958
1-Benzyl-4-chloro-6,7-dimethoxyphthalazine	—	(*E* 526)
4-Benzyl-6-chloro-2-(2-dimethylaminoethyl)-1(2*H*)-phthalazinone	—	(*E* 433)
2-Benzyl-7-chloro-1-oxo-1,2-dihydro-6-phthalazinesulfonamide	193–193, IR, MS, NMR	274
3-Benzyl-7-chloro-4-oxo-3,4-dihydro-6-phthalazinesulfonamide	236–237, IR, MS, NMR	274
2-Benzyl-7-chloro-1-oxo-1,2-dihydro-6-phthalazinesulfonyl chloride	153–156	274
N-Benzyl-4-chloro-1-oxo-2-phenyl-1,2-dihydro-6-phthalazinesulfonamide	—	260
4-Benzyl-5-chloro-2-phenyl-1(2*H*)-phthalazinethione	233	646
4-Benzyl-8-chloro-2-phenyl-1(2*H*)-phthalazinethione	85	646

TABLE A.2. (*Continued*)

Phthalazine	Melting Point (°C) etc.	Reference(s)
4-Benzyl-5-chloro-2-phenyl-1(2*H*)-phthalazinone	247	646
4-Benzyl-8-chloro-2-phenyl-1(2*H*)-phthalazinone	219	646
1-Benzyl-4-chlorophthalazine	150, NMR	(*H* 181; *E* 524) 778
4-Benzyl-5-chloro-1(2*H*)-phthalazinethione	173	646
4-Benzyl-8-chloro-1(2*H*)-phthalazinethione	231	646
4-Benzyl-5-chloro-1(2*H*)-phthalazinone	223	646
4-Benzyl-8-chloro-1(2*H*)-phthalazinone	211	646
4-Benzyl-2-(*N*,*N*-diethylcarbamoylmethyl)-1(2*H*)-phthalazinone	—	(*E* 425)
4-Benzyl-5,8-diiodo-2-phenyl-1(2*H*)-phthalazinone	221	647
4-Benzyl-5,8-diiodo-1(2*H*)-phthalazinone	201	647
4-Benzyl-6,7-diiodo-1(2*H*)-phthalazinone	185	648
1-Benzyl-6,7-dimethoxyphthalazine	—	(*H* 74; *E* 353)
4-Benzyl-2-(2-dimethylaminoethyl)-1(2*H*)-phthalazinone	—	(*E* 424)
1-Benzyl-4-dimethylaminophthalazine	129–130, IR, NMR	70, 689
1-Benzyl-4-(3-dimethylaminopropyl) phthalazine	—	(*E* 357)
4-Benzyl-2-(3-dimethylaminopropyl)-1(2*H*)-phthalazinone	—	(*E* 427)
N-Benzyl-1,4-dioxo-2-phenyl-1,2,3,4-tetrahydro-6-phthalazinesulfonamide	—	360
N-Benzyl-1,4-dioxo-1,2,3,4-tetrahydro-6-phthalazinesulfonamide	—	260
2-Benzyl-4-ethoxycarbonylmethyl-1(2*H*)-phthalazinone	—	(*E* 431)
4-Benzyl-2-ethoxycarbonylmethyl-1(2*H*)-phthalazinone	126, IR, NMR	676
4-Benzyl-2-(3-ethoxycarbonylpropyl)-1(2*H*)-phthalazinimine	HBr: 188–190, NMR	958
1-Benzyl-4-ethoxyphthalazine	—	(*H* 84)
2-Benzyl-4-ethoxy-1(2*H*)-phthalazinone	—	(*E* 492)
4-Benzyl-2-ethyl-1(2*H*)-phthalazinone	—	(*H* 101)
1-Benzyl-4-(*N*′-formylhydrazino)phthalazine	122–125, NMR	528
2-Benzyl-4-hydrazinocarbonylmethyl-1(2*H*)-phthalazinone	—	(*E* 431)
4-Benzyl-2-hydrazinocarbonylmethyl-1(2*H*)-phthalazinone	195–197, IR	676
1-(*N*-Benzylhydrazino)-4-methylphthalazine	—	(*E* 617)
1-Benzyl-4-hydrazinophthalazine	145–146, IR, NMR	(*E* 612) 528
2-Benzyl-4-hydrazino-1(2*H*)-phthalazinone	—	(*E* 618)
1-Benzyl-4-hydroxyaminophthalazine	—	(*E* 592)
2-Benzyl-8-hydroxy-7-methoxy-1(2*H*)-phthalazinone	—	(*H* 87)
2-Benzyl-8-hydroxy-4-methyl-1(2*H*)-phthalazinone	—	(*E* 434)
1-Benzylidenehydrazino-4-chlorophthalazine	174–175	275
1-Benzylidenehydrazino-4-phenylphthalazine	160–161	661
1-Benzylidenehydrazinophthalazine	—	(*E* 621)
5-Benzylidenehydrazino-1,4(2*H*,3*H*)-phthalazinedione	—	(*H* 148)

TABLE A.2. (*Continued*)

Phthalazine	Melting Point (°C) etc.	Reference(s)
4-Benzylidenehydrazino-1(2*H*)-phthalazinone	—	(*E* 625)
4-Benzylidene-3-phenyl-3,4-dihydro-1(2*H*)-phthalazinone	—	182
1-Benzyl-4-iodophthalazine	—	(*H* 181)
2-Benzyl-4-isopentyloxy-1(2*H*)-phthalazinone	—	(*E* 492)
1-Benzyl-4-isopropoxyphthalazine	—	(*E* 437)
1-Benzyl-4-isopropylaminophthalazine	—	(*E* 581)
1-Benzyl-4-methoxyphthalazine	211–213, IR, NMR	351, 525
1-Benzyl-7-methoxyphthalazine	—	(*H* 73)
2-Benzyl-4-methyl-8-nitro-1(2*H*)-phthalazinone	—	(*E* 434)
2-Benzyl-4-methyl 1(2*H*)-phthalazinone	—	(*E* 430)
4-Benzyl-2-methyl-1(2*H*)-phthalazinone	141–143 or 146–148, IR, NMR	(*H* 101) 119, 351, 689, 863, 864, 926
1-Benzyl-5-nitrophthalazine	233	330, 701
2-Benzyl-6/7-nitro-1,4(2*H*,3*H*)-phthalazinedione	—	(*H* 164)
4-Benzyl-7-nitro-1(2*H*)-phthalazinethione	200	641
4-Benzyl-6-nitro-1(2*H*)-phthalazinone	235	642
4-Benzyl-7-nitro-1(2*H*)-phthalazinone	206	522, 642
3-Benzyl-4-oxo-3,4-dihydro-1-phthalazinecarboxamide	—	(*E* 648)
3-Benzyl-4-oxo-3,4-dihydro-1-phthalazinecarboxylic acid	—	(*E* 646)
3-Benzyl-4-oxo-3,4-dihydro-5-phthalazinecarboxylic acid	168–171, IR, NMR	623
4-Benzyloxy-2-phenyl-1(2*H*)-phthalazinone	—	(*E* 495)
1-Benzyl-4-phenoxyphthalazine	162–164, NMR	(*H* 84) 958
1-Benzyl-3-phenyl-4-phenylhydrazono-3,4-dihydrophthalazine	156–157, IR, NMR	70, 689
1-Benzyl-4-phenylphthalazine 2-oxide	—	(*H* 373)
1-Benzyl-4-phenylphthalazine 3-oxide	—	(*E* 374)
1-Benzyl-2-phenylphthalazin-2-ium-4-olate	214–216, IR, UV	698
2-Benzyl-4-phenyl-1(2*H*)-phthalazinone	175–178	119
4-Benzyl-2-phenyl-1(2*H*)-phthalazinone	170	(*H* 101) 698
4-Benzyl-1-phthalazinamine	152–154, NMR; 3-$EtO_2C(CH_2)_3Br$: 192–194, NMR; 3-$HO_2C(CH_2)_3Br$: 210–212, NMR	958
1-Benzylphthalazine	—	(*H* 73; *E* 665)
4-Benzyl-1-phthalazinecarbonitrile	186–187, IR, MS, NMR	188, 399
2-Benzyl-1,4(2*H*,3*H*)-phthalazinedione	205	(*H* 159; *E* 480) 498
6-Benzyl-1,4(2*H*,3*H*)-phthalazinedione	276–277	704
4-Benzyl-1(2*H*)-phthalazinethione	156	641
2-Benzylphthalazin-2-ium-4-olate	—	(*E* 660)
2-Benzyl-1(2*H*)-phthalazinone	106–107, NMR	(*E* 399) 52
4-Benzyl-1(2*H*)-phthalazinone	196 to 201, IR; 3-PhBr: 220–225, NMR	(*H* 80) 182, 253, 296, 348, 351, 507, 642, 698, 778, 863, 864
4-Benzyl-2-piperidinomethyl-1(2*H*)-phthalazinone	—	348

TABLE A.2. (*Continued*)

Phthalazine	Melting Point (°C) etc.	Reference(s)
4-Benzyl-5,6,7,8-tetrabromo-2-phenyl-1(2*H*)-phthalazinone	291	643
4-Benzyl-5,6,7,8-tetrabromo-1(2*H*)-phthalazinone	285	643
4-Benzyl-5,6,7,8-tetrachloro-1(2*H*)-phthalazinone	204	643
4-Benzyl-5,6,7,8-tetraiodo-1(2*H*)-phthalazinone	—	207
4-Benzylthio-2-phenyl-1(2*H*)-phthalazinethione	—	(*E* 556)
4-Benzylthio-2-phenyl-1(2*H*)-phthalazinone	—	(*E* 557)
4-Benzylthio-1-phthalazinecarboxamide	—	(*E* 650)
4-Benzylthio-1-phthalazinecarboxylic acid	—	(*E* 650)
3-Benzyl-4-thioxo-3,4-dihydro-1-phthalazinecarboxylic acid	—	(*E* 649)
4-Benzyl-5,7,8-triiodo-2-phenyl-1(2*H*)-phthalazinone	145	648
4-Benzyl-5,7,8-triiodo-1(2*H*)-phthalazinone	135	648
1,4-Bis(*N*′-acetylhydrazino)phthalazine	—	(*E* 620)
1,4-Bis(benzylamino)phthalazine	122, MS, NMR, UV	133
1,4-Bis(benzylidenehydrazino)phthalazine	150	(*E* 627) 501
1,4-Bis(3-dimethylaminopropyl)phthalazine	—	(*E* 357)
1,4-Bis(ethylamino)phthalazine	—	(*E* 589)
6,7-Bis(hydroxymethyl)phthalazine	254–256, IR	584
4,6-Bis(hydroxymethyl)-1(2*H*)-phthalazinone	200–201, IR, NMR	404
1,4-Bis(isopropylidenehydrazino)phthalazine	—	(*E* 626)
1,4-Bis(methylamino)phthalazine	215–216, IR, NMR, UV	742
2,3-Bis(methylthiomethyl)-1,4(2*H*,3*H*)-phthalazinedione	—	365
1,4-Bis(methylthio)phthalazine	152–153, NMR	(*E* 553)
2-(4-Bromobutyl)-4-methyl-1(2*H*)-phthalazinone	166–172, NMR	280
2-(2-Bromoethyl)-4-chloro-6,7-dimethoxy-1(2*H*)-phthalazinone	176–178, IR, NMR	298
2-(2-Bromoethyl)-4-chloro-1(2*H*)-phthalazinone	115–116, IR, NMR	198
2-(2-Bromoethyl)-6,7-dimethoxy-1,4(2*H*,3*H*)-phthalazinedione	213–216, IR, NMR	198
2-(2-Bromoethyl)-4-methyl-1(2*H*)-phthalazinone	101–102, IR, NMR	298
2-(2-Bromoethyl)-4-phenyl-1(2*H*)-phthalazinone	119–121, IR, NMR	298
2-(2-Bromoethyl)-1,4(2*H*,3*H*)-phthalazinedione	176–179, IR, NMR	298
2-(2-Bromoethyl)-1(2*H*)-phthalazinone	73–74, IR, NMR	298
7-Bromo-4-hydroxymethyl-1(2*H*)-phthalazinone	222–223	574
1-Bromo-4-methoxyphthalazine	—	(*E* 532)
2-Bromomethyl-4-ethoxycarbonylmethyl-1(2*H*)-phthalazinone	96, NMR	68
1-Bromomethyl-4-methylphthalazine	60, IR, NMR	49
1-Bromomethyl-4-phenylphthalazine	114, IR, NMR	49
2-Bromo-4-nitromethyl-1(2*H*)-phthalazinone	—	(*E* 420)
4-Bromo-7-nitro-1(2*H*)-phthalazinone	237–239	908
7-Bromo-2-phenyl-1(2*H*)-phthalazinone	157–158, MS, NMR	411
1-Bromophthalazine	—	(*E* 531)
5-Bromophthalazine	118–119, NMR	208, 857
6-Bromophthalazine	110	80, 208
5-Bromo-1,4(2*H*,3*H*)-phthalazinedione	—	(*H* 147)

TABLE A.2. (*Continued*)

Phthalazine	Melting Point (°C) etc.	Reference(s)
6-Bromo-1,4(2*H*,3*H*)-phthalazinedione	—	(*H* 149)
4-Bromo-1(2*H*)-phthalazinone	—	(*H* 181)
6-Bromo-5,7,8-trimethyl-1,4(2*H*,3*H*)-phthalazinedione	215–216, NMR	114
1-(But-2-enylidenehydrazino)phthalazine	*E*: 111–114, IR, NMR, UV; *Z*: 83–85, IR, NMR	1026
2-(But-1-enyl)-1,4(2*H*,3*H*)-phthalazinedione	170–173	98
4-Butoxy-2-butyl-1(2*H*)-phthalazinone	—	(*E* 491)
1-Butoxy-4-isopropylphthalazine	—	(*E* 437)
1-Butoxy-4-phenylphthalazine	HCl: 245, IR, NMR	672
4-Butoxy-2-phenyl-1(2*H*)-phthalazinone	—	(*E* 554)
1-Butoxyphthalazine	—	(*E* 135)
6-Butoxy-1,4(2*H*,3*H*)-phthalazinedione	—	(*E* 475)
1-Butylaminophthalazine	—	(*E* 577)
6-Butylamino-1,4(2*H*,3*H*)-phthalazinedione	203–205, IR	313
4-Butylamino-1(2*H*)-phthalazinone	—	196
N-Butyl-4-chloro-1-oxo-2-phenyl-1,2-dihydro-6-phthalazinesulfonamide	—	260
N-Butyl-1,4-dioxo-2-phenyl-1,2,3,4-tetrahydro-6-phthalazinesulfonamide	—	260
N-Butyl-1,4-dioxo-1,2,3,4-tetrahydro-6-phthalazinesulfonamide	—	260
1-(*N*′-Butylguanidino)phthalazine	125–126, IR, MS, NMR	225
1-Butyl-4-hydroxyaminophthalazine	—	(*E* 591)
Butyl 1-hydroxymethyl-4-oxo-3,4-dihydro-6-phthalazinecarboxylate	144–146	404
2-Butyl-4-isopropyl-1(2*H*)-phthalazinone	—	(*E* 419)
4-*sec*-Butyl-2-isopropyl-1(2*H*)-phthalazinone	—	(*E* 419)
2-Butyl-4-phenyl-1(2*H*)-phthalazinone	220, IR, NMR	672
2-*tert*-Butyl-4-phenyl-1(2*H*)-phthalazinone	139–141	119
6-*tert*-Butyl-1,4-phthalazinediamine	—	(*E* 589)
2-Butyl-1(2*H*)-phthalazinone	liq, IR, NMR	47
4-Butyl-1(2*H*)-phthalazinone	152, IR, NMR	(*E* 404) 1015
4-*sec*-Butyl-1(2*H*)-phthalazinone	—	(*E* 404)
1-*tert*-Butylsulfinyl-4-methoxyphthalazine	147, IR, NMR	1010
1-Butylsulfonyl-4-phenylphthalazine	—	(*E* 558)
1-*tert*-Butylthio-4-methoxyphthalazine	61, IR, NMR	1010
1-Butylthio-4-phenylphthalazine	—	(*E* 553)
1-Butyrylmethyl-4-phenylphthalazine	—	(*E* 359)
3-Carbamoyl-6,7-dimethoxy-4-oxo-3,4-dihydro-1-phthalazinecarboxylic acid	—	(*H* 102)
4-Carbamoylmethyl-2-ethyl-1(2*H*)-phthalazinone	—	(*E* 431)
1-Carbamoylmethyl-4-hydroxyaminophthalazine	—	(*E* 591)
2-Carbamoylmethyl-4-methyl-1(2*H*)-phthalazinone	245–246	649
4-Carbamoylmethyl-2-methyl-1(2*H*)-phthalazinone	—	(*E* 431)
2-Carbamoylmethyl-1,4(2*H*,3*H*)-phthalazinedione	—	(*E* 479)
4-Carbamoylmethyl-1(2*H*)-phthalazinone	—	(*E* 411)

TABLE A.2. (*Continued*)

Phthalazine	Melting Point (°C) etc.	Reference(s)
3-(4-Carboxybutyl)-4-imino-3,4-dihydro-1-phthalazinamine	HBr: 248–250, IR, NMR	947
1-(4-Carboxybutyl)-4-phenylphthalazine	—	(*E* 359)
2-(4-Carboxybutyl)-4-phenyl-1(2*H*)-phthalazinimine	HBr: 233–235, IR, NMR	947
1-(2-Carboxyethyl)-4-methylphthalazine	—	(*E* 439)
2-(2-Carboxyethyl)-4-methyl-1(2*H*)-phthalazinone	—	(*E* 437)
2-(2-Carboxyethyl)-4-phenyl-1(2*H*)-phthalazinone	172–173	(*E* 427) 650
4-(2-Carboxyethyl)-2-phenyl-1(2*H*)-phthalazinone	—	(*H* 102)
2-(2-Carboxyethyl)-1,4(2*H*,3*H*)-phthalazinedione	biol	(*E* 480) 237
2-(2-Carboxyethyl)-1(2*H*)-phthalazinone	136–138	(*E* 399) 946
2-(2-Carboxy-1-methylethyl)-1,4(2*H*,3*H*)-phthalazinedione	—	(*E* 480)
2-(2-Carboxy-1-methylethyl)-1(2*H*)-phthalazinone	112–114	946
4-Carboxymethyl-2-ethyl-1(2*H*)-phthalazinone	—	(*E* 431)
1-Carboxymethyl-4-hydrazinophthalazine	—	(*E* 613)
1-Carboxymethyl-4-hydroxyaminophthalazine	—	(*E* 591)
2-Carboxymethyl-4-methyl-1(2*H*)-phthalazinone	253–255, NMR	649, 773
4-Carboxymethyl-2-methyl-1(2*H*)-phthalazinone	193–195, NMR	(*E* 430) 773
4-Carboxymethyl-2-phenyl-1(2*H*)-phthalazinone	—	(*H* 102) 209
2-Carboxymethyl-1,4(2*H*,3*H*)-phthalazinedione	—	(*E* 479)
2-Carboxymethyl-1(2*H*)-phthalazinone	233–235, NMR	773
4-Carboxymethyl-1(2*H*)-phthalazinone	164–165, NMR, pK_a	(*E* 411) 68, 169, 773
1-(5-Carboxypentyl)-4-phenylphthalazine	—	(*E* 359)
3-(3-Carboxypropyl)-4-imino-3,4-dihydro-1-phthalazinamine	HBr: 282–283, IR, NMR	947
2-(3-Carboxypropyl)-4-methyl-1(2*H*)-phthalazinimine	HBr: 266–267, IR, NMR	947
2-(3-Carboxypropyl)-4-phenyl-1(2*H*)-phthalazinimine	HBr: 249–251, IR, NMR	947
2-(2-Carboxypropyl)-1,4(2*H*,3*H*)-phthalazinedione	—	(*E* 480)
2-(2-Carboxypropyl)-1(2*H*)-phthalazinone	118–120	946
2-(4-Chlorobutyl)-1(2*H*)-phthalazinone	crude: liq, NMR	280
4-Chloro-2-(2-chloroethyl)-6,7-dimethoxy-1(2*H*)-phthalazinone	171–172, IR	298
4-Chloro-2-(2-chloroethyl)-1(2*H*)-phthalazinone	107–108, IR	298
4-Chloro-2-chloromethyl-1(2*H*)-phthalazinone	—	(*E* 529)
1-Chloro-4-cyanomethylphthalazine	—	(*E* 524)
4-Chloro-*N*-cyclohexyl-1-oxo-2-phenyl-1,2-dihydro-6-phthalazinesulfonamide	—	260
1-Chloro-4-diethylaminomethyl-7-methoxyphthalazine	—	195
1-Chloro-4-diethylaminomethylphthalazine	—	195
6-Chloro-1,4-dimethoxy-7-methyl phthalazine	crude, NMR	311
1-Chloro-6,7-dimethoxy-4-phenylphthalazine	—	(*E* 526)
1-Chloro-7,8-dimethoxyphthalazine	—	(*H* 180)

TABLE A.2. (*Continued*)

Phthalazine	Melting Point (°C) etc.	Reference(s)
6-Chloro-1,4-dimethoxyphthalazine	137–138, IR, NMR	311
1-Chloro-4-dimethylaminomethyl-7-methoxyphthalazine	—	195
1-Chloro-4-dimethylaminomethylphthalazine	—	195
1-Chloro-4-dimethylaminophthalazine	101, dip, NMR, UV, xl st; H_2SO_4: 141–147, IR	728, 740, 742
1-Chloro-4-(1,3-dimethylbut-2-enylidenehydrazino)phthalazine	138, IR, NMR, UV	738
1-Chloro-4-(1,3-dimethylbut-2-enylidenehydrazono)-3-methyl-3,4-dihydrophthalazine	94–95, IR, NMR, UV	738
1-Chloro-4-dimethylhydrazono-3-methyl-3,4-dihydrophthalazine	67–70, IR, NMR, UV	728
1-Chloro-4-ethoxyphthalazine	—	(*E* 525)
1-Chloro-4-(*N*′-ethylidene-*N*-methylhydrazino)phthalazine	145–147, NMR, UV	731
2-(2-Chloroethyl)-3-methyl-1,4(2*H*,3*H*)-phthalazinedione	118–120, IR, NMR, UV	745
2-(2-Chloroethyl)-4-methyl-1(2*H*)-phthalazinone	110–111	737
2-(2-Chloroethyl)-5-nitro-1(2*H*)-phthalazinone	—	(*E* 402)
2-(2-Chloroethyl)-7-nitro-1(2*H*)-phthalazinone	—	(*E* 402)
2-(2-Chloroethyl)-8-nitro-1(2*H*)-phthalazinone	—	(*E* 402)
2-(2-Chloroethyl)-4-phenyl-1(2*H*)-phthalazinone	—	(*E* 423)
1-Chloro-4-ethylphthalazine	—	(*H* 181; *E* 524)
2-(2-Chloroethyl-1(2*H*)-phthalazinone	—	(*E* 199)
1-Chloro-4-ethylthiophthalazine	—	(*E* 553)
1-Chloro-5/8-fluoro-4-methoxyphthalazine	—	(*E* 527)
1-Chloro-6/7-fluoro-4-methoxyphthalazine	—	(*E* 527)
1-Chloro-4-fluorophthalazine	NMR	887, 891
1-Chloro-5-fluorophthalazine	—	(*E* 521)
1-Chloro-6-fluorophthalazine	—	(*E* 521)
1-Chloro-8-fluorophthalazine	—	(*E* 521)
4-Chloro-6/7-fluoro-1(2*H*)-phthalazinone	—	(*E* 529)
1-Chloro-5/8-hydrazino-4-methoxyphthalazine	—	(*E* 527)
1-Chloro-6/7-hydrazino-4-methoxyphthalazine	—	(*E* 528)
1-Chloro-4-hydrazinophthalazine	200; PhNHN=: 174–175	(*E* 612) 222, 276, 593, 728
5-Chloro-1-hydrazinophthalazine	—	(*E* 611)
5-Chloro-4-hydrazinophthalazine	—	(*E* 611)
6-Chloro-1-hydrazinophthalazine	—	(*E* 611)
6-Chloro-4-hydrazinophthalazine	—	(*E* 611)
1-Chloro-4-hydrazono-3-methyl-3,4-dihydrophthalazine	106–108, IR, NMR, UV	728, 729
6-Chloro-7-(1-hydroxyethyl)-1,4-dimethoxyphthalazine	liq, NMR	311
4-Chloro-2-(2-hydroxyethyl)-6,7-dimethoxy-1(2*H*)-phthalazinone	178–180	298
4-Chloro-2-(2-hydroxyethyl)-1(2*H*)-phthalazinone	127–129, IR	298
6-Chloro-7-iodo-1,4-dimethoxyphthalazine	127–128, NMR	311

TABLE A.2. (*Continued*)

Phthalazine	Melting Point (°C) etc.	Reference(s)
1-Chloro-4-isobutylphthalazine	—	(*H* 181)
1-Chloro-4-isopropylidenehydrazinophthalazine	—	(*E* 624)
1-Chloro-4-(*N*′-isopropylidene-*N*-methylhydrazino)phthalazine	108–109, NMR, UV	731
1-Chloro-4-isopropylphthalazine	—	(*E* 524)
1-Chloro-7-methoxy-4-morpholinomethylphthalazine	—	195
1-Chloro-6-methoxy-4-phenylphthalazine	—	(*E* 526)
1-Chloro-7-methoxy-4-phenylphthalazine	—	(*E* 526)
1-Chloro-4-methoxyphthalazine	107 to 110, NMR, UV	(*E* 525) 120, 728, 1010
1-Chloro-7-methoxyphthalazine	—	(*H* 180; *E* 521)
6-Chloro-1-methoxyphthalazine	—	(*H* 84)
6-Chloro-4-methoxyphthalazine	—	(*H* 84)
1-Chloro-4-methoxyphthalazine 3-oxide	—	(*E* 373)
6-Chloro-7-methoxy-4-piperidinimethylphthalazine	—	195
1-Chloro-4-methylaminophthalazine	223–224, dip, IR, NMR, UV; H_2SO_4: 296–298	(*E* 581) 742
1-Chloro-4-(*N*-methylhydrazino)phthalazine	141–142, IR, NMR, UV	728, 731, 734
4-Chloromethyl-7-methoxy-1(2*H*)-phthalazinone	—	195
1-Chloro-4-(*N*-methyl-*N*′-methylenehydrazino)phthalazine	148, NMR, UV	731
1-Chloro-3-methyl-4-methylimino-3,4-dihydrophthalazine	80–81, dip, IR, NMR, UV; H_2SO_4: 230–233; F_2CCO_2H: 137–140	742
2-Chloromethyl-4-phenyl-1(2*H*)-phthalazinone	—	(*E* 420)
1-Chloro-4-methyl-7-nitrophthalazine	—	(*E* 526)
7-Chloro-2-methyl-1-oxo-1,2-dihydro-6-phthalazinesulfonamide	280, IR, MS, NMR	274
7-Chloro-3-methyl-4-oxo-3,4-dihydro-6-phthalazinesulfonamide	257–259, IR, MS, NMR	274
7-Chloro-2-methyl-1-oxo-1,2-dihydro-6-phthalazinesulfonyl chloride	189	274
1-Chloro-3-methyl-4-phenylimino-3,4-dihydrophthalazine	113–114, dip, IR, NMR, UV; H_2SO_4: 207–212	742
2-Chloromethyl-4-phenyl-1(2*H*)-phthalazinone	—	(*E* 420)
7-Chloromethyl-2-phenyl-1(2*H*)-phthalazinone	119–120, MS, NMR	411
1-Chloro-4-methylphthalazine	—	(*H* 180; *E* 524)
5-Chloro-2-methyl-1,4(2*H*,3*H*)-phthalazinedione	—	(*E* 485)
4-Chloro-2-methyl-1(2*H*)-phthalazinimine	100–104 or 116–119, dip, IR, NMR, UV; H_2SO_4: 291, IR, UV	728, 742
2-Chloromethyl-1(2*H*)-phthalazinone	—	(*E* 398)
4-Chloromethyl-1(2*H*)-phthalazinone	—	195

TABLE A.2. *(Continued)*

Phthalazine	Melting Point (°C) etc.	Reference(s)
4-Chloro-2-methyl-1(2*H*)-phthalazinone	127–128 or 128–130, NMR, UV	(*E* 529) 685, 728
6-Chloro-4-methyl-1(2*H*)-phthalazinone	294–297, IR, NMR	881
1-Chloro-4-methylthiophthalazine	123–125, NMR	512
1-Chloro-4-morpholinomethylphthalazine	—	195
1-Chloro-4-morpholinophthalazine	149–152	271
1-Chloro-7-nitrophthalazine	155–157	(*E* 521) 908
4-Chloro-5-nitro-1(2*H*)-phthalazinone	—	(*E* 529)
7-Chloro-1-oxo-1,2-dihydro-6-phthalazinesulfonamide	>300, IR, MS, NMR	274
7-Chloro-4-oxo-3,4-dihydro-6-phthalazinesulfonamide	>310, IR, MS, NMR	274
7-Chloro-1-oxo-1,2-dihydro-6-phthalazinesulfonyl chloride	225	274
5-Chloro-4-oxo-3-phenyl-3,4-dihydro-1-phthalazinecarboxylic acid	—	(*H* 102)
8-Chloro-4-oxo-3-phenyl-3,4-dihydro-1-phthalazinecarboxylic acid	—	(*H* 102)
1-Chloro-4-phenylphthalazine	158 or 160–161, NMR	(*H* 181; *E* 524) 282, 409, 650, 938
6-Chloro-3-phenyl-1,4(2*H*,3*H*)-phthalazinedione	—	(*E* 485)
4-Chloro-2-phenyl-1(2*H*)-phthalazinone	—	(*E* 530)
7-Chloro-2-phenyl-1(2*H*)-phthalazinone	160–161, IR, MS, NMR	411
4-Chloro-1-phthalazinamine	221–223, dip, IR, NMR, UV; H_2SO_4: 222, IR, NMR, UV	(*E* 476) 742
1-Chlorophthalazine	115 or 117, NMR, st, UV	(*H* 180; *E* 523) 282, 728, 734, 898
5-Chlorophthalazine	128	(*E* 335) 80, 208
6-Chlorophthalazine	139	(*E* 335) 80, 208
5-Chloro-1,4(2*H*,3*H*)-phthalazinedione	—	(*H* 147)
6-Chloro-1,4(2*H*,3*H*)-phthalazinedione	—	(*H* 149; *E* 475)
4-Chloro-1(2*H*)-phthalazinethione	222–224	(*E* 548) 201, 908
4-Chloro-1(2*H*)-phthalazinone	274, IR, NMR, UV	(*H* 79; *E* 529) 728, 734
5-Chloro-1(2*H*)-phthalazinone	—	(*E* 397)
6-Chloro-1(2*H*)-phthalazinone	—	(*H* 79; *E* 397)
7-Chloro-1(2*H*)-phthalazinone	—	(*H* 79; *E* 397)
8-Chloro-1(2*H*)-phthalazinone	—	(*E* 397)
1-Chloro-4-piperidinomethylphthalazine	—	195
1-Chloro-4-piperidinophthalazine	—	(*E* 582)
2-(3-Chloropropyl)-4-phenyl-1(2*H*)-phthalazinone	—	(*E* 426)
1-Chloro-4-propylphthalazine	—	(*H* 181)
1-Chloro-4-styrylphthalazine	103–105, IR, MS, NMR, UV	(*E* 525) 135
5-Chloro-6,7,8-trimethyl-1,4(2*H*,3*H*)-phthalazinedione	231–233, NMR	114
6-Chloro-5,7,8-trimethyl-1,4(2*H*,3*H*)-phthalazinedione	211–213, NMR	114

TABLE A.2. (*Continued*)

Phthalazine	Melting Point (°C) etc.	Reference(s)
1-Cyanoaminophthalazine	217–219	225
2-(2-Cyanoethyl)-4-methoxycarbonylmethyl-1(2*H*)-phthalazinone	125–126, NMR	68
1-(2-Cyanoethyl)-4-methylphthalazine	—	(*E* 439)
2-(2-Cyanoethyl)-4-methyl-1(2*H*)-phthalazinone	—	(*E* 427)
2-(2-Cyanoethyl)-4-phenyl-1(2*H*)-phthalazinone	144–145	(*E* 427) 214, 650
1-(1-Cyanoethyl)phthalazine	200/0.7, IR, NMR	417
2-(2-Cyanoethyl)-1,4(2*H*,3*H*)-phthalazinedione	—	(*E* 480)
2-(2-Cyanoethyl)-1(2*H*)-phthalazinone	—	(*E* 399) 214
2-Cyanomethyl-4-ethoxycarbonylmethyl-1(2*H*)-phthalazinone	113–114 or 125, NMR	68, 784
1-Cyanomethyl-4-hydrazihophthalazine	—	(*E* 613)
4-Cyanomethyl-1(2*H*)-phthalazinone	—	(*E* 405)
2-(Cyclohex-1-enyl)-1,4(2*H*,3*H*)-phthalazinedione	148–150	98
N-Cyclohexyl-1,4-dioxo-2-phenyl-1,2,3,4-tetrahydro-6-phthalazinesulfonamide	—	260
N-Cyclohexyl-1,4-dioxo-1,2,3,4-tetrahydro-6-phthalazinesulfonamide	—	260
1-Cyclohexylphthalazine	—	609
1,4-Diacetamidophthalazine	—	(*E* 589)
5,8-Diacetamido-1,4(2*H*,3*H*)-phthalazinedione	—	(*H* 150)
2,3-Diacetyl-1,4(2*H*,3*H*)-phthalazinedione	—	230
5,6-Diamino-1,4(2*H*,3*H*)-phthalazinedione	74, IR, NMR	4
5,7-Diamino-1,4(2*H*,3*H*)-phthalazinedione	—	(*E* 475)
5,8-Diamino-1,4(2*H*,3*H*)-phthalazinedione	—	(*H* 150)
6,7-Diamino-1,4(2*H*,3*H*)-phthalazinedione	407, IR, NMR	(*H* 150) 4
1,4-Dianilinophthalazine	230–231, dip, IR, NMR, UV	(*H* 184; *E* 590) 742
5,8-Dianilino-1,4(2*H*,3*H*)-phthalazinedione	—	(*H* 150)
1,4-Dibenzamidophthalazine	303	685
6,7-Dibenzoyl-1,4-diphenylphthalazine	261, IR	16, 942
6,7-Dibenzyl-1,4-diphenylphthalazine	233, UV	16, 942
1,4-Dibromophthalazine	—	(*E* 531)
6,7-Dibromo-1,4(2*H*,3*H*)-phthalazinedione	>350	704
6,7-Dibromo-5,8-phthalazinequinone	176–180, IR, NMR	715
Di-*tert*-butyl 5-amino-4-oxo-3-phenyl-1-thiocarbamoyl-3,4-dihydro-6,7-phthalazinedicarboxylate	30, IR, NMR	553
6,7-Dichloro-1,4-dihydrazinophthalazine	—	(*E* 619)
1,4-Dichloro-6,7-diphenylphthalazine	—	122
6,7-Dichloro-4-ethoxycarbonylmethyl-1(2*H*)-phthalazinone	250, NMR	68, 908
k,4-Dichloro-6-fluorophthalazine	—	(*E* 531)
1,4-Dichloro-6-phenylphthalazine	150–153	(*E* 531) 908
6,7-Dichloro-2-phenylphthalazin-2-ium-4-olate	280, IR	594
1,4-Dichlorophthalazine	163–165, NMR, UV; 2-Me_2SO_4: crude	(*E* 531) 160, 222, 512, 728, 734, 1003
1,5-Dichlorophthalazine	—	(*E* 521)

TABLE A.2. (*Continued*)

Phthalazine	Melting Point (°C) etc.	Reference(s)
1,6-Dichlorophthalazine	—	(*H* 180; *E* 521)
1,7-Dichlorophthalazine	—	(*H* 180; *E* 521)
1,8-Dichlorophthalazine	—	(*E* 521)
1,4-Dichloro-6-phthalazinecarbonitrile	solid, NMR	285
1,4-Dichloro-6-phthalazinecarbonyl chloride	—	(*E* 531)
1,4-Dichloro-6-phthalazinecarboxylic acid	—	(*E* 531)
5,8-Dichloro-1,4(2*H*,3*H*)-phthalazinedione	—	(*H* 149)
6,7-Dichloro-1,4(2*H*,3*H*)-phthalazinedione	—	(*H* 150) 856
6,7-Dichloro-5,8-phthalazinequinone	223–225, IR, MS, NMR	275, 715, 856
1,4-Dichloro-6-phthalazinesulfonic acid	—	(*E* 531)
5-Diethylamino-1,6-dimethyl-4,7,8-triphenylphthalazine	193–194, MS, NMR, UV	128
2-(2-Diethylaminoethyl)-4-ethylthio-1(2*H*)-phthalazinone	—	(*E* 557)
2-(2-Diethylaminoethyl)-4-hydrazino-1(2*H*)-phthalazinone	—	(*E* 618)
1-(2-Diethylaminoethyl)-4-hydroxyaminophthalazine	—	(*E* 591)
2-(2-Diethylaminoethyl)-4-methoxy-1(2*H*)-phthalazinone	—	(*E* 492)
2-(2-Diethylaminoethyl)-4-methyl-1(2*H*)-phthalazinone	—	(*E* 426)
2-(2-Diethylaminoethyl)-4-phenyl-1(2*H*)-phthalazinone	—	(*E* 426)
2-(2-Diethylaminoethyl)-1(2*H*)-phthalazinethione	—	(*E* 554)
2-(2-Diethylaminoethyl)-1(2*H*)-phthalazinone	—	(*E* 399)
4-Diethylamino-7-methoxy-1(2*H*)-phthalazinone	—	195
1-Diethylaminomethyl-4-hydrazino-6-methoxyphthalazine	—	195
1-Diethylaminomethyl-4-hydrazinophthalazine	—	195
6-Diethylamino-8-methyl-1,4(2*H*,3*H*)-phthalazinedione	290–291, IR, NMR, UV	628
4-Diethylaminomethyl-1(2*H*)-phthalazinone	—	195
5-Diethylamino-6-methyl-1,4,7,8-tetraphenylphthalazine	208–209, MS, NMR, UV	128
Diethyl 5-amino-4-oxo-3-phenyl-3,4-dihydro-1,6-phthalazinedicarboxylate	145, IR, NMR	490
6-Diethylamino-1,4(2*H*,3*H*)-phthalazinedione	—	(*E* 472)
5-Diethylamino-1,4,6-trimethyl-7,8-diphenylphthalazine	204–205, MS, NMR, UV	128
Diethyl 5,7-bis(bromomethyl)-4-oxo-3,4-dihydro-1,6-phthalazinedicarboxylate	—	250
Diethyl 5-bromomethyl-7-methyl-4-oxo-3,4-dihydro-1,6-phthalazinedicarboxylate	—	250
2-(*N*,*N*-Diethylcarbamoylmethyl)-4-methyl-1(2*H*)-phthalazinone	—	(*E* 425)
2-(*N*,*N*-Diethylcarbamoylmethyl)-4-phenyl-1(2*H*)-phthalazinone	—	(*E* 425)

TABLE A.2. (*Continued*)

Phthalazine	Melting Point (°C) etc.	Reference(s)
Diethyl 5,7-dimethyl-8-nitro-4-oxo-3,4-dihydro-1,6-phthalazinedicarboxylate	200–202	425
Diethyl 5,7-dimethyl-4-oxo-3,4-dihydro-1,6-phthalazinedicarboxylate	159–161, UV	405, 783
Diethyl 5,7-dimethyl-4-oxo-3-phenyl-3,4-dihydro-1,6-phthalazinedicarboxylate	108–109, IR, NMR	956
N,N-Diethyl-1-1,4-dioxo-1,2,3,4-tetrahydro-2-phthalazinecarbothioamide	—	477
Diethyl 5,8-dioxo-2,3,5,8-tetrahydro-6,7-phthalazinedicarboxylate	151–153, IR, UV	179, 697
N,N-Diethyl-3-methyl-1,4-dioxo-1,2,3,4-tetrahydro-2-phthalazinecarbothioamide	—	477
1,4-Diethyl-6-methylphthalazine	—	352
Diethyl 4-oxo-3,4-dihydro-1,6-phthalazinedicarboxylate	—	574
Diethyl 4-oxo-3,4-dihydro-1,7-phthalazinedicarboxylate	202–203, NMR	404
1,4-Diethylphthalazine	80–85, NMR	(*E* 357) 352, 399, 908
Diethyl 1,4-phthalazinedicarboxylate	—	(*E* 644)
Diethyl 6,7-phthalazinedicarboxylate	81–82, IR, UV	584
2,4-Diethyl-1(2*H*)-phthalazinone	—	(*H* 100)
2-[2-(*N,N*-Diethylsulfamoyl)ethyl]-1,4(2*H*,3*H*)-phthalazinedione	—	(*E* 483)
5,8-Difluoro-1,4-dioxo-1,2,3,4-tetrahydro-2-phthalazinecarbothioamide	237–238, NMR	994
6,7-Difluoro-1,4-dioxo-1,2,3,4-tetrahydro-2-phthalazinecarbothioamide	215–216, NMR	994
5,8-Difluoro-1,4-dioxo-1,2,3,4-tetrahydro-2-phthalazinecarboxamide	278–279, NMR	994
6,7-Difluoro-1,4-dioxo-1,2,3,4-tetrahydro-2-phthalazinecarboxamide	261–263, NMR	994
1,4-Difluorophthalazine	138–139, NMR	723, 887, 889
1,4-Dihydrazinophthalazine	187–190, IR, UV; 2HCl: anal; H_2SO_4: Raman	(*E* 619) 211, 223, 685, 770, 804, 895
1,8-Dihydrazinophthalazine	—	(*E* 611)
5,8-Dihydroxy-1,4(2*H*,3*H*)-phthalazinedione	—	(*H* 150) 229
6,7-Diiodo-2-phenylphthalazin-2-ium-4-olate	269–270, IR	594
1,4-Diiodophthalazine	—	(*E* 532)
2,3-Diisopropyl-1,4(2*H*,3*H*)-phthalazinedione	101–103, IR, MS, NMR	154
1,4-Dimethoxycarbonyl-5,8-phthalazinequinone	163–166, IR, NMR	975
5,8-Dimethoxy-2,3-dimethyl-1,4(2*H*,3*H*)-phthalazinedione	246–248, IR, MS, NMR	715
6,7-Dimethoxy-2,4-diphenyl-1(2*H*)-phthalazinone	—	(*E* 434)
5,7-Dimethoxy-4-methyl-2-phenyl-1(2*H*)-phthalazinone	—	(*E* 433)
6,7-Dimethoxy-1-methylphthalazine	161–162, IR, NMR	21
5,8-Dimethoxy-2-methyl-1,4(2*H*,3*H*)-phthalazinedione	160–170, IR, NMR	715

TABLE A.2. (*Continued*)

Phthalazine	Melting Point (°C) etc.	Reference(s)
5,7-Dimethoxy-4-methyl-1(2*H*)-phthalazinone	—	(*E* 414)
6,7-Dimethoxy-4-methyl-1(2*H*)-phthalazinone	—	(*E* 414)
7,8-Dimethoxy-2-methyl-1(2*H*)-phthalazinone	132, IR, MS	(*H* 87) 488
7,8-Dimethoxy-5-nitro-2-phenyl-1(2*H*)-phthalazinone	—	(*H* 87)
7,8-Dimethoxy-5-nitro-1(2*H*)-phthalazinone	—	(*H* 79)
6,7-Dimethoxy-4-oxo-3-phenyl-3,4-dihydro-1-phthalazinecarboxylic acid	—	(*H* 102)
6,7-Dimethoxy-1-phenylphthalazine	167–168, IR, MS, NMR	(*H* 74; *E* 353) 902
6,7-Dimethoxy-2-phenyl-1(2*H*)-phthalazinone	227–228	(*H* 87) 270
6,7-Dimethoxy-4-phenyl-1(2*H*)-phthalazinone	—	(*E* 415)
7,8-Dimethoxy-2-phenyl-1(2*H*)-phthalazinone	175, IR	(*H* 87) 488
1,4-Dimethoxyphthalazine	93	908
1,7-Dimethoxyphthalazine	—	(*H* 84)
6,7-Dimethoxyphthalazine	198–200, IR, NMR	(*E* 335) 557, 948
5,8-Dimethoxy-1,4(2*H*,3*H*)-phthalazinedione	244–246, IR, MS, NMR	715
6,7-Dimethoxy-1,4(2*H*,3*H*)-phthalazinedione	317–320	(*E* 475) 705
7,8-Dimethoxy-1(2*H*)-phthalazinone	—	(*H* 79)
1-(2-Dimethylaminoethyl)-4-hydroxyaminophthalazine	—	(*E* 591)
2-(2-Dimethylaminoethyl)-4-phenyl-1(2*H*)-phthalazinone	—	(*E* 423)
2-(2-Dimethylaminoethyl)-1(2*H*)-phthalazinone	—	(*E* 399)
1-Dimethylamino-4-isopropylphthalazine	62–63, IR	689
2-(2-Dimethylamino-1-methylethyl)-1(2*H*)-phthalazinone	—	(*E* 399)
1-Dimethylaminomethyl-4-hydrazino-6-methoxyphthalazine	—	195
1-Dimethylaminomethyl-4-hydrazinophthalazine	—	195
4-Dimethylaminomethyl-7-methoxy-1(2*H*)-phthalazinone	—	195
2-Dimethylaminomethyl-4-methyl-1(2*H*)-phthalazinone	—	(*E* 420)
2-Dimethylaminomethyl-4-phenyl-1(2*H*)-phthalazinone	—	(*E* 421)
7-Dimethylaminomethyl-2-phenyl-1(2*H*)-phthalazinone	113–115, IR, MS, NMR	411
1-Dimethylamino-4-methylphthalazine	—	(*E* 585)
2-Dimethylaminomethyl-1(2*H*)-phthalazinone	—	(*E* 399)
4-Dimethylaminomethyl-1(2*H*)-phthalazinone	—	195
1-Dimethylamino-4-phenylphthalazine	—	(*E* 581)
7-Dimethylamino-2-phenyl-1(2*H*)-phthalazinone	140–141, IR, NMR	411
7-Dimethylamino-2-phenyl-1(2*H*)-phthalazinone ω-*N*-oxide	159–160, MS, NMR	411
1-Dimethylaminophthalazine	146, NMR; 3-MeI: 214 or 215, NMR; 3-PrI: 202–203, NMR; 3-BzCH$_2$Br: 205	62, 63, 939, 943
5-Dimethylamino-1,4(2*H*,3*H*)-phthalazinedione	—	(*E* 472)

TABLE A.2. (*Continued*)

Phthalazine	Melting Point (°C) etc.	Reference(s)
6-Dimethylamino-1,4(2*H*,3*H*)-phthalazinedione	—	(*E* 472)
1-(3-Dimethylaminopropyl)-4-phenylphthalazine	—	(*E* 358)
1-(3-Dimethylaminopropyl)phthalazine	—	(*E* 353)
1-(1,3-Dimethylbut-2-enylidenehydrazino)-phthalazine	132–133, IR, NMR, UV	249, 396, 738
1-(1,3-Dimethylbut-2-enylidenehydrazono)-2-methyl-1,2-dihydrophthalazine	43–44, IR, NMR, UV	738
1-(1,2-Dimethylbutylidenehydrazino)phthalazine	liq, NMR	396
2,3-Dimethyl-1,4-dioxo-1,2,3,4-tetrahydro-6-phthalazinecarboxylic acid	282–285	918
1-(*N*,*N*′-Dimethylhydrazino)-4-methylphthalazine	—	(*E* 617)
Dimethyl 5-hydroxy-8-oxo-5,6,7,8-tetrahydro-1,4-phthalazinedicarboxylate	135–137, IR, NMR	975
2,4-Dimethyl-5-nitro-1(2*H*)-phthalazinone	—	(*E* 434)
N,*N*-Dimethyl-4-oxo-3,4-dihydro-1-phthalazinecarboxamide	—	(*E* 647)
N,*N*-Dimethyl-4-oxo-3-phenyl-3,4-dihydro-6-phthalazinecarboxamide	150–152, IR, MS, NMR	411
1,4-Dimethylphthalazine	104–106 or 106–107, IR, NMR	(*E* 357) 21, 290, 352
5,7-Dimethylphthalazine	106, NMR, UV	80
Dimethyl 2,3-phthalazinedicarboxylate	176–177, IR, MS, NMR	116, 131
2,3-Dimethyl-1,4(2*H*,3*H*)-phthalazinedione	172–174 or 175–176, IR, MS, NMR	(*E* 499) 1, 149
6,7-Dimethyl-1,4(2*H*,3*H*)-phthalazinedione	>350	704
2,3-Dimethyl-1,4(2*H*,3*H*)-phthalazinedithione	188–189	172, 460
1,2-Dimethylphthalazin-2-ium-4-olate	—	(*E* 660)
2,4-Dimethyl-1(2*H*)-phthalazinone	110–112	(*H* 97; *E* 417) 447, 575
2,8-Dimethyl-1(2*H*)-phthalazinone	98–101, IR, NMR	622
2-(1,2-Dimethylprop-1-enyl)-3-methyl-1,4(2*H*,3*H*)-phthalazinedione	crude, NMR	563
2-(1,2-Dimethylpropyl)-1(2*H*)-phthalazinone	—	(*E* 404)
2,3-Dimethyl-4-thioxo-3,4-dihydro-1(2*H*)-phthalazinone	—	172
1,4-Dimorpholino-6-phenylphthalazine	193–195	395
1,4-Dimorpholinophthalazine	—	(*E* 590)
5,7-Dinitro-1,4(2*H*,3*H*)-phthalazinedione	—	(*H* 149)
5,7-Dinitro-1(2*H*)-phthalazinone	288–289, IR, NMR	715
1,4-Dioxo-1,2,3,4-tetrahydro-5-phthalazinecarbohydrazide	—	(*H* 149)
1,4-Dioxo-1,2,3,4-tetrahydro-6-phthalazinecarbonitrile	solid, NMR	285
1,4-Dioxo-1,2,3,4-tetrahydro-2-phthalazinecarbothioamide	211–213 or 307(?), NMR	499, 994
1,4-Dioxo-1,2,3,4-tetrahydro-2-phthalazinecarboxamide	285–287 or 320(?), NMR	(*H* 161) 499, 994
1,4-Dioxo-1,2,3,4-tetrahydro-6-phthalazinecarboxamide	solid, NMR	285
1,4-Dioxo-1,2,3,4-tetrahydro-6,7-phthalazinedicarboxylic acid	$2NH_2NH_3$: xl st	885

TABLE A.2. (*Continued*)

Phthalazine	Melting Point (°C) etc.	Reference(s)
1,4-Diphenoxyphthalazine	222, NMR, UV	728, 908
1,4-Diphenylphthalazine	190 to 196, NMR 2-PhClO$_4$: 249–251	(*H* 77; *E* 361) 80, 81, 125, 315, 322, 444, 445, 707, 908
2,3-Diphenyl-1,4(2*H*,3*H*)-phthalazinedione	174 or 175, IR, MS, NMR, UV	149, 150
2,3-Diphenyl-1,4(2*H*,3*H*)-phthalazinedithione	226, MS, NMR	295
1,4-Diphenylphthalazine 2-oxide	193–194	(*E* 374) 315
2,4-Diphenyl-1(2*H*)-phthalazinethione	—	(*E* 554)
1,4-Diphenylphthalazin-2-ium-2-benzimidate	—	177
1,4-Diphenyl-5(3*H*) phthalazinone	296–298, IR, MS, UV	691
2,4-Diphenyl-1(2*H*)-phthalazinone	166 or 168; 3-EtBF$_4$: 251–253; 3-EtClO$_4$: 236–237	(*H* 102) 446, 447, 612, 638
2,3-Diphenyl-4-thioxo-3,4-dihydro-1(2*H*)-phthalazinone	192, MS, NMR	296
1,4-Dipropylphthalazine	—	352
6-Ethoxycarbonyl-5,7-dimethyl-8-nitro-4-oxo-3,4-dihydro-1-phthalazinecarboxylic acid	212	525
6-Ethoxycarbonyl-5,7-dimethyl-4-oxo-3,4-dihydro-1-phthalazinecarboxylic acid	216–218, UV	405, 783
6-Ethoxycarbonyl-5,7-dimethyl-4-oxo-3-phenyl-3,4-dihydro-1-phthalazinecarboxylic acid	218–219, IR, NMR	956
]-(2-Ethoxycarbonylethyl)-1,4(2*H*,3*H*)-phthalazinedione	—	(*E* 480)
3-Ethoxycarbonylmethyl-*N*,*N*-diethyl-1,4-dioxo-1,2,3,4-tetrahydro-2-phthalazine-carbothioamide	—	477
4-Ethoxycarbonylmethyl-2-ethyl-1(2*H*)-phthalazinone	—	(*E* 431)
1-Ethoxycarbonylmethyl-4-hydrazinophthalazine	—	(*E* 613)
4-Ethoxycarbonylmethyl-2-hydroxymethyl-1(2*H*)-phthalazinone	113–114	68
2-Ethoxycarbonylmethyl-4-methyl-1(2*H*)-phthalazinone	93–94	649
2-Ethoxycarbonylmethyl-5-nitro-1(2*H*)-phthalazinone	—	(*E* 402)
2-Ethoxycarbonylmethyl-7-nitro-1(2*H*)-phthalazinone	—	(*E* 402)
2-Ethoxycarbonylmethyl-4-phenyl-1(2*H*)-phthalazinone	194	659
1-Ethoxycarbonylmethylphthalazine	—	(*E* 355)
2-Ethoxycarbonylmethyl-1,4(2*H*,3*H*)-phthalazinedione	—	(*E* 479)
2-Ethoxycarbonylmethyl-1(2*H*)-phthalazinone	—	(*E* 399)
4-Ethoxycarbonylmethyl-1(2*H*)-phthalazinone	—	(*E* 411) 68
3-(3-Ethoxycarbonylpropyl)-4-imino-3,4-dihydro-1-phthalazinamine	92–93, IR, NMR; HBr: 207–208, IR, NMR	947
2-(3-Ethoxycarbonylpropyl)-4-methyl-1(2*H*)-phthalazinimine	HBr: 172–173, IR, NMR	947

TABLE A.2. (*Continued*)

Phthalazine	Melting Point (°C) etc.	Reference(s)
2-(3-Ethoxycarbonylpropyl)-4-phenyl-1(2*H*)-phthalazinone	HBr: 159–160, IR, NMR	947
2-(3-Ethoxycarbonylpropyl)-1(2*H*)-phthalazinimine	HBr: 176–177, IR, NMR	947
1-Ethoxy-4-ethylphthalazine	—	(*H* 84; *E* 436)
1-Ethoxy-4-hydroxyaminophthalazine	—	(*E* 591)
1-Ethoxy-4-isobutylphthalazine	—	(*H* 84)
1-Ethoxy-4-isopropoxyphthalazine	—	(*E* 436)
1-Ethoxy-4-isopropyl-5,8-dimethylphthalazine	—	(*E* 436)
1-Ethoxy-4-isopropylphthalazine	—	(*E* 436)
1-Ethoxy-4-methylphthalazine	—	(*H* 84; *E* 436, 665)
6-Ethoxy-7-nitro-1,4(2*H*,3*H*)-phthalazinedione	319–320, IR, NMR, UV	48
1-Ethoxy-4-phenylphthalazine	2-$MeBF_4$: 137–141	(*E* 436) 447
4-Ethoxy-2-phenyl-1(2*H*)-phthalazinethione	—	(*E* 554)
4-Ethoxy-2-phenyl-1(2*H*)-phthalazinone	—	(*H* 170; *E* 492)
1-Ethoxyphthalazine	—	(*H* 84; *E* 435, 665)
6-Ethoxy-1,4-phthalazinediamine	—	(*E* 589)
6-Ethoxy-1,4(2*H*,3*H*)-phthalazinedione	—	(*E* 475)
1-Ethoxyphthalazine 3-oxide	—	(*E* 372)
1-(2-Ethoxyvinyl)-4-phenylphthalazine	2-$MeClO_4$: 163–165; 2-$PhClO_4$: 123–124	461
Ethyl 8-acetamido-1-hydroxymethyl-5,7-dimethyl-4-oxo-3,4-dihydro-6-phthalazinecarboxylate	235–240	425
Ethyl 8-acetamido-4-oxo-3,4-dihydro-1-phthalazinecarboxylate	240–242	573
Ethyl 4-acetoxy-1-chloro-5,7-dimethyl-6-phthalazinecarboxylate	—	354
Ethyl 1-acetyl-5,7-dimethyl-4-oxo-3,4-dihydro-6-phthalazinecarboxylate	169–171	405
Ethyl 5-amino-6-benzoyl-4-oxo-3,7-diphenyl-3,4-dihydro-1-phthalazinecarboxylate	146, IR, NMR	380
Ethyl 5-amino-6-benzoyl-4-oxo-3-phenyl-3,4-dihydro-1-phthalazinecarboxylate	210, IR, NMR	380
Ethyl 5-amino-6-cyano-7-hydroxy-4-oxo-3-phenyl-3,4-dihydro-1-phthalazinecarboxylate	162, IR, NMR	526
Ethyl 5-amino-6-cyano-4-oxo-3,7-diphenyl-3,4-dihydro-1-phthalazinecarboxylate	271–272 or 275, IR, NMR	66, 489
Ethyl 5-amino-6-cyano-4-oxo-3-phenyl-3,4-dihydro-1-phthalazinecarboxylate	223 or 240, IR, NMR	379, 490
Ethyl 5-amino-6-cyano-4-oxo-2-phenyl-8-thioxo-2,3,4,8-tetrahydro-1-phthalazinecarboxylate	80, IR, NMR	520
Ethyl 5-amino-6-cyano-4-oxo-3-phenyl-8-thioxo-2,3,4,8-tetrahydro-1-phthalazinecarboxylate	170, IR, NMR	796
Ethyl 8-amino-5,7-dimethyl-1-morpholinomethyl-4-oxo-3,4-dihydro-6-phthalazinecarboxylate	204–206	425
Ethyl 1-amino-5,7-dimethyl-4-oxo-3,4-dihydro-6-phthalazinecarboxylate	218–220, NMR	573

TABLE A.2. (*Continued*)

Phthalazine	Melting Point (°C) etc.	Reference(s)
Ethyl 8-amino-5,7-dimethyl-4-oxo-3,4-dihydro-6-phthalazinecarboxylate	205–206	425
Ethyl 1-amino-5,7-dimethyl-4-oxo-3-phenyl-3,4-dihydro-6-phthalazinecarboxylate	173–174, NMR	573
Ethyl 1-(2-aminoethyl)-5,7-dimethyl-4-oxo-3,4-dihydro-6-phthalazinecarboxylate	143–144, IR, NMR, UV	405
Ethyl 8-amino-1-hydroxymethyl-5,7-dimethyl-4-oxo-3,4-dihydro-6-phthalazinecarboxylate	217–219, IR, NMR, UV	425
Ethyl 1-aminomethyl-5,7-dimethyl-4-oxo-3,4-dihydro-6-phthalazinecarboxylate	HCl: >200, anal	405
Ethyl 5-amino-6-nitro-4-oxo-3,7-diphenyl-3,4-dihydro-1-phthalazinecarboxylate	—	380
Ethyl 8-amino-4-oxo-3,4-dihydro-1-phthalazinecarboxylate	140–141	573
1-Ethylamino-4-phenylphthalazine	—	(*E* 581)
4-Ethylamino-1(2*H*)-phthalazinone	—	196
Ethyl 1-amino-3,5,7-trimethyl-4-oxo-3,4-dihydro-6-phthalazinecarboxylate	131–132, NMR	243, 573
Ethyl 8-amino-1,5,7-trimethyl-4-oxo-3,4-dihydro-6-phthalazinecarboxylate	181–182	425
Ethyl 8-amino-3,5,7-trimethyl-4-oxo-3,4-dihydro-6-phthalazinecarboxylate	178–179	425
Ethyl 1-benzyl-5,7-dimethyl-4-oxo-3,4-dihydro-6-phthalazinecarboxylate	—	262
Ethyl 3-benzyl-4-oxo-3,4-dihydro-5-phthalazinecarboxylate	147–149, IR, NMR	623
4-(1-Ethylbutyl)-2-isopropyl-1(2*H*)-phthalazinone	—	(*E* 419)
4-(1-Ethylbutyl)-2-methyl-1(2*H*)-phthalazinone	—	(*E* 417)
4-(1-Ethylbutyl)-1(2*H*)-phthalazinone	—	(*E* 455)
Ethyl 1-carbamoyl-5,7-dimethyl-4-oxo-3-phenyl-3,4-dihydro-6-phthalazinecarboxylate	229–230, NMR	956
Ethyl 1-carboxymethyl-5,7-dimethyl-4-oxo-3-phenyl-3,4-dihydro-6-phthalazinecarboxylate	178–180, MS, NMR	956
Ethyl 1-chloro-5,7-dimethyl-4-oxo-3,4-dihydro-6-phthalazinecarboxylate	biol	354
Ethyl 1-chloro-5,7-dimethyl-6-phthalazinecarboxylate	167, IR, NMR	421
Ethyl 4-chloro-5,7-dimethyl-6-phthalazinecarboxylate	solid, crude	783
Ethyl 1-chloroformyl-5,7-dimethyl-4-oxo-3-phenyl-3,4-dihydro-6-phthalazinecarboxylate	crude	956
Ethyl 8-chloro-1-hydroxymethyl-5,7-dimethyl-4-oxo-3,4-dihydro-6-phthalazinecarboxylate	202–203	425
Ethyl 1-chloromethyl-5,7-dimethyl-8-nitro-4-oxo-3,4-dihydro-6-phthalazinecarboxylate	207–209	425
Ethyl 1-chloromethyl-5,7-dimethyl-4-oxo-3,4-dihydro-6-phthalazinecarboxylate	187–189, NMR	405

TABLE A.2. (*Continued*)

Phthalazine	Melting Point (°C) etc.	Reference(s)
Ethyl 8-chloro-1,5,7-trimethyl-4-oxo-3,4-dihydro-6-phthalazinecarboxylate	188–189	425
Ethyl 8-cyano-5,7-dimethyl-4-oxo-3,4-dihydro-6-phthalazinecarboxylate	190–192	425
Ethyl 8-cyano-1-hydroxymethyl-5,7-dimethyl-4-oxo-3,4-dihydro-6-phthalazinecarboxylate	203–205	425
Ethyl 1-cyano-4-methoxy-5,7-dimethyl-6-phthalazinecarboxylate	solid	783
Ethyl 1-cyanomethyl-5,7-dimethyl-4-oxo-3,4-dihydro-6-phthalazinecarboxylate	193–194, IR, NMR	405
Ethyl 3-[*N*,*N*-diethyl(thiocarbamoyl)]-1,4-dioxo-1,2,3,4-tetrahydro-2-phthalazinecarboxylate	—	477
Ethyl dimethyl 5-amino-8-mercapto-4-oxo-3-phenyl-3,4-dihydro-1,6,7-phthalazinetricarboxylate	146, IR, NMR	380
Ethyl 5,7-dimethyl-1-morpholinomethyl-8-nitro-4-oxo-3,4-dihydro-6-phthalazinecarboxylate	164–166	425
Ethyl 5,7-dimethyl-3-nitro-1-nitroxymethyl-4-oxo-3,4-dihydro-6-phthalazinecarboxylate (revised structure)	126–128, IR, NMR, UV, xl st	415, 428
Ethyl 5,7-dimethyl-8-nitro-1-nitroxymethyl-4-oxo-3,4-dihydro-6-phthalazinecarboxylate	183–185, IR, NMR, UV	225
Ethyl 5,7-dimethyl-8-nitro-4-oxo-3,4-dihydro-6-phthalazinecarboxylate	210, IR, NMR	425
Ethyl 5,7-dimethyl-1-nitroxymethyl-4-oxo-3,4-dihydro-6-phthalazinecarboxylate	150–151, IR, MS, NMR, UV	415, 428
Ethyl 5,7-dimethyl-1-oxo-1,2-dihydro-6-phthalazinecarboxylate	190–192, IR, NMR	421
Ethyl 5,7-dimethyl-4-oxo-3,4-dihydro-6-phthalazinecarboxylate	171–172 or 177, MS, NMR	405, 783
Ethyl 5,7-dimethyl-4-oxo-1-phenyl-1,4-dihydro-6-phthalazinecarboxylate	—	262
Ethyl 5,7-dimethyl-4-oxo-3-phenyl-3,4-dihydro-6-phthalazinecarboxylate	138–140, NMR	270, 956
Ethyl 1-(2-ethoxycarbonylvinyl)-5,7-dimethyl-4-oxo-3,4-dihydro-6-phthalazinecarboxylate	170–172, NMR	424
Ethyl 1-ethoxymethyl-5,7-dimethyl-8-nitro-4-oxo-3,4-dihydro-6-phthalazinecarboxylate	150–152	425
Ethyl 4-ethylthio-1-phthalazinecarboxylate	—	(*E* 650)
Ethyl 8-fluoro-1-hydroxymethyl-5,7-dimethyl-4-oxo-3,4-dihydro-6-phthalazinecarboxylate	187–188, UV	425
Ethyl 1-formyl-5,7-dimethyl-4-oxo-3,4-dihydro-6-phthalazinecarboxylate	213–214, NMR	405
2-Ethyl-4-guanidinomethyl-1(2*H*)-phthalazinone	—	157
Ethyl 1-hydrazinocarbonyl-5,7-dimethyl-4-oxo-3,4-dihydro-6-phthalazinecarboxylate	249–252	573
2-Ethyl-4-hydrazinocarbonylmethyl-1(2*H*)-phthalazinone	—	(*E* 431)

TABLE A.2. (*Continued*)

Phthalazine	Melting Point (°C) etc.	Reference(s)
Ethyl 1-(2-hydrazinocarbonylvinyl)-5,7-dimethyl-4-oxo-3,4-dihydro-6-phthalazinecarboxylate	235–237, NMR	424
1-Ethyl-4-hydrazinophthalazine	—	(*E* 612)
1-Ethyl-4-hydroxyaminophthalazine	—	(*E* 591)
Ethyl 1-(1-hydroxyethyl)-5,7-dimethyl-4-oxo-3,4-dihydro-6-phthalazinecarboxylate	179–180, IR, NMR	405
Ethyl 1-(2-hydroxyethyl)-5,7-dimethyl-4-oxo-3,4-dihydro-6-phthalazinecarboxylate	143–145, IR, MS, NMR	405
Ethyl 1-hydroxymethyl-5,7-dimethyl-8-nitro-4-oxo-3,4-dihydro-6-phthalazinecarboxylate	177–179, IR, NMR, UV	425
Ethyl 1-hydroxymethyl-5,7-dimethyl-4-oxo-3,4-dihydro-6-phthalazinecarboxylate	173–175, IR, NMR, UV	405, 783
Ethyl 1-hydroxymethyl-5,7-dimethyl-4-oxo-3-phenyl-3,4-dihydro-6-phthalazinecarboxylate	145–146, IR, NMR, UV	956
Ethyl 1-(1-hydroxy-1-methylethyl)-5,7-dimethyl-4-oxo-3,4-dihydro-6-phthalazinecarboxylate	162–163, IR, MS, NMR	405
Ethyl 1-hydroxymethyl-3-methyl-4-oxo-3,4-dihydro-6-phthalazinecarboxylate	164–165, NMR	404
Ethyl 1-hydroxymethyl-4-oxo-3,4-dihydro-6-phthalazinecarboxylate	206–207, NMR	574
Ethyl 4-hydroxymethyl-1-oxo-1,2-dihydro-6-phthalazinecarboxylate	203–204, IR, NMR	404
Ethyl 1-hydroxymethyl-3,5,7-trimethyl-4-oxo-3,4-dihydro-6-phthalazinecarboxylate	144–145, MS, NMR	405
1-Ethylidenehydrazinophthalazine	—	(*E* 621)
1-Ethyl-4-iodophthalazine	—	(*H* 181)
1-Ethyl-4-isopropoxyphthalazine	—	(*E* 437)
Ethyl 1-[2-(*N*-isopropylcarbamoyl)vinyl]-5,7-dimethyl-4-oxo-3,4-dihydro-6-phthalazinecarboxylate	285–287, NMR	424
2-Ethyl-4-isopropyl-5,8-dimethyl-1(2*H*)-phthalazinone	—	(*E* 433)
2-Ethyl-4-isopropyl-1(2*H*)-phthalazinone	—	(*E* 418)
4-Ethyl-2-isopropyl-1(2*H*)-phthalazinone	—	(*E* 419)
Ethyl 4-methoxy-5,7-dimethyl-6-phthalazinecarboxylate	132, NMR	783
Ethyl 4-methoxy-5,7-dimethyl-6-phthalazinecarboxylate 2-oxide	186, NMR	783
Ethyl 2-methoxymethyl-1-oxo-4-phenyl-1,2-dihydro-6-phthalazinecarboxylate	189–190, NMR	1024
Ethyl 1-methoxymethyl-3,5,7-trimethyl-4-oxo-3,4-dihydro-6-phthalazinecarboxylate	105–106	405
1-Ethyl-4-methoxyphthalazine	—	(*H* 84; *E* 436)
Ethyl 3-methyl-4-oxo-3,4-dihydro-1-phthalazinecarboxylate	—	(*E* 645)
Ethyl 3-methyl-4-oxo-3,4-dihydro-5-phthalazinecarboxylate	103–104, IR, NMR	(*E* 402) 623
6-Ethyl-7-methyl-4-phenyl-1(2*H*)-phthalazinone	221–224, IR, NMR	598
2-Ethyl-3-methyl-1,4(2*H*,3*H*)-phthalazinedione	NMR, UV	457

TABLE A.2. (*Continued*)

Phthalazine	Melting Point (°C) etc.	Reference(s)
2-Ethyl-4-methyl-1(2*H*)-phthalazinone	—	(*H* 97)
4-Ethyl-2-methyl-1(2*H*)-phthalazinone	—	(*H* 100; *E* 417)
4-(1-Ethyl-2-methylpropyl)-1(2*H*)-phthalazinone	—	(*E* 405)
1-Ethyl-4-methylsulfonylphthalazine	143–144, UV	931
Ethyl 4-methylthio-1-phthalazinecarboxylate	—	(*E* 650)
Ethyl 6/7-nitro-1,4-dioxo-1,2,3,4-tetrahydro-2-phthalazinecarboxylate	—	(*H* 164)
Ethyl 8-nitro-4-oxo-3,4-dihydro-1-phthalazinecarboxylate	198–200, NMR	573
3-Ethyl-4-oxo-3,4-dihydro-1-phthalazinecarboxamide	—	(*E* 647)
Ethyl-4-oxo-3,4-dihydro-1-phthalazinecarboxylate	—	(*E* 645)
Ethyl 4-oxo-3,4-dihydro-5-phthalazinecarboxylate	169–171, IR	(*E* 397) 622, 623
3-Ethyl-4-oxo-3,4-dihydro-1-phthalazinecarboxylic acid	—	(*E* 645)
Ethyl 1-oxo-2-phenyl-1,2-dihydro-6-phthalazinecarboxylate	136–137, IR, MS, NMR	411
Ethyl 4-oxo-1-phenyl-3,4-dihydro-5-phthalazinecarboxylate	—	(*E* 415)
Ethyl 4-oxo-3-phenyl-3,4-dihydro-5-phthalazinecarboxylate	150–151, IR, NMR	(*E* 403) 623
Ethyl 4-oxo-3-phenyl-3,4-dihydro-6-phthalazinecarboxylate	166–167, IR, MS, NMR	411
1-Ethyl-4-phenoxyphthalazine	—	(*H* 84)
1-Ethyl-4-phenylphthalazine	—	(*E* 358)
2-Ethyl-3-phenyl-1,4(2*H*,3*H*)-phthalazinedione	109, NMR, UV	448
1-Ethyl-4-phenylphthalazine 2-oxide	—	(*E* 373)
2-Ethyl-4-phenyl-1(2*H*)-phthalazinone	—	(*H* 102)
4-Ethyl-2-phenyl-1(2*H*)-phthalazinone	—	(*H* 100)
1-Ethylphthalazine	pic: 175–177	(*H* 73) 931
4-Ethyl-1-phthalazinecarbonitrile	130, IR, MS, NMR	188, 399
2-Ethyl-1(2*H*)-phthalazinone	56, IR, NMR	(*H* 87; *E* 399) 399
4-Ethyl-1(2*H*)-phthalazinone	159 to 170, IR, NMR, UV	(*H* 79; *E* 404) 507, 931
2-(1-Ethylprop-1-enyl)-1,4(2*H*,3*H*)-phthalazinedione	129–130	98
7-Ethylsulfinyl-2-phenyl-1(2*H*)-phthalazinone	123–125, IR, MS, NMR	411
1-Ethylsulfonyl-4-methylphthalazine	128–130	649
1-Ethylsulfonyl-4-phenylphthalazine	—	(*E* 558)
7-Ethylsulfonyl-2-phenyl-1(2*H*)-phthalazinone	192–193, IR, MS, NMR	411
1-Ethylthio-4-methylphthalazine	75–77	649
1-Ethylthio-4-phenylphthalazine	—	(*E* 553)
4-Ethylthio-2-phenyl-1(2*H*)-phthalazinethione	—	(*E* 556)
7-Ethylthio-2-phenyl-1(2*H*)-phthalazinone	144–145, IR, MS, NMR	411
4-Ethylthio-1-phthalazinecarboxamide	—	(*E* 650)

TABLE A.2. (*Continued*)

Phthalazine	Melting Point (°C) etc.	Reference(s)
4-Ethylthio-1(2*H*)-phthalazinone	—	(*E* 553, 557)
Ethyl 4-thioxo-3,4-dihydro-1-phthalazinecarboxylate	—	(*E* 649)
3-Ethyl-4-thioxo-3,4-dihydro-1-phthalazinecarboxylic acid	—	(*E* 649)
1-Ethyl-5,7,8-trimethoxy-4-methylphthalazine	—	(*E* 357)
Ethyl 3,5,7-trimethyl-1-methylamino-4-oxo-3,4-dihydro-6-phthalazinecarboxylate	—	243
Ethyl 1,5,7-trimethyl-8-nitro-4-oxo-3,4-dihydro-6-phthalazinecarboxylate	216–217	425
Ethyl 3,5,7-trimethyl-8-nitro-4-oxo-3,4-dihydro-6-phthalazinecarboxylate	137–138	425
Ethyl 1,5,7-trimethyl-4-oxo-3,4-dihydro-6-phthalazinecarboxylate	190–192, MS, NMR	405
5-Fluoro-1,4-dioxo-1,2,3,4-tetrahydro-2-phthalazinecarbothioamide	220–222, NMR	994
5-Fluoro-1,4-dioxo-1,2,3,4-tetrahydro-2-phthalazinecarboxamide	288–290, NMR	994
5-Fluoro-1-hydrazinophthalazine	—	(*E* 611)
6-Fluoro-1-hydrazinophthalazine	—	(*E* 611)
6-Fluoro-4-hydrazinophthalazine	—	(*E* 611)
6-Fluoro-1-methylphthalazine	—	(*E* 436)
5-Fluoro-1,4(2*H*,3*H*)-phthalazinedione	—	(*E* 475)
6-Fluoro-1,4(2*H*,3*H*)-phthalazinedione	—	(*E* 475)
5-Fluoro-1(2*H*)-phthalazinone	—	(*E* 397)
6-Fluoro-1(2*H*)-phthalazinone	—	(*E* 397)
7-Fluoro-1(2*H*)-phthalazinone	—	(*E* 397)
8-Fluoro-1(2*H*)-phthalazinone	—	(*E* 397)
2-(2-Formylethyl)-1,4(2*H*,3*H*)-phthalazinedione	—	(*E* 479, 480)
1-(*N*′-Formylhydrazino)-4-hydrazinophthalazine	—	(*E* 620)
6-Formylmethyl-1,4(2*H*,3*H*)-phthalazinedione	—	(*E* 476)
1-Guanidinoamino-4-phenylphthalazine	225	658
4-Guanidinomethyl-2-methyl-1(2*H*)-phthalazinone	—	157
1-Guanidino-4-methylphthalazine	—	(*E* 584)
4-Guanidinomethyl-1(2*H*)-phthalazinone	—	157
1-Guanidinophthalazine	HCl: 241–243, IR, NMR	225
1,4,5,6,7,8-Hexachlorophthalazine	—	(*E* 531)
1,4,5,6,7,8-Hexafluorophthalazine	—	(*E* 532)
1,4,5,6,7,8-Hexaphenylphthalazine	249–250 or 251–252, UV	121, 691
2-Hexyl-4-hexyloxy-1(2*H*)-phthalazinone	—	(*E* 491)
2-Hexyl-4-isopropyl-1(2*H*)-phthalazinone	—	(*E* 420)
1-Hexyloxy-4-isopropyl-5,8-dimethylphthalazine	—	(*E* 437)
1-Hexyloxy-4-isopropylphthalazine	—	(*E* 437)
4-Hexyloxy-2-phenyl-1(2*H*)-phthalazinone	—	(*E* 493)
2-Hexyl-1,4(2*H*,3*H*)-phthalazinedione	—	(*E* 478)
1-(2-Hydrazinocarbonylethyl)-4-methylphthalazine	—	(*E* 439)

TABLE A.2. (*Continued*)

Phthalazine	Melting Point (°C) etc.	Reference(s)
1-Hydrazinocarbonylmethyl-4-hydroxyaminophthalazine	—	(*E* 591)
2-Hydrazinocarbonylmethyl-4-methyl-1(2*H*)-phthalazinone	220–221	649
4-Hydrazinocarbonylmethyl-2-methyl-1(2*H*)-phthalazinone	—	(*E* 431)
2-Hydrazinocarbonylmethyl-5-nitro-1(2*H*)-phthalazinone	—	(*E* 402)
2-Hydrazinocarbonylmethyl-7-nitro-1(2*H*)-phthalazinone	—	(*E* 402)
2-Hydrazinocarbonylmethyl-4-phenyl-1(2*H*)-phthalazinone	245, NMR	659
2-Hydrazinocarbonylmethyl-1,4(2*H*,3*H*)-phthalazinedione	—	(*E* 479)
2-Hydrazinocarbonylmethyl-1(2*H*)-phthalazinone	—	(*E* 399)
4-Hydrazinocarbonylmethyl-1(2*H*)-phthalazinone	—	(*E* 411)
1-Hydrazino-4-hydrazino carbonylmethylphthalazine	280	(*E* 613) 980
4-Hydrazino-1-hydroxymethyl-6(2*H*)-phthalazinone	—	195
1-Hydrazino-4-isobutylphthalazine	—	(*E* 612)
1-Hydrazino-7-methoxy-4-morpholinomethylphthalazine	—	195
1-Hydrazino-7-methoxyphthalazine	—	(*E* 611)
1-Hydrazino-7-methoxy-4-piperidino methylphthalazine	—	195
1-Hydrazino-4-methylphthalazine	112; HCl: 286	(*E* 612) 216
4-Hydrazino-2-methyl-1(2*H*)-phthalazinone	—	(*E* 618)
1-Hydrazino-4-morpholinomethylphthalazine	—	195
1-Hydrazino-4-phenylphthalazine	135–137	(*E* 612) 678
4-Hydrazino-2-phenyl-1(2*H*)-phthalazinone	—	(*E* 618)
1-Hydrazino-5-phthalazinamine	154–155	936
1-Hydrazinophthalazine (hydralazine)	170–173, IR, pK_a, NMR, UV; HCl: 172–174, biol	(*E* 611) 191, 222, 249, 593, 728, 908, 967
5-Hydrazino-1,4(2*H*,3*H*)-phthalazinedione	—	(*H* 148)
4-Hydrazino-1(2*H*)-phthalazinethione	—	(*E* 548)
4-Hydrazino-1(2*H*)-phthalazinone	269	(*E* 618) 199, 908
4-Hydrazino-6(2*H*)-phthalazinone	—	(*E* 611)
1-Hydrazino-4-piperidinomethylphthalazine	—	195
1-Hydrazino-4-propylphthalazine	—	(*E* 612)
1-Hydrazono-2-methyl-1,2-dihydrophthalazine	86–88, IR, NMR, UV	728, 729
1-Hydroxyamino-4-isobutylphthalazine	—	(*E* 591)
1-Hydroxyamino-4-isopropylphthalazine	—	(*E* 591)
1-Hydroxyamino-4-methoxycarbonyl methylphthalazine	—	(*E* 591)
1-Hydroxyamino-4-methoxymethylphthalazine	—	(*E* 591)
1-Hydroxyamino-4-methoxyphthalazine	—	(*E* 591)
1-Hydroxyamino-4-methylphthalazine	—	(*E* 491)

TABLE A.2. (*Continued*)

Phthalazine	Melting Point (°C) etc.	Reference(s)
1-Hydroxyamino-4-(2-morpholinoethyl)-phthalazine	—	(*E* 591)
1-Hydroxyamino-4-morpholinophthalazine	—	(*E* 592)
1-Hydroxyamino-4-phenoxyphthalazine	—	(*E* 591)
1-Hydroxyamino-4-phenylphthalazine	—	(*E* 592)
1-Hydroxyaminophthalazine	—	(*E* 591)
1-Hydroxyamino-4-propylphthalazine	—	(*E* 591)
5-Hydroxy-2,3-diphenyl-1,4(2*H*,3*H*)-phthalazinedione	—	(*E* 500)
6-Hydroxy-2,3-diphenyl-1,4(2*H*,3*H*)-phthalazinedione	—	(*E* 500)
2-(2-Hydroxyethyl)-6,7-dimethoxy-1(2*H*)-phthalazinone	267–270, IR	298
2-(2-Hydroxyethyl-3-methyl-1,4(2*H*,3*H*)-phthalazinedione	124–126, IR, NMR, UV	745
2-(2-Hydroxyethyl)-4-methyl-1(2*H*)-phthalazinone	152–153 or 154–156, IR	298, 737
2-(2-Hydroxyethyl)-5-nitro-1(2*H*)-phthalazinone	—	(*E* 402)
2-(2-Hydroxyethyl)-7-nitro-1(2*H*)-phthalazinone	—	(*E* 402)
2-(2-Hydroxyethyl)-8-nitro-1(2*H*)-phthalazinone	—	(*E* 402)
3-(2-Hydroxyethyl)-4-oxo-3,4-dihydro-5-phthalazinecarboxylic acid	165–168, IR, NMR	623
1-(1-Hydroxyethyl)-4-phenylphthalazine	101–102, IR, NMR	938
2-(2-Hydroxyethyl)-4-phenyl-1(2*H*)-phthalazinone	155–158, IR	(*E* 423) 298
1-(2-Hydroxyethyl)phthalazine	—	(*E* 354)
2-(2-Hydroxyethyl)-1,4(2*H*,3*H*)-phthalazinedione	206–207	453
2-(2-Hydroxyethyl)-1(2*H*)-phthalazinone	—	(*E* 399)
8-Hydroxy-7-methoxy-2-methyl-5-nitro-1(2*H*)-phthalazinone	—	(*H* 87)
8-Hydroxy-7-methoxy-2-methyl-1(2*H*)-phthalazinone	—	(*H* 87)
5-Hydroxy-8-methoxy-1,4(2*H*,3*H*)-phthalazinedione	248–250, IR, MS, NMR	715
8-Hydroxy-7-methoxy-1(2*H*)-phthalazinone	—	(*H* 79)
4-Hydroxymethyl-6,7-dimethyl-1(2*H*)-phthalazinone	253–255	574
4-(1-Hydroxy-1-methylethyl)-1(2*H*)-phthalazinone	212–213, IR, NMR	404, 574
1-Hydroxymethyl-3-methyl-4-oxo-3,4-dihydro-6-phthalazinecarboxylic acid	264–265	404
2-Hydroxymethyl-4-methyl-1(2*H*)-phthalazinone	—	(*E* 420)
1-Hydroxymethyl-4-oxo-3,4-dihydro-6-phthalazinecarbonitrile	226–227, IR	574
1-Hydroxymethyl-4-oxo-3,4-dihydro-6-phthalazinecarboxylic acid	298–300	574
2-Hydroxymethyl-4-phenyl-1(2*H*)-phthalazinone	—	(*E* 420)
4-Hydroxymethyl-2-phenyl-1(2*H*)-phthalazinone	166–168	(*E* 428) 574
6-Hydroxymethyl-2-phenyl-1(2*H*)-phthalazinone	151–153, NMR	411

TABLE A.2. (*Continued*)

Phthalazine	Melting Point (°C) etc.	Reference(s)
7-Hydroxymethyl-1-phenyl-1(2*H*)-phthalazinone	141–142, IR, MS, NMR	411
8-Hydroxy-4-methyl-2-phenyl-1(2*H*)-phthalazinone	—	(*E* 433)
4-Hydroxymethyl-1,7(2*H*,3*H*)-phthalazinedione	—	195
7-Hydroxy-8-methyl-5,6(2*H*,3*H*)-phthalazinedione	—	(*E* 335)
8-Hydroxy-2-methyl-1,7(2*H*,3*H*)-phthalazinedione	—	(*H* 87)
2-Hydroxymethyl-1(2*H*)-phthalazinone	—	(*E* 398)
4-Hydroxymethyl-1(2*H*)-phthalazinone	203	574
7-Hydroxy-6-methyl-5(3*H*)-phthalazinone	—	(*E* 335)
8-Hydroxy-4-methyl-1(2*H*)-phthalazinone	—	(*E* 414)
2-(2-Hydroxy-1-methylpropyl)-1,4(2*H*,3*H*)-phthalazinedione	*threo*: 139–141, IR, MS, NMR; *erythro*: 212–214, IR, MS, NMR	148
6-Hydroxy-7-nitro-1,4(2*H*,3*H*)-phthalazinedione	>350, IR, NMR, UV	48
5-Hydroxy-1,4(2*H*,3*H*)-phthalazinedione	pK_a, UV	(*H* 149) 805
6-Hydroxy-1,4(2*H*,3*H*)-phthalazinedione	pK_a, UV	(*H* 149; *E* 475) 805
8-Hydroxy-1,7(2*H*,3*H*)-phthalazinedione	—	(*H* 79)
2-(3-Hydroxypropyl)-4-phenyl-1(2*H*)-phthalazinone	—	(*E* 426)
4-Imino-3-methyl-3,4-dihydro-1-phthalazinamine	HCl: 312; pic: 252	685
1-Iodo-4-isobutylphthalazine	—	(*H* 181)
1-Iodo-4-methylphthalazine	—	(*E* 181)
1-Iodo-4-phenylphthalazine	—	(*H* 181)
1-Iodophthalazine	—	(*H* 181; *E* 532)
6-Iodo-1,4(2*H*,3*H*)-phthalazinedione	—	(*H* 149)
4-Isobutoxy-2-isobutyl-1(2*H*)-phthalazinone	—	(*E* 491)
4-Isobutoxy-2-phenyl-1(2*H*)-phthalazinone	—	(*E* 493)
6-Isobutoxy-1,4(2*H*,3*H*)-phthalazinedione	—	(*E* 475)
1-Isobutylamino-4-phenylphthalazine	—	(*E* 581)
1-Isobutylaminophthalazine	—	(*E* 577)
Isobutyl 1-hydroxymethyl-4-oxo-3,4-dihydro-6-phthalazinecarboxylate	160–162	404
1-Isobutyl-4-phenoxyphthalazine	—	(*H* 84)
1-Isobutylphthalazine	—	(*H* 73)
2-Isobutyl-1,4(2*H*,3*H*)-phthalazinedione	—	(*E* 478)
4-Isobutyl-1(2*H*)-phthalazinone	—	(*H* 79)
1-Isobutyrylmethyl-4-phenylphthalazine	—	(*E* 359)
4-Isocrotonoyloxy-1(2*H*)-phthalazinone	—	(*E* 486)
2-Isopentyl-4-isopentyloxy-1(2*H*)-phthalazinone	—	(*E* 491)
4-Isopentyloxy-2-phenyl-1(2*H*)-phthalazinone	—	(*E* 493)
2-Isopentyl-1,4(2*H*,3*H*)-phthalazinedione	—	(*E* 478)
2-Isopropenyl-1,4(2*H*,3*H*)-phthalazinedione	149–152	98
1-Isopropoxy-4-isopropyl-5,8-dimethylphthalazine	—	(*E* 437)
1-Isopropoxy-4-isopropylphthalazine	—	(*E* 437)
4-Isopropoxy-2-isopropyl-1(2*H*)-phthalazinone	—	(*E* 491)

TABLE A.2. (*Continued*)

Phthalazine	Melting Point (°C) etc.	Reference(s)
1-Isopropoxy-4-methoxy-5,8-dimethylphthalazine	—	(*E* 436)
1-Isopropoxy-4-methylphthalazine	—	(*E* 437)
4-Isopropoxy-2-methyl-1(2*H*)-phthalazinone	—	(*E* 493)
1-Isopropoxy-4-phenylphthalazine	—	(*E* 437)
1-Isopropoxyphthalazine	—	(*E* 435)
6-Isopropoxy-1,4(2*H*,3*H*)-phthalazinedione	—	(*E* 475)
1-Isopropoxyphthalazine 3-oxide	—	(*E* 372)
1-Isopropylamino-4-phenylphthalazine	—	(*E* 581)
1-Isopropylamino-5-phthalazinamine	299–300	936
2-[2-(*N*-Isopropylcarbamoyl)ethyl]-3-methyl-1,4(2*H*,3*H*)-phthalazinedione	146–147	918
Isopropyl 4-chloro-1-oxo-1,4-dihydro-2-phthalazinecarboxylate	—	(*E* 529)
1-(*N*′-Isopropylhydrazino)phthalazine	—	(*E* 615)
Isopropyl 1-hydroxymethyl-5,7-dimethyl-4-oxo-3,4-dihydro-6-phthalazinecarboxylate	177–179	405
Isopropyl 1-hydroxymethyl-4-oxo-3,4-dihydro-6-phthalazinecarboxylate	212–213, NMR	404
1-(Isopropylidenehydrazino)phthalazine	112–114, NMR, UV	(*E* 621) 396
1-(*N*′-Isopropylidene-*N*-methylhydrazino)phthalazine	liq, NMR	396
N′-Isopropylidene-4-thioxo-3,4-dihydro-1-phthalazinecarbohydrazide	—	(*E* 649)
1-Isopropyl-4-methoxyphthalazine	—	(*E* 436, 666)
2-Isopropyl-3-methyl-1,4(2*H*,3*H*)-phthalazinedione	NMR	457
2-Isopropyl-4-methyl-1(2*H*)-phthalazinone	—	(*E* 419)
1-Isopropyl-4-methylsulfonylphthalazine	124–125, UV	931
N-Isopropyl-4-oxo-3,4-dihydro-1-phthalazinecarboxamide	—	(*E* 647)
1-Isopropyl-4-pentyloxyphthalazine	—	(*E* 437)
4-Isopropyl-2-pentyl-1(2*H*)-phthalazinone	—	(*E* 420)
2-Isopropyl-3-phenyl-1,4(2*H*,3*H*)-phthalazinedione	122–123, NMR, UV	451
1-Isopropyl-4-phenylphthalazine 2-oxide	—	(*E* 373)
1-Isopropyl-4-phenylphthalazine 3-oxide	—	(*E* 374)
2-Isopropyl-4-phenyl-1(2*H*)-phthalazinone	137–138	119
4-Isopropyl-2-phenyl-1(2*H*)-phthalazinone	—	(*E* 428)
1-Isopropylphthalazine	165/0.35, MS, NMR; pic: 159–160	399, 417
4-Isopropyl-1-phthalazinecarbonitrile	145–146, IR, NMR	188, 399
2-Isopropyl-1,4(2*H*,3*H*)-phthalazinedione	—	(*E* 478)
2-Isopropyl-1(2*H*)-phthalazinone	liq, IR, NMR	399
4-Isopropyl-1(2*H*)-phthalazinone	156–157, IR, NMR, UV	(*E* 404) 931, 1015
1-Isopropyl-4-propoxyphthalazine	—	(*E* 436)
4-Isopropyl-2-propyl-1(2*H*)-phthalazinone	—	(*E* 418)
4-Isopropyl-2,5,8-trimethyl-1(2*H*)-phthalazinone	—	(*E* 433)
2-(Isopropylvinyl)-3-methyl-1,4(2*H*,3*H*)-phthalazinedione	230, NMR	563

TABLE A.2. (*Continued*)

Phthalazine	Melting Point (°C) etc.	Reference(s)
1-(2-Mercaptoethyl)phthalazine	—	(*E* 355)
1-(2-Mercapto-1,1,2-trimethylpropyl)phthalazine	solid, MS, NMR	417
4-Methacryloyloxy-1(2*H*)-phthalazinone	—	(*E* 486)
2-Methacryloyl-1,4(2*H*,3*H*)-phthalazinedione	—	(*E* 479)
4-Methanesulfonyloxy-1(2*H*)-phthalazinone	—	(*E* 488)
1-(1-Methoxycarbonylethyl)phthalazine	240/0.7, IR, NMR	417
4-Methoxycarbonylmethyl-2-methyl-1(2*H*)-phthalazinone	—	(*E* 431)
5-Methoxycarbonyl-2-methylphthalazin-2-ium-4-olate	305, IR, NMR; TsOH: 205–207, IR, NMR	623
4-Methoxycarbonylmethyl-1(2*H*)-phthalazinone	—	(*E* 411)
6-Methoxy-7,8-dimethyl-1,4(2*H*,3*H*)-phthalazinedione	203, NMR	114
2-(2-Methoxyethyl)-1,4(2*H*,3*H*)-phthalazinedione	146–148, NMR	949
8-Methoxy-1-methyl-5-nitro-1,7(2*H*,3*H*)-phthalazinedione	—	(*H* 87)
4-Methoxy-2-methyl-5-nitro-1(2*H*)-phthalazinone	—	(*E* 173)
4-Methoxy-2-methyl-8-nitro-1(2*H*)-phthalazinone	—	(*H* 173)
2-Methoxymethyl-1-oxo-4-phenyl-1,2-dihydro-6-phthalazinecarbonitrile	—	1024
1-Methoxy-4-methylphthalazine	35–36	(*H* 84; *E* 436) 649
1-Methoxy-4-methylphthalazine 3-oxide	—	(*E* 373)
4-Methoxy-2-methyl-1(2*H*)-phthalazinone	—	(*H* 170; *E* 491)
6-Methoxy-4-methyl-1(2*H*)-phthalazinone	247–251, IR, NMR	881
7-Methoxy-6-methyl-5(3*H*)-phthalazinone	—	(*E* 335)
8-Methoxy-7-methyl-6(2*H*)-phthalazinone	—	(*E* 335)
7-Methoxy-4-morpholinomethyl-1(2*H*)-phthalazinone	—	195
1-Methoxy-5-nitrophthalazine	207–210	908
6-Methoxy-7-nitro-1,4(2*H*,3*H*)-phthalazinedione	304–305, IR, NMR, UV	48
6-Methoxy-4-oxo-3,4-dihydro-1-phthalazinecarboxylic acid	—	(*H* 80)
1-Methoxy-4-phenylphthalazine	121–122, IR, NMR	(*E* 436, 665) 311
5-Methoxy-1-phenylphthalazine	—	(*H* 74)
6-Methoxy-1-phenylphthalazine	—	(*E* 353)
6-Methoxy-4-phenylphthalazine	—	(*H* 74; *E* 353)
4-Methoxy-2-phenyl-1(2*H*)-phthalazinone	—	(*H* 170; *E* 492)
6-Methoxy-4-phenyl-1(2*H*)-phthalazinone	—	(*E* 414)
7-Methoxy-2-phenyl-1(2*H*)-phthalazinone	168–171, IR, NMR	411
7-Methoxy-4-phenyl-1(2*H*)-phthalazinone	—	(*E* 414)
4-Methoxy-1-phthalazinamine	163–164	(*E* 576) 935
1-Methoxyphthalazine	59, pK_a, UV; 3-TsOMe: 155, IR, NMR, UV	(*H* 84; *E* 395, 435, 665) 681, 690
6-Methoxyphthalazine	—	557, 888
6-Methoxy-1,4(2*H*,3*H*)-phthalazinedione	—	(*E* 475)
7-Methoxy-5,6(1*H*,2*H*)-phthalazinedione	—	(*E* 335)
1-Methoxyphthalazine 3-oxide	—	(*E* 372)
4-Methoxy-1(2*H*)-phthalazinone	—	(*E* 486)

TABLE A.2. (*Continued*)

Phthalazine	Melting Point (°C) etc.	Reference(s)
7-Methoxy-1(2*H*)-phthalazinone	—	(*H* 79)
7-Methoxy-4-piperidinomethyl-1(2*H*)-phthalazinone	—	195
6-Methoxy-5,7,8-trimethyl-1,4(2*H*,3*H*)-phthalazinedione	162–163, NMR	114
5-(*N*-methylacetamido)-1,4(2*H*,3*H*)-phthalazinedione	—	(*H* 148)
1-Methylamino-7-nitrophthalazine	—	(*E* 580)
1-Methylamino-4-phenylphthalazine	—	(*E* 581)
1-Methylaminophthalazine	178, NMR; 3-MeI: 240–241, NMR; 3-PrI: 203, NMR	62, 939
5-Methylamino-1,4(2*H*,3*H*)-phthalazinedione	—	(*H* 148)
4-Methylamino-1(2*H*)-phthalazinone	—	(*E* 586)
Methyl 2-benzyl-1-oxo-1,2-dihydro-6-phthalazinecarboxylate	156–157, IR, MS, NMR	1036
Methyl 3-benzyl-4-thioxo-3,4-dihydro-1-phthalazinecarboxylate	—	(*E* 649)
1-[(2-Methylbut-2-enylidene) hydrazino]phthalazine	145–148, NMR, UV	396
4-(1-Methylbutyl)-1(2*H*)-phthalazinone	—	(*E* 404)
1-[(3-Methylbutyryl)methyl]-4-phenylphthalazine	—	(*E* 359)
6-Methyl-1,4-dipropylphthalazine	—	352
1-Methylenehydrazinophthalazine	—	(*E* 621)
Methyl 3-ethyl-4-oxo-3,4-dihydro-1-phthalazinecarboxylate	—	(*E* 645)
Methyl 3-ethyl-4-thioxo-3,4-dihydro-1-phthalazinecarboxylate	—	(*E* 649)
1-(*N*-Methylhydrazino)phthalazine	89–91, IR, NMR, UV	(*E* 615) 734
1-(*N*′-Methylhydrazino)phthalazine	102–108, IR, NMR, UV	728
Methyl 1-hydroxymethyl-5,7-dimethyl-4-oxo-3,4-dihydro-6-phthalazinecarboxylate	204–205, NMR	405
Methyl 1-hydroxymethyl-4-oxo-3,4-dihydro-6-phthalazinecarboxylate	224–225	404
Methyl 2-methoxymethyl-1-oxo-4-phenyl-1,2-dihydro-6-phthalazinecarboxylate	—	1024
4-Methyl-2-(2-methylallyl)-1(2*H*)-phthalazinone	liq, NMR	465
2-Methyl-5/8-methylamino-1,4(2*H*,3*H*)-phthalazinedione	—	(*H* 165)
1-Methyl-4-(*N*-methylhydrazino)phthalazine	—	(*E* 617)
Methyl 3-methyl-4-oxo-3,4-dihydro-1-phthalazinecarboxylate	—	(*E* 645)
Methyl 3-methyl-4-oxo-3,4-dihydro-5-phthalazinecarboxylate	125–126, IR, NMR	623
2-Methyl-3-methylsulfonylmethyl-1,4(2*H*,3*H*)-phthalazinedione	—	263
1-Methyl-4-methylsulfonylphthalazine	160–161, UV	931

TABLE A.2. (*Continued*)

Phthalazine	Melting Point (°C) etc.	Reference(s)
1-Methyl-4-methylthiophthalazine	109, UV	931
Methyl 3-methyl-4-thioxo-3,4-dihydro-1-phthalazinecarboxylate	—	(*E* 649)
Methyl 3-(2-morpholinoethyl)-4-oxo-3,4-dihydro-5-phthalazinecarboxylate	93–96, IR, NMR	623
Methyl 6-nitro-4-oxo-3-phenyl-3,4-dihydro-1-phthalazinecarboxylate	—	(*H* 102)
4-Methyl-8-nitro-2-phenyl-1(2*H*)-phthalazinone	—	(*E* 433)
4-Methyl-7-nitro-1-phthalazinamine	—	(*E* 575)
1-Methyl-5-nitrophthalazine	195–197	(*E* 665) 175, 330, 701
2-Methyl-5-nitro-1,4(2*H*,3*H*)-phthalazinedione	—	(*H* 164)
2-Methyl-6/7-nitro-1,4(2*H*,3*H*)-phthalazinedione	—	(*H* 164)
2-Methyl-8-nitro-1,4(2*H*,3*H*)-phthalazinedione	—	(*H* 164; *E* 485)
2-Methyl-7-nitro-1(2*H*)-phthalazinethione	—	(*E* 555)
2-Methyl-5-nitro-1(2*H*)-phthalazinone	—	(*E* 402)
2-Methyl-7-nitro-1(2*H*)-phthalazinone	—	(*E* 402)
2-Methyl-8-nitro-1(2*H*)-phthalazinone	—	(*E* 402)
4-Methyl-5-nitro-1(2*H*)-phthalazinone	273–275	(*E* 414) 908
4-Methyl-7-nitro-1(2*H*)-phthalazinone	—	(*E* 414)
4-Methyl-8-nitro-1(2*H*)-phthalazinone	293	(*E* 414) 908
3-Methyl-4-oxo-3,4-dihydro-1-phthalazinecarbohydrazide	—	(*E* 648)
3-Methyl-4-oxo-3,4-dihydro-1-phthalazinecarbonyl azide	—	(*E* 648)
3-Methyl-4-oxo-3,4-dihydro-1-phthalazinecarboxamide	—	(*E* 647)
4-Methyl-1-oxo-1,2-dihydro-2-phthalazinecarboxamide	—	(*E* 422)
N-Methyl-4-oxo-3,4-dihydro-1-phthalazinecarboxamide	—	(*E* 647)
Methyl 4-oxo-3,4-dihydro-5-phthalazinecarboxylate	210–212, IR, NMR	(*H* 397; *E* 646) 622
3-Methyl-4-oxo-3,4-dihydro-1-phthalazinecarboxylic acid	235–237, NMR	(*E* 645) 773
3-Methyl-4-oxo-3,4-dihydro-5-phthalazinecarboxylic acid	228–231, IR, NMR	623
Methyl 4-oxo-3-phenyl-3,4-dihydro-1-phthalazinecarboxylate	—	(*H* 102; *E* 646)
2-Methyl-1-oxo-7-phenylthio-1,2-dihydro-6-phthalazinesulfonamide	278–282, IR, MS, NMR	274
Methyl 4-oxo-3-(prop-2-ynyl)-3,4-dihydro-5-phthalazinecarboxylate	162–164, IR, NMR	623
2-(1-Methylpentyl)-1(2*H*)-phthalazinone	—	(*E* 405)
1-Methyl-4-phenoxyphthalazine	—	(*E* 440)
1-Methyl-4-phenylphthalazine	123–125, NMR, UV; 2-$MeClO_4$: 224–225; 2-PhI: 218–219; 2-$PhClO_4$: 212–213	(*E* 358, 665) 228, 290, 353, 449, 454, 456, 461

TABLE A.2. (*Continued*)

Phthalazine	Melting Point (°C) etc.	Reference(s)
2-Methyl-3-phenyl-1,4(2*H*,3*H*)-phthalazinedione	126–127, NMR, UV	448
1-Methyl-4-phenylphthalazine 2-oxide	—	(*E* 373)
1-Methyl-4-phenylphthalazine 3-oxide	181–183	(*E* 374) 938
4-Methyl-2-phenyl-1(2*H*)-phthalazinethione	—	(*E* 554)
1-Methyl-2-phenylphthalazin-2-ium-4-olate	—	(*H* 122)
2-Methyl-4-phenyl-1(2*H*)-phthalazinone	155 to 167, IR	(*H* 102; *E* 417) 119, 447, 726
4-Methyl-2-phenyl-1(2*H*)-phthalazinone	—	(*H* 97; *E* 428)
1-Methyl-4-phenylsulfonylphthalazine	—	216
1-Methyl-4-phenylthiophthalazine	130	33, 216
Methyl 3-phenyl-4-thioxo-3,4-dihydro-1-phthalazinecarboxylate	—	(*E* 649)
4-Methyl-1-phthalazinamine	200–202; 3-$EtO_2C(CH_2)_3Br$: 178–179, IR, NMR	(*E* 575) 508, 947
4-Methyl-1-phthalazinamine 2-oxide	218–220, NMR	508
1-Methylphthalazine	70–72, MS, NMR; pic: 203–205; 2-BuI: 140, IR, NMR; 2-MeI: 211, IR, NMR	(*H* 73; *E* 353, 665) 35, 47, 456, 610, 908, 931
5-Methylphthalazine	—	208
6-Methylphthalazine	72	80, 208
4-Methyl-1-phthalazinecarbonitrile	149–151, IR, MS, UV	188, 930
4-Methyl-1-phthalazinecarboxamide	228–230, IR, UV	930
4-Methyl-1,7-phthalazinediamine	—	(*E* 576)
2-Methyl-1,4(2*H*,3*H*)-phthalazinedione	238 to 242, IR, NMR	1, 105, 918
5-Methyl-1,4(2*H*,3*H*)-phthalazinedione	—	(*H* 149)
6-Methyl-1,4(2*H*,3*H*)-phthalazinedione	biol	(*E* 475) 237
2-Methyl-1(2*H*)-phthalazinethione	—	(*E* 544, 554) 171
4-Methyl-1(2*H*)-phthalazinethione	239 to 243, UV	(*E* 548) 33, 649, 931
2-Methyl-1(2*H*)-phthalazinimine	HI: 214–215, NMR	62
4-Methylphthalazin-2-ium-2-ethoxycarbonylmethylate	pK_a	194
2-Methylphthalazin-2-ium-4-olate	pK_a, UV; TsOH: 196, IR, NMR, UV	(*E* 660) 681
2-Methylphthalazin-2-ium-4-thiolate	—	171
2-Methyl-1(2*H*)-phthalazinone	111 to 116, IR, pK_a, NMR, UV; 3-TsOMe: 224–225, IR, NMR, UV	(*H* 87; *E* 388, 398) 303, 603, 623, 681, 690, 703, 728
4-Methyl-1(2*H*)-phthalazinone	219 to 231, IR, NMR	(*H* 79; *E* 404) 24, 33, 575, 773, 881, 908, 931
5-Methyl-1(2*H*)-phthalazinone	214–216, IR, NMR	622
8-Methyl-1(2*H*)-phthalazinone	191–192, IR, NMR	622
4-Methyl-1(2*H*)-phthalazinone 3-oxide	—	(*E* 373)
1-(1-Methylpropylidenehydrazino)phthalazine	—	(*E* 622)

TABLE A.2. (*Continued*)

Phthalazine	Melting Point (°C) etc.	Reference(s)
2-(2-Methylsulfinylethyl)-1,4(2*H*,3*H*)-phthalazinedione	199–200	949
1-Methylsulfinylphthalazine	105	(*E* 558) 908
2-(2-Methylsulfonylethyl)-1,4(2*H*,3*H*)-phthalazinedione	235–236	949
2-Methylsulfonylmethyl-1,4(2*H*,3*H*)-phthalazinedione	—	263
1-Methylsulfonyl-4-phenylphthalazine	210–212	(*E* 558) 931
1-Methylsulfonylphthalazine	156	(*E* 558) 908
2-(2-Methylthioethyl)-1,4(2*H*,3*H*)-phthalazinedione	151–153, NMR	949
2-Methylthiomethyl-1,4(2*H*,3*H*)-phthalazinedione	—	365
1-Methylthio-4-phenylphthalazine	—	(*E* 553)
1-Methylthiophthalazine	75–77	(*E* 544, 549) 171, 908
4-Methylthio-1-phthalazinecarboxamide	—	(*E* 650)
4-Methylthio-1-phthalazinecarboxylic acid	—	(*E* 650)
3-Methyl-4-thioxo-3,4-dihydro-1-phthalazinecarboxylic acid	—	(*E* 649)
Methyl 4-thioxo-3,4-dihydro-1-phthalazinecarboxylate	—	(*E* 649)
4-Methyl-2-vinyl-1(2*H*)-phthalazinone	141–142	737
4-Morpholinocarbonyl-2-phenyl-1(2*H*)-phthalazinone	177, IR, NMR	700
6-Morpholino-1,4-diphenylphthalazine	250–252, NMR	300
2-Morpholinomethyl-1,4(2*H*,3*H*)-phthalazinedione	—	(*E* 478)
4-Morpholinomethyl-1(2*H*)-phthalazinone	—	195
1-Morpholino-4-phenylphthalazine	—	(*E* 582)
4-Morpholino-2-phenyl-1(2*H*)-phthalazinone	—	(*E* 587)
2-[2-(Morpholinosulfonyl)ethyl]-1,4(2*H*,3*H*)-phthalazinedione	—	(*E* 483)
5-Nitro-1,4-dioxo-1,2,3,4-tetrahydro-2-phthalazinecarbothioamide	223–225, NMR	994
5-Nitro-1,4-dioxo-1,2,3,4-tetrahydro-2-phthalazinecarboxamide	279–281, NMR	994
5-Nitro-2,3-diphenyl-1,4(2*H*,3*H*)-phthalazinedione	—	(*E* 500)
6-Nitro-2,3-diphenyl-1,4(2*H*,3*H*)-phthalazinedione	—	(*E* 500)
4-Nitromethyl-1(2*H*)-phthalazinone	—	(*E* 405)
2-Nitro-4-nitroxymethyl-1(2*H*)-phthalazinone	92–93, MS, NMR, UV	428
5-Nitro-1-phenylphthalazine	188–189 or 256	175, 275, 330, 701
6-Nitro-2/3-phenyl-1,4(2*H*,3*H*)-phthalazinedione	—	(*H* 164)
5-Nitro-2-phenyl-1(2*H*)-phthalazinone	154–155, IR, MS, NMR	411
5-Nitro-4-phenyl-1(2*H*)-phthalazinone	—	(*E* 414)
6-Nitro-2-phenyl-1(2*H*)-phthalazinone	202–203, MS, NMR	411
7-Nitro-2-phenyl-1(2*H*)-phthalazinone	172–173, MS, NMR	(*H* 87) 411
8-Nitro-2-phenyl-1(2*H*)-phthalazinone	182–183, IR, MS, NMR	411

TABLE A.2. (*Continued*)

Phthalazine	Melting Point (°C) etc.	Reference(s)
7-Nitro-1-phthalazinamine	—	(*E* 575)
8-Nitro-5-phthalazinamine	>300, IR, NMR	856
5-Nitrophthalazine	187–188 or 188–189, IR, NMR	(*E* 335) 275, 715, 856, 908
6-Nitrophthalazine	—	254
5-Nitro-1,4(2*H*,3*H*)-phthalazinedione	pK_a, UV	(*H* 147; *E* 472) 203, 546, 805
6-Nitro-1,4(2*H*,3*H*)-phthalazinedione	pK_a, UV	(*H* 149) 805
5-Nitro-1(2*H*)-phthalazinone	263–265	(*E* 397) 908
7-Nitro-1(2*H*)-phthalazinone	—	(*E* 397)
8-Nitro-1(2*H*)-phthalazinone	253	(*E* 397) 908
4-Nitroxymethyl-1(2*H*)-phthalazinone	194–195, IR, MS, NMR, UV	428
4-Oxo-3,4-dihydro-5-phthalazinecarbaldehyde	252–254	(*E* 397) 622
4-Oxo-3,4-dihydro-1-phthalazinecarbohydrazide	—	(*E* 648)
4-Oxo-3,4-dihydro-5-phthalazinecarbonitrile	274, IR, NMR	615, 908
1-Oxo-1,2-dihydro-2-phthalazinecarbothioanilide	—	258
4-Oxo-3,4-dihydro-1-phthalazinecarboxamide	—	(*E* 647)
4-Oxo-3,4-dihydro-1-phthalazinecarboxanilide	322–323, IR, MS	484
4-Oxo-3,4-dihydro-1-phthalazinecarboxylic acid	230–231 or 232, NMR	(*H* 80; *E* 645) 773, 908
4-Oxo-3,4-dihydro-5-phthalazinecarboxylic acid	303–306, IR	(*E* 397) 622
4-Oxo-3,4-dihydro-1,7-phthalazinedicarboxylic acid	>280	404
1-Oxo-2-phenyl-1,2-dihydro-6-phthalazinecarbaldehyde	—	(*H* 87)
4-Oxo-3-phenyl-3,4-dihydro-1-phthalazinecarbohydrazide	—	(*E* 648)
4-Oxo-3-phenyl-3,4-dihydro-6-phthalazinecarbonitrile	232–233, IR, NMR	411
4-Oxo-3-phenyl-3,4-dihydro-1-phthalazinecarboxamide	—	(*E* 647)
4-Oxo-3-phenyl-3,4-dihydro-6-phthalazinecarboxamide	243–245, IR, MS, NMR	411
1-Oxo-2-phenyl-1,2-dihydro-6-phthalazinecarboxylic acid	crude	(*H* 87) 411
4-Oxo-1-phenyl-3,4-dihydro-5-phthalazinecarboxylic acid	197–198, IR, NMR	(*E* 414) 623
4-Oxo-3-phenyl-3,4-dihydro-1-phthalazinecarboxylic acid	222–223	(*H* 102; *E* 645) 574
4-Oxo-3-phenyl-3,4-dihydro-5-phthalazinecarboxylic acid	—	(*E* 403)
4-Oxo-3-phenyl-3,4-dihydro-6-phthalazinecarboxylic acid	crude	411
4-Oxo-7-phenylthio-3,4-dihydro-6-phthalazinesulfonamide	310, IR, MS, NMR	274
4,5,6,7,8-Pentachloro-1(2*H*)-phthalazinone	—	(*E* 529)
1,5,6,7,8-Pentafluoro-4-methoxyphthalazine	—	(*E* 532)
4,5,6,7,8-Pentafluoro-1(2*H*)-phthalazinone	—	(*E* 532)

TABLE A.2. (*Continued*)

Phthalazine	Melting Point (°C) etc.	Reference(s)
4-Pentylamino-1(2*H*)-phthalazinone	—	196
4-Pentyl-1(2*H*)-phthalazinone	129–130, IR, NMR	1015
4-Phenethyl-1(2*H*)-phthalazinone	146–147, IR, NMR	1015
1-Phenoxy-4-phenylphthalazine	167–168	650
4-Phenoxy-2-phenyl-1(2*H*)-phthalazinone	—	(*E* 491)
1-Phenoxyphthalazine	—	(*E* 435)
1-Phenoxyphthalazine 3-oxide	—	(*E* 372)
1-(*N*-Phenylhydrazino)phthalazine	—	(*E* 615)
1-Phenyl-4-phenylthiophthalazine	162–164	650
4-Phenyl-1-phthalazinamine	201	(*E* 576) 935
1-Phenylphthalazine	141 to 149, NMR, UV; 2-$PhCH_2I.H_2O$: 176–177, NMR	(*H* 74; *E* 353, 665) 80, 119, 290, 353, 908
5-Phenylphthalazine	—	(*E* 335)
6-Phenylphthalazine	139	(*E* 335) 80
4-Phenyl-1-phthalazinecarbonitrile	183–184 or 185–186, MS	(*E* 644) 9, 188, 587, 591
4-Phenyl-1-phthalazinecarboxamide	—	(*E* 644)
6-Phenyl-1,4-phthalazinediamine	—	(*E* 589)
2-Phenyl-1,4(2*H*,3*H*)-phthalazinedione	211 or 214–216, dip, IR	(*H* 159; *E* 481) 103, 498, 549, 675
6-Phenyl-1,4(2*H*,3*H*)-phthalazinedione	—	(*E* 476)
2-Phenyl-1,4(2*H*,3*H*)-phthalazinedithione	—	(*E* 556) 859
1-Phenylphthalazine 2-oxide	213, NMR	938
1-Phenylphthalazine 3-oxide	168–175 or 181–183, NMR	(*E* 372) 938
2-Phenyl-1(2*H*)-phthalazinethione	138	(*E* 554) 712
4-Phenyl-1(2*H*)-phthalazinethione	—	(*E* 548)
1-Phenylphthalazin-2-ium-2-benzamidate	245–246, UV	5
4-Phenylphthalazin-2-ium-2-benzamidate	203–204, NMR, UV	5
2-Phenylphthalazin-2-ium-4-olate	204–205, IR, NMR	(*H* 120; *E* 660) 696
2-Phenyl-1(2*H*)-phthalazinone	104–105 or 109, IR, NMR	(*H* 87; *E* 398) 411, 437
4-Phenyl-1(2*H*)-phthalazinone	232 to 247, IR, NMR, UV	(*H* 80; *E* 405) 119, 253, 278, 282, 296, 357, 507, 519, 537, 638, 669, 707, 708, 764, 935, 963
1-Phenyl-4-piperidinophthalazine	—	(*E* 582)
1-Phenyl-4-(1-propionylethyl)phthalazine	—	(*E* 359)
1-Phenyl-4-propionylmethylphthalazine	—	(*E* 359)
1-Phenyl-4-propoxyphthalazine	—	(*E* 436)
2-Phenyl-4-propoxy-1(2*H*)-phthalazinethione	—	(*E* 554)
2-Phenyl-4-propoxy-1(2*H*)-phthalazinone	—	(*E* 492)
1-Phenyl-4-propylaminophthalazine	—	(*E* 581)
2-Phenyl-3-propyl-1,4(2*H*,3*H*)-phthalazinedione	NMR	448
1-Phenyl-4-styrylphthalazine	3-$MeClO_4$: 122–124; 3-$PhClO_4$: 265–266	449, 461
2-Phenyl-4-styryl-1(2*H*)-phthalazinone	—	(*E* 429)

TABLE A.2. (*Continued*)

Phthalazine	Melting Point (°C) etc.	Reference(s)
4-Phenyl-2-(2-thiocarbamoylethyl)-1(2*H*)-phthalazinone	194–195, IR, NMR	214
4-Phenylthio-1(2*H*)-phthalazinone	—	(*E* 557)
1-Phenyl-4-thiosemicarbazidophthalazine	275, IR	658
3-Phenyl-4-thioxo-3,4-dihydro-1-phthalazinecarboxylic acid	—	(*E* 649)
3-Phenyl-4-thioxo-3,4-dihydro-1(2*H*)-phthalazinone	—	(*E* 554) 859(?)
1-Phthalazinamine	209–210, NMR; 3-MeI: 240–242, NMR; 3-PrI: 227, NMR; 3-BzCH$_2$Br: 252–255	(*H* 184; *E* 575) 62, 947
5-Phthalazinamine	223–224	275
Phthalazine	87 to 94, NMR, pK_a, xl st; TsOH: 158–160; 2-PhCH$_2$Cl: 175–178; 2-PhClO$_4$: 213–214, NMR; HBF$_4$: xl st; 2-MeI: 243–244, IR; 2-BuBr: IR, NMR; 2-AcCH$_2$CH$_2$I: 160–164; and many others	(*H* 70; *E* 324, 334, 661) 47, 56, 80, 90, 117, 152, 275, 277, 372, 374, 613, 629, 714, 728, 734, 762, 770, 813, 847, 853, 854, 908
1-Phthalazinecarbonitrile	156–157, MS	(*E* 644) 188, 930
1-Phthalazinecarboxamide	solid, anal, IR	435
1-Phthalazinecarboxylic acid	HCl: 198–200	(*E* 644) 908
1,4-Phthalazinediamine	254–255, IR, UV; HCl: 220; 2HCl: 226; pic: 302–303	(*E* 589) 176, 685, 908
1,7-Phthalazinediamine	—	(*E* 575)
5,8-Phthalazinediamine	249–250, IR, NMR	856
1,4-Phthalazinedicarbonitrile	—	(*E* 644)
6,7-Phthalazinedicarboxylic acid	>330, IR	584
6,7-Phthalazinedicarboxylic anhydride	200–260, IR	584
1,4(2*H*,3*H*)-Phthalazinedione	333 to 341, biol, IR, pK_a, NMR, st, UV	(*H* 147; *E* 472) 98, 105, 182, 237, 283, 498, 653, 685, 760, 805, 851, 875
1,7(2*H*,3*H*)-Phthalazinedione	—	(*E* 397)
5,6(2*H*,3*H*)-Phthalazinedione	—	(*H* 71)
1,4(2*H*,3*H*)-Phthalazinedithione	262–264	(*E* 548) 295, 512, 908
Phthalazine 2-oxide	143, NMR; pic: 153	152, 760, 908
5,8-Phthalazinequinone	>300, IR, NMR	715
1(2*H*)-Phthalazinethione	169–170 or 258–262, IR, st	(*E* 544, 548) 143, 171, 908
Phthalazin-2-ium-2-benzamidate	203–204, NMR, UV	5, 177

TABLE A.2. (*Continued*)

Phthalazine	Melting Point (°C) etc.	Reference(s)
1(2*H*)-Phthalazinone	182 to 190, IR, MS, NMR, pK_a, st, UV; 3-TsOMe: 196, IR	(*H* 79; *E* 388, 397) 193, 282, 305, 560, 593, 681, 685, 690, 707, 728, 734, 754, 760, 773, 908, 921
6(2*H*)-Phthalazinone	—	(*E* 335)
1(2*H*)-Phthalazinone 3-oxide	—	(*E* 372)
2-Piperidinomethyl-1,4(2*H*,3*H*)-phthalazinedione	—	(*E* 478)
4-Piperidinomethyl-1(2*H*)-phthalazinone	—	195
6-Piperidino-1,4(2*H*,3*H*)-phthalazinedione	—	(*E* 472)
1-(Prop-2-enylidenehydrazino)phthalazine	*E*: 61–63, IR, NMR, UV; *Z*: crude, NMR	1026
5-Propionamido-1,4(2*H*,3*H*)-phthalazinedione	—	(*E* 473)
1-Propoxyphthalazine	—	(*E* 665)
6-Propoxy-1,4(2*H*,3*H*)-phthalazinedione	—	(*E* 475)
1-Propoxyphthalazine 3-oxide	—	(*E* 372)
4-Propoxy-2-propyl-1(2*H*)-phthalazinone	—	(*E* 491)
1-Propylaminophthalazine	109, NMR; 3-MeI: 196–198, NMR; 3-PrI: 189 or 190, NMR	(*E* 577) 62, 939
4-Propylidenehydrazino-1(2*H*)-phthalazinone	—	(*E* 625)
2-Propyl-1,4(2*H*,3*H*)-phthalazinedione	—	(*E* 478)
4-Propyl-1(2*H*)-phthalazinone	—	(*H* 79; *E* 404)
1-Styrylphthalazine	—	(*H* 75)
4-Styryl-1(2*H*)-phthalazinone	262	649
2-(2-Sulfoethyl)-1,4(2*H*,3*H*)-phthalazinedione	—	(*E* 483)
5,6,7,8-Tetrachloro-2-phenyl-1,4(2*H*,3*H*)-phthalazinedione	266	643
5,6,7,8-Tetrachloro-2-phenyl-1(2*H*)-phthalazinone	235–236, IR, UV	288
1,4,6,7-Tetrachlorophthalazine	—	(*E* 531)
5,6,7,8-Tetrachloro-1,4(2*H*,3*H*)-phthalazinedione	291	(*H* 150), 643
5,6,7,8-Tetrachloro-1(2*H*)-phthalazinone	270, IR, NMR	288
1,4,6,7-Tetramethylphthalazine	—	(*E* 257)
5,6,7,8-Tetramethyl-1,4(2*H*,3*H*)-phthalazinedione	225–227, NMR	114
1,4,5,8-Tetraphenylphthalazine	242–244	(*E* 362) 121
5,6,7,8-Tetraphenylphthalazine	—	(*E* 335, 666)
4,5,6,7-Tetraphenyl-1(2*H*)-phthalazinone	—	(*E* 415)
2-(2-Thiocarbamoylethyl)-1(2*H*)-phthalazinone	159–161	214
1-Thiocyanatophthalazine	—	(*E* 551)
4-Thioxo-3,4-dihydro-1-phthalazinecarbohydrazide	—	(*E* 649)
4-Thioxo-3,4-dihydro-1-phthalazinecarboxamide	—	(*E* 649)
4-Thioxo-3,4-dihydro-1-phthalazinecarboxylic acid	—	(*E* 649)
4-Thioxo-3,4-dihydro-1(2*H*)-phthalazinone	—	(*E* 548)
5,6,8-Triacetoxyphthalazine	280–282, IR, NMR	715
1,6,7-Trichloro-4-methoxyphthalazine	—	(*E* 536)

TABLE A.2. (*Continued*)

Phthalazine	Melting Point (°C) etc.	Reference(s)
1,4,5-Trichlorophthalazine	—	(*E* 531)
1,4,6-Trichlorophthalazine	solid, NMR	285
3-*N*,*N*-Triethyl-1,4-dioxo-1,2,3,4-tetrahydro-2-phthalazinecarbothioamide	—	477
5,6,7-Trimethoxy-1,4(2*H*,3*H*)-phthalazinedione	—	(*E* 475)
1,4,6-Trimethylphthalazine	—	352
5,6,7-Trimethyl-1,4(2*H*,3*H*)-phthalazinedione	185–188, NMR	114
5,6,8-Trimethyl-1,4(2*H*,3*H*)-phthalazinedione	188–189, NMR	114
1-(Trimethylsiloxy)phthalazine	crude	986
1,4,6-Triphenylphthalazine	176–178, NMR	300, 908

References

Information was obtained from each original publication except where an additional reference to *Chemical Abstracts* is included. Abbreviations for journal titles are those recommended in the *Chemical Abstracts Service Source Index* (1994) and its supplements.

1. J. Nematollahi, *J. Heterocycl. Chem.*, **1972**, *9*, 963.
2. R. L. Zey, *J. Heterocycl. Chem.*, **1972**, *9*, 1177.
3. G. Axelrad and D. M. Lamontanaro, *J. Heterocyclic Chem.*, **1972**, *9*, 1407.
4. R. L. Williams and S. W. Shalaby, *J. Heterocycl. Chem.*, **1973**, *10*, 891.
5. Y. Tamura, Y. Miki, J. Minamikawa, and M. Ikeda, *J. Heterocycl. Chem.*, **1974**, *11*, 675.
6. G. Winters, V. Aresi, and G. Nathansohn, *J. Heterocycl. Chem.*, **1974**, *11*, 997.
7. G. Cignarella, F. Savelli, R. Cerri, and P. Sanna, *J. Heterocycl. Chem.*, **1974**, *11*, 1049.
8. Y. Tamura, Y. Miki, and M. Ikeda, *J. Heterocycl. Chem.*, **1975**, *12*, 119.
9. Y. Tamura, Y. Miki, K. Nakamura, and M. Ikeda, *J. Heterocycl. Chem.*, **1976**, *13*, 23.
10. B. Šket, M. Zupan, and A. Pollak, *J. Heterocycl. Chem.*, **1976**, *13*, 671.
11. A. Walser and G. Zenchoff, *J. Heterocycl. Chem.*, **1976**, *13*, 907.
12. R. P. Brundage and G. Y. Lesher, *J. Heterocycl. Chem.*, **1976**, *13*, 1085.
13. D. T. Connor, P. A. Young, and M. von Strandtnann, *J. Heterocycl. Chem.*, **1977**, *14*, 143.
14. M. B. Hocking, *J. Heterocycl. Chem.*, **1977**, *14*, 829.
15. D. T. Connor, P. A. Young, and M. von Standtmann, *J. Heterocycl. Chem.*, **1978**, *15*, 115.
16. L. Lepage and Y. Lepage, *J. Heterocycl. Chem.*, **1978**, *15*, 793.
17. T. Yamazaki, R. E. Draper, and R. N. Castle, *J. Heterocycl. Chem.*, **1978**, *15*, 1039.
18. G. R. Meyer, C. A. Kellert, and R. W. Ebert, *J. Heterocycl. Chem.*, **1979**, *16*, 461.
19. J.-M. Cosmao, N. Collignon, and G. Quéguiner, *J. Heterocycl. Chem.*, **1979**, *16*, 973.
20. I. Maeba and R. N. Castle, *J. Heterocycl. Chem.*, **1980**, *17*, 407.
21. D. Bhattacharjee and F. D. Popp, *J. Heterocycl. Chem.*, **1980**, *17*, 433.
22. D. Bhattacharjee and F. D. Popp, *J. Heterocycl. Chem.*, **1980**, *17*, 1035.
23. D. Bhattacharjee and F. D. Popp, *J. Heterocycl. Chem.*, **1980**, *17*, 1211.
24. T. L. Lemke and G. R. Parker, *J. Heterocycl. Chem.*, **1980**, *17*, 1519.
25. A. Amer and H. Zimmer, *J. Heterocycl. Chem.*, **1981**, *18*, 1625.

Cinnolines and Phthalazines: Supplement II, The Chemistry of Heterocyclic Compounds, Volume 64, by D.J. Brown

26. A. Compagnini, A. Lo-Vullo, U. Chiacchio, A. Corsaro, and G. Purrello, *J. Heterocycl. Chem.*, **1982**, *19*, 641.

27. R. A. Coburn and D. Gala, *J. Heterocycl. Chem.*, **1982**, *19*, 757.

28. L. Baiocchi and M. Giannangeli, *J. Heterocycl. Chem.*, **1983**, *20*, 225.

29. D. V. Bautista, G. Bullock, F. W. Hartstock, and L. K. Thompson, *J. Heterocycl. Chem.*, **1983**, *20*, 345.

30. G. Gelin and R. Dolmazon, *J. Heterocycl. Chem.*, **1983**, *20*, 543.

31. A. Amer and H. Zimmer, *J. Heterocycl. Chem.*, **1983**, *20*, 1231.

32. A. Counotte-Potman and H. C. van der Plas, *J. Heterocycl. Chem.*, **1983**, *20*, 1259.

33. M. Z. A. Badr, H. A. Sherief, G. M. El-Naggar, and S. A. Mahgoub, *J. Heterocycl. Chem.*, **1984**, *21*, 471.

34. A. D. Dunn, *J. Heterocycl. Chem.*, **1984**, *21*, 961.

35. J. Kant and F. D. Popp, *J. Heterocycl. Chem.*, **1985**, *22*, 1065.

36. S. Kaban, *J. Heterocycl. Chem.*, **1986**, *23*, 13.

37. R. L. Zey, *J. Heterocycl. Chem.*, **1987**, *24*, 1261.

38. R. L. Zey, *J. Heterocycl. Chem.*, **1988**, *25*, 847.

39. J. Esquerra and J. Alvaraz-Builla, *J. Heterocycl. Chem.*, **1988**, *25*, 917.

40. R. G. Sutherland, A. S. Abd-El-Aziz, A. Piórko, U. S. Gill, and C. C. Lee, *J. Heterocycl. Chem.*, **1988**, *25*, 1107.

41. M. A. Atfah, *J. Heterocycl. Chem.*, **1989**, *26*, 717.

42. R. L. Zey, G. Richter, and H. Randa, *J. Heterocycl. Chem.*, **1989**, *26*, 1437.

43. E. Licandro, S. Maiorana, A. Papagni, D. Tarallo, A. M. Z. Slawin, and D. J. Williams, *J. Heterocyclic Chem.*, **1990**, *27*, 1103.

44. A. Santagati, M. Santagati, and F. Russo, *J. Heterocycl. Chem.*, **1991**, *28*, 545.

45. B. Singh, *J. Heterocycl. Chem.*, **1991**, *28*, 881.

46. S. Shilcrat, I. Lantos, M. McGuire, L. Pridgen, L. Davis, D. Eggleston, D. Staiger, and L. Webb, *J. Heterocycl. Chem.*, **1993**, *30*, 1663.

47. J. M. Ruxer, J. Mauger, D. Bénard, and C. Lachoux, *J. Heterocycl. Chem.*, **1995**, *32*, 643.

48. L. R. Coswell and G. Cavasos, *J. Heterocycl. Chem.*, **1995**, *32*, 907.

49. S. Nan'ya, H. Ishida, K. Kanie, N. Ito, and Y. Butsugan, *J. Heterocycl. Chem.*, **1995**, *32*, 1299.

50. F. Saczewski and M. Gdaniec, *J. Heterocycl. Chem.*, **1998**, *35*, 707.

51. T. Billert, R. Beckert, P. Fehling, M. Döring, J. Brandenburg, H. Görls, and P. Langer, *J. Heterocycl. Chem.*, **1999**, *36*, 627.

52. M. J. Kornet and G. Shackleford, *J. Heterocycl. Chem.*, **1999**, *36*, 1095.

53. J. Daunis, M. Guerret-Rigail, and R. Jacquier, *Bull. Soc. Chim. Fr.*, **1972**, 1994.

54. J. Daunis, M. Guerret-Rigail, and R. Jacquier, *Bull. Soc. Chim. Fr.*, **1972**, 3198.

55. J. Schreiber, C.-G. Wermuth, and A. Meyer, *Bull. Soc. Chim. Fr.*, **1973**, 625.

56. A. Le Berre and A. Delacroix, *Bull. Soc. Chim. Fr.*, **1973**, 640.

57. A. Le Berre, A. Étienne, and J. Coquelin, *Bull. Soc. Chim. Fr.*, **1973**, 2266.

58. A. Le Berre and A. Delacroix, *Bull. Soc. Chim. Fr.*, **1973**, 2404.

59. J. Berlot and J. Renault, *Bull. Soc. Chim. Fr.*, **1973**, 3175.

60. G. Cauquis, B. Chabaud, and M. Genies, *Bull. Soc. Chim. Fr.*, **1973**, 3487.

61. G. Cauquis, B. Chabaud, and M. Genies, *Bull. Soc. Chim. Fr.*, **1975**, 583.

62. A. Guingant and J. Renault, *Bull. Soc. Chim. Fr.*, **1975**, 2246.

63. A. Guingant and J. Renault, *Bull. Soc. Chim. Fr.*, **1976**, 291.

64. R. Hazard and A. Tallec, *Bull. Soc. Chim. Fr.*, **1976**, 433.

65. G. Maghioros, G. Schlewer, and C.-G. Wermuth, *Bull. Soc. Chim. Fr.*, **1985**, 865.

66. F. F. Abdel-Latif, *Bull. Soc. Chim. Fr.*, **1990**, 129.
67. M. H. Elnagdi and A. W. Erian, *Bull. Soc. Chim. Fr.*, **1995**, *132*, 920.
68. B. L. Mylari, E. R. Larson, T. A. Beyer, W. J. Zembrowski, E. E. Aldinger, M. F. Dee, T. W. Siegel, and D. H. Singleton, *J. Med. Chem.*, **1991**, *34*, 108.
69. G. Cauquis and M. Genies, *Tetrahedron Lett.*, **1971**, 3959.
70. G. V. Boyd, *Tetrahedron Lett.*, **1972**, 2711.
71. G. Cauquis, B. Chabaud, and M. Genies, *Tetrahedron Lett.*, **1974**, 2389.
72. D. Johnston and D. M. Smith, *Tetrahedron Lett.*, **1975**, 1121.
73. E. Toja, A. Omodei-Salè, and G. Nathansohn, *Tetrahedron Lett.*, **1976**, 111.
74. N. Dennis, A. R. Katritzky, E. Lunt, M. Ramaiah, R. L. Harlow, and S. H. Simonson, *Tetrahedron Lett.*, **1976**, 1569.
75. S. Kanoktanaporn and J. A. H. MacBride, *Tetrahedron Lett.*, **1977**, 1817.
76. J.-P. Anselme, *Tetrahedron Lett.*, **1977**, 3615.
77. T. Durst and L. Tetreault-Ryan, *Tetrahedron Lett.*, **1978**, 2353.
78. B. I. Buzykin, A. P. Stolyarov, and N. N. Bystrykh, *Tetrahedron Lett.*, **1980**, *21*, 209.
79. M. Kuzuya, F. Miyake, and T. Okuda, *Tetrahedron Lett.*, **1980**, *21*, 2729.
80. S. K. Robev, *Tetrahedron Lett.*, **1981**, *22*, 345.
81. J. T. Sharp and C. E. D. Skinner, *Tetrahedron Lett.*, **1986**, *27*, 869.
82. F. P. J. T. Rutjes, H. Hiemstra, H. H. Mooiweer, and W. N. Spekamp, *Tetrahedron Lett.*, **1988**, *29*, 6975.
83. M. G. Hutchings and D. P. Devonald, *Tetrahedron Lett.*, **1989**, *30*, 3715.
84. W. L. F. Armarego, in *Physical Methods in Heterocyclic Chemistry*, A. R. Katritzky, ed., Academic Press, London, 1971, Vol. 3, pp. 67–222.
85. A. S. Kiselyov, *Tetrahedron Lett.*, **1995**, *36*, 1383.
86. M. S. South, T. L. Jakuboski, M. D. Westmeyer, and D. R. Dukesherer, *Tetrahedron Lett.*, **1996**, *37*, 1351.
87. A. F. Parsons and R. M. Pettifer, *Tetrahedron Lett.*, **1996**, *37*, 1667.
88. I. Gillies and C. W. Rees, *Tetrahedron Lett.*, **1996**, *37*, 4065.
89. L. Bourel, A. Tartar, and P. Melnyk, *Tetrahedron Lett.*, **1996**, *37*, 4145.
90. A. Ruiz, P. Rocca, F. Marsaia, A. Godard, and G. Quéguiner, *Tetrahedron Lett.*, **1997**, *38*, 6205.
91. Y. Matsubara, A. Horikawa, and Z. Yoshida, *Tetrahedron Lett.*, **1997**, *38*, 8199.
92. T. J. Sparey and T. Harrison, *Tetrahedron Lett.*, **1998**, *39*, 5873.
93. A. S. Kiselyav and C. Dominguez, *Tetrahedron Lett.*, **1999**, *40*, 5111.
94. S. Bräse, S. Dahmen, and J. Heuts, *Tetrahedron Lett.*, **1999**, *40*, 6201.
95. H. Feuer and J. P. Lawrence, *J. Org. Chem.*, **1972**, *37*, 3662.
96. H. Alper, *J. Org. Chem.*, **1972**, *37*, 3972.
97. J. A. Zoltewicz, T. M. Oestreich, J. K. O'Halloran, and L. S. Helmick, *J. Org. Chem.*, **1973**, *38*, 1949.
98. H. W. Heine, R. Henrie, L. Heitz, and S. R. Kovvali, *J. Org. Chem.*, **1974**, *39*, 3187.
99. A. L. Johnson, *J. Org. Chem.*, **1976**, *41*, 836.
100. J. A. Zoltewicz, L. S. Helmick, and J. K. O'Halloran, *J. Org. Chem.*, **1976**, *41*, 1308.
101. W. J. Scott, W. J. Bover, K. Bratin, and P. Zuman, *J. Org. Chem.*, **1976**, *41*, 1952.
102. S. Kasina and J. Nematollahi, *J. Org. Chem.*, **1977**, *42*, 159.
103. W. H. Pirkle and P. L. Gravel, *J. Org. Chem.*, **1977**, *42*, 1367.
104. M. J. Haddadin, S. J. Firson, and B. S. Nader, *J. Org. Chem.*, **1979**, *44*, 629.
105. H. W. Heine, L. S. Lehman, A. P. Glaze, and A. W. Douglas, *J. Org. Chem.*, **1980**, *45*, 1317.

106. D. H. Bown and J. S. Bradshaw, *J. Org. Chem.*, **1980**, *45*, 2320.

107. W. E. McEwen, I. C. Wang-Huang, C. P. Cartaya-Marin, F. McCarty, E. Marmugi-Segnini, C. M. Zepp, and J. J. Lubinkowski, *J. Org. Chem.*, **1982**, *47*, 3098.

108. B. Mylari, T. A. Beyer, P. J. Scott, C. E. Aldinger, M. F. Dee, T. W. Siegel, and W. J. Zembrowski, *J. Med. Chem.*, **1992**, *35*, 457.

109. A. Melikian, G. Schlewer, J.-P. Chambon, and C.-G. Wermuth, *J. Med. Chem.*, **1992**, *35*, 4092.

110. C. A. Bruynes and T. K. Jurriens, *J. Org. Chem.*, **1982**, *47*, 3966.

111. D. L. Boger, R. S. Coleman, J. S. Panek, and D. Yohannes, *J. Org. Chem.*, **1984**, *49*, 4405.

112. K. T. Potts, K. G. Bordeaux, W. B. Kuehnling, and R. L. Salsbury, *J. Org. Chem.*, **1985**, *50*, 1677.

113. S. F. Nelsen and N. P. Yumibe, *J. Org. Chem.*, **1985**, *50*, 4749.

114. T. Keumi, T. Morita, K. Teramoto, N. Takahashi, H. Yamamoto, K. Ikeno, M. Hanaki, T. Inagaki, and H. Kitajima, *J. Org. Chem.*, **1986**, *51*, 3439.

115. A. Hosomi, S. Hayashi, K. Hoashi, S. Kohra, and Y. Tominaga, *J. Org. Chem.*, **1987**, *52*, 4423.

116. S. C. Benson, J. L. Gross, and J. K. Snyder, *J. Org. Chem.*, **1990**, *55*, 3257.

117. M. T. Nguyen, L. F. Clarke, and A. F. Hegarty, *J. Org. Chem.*, **1990**, *55*, 6177.

118. B. L. Mylari and W. J. Zembrowski, *J. Org. Chem.*, **1991**, *56*, 2587.

119. R. E. Johnson, J. T. Hane, D. C. Schlegel, R. B. Perni, J. L. Herrmann, C. J. Opalka, P. M. Carabateas, J. H. Ackerman, J. Swestock, N. C. Birsner, and J. H. Tatlock, *J. Org. Chem.*, **1991**, *56*, 5218.

120. M. Kočever, P. Mihorko, and S. Polanc, *J. Org. Chem.*, **1995**, *60*, 1466.

121. K. P. Chan and A. S. Hay, *J. Org. Chem.*, **1995**, *60*, 3131.

122. H. Becker, S. B. King, M. Taniguchi, K. P. M. Vanhessche, and K. B. Sharpless, *J. Org. Chem.*, **1995**, *60*, 3940.

123. M. S. South, T. L. Jakuboski, M. D. Westmeyer, and D. R. Dukesherer, *J. Org. Chem.*, **1996**, *61*, 8921.

124. L. Brnati, G. Calestani, D. Nanni, and P. Spagnolo, *J. Org. Chem.*, **1998**, *63*, 4679.

125. R. Appel and P. Volz, *Chem. Ber.*, **1975**, *108*, 623.

126. H. Wamhoff and C. Wald, *Chem. Ber.*, **1977**, *110*, 1716.

127. U. Wolf, W. Sucrow, and H.-J. Vetter, *Chem. Ber.*, **1979**, *112*, 3237.

128. T. Eicher, T. Pfister, and M. Urban, *Chem. Ber.*, **1980**, *113*, 424.

129. H. Stetter and A. Mertens, *Chem. Ber.*, **1981**, *114*, 2479.

130. J. K. Rumiński and K. D. Przewoska, *Chem. Ber.*, **1982**, *115*, 3436.

131. R. Dhar, W. Hühnermann, T. Kämchen, W. Overhue, and G. Seitz, *Chem. Ber.*, **1983**, *116*, 97.

132. J. K. Rumiński, *Chem. Ber.*, **1983**, *116*, 970.

133. H. Verbrüggen and K. Krolikiewicz, *Chem. Ber.*, **1984**, *117*, 1523.

134. U. Dittrich and H.-F. Grützmacher, *Chem. Ber.*, **1985**, *118*, 4404.

135. G. Kaupp, H. Frey, and G. Behmann, *Chem. Ber.*, **1988**, *121*, 2135.

136. R. Kluge, L. Omelka, M. Reinhardt, and M. Schulz, *Chem. Ber.*, **1992**, *125*, 2075.

137. W. M. Horspool, J. R. Kershaw, and A. W. Murray, *J. Chem. Soc., Chem. Commun.*, **1973**, 345.

138. T. Igeta, T. Nakai, and T. Tsuchiya, *J. Chem. Soc., Chem. Commun.*, **1973**, 622.

139. K. Nagarajan and R. K. Shah, *J. Chem. Soc., Chem. Commun.*, **1973**, 926.

140. A. A. Sandison and G. Tennant, *J. Chem. Soc., Chem. Commun.*, **1974**, 752.

141. C. W. Rees, R. W. Stephenson, and R. C. Storr, *J. Chem. Soc., Chem. Commun.*, **1974**, 941.

142. F. McCapra and P. D. Leeson, *J. Chem. Soc., Chem. Commun.*, **1979**, 114.

143. K. T. Potts and P. Murphy, *J. Chem. Soc., Chem. Commun.*, **1984**, 1348.

144. R. Grigg, H. O. N. Gunaratne, and V. Sridharan, *J. Chem. Soc., Chem. Commun.*, **1985**, 1183.

145. R. N. Butler, A. M. Gillan, P. McArdle, and D. Cunningham, *J. Chem. Soc., Chem. Commun.*, **1987**, 1016.

146. M. Scobie and G. Tennant, *J. Chem. Soc., Chem. Commun.*, **1993**, 1756.
147. M. Scobie and G. Tennant, *J. Chem. Soc., Chem. Commun.*, **1994**, 2451.
148. L. Hoesch and A. S. Dreiding, *Helv. Chim. Acta*, **1975**, *58*, 1995.
149. L. Hoesch, M. Karpf, E. Dunkelblum, and A. S. Dreiding, *Helv. Chim. Acta*, **1977**, *60*, 816.
150. C. Leuenberger, M. Karpf, L. Hoesch, and A. Dreiding, *Helv. Chim. Acta*, **1977**, *60*, 831.
151. H. Link, K. Bernauer, S. Chaloupka, H. Heimgartner, and H. Schmid, *Helv. Chim. Acta*, **1978**, *61*, 2112.
152. W. Städeli, W. von Philipsborn, A. Wick, and I. Kompiš, *Helv. Chim. Acta*, **1980**, *63*, 504.
153. L. Hoesch, *Helv. Chim. Acta*, **1981**, *64*, 38.
154. C. Leuenberger, L. Hoesch, and A. S. Dreiding, *Helv. Chim. Acta*, **1981**, *64*, 1219.
155. A. Sammour, M. I. B. Selim, and M. W. Osman, *UAR J. Chem.*, **1971**, *14*, 305; *Chem. Abstr.*, **1972**, *77*, 152090.
156. N. Bregant, I. Perina, and K. Balenovic, *Bull. Sci., Cons. Acad. Sci. Arts RSF Yugoslavie, Sect. A*, **1972**, *17*(5–6), 148; *Chem. Abstr.*, **1972**, *77*, 152092.
157. A. Kost, A. Puodziunas, and A. Lubas, *Sin. Izuch. Fiziol. Aktiv. Veshchestv. Mater. Konf.*, **1971**, 61; *Chem. Abstr.*, **1973**, *79*, 78716.
158. A. Kost, A. Puodziunas, M. T. Grigoryan, and V. G. Afrikyan, *Sin. Geterotsikl. Soedin.*, **1972**(9), 15; *Chem. Abstr.*, **1973**, *79*, 126427.
159. A. Puodziunas, A. Kost, and A. Lubas, *Khim.-Farm. Zh.*, **1973**, *7*, 25; *Chem. Abstr.*, **1973**, *79*, 137066.
160. D. Twomey, *Proc. Roy. Ir. Acad., Sect. B*, **1974**, *74*, 37; *Chem. Abstr.*, **1974**, *81*, 3864.
161. M. Petrovanu, A. Saucic, and I. Zugravescu, *Rev. Roum. Chim.*, **1974**, *19*, 437; *Chem. Abstr.*, **1974**, *81*, 12723.
162. S. Yarnal and V. V. Badiger, *J. Karnatak Univ.*, **1973**, *18*, 25; *Chem. Abstr.*, **1975**, *82*, 43302.
163. V. A. Chuiguk and G. M. Pakholkov, *Ukr. Khim. Zh.* (*Russ. ed.*), **1974**, *40*, 1173; *Chem. Abstr.*, **1975**, *82*, 43319.
164. E. Domagalina and J. Ochynska, *Pol. J. Pharmacol. Pharm.*, **1974**, *26*, 473; *Chem. Abstr.*, **1975**, *82*, 57633.
165. M. Petrovanu, M. Caprosu, and I. Zugravescu, *An. Stiint. Univ. "Al. I. Cuza" Iasi, Sect. lc*, **1974**, *20*, 183; *Chem. Abstr.*, **1975**, *83*, 9954.
166. J. Tulecki, L. Senczuk, and D. Popiel, *Ann. Pharm.* (*Poznan*), **1975**, *11*, 101; *Chem. Abstr.*, **1975**, *83*, 147443.
167. M. F. Brana and J. L. Soto, *An. Quim.*, **1974**, *70*, 970; *Chem. Abstr.*, **1975**, *83*, 178966.
168. M. Ishikawa and Y. Eguchi, *Iyo Kizai Kenkyusho Hokoku, Tokyo Ika Shika Daigaku*, **1974**, *8*, 9; *Chem. Abstr.*, **1976**, *84*, 17259.
169. I. Sulekiene, A. Puodziunas, L. Zukauskaite, and A. Stankevicius, *Mater. Nauchn. Konf. Kaunas, Med. Inst., 22nd*, **1972**, 129; *Chem. Abstr.*, **1976**, *84*, 73461.
170. M. Caprosu, M. Petrovanu, and J. Zugravescu, *Bul. Inst. Politech. Iasi, Sect. 2*, **1975**, *21*, 61; *Chem. Abstr.*, **1976**, *84*, 43964.
171. K. Grabliauskas, *Mater. Nauchn. Konf. Kaunas, Med. Inst., 22nd*, **1972**, 131; *Chem. Abstr.*, **1976**, *84*, 58116.
172. K. Grabliauskas, *Mater. Nauchn. Konf. Kaunas, Med. Inst., 22nd*, **1972**, 133; *Chem. Abstr.*, **1976**, *84*, 59389.
173. K. Grabliauskas, *Mater. Nauchn. Konf. Kaunas, Med. Inst., 22nd*, **1972**, 130; *Chem. Abstr.*, **1976**, *84*, 59356.
174. J. Strumillo, *Acta Pol. Pharm.*, **1975**, *32*, 287; *Chem. Abstr.*, **1976**, *84*, 150580.
175. Y. Maki, T. Furuta, M. Kuzuya, and M. Suzuki, *Hukusokan Kagaku Toronkai Koen Yoshishu, 8th*, **1975**, 19; *Chem. Abstr.*, **1976**, *84*, 163839.

176. K. Mitsuhashi, N. Katou, N. Kembou, and T. Suzuki, *Seikei Daigaku Kogakubu Kogaku Hokoku*, **1976**, *21*, 1519; *Chem. Abstr.*, **1976**, *84*, 180170.

177. K. Lempert and K. Zauer, *Acta Chim. Acad. Sci. Hung.*, **1976**, *88*, 81; *Chem. Abstr.*, **1976**, *85*, 21266.

178. K. Kormendy, *Acta Chim. Acad. Sci. Hung.*, **1976**, *88*, 129; *Chem. Abstr.*, **1976**, *85*, 78068.

179. G. Adembri, S. Chimichi, F. De Sio, R. Nisi, and M. Scotton, *Chim. Ind.* (*Milan*), **1976**, *58*, 217; *Chem. Abstr.*, **1976**, *85*, 94300.

180. A. I. El-Sebai, M. Ragab, R. Soliman, and M. Gabr, *Pharmazie*, **1976**, *31*, 436; *Chem. Abstr.*, **1977**, *86*, 5391.

181. W. L. F. Armarego, in *MTP International Review of Science: Organic Chemistry; Heterocyclic Compounds*, D. H. Hey and K. Schofield, eds., Butterworth, London, Series 1, Vol. 4, 1973, p. 138; Series 2, Vol. 4, **1975**, p. 136.

182. S. M. A. E. Omran, M. R. M. Salem, and N. S. Harb, *Egypt. J. Chem.*, **1974**, *17*, 731; *Chem. Abstr.*, **1977**, *86*, 120345.

183. W. I. Award, F. G. Baddar, S. M. A. R. Omran, and S. Antonious, *Egypt. J. Chem.*, **1974**, *17*, 555; *Chem. Abstr.*, **1977**, *86*, 189825.

184. S. Baloniak, W. Kryk, and J. Szuscicka, *Acta Pol. Pharm.*, **1976**, *33*, 329; *Chem. Abstr.*, **1977**, *87*, 53188.

185. T. Kametani, K. Kigasawa, M. Hiiragi, N. Wagatsuma, T. Uryu, and K. Araki, *J. Med. Chem.*, **1973**, *16*, 301.

186. K. Nagarajan, J. David, and R. K. Shah, *J. Med. Chem.*, **1976**, *19*, 508.

187. M. Lora-Tamayo, B. Marco, and P. Navarro, *An. Quim.*, **1976**, *72*, 914; *Chem. Abstr.*, **1977**, *87*, 152108.

188. M. Uchida, E. Oishi, T. Higashino, and E. Hayashi, *Shitsuryo Bunseki*, **1977**, *25*, 175; *Chem. Abstr.*, **1978**, *88*, 21569.

189. M. Caprosu, I. Druta, and M. Petrovanu, *Bul. Inst. Polyteh. Iasi, Sect. 2*, **1977**, *23*, 71; *Chem. Abstr.*, **1978**, *88*, 50770.

190. K. Komendy, *Acta Chim. Acad. Sci. Hung.*, **1977**, *94*, 373; *Chem. Abstr.*, **1978**, *88*, 136550.

191. M. Valasco, H. Bartoncini, E. Romero, A. Urbina-Quintane, J. Guevara, and O. Hernandez-Pieretti, *Eur. J. Clin. Pharmacol.*, **1978**, *13*, 317; *Chem. Abstr.*, **1978**, *89*, 209243.

192. S. Baloniak, H. Blaszczak, M. Filczewski, A. Mroczkiewicz, K. Oledzka, M. Szymanska-Kosmala, and I. Zyczynska-Baloniak, *Acta Pol. Pharm.*, **1978**, *35*, 161; *Chem. Abstr.*, **1979**, *90*, 22943.

193. V. Basiunas, *Mater. Mezhvuz. Nauchn. Konf. Med. Inst., 25th*, **1976**, 58; *Chem. Abstr.*, **1979**, *90*, 38856.

194. M. Caprosu, R. Mocanu, and M. Petrovanu, *Rev. Chim.* (*Bucharest*), **1978**, *29*, 1087; *Chem. Abstr.*, **1979**, *90*, 151395.

195. Y. Eguchi, M. Nakajima, S. Kaneko, and M. Ishikawa, *Iyo Kizai Kenkyusho Hokoku* (*Tokyo Ika Shika Daigaku*), **1978**, *12*, 41; *Chem. Abstr.*, **1979**, *91*, 91579.

196. K. Kormendy, *Acta Chim. Acad. Sci. Hung.*, **1978**, *98*, 303; *Chem. Abstr.*, **1979**, *91*, 174384.

197. K. Kormendy, *Acta Chim. Acad. Sci. Hung.*, **1979**, *99*, 81; *Chem. Abstr.*, **1979**, *91*, 211257.

198. W. I. Awad, M. F. Ismail, and N. G. Kandila, *Egypt. J. Chem.*, **1976**, *19*, 761; *Chem. Abstr.*, **1980**, *92*, 6332.

199. K. Komendy, K. A. Juhasz, and E. Lemberkovics, *Acta Chim. Acad. Sci. Hung.*, **1979**, *102*, 39; *Chem. Abstr.*, **1980**, *92*, 198335.

200. R. Solimam and H. Mokhtar, *Pharmazie*, **1979**, *34*, 297; *Chem. Abstr.*, **1980**, *92*, 76431.

201. D. Twomey, *Proc. R. Ir. Acad., Sect. B*, **1979**, *79*(3), 29; *Chem. Abstr.*, **1980**, *92*, 94311.

202. A. M. Islam, I. B. Nammout, A. A. El-Maghraby, and F. M. Aly, *Egypt. J. Chem.*, **1977**, *20*, 25; *Chem. Abstr.*, **1980**, *93*, 7937.

203. K. K. Zauer, *Sint. Geterotsikl. Soedin.*, **1979**, *11*, 39; *Chem. Abstr.*, **1981**, *94*, 15664.

204. M. Gabr, A. A. B. Hazzaa, and M. A. Khalil, *Sci. Pharm.*, **1980**, *48*, 141; *Chem. Abstr.*, **1981**, *94*, 47249.

205. K. Kormendy and F. Ruff, *Acta Chim. Acad. Sci Hung.*, **1981**, *106*, 155; *Chem. Abstr.*, **1981**, *95*, 97698.

206. H. M. Feid-Allah and R. Soliman, *Pharmazie*, **1981**, *36*, 471; *Chem. Abstr.*, **1981**, *95*, 169112.

207. A. M. Islam and S. M. M. El-Shafie, *Egypt. J. Chem.*, **1979**, *22*, 135; *Chem. Abstr.*, **1981**, *95*, 187182.

208. S. Robev, *Dokl. Bolg. Akad. Nauk*, **1981**, *34*, 799; *Chem. Abstr.*, **1982**, *96*, 35182.

209. A. Mroczkiewicz, *Acta Pol. Pharm.*, **1981**, *38*, 263; *Chem. Abstr.*, **1982**, *96*, 181225.

210. M. Lora-Tamayo, P. Navarro, D. Romero, and J. L. Soto, *An. Quim., Ser. C*, **1981**, *77*, 296; *Chem. Abstr.*, **1982**, *97*, 23723.

211. S. Gao, *Yaoxuc Tongbao*, **1982**, *17*, 501; *Chem. Abstr.*, **1982**, *97*, 182335.

212. A. A. Asselin, L. G. Humber, T. A. Dobson, J. Komlossy, and R. R. Martel, *J. Med. Chem.*, **1976**, *19*, 787.

213. D. Holland, G. Jones, P. W. Marshall, and G. D. Tringham, *J. Med. Chem.*, **1976**, *19*, 1225.

214. T. Yamada, Y. Nobuhara, A. Yamaguchi, and M. Ohki, *J. Med. Chem.*, **1982**, *25*, 975.

215. M. Petrovanu, M. Caprosu, E. Stefanescu, and N. Stavri, *Rev. Med.-Chir.*, **1982**, *86*, 141; *Chem. Abstr.*, **1983**, *98*, 16631.

216. M. Z. A. Badr, H. A. H. El-Sherief, G. M. El-Naggar, and S. A. Mahgoub, *Bull. Fac. Sci. Assiut Univ.*, **1982**, *11*, 25; *Chem. Abstr.*, **1983**, *98*, 53807.

217. G. Heinisch and M. Koch, *Sci. Pharm.*, **1982**, *50*, 246; *Chem. Abstr.*, **1982**, *50*, 246.

218. G. V. Shatalov, S. A. Gridchin, G. V. Kovalev, B. I. Mikhant'ev, and S. F. Gofman, *Izv. Vyssh. Uchebn. Zavid., Khim. Khim. Tekhnol.*, **1982**, *25*, 1179; *Chem. Abstr.*, **1983**, *98*, 89286.

219. P. V. Padmanabhan, S. R. Ramadas, B. Varghese, and S. Srinivasan, *Cryst. Struct. Commun.*, **1982**, *11*, 1277; *Chem. Abstr.*, **1983**, *98*, 207886.

220. S. Robev and N. Boyadzhieva, *Dokl. Bolg. Akad. Nauk*, **1982**, *35*, 1587; *Chem. Abstr.*, **1983**, *99*, 5578.

221. K. Kormendy and F. Ruff, *Acta Chim. Hung.*, **1983**, *112*, 65; *Chem. Abstr.*, **1983**, *99*, 158354.

222. T. S. Chou and W. J. Cheng, *Bull. Inst. Chem., Acad. Sin.*, **1983**, *30*, 7; *Chem. Abstr.*, **1984**, *100*, 85645.

223. S. Gao, *Yiyao Gongye*, **1983**(10), 41; *Chem. Abstr.*, **1984**, *100*, 156556.

224. D. M. Stout, W. L. Matier, C. Barcelon-Yang, R. D. Reynolds, and B. S. Brown, *J. Med. Chem.*, **1983**, *26*, 808.

225. C. B. Chapleo, J. C. Doxey, and P. L. Myers, *J. Med. Chem.*, **1982**, *25*, 821.

226. A. Arcoleo, G. Fontana, and G. Giammona, *J. Chem. Eng. Data*, **1984**, *29*, 354; *Chem. Abstr.*, **1984**, *101*, 54861.

227. Y. M. Volovenko, V. A. Litenko, and F. S. Babichev, *Dopev. Akad. Nauk Ukr. RSR, Ser. B*, **1984**(5), 33; *Chem. Abstr.*, **1984**, *101*, 130649 (1984).

228. S. Nan'ya, Y. Suzuki, and E. Maekawa, *Nagoya Kogyo Daigaku Gakuho*, **1982**, *34*, 157; *Chem. Abstr.*, **1984**, *102*, 26346.

229. M. Lora-Tamayo, B. Lopez, P. Navarro, L. Palacios, and F. Gomez-Contreras, *An. Quim., Ser. C*, **1983**, *79*, 138; *Chem. Abstr.*, **1984**, *102*, 45862.

230. T. Otto, K. Galewicz, and L. Strzemecka, *Acta Pol. Pharm.*, **1984**, *41*, 333; *Chem. Abstr.*, **1984**, *102*, 166725.

231. S. G. Kon'kova, A. A. Safaryan, and A. N. Akopyan, *Arm. Khim. Zh.*, **1984**, *37*, 572; *Chem. Abstr.*, **1985**, *103*, 37438.

232. M. Caprosu, M. Ungureanu, C. Radu, and M. Petrovanu, *Rev. Med.-Chir.*, **1984**, *88*, 149; *Chem. Abstr.*, **1985**, *103*, 84883.

233. Y. H. Ahn and Y. Y. Lee, *Taehan Hwahakhue Chi*, **1985**, *29*, 61; *Chem. Abstr.*, **1985**, *103*, 87830.
234. X. Lu, *Yiyao Gongye*, **1985**, *16*, 167; *Chem. Abstr.*, **1985**, *103*, 104906.
235. K. Kormendy, F. Ruff, and I. Kovesti, *Acta Chim. Hung.*, **1984**, *117*, 371; *Chem. Abstr.*, **1985**, *103*, 177767.
236. S. M. Yarnal and V. V. Badiger, *J. Karnatak Univ., Sci.*, **1984**, *29*, 82; *Chem. Abstr.*, **1986**, *104*, 224864.
237. A. R. K. Murthy, I. H. Hall, J. M. Chapman, K. A. Rhyne, and S. D. Wyrick, *Pharm. Rev.*, **1986**, *3*, 93; *Chem. Abstr.*, **1986**, *105*, 91100.
238. R. S. Vartanyan, L. O. Avetyan, S. A. Karamyan, and S. A. Vartanyan, *Arm. Khim. Zh.*, **1985**, *38*, 743; *Chem. Abstr.*, **1986**, *105*, 133826.
239. S. C. Shin, H. C. Wang, W. Y. Lee, and Y. Y. Lee, *Bull. Korean Chem. Soc.*, **1985**, *6*, 323; *Chem. Abstr.*, **1986**, *105*, 133854.
240. Y. Y. Lee, H. H. Kim, and Y. M. Goo, *Taehan Hwahakhoe Chi*, **1986**, *30*, 243; *Chem. Abstr.*, **1987**, *106*, 18331.
241. A. A. Afify, M. A. Sayed, A. El-Farargy, and G. H. Tamam, *J. Chem. Soc. Pak.*, **1986**, *8*, 53; *Chem. Abstr.*, **1987**, *106*, 156382.
242. V. Brizzi, S. Masi, and G. Corbini, *Boll. Chim. Farm.*, **1986**, *125*, 154; *Chem. Abstr.*, **1987**, *107*, 58958.
243. Y. Eguchi and M. Ishikawa, *Iyo Kizai Konkyusho Hokoku* (*Tokyo Ika Shika Daigaku*), **1985**, *19*, 39; *Chem. Abstr.*, **1987**, *107*, 198214.
244. R. K. Bansal, S. K. Jain, and R. C. Sarkar, *Proc. Natl. Acad. Sci., India; Sect. A*, **1986**, *56*, 384; *Chem. Abstr.*, **1988**, *108*, 131719.
245. M. Caprosu, M. Ungureanu, N. Stavri, and M. Petrovanu, *Rev. Med.-Chir.*, **1986**, *90*, 719; *Chem. Abstr.*, **1988**, *108*, 150400.
246. M. Z. A. Badr, S. A. Mahgoub, A. M. Fahmy, and F. F. A. El-Latif, *Acta Chim. Hung.*, **1987**, *124*, 413; *Chem. Abstr.*, **1988**, *108*, 167406.
247. B. Abd El-Fattah, E. Roeder, M. I. Al-Ashmawi, and S. El-Feky, *Egypt. J. Pharm. Sci.*, **1987**, *28*, 371; *Chem. Abstr.*, **1988**, *108*, 204579.
248. B. Abd El-Fattah, M. I. Al-Ashmawi, S. El-Feky, and E. Roeder, *Egypt. J. Pharm. Sci.*, **1987**, *28*, 383; *Chem. Abstr.*, **1988**, *108*, 204580.
249. J. Jiao, W. Wang, and L. Gao, *Yiyao Gongyo*, **1988**, *19*, 29; *Chem. Abstr.*, **1988**, *109*, 110360.
250. Y. Eguchi, Y. Hasegawa, E. Katano, Y. Komoda, and M. Ishikawa, *Iyo Kizai Kenkyusho Hokoku* (*Tokyo Ika Shika Daigaku*), **1987**, *21*, 27; *Chem. Abstr.*, **1989**, *110*, 8145.
251. K. Kormendy, F. Ruff, and I. Kovesdi, *Acta Chim. Hung.*, **1988**, *125*, 99; *Chem. Abstr.*, **1989**, *110*, 57601.
252. J. Shen, T. Sun, R. Chen, and X. Xu, *Huaxue Shiji*, **1988**, *10*, 178; *Chem. Abstr.*, **1989**, *110*, 154249.
253. A. M. A. El-Khamry, A. Y. Soliman, A. A. Afify, and M. A. Sayed, *Orient. J. Chem.*, **1988**, *4*, 318; *Chem. Abstr.*, **1989**, *110*, 212739.
254. V. N. Lysitsyn and A. G. Borovko, *Tr. Inst.-Mosk. Khim.-Tekhnol. Inst. im D. I. Mendelaeva*, **1986**, *141*, 14; *Chem. Abstr.*, **1989**, *110*, 212737.
255. S. C. Shin and Y. Y. Lee, *Bull. Korean Chem. Soc.*, **1988**, *9*, 359; *Chem. Abstr.*, **1989**, *110*, 212774.
256. B. Abdel-Fattah, M. I. Al-Ashmawi, S. El-Feky, and E. Roder, *Egypt. J. Pharm. Sci.*, **1988**, *29*, 259; *Chem. Abstr.*, **1989**, *111*, 7320.
257. M. A. El-Hashash, S. I. El-Nagdy, and A. Y. Soliman, *Egypt. J. Chem.*, **1986**, *29*, 529; *Chem. Abstr.*, **1989**, *111*, 7325.
258. B. Chaturvadi, N. Tiwari, and ?. Nizamuddin, *Indian J. Pharm. Sci.*, **1988**, *50*, 316; *Chem. Abstr.*, **1989**, *111*, 70577.

259. E. M. M. Kassem, M. M. Kamel, A. A. Makhlouf, and M. T. Omar, *Pharmazie*, **1989**, *44*, 62; *Chem. Abstr.*, **1989**, *111*, 194702.

260. T. Dabhi and A. R. Parikh, *J. Inst. Chem.* (*India*), **1988**, *60*, 214; *Chem. Abstr.*, **1990**, *112*, 35783.

261. K. R. Desai and P. H. Desai, *J. Inst. Chem.* (*India*), **1988**, *60*, 227; *Chem. Abstr.*, **1990**, *112*, 35784.

262. Y. Eguchi, Y. Hasegawa, and M. Ishikawa, *Iyo Kizai Kenkyusho hokoku* (*Tokyo Ika Shika Daigaku*), **1988**, *22*, 47; *Chem. Abstr.*, **1990**, *112*, 55748.

263. W. Czuba and D. Sedzik-Hibner, *Pol. J. Chem.*, **1989**, *63*, 113; *Chem. Abstr.*, **1990**, *112*, 178845.

264. A. A. El-Sawy, S. G. Donia, S. A. Essawy, and A. N. F. Eissa, *J. Chem. Soc. Pak.*, **1989**, *11*, 111; *Chem. Abstr.*, **1990**, *112*, 198274.

265. S. Kishimoto, T. Okushi, and T. Hirashima, *Kagaku to Kogyo,* (*Osaka*), **1990**, *64*, 40; *Chem. Abstr.*, **1990**, *113*, 40487.

266. S. C. Shin, K. H. Kang, Y. Y. Lee, and Y. M. Goo, *Bull. Korean Chem. Soc.*, **1990**, *11*, 22; *Chem. Abstr.*, **1990**, *113*, 40633.

267. N. G. Kandile, *Acta Chim. Hung.*, **1989**, *126*, 533; *Chem. Abstr.*, **1990**, *113*, 78306.

268. F. A. Yassin, B. E. Bayoumy, and A. F. El-Farargy, *Bull. Korean Chem. Soc.*, **1990**, *11*, 7; *Chem. Abstr.*, **1990**, *113*, 132113.

269. B. E. Bayoumy, S. R. El-Feky, and M. El-Mobayed, *J. Chem. Soc. Pak.*, **1989**, *11*, 302; *Chem. Abstr.*, **1990**, *113*, 191266.

270. A. Sugimoto, H. Tanaka, Y. Eguchi, S. Ito, Y. Takashima, and M. Ishikawa, *J. Med. Chem.*, **1984**, *27*, 1300.

271. E. Ballasio, A. Campi, N. Di Mola, and E. Baldoli, *J. Med. Chem.*, **1984**, *27*, 1077.

272. B. S. Ross and R. A. Wiley, *J. Med. Chem.*, **1985**, *28*, 870.

273. I. Sircar, B. L. Duell, G. Bobowski, J. A. Bristol, and D. B. Evans, *J. Med. Chem.*, **1985**, *28*, 1405.

274. S. Cherkez, J. Herzic, and H. Yellin, *J. Med. Chem.*, **1986**, *29*, 947.

275. I. A. Shaikh, F. Johnson, and A. P. Grollman, *J. Med. Chem.*, **1986**, *29*, 1329.

276. G. Tarzia, E. Occelli, E. Toja, D. Barone, N. Corsico, L. Gallico, and F. Luzzani, *J. Med. Chem.*, **1988**, *31*, 1115.

277. S. J. Dominianni and T. T. Yen, *J. Med. Chem.*, **1989**, *32*, 2301.

278. M. Yamaguchi, K. Kamei, T. Koga, M. Akima, T. Kuroki, and N. Ohi, *J. Med. Chem.*, **1993**, *36*, 4052.

279. M. Yamaguchi, K. Kamei, T. Koga, M. Akima, A. Maruyama, T. Kuroki, and N. Ohi, *J. Med. Chem.*, **1993**, *36*, 4061.

280. M. H. Norman, G. C. Rigdon, F. Navas, and B. R. Cooper, *J. Med. Chem.*, **1994**, *37*, 2552.

281. G. W. Rewcastle, W. A. Denny, A. J. Bridges, H. Zhou, D. R. Cody, A. McMichael, and D. W. Fry, *J. Med. Chem.*, **1995**, *38*, 3482.

282. Y. Rival, R. Hoffmann, B. Didier, V. Rybaltchenko, J.-J. Bourguignon, and C.-G. Wermuth, *J. Med. Chem.*, **1998**, *41*, 311.

283. S. F. Brady, K. Stauffer, W. C. Lumma, G. M. Smith, H. G. Ramjit, S. D. Lewis, B. J. lucas, S. J. Gardell, E. A. Lyle, S. D. Appleby, J. J. Cook, M. A. Holahan, M. T. Stranieri, J. J. Lynch, J. H. Lin, J.-W. Chen, K. Vastag, A. M. Naylor-Olsen, and J. P. Vacca, *J. Med. Chem.*, **1998**, *41*, 401.

284. C. Altomare, S. Cellamare, L. Summo, M. Catto, and A. Carotti, *J. Med. Chem.*, **1998**, *41*, 3812.

285. N. Watanabe, Y. Kabasawa, Y. Takase, M. Matsukura, K. Miyazaki, H. Ishihara, K. Kodama, and H. Adachi, *J. Med. Chem.*, **1998**, *41*, 3367.

286. T. Uketa, M. Sugahara, Y. Terakawa, T. Kuroda, K. Wada, A. Nakata, Y. Ohmachi, H. Kikkawa, K. Ikezawa, and K. Naito, *J. Med. Chem.*, **1999**, *42*, 1088.

287. M. S. South and T. L. Jakuboski, *Tetrahedron Lett.*, **1995**, *36*, 5703.

288. A. Roedig, G. Bonse, E. M. Ganns, and H. Heinze, *Tetrahedron*, **1977**, *33*, 2437.

289. R. G. Guy, F. J. Swinbourne, and T. McD. McClymont, *Tetrahedron*, **1978**, *34*, 941.

290. R. K. Anderson, S. D. Carter, and G. W. H. Cheeseman, *Tetrahedron*, **1979**, *35*, 2463.
291. D. E. Ames and D. Bull, *Tetrahedron*, **1981**, *37*, 2489.
292. C. B. Kanner and U. K. Pandit, *Tetrahedron*, **1981**, *37*, 3513.
293. D. E. Ames and D. Bull, *Tetrahedron*, **1982**, *38*, 383.
294. J. Gronowska and H. M. Mokhtar, *Tetrahedron*, **1982**, *38*, 1657.
295. S. Scheibye, A. A. El-Barbary, S.-O. Lawesson, H. Fritz, and G. Rihs, *Tetrahedron*, **1982**, *38*, 3753.
296. M. F. Ismail, F. H. El-Bassiouny, and H. A. Young, *Tetrahedron*, **1984**, *40*, 2983.
297. M. H. Elnagdi, F. M. Abdelrazek, N. S. Ibrahim, and A. W. Erian, *Tetrahedron*, **1989**, *45*, 3597.
298. A. Csámpai, K. Körmendy, P. Sohár, and F. Ruff, *Tetrahedron*, **1989**, *45*, 5539.
299. A. Csámpai, K. Körmendy, R. Sohár, and F. Ruff, *Tetrahedron*, **1990**, *46*, 6895.
300. N. Haider, *Tetrahedron*, **1991**, *47*, 3959.
301. A. Csámpai, K. Körmendy, and F. Ruff, *Tetrahedron*, **1991**, *47*, 4457.
302. F. P. J. T. Rutjes, N. M. Teerhuis, H. Hiemstra, and W. N. Speckamp, *Tetrahedron*, **1993**, *49*, 8605.
303. F. Fariña, M. V. Martín, and M. Romañach, *Tetrahedron*, **1994**, *50*, 5169.
304. T. Itoh, H. Hasegawa, K. Nagata, M. Okada, and A. Ohsawa, *Tetrahedron*, **1994**, *50*, 13089.
305. G. Grassi, F. Risitano, and F. Foti, *Tetrahedron*, **1995**, *51*, 11855.
306. H. Al-Awadhi, F. Al-Omran, M. H. Elnagdi, L. Infantes, C. Foces-Foces, N. Jagerovic, and J. Elguero, *Tetrahedron*, **1995**, *51*, 12745.
307. A. Turck, N. Plé, V. Tallon, and G. Quéguiner, *Tetrahedron*, **1995**, *51*, 13045.
308. A. Szabó, A. Csámpai, K. Körmendy, and Z. Böcskei, *Tetrahedron*, **1997**, *53*, 7021.
309. I. Torrini, G. P. Zecchini, M. P. Paradisi, G. Lucente, G. Mastropietro, E. Gavuzzo, F. Mazza, and G. Pochetti, *Tetrahedron*, **1998**, *54*, 165.
310. I. Torrini, G. P. Zecchini, M. P. Paradisi, G. Mastropietro, G. Lucente, E. Gavuzzo, and F. Mazza, *Tetrahedron*, **1999**, *55*, 2077.
311. V. G. Chapoulaud, I. Salliot, N. Plé, A. Turck, and Quéguiner, *Tetrahedron*, **1999**, *55*, 5389.
312. T. Böhm, K. Elender, P. Riebel, T. Troll, A. Weber, and J. Sauer, *Tetrahedron*, **1999**, *55*, 9515.
313. R. B. Brundrett, D. F. Roswell, and E. H. White, *J. Am. Chem. Soc.*, **1972**, *94*, 7536.
314. C. R. Flynn and J. Michl, *J. Am. Chem. Soc.*, **1973**, *95*, 5802.
315. K. B. Tomer, N. Harrit, I. Rosenthal, O. Buchardt, P. L. Kumier, and D. Creed, *J. Am. Chem. Soc.*, **1973**, *95*, 7402.
316. R. B. Brundrett and E. H. White, *J. Am. Chem. Soc.*, **1974**, *96*, 7497.
317. J. A. Zoltewicz and J. H. O'Halloran, *J. Am. Chem. Soc.*, **1975**, *97*, 5531.
318. W. H. Hersh and R. G. Bergman, *J. Am. Chem. Soc.*, **1983**, *105*, 5846.
319. J. Lind, G. Merényi, and T. E. Eriksen, *J. Am. Chem. Soc.*, **1983**, *105*, 7655.
320. C. R. Flynn and J. Michl, *J. Am. Chem. Soc.*, **1974**, *96*, 3280.
321. M. J. Cook, A. R. Katritzky, and A. D. Page, *J. Am. Chem. Soc.*, **1977**, *99*, 165.
322. Y. Cohen, A. Y. Meyer, and M. Rabinovitz, *J. Am. Chem. Soc.*, **1986**, *108*, 7039.
323. G. Merényi, J. Lind, and T. E. Eriksen, *J. Am. Chem. Soc.*, **1986**, *108*, 7716.
324. A. L. G. Beckwith, S. Wang, and J. Warkentin, *J. Am. Chem. Soc.*, **1987**, *109*, 5289.
325. S. F. Wang, L. Mathew, and J. Wartentin, *J. Am. Chem. Soc.*, **1988**, *110*, 7235.
326. R. Meissner, X. Garcias, S. Mecozzi, and J. Rebek, *J. Am. Chem. Soc.*, **1997**, *119*, 77.
327. J. Kang, G. Hilmersson, J. Santamaria, and J. Rebek, *J. Am. Chem. Soc.*, **1998**, *120*, 3650.
328. M. Suárez, S.-M. Lehn, S. C. Zimmerman, A. Skoulios, and B. Heinrich, *J. Am. Chem. Soc.*, **1998**, *120*, 9526.
329. G. Fischer and M. Puza, *Synthesis*, **1973**, 218.
330. Y. Maki and T. Furuta, *Synthesis*, **1976**, 263.

331. B. C. Uff and R. S. Budhram, *Synthesis*, **1978**, 206.
332. Y. Kikugawa, K. Saito, and S. Yamada, *Synthesis*, **1978**, 447.
333. D. E. Ames, O. T. Leung, and A. G. Singh, *Synthesis*, **1983**, 52.
334. P. Del Buttero, E. Licandro, S. Maiorana, and A. Papagni, *Synthesis*, **1986**, 1059.
335. C. Baldoli, E. Licandro, S. Maiorana, E. Merita, and A. Papagni, *Synthesis*, **1987**, 288.
336. A. M. Bernard, M. T. Cocco, C. Congiu, V. Onnis, and P. P. Piras, *Synthesis*, **1998**, 317.
337. F. Csende, Z. Szabó, G. Bernáth, and G. Stájer, *Synthesis*, **1995**, 1240.
338. D. E. Ames, D. Bull, and C. Takundwa, *Synthesis*, **1981**, 364.
339. K. D. Deodhar and S. D. Deval, *Synthesis*, **1983**, 421.
340. J. Sauer, D. K. Heldmann, J. Hetzenegger, J. Krauthan, H. Sichert, and J. Schuster, *Eur. J. Org. Chem.*, **1998**, 2885.
341. B. A. Frontana-Uriba, C. Moinet, and L. Toupet, *Eur. J. Org. Chem.*, **1999**, 419.
342. M. A. Salem, A. M. El-Gendy, and S. I. El-Nagdy, *Rev. Roum. Chim.*, **1989**, *34*, 1963; *Chem. Abstr.*, **1991**, *114*, 6413.
343. S. Robev, *Dokl. Bolg. Akad. Nauk*, **1989**, *42*, 47; *Chem. Abstr.*, **1991**, *114*, 42695.
344. M. A. Badawy, K. Abou-Hadeed, and S. A. Abdel-Hady, *Sulfur Lett.*, **1990**, *12*, 1; *Chem. Abstr.*, **1991**, *114*, 122235.
345. H. Horn, E. Morgenstern, and K. Unverferth, *Pharmazie*, **1990**, *45*, 724; *Chem. Abstr.*, **1991**, *114*, 122240.
346. S. A. H. El-Feky, Z. K. M. Abd El-Samii, M. I. Jaeda, and A. A. I. El-Zurkany, *Mansoura J. Pharm. Sci.*, **1990**, *6*, 30; *Chem. Abstr.*, **1991**, *114*, 164133.
347. E. M. M. Kassem, M. M. Kamel, and N. Y. Khir-Eldin, *Pak. J. Sci. Ind. Res.*, **1990**, *33*, 315; *Chem. Abstr.*, **1991**, *114*, 185415.
348. A. Y. Soliman, H. M. Bakeer, M. A. Sayed, I. Islam, and A. A. Mohamed, *Chin. J. Chem.*, **1990**, 549; *Chem. Abstr.*, **1991**, *114*, 247220.
349. E. Roeder and B. Abd El-Fattah, *Chung-hua Yao Hsueh Tsa Chih*, **1991**, *43*, 49; *Chem. Abstr.*, **1991**, *115*, 29230.
350. M. A. E. Shaban, M. A. M. Taha, and A. Z. Nasr, *Pharmazie*, **1991**, *46*, 105; *Chem. Abstr.*, **1991**, *115*, 29231.
351. M. A. Sayed, I. Islam, A. A. Mohamed, A. Y. Soliman, and H. M. Bakeer, *Chin. J. Chem.*, **1991**, *9*, 45; *Chem. Abstr.*, **1991**, *115*, 49575.
352. S. Bobev and Z. Kostadinova, *Dokl. Bolg. Akad. Nauk*, **1990**, *43*, 65; *Chem. Abstr.*, **1991**, *115*, 49577.
353. S. Robev, Z. Kostadinova, and M. Dicheva, *Dokl. Bolg. Akad. Nauk*, **1990**, *43*, 53; *Chem. Abstr.*, **1991**, *115*, 49578.
354. S. Ito, H. Azuma, N. Funayama, T. Kubota, S. Sekizaki-Ohyama, and M. Ishikawa, *Iyo Kizai Kenkyusho Hokoku, (Tokyo Ika Shika Daigaku)*, **1990**, *24*, 41; *Chem. Abstr.*, **1991**, *115*, 49579.
355. S. El-Feky, M. I. Ashmawi, E. Roeder, and B. Abd El-Fattah, *Chung-hua Yao Hsueh Tsa Chih*, **1991**, *43*, 139; *Chem. Abstr.*, **1991**, *115*, 71516.
356. B. Abd El-Fattah, M. I. Al-Ashmawi, S. El-Feky, and E. Roeder, *Chung-hua Yao Hsueh Tsa Chih*, **1991**, *43*, 129; *Chem. Abstr.*, **1991**, *115*, 92191.
357. M. F. Ismail and N. G. Kandile, *Acta Chim. Hung.*, **1991**, *128*, 251; *Chem. Abstr.*, **1991**, *115*, 92197.
358. M. Caprosu, E. Stefanescu, M. Petrovanu, and C. Radu, *Rev. Med.-Chir.*, **1989**, *93*, 357; *Chem. Abstr.*, **1991**, *115*, 136031.
359. J. Matsumoto and T. Miyamoto, *Quinolones*, **1989**, 109; *Chem. Abstr.*, **1991**, *115*, 228109.
360. S. Groszkowski and A. Stanczak, *Acta Pol. Pharm.*, **1989**, *46*, 320; *Chem. Abstr.*, **1991**, *115*, 232158.

361. M. M. Maslova, N. B. Marchenko, and R. G. Glushkov, *Khim.-Farm. Zh.*, **1991**, *25*, 62; *Chem. Abstr.*, **1991**, *115*, 279937.

362. M. Caprosu, E. Stefanescu, M. Ungureanu, and M. Petrovanu, *Rev. Roum. Chim.*, **1991**, *36*, 657; *Chem. Abstr.*, **1992**, *116*, 235554.

363. M. A. El-Safty, E. El-Bahaei, M. A. El-Hashash, A. El-Faragy, and F. Yasin, *Rev. Roum. Chim.*, **1991**, *36*, 187; *Chem. Abstr.*, **1992**, *116*, 235555.

364. A. F. A. Harb, *Bull. Fac. Sci. Assiut Univ.*, **1991**, *20*, 65; *Chem. Abstr.*, **1992**, *116*, 235556.

365. W. Czuba and D. Sedzik-Hibner, *Univ. Iagellon. Acta Chim.*, **1991**, *35*, 57; *Chem. Abstr.*, **1992**, *117*, 48458.

366. I. M. El-Deen, *Pak. J. Sci. Ind. Res.*, **1991**, *34*, 483; *Chem. Abstr.*, **1992**, *117*, 69751.

367. M. S. Amine, *Asian J. Chem.*, **1992**, *4*, 865; *Chem. Abstr.*, **1992**, *117*, 191787.

368. J. L. Mokrosz, B. Duszyńska, S. Charakchieva-Minol, A. J. Bojarski, M. J. Mokrosz, R. L. Wydra, L. Janda, and L. Strekowski, *Eur. J. Med. Chem.*, **1996**, *31*, 973.

369. G. Strappaghetti, S. Corsano, R. Barbaro, A. Lucacchini, G. Giannicccini, and L. Betti, *Eur. J. Med. Chem.*, **1998**, *33*, 501.

370. S. Kanoktanaporn and J. A. H. MacBride, *J. Chem. Res.*, **1980**, *Synop.* 206, *Minipr.* 2941.

371. A. Katritzky, N. E. Grzeskowiak, T. Siddiqui, C. Jayaram, and S. N. Vassilatos, *J. Chem. Res.*, **1982**, *Synop.* 26, *Minipr.*, 528.

372. R. N. Butler and J. P. James, *J. Chem. Res.*, **1982**, *Synop.* 248.

373. M. T. Williams, *J. Chem. Res.*, **1986**, *Synop.* 136, *Minipr.* 1185.

374. B. C. Uff, R. S. Budhram, G. Ghaem-Maghami, A. S. Mallard, V. Harutunian, S. Calinghen, N. Choudhury, J. Kant, and F. D. Popp, *J. Chem. Res.*, **1986**, *Synop.* 206, *Minipr.* 1901.

375. R. N. Butler, A. M. Gillan, J. P. James, P. McArdle, and D. Cunningham, *J. Chem. Res.*, **1989**, *Synop.* 62, *Minipr.* 523.

376. B. C. Uff, D. L. W. Burford, and Y.-P. Ho, *J. Chem. Res.*, **1989**, *Synop.* 386.

377. B. C. Uff, Y.-P. Ho, F. Hussain, and M. S. Haji, *J. Chem. Res.*, **1989**, *Synop.* 24.

378. M. H. Elnagdi, A. W. Erian, K. U. Sakek, and M. A. Selim, *J. Chem. Res.*, **1990**, *Synop.* 148, *Minipr.* 1124.

379. M. H. Elnagdi, S. A. S. Shozlan, F. M. Abdelrazek, and M. A. Selim, *J. Chem. Res.*, **1991**, *Synop.* 116, *Minipr.* 1021.

380. M. H. Elnagdi, A. M. Negm, E. M. Hassan, and A. El-Boreiy, *J. Chem. Res.*, **1993**, *Synop.* 130.

381. F. A. Abu-S-anab, B. Wakefield, F. Al-Omran, N. M. Abdel-Khalek, and M. H. Elnagdi, *J. Chem. Res.*, **1995**, *Synop.* 488, *Minipr.* 2924.

382. R. N. Butler, D. M. Farrell, and C. S. Pyne, *J. Chem. Res.*, **1996**, *Synop.* 418.

383. A. A. Nada, A. W. Erian, N. R. Mohamed, and A. M. Mahran, *J. Chem. Res.*, **1997**, *Synop.* 236, *Minipr.* 1576.

384. I. S. A. Hafiz, M. E. E. Rashad, M. A. E. Mahfouz, and M. H. Elnagdi, *J. Chem. Res.*, **1998**, *Synop.* 690, *Minipr.* 2946.

385. A. Arcoleo, M. C. Natoll, and M. L. Marino, *Synth. Commun.*, **1975**, *5*, 207.

386. E. Bellasio, *Synth. Commun.*, **1976**, *6*, 85.

387. H. A. Dowlatshahi, *Synth. Commun.*, **1985**, *15*, 1095.

388. S. F. Vasilevsky, E. V. Tretyakov, and H. D. Verkruijsse, *Synth. Commun.*, **1994**, *24*, 1733.

389. T. Klindert and G. Seitz, *Synth. Commun.*, **1996**, *26*, 2587.

390. W. M. Abdou, *Synth. Commun.*, **1997**, *27*, 3599.

391. J. Z. Brzeziński, J. Epsztajn, A. D. Bakalarz, A. Lajszczak, and Z. Malinowski, *Synth. Commun.*, **1999**, *29*, 457.

392. G. Tarzia, E. Occelli, N. Corsico, L. Gallico, F. Luzzani, and D. Barone, *Farmaco*, **1989**, *44*, 17; *Chem. Abstr.*, **1990**, *112*, 30473.

393. T. Kametani, T. Kohno, K. Kigasawa, M. Hiragi, K. Wakisaka, and T. Uryu, *Chem. Pharm. Bull.*, **1971**, *19*, 1794.

394. T. Jojima, H. Takeshiba, and T. Konotsune, *Chem. Pharm. Bull.*, **1972**, *20*, 2191.

395. Y. Oka, K. Omura, A. Miyake, K. Itoh, M. Tomimoto, N. Tada, and S. Yurugi, *Chem. Pharm. Bull.*, **1975**, *23*, 2239.

396. K. Ueno, R. Moroi, M. Kitagawa, K. Asano, and S. Miyazaki, *Chem. Pharm. Bull.*, **1976**, *24*, 1068.

397. T. Jojima, H. Takeshiba, and T. Kinoto, *Chem. Pharm. Bull.*, **1976**, *24*, 1581.

398. T. Jojima, H. Takeshiba, and T. Kinoto, *Chem. Pharm. Bull.*, **1976**, *24*, 1588.

399. E. Hayashi, M. Iinuma, I. Utsunomiya, C. Iijima, E. Ôishi, and T. Higashino, *Chem. Pharm. Bull.*, **1977**, *25*, 579.

400. K. Yamazaki, R. Moroi, and M. Sano, *Chem. Pharm. Bull.*, **1977**, *25*, 1147.

401. M. Uchida, T. Higashino, and E. Hayashi, *Chem. Pharm. Bull.*, **1977**, *25*, 2225.

402. S. Kamiya, S. Sueyoshi, M. Miyahara, K. Yanagmachi, and T. Nakashima, *Chem. Pharm. Bull.*, **1980**, *28*, 1485.

403. M. Somei, K. Kato, and S. Inoue, *Chem. Pharm. Bull.*, **1980**, *28*, 2515.

404. Y. Eguchi, A. Sugimoto, and M. Ishikawa, *Chem. Pharm. Bull.*, **1980**, *28*, 2763.

405. M. Ishikawa, Y. Eguchi, and A. Sugimoto, *Chem. Pharm. Bull.*, **1980**, *28*, 2770.

406. M. Kuzuya, T. Usui, S. Ito, F. Miyaki, S. Nozawa, and T. Okuda, *Chem. Pharm. Bull.*, **1980**, *28*, 3561.

407. H. Yamanaka, T. Shiraishi, T. Sakamoto, and H. Matsuda, *Chem. Pharm. Bull.*, **1981**, *29*, 1049.

408. H. Yamanaka, T. Shiraishi, and T. Sakamoto, *Chem. Pharm. Bull.*, **1981**, *29*, 1056.

409. M. Kuzuya, T. Usui, F. Miyaki, S. Ito, S. Nosawa, and T. Okuda, *Chem. Pharm. Bull.*, **1981**, *29*, 2856.

410. M. Kuzuya, F. Miyaki, and T. Okuda, *Chem. Pharm. Bull.*, **1983**, *31*, 791.

411. A. Sugimoto, K. Sakamoto, Y. Fujino, Y. Takashima, and M. Ishikawa, *Chem. Pharm. Bull.*, **1985**, *33*, 2809.

412. T. Higashino, M. Takemoto, K. Tanji, C. Iijima, and E. Hayashi, *Chem. Pharm. Bull.*, **1985**, *33*, 4193.

413. E. Sato, Y. Ikeda, and Y. Kanaoka, *Chem. Pharm. Bull.*, **1987**, *35*, 3641.

414. T. Miyamoto and J. Matsumoto, *Chem. Pharm. Bull.*, **1988**, *36*, 1321.

415. S. Ito, Y. Komoda, S. Sekizaki, H. Azuma, and M. Ishikawa, *Chem. Pharm. Bull.*, **1988**, *36*, 2669.

416. T. Miyamoto and J. Matsumoto, *Chem. Pharm. Bull.*, **1989**, *37*, 93.

417. E. Sato, Y. Ikeda, and Y. Kanaoka, *Chem. Pharm. Bull.*, **1990**, *38*, 1205.

418. K. Sasamoto and Y. Ohkura, *Chem. Pharm. Bull.*, **1990**, *38*, 1323.

419. I. Takeuchi, Y. Hamada, K. Hatano, Y. Kurono, and T. Yashiro, *Chem. Pharm. Bull.*, **1990**, *38*, 1504.

420. J. M. Chen, S. L. Xu, Z. Wawrzak, G. S. Basarab, and D. B. Jordan, *Biochemistry*, **1998**, *37*, 17735.

421. Y. Nomoto, H. Obasi, H. Takai, M. Terqnishi, J. Nakamura, and K. Kubo, *Chem. Pharm. Bull.*, **1990**, *38*, 2179.

422. E. Oishi, N. Taido, K. Iwamoto, A. Miyashita, and T. Higashino, *Chem. Pharm. Bull.*, **1990**, *38*, 3268.

423. H. Arakawa, J. Ishida, M. Yamaguchi, and M. Nakamura, *Chem. Pharm. Bull.*, **1990**, *38*, 2491.

424. Y. Eguchi, F. Sasaki, Y. Takashima, M. Nakajima, and M. Ishikawa, *Chem. Pharm. Bull.*, **1991**, *39*, 795.

425. Y. Eguchi and M. Ishikawa, *Chem. Pharm. Bull.*, **1991**, *39*, 1846.
426. Y. Eguchi, Y. Sato, S. Sekizaki, and M. Ishikawa, *Chem. Pharm. Bull.*, **1991**, *39*, 2009.
427. L. Compayo, F. Gomez-Contretas, and P. Navarro, *Chem. Pharm. Bull.*, **1992**, *40*, 2905.
428. S. Ito, K. Yamaguchi, and Y. Komoda, *Chem. Pharm. Bull.*, **1992**, *40*, 3327.
429. K. Iwamoto, S. Suzuki, E. Bishi, K. Tanji, A. Miyashita, and T. Higashino, *Chem. Pharm. Bull.*, **1994**, *42*, 413.
430. M. Yamaguchi, T. Koga, K. Kamei, M. Akima, T. Kuroki, M. Hamana, and N. Ohi, *Chem. Pharm. Bull.*, **1994**, *42*, 1601.
431. T. Itoh, H. Hasegawa, K. Nagata, Y. Matsuya, M. Okada, and A. Ohsawa, *Chem. Pharm. Bull.*, **1994**, *42*, 1768.
432. M. Yamaguchi, T. Koga, K. Kamei, M. Akima, N. Maruyama, T. Kuroki, M. Hamana, and N. Ohi, *Chem. Pharm. Bull.*, **1994**, *42*, 1850.
433. Y. Satoh, T. Matsuo, H. Sogabe, H. Itoh, T. Tada, T. Kinoshita, K. Yoshida, and T. Takaya, *Chem. Pharm. Bull.*, **1994**, *42*, 2071.
434. M. Yamaguchi, N. Maruyama, T. Koga. K. Kamei, M. Akima, T. Kuroki, M. Hamana, and N. Ohi, *Chem. Pharm. Bull.*, **1995**, *43*, 332.
435. K. Iwamoto, S. Suzuki, E. Oishi, K. Tanji, A. Miyashita, and T. Higashino, *Chem. Pharm. Bull.*, **1995**, *43*, 679.
436. H. Harada, T. Morie, Y. Hirokawa, H. Terauchi, I. Fujiwara, N. Yoshida, and S. Kato, *Chem. Pharm. Bull.*, **1995**, *43*, 1912.
437. M. Sugahara and T. Ukita, *Chem. Pharm. Bull.*, **1997**, *45*, 719.
438. Y. Suzuki, Y. Takamura, K. Iwamoto, T. Higashima, and A. Miyashita, *Chem. Pharm. Bull.*, **1998**, *46*, 199.
439. H. Horino, T. Mimura, K. Kagechika, M. Ohta, H. Kubo, and M. Kitagawa, *Chem. Pharm. Bull.*, **1998**, *46*, 602.
440. B. I. Buzykin, N. N. Bystrykh, and P. Kitaev, *Zh. Org. Khim.*, **1975**, *11*, 1570.
441. O. N. Chupakhin, E. O. Sidorov, and E. G. Kovalev, *Zh. Org. Khim.*, **1977**, *13*, 204.
442. K. N. Zelenin, Z. M. Matveeva, and V. N. Verbov, *Zh. Org. Khim.*, **1977**, *13*, 1340.
443. K. N. Zelenin, V. N. Verbov, and Z. M. Matveeva, *Zh. Org. Khim.*, **1979**, *15*, 2370.
444. N. V. L'vova, I. V. Samartseva, and L. A. Pavlova, *Zh. Org. Khim.*, **1980**, *16*, 2340.
445. N. V. L'vova, I. V. Samartseva, and L. A. Pavlova, *Zh. Org. Khim.*, **1981**, *17*, 2235.
446. N. V. L'vova, I. V. Samartseva, and L. A. Pavlova, *Zh. Org. Khim.*, **1982**, *18*, 1071.
447. N. F. Volynets, I. V. Samartseva, I. V. Khramova, and L. A. Pavlova, *Zh. Org. Khim.*, **1982**, *18*, 1533.
448. M. A. Kuznetsov and A. A. Suvarov, *Zh. Org. Khim.*, **1983**, *19*, 2577.
449. I. G. Ovchinnikova, I. V. Samartseva, and L. A. Pavlova, *Zh. Org. Khim.*, **1984**, *20*, 2248.
450. V. N. Lisitsyn, A. G. Borovkov, A. B. Kudryavtsev, and M. Mansur, *Zh. Org. Khim.*, **1985**, *21*, 1556.
451. A. A. Suvorov, M. A. Kuznetsov, and V. V. Zuev, *Zh. Org. Khim.*, **1985**, *21*, 2126.
452. V. Y. Komarov, N. F. Volynets, I. V. Samartseva, L. A. Pavlova, and B. I. Ionin, *Zh. Org. Khim.*, **1985**, *21*, 2445.
453. V. V. Shevchenko, G. A. Vasil'euskaya, N. S. Klimenko, S. N. Loshkareva, and T. S. Khramova, *Zh. Org. Khim.*, **1986**, *22*, 74.
454. I. G. Ovchinnikova, I. V. Samartseva, and L. A. Pavlova, *Zh. Org. Khim.*, **1986**, *22*, 1089.
455. I. G. Ovchinnikova, I. V. Samartseva, and L. A. Pavlova, *Zh. Org. Khim.*, **1986**, *22*, 2238.
456. Y. V. Shurukhin, N. A. Klyuev, and I. I. Grandberg, *Zh. Org. Khim.*, **1987**, *23*, 1063.
457. A. A. Suvorov, M. Y. Malov, and M. A. Kuznetsov, *Zh. Org. Khim.*, **1987**, *23*, 1962.
458. I. G. Ovchinnikova, I. V. Samartseva, and L. A. Pavlova, *Zh. Org. Khim.*, **1987**, *23*, 2607.
459. S. V. Yakovlev and L. A. Pavlova, *Zh. Org. Khim.*, **1988**, *24*, 2433.

460. N. M. Przheval'skii and V. N. Drozd, *Zh. Org. Khim.*, **1990**, *16*, 214.

461. I. G. Ovchinnikova, I. V. Samartseva, and L. A. Pavlova, *Zh. Org. Khim.*, **1990**, *26*, 1330.

462. I. G. Ovchinnikova, I. V. Samartseva, and L. A. Pavlova, *Zh. Org. Khim.*, **1990**, *26*, 2619.

463. D. A. Oparin, V. S. Yakovlev, D. B. Motovilin, and V. A. Galishev, *Zh. Org. Khim.*, **1992**, *28*, 1317.

464. D. A. Oparin, *Zh. Org. Khim.*, **1993**, *29*, 418.

465. V. N. Veshchula, A. N. Vdevichenko, A. V. Krivoruchko, and V. I. Dulenko, *Zh. Org. Khim.*, **1996**, *32*, 591.

466. O. P. Shvaika, N. A. Kovach, O. N. Chupakhin, and V. N. Charushin, *Zh. Org. Khim.*, **1996**, *32*, 941.

467. E. M. M. Kassem, *Proc. Pak. Acad. Sci.*, **1991**, *28*, 225; *Chem. Abstr.*, **1993**, *118*, 59640.

468. S. Elkousy, A. M. El-Torgoman, M. R. Abdel-Azim, and H. A. El-Fahham, *Egypt. J. Pharm. Sci.*, **1991**, *32*, 851; *Chem. Abstr.*, **1993**, *118*, 59657.

469. A. A. El-Sawy, S. G. Donia, S. A. Essawy, and A. M. F. Eissa, *Egypt. J. Chem.*, **1990**, *33*, 13; *Chem. Abstr.*, **1993**, *118*, 124476.

470. F. A. Yassin, B. E. Bayoumy, M. A. El-Safty, and A. F. El-Farargy, *Egypt. J. Chem.*, **1990**, *33*, 199; *Chem. Abstr.*, **1993**, *118*, 124477.

471. S. El-Bahaie, M. A. El-Safty, and F. Yassin, *Egypt. J. Chem.*, **1990**, *33*, 381; *Chem. Abstr.*, **1993**, *118*, 147524.

472. B. E. Bayoumy, S. A. El-Feky, and M. El-Mobayed, *Egypt. J. Chem.*, **1990**, *33*, 267; *Chem. Abstr.*, **1993**, *118*, 169038.

473. A. F. El-Kafrawy, S. A. Shiba, E. S. Donia, and M. A. El-Hashash, *J. Chem. Soc. Pak.*, **1992**, *14*, 132; *Chem. Abstr.*, **1993**, *118*, 169058.

474. M. E. Hassan and M. A. Parsy, *Aswan Sci. Technol. Bull.*, **1992**, *13*, 33; *Chem. Abstr.*, **1993**, *119*, 28083.

475. W. Kwapiszwski and A. Stanczak, *Acta Pol. Pharm.*, **1992**, *49*, 53; *Chem. Abstr.*, **1994**, *120*, 217502.

476. W. Kwapiszewski, A. Stanczak, and S. Groszkowski, *Acta Pol. Pharm.*, **1992**, *49*, 47; *Chem. Abstr.*, **1994**, *120*, 217501.

477. T. Otto-Jaworska, *Acta Pol. Pharm.*, **1992**, *49*, 81; *Chem. Abstr.*, **1994**, *120*, 298568.

478. O. Morgenstern, M. Meusel, and P. H. Richter, *Pharmazie*, **1994**, *49*, 489; *Chem. Abstr.*, **1994**, *121*, 108710.

479. O. Morgenstern, M. Meusel, S. Denke, J. Vepsaelaeinen, and P. H. Richter, *Pharmazie*, **1994**, *49*, 419; *Chem. Abstr.*, **1994**, *121*, 157602.

480. M. L. Główka, I. Iwanicka, and A. Stańczak, *J. Chem. Crystallogr.* **1994**, *24*, 397.

481. N. Yoshikuni, *Talanta*, **1992**, *39*, 805.

482. S. Rádl, L. Kovárová, P. Hazky, V. Vosátka, O. Königová, J. Proška, and I. Krejci, *Arch. Pharm. (Weinheim)*, **1999**, *332*, 208.

483. E. S. H. El-Ashry, A. A.-H. Abdel-Rahman, N. Rashed, and H. A. Rasheed, *Arch. Pharm. (Weinheim)*, **1999**, *332*, 327.

484. S. Petersen and H. Heitzer, *Liebigs Ann. Chem.*, **1978**, 283.

485. R. Albrecht, *Liebigs Ann. Chem.*, **1978**, 617.

486. C. Rüchardt and V. Hassmann, *Liebigs Ann. Chem.*, **1980**, 908.

487. K. Gewald, O. Calderon, H. Schäfer, and U. Hain, *Liebigs Ann. Chem.*, **1984**, 1390.

488. L. Somogyi, *Liebigs Ann. Chem.*, **1985**, 1679.

489. M. H. Elnagdi, N. S. Ibrahim, K. U. Sadek, and M. S. Mohamed, *Liebigs Ann. Chem.*, **1988**, 1005.

490. M. H. Elnagdi, A. M. Negm, and A. W. Erian, *Liebigs Ann. Chem.*, **1989**, 1255.

491. R. Hoferichter and G. Seitz, *Liebigs Ann. Chem.*, **1992**, 1153.

492. S. F. Vasilevsky and E. V. Tretyakov, *Liebigs Ann. Chem.*, **1995**, 775.

493. V. J. Arán, M. Flores, P. Muñoz, J. R. Ruiz, P. Sánchez-Verdú, and M. Stud, *Liebigs Ann. Chem.*, **1995**, 817.

494. B. Wünsch, S. Nerdinger, and G. Höfner, *Liebigs Ann. Chem.*, **1995**, 1303.

495. M. Knaack, I. Fleischhauer, P. Emig, B. Kutscher, and A. Müller, *Liebigs Ann. Chem.*, **1996**, 1477.

496. F. Saczewski and M. Gdaniec, *Liebigs Ann. Chem.*, **1996**, 1673.

497. P. S. Fernandes, V. V. Nadkarny, G. A. Jabbar, and R. P. Jani, *J. Indian Chem. Soc.*, **1975**, *52*, 546.

498. A. C. Desai and C. M. Desai, *J. Indian Chem. Soc.*, **1979**, *56*, 324.

499. A. C. Desai and C. M. Desai, *J. Indian Chem. Soc.*, **1980**, *57*, 757.

500. N. Thankarajan, K. Krishnankutty, and T. K. K. Srinivasan, *J. Indian Chem. Soc.*, **1986**, *63*, 977.

501. H. Joshi, P. Upadhyay, and A. J. Baxi, *J. Indian Chem. Soc.*, **1990**, *63*, 779.

502. R. G. Menon and E. Purushothaman, *J. Indian Chem. Soc.*, **1993**, *70*, 533.

503. R. G. Menon and E. Purushothaman, *J. Indian Chem. Soc.*, **1995**, *72*, 731.

504. M. Z. A. Badr, S. A. Mahgoub, F. M. Atta, and O. S. Moustafa, *J. Indian Chem. Soc.*, **1997**, *74*, 30.

505. R. G. Menon and E. Purushothaman, *J. Indian Chem. Soc.*, **1997**, *74*, 123.

506. J. R. Honner, P. R. Nott, and B. K. Selinger, *Aust. J. Chem.*, **1974**, *27*, 1613.

507. W. I. Awad, M. F. Ismail, and N. G. Kandile, *Aust. J. Chem.*, **1975**, *28*, 1621.

508. L. W. Deady and M. S. Stanborough, *Aust. J. Chem.*, **1981**, *34*, 1295.

509. P. A. Marshall, B. A. Mooney, R. H. Prager, and A. D. Ward, *Aust. J. Chem.*, **1981**, *34*, 2619.

510. D. B. Paul, *Aust. J. Chem.*, **1984**, *37*, 893.

511. D. B. Paul, *Aust. J. Chem.*, **1984**, *37*, 1001.

512. G. B. Barlin and S. J. Ireland, *Aust. J. Chem.*, **1985**, *38*, 1685.

513. G. B. Barlin, *Aust. J. Chem.*, **1986**, *39*, 1803.

514. M. Kač, F. Kovač, B. Stanovnik, and M. Tišler, *Gazz. Chim. Ital.*, **1975**, *105*, 1291.

515. A. Alemagna, P. Del Buttero, E. Licandro, and S. Maiorana, *Gazz. Chim. Ital.*, **1981**, *111*, 285.

516. M. T. Cocco, A. Maccioni, and A. Plumitallo, *Gazz. Chim. Ital.*, **1984**, *114*, 521.

517. N. Bonanomi and L. Baiocchi, *Gazz. Chim. Ital.*, **1989**, *119*, 445.

518. T. Benincori, R. Fusco, and F. Sannicolo, *Gazz. Chim. Ital.*, **1990**, *120*, 635.

519. M. F. Ismail, E. I. Enayat, F. A. A. El-Bassiouny, and H. A. Younes, *Gazz. Chim. Ital.*, **1990**, *120*, 677.

520. R. M. Mohareb, A. M. El-Torgoman, S. I. Aziz, S. M. El-Kousy, and M. Riad, *Gazz. Chim. Ital.*, **1992**, *122*, 503.

521. A. M. Islam, I. B. Hannout, L. M. Souka, A. M. Naser, A. A. El-Maghraby, and I. E. Islam, *J. Prakt. Chem.*, **1973**, *315*, 1025.

522. A. M. Islam, I. B. Hannout, and A. M. El-Sharief, *J. Prakt. Chem.*, **1975**, *317*, 567.

523. M. F. Islam, F. A. Bassiouny, E. I. Enayat, A. M. Kaddah, and H. A. Younes, *J. Prakt. Chem.*, **1985**, *327*, 177.

524. M. M. Abbasi, M. Abou-Sekkina, Y. Hafez, and H. H. Zoorob, *J. Prakt. Chem.*, **1986**, *328*, 932.

525. A. H. Bedair and R. Q. Lamphon, *J. Prakt. Chem.*, **1987**, *329*, 675.

526. A. A. El-Bannany, A. A. El-Assar, and M. H. El-Nagdy, *J. Prakt. Chem.*, **1989**, *331*, 726.

527. A. R. Katritzky, V. Feygelman, G. Musumarra, P. Barczynski, and M. Szafran, *J. Prakt. Chem.*, **1990**, *332*, 870.

528. S. El-Feky and B. E. Bayoumy, *J. Prakt. Chem.*, **1990**, *332*, 1041.

529. M. Makosza and Z. Wróbel, *J. Prakt. Chem./Chem. Ztg.*, **1992**, *334*, 131.

530. A. H. H. Elghandour, A. H. M. Hussein, M. H. Elnagdi, A. F. A. Harb, and S. M. Metwally, *J. Prakt. Chem./Chem. Ztg.*, **1992**, *234*, 723.

531. J. Engel, A. Bork, I. Fleischhauer, B. Kutscher, and M. Lieflånder, *J. Prakt. Chem./Chem. Ztg.*, **1994**, *336*, 207.

532. T. Klindert, P. von Hagel, L. Baumann, and G. Seitz, *J. Prakt. Chem./Chem. Ztg.*, **1997**, *339*, 623.

533. G. Zacharias and H. Junek, *Z. Naturforsch. B, Chem. Sci.*, **1976**, *31*, 371.

534. M. H. Elnagdi, F. A. M. Abdul-Aal, E. A. A. Hafez, and Y. M. Yassin, *Z. Naturforsch. B, Chem. Sci.*, **1989**, *44*, 683.

535. M. H. Elnagdi, F. A. M. Abdul-Aal, N. M. Taha, and Y. M. Yassin, *Z. Naturforsch. B, Chem. Sci.*, **1990**, *45*, 389.

536. J. Laue, G. Seitz, and H. Wassmuth, *Z. Naturforsch. B, Chem. Sci.*, **1996**, *51*, 348.

537. H. A. Y. Derbala, *Afinidad*, **1994**, *51*, 383; *Chem. Abstr.*, **1995**, *122*, 81061.

538. F. M. A. Soliman, L. M. Souka, and N. F. A. El-Gaffar, *Rev. Roum. Chim.*, **1994**, *39*, 827; *Chem. Abstr.*, **1995**, *122*, 81270.

539. M. L. Główka, A. Olczak, and A. Stańczak, *Pol. J. Chem.*, **1994**, *68*, 1631; *Chem. Abstr.*, **1995**, *122*, 227350.

540. H. Kisch, F. Knoch, G. Halbritter, and D. Fenske, *Z. Krystallogr.*, **1995**, *210*, 222.

541. M. F. Ismail, O. E. Mustafa, S. A. Emara, and H. A. M. Sallam, *Egypt. J. Chem.*, **1994**, *37*, 89; *Chem. Abstr.*, **1995**, *123*, 143775.

542. F. K. Mohamed, M. R. Mahmoud, and M. El-Komy, *Egypt. J. Chem.*, **1994**, *37*, 433; *Chem. Abstr.*, **1995**, *123*, 228809.

543. M. A. El-Hashash, F. M. A. Soliman, L. Souka, and N. Abdel-Ghaffar, *Rev. Roum. Chim.*, **1995**, *40*, 173; *Chem. Abstr.*, **1996**, *124*, 8733.

544. A. F. El-Farargy, S. A. El-Rahman, A. E.-F. Z. Haikal, E. A. El-Ghani, and A. A. Shahin, *Zagazig J. Pharm. Sci.*, **1995**, *4*, 76; *Chem. Abstr.*, **1996**, *124*, 86918.

545. M. F. Ismail, O. E. A. Mustafa, S. A. Emara, and H. A. Sallam, *Egypt. J. Chem.*, **1993**, *36*, 497; *Chem. Abstr.*, **1996**, *124*, 202144.

546. V. Nenzel, *Bull. Union Physiciens*, **1995**, *89*, 1819; *Chem. Abstr.*, **1996**, *124*, 232361.

547. A. M. Hussein, A. A. Atalla, I. S. Abdel-Hafez, and M. H. Elnagdi, *Pol. J. Chem.*, **1996**, *70*, 589; *Chem. Abstr.*, **1996**, *125*, 58421.

548. M. G. Assy and F. M. A.-E. Motti, *Egypt. J. Chem.*, **1996**, *39*, 581; *Chem. Abstr.*, **1996**, *125*, 328539.

549. S. Baloniak, A. Mroczkiewicz, and A. Katrusiak, *Pol. J. Chem.*, **1996**, *70*, 1274; *Chem. Abstr.*, **1997**, *126*, 47171.

550. M. L. Główka and A. Stańczak, *Pol. J. Chem.*, **1996**, *70*, 1512; *Chem. Abstr.*, **1997**, *126*, 111328.

551. D. A. Oparin and A. A. Solodunov, *Vesti Akad. Navuk Belarusi, Ser. Khim. Navuk*, **1996**, 121; *Chem. Abstr.*, **1997**, *126*, 117940.

552. W. J. Coates, *Compr. Heterocycl. Chem. II*, **1996**, *6*, 1-91 & 1177-1307.

553. A. M. Hussein, A. A. Atalla, I. S. Abdel-Hafez, and M. H. Elnagdi, *Heteroat. Chem.*, **1997**, *8*, 29.

554. A. Stańczak, W. Lewgowd, W. Pakulska, and A. Szadowska, *Pharmazie*, **1997**, *52*, 91; *Chem. Abstr.*, **1997**, *126*, 171546.

555. M. L. Główka, A. Olczak, D. Martynowski, and A. Staszewska, *Pol. J. Chem.*, **1997**, *71*, 170; *Chem. Abstr.*, **1997**, *126*, 238015.

556. Y. N. Manohara and P. V. Nayak, *Synth. React. Inorg. Met. Org. Chem.*, **1997**, *27*, 589; *Chem. Abstr.*, **1997**, *126*, 324463.

557. C. S. Lee, S. H. Kim, and C.-H. Lee, *J. Korean Chem. Soc.*, **1997**, *41*, 677; *Chem. Abstr.*, **1998**, *128*, 88869.

558. G. T. Crisp, Y.-L. Jiang, and E. R. T. Tiekink, *Z. Krystallogr.—New Cryst. Struct.*, **1998**, *213*, 391.

559. J. B. Sutherland, J. P. Freeman, and A. J. Williams, *Appl. Microbiol. Biotechnol.*, **1998**, *49*, 445.

560. J. B. Sutherland, J. P. Freeman, and A. J. Williams, *Appl. Microbiol. Biotechnol.*, **1998**, *49*, 445.

561. R. D. Rogers, C. V. K. Sharma, and D. R. Whitcomb, *Cryst. Eng.*, **1998**, *1*, 255; *Chem. Abstr.*, **1999**, *130*, 345298.
562. Y. N. Manohara and P. V. Nayak, *Orient. J. Chem.*, **1998**, *15*, 123; *Chem. Abstr.*, **1999**, *131*, 138315.
563. G. Baccolini, C. Boga, and V. Negri, *Heyeroat. Chem.*, **1999**, *10*, 291.
564. B. López, M. Lora-Tamayo, P. Navarro, and J. L. Soto, *Heterocycles*, **1974**, *2*, 649.
565. I. Matsuura, *Heterocycles*, **1975**, *3*, 381.
566. H. Yamanaka, T. Shiraishi, and T. Sakamoto, *Heterocycles*, **1975**, *3*, 1069.
567. H. Yamanaka, T. Shiraishi, and T. Sakamoto, *Heterocycles*, **1975**, *3*, 1075.
568. B. C. Uff and R. S. Budhram, *Heterocycles*, **1977**, *6*, 1789.
569. D. Bhattacharjee and F. D. Popp, *Heterocycles*, **1977**, *6*, 1905.
570. C. B. Kanner and U. K. Pandit, *Heterocycles*, **1978**, *9*, 1381.
571. M. Quinteiro, C. Seoane, and J. L. Soto, *Heterocycles*, **1978**, *9*, 1771.
572. T. Kametani, K. Kigasawa, M. Hiiragi, H. Ishimaru, N. Wagatsuma, T. Kohagisawa, and T. Nakamura, *Heterocycles*, **1980**, *14*, 449.
573. M. Ishikawa, Y. Eguchi, and A. Sugimoto, *Heterocycles*, **1981**, *16*, 21.
574. M. Ishikawa and Y. Eguchi, *Heterocycles*, **1981**, *16*, 25.
575. Z.-T. Jin, K. Imafuku, and H. Matsumura, *Heterocycles*, **1981**, *16*, 935.
576. R. Prager, A. D. Ward, P. Marshall, and B. Mooney, *Heterocycles*, **1982**, *18*, 327.
577. S. Gelin and R. Dolmazon, *Heterocycles*, **1983**, *20*, 61.
578. Y. Ikimi, K. Matsumoto, and T. Uchida, *Heterocycles*, **1983**, *20*, 1009.
579. F. D. Popp and J. Kant, *Heterocycles*, **1985**, *23*, 2193.
580. H. Zimmer and A. Amer, *Heterocycles*, **1987**, *26*, 1177.
581. A. Amer, N. Rashed, M. M. A. Rahman, and H. Zimmer, *Heterocycles*, **1987**, *26*, 1277.
582. Y. Tominaga, Y. Ichihara, and A. Hosomi, *Heterocycles*, **1988**, *27*, 2345.
583. C. F. Tómez-Contreras, M. Lora-Tamayo, and A. M. Sanz, *Heterocycles*, **1989**, *28*, 791.
584. F. De Sio, S. Chimichi, R. Nesi, and L. Cecchi, *Heterocycles*, **1983**, *20*, 1279.
585. D. Villemin and D. Goussu, *Heterocycles*, **1989**, *29*, 1255.
586. H. Yamanaka and S. Ohba, *Heterocycles*, **1990**, *31*, 895.
587. A. Miyashita, T. Kawashima, C. Iijima, and T. Higashino, *Heterocycles*, **1992**, *33*, 211.
588. T. Itoh, H. Hasegawa, K. Nagata, Y. Matsuya, and A. Ohsawa, *Heterocycles*, **1994**, *37*, 709.
589. G. Stájer, A. E. Szabó, G. Bernáth, and P. Sohár, *Heterocycles*, **1994**, *38*, 1061.
590. N. Haider, K. Mereiter, and R. Wanko, *Heterocycles*, **1994**, *38*, 1845.
591. A. Miyashita, Y. Suzuki, K. Ohta, and T. Higashino, *Heterocycles*, **1994**, *39*, 345.
592. S. Lehnhoff and I. Ugi, *Heterocycles*, **1995**, *40*, 801.
593. I. Butula, A. Ruždija, and J. Kalmar, *Croat. Chem. Acta*, **1979**, *52*, 43.
594. N. Celebi, I. Kayalidere, and L. Türker, *Bull. Soc. Chim. Belg.*, **1995**, *104*, 595.
595. S. Nan'ya, H. Tôyama, and E. Maekawa, *Nippon Kagaku Kaishi,* **1982**, 706.
596. T. Itoh, H. Hasegawa, K. Nagata, and A. Ohsawa, *Synlett*, **1994**, 557.
597. C. Saturnino, M. Abarghas, M. Schmitt, C.-G. Wermuth, and J.-J. Bourquignon, *Heterocycles*, **1995**, *41*, 1491.
598. N. Haider, *Heterocycles*, **1995**, *41*, 2519.
599. K. Iwamoto, S. Suzuki, E. Oishi, A. Miyashita, and T. Higashino, *Heterocycles*, **1996**, *43*, 199.
600. F. Csende, G. Bernáth, Z. Böcskei, P. Sohár, and G. Stájer, *Heterocycles*, **1997**, *45*, 323.
601. A. Miyashita, Y. Suzuki, K. Iwamoto, E. Oishi, and T. Higashino, *Heterocycles*, **1998**, *49*, 405.
602. S. Sommer, *Chem. Lett.*, **1977**, 583.

603. Y. Maki, M. Kawamura, H. Okamoto, M. Suzuki, and K. Kaji, *Chem. Lett.*, **1977**, 1005.

604. E. Somei and K. Ura, *Chem. Lett.*, **1978**, 707.

605. M. Somei and Y. Kurizuka, *Chem. Lett.*, **1979**, 127.

606. R. Sano and H. Inoue, *Chem. Lett.*, **1984**, 1901.

607. E. Sato, Y. Ikeda, and Y. Kanaoka, *Chem. Lett.*, **1987**, 273.

608. H. Uno, S. Okada, Y. Shiraisi, K. Shimokawa, and H. Suzuki, *Chem. Lett.*, **1988**, 1165.

609. H. Togo, M. Aoki, and M. Yokoyama, *Chem. Lett.*, **1991**, 1691.

610. S. Wake, Y. Takayama, Y. Otsuji, and E. Imoto, *Bull. Chem. Soc. Jpn.*, **1974**, *47*, 1257.

611. A. M. Islam, A. M. Khalil, and A. A. El-Maghraby, *Bull. Chem. Soc. Jpn.*, **1974**, *47*, 1274.

612. A. F. Fahmy and N. F. Aly, *Bull. Chem. Soc. Jpn.*, **1976**, *49*, 1391.

613. A. Tamaoki and K. Takahashi, *Bull. Chem. Soc. Jpn.*, **1986**, *59*, 1650.

614. H. Inoue, T. Sakurai, K. Tomiyama, R. Sano, M. Oshio, T. Hoshi, and J. Okubo, *Bull. Chem. Soc. Jpn.*, **1988**, *61*, 893.

615. R. Sato, H. Endoh, A. Aba, S. Yamaichi, T. Goto, and M. Saito, *Bull. Chem. Soc. Jpn.*, **1990**, *63*, 1160.

616. Z. Vejdělek, J. Metyš, J. Hallobek, M. Buděšinský, E. Svátek, O. Matoušová, and M. Protiva, *Collect. Czech. Chem. Commun.*, **1988**, *53*, 132.

617. S. Rádl, J. Moural, and R. Bendová, *Collect. Czech. Chem. Commun.*, **1990**, *55*, 1311.

618. S. El-Kousy, I. El-Sakka, A. N. El-Torgoman, H. Roshdy, and M. H. Elnagdi, *Collect. Czech. Chem. Commun.*, **1990**, *55*, 2977.

619. M. S. K. Youssef, A. M. K. El-Dean, M. S. Abbady, and K. M. Hassan, *Collect. Czech. Chem. Commun.*, **1991**, *56*, 1768.

620. M. S. Abbady and A. E.-A. M. Gaber, *Collect. Czech. Chem. Commun.*, **1992**, *57*, 1149.

621. A. El-Farargy, F. Yassin, and T. Hafez, *Collect. Czech. Chem. Commun.*, **1993**, *58*, 1937.

622. J. E. Francis, K. J. Doebel, P. M. Schutte, E. C. Savarese, S. E. Hopkins, and E. F. Bachmann, *Can. J. Chem.*, **1979**, *57*, 3320.

623. J. E. Francis, K. J. Doebel, P. M. Schutte, E. F. Bachmann, and R. E. Detlefsen, *Can. J. Chem.*, **1982**, *60*, 1214.

624. D. V. Bautista, J. C. Dewan, and L. K. Thompson, *Can. J. Chem.*, **1982**, *60*, 2583.

625. G. Bullock, F. W. Hartstock, and L. K. Thompson, *Can. J. Chem.*, **1983**, *61*, 57.

626. L. K. Thompson, F. W. Hartstock, P. Robichaud, and A. W. Hanson, *Can. J. Chem.*, **1984**, *62*, 2755.

627. K. Vaughan, R. J. LaFrance, Y. Tang, and D. L. Hooper, *Can. J. Chem.*, **1985**, *63*, 2455.

628. A. Bélanger, P. Brassard, D. Laquerre, and Y. Mérand, *Can. J. Chem.*, **1987**, *65*, 1392.

629. C. Huiszoon, B. W. van de Waal, A. B. van Egmond, and S. Harkema, *Acta Crystallogr., Sect. B*, **1972**, *28*, 3415.

630. G. J. van Hummel, D. M. W. van den Ham, and C. Huiszoon, *Acta Crystallogr., Sect. B*, **1979**, *35*, 516.

631. G. Yan and P. Coopens, *Acta Crystallogr., Sect. C*, **1987**, *43*, 1610.

632. M. L. Główka and I. Iwanicka, *Acta Crystallogr., Sect. C*, **1991**, *47*, 196.

633. M. L. Główka, I. Iwanicka, and A. Stańczak, *Acta Crystallogr., Sect. C*, **1991**, *47*, 668.

634. M. L. Główka, I. Iwanicka, and A. Olczak, *Acta Crystallogr., Sect. C*, **1994**, *50*, 1770.

635. A. N. Kaushal and K. S. Narang, *Indian J. Chem.*, **1972**, *10*, 675.

636. D. R. Sinha and A. B. Lal, *Indian J. Chem.*, **1973**, *11*, 126.

637. S. N. Bannore and J. L. Bose, *Indian J. Chem.*, **1973**, *11*, 631.

638. A. F. Fahmy and M. A. Elkomy, *Indian J. Chem.*, **1975**, *13*, 652.

639. A. B. Sakla, G. Aziz, S. A. El-Sayed, and N. S. Ibrahim, *Indian J. Chem., Sect. B*, **1976**, *14*, 742.

640. V. P. Arya and S. J. Shenoy, *Indian J. Chem., Sect. B*, **1976**, *14*, 766.
641. L. B. Hannout, A. M. Islam, L. M. Souka, A. A. El-Maghraby, and N. M. Taha, *Indian J. Chem., Sect. B*, **1976**, *14*, 868.
642. A. M. Islam, I. B. Hannout, N. M. Taha, and A. A. El-Maghraby, *Indian J. Chem., Sect. B*, **1977**, *15*, 58.
643. I. B. Hannout, A. M. Islam, A. A. El-Maghraby, and S. A. Ahmed, *Indian J. Chem., Sect. B*, **1977**, *15*, 112.
644. H. Jahine, A. Sayed, H. A. Zaher, and O. Sherif, *Indian J. Chem., Sect. B*, **1977**, *15*, 250.
645. M. A. Elkasaby, M. Elkady, and N. A. N. Eldin, *Indian J. Chem., Sect. B*, **1977**, *15*, 436.
646. A. M. Islam, A. M. S. El-Sherief, and F. A. Ead, *Indian J. Chem., Sect. B*, **1978**, *16*, 50.
647. A. M. Islam, A. M. S. El-Sharief, and A. H. Bedear, *Indian J. Chem., Sect. B*, **1978**, *16*, 491.
648. A. M. Islam, A. M. S. El-Sharief, and A. H. Bedair, *Indian J. Chem., Sect. B*, **1978**, *16*, 593.
649. H. Jahine, H. A. Zaher, Y. Akhnookh, and Z. El-Gendy, *Indian J. Chem., Sect. B*, **1978**, *16*, 689.
650. J. R. Merchant, S. D. Kulkarni, and M. S. Venkatesh, *Indian J. Chem., Sect. B*, **1980**, *19*, 914.
651. R. K. Bansal, S. K. Sharma, and G. Bhagchandani, *Indian J. Chem., Sect. B*, **1982**, *21*, 149.
652. N. A. Shams, A. M. Kaddah, and A. H. Moustafa, *Indian J. Chem., Sect. B*, **1982**, *21*, 317.
653. M. F. Ismail and N. G. Kamdile, *Indian J. Chem., Sect. B*, **1982**, *21*, 462.
654. J. S. Singh, V. Sardana, P. C. Jain, and N. Anand, *Indian J. Chem., Sect. B*, **1983**, *22*, 1083.
655. K. D. Deodhar and S. D. Deval, *Indian J. Chem., Sect. B*, **1985**, *24*, 210.
656. K. Nagarajan, R. K. Shah, and S. J. Shenoy, *Indian J. Chem., Sect. B*, **1986**, *25*, 697.
657. N. Thankarajan and K. Krishnankutty, *Indian J. Chem., Sect. A*, **1987**, *26*, 612.
658. Z. El-Gendy, R. M. Abdel-Rahman, and M. S. Abdel-Malik, *Indian J. Chem., Sect. B*, **1989**, *28*, 479.
659. Z. El-Gendy and R. M. Abdel-Rahman, *Indian J. Chem., Sect. B*, **1989**, *28*, 647.
660. Z. El-Gendy and R. M. Abdel-Rahman, *Indian J. Chem., Sect. B*, **1989**, *28*, 654.
661. M. Razvi, T. Ramalingam, and P. B. Sattur, *Indian J. Chem., Sect. B*, **1989**, *28*, 695.
662. M. Razvi, T. Ramalingam, and P. B. Sattur, *Indian J. Chem., Sect. B*, **1990**, *29*, 399.
663. M. S. Abbady, S. M. Radwan, and E. A. Bakhite, *Indian J. Chem., Sect. B*, **1993**, *32*, 1281.
664. Z. El-Gendy, *Indian J. Chem., Sect. B*, **1994**, *33*, 326.
665. S. Abdel-Rahman, G. A. Latif, M. G. Assy, and A. A. Shahin, *Indian J. Chem., Sect. B*, **1994**, *33*, 491.
666. R. M. Fikry, F. A. Yassin, and E. K. Mohamed, *Indian J. Chem., Sect. B*, **1994**, *33*, 742.
667. L. V. G. Nargund, V. V. Badiger, and S. M. Yarnal, *Indian J. Chem., Sect. B*, **1994**, *33*, 764.
668. F. K. Mohamed, M. R. Mahmoud, and M. El-Komy, *Indian J. Chem., Sect. B*, **1994**, *33*, 769.
669. H. A. Y. Derbala, *Indian J. Chem., Sect. B*, **1994**, *33*, 779.
670. L. V. G. Nargund, Y. N. Manohara, and V. Hariprasad, *Indian J. Chem., Sect. B*, **1994**, *33*, 1001.
671. A. A. F. Wasfy, *Indian J. Chem., Sect. B*, **1994**, *33*, 1098.
672. S. A. El-Abbady and S. M. Agami, *Indian J. Chem., Sect. B*, **1995**, *34*, 504.
673. M. G. Assy and F. M. A.-E. Motti, *Indian J. Chem., Sect. B*, **1996**, *35*, 608.
674. M. K. A. Ibrahim, A. A. H. Eighandour, and B. El-Badry, *Indian J. Chem., Sect. B*, **1997**, *36*, 612.
675. E. M. Kassim, K. N. Abd-El-Nour, A. I. El-Diwany, and T. H. Abou-Aiad, *Indian J. Chem., Sect. B*, **1996**, *35*, 692.
676. El-S. H. El-Tamaty, M. E. Abdel-Fattah, and I. M. El-Deen, *Indian J. Chem., Sect. B*, **1996**, *35*, 1067.
677. R. G. Menon and E. Purushothaman, *Indian J. Chem., Sect. B*, **1996**, *35*, 1185.
678. Z. El-Gendy, *Indian J. Chem., Sect. B*, **1998**, *37*, 342.

679. C. O. Kangani and H. E. Master, *Indian J. Chem., Sect. B*, **1998**, *37*, 778.

680. M. Razvi and T. Ramalingam, *Indian J. Chem., Sect. B*, **1992**, *31*, 788; *Chem. Abstr.*, **1993**, *118*, 180723.

681. M. J. Cook, A. R. Katritzky, A. D. Page, and M. Ramaiah, *J. Chem. Soc., Perkin Trans. 2*, **1977**, 1184.

682. T. C. Woon, R. McDonald, S. K. Mandal, L. K. Thompson, S. P. Connors, and A. W. Addison, *J. Chem. Soc., Dalton Trans.*, **1986**, 2381.

683. M. A. V. Ribero da Silva, M. A. R. Matos, and V. M. F. Moraid, *J. Chem. Soc., Faraday Trans.*, **1995**, *91*, 1907.

684. A. R. Katritzky, B. Terem, E. V. Scriven, S. Clementi, and H. O. Tarhan, *J. Chem. Soc., Perkin Trans. 2*, **1975**, 1600.

685. J. A. Elvidge and A. P. Redman, *J. Chem. Soc., Perkin Trans. 1*, **1972**, 2820.

686. T. Kametani, K. Kigasawa, M. Hiiragi, H. Ishimaru, T. Uryu, and S. Haga, *J. Chem. Soc., Perkin Trans. 1*, **1973**, 471.

687. C. W. Rees and A. A. Sale, *J. Chem. Soc., Perkin Trans. 1*, **1973**, 545.

688. R. D. Chambers, M. Clark, J. A. H. McBride, W. K. R. Musgrave, and K. C. Srivastava, *J. Chem. Soc., Perkin Trans. 1*, **1974**, 125.

689. G. V. Boyd, *J. Chem. Soc., Perkin Trans. 1*, **1973**, 1731.

690. N. Dennis, A. R. Katritzky, and M. Ramaiah, *J. Chem. Soc., Perkin Trans. 1*, **1975**, 1506.

691. T. L. Gilchrist, G. E. Gymer, and C. W. Rees, *J. Chem. Soc., Perkin Trans. 1*, **1975**, 1747.

692. R. F. Cookson and R. E. Rodway, *J. Chem. Soc., Perkin Trans. 1*, **1975**, 1854.

693. D. E. Ames, S. Chandrasekhar, and R. Simpson, *J. Chem. Soc., Perkin Trans. 1*, **1975**, 2035.

694. D. Johnston and D. M. Smith, *J. Chem. Soc., Perkin Trans. 1*, **1976**, 399.

695. D. E. Ames and C. J. A. Byrne, *J. Chem. Soc., Perkin Trans. 1*, **1976**, 592.

696. N. Dennis, A. R. Katritzky, and M. Ramaiah, *J. Chem. Soc., Perkin Trans. 1*, **1976**, 2281.

697. G. Adembri, S. Chimichi, and R. Nesi, *J. Chem. Soc., Perkin Trans. 1*, **1977**, 1020.

698. V. Scartoni, I. Marelli, A. Marsili, and S. Catalano, *J. Chem. Soc., Perkin Trans. 1*, **1977**, 2332.

699. M. G. Barlow, R. H. Haszeldine, and A. Pickett, *J. Chem. Soc., Perkin Trans. 1*, **1978**, 378.

700. G. V. Boyd, R. L. Monteil, P. F. Lindley, and M. M. Mahmoud, *J. Chem. Soc., Perkin Trans. 1*, **1978**, 1351.

701. Y. Maki, T. Furuta, and M. Suzuki, *J. Chem. Soc., Perkin Trans. 1*, **1979**, 553.

702. L. Garanti and G. Zecchi, *J. Chem. Soc., Perkin Trans. 1*, **1979**, 1195.

703. Y. Maki, M. Suzuki, T. Furuta, M. Kawamura, and M. Kuzuya, *J. Chem. Soc., Perkin Trans. 1*, **1979**, 1199.

704. K. J. Gould, N. P. Hacker, J. F. W. McOmie, and D. H. Perry, *J. Chem. Soc., Perkin Trans. 1*, **1980**, 1834.

705. O. Abou-Teim, R. B. Jansen, J. F. W. McOmie, and D. H. Perry, *J. Chem. Soc., Perkin Trans. 1*, **1980**, 1841.

706. M. F. Aldersley, F. M. Dean, and R. Nayyir-Mazhir, *J. Chem. Soc., Perkin Trans. 1*, **1983**, 1753.

707. C. J. Wharton and R. Wrigglesworth, *J. Chem. Soc., Perkin Trans. 1*, **1985**, 809.

708. S. P. Breukelman, G. D. Meakins, and A. M. Roe, *J. Chem. Soc., Perkin Trans. 1*, **1985**, 1527.

709. D. Johnston, D. M. Smith, T. Shepherd, and D. Thompson, *J. Chem. Soc., Perkin Trans. 1*, **1987**, 495.

710. C. Glidewell, T. Shepherd, and D. M. Smith, *J. Chem. Soc., Perkin Trans. 1*, **1987**, 507.

711. J. W. Barton and N. D. Pearson, *J. Chem. Soc., Perkin Trans. 1*, **1987**, 1541.

712. R. N. Butler, A. M. Gillan, F. A. Lysaght, P. McArdle, and D. Cunningham, *J. Chem. Soc., Perkin Trans. 1*, **1990**, 555.

713. J. R. Russell, C. D. Garner, and J. A. Joule, *J. Chem. Soc., Perkin Trans. 1*, **1992**, 409.
714. R. N. Butler and C. S. Pyne, *J. Chem. Soc., Perkin Trans. 1*, **1992**, 3387.
715. J. Parrick and R. Ragunathan, *J. Chem. Soc., Perkin Trans. 1*, **1993**, 211.
716. A. W. Bridge, M. B. Hursthouse, C. W. Lehmann, D. J. Lythgoe, and C. G, Newton, *J. Chem. Soc., Perkin Trans. 1*, **1993**, 1839.
717. V. J. Arán, J. L. Asensio, J. Molina, P. Muñoz, J. R. Ruiz, and M. Stud, *J. Chem. Soc., Perkin Trans. 1*, **1997**, 2229.
718. M. Kuzuya, F. Miyake, and T. Okuda, *J. Chem. Soc., Perkin Trans. 2*, **1984**, 1465.
719. M. Kuzuya, F. Miyake, and T. Okuda, *J. Chem. Soc., Perkin Trans. 2*, **1984**, 1471.
720. S. Sommer, *Angew. Chem.*, **1976**, *88*, 449.
721. G. Seitz, R. Hoferichter, and R. Mohr, *Angew. Chem.*, **1987**, *99*, 345.
722. G. Jalbritter, F. Knoch, A. Wolski, and H. Kisch, *Angew. Chem.*, **1994**, *106*, 1676.
723. D. M. W. van den Ham, G. F. S. Harrison, A. Spaans, and D. van der Meer, *Recl. Trav. Chim. Pays-Bas*, **1975**, *94*, 168.
724. H. C. van der Plas, D. J. Buurman, and C. M. Vos, *Recl. Trav. Chim. Pays-Bas*, **1978**, *97*, 50.
725. J. H. Wolsink, A. Spaans, D. M. W. van den Ham, and D. Feil, *Recl., J. R. Neth. Chem. Soc.*, **1982**, *101*, 141.
726. R. Valter, A. E. Batse, and S. P. Valter, *Khim. Geterotsikl. Soedin.*, **1973**, 1124.
727. A. N. Kost, M. A. Yurovskaya, and N. M. Thao, *Khim. Geterotsikl. Soedin.*, **1975**, 1512.
728. B. I. Buzykin, N. N. Bystrykh, A. P. Stolyarov, S. A. Flegontov, V. V. Zverev, and Y. P. Kitaev, *Khim. Geterotsikl. Soedin.*, **1976**, 402.
729. B. I. Buzykin, N. N. Bystrykh, A. P. Stolyarov, and Y. P. Kitaev, *Khim. Geterotsikl. Soedin.*, **1977**, 1264.
730. K. N. Zelenin, V. N. Verbov, and Z. M. Matveeva, *Khim. Geterotsikl. Soedin.*, **1977**, 1658.
731. B. I. Buzykin, N. N. Bystrykh, A. P. Stolyarov, S. A. Flegontov, and Y. P. Kitaev, *Khim. Geterotsikl. Soedin.*, **1978**, 530.
732. Y. A. Sedov and M. A. Chernova, *Khim. Geterotsikl. Soedin.*, **1978**, 545.
733. B. I. Buzykin, N. N. Bystrykh, A. P. Stolyarov, L. V. Belova, and Y. P. Kitaev, *Khim. Geterotsikl. Soedin.*, **1978**, 699.
734. Y. H. Ivanova, B. I. Buzykin, and N. N. Bystrykh, *Khim. Geterotsikl. Soedin.*, **1979**, 541.
735. V. N. Eraksina, T. M. Ivanova, T. A. Babushkina, A. M. Vasil'ev, and N. N. Suvorov, *Khim. Geterotsikl. Soedin.*, **1979**, 916.
736. K. H. Zelenin and V. N. Verbov, *Khim. Geterotsikl. Soedin.*, **1979**, 1547.
737. G. V. Shatalov, S. A. Gridchin, and B. I. Mikhant'ev, *Khim. Geterotsikl. Soedin.*, **1980**, 394.
738. N. N. Bystrykh, B. I. Buzykin, A. P. Stolyarov, S. A. Flegontov, and Y. P. Kitaev, *Khim. Geterotsikl. Soedin.*, **1981**, 678.
739. O. N. Chupakhin, V. N. Charushin, L. M. Naumova, A. I. Rezvukhin, and N. A. Klyuev, *Khim. Geterotsikl. Soedin.*, **1981**, 1549.
740. L. A. Litvinov, Y. T. Struchkov, N. N. Bystrykh, Y. P. Kitaev, and B. I. Buzykin, *Khim. Geterotsikl. Soedin.*, **1982**, 977.
741. S. S. Mochalov, A. N. Fedotov, E. A. Kupriyanova, and Y. S. Shabarov, *Khim. Geterotsikl. Soedin.*, **1983**, 688.
742. N. N. Bystrykh, B. I. Buzykin, A. P. Stolyarov, S. A. Flegontov, and Y. P. Kitaev, *Khim. Geterotsikl. Soedin.*, **1983**, 826.
743. L. N. Donchak, V. A. Kaminskii, and M. N. Tilichenko, *Khim. Geterotsikl. Soedin.*, **1986**, 1271.
744. V. A. Samsonov, L. B. Volodarskii, I. Y. Bagryanskaya, and Y. V. Gatilov, *Khim. Geterotsikl. Soedin.*, **1996**, 1055.

745. B. G. Pring and C.-G. Swahn, *Acta Chem. Scand.*, **1973**, *27*, 1891.
746. G. Ahlgren, B. Åkermark, J. Lewandowska, and R. Wahren, *Acta Chem. Scand., Ser. B*, **1975**, *29*, 524.
747. H. Lund and N. H. Nilsson, *Acta Chem. Scand., Ser. B*, **1976**, *30*, 5.
748. H. Alper, *J. Organomet. Chem.*, **1973**, *50*, 209.
749. G. Halbritter, F. Knoch, and H. Kisch, *J. Organomet. Chem.*, **1995**, *492*, 87.
750. U. Dürr, F. W. Heinemann, and H. Kisch, *J. Organomet. Chem.*, **1997**, *541*, 307.
751. U. Dürr, F. W. Heinemann, and H. Kisch, *J. Organomet. Chem.*, **1998**, *558*, 91.
752. H. Horino, T. Mimura, M. Ohta, H. Kubo, and M. Kitagawa, *Bioorg. Med. Chem. Lett.*, **1997**, *7*, 437.
753. W. Xu, R. Mohan, and M. M. Morrissey, *Bioorg. Med. Chem. Lett.*, **1998**, *8*, 1089.
754. Y. Chen, W. Hua, M. Lü, and J. Pan, *Magn. Reson. Chem.*, **1999**, *37*, 149.
755. P. Bruck, J. Tamás, and K. Körmendy, *Org. Mass Spectrom.*, **1974**, *9*, 335.
756. G. Cauquis, B. Chabaud, and J. Ulrich, *Org. Mass Spectrom.*, **1977**, *12*, 717.
757. P. van de Weijer, H. H. Thijsse, and D. van der Meer, *Org. Magn. Reson.*, **1976**, *8*, 187.
758. G. Stájer, F. Csende, G. Bernáth, P. Sohár, and J. Szúnyog, *Monatsh. Chem.*, **1994**, *125*, 933.
759. N. Öcal, Z. Turgut, and S. Kaban, *Monatsh. Chem.*, **1999**, *130*, 915.
760. L. Stefaniak, *Spectrochim. Acta, Part A*, **1976**, *32*, 345.
761. W. M. F. Fabian, *J. Comput. Chem.*, **1991**, *12*, 17.
762. K. Woźniak, T. M. Krygoeski, B. Kariuki, W. Jones, and E. Grech, *J. Mol. Struct.*, **1990**, *240*, 119.
763. G. Fischer, *J. Mol. Spectrosc.*, **1974**, *49*, 201.
764. F. G. Baddar, A. F. M. Fahmy, and N. F. Aly, *J. Chem. Soc., Perkin Trans. 1*, **1973**, 2448.
765. W. J. Guilford, T. G. Patterson, R. O. Vega, L. Fang, Y. Liang, H. A. Lewis, and J. N. Labovitz, *J. Agric. Food Chem.*, **1992**, *40*, 2026.
766. F. A. Neugebauer and H. Weger, *J. Phys. Chem.*, **1978**, *82*, 1152.
767. S. Kaban and J. G. Smith, *Organometallics*, **1983**, *2*, 1351.
768. G. Wu, A. L. Rheingold, and R. F. Heck, *Organometallics*, **1986**, *5*, 1922.
769. G. Wu, A. L. Rheingold, and R. F. Heck, *Organometallics*, **1987**, *6*, 2386.
770. S. D. Carter and G. W. H. Cheeseman, *Org. Prep. Proced. Int.*, **1974**, *6*, 67.
771. S. Ruchirawat and P. Thopchumrune, *Org. Prep. Proced. Int.*, **1980**, *12*, 263.
772. M. Lora-Tamayo, B. Marco, and P. Navarro, *Org. Prep. Proced. Int.*, **1976**, *8*, 45; *Chem. Abstr.*, **1976**, *85*, 32943.
773. S. Spassov, I. A. Atanassova, and M. A. Halmova, *Magn. Reson. Chem.*, **1985**, *23*, 795.
774. E. V. Tratyakov and S. F. Vasilevsky, *Russ. Chem. Bull.*, **1998**, *47*, 1233; *Chem. Abstr.*, **1998**, *129*, 230685.
775. B. A. Rusin and A. N. Leksin, *Izv. Akad. Nauk SSSR, Ser. Khim.*, **1982**, 2685.
776. B. A. Rusin, A. N. Leksin, A. S. Shalomeer, and M. P. Seminova-Zhukova, *Izv. Akad. Nauk SSSR, Ser. Khim.*, **1984**, 305.
777. B. A. Rusin and A. H. Leksin, *Izv. Akad. Nauk SSSR, Ser. Khim.*, **1984**, 314.
778. S. F. Vasilevskii, A. V. Pozdnyakov, and M. S. Shvartsberg, *Izv. Akad. Nauk SSSR, Ser. Khim.*, **1985**, 1367.
779. C. Tamborski, U. D. G. Prabhu, and K. C. Eapen, *J. Fluorine Chem.*, **1985**, *28*, 139.
780. M.-J. Chen, C.-S. Chi, and Q.-Y. Chen, *J. Fluorine Chem.*, **1990**, *49*, 99.
781. Y. V. Burgart, A. S. Fokin, O. G. Kuzueva, O. N. Chupakhin, and V. I. Saloutin, *J. Fluorine Chem.*, **1998**, *92*, 101.
782. S. W. Longworth, D. C. H. Bigg, D. F. White, and J. Burns, *J. Labelled Compd. Radiopharm.*, **1974**, *10*, 423.

783. H. Abuki and H. Miyazaki, *J. Labelled Compd. Radiopharm.*, **1981**, *18*, 889.

784. B. L. Mylari and J. J. Zembrowski, *J. Labelled Compd. Radiopharm.*, **1991**, *29*, 143.

785. E. Lippmann and S. Ungethüm, *Z. Chem.*, **1973**, *13*, 343.

786. W. Jugelt and S. Schwartner, *Z. Chem.*, **1984**, *24*, 149.

787. M. Reinhardt, R. Kluge, and M. Schulz, *Z. Chem.*, **1989**, *29*, 421.

788. H. R. Griffin, M. B. Hocking, and D. G. Lowery, *Chem. Ind.* (*London*), **1974**, 829.

789. A. Arcoleo, M. Marino, and G. Giammona, *Chem. Ind.*, (*London*), **1976**, 651.

790. K. Imae, S. Iimura, T. Hasegawa, T. Okita, M. Hirano, H. Kamachi, and H. Kamei, *J. Antibiot.*, **1993**, *46*, 840.

791. S. Iimura, K. Imae, T. Hasegawa, T. Okita, M. Tamaoka, S. Murata, H. Kamachi, and H. Kamei, *J. Antibiot.*, **1993**, *46*, 850.

792. P. Sohár, G. Stájer, A. E. Szabó, and G. Bernáth, *J. Mol. Struct.*, **1996**, *382*, 187.

793. M. Pulici and G. Sello, *J. Mol. Struct.* (*Theochem*), **1993**, *288*, 245.

794. J. M. Ames and G. MacLeod, *Phytochemistry*, **1990**, *29*, 1201.

795. S. M. Radwan, M. S. Abbady, and R. A. Ahmed, *Phosphorus, Sulfur, Silicon Relat. Elem.*, **1991**, *63*, 363.

796. M. H. Elnagdi, R. M. Mohareb, F. A.-E. Abd-Elaal, and H. A. Mohamed, *Phosphorus, Sulfur, Silicon Relat. Elem.*, **1993**, *82*, 195.

797. I. Maeba, T. Ishikawa, and H. Furukawa, *Carbohydr. Res.*, **1985**, *140*, 332.

798. J. E. Ridley and M. C. Zerner, *J. Mol. Spectrosc.*, **1974**, *50*, 457.

799. L. K. Thompson, T. C. Woon, D. B. Murphy, E. J. Gabe, F. L. Lee, and Y. Le Page, *Inorg. Chem.*, **1985**, *24*, 4719.

800. L. Chen, L. K. Thompson, S. S. Tandon, and J. N. Bridson, *Inorg. Chem.*, **1993**, *32*, 4063.

801. Y. Zang, L. K. Thompson, J. N. Bridson, and M. Bubenik, *Inorg. Chem.*, **1995**, *34*, 5870.

802. F. W. Hartstock and L. K. Thompson, *Inorg. Chim. Acta*, **1983**, *72*, 227.

803. L. K. Thompson, F. W. Hartstock, L. Rosenberg, and T. C. Woon, *Inorg. Chim. Acta*, **1985**, *97*, 1.

804. D. Attanasio, G. Dessy, and V. Fares, *Inorg. Chim. Acta*, **1985**, *104*, 99.

805. I. E. Kalinichenko, A. T. Pilipenko, and A. V. Barovskii, *Zh. Obshch. Khim.*, **1978**, *48*, 334.

806. S. Basu, K. A. McLaughlan, and A. J. D. Ritchie, *Chem. Phys.*, **1983**, *79*, 95.

807. M. H. Palmer, R. O. Gould, A. J. Blake, J. A. S. Smith, and D. Stephenson, *Chem. Phys.*, **1987**, *112*, 213.

808. M. H. Palmer, A. J. Blake, M. M. P. Khurshid, and J. A. S. Smith, *Chem. Phys.*, **1992**, *168*, 41.

809. M. Lora-Tamayo, B. Marco, and C. Sender, *Org. Prep. Proced. Int.*, **1978**, *10*, 298.

810. A. H. H. Elghandour, M. K. A. Ibrahim, B. El-Badry, and H. K. Waly, *Phosphorus, Sulfur, Silicon Relat. Elem.*, **1994**, *88*, 147; *Chem. Abstr.*, **1996**, *122*, 81263.

811. A. A. Nada, N. K. El-Din, S. T. Gaballah, and M. F. Zayed, *Phosphorus, Sulfur, Silicon Relat. Elem.*, **1996**, *119*, 27; *Chem. Abstr.*, **1997**, *127*, 220723.

812. E. J. Alford and K. Schofield, *J. Chem. Soc.*, **1952**, 2102.

813. H. J. S. Machado and A. Hinchliffe, *J. Mol. Struct.* (*Theochem*), **1995**, *339*, 255.

814. S. Bräse, C. Gil, and K. Knepper, *Bioorg. Med. Chem.*, **2002**, *10*, 2415.

815. A. M. Amer, M. A. El-Bermaui, A. F. S. Ahmed, and S. M. Soliman, *Monatsh. Chem.*, **1999**, *130*, 1409.

816. D. B. Kimball, T. J. R. Weakley, R. Herges, and M. M. Haley, *J. Am. Chem. Soc.*, **2002**, *124*, 13463.

817. D. B. Kimball, T. J. R. Weakley, and M. M. Haley, *J. Org. Chem.*, **2002**, *67*, 6395.

818. A. Kumar, N. A. Al-Awadi, M. H. Elnagdi, Y. A. Ibrahim, and K. Kaul, *Int. J. Chem. Kinet.*, **2001**, *33*, 402.

819. D. B. Kimball, A. G. Hayes, and M. M. Haley, *Org. Lett.*, **2000**, *2*, 3825.

820. D. H. Jeong, N. H. Jang, J. S. Suh, and M. Moskovits, *J. Phys. Chem. Sect. B*, **2000**, *104*, 3594.

821. A. Leprêtre, A. Turck, N. Plé, and G. Quéguiner, *Tetrahedron*, **2000**, *56*, 3709.

822. V. G. Chapoulaud, N. Plé, A. Turck, and G. Quéguiner, *Tetrahedron*, **2000**, *56*, 5499.

823. V. T. Abaev, A. V. Gutnov, A. V. Butin, and V. E. Zavodnik, *Tetrahedron*, **2000**, *56*, 8933.

824. N. A. Al-Awadi, M. H. Elnagdi, Y. A. Ibrahim, K. Kaul, and A. Kumar, *Tetrahedron*, **2001**, *57*, 1609.

825. Y. A. Ibrahim, N. A. Al-Awadi, and K. Kaul, *Tetrahedron*, **2001**, *57*, 7377.

826. M. R. Lentz, P. E. Fanwick, and J. P. Rothwell, *Chem. Commun.*, **2002**, 2482.

827. B.-X. Wang, Y.-Y. Sun, G,-D. Li, X.-R. Wang, and H.-W. Hu, *Acta Chim. Sinica*, **2002**, *60*, 1883.

828. I. F. Hennequin, A. P. Thomas, C. Johnstone, E. S. E. Stokes, P. A. Plé, J.-J. M. Lohmann, D. J. Ogilvie, M. Dukes, S. A. Wedge, J. O. Curwen, J. Kendrew, and C. Lambert-van der Brempt, *J. Med. Chem.*, **1999**, *42*, 5369.

829. Y. Murakami, H. Yokoo, Y. Yokoyama, and T. Watanabe, *Chem. Pharm. Bull.*, **1999**, *47*, 791.

830. A. M. Amer, *Monatsh. Chem.*, **2001**, *132*, 859.

831. G. Cirrincione, A. M. Almerico, P. Barraja, P. Diana, A. Lauria, A. Passannanti, C. Musiu, A. Pani, P. Murtas, C. Minnei, M. E. Marongiu, and P. La Colla, *J. Med. Chem.*, **1999**, *42*, 2561.

832. P. Tapolcsányi, G. Krajsovszky, R. Andró, P. Lipcsey, G. Horváth, P. Mátyus, Z. Riedl, G. Hajós, B. U. W. Maes, and G. L. F. Lumière, *Tetrahedron*, **2002**, *58*, 10137.

833. I. Constantino, G. Rastelli, G. Cignarella, and D. Barlocco, *Farmaco*, **2000**, *55*, 544.

834. P. Barraja, P. Diana, A. Lauria, A. Passannanti, A. M. Almerico, C. Minnei, S. Longu, D. Congiu, C. Musiu, and P. La Colla, *Bioorg. Med. Chem.*, **1999**, *7*, 1591.

835. B. El-Saleh, M. M. Abdel-Khalik, E. Darwich, O. A.-M. Salah, and M. H. Elnagdi, *Heteroat. Chem.*, **2002**, *13*, 141.

836. R. N. Butler, A. G. Coyne, P. McArdle, D. Cunningham, and L. A. Burke, *J. Chem. Soc., Perkin Trans. 1*, **2001**, 1391.

837. G. Giorgi, F. Ponticelli, L. Savini, L. Chiasserini, and C. Pellerano, *J. Chem. Soc., Perkin Trans. 2*, **2000**, 2259.

838. R. N. Butler, A. G. Coyne, and L. A. Burke, *J. Chem. Soc., Perkin Trans. 2*, **2001**, 1781.

839. R. N. Butler, A. G. Coyne, W. J. Cunningham, and L. A. Burke, *J. Chem. Soc., Perkin Trans. 2*, **2002**, 1807.

840. H. D. Joshi, P. S. Upadhyay, and A. J. Baxi, *Indian J. Chem., Sect. B*, **2000**, *39*, 967.

841. P. Upadhyay, H. Joshi, J. Upadhyay, A. J. Baxi, and A. R. Parikh, *Indian J. Chem., Sect. B*, **2001**, *40*, 500.

842. T. Ghafourian and M. R. Rashidi, *Chem. Pharm. Bull.*, **2001**, *49*, 1066.

843. Y. Murata, N. Kato, and K. Komatsu, *J. Org. Chem.*, **2001**, *66*, 7235.

844. A. M. Barrios and S. J. Lippard, *J. Am. Chem. Soc.*, **1999**, *121*, 11751.

845. A. M. Barrios and S. J. Lippard, *J. Am. Chem. Soc.*, **2000**, *122*, 9172.

846. J. Kuzelka, B. Spingler, and S. J. Lippard, *Inorg. Chim. Acta*, **2002**, *337*, 212.

847. J. Zhou, Y. Hu, and H. Hu, *J. Heterocycl. Chem.*, **2000**, *37*, 1165.

848. G. Lukács and G. Simig, *J. Heterocycl. Chem.*, **2002**, *39*, 989.

849. M. Hashimoto, T. Nakamura, K. Rsukagashi, R. Nakajima, and K. Kondo, *Bull. Chem. Soc. Jpn.*, **1999**, *72*, 2673.

850. B. I. Buzykin, V. V. Yanilkin, V. I. Morozov, N. I. Maksymyuk, R. M. Eliseenkova, and N. V. Nastapova, *Mendeleev Commun.*, **2000**, 34.

851. A. H. M. Elwahy, M. M. Ahmed, and M. El-Sadek, *J. Chem. Res.*, **2001**, *Synop.* 175, *Minipr.* 552.

852. J. S. Suh, D. H. Jeang, and M. S. Lee, *J. Raman Spectrosc.*, **1999**, *30*, 595.

853. F. Drumitrascu, C. I. Mitan, C. Draghici, M. T. Caproiu, D. Caprau, and D. Dumitrescu, *Rev. Roum. Chim.*, **2002**, *47*, 309.

854. F. Dumitrascu, C. I. Mitan, D. Caprau, C. Draghici, M. T. Caproiu and D. Dumitrescu, *Rev. Chim. (Bucharest)*, **2002**, *53*, 736.

855. F. Dumitrascu, C. I. Mitan, C. Draghici, M. T. Caproiu, and A. Niculae, *Rev. Roum. Chim.*, **2002**, *47*, 179.

856. J. S. Kim, K. J. Shin, D. C. Kim, Y. H. Kabg, D. J. Kim, K. H. Yoo, and S. W. Park, *Bull. Korean Chem. Soc.*, **2002**, *23*, 1425.

857. A. K. Karim, M. Armengol, and J. A. Joule, *Heterocycles*, **2001**, *55*, 2139.

858. M. I. H. Helaleh, M. Kumemura, S. Fujii, and T. Korenaga, *Analyst*, **2001**, *126*, 104.

859. A. A. Nada, N. K. El-Din, S. T. Gab-Allah, and M. F. Zayed, *Phosphorus, Sulfur, Silicon Relat. Elem.*, **2000**, *156*, 213.

860. F. Badea, E. Iordache, A. Maghea, A. Simion, I. Costea, S. Iordache, and C. Simion, *Rev. Roum. Chim.*, **1999**, *44*, 585.

861. M. Napoletano, G. Norcini, F. Pellacini, F. Marchini, G. Morazzoni, P. Ferlenga, and L. Pradella, *Bioorg. Med. Chem. Lett.*, **2000**, *10*, 2235.

862. M. Napoletano, G. Norcini, F. Pellacini, F. Marchini, G. Morazzoni, P. Ferlenga, and L. Pradella, *Bioorg. Med. Chem. Lett.*, **2001**, *11*, 33.

863. E. del Olmo, M. G. Armas, J. L. López-Pérez, V. Muñoz, E. Deharo, and A. S. Feliciano, *Bioorg. Med. Chem. Lett.*, **2001**, *11*, 2123.

864. F. del Olmo, M. G. Armas, J. L. López-Pérez, G. Ruiz, F. Vargas, A. Giménez, E. Deharo, and A. S. Feliciano, *Bioorg. Med. Chem. Lett.*, **2001**, *11*, 2755.

865. M. Napoleyano, G. Norcini, F. Pellacini, F. Marchini, G. Morazzoni, R. Fattori, P. Ferlenga, and L. Pradella, *Bioorg. Med. Chem. Lett.*, **2002**, *12*, 5.

866. Y. T. Chen and W. Hua, *Spectrochim. Acta, Part A*, **2000**, *56*, 1045.

867. A. Yatani, M. Fujii, Y. Nakao, S. Kashino, M. Kinoshita, W. Mori, and S. Suzuki, *Inorg. Chim. Acta*, **2001**, *316*, 127.

868. S. Marcaccini, R. Pepino, C. Polo, and M. C. Pozo, *Synthesis*, **2001**, 85.

869. M. van der Mey, H. Boss, A. Hatzelmann, I. J. van der Laan, G. K. Sterk, and H. Timmerman, *J. Med. Chem.*, **2002**, *45*, 2520.

870. G. Bold, K.-H. Altmann, J. Frei, M. Lang, P. W. Manley, P. Traxler, B. Wietfeld, J. Brüggen, E. Buchdunger, R. Cozens, S. Ferrari. P. Furet, F. Hofmann, G. Martiny-Baron, J. Mestan, J. Rösel, M. Sills, D. Stover, F. Acemoglu, E. Boss, R. Emmenegger, L. Lässer, E. Masso, R. Roth, C. Schlachter, W. Vetterli, D. Wyss, and J. M. Wood, *J. Med. Chem.*, **2000**, *43*, 2310.

871. N. Watanabe, H. Adachi, Y. Takase, H. Ozaki, M. Matsukura, K. Miyazaki, K. Ishibashi, H. Ishihara, K. Kodama, M. Nishino, M. Kakiki, and Y. Kabasawa, *J. Med. Chem.*, **2000**, *43*, 2523.

872. M. van der Mey, A. Hatzelmann, I. J. van der Laan, G. J. Sterk, U. Thibaut, and H. Timmerman, *J. Med. Chem.*, **2001**, *44*, 2511.

873. M. van der Mey, A. Hatzelmann, G. P. M. van Klink, T. J. van der Laan, G. J. Sterk, U. Thibaut, W. R. Ulrich, and H. Timmermann, *J. Med. Chem.*, **2001**, *44*, 2523.

874. G. R. Madhavan, R. Chakrabarti, S. K. B. Kumar, P. Misra, R. N. V. S. Mamidi, V. Balraju, K. Kasiram, R. K. Babu, J. Suresh, B. B. Lohray, V. B. Lohray, J. Iqbal, and R. Rajagopalan, *Eur. J. Med. Chem.*, **2001**, *36*, 627.

875. R. Sivakumar, S. K. Gnanasam, S. Ramachandran, and J. T. Leonard, *Eur. J. Med. Chem.*, **2002**, *37*, 793.

876. N. Watanabe, T. Tokumura, and T. Nakamura, *J. Labelled Compd. Radiopharm.*, **2001**, *44*, 843.

877. J. Sinkonen, V. Ovcharenko, K. N. Zelenin, I. P. Bezhan, B. A. Chakchir, F. Al-Assar, and K. Pihlaja, *Eur. J. Org. Chem.*, **2002**, 2046.

878. J. Sinkonen, V. Ovcharenko, K. N. Zelenin, I. P. Bezhan, B. A. Chakchir, F. Al-Assar, and K. Pihlaja, *Eur. J. Org. Chem.*, **2002**, 3447.

879. R. N. Warrener, D. N. Butler, L. Liu, D. Margetic, and R. A. Russell, *Chem. Eur. J.*, **2001**, *7*, 3406.

880. S. Guery, I. Parrot, Y. Rival, and C.-G. Wermuth, *Synthesis*, **2001**, 699.

881. J. Epsztajn, Z. Malinowski, J. Z. Brzeziñki, and M. Karzatka, *Synthesis*, **2001**, 2085.

882. A. M. Barrios and S. J. Lippard, *Inorg. Chem.*, **2001**, *40*, 1060.

883. A. L. Rose and T. D. Waite, *Anal. Chem.*, **2001**, *73*, 5909.

884. S. Mavel, I. Thery, and A. Gueiffier, *Arch. Pharm.* (*Weinheim*), **2002**, *335*, 7.

885. A. C. Benniston, D. S. Yufit, and J. A. K. Howard, *Acta Crystallogr., Sect. C*, **1999**, *55*, 1535.

886. D. J. Berg, J. M. Boncella, and R. A. Andersen, *Organometallics*, **2002**, *21*, 4622.

887. M. Darabantu, T. Lequeux, J.-C. Pommelet, N. Plé, A. Turck, and L. Toupet, *Tetrahedron Lett.*, **2000**, *41*, 6763.

888. P. G. Tsoungas and M. Searcey, *Tetrahedron Lett.*, **2001**, *42*, 6589.

889. F. Dumitraşcu, C. I. Mitan, C. Drăghici, M. T. Căproiu, and D. Răileanu, *Tetrahedron Lett.*, **2001**, *42*, 8379; **2003**, *44*, 9385.

890. H. Buff and U. Kuckländer, *Tetrahedron*, **2000**, *56*, 5137.

891. M. Darabantu, T. Lequeux, J.-C. Pommelet, N. Plé, and A. Turck, *Tetrahedron*, **2001**, *57*, 739.

892. A. S. Amarasekara and S. Chandrasekara, *Org. Lett.*, **2002**, *4*, 773.

893. T. Z. Wang, L. Cheng, T. Zhang, H. Yuan, and S. Mao, *Magn. Reson. Chem.*, **2002**, *40*, 738.

894. M. van der Mey, H. Boss, D. Couwenberg, A. Hatzelmann, G. J. Sterk, K. Goubitz, H. Schenk, and H. Timmerman, *J. Med. Chem.*, **2002**, *45*, 2526.

895. A. Pavel, D. Moigno, S. Cîntă, and W. Kiefer, *J. Phys. Chem. A*, **2002**, *106*, 3337.

896. D. E. Lynch and I. McClenaghan, *Acta Crystalogr., Sect. E*, **2002**, *58*, 01051.

897. L. Türker, *J. Mol. Struct.* (*Theochem*), **2002**, *588*, 165.

898. M. K. Award, *J. Mol. Struct.* (*Theochem*), **2001**, *542*, 139.

899. S. K. Kundu, A. Pramanik, and A. Patra, *Synlett*, **2002**, 823.

900. S. Janelli and M. Carcelli, *Z. Krystallogr.—New Cryst. Struct.*, **2002**, *217*, 203.

901. K. H. Park, K. Jun, S. R. Shin, and S. W. Oh, *Synth. Commun.*, **1999**, *29*, 583.

902. S. Shashikanth, S. K. Ahmed, G. L. Hegde, and K. M. L. Rai, *Synth. Commun.*, **1999**, *29*, 3503.

903. R. N. Hunston, J. Parrick, and C. J. G. Shaw, in *Rodd's Chemistry of Carbon Compounds*, 2nd ed., Vol. IV, Part I/J, M. F. Ansell, ed., Elsevier, Amsterdam, 1989, p. 1.

904. J. Parrick, C. J. G. Shaw, and L. K. Mehta, in *Rodd's Chemistry of Carbon Compounds*, 2nd ed., 1st suppl., Vol. IV, Part I/J, M. F. Ansell, ed., Elsevier, Amsterdam, 1995, p. 1.

905. J. Parrick, C. J. G. Shaw, and L. K. Mehta, in *Rodd's Chemistry of Carbon Compounds*, 2nd ed., 2nd suppl., Vol. IV, Part I/J, M. Sainsbury, ed., Elsevier, Amsterdam, 1999, p. 1.

906. J. C. E. Simpson, *Condensed Pyridazine and Pyrazine Rings*, (*Cinnolines, Phthalazines, and Quinoxalines*) Interscience, New York, 1953, pp. 1–190.

907. G. M. Singerman and N. R. Patel, in *Condensed Pyridazines Including Cinnolines and Phthalazines*, R. N. Castle, ed., Wiley, New York, 1973, pp. 1–740.

908. B. Stanovnik, in *Methods of Organic Chemistry* (*Houben-Weyl*), 4th ed., suppl. Vol. E9a, E. Schaumann, ed., Thieme, Stuttgart, 1997, pp. 683–789.

909. T. J. Kress, *Prog. Heterocycl. Chem.*, **1989**, *1*, 243.

910. T. J. Kress and D. L. Narie, *Prog. Heterocycl. Chem.*, **1990**, *2*, 185; **1991**, *3*, 205; **1992**, *4*, 186.

911. D. T. Hurst, *Prog. Heterocycl. Chem.*, **1993**, *5*, 220.

912. G. Heinisch and B. Matuszczak, *Prog. Heterocycl. Chem.*, **1994**, *6*, 231; **1995**, *7*, 226.

913. M. P. Groziak, *Prog. Heterocycl. Chem.*, **1996**, *8*, 231; **1997**, *9*, 249; **1998**, *10*, 251; **1999**, *11*, 256; **2003**, *15*, 306.

914. B. R. Lahue and J. K. Snyder, *Prog. Heterocycl. Chem.*, **2000**, *12*, 263.
915. B. R. Lahue, G. H. C. Woo, and J. K. Snyder, *Prog. Heterocycl. Chem.*, **2001**, *13*, 261.
916. G. H. C. Woo and J. K. Snyder, *Prog. Heterocycl. Chem.*, **2002**, *14*, 279.
917. H. J. Barber and E. Lunt, *J. Chem. Soc.*, **1965**, 1468.
918. J. A. Beisler, G. W. Peng, and J. S. Driscoll, *J. Pharm. Sci.*, **1977**, *66*, 849.
919. E. Bellasio and E. Arrigoni-Martelli, *Farmaco, Ed. Sci.*, **1972**, *27*, 627.
920. E. Bellasio, *Farmaco, Ed. Sci.*, **1974**, *29*, 210.
921. E. Bellasio and G. Tuan, *Farmaco, Ed. Sci.*, **1975**, *30*, 343.
922. G. Winters, N. Di Mola, E. Oppici, and G. Nathansohn, *Farmaco, Ed. Sci.*, **1975**, *30*, 620.
923. G. Seitz and W. Overhue, *Arch. Pharm.* (*Weinheim*), **1979**, *312*, 452.
924. S. Groszkowski and B. Wesołowska, *Arch. Pharm.* (*Weinheim*), **1981**, *314*, 880.
925. S. Groszkowski and B. Wesołowska, *Arch. Pharm.* (*Weinheim*), **1982**, *135*, 136.
926. J. Dusemund, *Arch. Pharm.* (*Weinheim*), **1982**, *315*, 925.
927. S. Groszkowski, B. Wesołowska, and J. Wrona, *Arch. Pharm.* (*Weinheim*), **1985**, *318*, 678.
928. E. Schenker and R. Salzmann, *Arzneim.-Forsch.*, **1979**, *29*, 1835.
929. M. Hieda, K. Ōmura, and S. Yurugi, *Yakugaku Zasshi*, **1972**, *92*, 1327.
930. E. Hayashi, T. Higashino, and I. Watanabe, *Yakugaku Zasshi*, **1973**, *93*, 409.
931. E. Ôishi, K. Ôsumi, and E. Hayashi, *Yakugaku Zasshi*, **1974**, *94*, 672.
932. E. Ôishi, *Yakugaku Zasshi*, **1974**, *94*, 746.
933. E. Hayashi and I. Utsunomiya, *Yakugaku Zasshi*, **1974**, *94*, 1015.
934. E. Hayashi and I. Utsunomiya, *Yakugaku Zasshi*, **1974**, *94*, 1159.
935. C. Iijima, T. Morikawa, and E. Hayashi, *Yakugaku Zasshi*, **1975**, *95*, 784.
936. T. Watanabe, F. Hamaguchi, and S. Ohki, *Yakugaku Zasshi*, **1976**, *96*, 721.
937. E. Hayashi, N. Shimada, and A. Miyashita, *Yakugaku Zasshi*, **1976**, *96*, 1370.
938. E. Ōishi, Y. Kawamura, D. Kojima, and E. Hayashi, *Yakugaku Zasshi*, **1977**, *97*, 1082.
939. A. Guingant and J. Renault, *C. R. Acad. Sci., Ser. C*, **1972**, *275*, 705.
940. M. Lamant, *C. R. Acad. Sci., Ser. C*, **1973**, *277*, 319.
941. H. Lund and J. Simonet, *C. R. Acad. Sci., Ser. C*, **1973**, *177*, 1387.
942. L. Lepage and Y. Lepage, *C. R. Acad. Sci., Ser. C*, **1974**, *278*, 541.
943. A. Guingant and J. Renault, *C. R. Acad. Sci., Ser. C*, **1974**, *279*, 49.
944. A. Guingant and J. Renault, *C. R. Acad. Sci. Ser. C*, **1974**, *279*, 121.
945. A. Guingant and J. Renault, *C. R. Acad. Sci., Ser. C*, **1974**, *279*, 179.
946. A. Guingant and J. Renault, *C. R. Acad. Sci., Ser. C*, **1974**, *279*, 209.
947. D. Catarzi, L. Cecchi, G. Filacchioni, A. Bartolini, R. Carpenedo, A. Galli, and F. Mori, *Drug Des. Discovery*, **1993**, *10*, 23.
948. D. Bhattacharjee and F. D. Popp, *J. Pharm. Sci.*, **1980**, *69*, 120.
949. J. M. Chapman, J. W. Sowell, G. Abdalla, I, H. Hall, and O. T. Wang, *J. Pharm. Sci.*, **1989**, *78*, 903.
950. L. V. G. Nargund, V. V. Badiger, and S. M. Yarnal, *J. Pharm. Sci.*, **1992**, *81*, 365.
951. G. Sheffler, J. Engel, B. Kutscher, W. S. Sheddrick, and P. Bell, *Arch. Pharm.* (*Weinheim*), **1988**, *321*, 205.
952. C. Terán, E. Raviña, L. Santana, N. Garcia-Dominguez, G. Garcia-Mera, J. A. Fontenla, F. Orallo, and J. M. Calleja, *Arch. Pharm.* (*Weinheim*), **1989**, *322*, 331.
953. X.-G. Yang and G. Seitz, *Arch. Pharm.* (*Weinheim*), **1992**, *325*, 559.
954. M. J. Mokrosz, *Arch. Pharm.* (*Weinheim*), **1993**, *326*, 39.

955. P. Mátyus, E. Kasztreiner, E. Diesler, Á. Behr, I. Varga, J. Kosáry, G. Rabloczky, and L. Jaszlits, *Arch. Pharm. (Weinheim)*, **1994**, *327*, 543.

956. A. Sugimoto, S. Ito, Y. Eguchi, H. Tanaka, Y. Takashima, and M. Ishikawa, *Eur. J. Med. Chem.*, **1984**, *19*, 223.

957. F. D. Popp, *Eur. J. Med. Chem.*, **1989**, *29*, 313.

958. V. Colotta, L. Cecchi, D. Catarzi, G. Filacchioni, A. Galli, and F. Mori, *Eur. J. Med. Chem.*, **1994**, *29*, 95.

959. C. T. Bahner, L. M. Rives, S. W. McGaha, D. Rutledge, D. Ford, E. Gooch, D. Westberry, D. Ziegler, and R. Ziegler, *Arzneim.-Forsch.*, **1981**, *31*, 404.

960. J. Hasegawa, Y. Tomono, M. Tanaka, T. Fujita, and K. Sugiyama, *Arzneim.-Forsch.*, **1981**, *31*, 1215.

961. H. Fischer, H. Möller, M. Budnowski, G. Atassi, P. Dumont, J. Venditti, and O. C. Yoder, *Arzneim.-Forsch.*, **1984**, *34*, 663.

962. T. Kametani, K. Kigasawa, M. Hiragi, H. Ishimaru, N. Wagatsuma, T. Kohagizawa, and T. Nakamura, *Yakugaku Zasshi*, **1980**, *100*, 641.

963. E. Ōishi, H. Yamamoto, and E. Hayashi, *Yakugaku Zasshi*, **1981**, *101*, 1042.

964. T. Kametani, K. Kigasawa, M. Hiragi, N. Wagatsuma, T. Kohagizawa, T. Uryu, and T. Nakamura, *Yakugaku Zasshi*, **1982**, *102*, 173.

965. E. Ōishi and E. Hayashi, *Yakugaku Zasshi*, **1983**, *103*, 34.

966. K.-D. Gundmann and F. McCapra, *Chemiluminescence in Organic Chemistry*, Springer-Verlag, Berlin, 1987.

967. W. L. F. Armarego and C. L. L. Chai, *Purification of Laboratory Chemicals*, 5th ed., Elsevier, London, 2003.

968. M. Kuzuya, T. Usai, F. Miyake, K. Kamiya, and T. Okuda, *Chem. Pharm. Bull.*, **1982**, *30*, 708.

969. M. M. Abbasi, M. Abou-Sekkina, Y. Hafez, and H. H. Zoorob, *J. Prakt. Chem.*, **1986**, *328*, 932.

970. O. Pummerer, *Ber. Dtsch. Chem. Ges.*, **1909**, *42*, 2282.

971. H. E. Baumgarten, *J. Am. Chem. Soc.*, **1955**, *77*, 5109.

972. K. M. Shubin, V. A. Kuznetsov, and V. A. Galishev, *Tetrahedron Lett.*, **2004**, *45*, 1407.

973. M. A.-M. Gomaa, *Tetrahedron Lett.*, **2003**, *44*, 3493.

974. M. Pal, R. B. Venkateswara, K. Parasuraman, and K. R. Yeleswarapu, *J. Org. Chem.*, **2003**, *68*, 6806.

975. G. Özer, N. Saraçoğlu, and M. Balci, *J. Org. Chem.*, **2003**, *68*, 7009.

976. M. C. Caprosu, G. N. Zbancioc, C. C. Moldoveanu, and I. I. Mangalagiu, *Collect. Czech. Chem. Commun.*, **2004**, *69*, 426.

977. P. Tapolcsányi, B. U. W. Maes, K. Monsieurs, G. L. F. Lumière, Z. Riedl, G. Hajós, B. van den Driessche, R. A. Dommisse, and P. Mátyus, *Tetrahedron*, **2003**, *59*, 5919.

978. A.-Z. A. Elassar and Y. El-Kholy, *Heteroatom Chem.*, **2003**, *14*, 427.

979. T. Mitsumori, M. Bendikov, J. Sedó, and F. Wudl, *Chem. Mater.*, **2003**, *15*, 3759.

980. A. Fogain-Ninkam, A. Daich, B. Decroix, and P. Netchitailo, *Eur. J. Org. Chem.*, **2003**, 4273.

981. M. L. Głowka, D. Martynowski, A. Olczak, J. Bojarska, M. Srczesio, and K. Kozlowska, *J. Mol. Struct.*, **2003**, *658*, 43.

982. N. M. Rageh, *Spectrochim. Acta, Part A*, **2004**, *603*, 103.

983. U. Siemeling, T. Türk, U. Vorfeld, and H. Fink, *Monatsh. Chem.*, **2003**, *134*, 419.

984. A. A. Aly, *Org. Biomol. Chem.*, **2003**, *1*, 756.

985. M. van der Mat, K. M. Rommelé, H. Boss, A. Hatzelmann, M. van Slingerland, G. J. Sterk, and H. Timmerman, *J. Med. Chem.*, **2003**, *46*, 2008.

986. A. Z. Haikal, E. S. H. E. Ashry, and J. Banoub, *Carbohydr. Res.*, **2003**, *338*, 2291.

987. M. Johnsen, K. Rehse, H. Pertz, J. P. Stasch, and E. Bischoff, *Arch. Pharm. Med. Chem.*, **2003**, *336*, 591.

988. M. R. Lentz, J. S. Vilardo, M. A. Lockwood, P. F. Fanwick, and J. P. Rothwell, *Organometallics*, **2004**, *23*, 329.

989. G. V. Kalayda, S. Komeda, K. Ikeda, T. Sato, M. Chikuma, and J. Reedijk, *Eur. J. Inorg. Chem.*, **2003**, 4347.

990. J. Kuzolka, J. R. Farrell, and S. J. Lippard, *Inorg. Chem.*, **2003**, *42*, 8652.

991. N. V. Nastapova, V. V. Vanilkin, R. M. Eliseenkova, V. I. Morozov, E. I. Strunskaya, Z. A. Bredikhina, A. A. Bredikhin, and B. I. Buzykin, *Russ. J. Electrochem.*, **2003**, *39*, 1166.

992. K. Mogilaiah, D. S. Chowdary, P. R. Reddy, and N. V. Reddy, *Synth. Commun.*, **2003**, *33*, 127.

993. N. A. Buluchevskaya, A. V. Gulevskaya, and A. F. Pozharskii, *Chem. Heterocycl. Compd.*, **2003**, *39*, 87.

994. M. C. Cardia, S. Distinto, E. Maccioni, L. Bobsignore, and A. de Logu, *J. Heterocycl. Chem.*, **2003**, *40*, 1011.

995. S. K. Kundu, P. A. Mazumdar, A. K. Das, V. Bertolasi, and A. Pramanak, *J. Chem. Res.*, **2003**, *Synop.* 574.

996. A. A. F. Wasfy, *J. Chem. Res.*, **2003**, *Synop. 457*, *Minipr.* 835.

997. Y. Cheng, B. Ma, and F. Wudl, *J. Mater. Chem.*, **1999**, *9*, 2183.

998. O. Sato, K. Tsurumaki, S. Tamaru, and J. Tsunetsugu, *Heterocycles*, **2004**, *62*, 535.

999. A. Roedig, G. Bonse, R. Helm, and R. Kohlhaupt, *Chem. Ber.*, **1971**, *104*, 3378.

1000. A. Hirsch and D. Orphanos, *J. Heterocycl. Chem.*, **1965**, *2*, 726.

1001. L. A. Carpino, *J. Am. Chem. Soc.*, **1963**, *85*, 2144.

1002. F. D. Popp, J. M. Wefer, and C. W. Klinowski, *J. Heterocycl. Chem.* **1968**, *5*, 879.

1003. W. Amberg, Y. L. Bennani, R. K. Chadha, G. A. Crispino, W. D. Davis, J. Hartung, K.-S. Jeong, Y. Ogino, T. Shibata, and K. B. Sharpless, *J. Org. Chem.*, **1993**, *58*, 844.

1004. B. A. Rusin, A. N. Leksin, and A. L. Roshchin, *Zh. Org. Khim.*, **1980**, *16*, 209.

1005. J. Morley, *J. Chem. Soc.*, **1951**, 1971.

1006. A. Hassner and C. Stumer, *Organic Syntheses Based on Name Reactions*, 2nd ed., Pergamon, Oxford, 2002.

1007. G. B. Barlin and C. Y. Yap, *Aust. J. Chem.*, **1977**, *30*, 2319.

1008. J. Druey and A. Marxer, *J. Med. Pharm. Chem.*, **1959**, *1*, 1.

1009. B. Radziszewski, *Ber. Dtsch. Chem. Ges.*, **1885**, *18*, 355.

1010. N. Le Fur, L. Mojovic, A. Turck, N. Plé, G. Quéguiner, V. Reboul, S. Perrio, and P. Metzner, *Tetrahedron*, **2004**, *60*, 7983.

1011. A. M. Amer, M. M. El-Mobayed, and S. Asker, *Monatsh. Chem.*, **2004**, *135*, 595.

1012. A. L. Ruchelman, S. K. Singh, A. Ray, X. Wu, J.-M. Yang, N. Zhou, A. Liu, L. F. Liu, and E. J. La Voie, *Biorg. Med. Chem.*, **2004**, *12*, 795.

1013. S. K. Kundu and A. Pramani, *Indian J. Chem., Sect. B*, **2004**, *43*, 595.

1014. S. Al-Mousawi, E. John, and N. Al-Kandery, *J. Heterocycl. Chem.*, **2004**, *41*, 381.

1015. T. G. Chun, K. S. Kim, S. Lee, T.-S. Jeong, H.-Y. Lee, Y. H. Kim, and W. S. Lee, *Synth. Commun.*, **2004**, *34*, 1301.

1016. J. S. Kim, H.-J. Lee, M.-E. Suh, H.-Y. P. Choo, S. K. Lee, H. J. Park, C. Kim, S. W. Park, and C. O. Lee, *Bioorg. Med. Chem.*, **2004**, *12*, 3683.

1017. H.-J. Lee, J. S. Kim, S.-Y. Park, M.-E. Suh, H. J. Kim, E.-K. Seo, and C.-O. Lee, *Bioorg. Med. Chem.*, **2004**, *12*, 1623.

1018. L. J. Street, F. Sternfeld, R. A. Jelley, A. J. Reeve, R. W. Carling, K. W. Moore, R. M. McKernan, B. Sohal, A. Pike, G. R. Dawson, F. A. Bromidge, K. A. Wafford, G. R. Seabrook, S. A. Thompson,

G. Marshall, G. V. Pillai, J. S. Castro, J. R. Atack, and A. M. MacLeod, *J. Med. Chem.*, **2004**, *47*, 3642.

1019. R. W. Carling, K. W. Moore, L. J. Street, D. Wild, C. Isted, P. D. Leeson, S. Thomas, D. O'Connor, R. M. McKernan, K. Quirk, S. M. Cook, J. R. Atack, K. A. Wafford, S. A. Thompson, G. R. Dawson, P. Ferris, and J. L. Castro, *J. Med. Chem.*, **2004**, *47*, 1807.
1020. A. D. Lebsack, J. Gunzner, B. Wang, R. Pracitto, H. Schaffhauser, A. Santini, J. Aiyar, R. Bezverkov, B. Munoz, W. Liu, and S. Venkatraman, *Bioorg. Med. Chem. Lett.*, **2004**, *14*, 2463.
1021. J. Kuzelka, S. Mukhopadhyay, B. Springler, and S. J. Lippard, *Inorg. Chem.*, **2004**, *43*, 1751.
1022. A. I. Siriwardana, I. Nakamura, and Y. Yamamoto, *J. Org. Chem.*, **2004**, *69*, 3202.
1023. L. K. Thompson, V. Niel, H. Grove, D. O. Miller, M. J. Newlands, P. E. Bird, W. A. Wickramasinghe, and A. B. P. Lever, *Polyhedron*, **2004**, *23*, 1175.
1024. A. Coelho, E. Sotelo, H. Novoa, O. M. Peeters, N. Blaton, and E. Raviña, *Tetrahedron Lett.*, **2004**, *45*, 8459.
1025. P. G. Plieger, A. J. Downard, B. Moubaraki, K. S. Murray, and S. Brooker, *J. Chem. Soc., Dalton Trans.*, **2004**, 2157.
1026. L. M. Kaminskas, S. O. Pyke, and P. C. Burcham, *Org. Biomol. Chem.*, **2004**, *2*, 2578.
1027. S. A. A. El-Maksoud, *Electrochim. Acta*, **2004**, *49*, 4205.
1028. M. I. H. Helaleh, A. Al-Omair, and K. Tanaka, *J. Anal. Chem.*, **2004**, *59*, 560.
1029. Ö. Çelik, S. Ide, M. Kurt, and S. Yurdakul, *Acta Crystallogr., Sect. E*, **2004**, *60*, M 424.
1030. Ö. Çelik, S. Ide, M. Kurt, and S. Yurdakul, *Acta Crystallogr., Sect. E*, **2004**, *60*, M 1134.
1031. I. Hausmann, M. H. Klingele, V. Lozan, G. Steinfeld, D. Siebert, Y. Journaux, J. J. Girerd, and B. Kersting, *Chem.—Eur. J.*, **2004**, *10*, 1716.
1032. D. S. Dogruer, E. Kupeli, E. Yesilada, and M. F. Sahin, *Arch. Pharm.*, **2004**, *337*, 303.
1033. R. Jiménez, A. M. Sanz, F. Gómez-Contreras, M. C. Cano, M. I. R. Yunta, M. Pardo, and L. Campayo, *Heterocycles*, **2004**, *63*, 1299.
1034. L. Sereni, M. Tató, F. Sola, and W. K.-D. Brill, *Tetrahedron*, **2004**, *60*, 8561.
1035. H.-R. Bjørsvik, R. R. González, and L. Liguori, *J. Org. Chem.*, **2004**, *69*, 7720.
1036. A. Coelho, E. Satelo, H. Novoa, O. M. Peeters, N. Blaton, and E. Raviña, *Tetrahedron*, **2004**, *60*, 12177.
1037. A. Gomtsyan, E. K. Rayburt, R. G. Schmidt, G. Y. Zheng, R. J. Perner, S. Didomenico, J. R. Koenig, S. Turner, T. Jinkerson, I. Drizin, S. M. Hannick, B. S. Macri, H. A. McDonald, P. Honore, C. T. Wismer, K. C. Marsh, J. Wetter, K. D. Stewart, T. Oie, M. F. Jarvis, C. J. Surowy, C. R. Faltynek, and C.-H. Lee, *J. Med. Chem.*, **2005**, *48*, 744.
1038. P. Sohár, A. Csámpai, G. Magyarfaivi, A. E. Szabó, and G. Stájer, *Monatsh. Chem.*, **2004**, *135*, 1519.
1039. S. Demirayak, A. C. Karaburum, and R. Beis, *Eur. J. Med. Chem.*, **2004**, *39*, 1089.
1040. J. Epsztajn and Z. Malinowski, *Synth. Commun.*, **2005**, *35*, 179.
1041. G. Özer, N. Saracoglu, A. Menzek, and M. Balci, *Tetrahedron*, **2005**, *61*, 1545.

Index

This index covers the text but neither the Appendix (Tables of Simple Cinnolines and Simple Phthalazines) nor the Glance Indices (appended to Chapters 1 and 8).

The page number(s) following each primary entry refer to synthesis or general information. Although each number indicates that the subject is treated on that page (and possibly also on subsequent pages), the actual words of the primary entry may appear only in an abbreviated form in the text.

Some unusual terms have been employed extensively as succinct secondary entries. For example, the term *alkanelysis* has been used to indicate the direct replacement of an appropriate leaving group by an alkyl substituent, thus mimicking conventional terms such as *aminolysis*, *alcoholysis*, and the like.